Undergraduate Texts in Physics

Undergraduate Texts in Physics (UTP) publishes authoritative texts covering topics encountered in a physics undergraduate syllabus. Each title in the series is suitable as an adopted text for undergraduate courses, typically containing practice problems, worked examples, chapter summaries, and suggestions for further reading. UTP titles should provide an exceptionally clear and concise treatment of a subject at undergraduate level, usually based on a successful lecture course. Core and elective subjects are considered for inclusion in UTP.

UTP books will be ideal candidates for course adoption, providing lecturers with a firm basis for development of lecture series, and students with an essential reference for their studies and beyond.

Simon Mochrie • Claudia De Grandi

Introductory Physics for the Life Sciences

Simon Mochrie
Departments of Physics and Applied
Physics
Yale University
New Haven, CT, USA

Claudia De Grandi
Department of Physics and Astronomy
University of Utah
Salt Lake City, UT, USA

ISSN 2510-411X   ISSN 2510-4128   (electronic)
Undergraduate Texts in Physics
ISBN 978-3-031-05807-3   ISBN 978-3-031-05808-0   (eBook)
https://doi.org/10.1007/978-3-031-05808-0

This Springer imprint is published by the registered company Springer Nature Switzerland AG
The registered company address is: Gewerbestrasse 11, 6330 Cham, Switzerland

"My fingers," said Elizabeth, "do not move over physics in the masterly manner which I see so many women's do. They have not the same force or rapidity, and do not produce the same expressions. But then I have always supposed it to be my own fault – because I would not take the trouble of practising. It is not that I do not believe my fingers as capable as any other woman's of superior execution." Elizabeth Bennett in Jane Austen's Pride and Prejudice talking about physics.

Preface

In a seminal 2003 article, Bialek and Botstein [1] wrote:

> Dramatic advances in biological understanding, coupled with equally dramatic advances in experimental techniques and computational analyses, are transforming the science of biology. The emergence of new frontiers of research in functional genomics, molecular evolution, intracellular and dynamic imaging, systems neuroscience, complex diseases, and the system-level integration of signal-transduction and regulatory mechanisms requires an ever-larger fraction of biologists to confront deeply quantitative issues that connect to ideas from the more mathematical sciences. At the same time, increasing numbers of physical scientists and engineers are recognizing that exciting frontiers of their own disciplines lie in the study of biological phenomena. Characteristic of this new intellectual landscape is the need for strong interaction across traditional disciplinary boundaries.

Recent national reports [2–5] have also highlighted the increasing importance of quantitative skills for students who are planning biomedical careers, and the need to modify and augment undergraduate biology and premedical education accordingly.

The two-semester Introductory Physics for Life Sciences (IPLS) sequence, currently required for premedical students and biological science majors, is where these students encounter quantitative and mathematical descriptions of the natural world for the first time. Biology and premedical students are usually skeptical about the relevance of physics and mathematics to their academic and professional goals. Often, their skepticism is reinforced by the topics presented in introductory physics, which generally owe more to tradition than to what is meaningful to biologists. The disconnect between traditional introductory physics and the needs of today's biology undergraduates is compounded by the fact that many physics instructors feel unqualified or have too little time to develop a biologically focused physics course. However, motivated by national calls for reforming biomedical education [2–5] and inspired by backward design approaches to curriculum reform [6], physics departments are increasingly receptive to IPLS curriculum changes, even if it means putting aside cherished traditional topics.

It seems then that the moment is right to transform IPLS into a course that is understood to be essential to every biologist's undergraduate education. To achieve such a transformation, IPLS students and instructors critically need appropriate curricular materials, including suitable textbooks and homework problems. To this end, this book is the text for a two-semester, calculus-based introductory physics course, designed for life science and pre-medical students and their instructors, that re-imagines the IPLS syllabus to incorporate myriad biologically and medically relevant topics. Our purposes in writing this book are to:

- Introduce biological science majors and future clinicians to a set of physical and mathematical principles and tools that will enable a deeper scientific understanding of biological systems, including the human body.
- Demonstrate the application of physics and mathematics to the life sciences and medicine, via a number of highly relevant examples.
- Transform introductory physics for the life sciences into an engaging and exciting subject that is understood to be essential to every biologist's undergraduate education.
- Provide curricula materials, including problems, that will enable every physics faculty member to teach a biologically meaningful version of IPLS.

The book comprises 19 chapters. Many would not be far out of place in a more traditional introductory physics class, including chapters on Newton's laws, simple harmonic motion, waves, and electricity and magnetism. However, chapters on random walks, diffusion, rates of change, statistical mechanics, fluid mechanics, and biologic stand outside the traditional curriculum. In developing these chapters, we owe a special debt to [7–9]. A version of IPLS using these materials has been taught at Yale University since Fall 2010 [10–12]. As of Spring 2022, approximately 1500 students have taken the sequence. The majority of these students are biological science majors (64%), with significant numbers of psychology and cognitive science majors (14%). Eighty-one percent identify themselves as premedical students. There have been roughly equal numbers of sophomores and juniors in the class, with fewer seniors, and essentially no first years. Students generally arrive in the class possessing considerable biological and chemical sophistication, as a result of the prior science classes. Almost all take the class in order to fulfill the physics requirement for medical school and/or the physics requirement of their major.

A key issue for any IPLS course is the level of mathematics to use. Almost all students who have taken the class at Yale had previously taken a first course in calculus, either at Yale or in high school. In fact, the class seeks to further develop students' mathematical and analytical skills. Thus, we include probability, simple differential equations, simple linear algebra, and complex numbers, for example. These mathematical topics are introduced and exploited as necessary, and earn their place in the curriculum. On the other hand, we avoid multi-dimensional integrals as far as possible, because the students find them challenging and the pay-off seems not worth it. We use Wolfram Alpha (http://www.wolframalpha.com[1]) to facilitate mathematical manipulations as much as possible, including the solution of systems of algebraic equations, the evaluation of derivatives and integrals, and the numerical solution of differential equations. Using Wolfram Alpha empowers students to

---

[1] http://www.wolframalpha.com

carry out more sophisticated mathematics than otherwise, and is a handy skill for their future careers. Because computational approaches constitute an essential aspect of how research is now carried out, both in the physical and life sciences, the course includes a number of simulations and visualizations implemented as Wolfram Demonstrations.

We would like to thank Rona Ramos, Sarah Demers, Daisuke Nagai, Sidney Cahn, and Alison Sweeney for their advice and help. We would especially like to express our deep gratitude to the graduate student teaching assistants, the undergraduate peer tutors, and, especially, all of the students who have been involved in our classes over the last decade. SM would like to thank the NSF for research support. Finally, SM would like to thank Lynne, Gemma, and James for all their encouragement and support.

New Haven, CT, USA   Simon Mochrie
Salt Lake City, UT, USA   Claudia De Grandi

References

1. W. Bialek, D. Botstein, Introductory science and mathematics education for 21st-century biologists. Science **303**, 788–790 (2004)
2. National Research Council, *BIO2010: Transforming Undergraduate Education for Future Research Biologists* (National Academies Press, Washington, 2003)
3. Association of American Medical Colleges (AAMC) and the Howard Hughes Medical Institute (HHMI), *Scientific Foundations for Future Physicians* (2009)
4. American Association for the Advancement of Science (AAAS), *Vision and Change in Undergraduate Biology Education* (2011)
5. National Research Council, *Convergence: Transdisciplinary Integration of Life Sciences, Physical Sciences, Engineering and Beyond* (National Academies Press, Washington, 2014)
6. G. Wiggins, J. McTighe, *Understanding by Design* (Merrill Education/Prentice Hall, Hoboken, 2005)
7. H.C. Berg, *Random Walks in Biology* (Princeton University Press, Princeton, 1993)
8. P.C. Nelson, *Biological Physics: Energy, Information, Life* (W. H. Freeman, New York, 2004)
9. R. Phillips, J. Kondev, J. Theriot, *Physical Biology of the Cell* (Garland Science, New York, 2008)
10. S.G.J. Mochrie, Vision and change in introductory physics for the life sciences. Amer. J. Phys. **84**, 542–551 (2016)
11. C. De Grandi, R. Ramos, S.G.J. Mochrie, Assessment of strategies to build a welcoming STEM classroom environment for all students, in *Proceedings of the Physics Education Research Conference 2018* (2019)
12. C. De Grandi, S. Mochrie, R. Ramos, Pedagogical strategies to increase students' engagement and motivation, in *Concepts, Strategies and Models to Enhance Physics Teaching and Learning* (Springer, Cham, 2019)

Electronic Supplementary Material

Chapters 1, 2, 5, 7, 11, and 16 are supplemented by Mathematica demonstrations which the reader can download by following the link in the footnote at the start of the chapter and then navigating to the "Electronic Supplementary Material" section of the page. Chapters 1, 5, 7, 8, 11, 12, and 16 are supplemented by Mathematica demonstrations that can be downloaded from the Wolfram site from the links provided at the relevant point in the text.

Contents

Vectors and Kinematics

1

Nothing in life is to be feared, it is only to be understood. Now is the time to understand more, so that we may fear less.

Marie Curie

The good thing about science is that it is true whether or not you believe in it.

Neil deGrasse Tyson

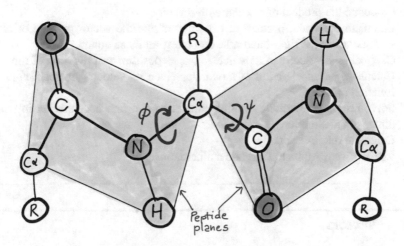

Electronic Supplementary Material The online version of this article (https://doi.org/10.1007/978-3-031-05808-0_1) contains supplementary material, which is available to authorized users.

1.1   Introduction

In this chapter, we will discuss the tools needed to describe the motion of an object, such as its position, velocity, and acceleration. These tools are part of the branch of physics called kinematics. In order to be able to describe the motion of an object in three-dimensional space, we will introduce vectors and related vector operations. Vectors will also prove to be helpful tools to describe relationships between different parts of geometrical objects, as we will see in the context of certain intriguing structures called "tensegrities." In addition, in preparation for Chap. 2. *Force and Momentum: Newton's Laws and How to Apply Them*, which focusses on how the net force on an object leads to its acceleration, the second part of this chapter will focus on the facts that acceleration is rate of change of velocity and velocity is rate of change of position, permitting in principle a complete description of an object's translational behavior, given the net force on the object and Newton's laws.

1.2   Your Learning Goals for This Chapter

By the end of this chapter you should be able to:

- Understand what a vector is and be able to resolve a vector into components
- Add and subtract vectors
- Calculate the product of a scalar with a vector
- Calculate the scalar product of two vectors and the vector product of two vectors, using the right-hand rule, and interpret these quantities
- Calculate the velocity of a particle given its position as a function of time
- Calculate the acceleration of a particle, given its velocity or position as a function of time
- Solve problems involving motion subject to constant acceleration in one-dimensional situations
- Calculate relative velocity
- Use Wolfram Alpha to facilitate mathematics

1.3   Vectors

From a geometrical point of view, a vector is a directed line segment with both a magnitude "or length" and a direction. We denote a vector with an arrow on top, $\vec{a}$, or using boldface, **a** (Fig. 1.1). Either of these is "the vector a." The length or magnitude of **a** is denoted $|\mathbf{a}|$ or simply a.

Consider two vectors **a** and **b** (Fig. 1.2). If **a** and **b** have the same length and the same direction, they are equal: **a** = **b**. It is important to note that it does not matter where a vector "starts," i.e., where the arrow representing the vector is located; a

Fig. 1.1 A vector **a** is characterized by a length and a direction

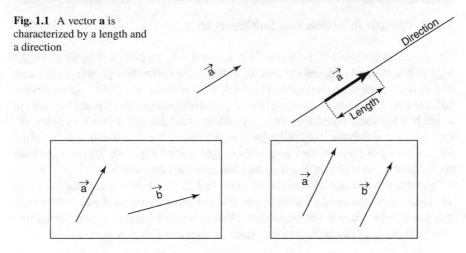

Fig. 1.2 Two vectors **a** and **b**: on the left the vectors are different **a** $\neq$ **b**; on the right the vectors are equal **a** = **b**

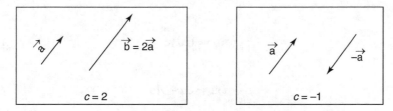

Fig. 1.3 Multiplication of a vector by a scalar c. Left side: $c = 2$. Right side: $c = -1$

vector can be translated to different points in space and, as long as it keeps the same length and direction, it is still the same vector. That is why on the right side of Fig. 1.2 the two vectors **a** and **b** are equal even if they are located at different points.

1.3.1 Multiplication of a Vector by a Scalar

Consider a vector **a**. Multiplication of **a** by a number c gives another vector **b** that has the same direction, i.e., is parallel, to **a**, but a different length, i.e.,

$$\mathbf{b} = c\mathbf{a}. \tag{1.1}$$

c is a number, a.k.a. a "scalar." A scalar has a magnitude and a sign ($\pm$), but not a direction. See Fig. 1.3 for two examples: $c = 2$, doubling a vector: **b** = 2**a**, and $c = -1$, creating the opposite vector: **b** = −**a**, which is in the opposite direction to **a**.

1.3.2   Vector Addition and Subtraction

The sum of two vectors **a** and **b** is a third vector **a** + **b**. To draw **a** + **b**, we place the start of **b** at the end of **a**, the sum vector **a** + **b** is the vector that goes from the start of **a** to the end of **b**, see left side of Fig. 1.4. This is called the *tail-to-tip method* to add vectors. Alternatively, you can use the *parallelogram method*: place the start of **a** and **b** at the same point, then draw a parallelogram that has **a** and **b** as sides, the sum vector **a** + **b** is the longest diagonal of the parallelogram, starting at the initial place where you drew the two vectors (see right side of Fig. 1.4). The two methods are equivalent, as shown in Fig. 1.4, they bring to the same result.

The order in a vector addition is not important: if you draw **b** + **a** (either with the tail-to-tip or parallelogram method), you find the same vector as if you draw **a** + **b**. It follows that **a** + **b** = **b** + **a**, i.e., vector addition is commutative. Vector subtraction may be thought of as addition of the opposite vector: **a** − **b** = **a** + (−**b**).

Also

$$(\mathbf{a} + \mathbf{b}) + \mathbf{c} = \mathbf{a} + (\mathbf{b} + \mathbf{c}), \qquad (1.2)$$

$$c(d\mathbf{a}) = (cd)\mathbf{a}, \qquad (1.3)$$

$$(c + d)\mathbf{a} = c\mathbf{a} + d\mathbf{a}, \qquad (1.4)$$

and

$$c(\mathbf{a} + \mathbf{b}) = c\mathbf{a} + c\mathbf{b}, \qquad (1.5)$$

where c and d are both scalars.

Wolfram Alpha can add vectors nicely. For example, consider the two-components vectors (5,1) and (3,6) (components will be discussed shortly in 1.3.5),

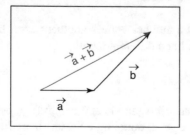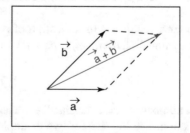

Fig. 1.4 Methods for vector addition. Left side: tail-to-tip method. Right side: parallelogram method

to add $(5, 1) + (3, 6)$, go to the Wolfram Alpha website: `http://www.wolfram alpha.com/`,[1] and in the entry box, type:

```
(5,1)+(3,6)
```

Then, hit the "=" button and behold the results.[2]

1.3.3 Products of Two Vectors

There are two different ways to calculate the product of two vectors. One way gives a scalar result. The other gives a vector.

Scalar Products of Two Vectors
The scalar or dot product of **a** and **b** is denoted **a** · **b** and is a scalar—hence the name. The value of **a** · **b** is

$$\mathbf{a} \cdot \mathbf{b} = ab \cos \theta, \tag{1.6}$$

where a and b are the lengths of **a** and **b**, respectively, and θ is the angle between the two vectors (see Fig. 1.5).

From the right side of Fig. 1.5, we see that we can interpret $b \cos \theta$ as the projection of the vector **b** onto the (direction of) **a**. So **a** · **b** is equal to the product of the length of **a** and the projection of **b** onto **a** (or vice versa). The scalar product, in other words, quantifies how much two vectors overlap:

- if two vectors are perpendicular, $\theta = \pi/2$ radians (90°), then the scalar product is zero: **a** · **b** $= ab \cos(\pi/2) = 0$.
- if two vectors are parallel, $\theta = 0$ radians (0°), the scalar product is maximum **a** · **b** $= ab \cos(0) = ab$.

Notice too that **a** · **a** $= a^2 \cos 0 = a^2$. Therefore, calculating the scalar product of a vector with itself yields the square of the vector's length. Then taking the square root gives the length of the vector $a = \sqrt{\mathbf{a} \cdot \mathbf{a}}$.

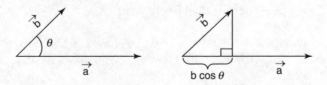

Fig. 1.5 Quantities involved in calculating the scalar product of **a** and **b**

---

[1] http://www.wolframalpha.com/

[2] http://www.wolframalpha.com/input/?i=%7B5%2C1%7D%2B%7B3%2C6%7D

Fig. 1.6 Cross product of two vectors. Right side: how to apply the right-hand rule to find the direction of $\mathbf{a} \times \mathbf{b}$

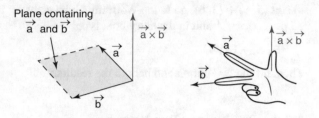

Vector or Cross Product

The vector or cross product of $\mathbf{a}$ and $\mathbf{b}$ is denoted $\mathbf{a} \times \mathbf{b}$ and is a vector (call it $\mathbf{c}$)—hence the name:

$$\mathbf{c} = \mathbf{a} \times \mathbf{b}. \tag{1.7}$$

The magnitude of $\mathbf{a} \times \mathbf{b}$, i.e., the magnitude of $\mathbf{c}$, is

$$|\mathbf{c}| = c = |\mathbf{a} \times \mathbf{b}| = ab\sin\theta, \tag{1.8}$$

where a and b are the lengths of $\mathbf{a}$ and $\mathbf{b}$, respectively, and θ is the angle between these two vectors. In this case, since the vector product involves $\sin\theta$, for parallel vectors, for which $\theta = 0$, their vector (cross) product equals zero.

The direction of $\mathbf{a} \times \mathbf{b}$ is shown in Fig. 1.6. Specifically, pick the plane that contains both $\mathbf{a}$ and $\mathbf{b}$. Then, $\mathbf{a} \times \mathbf{b}$ is perpendicular to that plane. But does $\mathbf{a} \times \mathbf{b}$ point up or down? This we must decide using the "right-hand rule." In Eq. 1.7, to determine the direction of $\mathbf{a} \times \mathbf{b}$, set your right-hand index finger along $\mathbf{a}$, set you right-hand middle finger along $\mathbf{b}$, then $\mathbf{a} \times \mathbf{b}$ lies along your thumb. This prescription leads to:

$$\mathbf{b} \times \mathbf{a} = -\mathbf{a} \times \mathbf{b}. \tag{1.9}$$

This means that the cross product is not commutative, in other words the order matters. Pay attention which vector goes first!

1.3.4   Example: Tetrahedral Bond Angles

In chemistry classes, we learn that in organic compounds, carbon is often tetrahedrally coordinated. Determine the angle, α, between tetrahedral bonds.

Answer: We define four unit vectors, e_1, e_2, e_3, and e_4, which are directed from a carbon atom along each of its tetrahedral bonds (Fig. 1.7). Note: a unit vector is a vector with length 1. The angle

(continued)

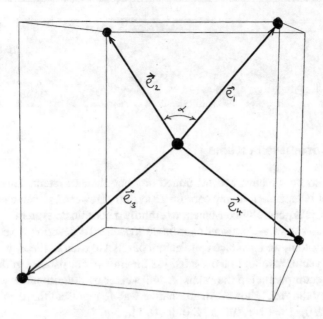

Fig. 1.7 Vectors for "Tetrahedral bond angles"

α is given by $\mathbf{e}_i \cdot \mathbf{e}_j = \cos \alpha$ for $i \neq j$. If we think about the vector sum of $\mathbf{e}_1$ and $\mathbf{e}_2$, we see that it is a vector that is directed vertically upwards in the figure. On the other hand, we can similarly see that the vector sum of $\mathbf{e}_3$ and $\mathbf{e}_4$ is directed vertically downwards. Therefore, the two vectors, $\mathbf{e}_1 + \mathbf{e}_2$ and $\mathbf{e}_3 + \mathbf{e}_4$ have opposite directions and equal lengths. It follows that the sum of all four vectors is zero:

$$\mathbf{e}_1 + \mathbf{e}_2 + \mathbf{e}_3 + \mathbf{e}_4 = \mathbf{0}. \tag{1.10}$$

Therefore the scalar product of one of them with their sum must equal zero:

$$\mathbf{e}_1 \cdot (\mathbf{e}_1 + \mathbf{e}_2 + \mathbf{e}_3 + \mathbf{e}_4) = 0, \tag{1.11}$$

i.e.,

$$1 + 3 \cos \alpha = 0, \tag{1.12}$$

(continued)

or

$$\alpha = \arccos\left(-\frac{1}{3}\right) \simeq 109.5°. \tag{1.13}$$

1.3.5 Coordinate Systems

Notice that so far we have not mentioned any coordinate system. This is because vectors exist independent of any coordinate system. However, to carry out concrete calculations, it is generally convenient to employ a coordinate system.

Figure 1.8 shows a Cartesian coordinate system. To describe a vector in this coordinate system, we introduce unit vectors parallel to each of the x, y, and z axes. $\hat{\mathbf{i}}$ is the unit vector parallel to the x-axis. $\hat{\mathbf{j}}$ is the unit vector parallel to the y-axis. $\hat{\mathbf{k}}$ is the unit vector parallel to the z-axis. A unit vector by definition has length 1 and is denoted by the "hat" shown. An alternative way to represent these unit vectors is via $\hat{\mathbf{i}} = (1, 0, 0)$, $\hat{\mathbf{j}} = (0, 1, 0)$, and $\hat{\mathbf{k}} = (0, 0, 1)$.

Correspondingly, we can express the vector $\mathbf{a}$ in terms of coordinates in two ways: Either as

$$\mathbf{a} = a_x\hat{\mathbf{i}} + a_y\hat{\mathbf{j}} + a_z\hat{\mathbf{k}}, \tag{1.14}$$

where a_x is the component of $\mathbf{a}$ along the x-axis, a_y is the component of $\mathbf{a}$ along the y-axis, and a_z is the component of $\mathbf{a}$ along the z-axis, or as

$$\mathbf{a} = (a_x, a_y, a_z), \tag{1.15}$$

see Fig. 1.9.

Fig. 1.8 A Cartesian coordinate system and the corresponding unit vectors $\hat{\mathbf{i}}$, $\hat{\mathbf{j}}$, and $\hat{\mathbf{k}}$

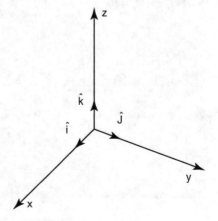

Fig. 1.9 A vector **a** relative to a particular Cartesian coordinate system and the corresponding components of **a** along the x, y, and z axes, namely a_x, a_y, and a_z, respectively

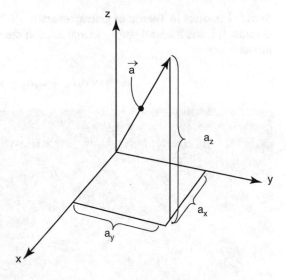

It is important to understand that

$$\mathbf{a} = \mathbf{b} \tag{1.16}$$

is actually three equations, one each of the components along x, y, and z:

$$a_x = b_x, \tag{1.17}$$

$$a_y = b_y, \tag{1.18}$$

and

$$a_z = b_z. \tag{1.19}$$

Right-Handed Coordinate Systems and the Right-Hand Rule

When you are sketching a coordinate system, after you have drawn the x and y axes, it might seem that there is dilemma, namely what is the direction of z? However, by unwavering convention, we resolve this dilemma by *always* using a "right-handed coordinate system" and the right-hand rule already discussed in the context of the cross product of two vectors (see right side Fig. 1.6).

To use the right-hand rule, point the index finger of your right hand along the positive x-direction, i.e., along $\hat{\mathbf{i}}$, arrange your middle finger along the positive y-direction, i.e., along $\hat{\mathbf{j}}$. Extend your thumb, so that it is perpendicular to both your index and middle finger. The right-hand rule is that your thumb now points along the positive z direction, i.e., along $\hat{\mathbf{k}}$. It will take you just a moment to convince yourself that the coordinate systems in Figs. 1.8 and 1.9 are both right-handed coordinate systems.

Scalar Product in Terms of Components

Because $\hat{\mathbf{i}}, \hat{\mathbf{j}}$, and $\hat{\mathbf{k}}$ are all perpendicular to each other, one can show, using the scalar product, that

$$\mathbf{a} \cdot \mathbf{b} = a_x b_x + a_y b_y + a_z b_z. \tag{1.20}$$

1.3.6   Example: A Trigonometric Identity

(a) Sketch the two vectors

$$\mathbf{v} = (\cos \alpha, \sin \alpha) \tag{1.21}$$

and

$$\mathbf{w} = (\cos \beta, \sin \beta), \tag{1.22}$$

where α and β are two arbitrary angles such that: $0 < \alpha < \beta < \frac{\pi}{2}$.

Answer: The required sketch is shown in Fig. 1.10.

(b) Calculate the length of $\mathbf{v}$ and $\mathbf{w}$, namely v and w, respectively.

Answer:

$$v = \sqrt{\cos^2 \alpha + \sin^2 \alpha} = 1. \tag{1.23}$$

Similarly $w = 1$. Both $\mathbf{v}$ and $\mathbf{w}$ are unit vectors.

(c) Calculate the scalar product of $\mathbf{v}$ and $\mathbf{w}$.

Answer: We have that

$$\mathbf{v} \cdot \mathbf{w} = \cos \alpha \cos \beta + \sin \alpha \sin \beta. \tag{1.24}$$

(d) Determine the angle, θ, between $\mathbf{v}$ and $\mathbf{w}$ in two different ways and hence establish a trigonometric identity for $\cos(\beta - \alpha)$.

Answer: We also have that

$$\mathbf{v} \cdot \mathbf{w} = vw \cos \theta. \tag{1.25}$$

(continued)

Fig. 1.10 The vectors **v** and
w of Sect. 1.3.6

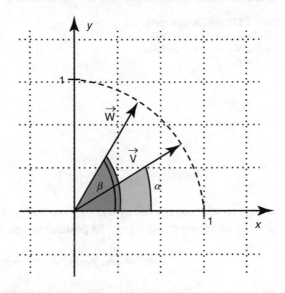

Using the answers to (b) and (c) in Eq. 1.25, it follows that

$$\cos\theta = \cos\alpha\cos\beta + \sin\alpha\sin\beta. \tag{1.26}$$

On the other hand, from inspection of Fig. 1.10, it is clear that

$$\theta = \beta - \alpha. \tag{1.27}$$

Combining these two results (Eqs. 1.26 and 1.27), we find the desired trigonometric identity:

$$\cos(\beta - \alpha) = \cos\alpha\cos\beta + \sin\alpha\sin\beta. \tag{1.28}$$

1.3.7 Position Vector and Displacement Vector

It is often convenient to define an object's *position vector*, which is the vector from the origin of the coordinate system to the object. Thus, if object 1 is at coordinates (x_1, y_1, z_1) its position vector is

$$\mathbf{r}_1 = (x_1, y_1, z_1) = x_1\hat{\mathbf{i}} + y_1\hat{\mathbf{j}} + z_1\hat{\mathbf{k}}, \tag{1.29}$$

Fig. 1.11 The displacement
vector $\mathbf{r}_2 - \mathbf{r}_1$ goes from the
end of vector $\mathbf{r}_1$ to the end of
vector $\mathbf{r}_2$

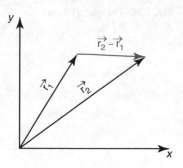

which we will also frequently write simply as (x_1, y_1, z_1). If another object—object
2—has coordinates (x_2, y_2, z_2), its position vector is

$$\mathbf{r}_2 = (x_2, y_2, z_2) = x_2\hat{\mathbf{i}} + y_2\hat{\mathbf{j}} + z_2\hat{\mathbf{k}}, \qquad (1.30)$$

or (x_2, y_2, z_2). Then, the vector that goes from object 1 to object 2 is $\mathbf{r}_2 - \mathbf{r}_1$. This
vector is the displacement vector from object 1 to object 2, and it is shown for a
two-dimensional case in Fig. 1.11.

Displacement vectors and position vectors are measured in units of length. The
SI unit for displacement and position is the meter: m. SI units are standard units
(International System of Units) that we will commonly use in this book.

1.3.8 Example: Tensegrity Geometry

In this example, we start discussing a cool class of structures, called "tensegrities." A tensegrity (tensional integrity) structure, or simply a tensegrity,
consists of a number of struts, which are compressed, connected together by
cables, which are stretched. Tensegrities are self-supporting and more-or-less
rigid. A number of tensegrities are works of art, including the Needle Tower at
the Hirschhorn Museum in Washington, DC. Tensegrities have also become a
focus of recent robotics research.[3] Interestingly, they have also been proposed
as models for a number of biological structures, including the human spine
and the cytoskeleton.

The simplest example of a tensegrity is shown in Fig. 1.12. In this chapter,
our goal is to analyze the geometry of this tensegrity, which consists of two
equally sized equilateral triangles of cables (shown red in the schematic on
the right), one in the plane $z = 0$ and one in the plane $z = 2$. These two

[3]https://royalsocietypublishing.org/doi/10.1098/rsif.2014.0520

(continued)

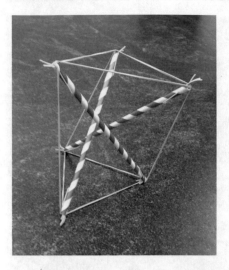 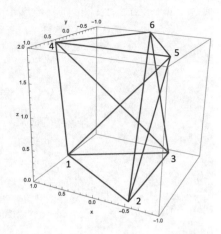

Fig. 1.12 Left: A threefold rotationally symmetric prismatic "tensegrity," built from drinking straws, rubber bands, and paper clips. Right: Schematic of a threefold rotationally symmetric prismatic "tensegrity," including a coordinate system

triangles are rotated relative to each other by an angle α and are connected to each other by additional cables (also red) and struts (blue). The center of the $z = 0$-plane triangle is at the origin of the coordinate system, so that the position vectors of the vertices of that triangle are

$$\mathbf{r}_1 = (1, 0, 0), \tag{1.31}$$

$$\mathbf{r}_2 = (\cos \frac{2\pi}{3}, \sin \frac{2\pi}{3}, 0), \tag{1.32}$$

$$\mathbf{r}_3 = (\cos \frac{4\pi}{3}, \sin \frac{4\pi}{3}, 0). \tag{1.33}$$

The expressions for $\mathbf{r}_2$ and $\mathbf{r}_3$ follow from the observation that these vectors are obtained by rotating a unit vector, originally along the x-axis, by $\frac{2\pi}{3}$ and $\frac{4\pi}{3}$, respectively.

(a) What are the position vectors of the vertices of the triangle in the $z = 2$-plane?

(continued)

Answer: To go from vertex 1, 2, and 3 to vertex
4, 5, and 5, there must occur a further rotation
of α in the xy-plane and a translation of 2
along the z-direction. Thus,

$$\mathbf{r}_4 = (\cos\alpha, \sin\alpha, 2), \tag{1.34}$$

$$\mathbf{r}_5 = (\cos(\alpha + \frac{2\pi}{3}), \sin(\alpha + \frac{2\pi}{3}), 2), \tag{1.35}$$

$$\mathbf{r}_6 = (\cos(\alpha + \frac{4\pi}{3}), \sin(\alpha + \frac{4\pi}{3}), 2). \tag{1.36}$$

(b) What is the length, ℓ_{14}, of the cable that runs from vertex 1 to vertex 4?
Does your answer make sense for $\alpha = 0$?

Answer: The vector from vertex 1 to vertex 4 is
$\mathbf{r}_4 - \mathbf{r}_1$. The length of this vector is the length of
the cable from vertex 1 to vertex 4. This vector
is

$$\mathbf{r}_4 - \mathbf{r}_1 = (\cos\alpha - 1, \sin\alpha, 2). \tag{1.37}$$

Its length is

$$\ell_{14} = \sqrt{(\mathbf{r}_4 - \mathbf{r}_1)\cdot(\mathbf{r}_4 - \mathbf{r}_1)} = \sqrt{(\cos\alpha - 1)^2 + \sin^2\alpha + 4} = \sqrt{6 - 2\cos\alpha}, \tag{1.38}$$

where we used the trigonometric identity,
$\cos^2\alpha + \sin^2\alpha = 1$, valid for any α. For $\alpha = 0$,
$\cos\alpha = 1$, and this calculation gives $\sqrt{4} = 2$ for
the length of the cable. At the same time, for
$\alpha = 0$, vertex 4 is directly over vertex 1 and so
the distance between them is indeed 2. So our
answer does make sense for $\alpha = 0$.

(c) What is the length, ℓ_{12}, of the cable that runs from vertex 1 to vertex 2?

Answer: Similarly to (b), the vector from vertex
1 to vertex 2 is

$$\mathbf{r}_2 - \mathbf{r}_1 = \left(\cos\frac{2\pi}{3} - 1, \sin\frac{2\pi}{3}, 0\right), \tag{1.39}$$

(continued)

with length

$$\ell_{12} = \sqrt{(\mathbf{r}_2 - \mathbf{r}_1) \cdot (\mathbf{r}_2 - \mathbf{r}_1)} = \sqrt{\left(\cos\frac{2\pi}{3} - 1\right)^2 + \sin^2\frac{2\pi}{3}}$$

$$= \sqrt{2 - 2\cos\frac{2\pi}{3}} = \sqrt{3}, \qquad (1.40)$$

since $\cos\frac{2\pi}{3} = -\frac{1}{2}$.

(d) What is the angle, β, between the cable that runs from vertex 1 to vertex 4 and the cable that runs from vertex 1 to vertex 2?

Answer: To calculate β, we can use the scalar product. Specifically, we have that

$$(\mathbf{r}_2 - \mathbf{r}_1) \cdot (\mathbf{r}_4 - \mathbf{r}_1) = \ell_{12}\ell_{14}\cos\beta. \qquad (1.41)$$

Therefore, using

$$(\mathbf{r}_2 - \mathbf{r}_1) \cdot (\mathbf{r}_4 - \mathbf{r}_1) = \frac{1}{2}(3 - 3\cos\alpha + \sqrt{3}\sin\alpha), \qquad (1.42)$$

we find

$$\beta = \arccos\frac{\frac{1}{2}(3 - 3\cos\alpha + \sqrt{3}\sin\alpha)}{\ell_{12}\ell_{14}} = \arccos\frac{3 - 3\cos\alpha + \sqrt{3}\sin\alpha}{2\sqrt{3}\sqrt{6 - 2\cos\alpha}}.$$
$$(1.43)$$

For $\alpha = 0$, we find $\cos\beta = 0$, i.e., $\beta = \frac{\pi}{2}$, which is sensible.

1.4   Kinematics

1.4.1   Motion in One-Dimension

We now consider motion along a line, i.e., along the x-axis only. Then, the mean velocity ($\bar{v}$) of the particle between times t_1 and t_2 is:

$$\bar{v} = \frac{x_2 - x_1}{t_2 - t_1}. \qquad (1.44)$$

The bar on top of the v denotes a mean value. If we envision the particle's position to be a function of time, we can re-write Eq. 1.44:

$$\bar{v} = \frac{x(t_2) - x(t_1)}{t_2 - t_1}. \tag{1.45}$$

Now, let us consider the case that t_2 is actually only slightly different from t_1. To this end, we write $t_1 = t$ and $t_2 = t + \Delta t$, where Δt is a small time increment. Then Eq. 1.45 becomes:

$$\bar{v} = \frac{x(t + \Delta t) - x(t)}{t + \Delta t - t} = \frac{x(t + \Delta t) - x(t)}{\Delta t}. \tag{1.46}$$

In Eq. 1.46, $\bar{v}$ is the mean velocity between t and $t + \Delta t$. As Δt gets progressively smaller, $\bar{v}$ approaches the (instantaneous) velocity at time t, namely $v(t)$. We can write this mathematically in terms of the limit as $\Delta t \to 0$:

$$v(t) = \lim_{\Delta t \to 0} \frac{x(t + \Delta t) - x(t)}{\Delta t}. \tag{1.47}$$

Recall from calculus, that the right-hand side of Eq. 1.47 is exactly the derivative of $x(t)$ with respect to t, i.e.,

$$v(t) = \lim_{\Delta t \to 0} \frac{x(t + \Delta t) - x(t)}{\Delta t} = \frac{dx}{dt} = \dot{x}(t) = x'(t), \tag{1.48}$$

where $\dot{x}$ (x dot) and x' are alternative notations for dx/dt.

Thus, we see that for 1D motion *the velocity is the derivative of the position with respect to time*. This is one key result from this chapter.

The mean acceleration ($\bar{a}$) of the particle between times t_1 and t_2 is:

$$\bar{a} = \frac{v_2 - v_1}{t_2 - t_1}. \tag{1.49}$$

We can repeat the same steps done above to find the limit in which t_2 is only slightly different from t_1, and with our calculus knowledge we find that:

$$a(t) = \frac{dv}{dt} = \dot{v}(t) = v'(t). \tag{1.50}$$

This is another key result: *the acceleration is the derivative of the velocity with respect to time*. Equations 1.48 and 1.50 together imply that

$$a(t) = \frac{d^2x}{dt^2} = x''(t). \tag{1.51}$$

The acceleration is also the *second derivative of the position with respect to time*. To summarize: the velocity describes the rate of change of the position of an object as a function of time; the acceleration describes the rate of change of the velocity of an object as a function of time. For instance, zero velocity means that the object is not moving; zero acceleration means that the object may still be moving, but with a constant velocity.

The SI units for velocity are meter over seconds: $\frac{m}{s}$, or m/s.

The SI units for acceleration are meter over seconds squared: $\frac{m}{s^2}$, or m/s².

1.4.2 Motion in Two or Three Dimensions

In two or three dimensions, the mean velocity between times t_1 and t_2 is:

$$\bar{\mathbf{v}} = \frac{\mathbf{r}(t_2) - \mathbf{r}(t_1)}{t_2 - t_1}. \tag{1.52}$$

Again setting $t_1 = t$ and $t_2 = t + \Delta t$, we quickly find

$$\mathbf{v}(t) = \lim_{\Delta t \to 0} \frac{\mathbf{r}(t + \Delta t) - \mathbf{r}(t)}{\Delta t}. \tag{1.53}$$

To see what this really means, it is convenient to express $\mathbf{r}(t)$ in terms of its components along $\hat{\mathbf{i}}$, $\hat{\mathbf{j}}$, and $\hat{\mathbf{k}}$, i.e.,

$$\mathbf{v}(t) = \lim_{\Delta t \to 0} \frac{(x(t + \Delta t) - x(t))\hat{\mathbf{i}} + (y(t + \Delta t) - y(t))\hat{\mathbf{j}} + (z(t + \Delta t) - z(t))\hat{\mathbf{k}}}{\Delta t}, \tag{1.54}$$

i.e.,

$$\mathbf{v}(t) = \lim_{\Delta t \to 0} \frac{(x(t + \Delta t) - x(t))}{\Delta t}\hat{\mathbf{i}} + \lim_{\Delta t \to 0} \frac{y(t + \Delta t) - y(t)}{\Delta t}\hat{\mathbf{j}} + \lim_{\Delta t \to 0} \frac{z(t + \Delta t) - z(t)}{\Delta t}\hat{\mathbf{k}}, \tag{1.55}$$

i.e.,

$$\mathbf{v}(t) = \frac{dx}{dt}\hat{\mathbf{i}} + \frac{dy}{dt}\hat{\mathbf{j}} + \frac{dz}{dt}\hat{\mathbf{k}} = v_x\hat{\mathbf{i}} + v_y\hat{\mathbf{j}} + v_z\hat{\mathbf{k}}, \tag{1.56}$$

where $dx/dt = v_x$ is the component of the (vector) velocity along the x-direction, $dy/dt = v_y$ is the component of the velocity along the y-direction, and $dz/dt = v_x$ is the component of the velocity along the z-direction.

Equation 1.56 provides a prescription for how to determine the velocity, given the position vector, and is another key result of this chapter. Concerning the connection between the acceleration and the velocity, we may similarly show that

$$\mathbf{a}(t) = \frac{dv_x}{dt}\hat{\mathbf{i}} + \frac{dv_y}{dt}\hat{\mathbf{j}} + \frac{dv_z}{dt}\hat{\mathbf{k}} = a_x\hat{\mathbf{i}} + a_y\hat{\mathbf{j}} + a_z\hat{\mathbf{k}}, \tag{1.57}$$

which is our fourth key result, in addition to our previous three key results, namely Eqs. 1.48, 1.51, and 1.56.

1.4.3   Example: Motion in a Circle at Constant Angular Velocity

Motion around a circular path at a constant rate, i.e., at a constant *angular velocity*, provides an instructive example of some of these ideas. Consider, then, a particle that moves around the circumference of a circle of radius r, lying in the xy-plane, at a constant angular velocity, ω. What is the velocity and acceleration of such a particle? Describe how the particle's acceleration is related to its position vector?

First, we figure out how to express the circular motion mathematically. To this end, suppose that at time t, the position vector of the particle makes an angle $\theta(t)$ with the x-axis. Then, the position vector of the particle at time t is:

$$\mathbf{r}(t) = r\cos\theta(t)\mathbf{i} + r\sin\theta(t)\mathbf{j}. \tag{1.58}$$

In order for the particle to go around at a constant rate, we must have $\theta(t) = \omega t$, where ω (Greek omega) is called the *angular velocity*. Thus we have:

$$\mathbf{r}(t) = r\cos\omega t\mathbf{i} + r\sin\omega t\mathbf{j}, \tag{1.59}$$

for motion around a circle in the xy plane at a constant rate.

Now what about the velocity? This is given by Eq. 1.55:

$$\mathbf{v}(t) = -r\omega\sin\omega t\hat{\mathbf{i}} + r\omega\cos\omega t\hat{\mathbf{j}}, \tag{1.60}$$

where we used:

$$\frac{d}{dt}\sin\omega t = \omega\cos\omega t \tag{1.61}$$

and

$$\frac{d}{dt}\cos\omega t = -\omega\sin\omega t. \tag{1.62}$$

Notice that using Eqs. 1.59 and 1.60, we can see that $\mathbf{r} \cdot \mathbf{v} = 0$, which immediately tells us that $\mathbf{v}(t)$ and $\mathbf{r}(t)$ are perpendicular for all times. By

(continued)

reflecting on it you will convince yourself that this is what we expect for motion in a circle, so that is good :)

Now what about the acceleration? Using Eq. 1.60, the acceleration is given by Eq. 1.57:

$$\mathbf{a}(t) = -r\omega^2 \cos \omega t \hat{\mathbf{i}} - r\omega^2 \sin \omega t \hat{\mathbf{j}} = -\omega^2 (r \cos \omega t \hat{\mathbf{i}} + r \sin \omega t \hat{\mathbf{j}}) = -\omega^2 \mathbf{r}(t).$$

(1.63)

Comparing Eqs. 1.59 and 1.63, we can see that these two vectors differ by a factor $-\omega^2$ which is a (negative) scalar. Therefore, these two vectors, $\mathbf{r}(t)$ and $\mathbf{a}(t)$ always point in exactly opposite directions!

1.4.4 Motion Subject to a Constant Acceleration in One Dimension

The special case of motion subject to constant acceleration includes the very important cases of motion near the Earth's surface subject to gravity only, and also motion of a charged particle in a constant electric field. In particular, the acceleration due to gravity near the Earth's surface is $9.8 \, \mathrm{ms}^{-2}$.

First, we consider the motion of a particle in 1D, along the x axis with constant acceleration a. What is the corresponding velocity as a function of time? What is the corresponding position as a function of time?

One very convenient way to figure this sort of thing out—and a method that we will use often in this book—is to hypothesize, or guess, a solution and then demonstrate that our solution works. Some folks are uncomfortable with this procedure. Somehow, they think that it is unscientific or unmathematical. But, in fact, this procedure is a perfectly valid route for finding the correct solution, and we will use it repeatedly. In fact, guessing or hypothesizing an answer is central to the scientific method, as eloquently described by Richard Feynman, who was awarded the 1965 Physics Nobel Prize, in

http://www.youtube.com/watch?v=b240PGCMwV0.[4]

Here, let us guess that the velocity is given by

$$v(t) = v_0 + at,$$

(1.64)

where v_0 is a constant, independent of t, and a is the constant acceleration. On the other hand, Eq. 1.51 informs us that

$$a(t) = \frac{d}{dt}(v_0 + at) = a,$$

(1.65)

---

[4] http://www.youtube.com/watch?v=b240PGCMwV0

which is the desired constant acceleration, and we have thus found the appropriate $v(t)$ in the case of 1D motion subject to constant acceleration.

Similarly, we guess that the position is given by

$$x(t) = x_0 + v_0 t + \frac{1}{2}at^2. \tag{1.66}$$

Here we may apply Eq. 1.48 with the result that

$$v(t) = \frac{d}{dt}\left(x_0 + v_0 t + \frac{1}{2}at^2\right) = v_0 + at, \tag{1.67}$$

which is just the result that we found for the velocity in the case of constant acceleration, implying that this solution for the position works too. Equations 1.64 and 1.66 represent two additional key results of this chapter.

Although Eqs. 1.64 and 1.66 represent the velocity and position of a particle subject to a constant acceleration a, we did not yet specify what are v_0 and x_0. To do this, let us specialize, Eq. 1.64 to $t = 0$, then Eq. 1.64 becomes $v(0) = v_0$ and we see that we must interpret v_0 as the value of the particle's velocity at $t = 0$, i.e., its initial velocity. Similarly, by considering Eq. 1.66 at $t = 0$, we discover that x_0 is the position of the particle at $t = 0$. If we know these *initial conditions*, the subsequent motion of the particle is completely defined.

1.4.5   Motion Subject to Constant Acceleration in Two and Three Dimensions

Generalization of the discussion of the last section to multiple dimensions is obtained by utilizing the appropriate corresponding vectors with the results that

$$\mathbf{v}(t) = \mathbf{v}_0 + \mathbf{a}t, \tag{1.68}$$

$$\mathbf{r}(t) = \mathbf{r}_0 + \mathbf{v}_0 t + \frac{1}{2}\mathbf{a}t^2, \tag{1.69}$$

for the velocity and position vector, respectively, where $\mathbf{a}$ is the (vector) acceleration, $\mathbf{r}_0$ is the initial position vector of the particle, and $\mathbf{v}_0$ is its initial velocity. The equations Eqs. 1.68 and 1.69 are the *kinematic equations for a motion at constant acceleration*. These equations, together with their 1D version Eqs. 1.64 and 1.66, are incredibly important. You will find yourself apply them more often than you think, since often times objects can be described by a constant acceleration. Be aware that these equations are valid *only* if the acceleration is constant, i.e., if it does not change in time.

1.4.6 Example: Trajectory of Projectile Motion

The trajectory of a particle is its path in space, e.g., for the circular motion example described above, it is a circle; more generally for a motion confined to the xz-plane, the trajectory is the behavior of the particle's z coordinate, $z(t)$, versus its x coordinate, $x(t)$. The goal of this example is to determine the trajectory of a particle near the surface of the Earth, where it is subject to a constant acceleration. $(0, 0, -g)$, after it is launched with initial velocity $\mathbf{v}_0 = (u, 0, w)$ from $\mathbf{r}_0 = (0, 0, 0)$.

(a) What is the particle's position vector at time t?

    Answer: Equation 1.69 informs us that

$$(x, y, z) = (0, 0, 0) + (ut, 0, wt) + \frac{1}{2}(0, 0, -g)t^2. \tag{1.70}$$

(b) Eliminate t to obtain z as a function of x.

    Answer: We have

$$x = ut, \tag{1.71}$$

and

$$z = wt - \frac{1}{2}gt^2. \tag{1.72}$$

    Equation 1.71 implies that $t = \frac{x}{u}$, which we can substitute into Eq. 1.72 with the result that

$$z - \frac{w}{u}x - \frac{g}{2u^2}x^2. \tag{1.73}$$

(c) What curve describes the trajectory?

    Answer: z is a quadratic function of x. This curve, namely the trajectory, is a parabola.

(d) Where does the particle return to the height from which it was launched?

    Answer: The launch height was $z = 0$. From the answer to (b), we have

(continued)

$$z = \frac{w}{u}x - \frac{g}{2u^2}x^2 = \frac{gx}{2u^2}\left(\frac{2uw}{g} - x\right). \qquad (1.74)$$

We can see that $z = 0$ for $x = 0$, which is the launch point, but also for $x = \frac{2uw}{g}$, which therefore is the location in x at which the particle returns to its launch height. Therefore we know the launch range is:

$$R = \frac{2uw}{g}.$$

(e) What is the maximum height that the particle achieves?

Answer: The particle achieves its maximum height at the halfway point, i.e., for $x = \frac{uw}{g}$, where the corresponding value of z is

$$z = \frac{uw}{g}\frac{g}{2u^2}\left(\frac{2uw}{g} - \frac{uw}{g}\right) = \frac{w^2}{2g}. \qquad (1.75)$$

Therefore the maximum height reached by the particle is:

$$h_{MAX} = \frac{w^2}{2g}. \qquad (1.76)$$

From this result, it is important to notice that the maximum height achieved is independent of the horizontal component of the initial velocity (u).

1.5   Frames of Reference and Relative Velocity

A frame of reference is a system of coordinates with respect to which we define the position of an object. Coordinate systems can be defined in any way we want, that is why there are infinite reference frames, therefore it is incredibly important to specify the one we use every time. For instance in Fig. 1.13 the position of the object at location P is represented by a vector $\mathbf{r}$ in the reference frame $S(x, y, z)$, while it is represented with a different vector $\mathbf{r}'$ according to the reference frame $S'(x', y', z')$.

Fig. 1.13 Two different reference frames, S and S', to represent the position of point P

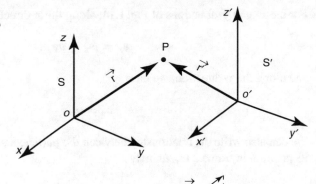

Fig. 1.14 Top: In frame 1, a particle moves with velocity, $\mathbf{v} = (v_x, v_y)$. Bottom: In frame 2, which is moving with velocity $\mathbf{v}_F = (v_F, 0)$ relative to frame 1, the particle moves with velocity $\mathbf{w}$. In this situation, $\mathbf{w}$ and $\mathbf{v}$ are related via $\mathbf{v} = \mathbf{w} + \mathbf{v}_F$ or equivalently $\mathbf{w} = \mathbf{v} - \mathbf{v}_F$

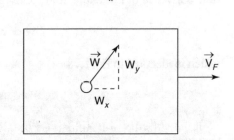

Some frames of reference may be moving with respect to each other, e.g., a train is a moving frame with respect to the frame of the train station, which is not moving.

Often, their motion depends on the velocity of two objects with respect to each other i.e. on their relative velocity. For example, the frictional force between two objects in contact depends on the relative velocity of the two objects, not on their velocities observed in an arbitrary "frame of reference" (or simply "frame"). To elucidate relative velocity, which is a nice application of vector addition and subtraction, we consider the situation shown in Fig. 1.14. In frame 1 (the laboratory frame), which has a velocity of zero, the particle has velocity $\mathbf{v} = (v_x, v_y, 0)$. In frame 2, which has a velocity $\mathbf{v}_F = (v_F, 0, 0)$, the same particle moves with velocity $\mathbf{w} = (w_x, w_y, 0)$. We pick the x-axis to be parallel to the relative velocity for convenience.

The relationship among these velocities is that

$$\mathbf{v} = \mathbf{w} + \mathbf{v}_F, \tag{1.77}$$

that is, we add $\mathbf{v}_F$ to $\mathbf{w}$ to get $\mathbf{v}$. It follows that:

$$\mathbf{w} = \mathbf{v} - \mathbf{v}_F. \tag{1.78}$$

For the x- and y-directions of Fig. 1.14, along the x-direction, we have

$$v_x = w_x + v_F, \tag{1.79}$$

and along the y-direction, we have

$$v_y = w_y. \tag{1.80}$$

We can also write the relationship between the particle's position in frame 1 ($\mathbf{r}$) and its position in frame 2 ($\mathbf{s}$), namely

$$\mathbf{s} = \mathbf{r} - \mathbf{v}_F t. \tag{1.81}$$

(We pick the position vectors in the two frames to coincide at $t = 0$.) Taking the time derivative of this expression yields

$$\mathbf{w} = \mathbf{v} - \mathbf{v}_F, \tag{1.82}$$

which is identical to Eq. 1.78.

1.5.1   Example: Relative Position

Object 1 has a constant velocity $(v_1, w_1) = v_1\hat{\mathbf{i}} + w_1\hat{\mathbf{j}}$ relative to a stationary observer, and object 2 has a constant velocity (v_2, w_2), relative to the same stationary observer. The acceleration is zero.

(a) What is the velocity of object 2 relative to a frame in which object 1 is stationary, i.e., what is the velocity of object 2 relative to object 1?

Answer:

$$(v_2 - v_1, w_2 - w_1). \tag{1.83}$$

(b) At $t = 0$, the position vector of object 1 is $(x_1, 0)$ and the position vector of object 2 is $(x_2, 0)$. What is the position vector of object 2 relative to object 1 at time t?

Answer:

$$(x_2 - x_1, 0) + (v_2 - v_1, w_2 - w_1, 0)t = (x_2 + v_2 t - x_1 - v_1 t, w_2 t - w_1 t,). \tag{1.84}$$

(continued)

(c) How far apart are object 1 and object 2 at time t?

Answer:

$$\sqrt{(x_2 + v_2t - x_1 - v_1t)^2 + (w_2t - w_1t)^2}. \tag{1.85}$$

1.6   Chapter Outlook

Many of the quantities that we will encounter in this book are vectors, including

- acceleration, which we will discuss further in Chap. 2. *Force and Momentum: Newton's Laws and How to Apply Them*,
- force, which we will also discuss in Chap. 2. *Force and Momentum: Newton's Laws and How to Apply Them*. Chapter 9. *Fluid Mechanics: Laminar Flow, Blushing, and Murray's Law*, Chap. 10. *Oscillations and Resonance*, and Chap. 11. *Wave Equations: Strings and Wind*.
- relative velocity, which we discussed in this chapter. Understanding relative velocity is essential for understanding frictional forces in Chap. 2. *Force and Momentum: Newton's Laws and How to Apply Them*,
- momentum, which we will also discuss in Chap. 2. *Force and Momentum: Newton's Laws and How to Apply Them*, and which plays a pivotal role in Chap. 9. *Fluid Mechanics: Laminar Flow, Blushing, and Murray's Law*,
- eigenvectors, which we will first encounter in Chap. 7. *Rates of Change: Drugs, Infections, and Weapons of Mass Destruction*, and which are also essential for Chap. 10. *Oscillations and Resonance*, and Chap. 11. *Waves: Equations Strings and Wind*, and
- electric and magnetic fields, which we will encounter in Chap. 12. *Gauss's Law: Charges and Electric Fields*, in Chap. 17. *Magnetic Fields and Ampere's Law*, in Chap. 18. *Faraday's Law and Electromagnetic Induction*, and in Chap. 19. *Maxwell's Equations and Then There Was Light*.

The material on vectors in this chapter is a prerequisite for all of these later chapters. The tensegrity vectors, which were a focus in this chapter, will be relevant in the next chapter, when we will introduce tension and compression forces.

Finally, the material in this chapter on rates of change foreshadows a discussion of how to model a number of interesting processes in Chap. 7. *Rates of Change: Drugs, Infections, and Weapons of Mass Destruction*, which often involves consideration of the rates of change of the quantities of interest.

1.7   Problems

Problem 1: Vectors
Consider the two vectors:

$$\mathbf{A} = A_x\mathbf{i} + A_y\mathbf{j} + A_z\mathbf{k} \tag{1.86}$$

and

$$\mathbf{B} = B_x\mathbf{i} + B_y\mathbf{j} + B_z\mathbf{k}, \tag{1.87}$$

where boldface indicates a vector quantity.

(a) Write the components of $\mathbf{C} = \mathbf{A} - \mathbf{B}$ in terms of the components of $\mathbf{A}$ and $\mathbf{B}$.
(b) For $(A_x, A_y, A_z) = (1, 5, 0)$ and $(B_x, B_y, B_z) = (4, 3, 0)$, calculate the components of $\mathbf{C}$.
(c) Given your answer to (b), find the components of the vector $\mathbf{D} = 3\mathbf{C}$.
(d) Calculate the scalar product of $\mathbf{A}$ and $\mathbf{B}$, namely $\mathbf{A} \cdot \mathbf{B}$.
(e) Calculate the scalar product of $3\mathbf{i} + 2\mathbf{j}$ and $-2\mathbf{i} + 4\mathbf{k}$, and determine the angle between these two vectors.
(f) Calculate the scalar product of $3\mathbf{i} + \mathbf{j} - 9\mathbf{k}$ and $2\mathbf{i} + 3\mathbf{j} + \mathbf{k}$, and determine the angle between these two vectors.
(g) Calculate the scalar product of $(1, 1, 1)$ and $(-1, -1, 1)$, and determine the angle between these two vectors.
(h) Calculate the scalar product of $a_x\mathbf{i} + a_y\mathbf{j}$ and $b_x\mathbf{i} + b_y\mathbf{j}$, and determine the angle between these two vectors.

Problem 2: Geometry of a Fourfold Rotationally Symmetric Tensegrity
A fourfold rotationally symmetric prismatic tensegrity is shown in Fig. 1.15. It consists of two equally sized squares of cables (red), one in the plane $z = 0$ and one in the plane $z = 2$, These two squares are rotated relative to each other by an angle α and are connected to each other by additional cables (also red) and struts (blue). The center of the $z = 0$-plane square is at the origin of the coordinate system, so that the position vectors of the vertices of the bottom square are

$$\mathbf{r}_1 = (1, 0, 0), \tag{1.88}$$

$$\mathbf{r}_2 = (\cos\frac{\pi}{2}, \sin\frac{\pi}{2}) = (0, 1, 0), \tag{1.89}$$

$$\mathbf{r}_3 = (\cos\pi, \sin\pi) = (-1, 0, 0), \tag{1.90}$$

$$\mathbf{r}_4 = (\cos\frac{3\pi}{2}, \sin\frac{3\pi}{2}) = (0, -1, 0). \tag{1.91}$$

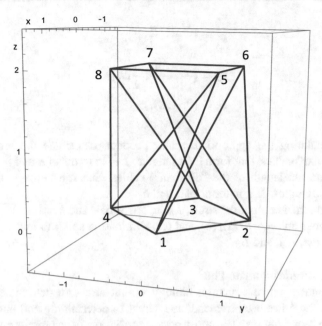

Fig. 1.15 Schematic of a fourfold rotationally symmetric prismatic "tensegrity," including a coordinate system

The goal of this problem is to analyze the geometry of this tensegrity.

(a) What are the position vectors of the vertices of the square in the $z = 2$-plane?
(b) What is the length, ℓ_{15}, of the cable that runs from vertex 1 to vertex 5?
(c) What is the length, ℓ_{16}, of the strut that runs from vertex 1 to vertex 6?
(d) What is the length, ℓ_{12}, of the cable that runs from vertex 1 to vertex 2?
(e) What is the angle, β, between the cable that runs from vertex 1 to vertex 5 and the cable that runs from vertex 1 to vertex 2?

Problem 3: Threefold Symmetric Tensegrity Revisited
A certain threefold prismatic tensegrity, similar to the tensegrity of Example 1.3.8, has

$$\mathbf{r}_1 = (a, 0, 0), \tag{1.92}$$

$$\mathbf{r}_2 = \left(-\frac{a}{2}, \frac{\sqrt{3}a}{2}, 0\right), \tag{1.93}$$

$$\mathbf{r}_3 = \left(-\frac{a}{2}, -\frac{\sqrt{3}a}{2}, 0\right), \tag{1.94}$$

$$\mathbf{r}_4 = \left(\frac{\sqrt{3}a}{2}, \frac{a}{2}, h\right), \tag{1.95}$$

$$\mathbf{r}_5 = \left(-\frac{\sqrt{3}a}{2}, \frac{a}{2}, h\right), \tag{1.96}$$

$$\mathbf{r}_6 = (0, -a, h). \tag{1.97}$$

(a) By calculating the appropriate scalar product, determine the length of the "triangular" cables, that form the triangles, L_T, in terms of a and h.
(b) Calculate the length of the "diagonal" cables, that run between the top and bottom triangles, L_D, in terms of a and h.
(c) Calculate the length of the struts, L_S, in terms of a and h.
(d) Using your answers to (a), (b), and (c), eliminate a and h to find a relationship among L_T, L_D, and L_S.

Problem 4: Ramachandran Plot[5]

A fundamental tenet of structural biology is that structure determines function. As a result, there has been tremendous interest in determining and understanding the structures of proteins. The most basic description of a protein is that it is a chain of amino acids, a.k.a. peptides. A portion of a peptide chain in a protein is shown schematically in Fig. 1.16. Chemistry classes tell us that properties of the electronic bonds among each O-C-N-Cα motif require that these atoms exhibit a planar configuration, highlighted by the yellow rectangles in Fig. 1.16. However, electronic bonding does not restrict the rotation of these rectangles about the N-Cα

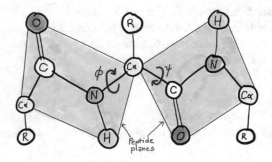
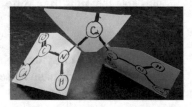

Fig. 1.16 Two representations of two neighboring peptide (amino acid) constituents of a protein, linked together at a so-called Cα carbon atom. Chemistry classes tell us that properties of the electronic bonds among O-C-N-Cα require that these atoms have a planar configuration within each peptide. However, electronic bonding does not restrict the rotation of neighboring peptide planes relative to each other. There are two angles, ϕ and ψ, which specify the relative rotation of neighboring peptides

---

[5] https://en.wikipedia.org/wiki/Ramachandran_plot

bond and about the Cα-C bond. As shown in the figure, these two rotations may be specified by giving the values of the two so-called dihedral angles, ϕ and ψ. The angle ϕ specifies the angle between the plane the yellow rectangle on the left and the plane defined by central Cα atom and its N-Cα bond on the left and its Cα-C bond on the right. The angle ψ specifies the angle between the plane of the yellow rectangle on the right and the plane of these N-Cα and Cα-C bonds.

In Fig. 1.17 shows a scatter plot of measurements of these angles, determined from actual protein structures in the Protein Data Bank.[6] Evidently, the measured dihedral angles overwhelmingly fall into three regions. Points that fall in the high-point-density region in the middle left of the plot correspond to a type of protein

Ramachandran Analysis of Protein Backbone Dihedral Angles

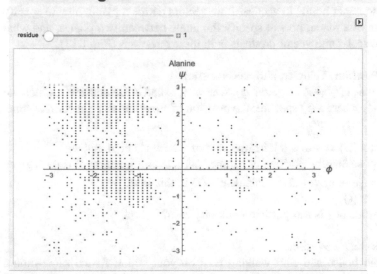

This Demonstration shows the distribution of backbone dihedral angles for each amino acid in a representative set of human proteins tabulated in *Mathematica* via the `ProteinData` command. Available conformations are restricted due to steric clashes of the side chains; Ramachandran analysis reveals that certain conformations are preferentially accessible to amino acids.

Contributed by: Daniel Barr

Fig. 1.17 Screenshot of the Wolfram Demonstration at http://demonstrations.wolfram.com/ RamachandranAnalysisOfProteinBackboneDihedralAngles/, showing the dihedral angles for alanine determined from experimentally measured protein structures (points). The possible angles and angle combinations are restricted by steric interferences. The demonstration can show the dihedral angles for any selected amino acid

---

[6] http://www.rcsb.org/pdb/home/home.do

secondary structure called α-helix. Points that fall in the high-point-density region at the top left in the plot correspond to a type of protein secondary structure called β-sheet. There are many fewer points in the third high-point-density region on the right of the figure. These structures correspond to a left-handed helical structure, which are uncommon, relative to α-helical structures, which are right-handed.

Based on the hypothesis that atoms in proteins behave like connected hard spheres, that cannot overlap each other, Ramachandran et al. [1] predicted the ranges of possible dihedral angles that could be realized in proteins. Their predictions, which were made in 1963 when only a few protein structures were known, conform well to the observed distribution. We may infer that protein secondary structure follows directly from the shape of the protein's constituent amino acids.

Given the definitions of ϕ and ψ given above, explain how to determine the values of ψ and ϕ for a protein, given the coordinates of its atomic constituents, which are available from the Protein Data Bank. Your explanation probably will include how to relate bonds to vectors, how vectors define a plane, how to specify the orientation of a plane, how to specify the angle between two planes, and what the role of the scalar and vector products is in these questions (Fig. 1.17).

Problem 5: Position, Velocity, and Acceleration

(a) Suppose that $x(t) = A\cos(\omega t + \phi)$, where A, ω and ϕ are constants. Calculate $x'(t) = \frac{dx}{dt}$, where $x(t)$ specifies the position of a particle as a function of time.

(b) Calculate $x''(t) = \frac{d^2x}{dt^2}$.

(c) Show that $x''(t) = -\omega^2 x(t)$, irrespective of the values of A and ϕ.

(d) Now suppose that the position of another particle as a function of time is given by $z(t) = \frac{D^2}{E} - 2Dt + Et^2$. Calculate $z'(t)$ in this case.

(e) Calculate $z''(t)$.

(f) At what value of t is this particle's velocity zero?

Problem 6: Relative Motion

You leave your house and start walking East on your street. At the same time a friend leaves a coffee shop, that is located a short distance East of your house, and he also starts walking East. You walk faster than your friend and eventually catch up with him. You both stop to chat for a little bit and then decide to walk back to the coffee shop together. Since you are chatting, both of you walk much slower now than before.

(a) Include on a single graph, a sketch representing the positions as a function of time of you and your friend, and your relative position, from the moment you left your house and he left the coffee shop, until the moment you reached the coffee shop together.

(b) Include on a second graph, the velocities as a function of time of you and your friend, and your relative velocity from the moment you left your house and he left the coffee shop, until the moment you reached the coffee shop together.

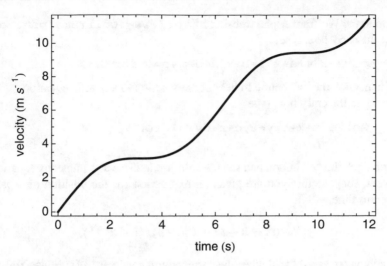

Fig. 1.18 Velocity versus time

Problem 7: On the River Bank
A boat, which can move at velocity V relative to the water, shuttles passengers between two locations on the same bank of a river. The distance between the two locations is d and the river flows at a velocity U. ($V > U$.)

(a) The boat starts off at Location 1, which is upstream of Location 2, so that the river flows from Location 1 toward Location 2. What is the velocity of the boat, relative to the river bank, as it travels from Location 1 to Location 2?
(b) How long does it take for the boat to travel from Location 1 to Location 2?
(c) What is the boat's velocity, relative to the river bank, on the way back from Location 2 to Location 1?
(d) How long does it take for the boat to travel from Location 2 back to Location 1?
(e) What is the total round trip time from Location 1 to Location 2 and back again? Does the flow of the river speed up or slow down the round trip time?

Problem 8: Position, Velocity, and Acceleration Revisited
An object moves along x as described by the velocity vs. time graph shown in Fig. 1.18. Sketch both the object's corresponding acceleration and position vs. time.

Problem 9: Wolfram Alpha
Wolfram Alpha facilitates mathematical manipulations of all sorts, including the solution of systems of algebraic equations, the evaluation of derivatives and integrals, the numerical solution of differential equations, and the creation of plots. As a web application, Wolfram Alpha is highly accessible, and its largely syntax-free input make it very easy to use successfully. This problem provides an introduction to Wolfram Alpha.

(a) Go to the Wolfram Alpha website: `http://www.wolframalpha.com/`[7]
(b) In the entry box, type:

```
solve x^4+a*x^3+b*x^2+c*x+d=0 for x
```

Then, click the "=" button to behold the solution to a quartic equation.
(c) Next, in the entry box, type

```
solve a*x+b*y=0,c*x+d*y=1 for x,y
```

and click the "=" button and see the solution to these simultaneous equations.
(d) Next, suppose that you are given an expression for the position of a particle versus time,

$$x(t) = b + ut + (w - u)\tau(1 - e^{-t/\tau}), \tag{1.98}$$

and you are asked to calculate the corresponding velocity, $v(t) = \frac{dx}{dt}$. You could carry out the derivative yourself. But type

```
d/dt(b+u*t+(w-u)*tau*(1-exp(-t/tau)))
```

and hit "=", and Wolfram Alpha will do it for you.
(e) Calculate the particle's acceleration, using Wolfram Alpha.
(f) Next, type

```
solve v'(t)=u/tau-v(t)/tau, v(0)=w
```

and hit "=".

In fact, v' is a shorthand for the derivative of v with respect to its argument, namely t, so actually you have just solved a differential equation.
(g) Briefly explain the meaning of this differential equation, assuming that v is the velocity of a particle, and that tau, u and w are constants.
(h) Finally, use Wolfram Alpha to plot your solution for tau $= 1$, u $= 1$, and w $= 2$ and for tau $= 3$, u $= 1$, and w $= 2$ from t $= 0$ to t $= 5$. Explain why the difference that you see between the two curves is sensible in view of how the value of tau is different in the two cases.

You should now be able to run Wolfram Alpha to facilitate all sorts of mathematics.

Problem 10: Wolfram Demonstrations
Computer visualizations and simulations can be extremely valuable in learning about abstract concepts in physics. Mathematica-based Wolfram Demonstrations are a convenient way to show visualizations and simulations, because they run directly

---

[7] http://www.wolframalpha.com/

in a web browser from anywhere you are studying. This problem introduces you to Wolfram Demonstrations.

- Check out the following movie,

  http://www.youtube.com/watch?v=G_nCBT9O0EI,[8]

  which shows the motion of a number of small Delrin beads—Delrin is a type of plastic—after each one is dropped into a beaker of canola oil. In this activity, you will exam the motion of one of these beads in more detail, using a Wolfram Demonstration.
- If you don't have access to a computer with Mathematica, you can install the CDF Player, located at

  http://www.wolfram.com/cdf-player/.[9]

- Next, download the following Wolfram Demonstration:

  FallingBead-Ch01.cdf

  Then, fire up the demonstration, showing a thin vertical slice from the movie.
- Hit the little + button to the right of the "frame number" bar to reveal the drop-down controls. Then, hit the-drop-down-control's triangular play button. You should see an animation of one of the falling beads. The demonstration tracks the bead's position and indicates where it thinks the bead is via the cross-hairs that you can see.
- Now, hit the "Toggle height" button only. This will "toggle on" an animated plot of the bead's height as a function of time. Be sure that you understand the connection between what you see in the movie and the animated plot. Explain in simple terms what you see happening as the bead moves through the fluid. (You may find it helpful to step through the movie.) Note the "units" that I used.
- Next, hit the "Toggle velocity" button, so that both the height plot and the velocity plot are visible. To create the velocity plot, the height at frame n was subtracted from the height at frame $n + 1$, and that difference was plotted at frame n.
- Note that the image in frame 13 is the same as the image in frame 12. This is because the camera missed the real image at frame 13, and so it "handed in" the previous frame. As a result, there is no apparent change in position between frames 12 and 13, but twice the usual change in position between frames 13 and 14. Please ignore this behavior.

---

[8] http://www.youtube.com/watch?v=G_nCBT9O0EI
[9] http://www.wolfram.com/cdf-player/

(a) Interpret and explain what you see in these plots. How is the slope of the height plot related to the velocity plot?
(b) Critique the procedure used to calculate the velocity.
(c) How would you calculate and plot the bead's acceleration as a function of time?
(d) Being able to read graphs is an important skill. Given that the functions encountered in Problem 9(d) mathematically describe the position versus time and the velocity versus time of the falling bead, *estimate* the experimental values of w, u, and τ. (Eq. 1.98 describes the bead's position versus time.)

You should now be able to run Wolfram Demonstrations.

Problem 11: Two Trains and a Bird
Two empty trains (1 and 2), each having a speed u are initially (i.e., $t = 0$) a distance d apart and are headed toward each other on the same track. A bird that can fly at a speed v (where $v > u$) flies from the front of train 1 directly toward train 2. On reaching train 2, the bird turns around and immediately flies back to train 1, and so forth. The goal of this problem is to calculate the total distance the bird flies before the trains reach each other.

(a) Calculate the time at which the trains crash.
(b) Calculate the total distance the bird flies before the trains crash.
(c) Sketch the position of the two trains and of the bird as a function of time.

Problem 12: Measuring g
A measurement of g can be made by throwing a ball upwards in an evacuated tube (to eliminate air resistance) past two level marks on the tube, and letting it return. (See Fig. 1.19.) Let the separation between the two level marks be h, and $t_4 - t_1$ and $t_3 - t_2$ be the time intervals between the two passages across the lower and upper level marks, respectively. Calculate g in terms of the measured quantities, namely h, $t_4 - t_1$, and $t_3 - t_2$.

Fig. 1.19 Sketch for "Measuring g"

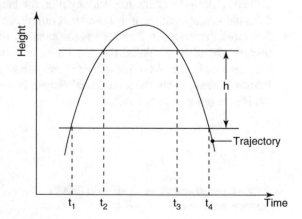

Problem 13: A Date on Michigan Avenue

Michigan Avenue is a major north-south street in Chicago. The famous Art Institute of Chicago is also along this avenue. You decide to meet a friend at the Art Institute, you both live on Michigan Avenue, but you live at a distance L south of the Art Institute, while your friend lives at a distance $2L$ north of the Art Institute. You both leave your own house at time $t = 0$ and walk with a speed v toward the Art Institute. (For your coordinate system, take an x-axis centered at the location of the Art Institute (i.e., the Art Institute is at $x = 0$) with the positive direction going North.)

(a) Include on a single graph, a sketch representing the positions as a function of time of: (i) you and (ii) your friend from time $t = 0$ to a time big enough so that you have both reached the Art Institute. Clearly label on your graph which one is your position $x_{you}(t)$ and which one your friend's position $x_{friend}(t)$.

On a separate graph sketch the relative position as a function of time of your friend as seen by you, from time $t = 0$ to a time big enough so that you have both reached the Art Institute. Label your graph with your friend's position relative to you with $x_{friend}^{(you)}(t)$.

(b) Include on a single graph, a sketch of the velocities as a function of time of: (i) you and (ii) your friend, from time $t = 0$ to a time big enough so that you have both reached the Art Institute. Clearly label on your graph which one is your velocity $v_{you}(t)$ and which one is your friend's velocity $v_{friend}(t)$.

On a separate graph sketch the relative velocity as a function of time of your friend as seen by you, from time $t = 0$ to a time big enough so that you have both reached the Art Institute. Label your graph with your friend's velocity relative to you with $v_{friend}^{(you)}(t)$.

Reference

1. G.N. Ramachandran, C. Ramakrishnan, V. Sasisekharan, Stereochemistry of polypeptide chain configurations. J. Molec. Biol. **7**, 95–99 (1963)

Force and Momentum: Newton's Laws and How to Apply Them

2

Nature and nature's laws lay hid in night; God said "Let Newton be and all was light."

A. Pope.[1]

We choose to go to the moon in this decade and do the other things, not because they are easy, but because they are hard, because that goal will serve to organize and measure the best of our energies and skills, because that challenge is one that we are willing to accept, one we are unwilling to postpone, and one which we intend to win, and the others, too.

President John F. 13 Kennedy, Speech at Rice University, September 12, 1962.[2]

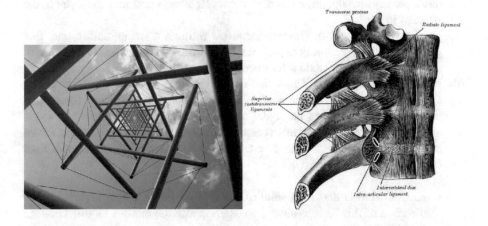

 Electronic Supplementary Material The online version of this article (https://doi.org/10.1007/978-3-031-05808-0_2) contains supplementary material, which is available to authorized users.

[1] http://en.wikipedia.org/wiki/Alexander_Pope

[2] https://www.youtube.com/watch?v=QXqlziZV63k

2.1   Introduction

Newton's laws, which are the subject of this chapter, represent a landmark in the history of science and philosophy in that they, for the first time, replaced *descriptions* of natural phenomena with *explanations in terms of cause and effect*. In particular, in the seventeenth century, Kepler's laws were able to describe planetary motion, but Newton's laws of motion and Newton's law of universal gravitation were able to explain planetary motion in terms of fundamental principles. Furthermore, Newton's laws are applicable far beyond planetary motion. They apply to all motions and lead to many testable predictions that agree extremely well with experiments. The goal of this chapter is for you to be able to apply the principles of classical mechanics. This material is foundational for all phenomena that involve motion, including all biologically and medically relevant phenomena that involve motion.

An essential aspect of Newton's laws is that any change in an object's velocity is the result of force. In this chapter, we will encounter a number of forces, including:

(a) Weight (W). $W = mg$, where m is the mass of an object, and g is the acceleration due to gravity at the Earth's surface. An object's weight corresponds to the attractive gravitational force between the Earth and the object in question.

(b) "Normal" force (N). This force is an example of a contact force that operates whenever two bodies come into contact. The normal force is the force that prevents two objects penetrating each other. Ultimately, the normal force originates in the atomic structure of matter and the electrostatic forces between electric charges. The "normal" force is called "normal" because "normal" can mean perpendicular, and in this case it does. It always acts perpendicular to the surfaces in contact.

(c) Tension (T). This is the force transmitted through a string, spring, etc. that permits the string or spring to pull. It too originates with the atomic structure of matter and the electrostatic forces between electric charges.

(d) Friction (f). Frictional forces are the forces that develop when an object slides on a surface, or would slide in the absence of frictional forces that prevent sliding, or when an object moves through a viscous liquid. Friction also originates with the atomic structure of matter and the electrostatic forces between electric charges. In this chapter, we will encounter three distinct frictional forces:

- Kinetic friction acts on a solid object that is sliding across a surface. The direction of this force is always so as to oppose the relative motion between the object and the surface.
- Static friction acts on a stationary object that is in contact with a surface and acts to oppose other forces and thus prevents the object from sliding across the surface.
- Fluid friction acts whenever an object is moving relative to a viscous fluid. The direction of fluid friction is also in the direction that opposes the relative motion. We will discuss the viscous frictional force exerted by a volume of fluid on a neighboring volume of fluid that has a different velocity, which

is critical for the human circulatory system, in Chap. 9. *Fluid Mechanics: Laminar Flow, Blushing, and Murray's Law.*

(e) The gravitational force between two masses (m_1 and m_2) separated by a distance r:

$$\mathbf{F} = -\frac{Gm_1m_2}{r^2}\hat{\mathbf{r}}, \tag{2.1}$$

where G is the gravitational constant, $\hat{\mathbf{r}}$ is the unit vector along the radial direction, from one mass to the other, and the minus sign indicates that the direction of the force is in the direction of decreasing r, which implies that gravity is an attractive force. In contrast to the normal force, friction, and tension, which are manifestations of the atomic structure of matter and electrostatic forces between charges, the gravitational force is said to be a fundamental force, because there is no underlying explanation of the gravitational force beyond Eq. 2.1. Near the surface of the Earth, Eq. 2.1 simplifies to the simpler case discussed above in (a). To see this simplification consider Eq. 2.1 for an object of mass m_1 on the surface of the Earth. In this case $m_2 = M_{\text{Earth}}$ and $r = R$ the radius of the Earth, calling $g = Gm_2/R^2$, we get to the weight of an object of mass m_1: $F = W = m_1 g$, which is intrinsically an attractive force, pulling downwards on any object on the surface of Earth.

(f) The electrostatic force on a charge, Q, in an electric field $\mathbf{E}$ is $\mathbf{F} = Q\mathbf{E}$. This is also a fundamental force. We will discuss electrostatic phenomenon in detail in *12 Gauss's law: Charges and electric fields* and *13 Electric potential, capacitors, and dielectrics.* Beyond gravitational and electromagnetic forces, two additional fundamental forces may be identified: the strong force and the weak force. However, the recent discovery of the Higgs boson[3] is compelling evidence that the electromagnetic and weak forces must be considered a single "electro-weak" force, that we now understand in detail. The strong force, or strong interactions, is/are primarily responsible for nuclear binding, which we will discuss in Chap. 3. *Energy, Work, Geckos, and ATP*, and therefore for the energy released in an atomic explosion, which we will discuss in Chap. 7. *Rates of Change: Drugs, Infections, and Weapons of Mass Destruction.*

This chapter also introduces the concept of momentum and the result that the rate of change of momentum of a collection of particles is equal to the net external force on the collection of particles. This material will be especially important in Chap. 9. *Fluid Mechanics: Laminar Flow, Blushing, and Murray's Law.* When there is no external force, the rate of change of the total momentum of a system of particles is zero. Thus, if there is no external force, the total momentum of a system of particles is constant, i.e., it is conserved.

In this chapter, we will also introduce a framework that we call the 7-step strategy, for always correctly solving problems that involve forces acting on a

---

[3] http://en.wikipedia.org/wiki/Higgs_boson

system. In this chapter, we will encounter a variety of situations (blocks, pulleys, strings, tensegrities, beads in fluids, DNA curtains, etc.). In every case, the 7-step strategy will enable us to correctly predict the motion of objects by carefully analyzing the forces acting on them and solving the appropriate equations of motion. Applying the 7-step strategy to a Newton's law problem guarantees a correct solution.

2.2   Your Learning Goals for This Chapter

By the end of this chapter you should be able to:

- Apply (via the 7-Step Strategy described below) Newton's laws to calculate the acceleration of objects of interest and/or other unknown quantities, and in particular:
  - Construct "force diagrams," properly including all Newton's Third Law forces.
  - Write the equations of motion, that is, Newton's Second Law, including resolving the components of vector quantities along coordinate axes.
  - Recognize "constraints" of the motion.
  - Solve the resultant equations in terms of the given quantities to find the desired unknown quantities.
- Solve Newton's-law problems involving: weight, normal forces, tension, pulleys, static and kinetic friction, fluid friction, and the gravitational force between masses.
- Interpret and use force-extension relationships, such as Hooke's law, the force-extension relation for rubber elasticity, etc.
- Determine the terminal velocity of an object moving through a viscous fluid under the influence of a constant force.
- Calculate the momentum of a particle and of a collection of particles.
- Apply Newton's second law expressed as the rate of change of the momentum of a system of particles equals the external force acting on that system.
- Explain the significance of the center of mass in the context of Newton's second law.
- Calculate the rate of change of momentum of a stream of particles hitting a surface, or of a fluid jet hitting a surface, and thus calculate the force on the surface.

2.3   Newton's Laws of Motion

Forces cause motion as prescribed by the following three Newton's laws.

1. If an object has zero total force on it, it has constant velocity.

2. $\mathbf{F} = m\mathbf{a}$ where $\mathbf{F} = \mathbf{F}_1 + \mathbf{F}_2 + \mathbf{F}_3 + \ldots = \Sigma_i \mathbf{F}_i$ is the total force on the object, equal to the vector sum of the individual constituent forces $\mathbf{F}_i$, $\mathbf{a}$ is the object's acceleration, and m is the mass of the object. Conceptually, Newton's Second Law specifies how the acceleration of an object responds to the total force on the object. From this point of view, it might be better written $\mathbf{a} = \frac{\mathbf{F}}{m}$.

3. If body b exerts a force $\mathbf{F}_a$ on body a, then body a exerts a force $\mathbf{F}_b$ on body b, where $\mathbf{F}_a = -\mathbf{F}_b$. The forces are equal in magnitudes: $|\mathbf{F}_a| = |\mathbf{F}_b|$, but opposite in direction.

It is important to recognize that Newton's First and Second Laws describes the behavior of an individual object, subject to a force, while Newton's Third Law describes how two objects interact. These laws hold in so-called inertial coordinate systems, that is, coordinate systems that are not accelerating. The Earth provides a reasonable approximation to an inertial coordinate system, but since the Earth is rotating about its axis and about the Sun, actually we are always accelerating a little. We will always take our coordinate systems to be inertial coordinate systems, but for motion on Earth this is an approximation, albeit usually a good one.

2.4 How to Apply Newton's Laws: 7-Step Strategy

We will encounter a class of problems in which it is necessary to calculate, the acceleration of one or more objects, or the tension in a piece of string, etc. To solve this class of problems, it is very helpful to follow a systematic procedure that invariably yields the correct answer. This procedure involves the following seven steps:

1. Make a sketch of your system.
2. Mentally partition the system into parts, each one of which can be treated as a "point mass."
3. Draw a force diagram for each point mass, representing each mass as a point, and drawing a force vector from the point, corresponding to each force acting on the mass. Only include forces acting directly on the mass in question and include Newton's third law forces properly, so that each force of a Newton's third law pair acts on a different mass in the opposite direction to its partner.
4. Introduce a coordinate system for each point mass.
5. Using your force diagrams as a guide, write the component force equations for each point mass, e.g., for an object of mass m, subject to two forces $\mathbf{F}_1$, $\mathbf{F}_2$, ... we have

$$F_{1x} + F_{2x} + \ldots = ma_x, \tag{2.2}$$

$$F_{1y} + F_{2y} + \ldots = ma_y, \tag{2.3}$$

in the x and y directions, respectively.

6. Look for any "constraints" of the motion in your problem and write the equations that express the constraints. Examples will show better what exactly is meant by a "constraint" of the motion. For instance, if an object is not moving, its acceleration is zero, so in this case the "constraint" is that $a = 0$. Instead if two objects are tied together via an inextensible string under tension, since the length of the string is fixed, the magnitude of the acceleration of the two objects must be equal to each other; the constraint in this case is that $a_1 = a_2$.

7. Substitute your constraint equations from step 6. into your equations in step 5. and then solve for the unknown quantities that the problem asks (for instance the acceleration). Make sure to give your answer in terms of the "given" quantities, i.e., the quantities that are known, as far as the problem is concerned.

We will call this the 7-Step Strategy. If you follow this strategy, you are guaranteed to achieve the correct answer to Newton's law problems. Therefore, you are urged to follow this procedure in all cases. We will now work a number of examples via this procedure.

2.5   Weight and Normal Forces

Weight is the gravitational force exerted by the Earth on a mass. A normal force is the contact force exerted by one object on a second that prevents the second object from penetrating the first. One meaning of "normal" is perpendicular, and this is the meaning here: normal forces are exerted between two objects in contact, directed perpendicular to their contact surfaces.

2.5.1   Example: Normal Forces Between Objects Sitting on a Table

An object (B) of mass m_B rests on a table. A second object (A) of mass m_A is sitting on top of B. Calculate the contact forces between A and B and between B and the table.

1. According to our procedure, we should make a sketch, which is shown in Fig. 2.1.
2. Then, we should mentally divide the system into parts, each of which can be treated as a point mass. In this case, object A and object B can each be treated as a point mass.
3. Next, we should draw a force diagram for each "point mass." To construct the force diagrams, we need to introduce the weight of each object, which is equal to the mass of the object multiplied by the acceleration due to gravity (g), and originates in the gravitational force exerted on the object by the Earth. We also need to introduce normal forces: a normal force

(continued)

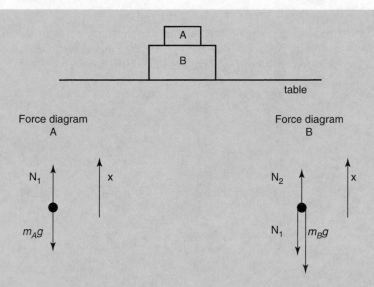

Fig. 2.1 Top: Sketch of the problem. Bottom: Force diagrams for object A and object B

upwards of magnitude N_1 is the force that object B exerts on object A. A normal force downwards of magnitude N_1 is the force that object A exerts on object B. A normal force upwards of magnitude N_2 is the force that the table exerts on object B. Object B is subject to compression, because N_1 downwards and N_2 downwards tend to compress it. The resultant force diagrams for each of the two masses, treated as point masses, are shown in Fig. 2.1.

4. Next, we should introduce a coordinate system. In this case, we only need to worry about the vertical direction. As also shown in the figure, we picked the positive x-direction to be upwards.

5. Using the force diagrams to guide us we can write the following two Newton's laws. For object A, we have

$$m_A a_A = -m_A g + N_1. \qquad (2.4)$$

For object B, we have

$$m_B a_B = -m_B g - N_1 + N_2. \qquad (2.5)$$

It is a direct consequence of Newton's third law that, if the normal force on object A from object B is N_1, then the normal force on object B from object A must be $-N_1$.

(continued)

6. The next step in our procedure is to apply "constraints" of the motion. In this case, we know that the two objects are at rest, therefore their acceleration must be zero, therefore the constraints are:

$$a_A = 0, \tag{2.6}$$

and

$$a_B = 0. \tag{2.7}$$

7. Then, we solve the problem. Using the constraints, Eqs. 2.4 and 2.5 become:

$$0 = -m_A g + N_1, \tag{2.8}$$

and

$$0 = -m_B g - N_1 + N_2. \tag{2.9}$$

Equation 2.8 implies that

$$N_1 = m_A g. \tag{2.10}$$

Then, using Eq. 2.8 in Eq. 2.9 yields

$$N_2 = m_B g + N_1 = (m_A + m_B)g. \tag{2.11}$$

Equations 2.10 and 2.11 are the solutions, we seek. Importantly, they are given in terms of "given" quantities, namely m_A, m_B, and g.

2.5.2 Example: The Normal Forces Between Three Masses Being Pushed Along

Three crates of masses m_1, m_2, and m_3 are touching each other in a line on a horizontal frictionless table as shown in Fig. 2.2, and they all have a horizontal acceleration a, as shown, as a result of an external force which acts on m_1 only.

(continued)

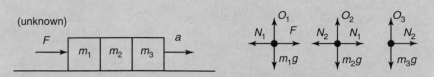

Fig. 2.2 Left: Sketch for the example of Sect. 2.5.2. Right: The corresponding force diagrams

(a) What are the "given quantities" in this problem?

Answer: The given quantities are: m_1, m_2, m_3, and a.

(b) Sketch a force diagram for each mass. Introduce and define the quantities/variables that you need to accomplish this part.

Answer: The force diagrams are shown on the right-hand side of Fig. 2.2. N_1 and N_2 are normal forces in the horizontal direction. O_1, O_2, and O_3 are normal forces in the vertical direction. In the vertical direction, since the vertical acceleration is zero, we have $O_1 = m_1g$, $O_2 = m_2g$, and $O_3 = m_3g$.

(c) Write the equations of motion in the horizontal direction, i.e., Newton's second law in the horizontal direction, for each mass.

Answer:

$$m_1a_1 = F - N_1, \qquad (2.12)$$

$$m_2a_2 = N_1 - N_2, \qquad (2.13)$$

and

$$m_3a_3 = N_2, \qquad (2.14)$$

where N_1 is the normal force between m_1 and m_2 and N_2 is the normal force between m_2 and m_3. Newton's third law demands that if the normal

(continued)

force on object 1 from object 2 is $-N_1$, then
the normal force on object 2 from object 1 must
be N_1, and if the normal force on object 2 from
object 3 is $-N_2$, then the normal force on object
2 from object 1 must be N_2.

(d) Write any constraints.

Answer: In this case we know the three creates
are touching each other and moving together
with the same acceleration a. Therefore, the
constraint of the motion is: $a_1 = a_2 = a_3 = a$.

(e) What external force (call it F) is needed to push all three masses so that
they accelerate with acceleration a?

Answer: We have

$$m_1 a = F - N_1, \tag{2.15}$$

$$m_2 a = N_1 - N_2, \tag{2.16}$$

and

$$m_3 a = N_2. \tag{2.17}$$

Adding these three equations together, we have

$$F = (m_1 + m_2 + m_3)a \tag{2.18}$$

which is also the answer.

(f) What is the force exerted by mass m_2 on mass m_3?

Answer: We are asked to find N_2, since a is
given, it is correct to express

$$N_2 = m_3 a. \tag{2.19}$$

(continued)

(g) What is the force exerted by mass m_1 on mass m_2?

Answer: Now we have to find N_1. Using

$$m_2 a = N_1 - N_2, \tag{2.20}$$

we have

$$N_1 = m_2 a + N_2 = (m_2 + m_3)a. \tag{2.21}$$

(h) What is the total force on m_2?

Answer: $F_2 = m_2 a$ by Newton's second law. Or $N_1 - N_2 = m_2 a$ the same, of course.

2.6 Newton's Universal Law of Gravitation

It is only appropriate to treat the gravitational force between two masses as a constant "weight," independent of height, when the height does not change appreciably. More generally, the gravitational force between two masses (m_1 and m_2) separated by a distance r is given by:

$$\mathbf{F} = -\frac{Gm_1 m_2}{r^2}\hat{\mathbf{r}}, \tag{2.22}$$

where G is the gravitational constant. This force is directed along the radial direction that means the line that connects the center of mass m_1 with the center of mass m_2; $\hat{\mathbf{r}}$ is the unit vector in the radial direction. The minus sign indicates that the direction of the force is in the direction of decreasing r, which implies that gravity is an attractive force. Equation 2.22 is Newton's Universal Law of Gravitation, and certainly represents a fundamental principle of Nature.

2.6.1 Example: Geostationary Orbits

To an excellent approximation, a satellite in orbit about the Earth rotates about the center of the Earth under the influence of this force. In this example, we analyze this motion in order to determine the radius, R, of a geostationary orbit, i.e., an orbit that remains over the same place on the Earth's surface

(continued)

all the time. Suppose that the Earth has mass m_1 and the satellite has mass m_2 with $m_1 \gg m_2$, and that the motion of the satellite is uniform circular motion. In Chap. 1. *Vectors and Kinematics*, we discussed motion in a circle of radius r about the origin at constant angular velocity ω, and we discovered in Eq. 1.63 that the acceleration and position vector are proportional but with a negative sign:

$$\mathbf{a} = -\omega^2 \mathbf{r}. \tag{2.23}$$

This result permits us to discuss uniform circular motion using Newton's laws. Specifically, we have for the acceleration in the radial direction that

$$a = -\omega^2 r, \tag{2.24}$$

where the negative sign indicates that the acceleration is opposite the direction of increasing r.

(a) Draw a sketch of this system and indicate on your sketch the coordinate system that you will use.

   The required sketch and coordinate system are shown in Fig. 2.3.

(b) What is the numerical value of the angular velocity of the satellite, ω, (in radians per second) for a geostationary orbit?

   Answer: 2π radians per day, i.e., $2\pi/24/60/60 = 7.3 \times 10^{-5}$ radians s^{-1}.

(c) What is the corresponding acceleration of the of the satellite in terms of ω?

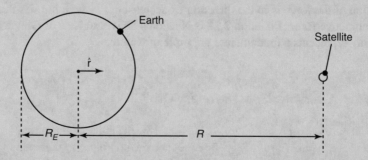

Fig. 2.3 Sketch for "Geostationary orbits"

(continued)

Answer:

$$- \omega^2 R \hat{\mathbf{r}}, \tag{2.25}$$

where $\hat{\mathbf{r}}$ is unit vector in the direction from the center of the Earth to the satellite.

(d) What is the force on the satellite?

Answer:

$$- \frac{Gm_1m_2}{R^2} \hat{\mathbf{r}}, \tag{2.26}$$

according to Newton's universal law of gravitation.

(e) Write Newton's Second Law as it applies to the satellite.

Answer:

$$- m_2\omega^2 R\hat{\mathbf{r}} = -\frac{Gm_1m_2}{R^2}\hat{\mathbf{r}}, \tag{2.27}$$

(f) Determine R for a geostationary orbit in terms of the Earth's radius (R_E).

Answer: Canceling $-m_2\hat{\mathbf{r}}$, we find

$$\omega^2 R = \frac{Gm_1}{R^2}, \tag{2.28}$$

i.e.,

$$R^3 = \frac{Gm_1}{\omega^2}. \tag{2.29}$$

It is convenient to express

$$g = \frac{Gm_1}{R_E^2}, \tag{2.30}$$

since $g = 9.8$ m s^{-2}. Therefore,

(continued)

$$\frac{R}{R_E} = \left(\frac{g}{R_E\omega^2}\right)^{\frac{1}{3}} = \left(\frac{9.8}{6.4 \times 10^6 \times (7.3 \times 10^{-5})^2}\right)^{\frac{1}{3}} = 6.6,$$

(2.31)

```
where we used R_E     =    6.4  ×  10^6 m. We see that
geostationary orbits are up at a radius of
R = 6.6R_E .
```

(g) Explain why geostationary orbits must be above the equator.

```
Answer: The satellite and the Earth must
have the same axis of rotation in order for
the satellite to remain geostationary. This
requirement places the orbit above the equator.
```

2.7   Tension and Compression

Tension is the force, exerted by a string, cable, or spring, that is itself being pulled. Tension opposes the external pulling force. The string, cable, or spring is then said to be "under tension." By contrast, when a rigid object is being squeezed, it is "under compression" and resists being compressed. Previously, we called this force, which resists compression, the "normal" force.

For many objects, there is a well-defined relationship between the length of the object (ℓ) and the tension. One important example is a linear relationship between the tension, T, and the length, ℓ:

$$T = k(\ell - L),$$

(2.32)

where L is the length of the object in equilibrium, i.e., in the absence of tension. The spring constant, k, specifies the stiffness of the object: the higher the value of k the stiffer is the object. Equation 2.32 is Hooke's Law, it is an example of a "force-versus-extension" relationship, which says that the tension (T) is linearly proportional to the change in length—the extension—of the object, $\ell - L$. We will discuss some of the consequences of Hooke's law in detail in Chap. 10. *Oscillations and Resonance*.

For small-enough extensions, almost every material/object obeys Hooke's law, which makes Hooke's law an important result. But it is important to remember that Hooke's law applies only for small-enough extensions. A more complete expression for the tension-extension relationship of a rubber band for example, which works over a wider range of extensions is

$$T = \frac{1}{3}k\left(\ell - \frac{L^3}{\ell^2}\right). \tag{2.33}$$

The factor of $\frac{1}{3}$ ensures that $T \simeq k(\ell - L)$ at the smallest extensions, in other words: in the limit of small extensions Eq. 2.33 becomes equal to Hooke's law Eq. 2.32.

2.7.1 Validity of Hooke's Law

(a) How can the statement that "almost every material/object follows Hooke's law" be correct? After all, in the rubber band and straw tensegrities shown in Fig. 2.6, is not the length of the straws fixed? Does not that immediately tell us that this statement is incorrect?

Answer: The statement is correct. For stiff
materials, such as the straws of Fig. 2.6 or
a table, k is necessarily large--that is what
"stiff" means--and therefore any change in
length is unnoticeably small, even for large
tensions or compressions.

(b) How would you demonstrate Eq. 2.32 from Eq. 2.33?

Answer: One way to do this is to use Wolfram
Alpha to plot both functions. (Here, for $\ell = 1$.)[4]

(c) For small enough extensions, Eq. 2.32 is correct both for $\ell > L$ (in which case we call the resultant force "tension"), as well as for $\ell < L$, in which case, the resultant force resists further compression. What is the connection between Hooke's law and the normal force?

Answer: A table, or straw, or any rigid object,
realizes Eq. 2.32, albeit with a large value
of k. The normal force is a manifestation
of Eq. 2.32 for large k and ℓ < L, resisting
compression. From this point of view, there
is not a major conceptual difference between
tension and the normal force. Nevertheless, it
is conventional to reserve "tension" to refer

---

[4]http://www.wolframalpha.com/input/?i=plot(x-1,1%2F3*(x-1%2Fx%5E2)+from+x%3D0.8+to+1.2

(continued)

to the force in cases that $\ell > L$ and the "normal
force" to refer to the force in cases that $\ell < L$.
Go figure! Just like everyone else, though, we
will respect this convention.

Only for sufficiently small compressional forces do long, thin objects (such as
straws) remain straight. At larger compressional forces, they buckle. We will
learn about buckling in Chap. 3. *Energy, Work, Geckos, and ATP*. (Rubber
bands buckle at very small compressional forces.)

Whenever a clinician palpates a patient or a woman carries out a breast self-
examination, they are effectively measuring stiffness—the ratio of the displacement
of the tissues in question to the force applied to those tissues. Figure 2.4 depicts
the elastic modulus of various tissues in the human body. Elastic modulus, E, is a
material property that depends only on the material and not on shape or size. But
the stiffness, k, of an object is set by its elastic modulus via $k = \frac{AE}{L}$, where A is its
cross-sectional area and L its length.

Within our bodies, each cell type develops and functions properly only in an
environment with the correct stiffness. The importance of environmental stiffness
for cell development is dramatically illustrated in experiments in which undifferen-
tiated stem cells were cultured in environments that differed only in stiffness [1].
In a soft environment, with a stiffness comparable to that of the brain, the stem
cells develop a phenotype characteristic of neurons; in matrices with the stiffness
of muscle, the stem cells develop a phenotype characteristic of myocytes (muscle
cells); in very stiff environments, the stem cells develop a phenotype characteristic
of osteocytes, which make bone.

2.7.2 Tensegrity Structures

In order to become more familiar with tension and compression, we will consider
a class of remarkable structures, called "tensegrities," which we already introduced
in the previous chapter (see for instance Example 1.3.8). A tensegrity (tensional
integrity) structure, or simply a tensegrity, consists of a number of compression
elements that are connected together by tension elements. According to the strictest
definition, a tensegrity's compression elements cannot contact each other, but the
tensegrity nevertheless is self-supporting and is more-or-less rigid under the action
of external forces. Interestingly, tensegrities may be a model for a number of
biological structures. In particular, as shown in Fig. 2.5, tensegrities may be a
valuable conceptual model for the human spine. At first sight, when we see a skeletal
human spine, we may be tempted to conceive of it as somehow analogous to a tower
of children's building blocks, with each vertebra supported by the one below. Such
a structure involves solely compression elements. However, you know that if you
try to tilt a tower of building blocks, it soon falls over. By contrast, the human spine

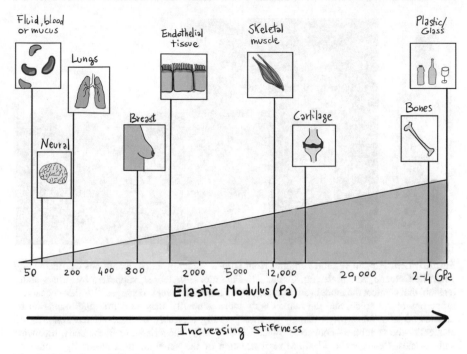

Fig. 2.4 Elastic modulus of various tissues within the human body

functions at any angle, and, moreover, is flexible. In the body, vertebrae are attached to each other by ligaments. This observation suggests that it may be appropriate to model the spine as a tensegrity structure, as shown in the center left panel of Fig. 2.4. This simplification is taken one step further in the center right panel of this figure, which shows a tensegrity structure containing stellated tetrahedra as its compression elements, each of which represents a vertebra. Finally, the monumental tensegrity depicted in the far right panel, which is Kenneth Snelson's *Needle Tower* takes the abstraction a step further. In this structure, the compression elements— the vertebrae—are simply struts, to which are attached cables under tension. This structure is actually comprised of multiple layers of a version of the simplest three-dimensional tensegrity, which is shown in Fig. 2.6.

The structure shown in Fig. 2.6 consists of three equal-length struts, shown in blue in the figure, that are under compression, and nine cables, shown in red in the figure, that are under tension. The bottom ends of the three struts form the vertices of one equilateral triangle. The top ends form the vertices of a second equilateral triangle (note that we already analyzed the geometry of this tensegrity using vectors in Example 1.3.8). Although these two triangles are the same size and have the same center, they are rotated relative to each other by an angle α. Importantly, this structure has three-fold rotational symmetry about the axis defined by the center of the two triangles. What do we mean when we say that the structure has three-fold rotational symmetry? It means that the structure is unchanged by rotations of integer multiples of $\frac{1}{3}$ of a complete turn about this axis, i.e., by rotations by integer

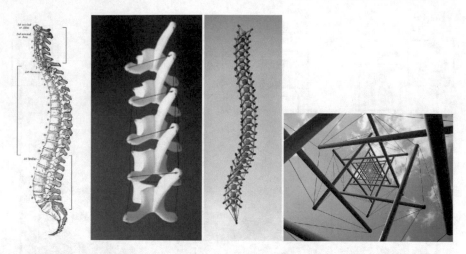

Fig. 2.5 Left: Human vertebral column from Gray's Anatomy https://commons.wikimedia.org/wiki/File:Colonne_vertebrale.png. Center left: Five model vertebrae, supported by strings under tension, that connect the model vertebrae together. This model seeks to suggest that the mechanical properties of the spine, and the human body more generally, may as a first approximation be understood as a "tensegrity," namely a self-supporting assembly of disconnected compression elements—the vertebrae—connected by tension elements—the strings, or in the body, ligaments and muscles. Center right: Idealized representation of the vertebrae as a tensegrity containing stellated tetrahedra as its compression elements and elastic strings as its tension elements. These three images, are copyright T. Flemons 2006, and are used with permission. Right: Another tensegrity structure that could be envisioned as a model for the spine, namely the *Needle Tower* https://en.wikipedia.org/wiki/Needle_Tower, created by Kenneth Snelson https://en.wikipedia.org/wiki/Kenneth_Snelson. One version of the needle tower is located at the Hirshhorn Museum and Sculpture Garden in Washington, DC. The photograph here shows the version at the Kroller-Muller Museum in the Netherlands. From https://commons.wikimedia.org/wiki/File:Kenneth_Snelson_Needle_Tower.JPG Licensed under the Creative Commons Attribution-Share Alike 2.5 Netherlands license

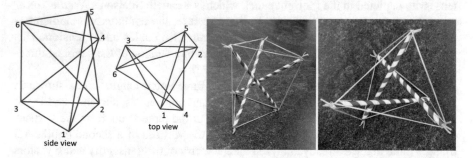

Fig. 2.6 The simplest three-dimensional tensegrity, consisting of three struts under compression (blue on left), and nine cables under tension (red on left). Left: Schematic representation of the tensegrity, showing our vertex labeling convention. Right: The tensegrity constructed using drinking straws, identical rubber bands, and paper clips

multiples of $\frac{2\pi}{3}$. As we will discover, the magnitude of α must have a particular value. However, α can be positive or negative, corresponding to a right-handed or left-handed tensegrity, respectively.

Beyond macroscopic skeletons, it has also been hypothesized that the cytoskeleton[5] forms a microscopic tensegrity [2], in which actin filaments[6] are cables under tension and microtubules[7] are struts under compression.

2.7.3 Example: Prismatic Tensegrity with Three-Fold Rotational Symmetry

To better understand how tensegrities can support themselves, here, we will seek to determine what is required for the "prismatic" tensegrity of Fig. 2.6 to be stable. This tensegrity has three-fold rotation symmetry about the axis that goes through the center of the two equilateral triangles at the top and bottom of the structure. The tension in the triangular cables, which connect vertices 1 and 2, 2 and 3, 3 and 1, 4 and 5, 5 and 6, and 6 and 4, is T; the tension in the diagonal cables, which connect vertices 1 and 4, 2 and 5, and 3 and 6, is D; and the tension in the struts, which connect vertices 1 and 5, 2 and 6, and 3 and 4, is S. In order for the cables to be under tension, T and D must be positive. In order for the struts to be in compression, S must be negative.

(a) Write an expression for the total vector force at vertex 1, $\mathbf{F}_1$, in terms of the tensions, T, D, and S, and the position vectors of vertices 1 through 5, namely $\mathbf{r}_1$, $\mathbf{r}_2$, $\mathbf{r}_3$, $\mathbf{r}_4$, and $\mathbf{r}_5$.

Answer: It is valuable to be guided by the procedure described above for solving Newton's law problems as far as possible. Figure 2.6 already constitutes a sketch of the system, which is the first step in this procedure. The second step is to partition the system into parts, each of which can be treated as a point mass. In this case, we will treat each vertex as a point mass. Third is to sketch a force diagram for each point mass. Since all six vertices are identical, it is only necessary to create the force diagram for one of them. Figure 2.7 shows the force diagram for vertex 1.

(continued)

Fig. 2.7 Force diagram for vertex 1. The vertex is pulled by the three cables and pushed by the strut

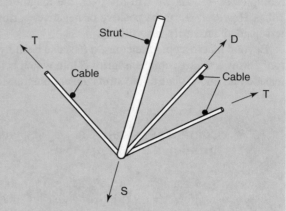

The problem asks for the total vector force at vertex 1, which equals the vector sum of the forces in the strut and three cables, that are attached to vertex 1. As shown in Fig. 2.7, the direction of each contribution is along the relevant cable or struct. Therefore, we can write:

$$\mathbf{F}_1 = T\hat{\mathbf{n}}_{21} + T\hat{\mathbf{n}}_{31} + D\hat{\mathbf{n}}_{41} + S\hat{\mathbf{n}}_{51}, \tag{2.34}$$

where

$$\hat{\mathbf{n}}_{21} = \frac{\mathbf{r}_2 - \mathbf{r}_1}{|\mathbf{r}_2 - \mathbf{r}_1|} \tag{2.35}$$

is a unit vector parallel to the cable from vertex 1 to vertex 2,

$$\hat{\mathbf{n}}_{31} = \frac{\mathbf{r}_3 - \mathbf{r}_1}{|\mathbf{r}_3 - \mathbf{r}_1|} \tag{2.36}$$

is a unit vector parallel to the cable from vertex 1 to vertex 3,

$$\hat{\mathbf{n}}_{41} = \frac{\mathbf{r}_4 - \mathbf{r}_1}{|\mathbf{r}_4 - \mathbf{r}_1|} \tag{2.37}$$

is a unit vector parallel to the cable from vertex 1 to vertex 4, and

(continued)

$$\hat{\mathbf{n}}_{51} = \frac{\mathbf{r}_5 - \mathbf{r}_1}{|\mathbf{r}_5 - \mathbf{r}_1|} \tag{2.38}$$

is a unit vector parallel to the strut from
vertex 1 to vertex 5.

(b) It turns out that it is convenient to express each force at vertex 1 in terms
of the tension in a cable/strut divided by the length of the cable/strut,
namely in terms of each cable's/strut's tension per unit length. The tension
per unit length in the triangular cables, which connect vertices 1 and 2,
2 and 3, 3 and 1, 4 and 5, 5 and 6, and 6 and 4, is t; the tension per unit
length in the diagonal cables, which connect vertices 1 and 4, 2 and 5, and
3 and 6, is d; and the tension per unit length in the struts, which connect
vertices 1 and 5, 2 and 6, and 3 and 4, is s. Express t, d, and s in terms of
T, D, and S and $\mathbf{r}_1$, $\mathbf{r}_2$, $\mathbf{r}_3$, $\mathbf{r}_4$, and $\mathbf{r}_5$. Thus, re-write your answer to (a)
in terms of t, d, and s and $\mathbf{r}_1$, $\mathbf{r}_2$, $\mathbf{r}_3$, $\mathbf{r}_4$, and $\mathbf{r}_5$.

Answer: The length of the triangular cables from
vertex 1 to vertex 2 is $L_T = |\mathbf{r}_2 - \mathbf{r}_1| = |\mathbf{r}_3 - \mathbf{r}_1|$;
the length of the diagonal cable from vertex 1
to vertex 4 is $L_D = |\mathbf{r}_4 - \mathbf{r}_1|$; and the length of the
strut from vertex 1 to vertex 5 is $L_S = |\mathbf{r}_5 - \mathbf{r}_1|$.
Therefore, the tensions per unit length are:

$$t = \frac{T}{L_T} = \frac{T}{|\mathbf{r}_2 - \mathbf{r}_1|} = \frac{T}{|\mathbf{r}_3 - \mathbf{r}_1|}, \tag{2.39}$$

$$d = \frac{D}{L_D} = \frac{D}{|\mathbf{r}_4 - \mathbf{r}_1|}, \tag{2.40}$$

and

$$s = \frac{S}{L_S} = \frac{S}{|\mathbf{r}_5 - \mathbf{r}_1|}, \tag{2.41}$$

and we can write:

$$\mathbf{F}_1 = t(\mathbf{r}_2 - \mathbf{r}_1) + t(\mathbf{r}_3 - \mathbf{r}_1) + d(\mathbf{r}_4 - \mathbf{r}_1) + s(\mathbf{r}_5 - \mathbf{r}_1), \tag{2.42}$$

which is the desired result.

(continued)

(c) If the tensegrity is stable, what is the value of the total force at each vertex of the structure? Explain why it is only necessary to consider the force at one vertex. What is the total force at vertex 1, namely $\mathbf{F}_1$?

```
Answer: In the context of our procedure, step
six is to look for any "constraints." In this
case, the "constraint" is that the acceleration
of each vertex is zero. It follows that the
total force at each vertex is zero. Moreover,
since all the vertices are the same, it is only
necessary to consider the force at one of them.
For example, F₁ = 0.
```

(d) Pick the origin of your coordinate system to be at the center of the bottom triangle with the z-axis vertically upwards, so that $\hat{\mathbf{k}}$ is the unit vector directed from the center of the lower triangle toward the center of the upper triangle, and $\hat{\mathbf{i}}$ is directed toward vertex 1. Then, use your answers to (b) and (c) to obtain an equation for d and s via $F_{1z} = 0$.

```
Answer: Steps four and five in our procedure
are to pick a coordinate system and write the
corresponding component force equations. As
instructed, we pick the origin of the coordinate
system to be at the center of the tensegrity's
lower equilateral triangle. We also pick vertex
1 to be located at x = a on the x axis. Thus,
```

$$\mathbf{r}_1 = (a, 0, 0). \tag{2.43}$$

```
From this point on, the coordinates of all of
the other vertices are fixed, and are the same
as given in Sec. 1.3.8 It follows that
```

$$\mathbf{r}_2 - \mathbf{r}_1 = \left(a \cos \frac{2\pi}{3} - a, a \sin \frac{2\pi}{3}, 0 \right), \tag{2.44}$$

$$\mathbf{r}_3 - \mathbf{r}_1 = \left(a \cos \left(-\frac{2\pi}{3} \right) - a, a \sin \left(-\frac{2\pi}{3} \right), 0 \right), \tag{2.45}$$

$$\mathbf{r}_4 - \mathbf{r}_1 = (a \cos \alpha - a, a \sin \alpha, h), \tag{2.46}$$

(continued)

$$\mathbf{r}_5 - \mathbf{r}_1 = \left(a \cos \left(\alpha + \frac{2\pi}{3} \right) - a, a \sin \left(\alpha + \frac{2\pi}{3} \right), h \right), \qquad (2.47)$$

and

$$\mathbf{r}_6 - \mathbf{r}_1 = \left(a \cos \left(\alpha - \frac{2\pi}{3} \right) - a, a \sin \left(\alpha - \frac{2\pi}{3} \right), h \right). \qquad (2.48)$$

Since $\mathbf{F}_1 = \mathbf{0}$, it must also be that $F_{1z} = \hat{\mathbf{k}} \cdot \mathbf{F}_1 = 0$. From these expressions for $\mathbf{r}_2 - \mathbf{r}_1$, $\mathbf{r}_3 - \mathbf{r}_1$, etc., it follows that

$$F_{1z} = 0 = t\hat{\mathbf{k}} \cdot (\mathbf{r}_2 - \mathbf{r}_1) + t\hat{\mathbf{k}} \cdot (\mathbf{r}_3 - \mathbf{r}_1) + d\hat{\mathbf{k}} \cdot (\mathbf{r}_4 - \mathbf{r}_1) + s\hat{\mathbf{k}} \cdot (\mathbf{r}_5 - \mathbf{r}_1), \qquad (2.49)$$

i.e.,

$$s + d = 0. \qquad (2.50)$$

(e) Use your answers to (b) and (c) to obtain a second equation for $t, d, s,$ and α via $F_{1x} = 0$.

Answer: In this case, we have that

$$F_{1x} = 0 = t\hat{\mathbf{i}} \cdot (\mathbf{r}_2 - \mathbf{r}_1) + t\hat{\mathbf{i}} \cdot (\mathbf{r}_3 - \mathbf{r}_1) + d\hat{\mathbf{i}} \cdot (\mathbf{r}_4 - \mathbf{r}_1) + s\hat{\mathbf{i}} \cdot (\mathbf{r}_5 - \mathbf{r}_1), \qquad (2.51)$$

i.e.,

$$-3t - d - s + d \cos \alpha + s \cos \left(\frac{2\pi}{3} + \alpha \right) = 0. \qquad (2.52)$$

(f) Use your answers to (b) and (c) to obtain a third equation for $t, d, s,$ and α via $F_{1y} = 0$.

Answer: Now, we have that

$$F_{1y} = 0 = t\hat{\mathbf{j}} \cdot (\mathbf{r}_2 - \mathbf{r}_1) + t\hat{\mathbf{j}} \cdot (\mathbf{r}_3 - \mathbf{r}_1) + d\hat{\mathbf{j}} \cdot (\mathbf{r}_4 - \mathbf{r}_1) + s\hat{\mathbf{j}} \cdot (\mathbf{r}_5 - \mathbf{r}_1), \qquad (2.53)$$

i.e.,

$$d \sin \alpha + s \sin(\frac{2\pi}{3} + \alpha) = 0. \qquad (2.54)$$

(continued)

(g) The answers to (d), (e), and (f) constitute three simultaneous equations for the three unknowns: $\frac{d}{t}$, $\frac{s}{t}$ and α. Solve these equations to find the values of $\frac{d}{t}$, $\frac{s}{t}$ and α in order that the tensegrity is stable.

Answer: Step seven in our procedure is to solve the problem. Equations 2.50, 2.52, and 2.54 can be solved using Wolfram Alpha[8] with the results that

$$\alpha = \frac{\pi}{6}, \tag{2.55}$$

$$\frac{d}{t} = \sqrt{3}, \tag{2.56}$$

and

$$\frac{s}{t} = -\sqrt{3}, \tag{2.57}$$

where we picked Wolfram Alpha's solution for which t and d are positive and s is negative, as required. These are the required conditions for the tensegrity of Fig. 2.6 to be stable. Remarkably, no matter what the dimensions and the tensions, the two triangles in this tensegrity are necessarily rotated relative to each other by $\frac{\pi}{6}$. In addition, the tension per unit length ratios must also be as given in Eqs. 2.56 and 2.57 in order for the structure to be stable. The negative sign in the expression for $\frac{s}{t}$ indicates that the tension in the strut is negative, i.e., that the struct is in compression, as required.

---

[8]http://www.wolframalpha.com/input/?i=Solve+%28s+%2B+d+%3D+0+%26+-3*t+-+d+-+s%2Bd*Cos%28a%29+%2B+s*Cos%282*pi%2F3%2Ba%29%3D0+%26+d*Sin%28a%29%2B+s*Sin%282*pi%2F3%2Ba%29%3D0+%29%2C+for+a%2C+s%2C+d

2.7.4 Pulleys

Tension also appears together with pulleys in a favorite class of Newton's law problems. Some of us struggle with this kind of problem every day at the gym, as suggested in Fig. 2.8 (groan).

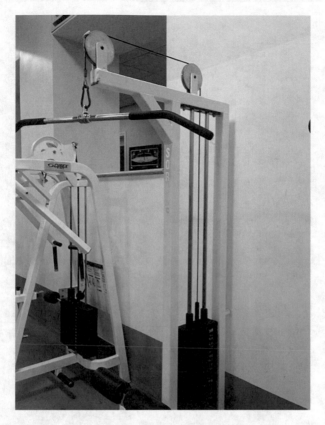

Fig. 2.8 A weight machine at the gym, comprising two pulleys with negligible mass and negligible friction, an effectively inextensible cable, and a mass. All that is needed is an external force, namely a person doing weight training

2.7.5 Example: Atwood's Machine

Figure 2.9 shows a simple, freely rotating pulley which supports a massless, inextensible string with masses m_1 and m_2 at each end. As one mass moves up the pulley rotates and the other mass moves down. This arrangement is often known as Atwood's[9] machine. Determine the accelerations of the masses.

---

[9]http://en.wikipedia.org/wiki/George_Atwood

(continued)

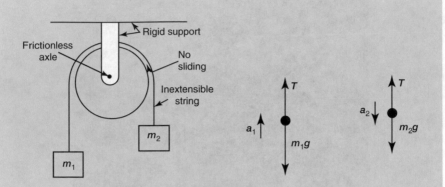

Fig. 2.9 Left: Two masses attached to one another by a *massless, inextensible* string that is drapped over a massless pulley. These assumptions—a massless, inextensible string and a massless pulley—permit us to only consider the two masses. Gravity acts downwards. There is no slippage between the string and the pulley, and there is no friction at the axle of the pulley. Right: The corresponding force diagrams and coordinate systems

To answer this question, we follow our procedure:

1. The left-hand side of Fig. 2.9 shows a sketch.
2. Each mass can be treated as a point mass.
3. The right-hand side of Fig. 2.9 shows the force diagrams.
4. The right-hand side of Fig. 2.9 also shows the coordinate system for each mass: a_1 is upwards and a_2 is downwards.
5. Using these force diagrams and coordinate systems, we have:

$$m_1 a_1 = T - m_1 g, \qquad (2.58)$$

and

$$m_2 a_2 = m_2 g - T, \qquad (2.59)$$

where T is the (as-yet-unknown) tension.
6. Writing the constraints, we have $a_1 = a_2 = a$, for a_1 upwards and a_2 downwards.
7. To solve these equations to determine the accelerations of the two masses, first, we add

$$m_1 a = T - m_1 g, \qquad (2.60)$$

(continued)

and

$$m_2 a = m_2 g - T, \tag{2.61}$$

with the result that

$$(m_1 + m_2)a = (m_2 - m_1)g. \tag{2.62}$$

Now, dividing by $m_1 + m_2$, we arrive at

$$a = \frac{m_2 - m_1}{m_1 + m_2}g, \tag{2.63}$$

which is the desired solution. If the two masses are equal, $m_1 = m_1$, the acceleration is zero, which is physically sensible. The solution is similarly sensible, when $m_1 > m_2$. In this case, the acceleration is negative, corresponding to mass m_1 falling and mass m_2 rising.

2.7.6 Example: Mechanical Advantage

Historically, pulleys have played an essential role in construction projects of all kinds from Roman aqueducts[10] to medieval cathedrals.[11] This is because of the mechanical advantage pulleys can provide—a suitable arrangement of pulleys can allow a modest force to be multiplied many fold—which enabled our forebears to lift massive blocks of stone to construct these sorts of monuments.

In this problem, we investigate mechanical advantage in the context of the arrangement of four pulleys shown at the left of Fig. 2.10. In this example, we consider and refer to this arrangement of four pulleys only. Specifically, we seek to determine what value of the weight $P = mg$ is necessary to hold weight $W = Mg$ stationary, and, more generally, what are the accelerations of the masses for a given m and M. We assume that all the pulleys are massless and the strings are massless and inextensible.

---

[10]https://en.wikipedia.org/wiki/Aqueduct_of_Segovia

[11]https://en.wikipedia.org/wiki/Chartres_Cathedral

(continued)

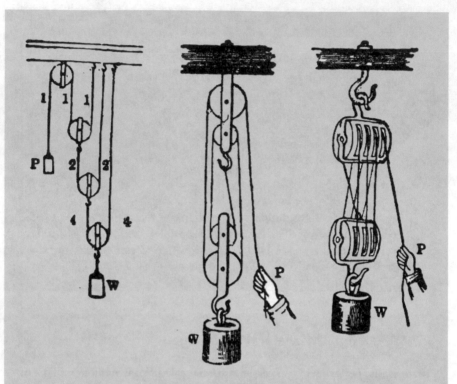

Fig. 2.10 A number of pulley arrangements that yield mechanical advantage. Public domain image from https://commons.wikimedia.org/wiki/File:NSRW_Block_and_tackle_arrangements.jpg

(a) Divide the pulley system into parts that can be treated as point masses.

> We pick mass m as the first point mass. We pick the second (center left) pulley as the second point mass. Even though this pulley is massless, it connects one string with another string. Therefore, applying Newton's second law to this pulley will establish the relationship between the tension in these two strings. Likewise, we pick the third (center right) pulley as the third point mass. We pick the fourth point mass to consist of mass M together with the fourth (rightmost) pulley, because these two objects necessarily move together.

(continued)

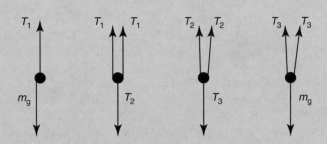

Fig. 2.11 Force diagrams for "Mechanical Advantage"

(b) Sketch the force diagrams for your point masses.

The force diagrams for these objects are shown in Fig. 2.11.

(c) Pick a coordinate system.

We chose the acceleration of both mass m and mass M, namely a_1 and a_4, respectively, to be positive upwards.

(d) Write Newton's second law for each point mass.

Using this coordinate system, we have

$$ma_1 = T_1 - mg, \tag{2.64}$$

$$0 = 2T_1 - T_2 \tag{2.65}$$

$$0 = 2T_2 - T_3 \tag{2.66}$$

and

$$Ma_4 = 2T_3 - Mg. \tag{2.67}$$

(e) Specify the constraints in this case.

When mass m moves up a certain distance, the center left pulley moves down one-half of that

(continued)

distance. It follows that $x_1 = -2x_2$ and therefore that $a_1 = -2a_2$. Similarly, when the center left pulley moves up a certain distance, the third, center right pulley moves up one-half that distance. It follows that $x_2 = 2x_3$ and therefore that $a_2 = 2a_3$. Similarly, when the center right pulley moves up a certain distance, the fourth, rightmost pulley moves up one-half that distance. It follows that $x_3 = 2x_4$ and therefore that $a_3 = 2a_4$. Eliminating a_2 and a_3, we find the constraint that we seek, namely $a_1 = -8a_4$

(f) Solve your collected equations to find a_1 and a_3 and how m and M are related in order for the accelerations to be zero.

Using $a_1 = -8a_4$ and $4T_1 = T_3$, we have

$$-8ma_4 = T_1 - mg, \tag{2.68}$$

and

$$Ma_4 = 8T_1 - Mg. \tag{2.69}$$

Subtracting 8 times the first equation from the second, we find

$$(M + 64m)a_4 = 8mg - Mg, \tag{2.70}$$

i.e.,

$$a_4 = -\frac{M - 8m}{M + 64m}g, \tag{2.71}$$

and

$$a_1 = \frac{8M - 64m}{M + 64m}g \tag{2.72}$$

which are the desired results. It follows that the accelerations are zero for $m = \frac{1}{8}M$. For $m > \frac{1}{8}M$, it is possible to raise the large

(continued)

mass M using a nearly eight times smaller
counterweight, m. This is mechanical advantage.

(g) By extrapolating from this arrangement, determine what is the minimum
    number of pulleys needed to be able to hold a mass $2^n m$, where n is an
    integer greater than 1?

    The first pulley here changes the direction of
    the force, but gives no mechanical advantage.
    So, here, four pulleys give a mechanical
    advantage of $8 = 2^3$. Extrapolating, we can expect
    n pulleys to give a mechanical advantage of 2^{n-1}.

2.8   Frictional Forces Between Solid Surfaces

Frictional forces come into play when two solid surfaces are in relative motion,
sliding passed each other, or are held stationary relative to each other by friction,
if the other forces in the problem are such that the two objects would slide past
each other in the absence of friction. How important frictional forces are in a given
situation is a property of the surfaces in contact. A simplified model for the frictional
forces between two solid objects in contact is as follows:

1. *Kinetic Friction.* If the objects are sliding past each other, then the magnitude
   of the frictional force is

$$F = \mu_K N, \qquad (2.73)$$

   where μ_K is the coefficient of kinetic friction, which is a property of the
   surfaces in contact, and N is the normal force. *The direction of the frictional
   force on an object is opposite to the direction of the velocity of the object relative
   to the velocity of the surface that is causing the friction.* Thus, if the surface
   originating the frictional force is stationary, the frictional force on the object is
   opposite to its velocity. However, if the velocity of the surface originating the
   frictional force is greater than the velocity of the object, then the direction of
   the frictional force is along the direction of the velocity of the surface causing
   the frictional force.
2. *Static Friction.* If the objects are not moving relative to each other, the
   magnitude of the frictional force (F) is such that

$$0 \le F \le \mu_S N, \qquad (2.74)$$

where μ_S is the coefficient of static friction, also a property of the surfaces, and N is the normal force between the two objects. It is generally true that sliding frictional forces are less or equal to the maximum-possible static (sticking) frictional forces, so that $\mu_S \geq \mu_K$.

Equations 2.74 and 2.73 represent an *approximate* description of how many solid surfaces in contact actually behave.

A material that is particularly well-known for its frictional properties is rosin,[12] which is used by pitchers in baseball, for example, to enhance their grip on the ball, and by violinists to improve the grip of the bow hair on the strings of the violin. Figure 2.12 shows the friction coefficients of rosin as a function of temperature, Evidently, for temperatures below about 30 °C, there is little difference between the coefficients of static (sticking) and kinetic (sliding) friction, which both have a value between 0.3 and 0.4. However, on raising the temperature, we see that between 30 and 40 °C, the static friction increases significantly, reaching about 0.7 by 40 °C and about 0.75 by 45 °C.

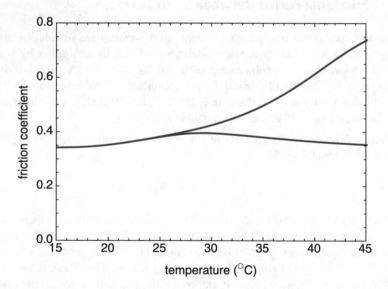

Fig. 2.12 Friction coefficients of violin rosin, measured by sliding a polished granite block down a temperature-controlled polished aluminum surface, coated with violin rosin. For temperatures below about 29 °C, rosin's coefficients of static and kinetic friction are the same, but, for temperatures above about 29 °C, rosin's coefficient of static friction (upper curve) is larger than its coefficient of kinetic friction (lower curve). Adapted from Ref. [3]

---

[12] http://en.wikipedia.org/wiki/Rosin

2.8.1   Example: Two Masses Pulled Along a Surface with Friction

As a first example problem involving friction, consider, as shown in Fig. 2.13, two blocks of mass m and $2m$, located on a flat plane, attached together via a massless, inextensible string. A given force F parallel to the surface is applied to the first mass only, which causes the masses to accelerate. The coefficient of kinetic friction between each mass and the surface is μ_K. The acceleration due to gravity is g, downwards. What is the acceleration of the two masses (a_1 and a_2 for mass m and $2m$, respectively) and what is the tension in the string between the two masses (T)?

(a) Draw two force diagrams, one for each mass, clearly showing all the relevant forces. (Call the normal force on m N_1, and the normal force on $2m$ N_2. Call the friction force on m f_1, and the friction force on $2m$ f_2.)

```
Answer: The required force diagrams are shown in
Fig. 2.14.
```

(b) Using a coordinate system, such that the x direction is positive, along the plane, parallel to F, apply Newton's Second Law to write *two* equations that describe the motion of the two masses in the x-direction.

```
Answer:
```

$$ma_1 = F - T - f_1 \tag{2.75}$$

```
and
```

$$2ma_2 = T - f_2. \tag{2.76}$$

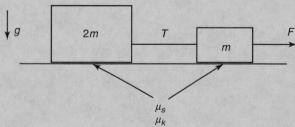

Fig. 2.13 Sketch for "Two masses pulled along a surface with friction"

(continued)

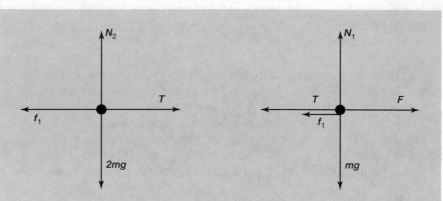

Fig. 2.14 Force diagrams for "Two masses pulled along a surface with friction"

(c) Given that the acceleration of each mass along the y direction—perpendicular to the plane—is zero, apply Newton's Second Law to write *two* equations that determine N_1 and N_2 in terms of the given quantities.

Answer:

$$0 = N_1 - mg \qquad (2.77)$$

and

$$0 = N_2 - 2mg. \qquad (2.78)$$

(d) How are a_1 and a_2 related?

Answer:

$$a_1 = a_2 = a, \qquad (2.79)$$

say.

(e) Write the relationships among f_1, f_2, N_1, and N_2 to solve your equations to find a_1, a_2, and T in terms of the given quantities, only.

Answer: For masses in motion, the desired relationships are:

$$f_1 = \mu_K N_1 \qquad (2.80)$$

(continued)

and

$$f_2 = \mu_K N_2. \tag{2.81}$$

Substituting these results into Eqs. 2.75 and 2.76, we have:

$$ma = F - T - \mu_K mg \tag{2.82}$$

and

$$2ma = T - 2\mu_K mg. \tag{2.83}$$

Adding these two equations together then yields

$$3ma = F - 3\mu_K mg, \tag{2.84}$$

i.e.,

$$a = \frac{F}{3m} - \mu_K g, \tag{2.85}$$

which is the desired solution for $a = a_1 = a_2$. Furthermore, using this expression for a in Eq. 2.83, we find

$$2m\left(\frac{F}{3m} - \mu_K g\right) = T - 2\mu_K mg. \tag{2.86}$$

The terms involving $\mu_K mg$ cancel, leaving

$$T = \frac{2}{3}F, \tag{2.87}$$

which is the desired expression for T.

2.8.2 Example: Stationary Block on an Inclined Plane with Friction

A block is on an inclined plane with friction, as sketched at the top of Fig. 2.15: A block of mass m is on a plane, which is inclined at an angle θ to the horizontal. The coefficient of static friction between the block and the plane is μ_S. The coefficient of kinetic friction between the block and the plane is μ_K. g is the acceleration due to gravity. The goal of this example is to show that an initially stationary block remains stationary for values of θ less than a threshold value and to find that threshold value of θ.

(a) Sketch the system of interest.

Answer: See the top of Fig. 2.15.

(b) Divide the system into parts, each of which can be treated as a point mass.

Answer: The block is the point mass in this problem.

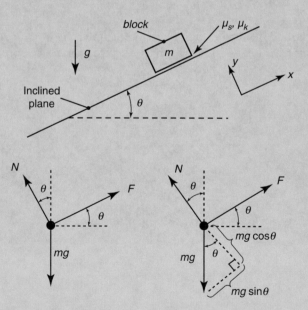

Fig. 2.15 Top: Sketch for "Stationary block on an inclined plane with friction" and "Moving block on an inclined plane with friction." Bottom left: The corresponding force diagram for the block. Bottom right: The relevant trigonometry

(continued)

(c) Draw a force diagram for each point mass.

    Answer: See the bottom of Fig. 2.15.

(d) Introduce a coordinate system.

    Answer: In this case, it is convenient to define the positive x direction to be parallel to inclined plane and to the frictional force, F, and the positive y direction parallel to the normal force, N, as shown in Fig. 2.15.

(e) Write the equations of motion.

    Answer: In the x-direction, we have

$$ma_x = F - mg \sin \theta. \tag{2.88}$$

    In the y-direction, we have

$$ma_y = N - mg \cos \theta. \tag{2.89}$$

(f) Write the equations that express the constraints of the motion.

    Answer: By assumption the block does not move, so the acceleration of the block along both x and y is zero:

$$a_x = 0 \tag{2.90}$$

    and

$$a_y = 0. \tag{2.91}$$

(g) Solve the problem.

    Answer: Equations 2.88, 2.89, 2.90, and 2.91 together imply that

$$N = mg \cos \theta \tag{2.92}$$

    and

(continued)

$$F = mg \sin \theta. \tag{2.93}$$

However, the static frictional force, F, must lie in the range

$$0 \leq F \leq \mu_S N. \tag{2.94}$$

Therefore, the *maximum* possible value of the static frictional force is $\mu_S N$.

On the other hand, Eq. 2.93 implies that the frictional force increases as θ increases, so that the frictional force exactly cancels the component of the gravitational along the plane, which tends to cause the block to accelerated down the slope. Then, the total force along the plane is then zero and the block is stationary.

With increasing θ, however, the calculated frictional force (Eq. 2.93) eventually becomes larger than $\mu_S N = \mu_S mg \cos \theta$. Beyond this point, Eq. 2.94 informs us that our calculation, which presumes a stationary block, becomes invalid and the block *must* slide. The condition that the required frictional force equals the maximum possible static frictional force is given by $mg \sin \theta = \mu_S mg \cos \theta$, that is, $\theta = \tan^{-1} \mu_S$. Therefore, we see that the block must slide if $\theta > \tan^{-1} \mu_S$, which is the desired solution.

2.8.3   Example: Moving Block on an Inclined Plane with Friction

Consider the situation shown in sketched at the top of Fig. 2.15: a block of mass m is on a plane, which is inclined at an angle θ to the horizontal. The coefficient of static friction between the block and the plane is μ_S. The coefficient of kinetic friction between the block and the plane is μ_K. g is the acceleration due to gravity. The goal of this example is to show that an initially moving block remains in motion for values of θ greater than a threshold value and to find that threshold value of θ.

(a) Sketch the system of interest, divide the system into parts, each of which can be treated as a point mass, draw a force diagram for each point mass, and introduce a coordinate system.

(continued)

Answer: The sketch, force diagram, and coordinate system are the same as in the previous example, and are shown Fig. 2.15.

(b) Write the equations of motion.

Answer: in the case of a moving block, Newton's laws lead to Eqs. 2.89 and 2.88, just as before.

(c) Write the equations that express the constraints of the motion.

In the case of a moving block, the constraint is

$$a_y = 0. \tag{2.95}$$

In addition, the frictional force is given by

$$F = \mu_K N, \tag{2.96}$$

which is how friction behaves when the surfaces are moving relative to each other.

(d) Solve the problem.

Using Eq. 2.95 and 2.96 in Eq. 2.89 and Eq. 2.88, we find that

$$ma_x = -mg(\sin\theta - \mu_K \cos\theta) = -mg\cos\theta(\tan\theta - \mu_K). \tag{2.97}$$

On physical grounds, the block cannot slide up the inclined plane. This observation implies that here a_x must be negative or zero. It follows from Eq. 2.97 that we must have that

$$\tan\theta \geq \mu_K. \tag{2.98}$$

When this condition is NOT satisfied, the premise of our calculation, namely that the block is moving, is invalid, i.e., it must be stationary.

2.8.4 Block on an Inclined Plane with Friction: Visualization of Results

It is instructive to visualize the results of the last two examples by plotting $\tan\theta$, μ_S, and μ_K versus θ, as shown in Fig. 2.16. In the case that the block does not slide, we found $F/N = \tan\theta$. But according to Eq. 2.74, the maximum possible valuable of F/N is μ_S. Both of these conditions are satisfied when the $\tan\theta$ curve is below the μ_S curve, implying that in the corresponding range of θ our calculation assuming a stationary block is valid, that is a stationary block is consistent with Newton's law. In the case that the block slides, we saw that our calculation was only sensible for $\tan\theta \geq \mu_K$, implying that only for $\tan\theta \geq \mu_K$ can the block slide. This implies that the range of values of θ in which the block slides is where the $\tan\theta$ curve is above the μ_K curve.

Since, in general, experiments show that $\mu_S \geq \mu_K$, we can now distinguish three regions of θ: for θ such that $\tan\theta \leq \mu_K$, only the solution for which the object is stationary is valid; for θ such that $\tan\theta \geq \mu_S$, only the solution for which the object is moving is valid; for θ such that $\mu_K \leq \tan\theta \leq \mu_S$, both solutions are valid. Therefore, within the range of inclination angles given by $\mu_K \leq \tan\theta \leq \mu_S$, we may expect that if the block is moving, it remains moving and if the block is stationary, it remains stationary. This behavior is just what is observed experimentally.

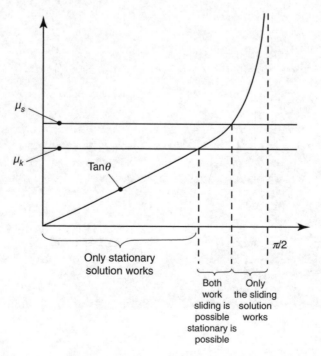

Fig. 2.16 $\tan\theta$, μ_S, and μ_K plotted versus θ. (μ_S and μ_K are both constants.) In the labeling of this plot, "works" means "is physically sensible"

2.9   Fluid Friction

When an object moves in a viscous fluid or when viscous fluid flows past an object, there is also a frictional force[13] on the object. In this case, our model for the frictional force is:

$$\mathbf{F}_f = -\zeta \mathbf{v}_R, \qquad (2.99)$$

where ζ is the friction coefficient of the object, and $\mathbf{v}_R$ is the *velocity of the object, relative to the fluid.* In terms of the velocity of the object, $\mathbf{v}$, and the velocity of the fluid flow, $\mathbf{v}_F$, $\mathbf{v}_R = \mathbf{v} - \mathbf{v}_F$. Equation 2.99 is believed to be appropriate at sufficiently low relative velocities. The sign in Eq. 2.99 is the same as for the case of friction between two solid objects sliding past each other, where, as noted previously, the direction of the frictional force on an object is opposite to the direction of the velocity of the object relative to the velocity of the contact surface causing the friction. The friction coefficient, ζ, depends both on the geometry and size of the object as well as the fluid viscosity, which is the fluid property that determines how easily a fluid flows. A thick, goopy fluid, such as canola oil or Karo syrup or honey, has a large viscosity. A thin liquid, which flows easily, such as water or gasoline (octane) or alcohol, has small viscosity. Viscosity is extremely important for an understanding of blood flow through the vascular system, and indeed for the physiology of the vascular system, as we will discuss in Chap. 9. *Fluid Mechanics: Laminar Flow, Blushing and Murray's Law.*

Here, we will examine two examples, in which fluid friction is important. In the first example, the object is stationary ($\mathbf{v} = \mathbf{0}$) and the fluid is flowing past the stationary object, which will be a strand of DNA. This implies that the velocity of the DNA relative to the fluid is $-\mathbf{v}_F$. Therefore, Eq. 2.99 implies that the force on the DNA is $+\zeta \mathbf{v}_F$, parallel to the flow velocity of the fluid. This force is comparable to the force that you feel when you stand in a fast flowing stream, when you feel a force parallel to the flow. In the second example, the object, which will be a sphere, is moving with velocity $\mathbf{v}$ under the influence of an external force through a stationary fluid ($\mathbf{v}_F = \mathbf{0}$). In this case, the velocity of the sphere relative to the fluid is simply $\mathbf{v}$, and therefore the force on the sphere is $= -\zeta \mathbf{v}$, which is in the opposite direction to the velocity of the sphere. This force is comparable to the force that you feel when you are swimming, when you feel a force opposite to the direction of your motion.

---

[13] http://en.wikipedia.org/wiki/Stokes%27_law

2.9.1　Example: "DNA Curtains"

DNA-protein interactions are tremendously important in biology. Eric Greene at Columbia University has developed an elegant experimental platform for examining DNA-protein interactions, called DNA curtains [4]. In brief, fluorescent DNA is functionalized with a reactive group at one end. In addition, a surface is patterned with lines of a complementary molecule that will react with the functionalized end of the DNA. When the surface is exposed to the functionalized DNA, the result is linear regions across the surface to which are attached DNA from its end. Next, a fluid is flowed past the DNA, stretching out the DNA, thus permitting the fluorescent DNA and any fluorescent proteins in solution that bind and interact with the DNA to be visualized via microscopy. In this way, it is possible to study these proteins' behavior on the DNA. This all becomes clearer, including the motivation for the name "DNA curtains" by looking at movies that show DNA curtains: `http://thegreenelab.cumc.columbia.edu/ECG%20Home.html`.[14]

However, the process of stretching out the DNA applies a tension to the DNA, and in order to be able to properly interpret experiments with DNA curtains, it is important to be aware of the possibility that the behavior may be different in a region where the tension in the DNA is low from a region where the tension in the DNA is large. Our goal in this example is to understand how the tension varies along the DNA, and thus place ourselves in a position to better interpret experiments carried out using DNA curtains.

We will represent the DNA as a string of beads, and may think of each bead as representing one base pair of the DNA. Fluid flows past the DNA at a constant velocity, exerting a frictional force on each bead in the direction of the flow, as discussed above. In addition to the frictional force exerted by the fluid on each element of the DNA, there are also the forces exerted by the neighboring elements of DNA, namely the tension in the connections between the beads. Our model is sketched in Fig. 2.17. Although this representation of DNA is simplified, it will nevertheless permit us to understand how the tension varies along the DNA. Force diagrams (free body diagrams) for the element of DNA at the free, unattached end (element 1) and for an interior element of DNA (element n) are shown at the bottom of Fig. 2.17. Following the 7-step Strategy from Section 2.4:

(a)　Sketch the system of interest.

```
Answer: See the top of Fig. 2.17.
```

---

[14]http://thegreenelab.cumc.columbia.edu/ECG%20Home.html

(continued)

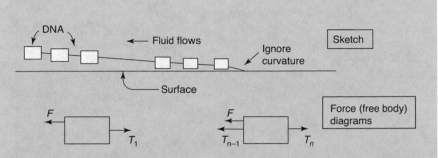

Fig. 2.17 Top: Our model for DNA curtains. Bottom: The corresponding force diagrams (free body diagrams) for the end element of DNA (element 1) and for an interior element of DNA (element n)

(b) Divide the "system" into parts, each of which can be treated as a point mass.

    Answer: Each DNA element - i.e., each bead - is a "point mass."

(c) Draw a force diagram for each "point mass."

    Answer: See the bottom of Fig. 2.17.

(d) Introduce a coordinate system.

    Answer: Positive x is parallel to the frictional force, i.e., horizontal and pointing to the left.

(e) Write the equations of motion.

    Answer: For bead n, we have

$$ma_n = F + T_{n-1} - T_n, \qquad (2.100)$$

    where F is the fluid frictional force on each bead, $-T_n$ is the tension exerted by bead $n + 1$ on bead n, and T_{n-1} is the tension exerted by bead $n - 1$ on bead n. For bead 1, at the free end of the DNA, we have

$$ma_1 = F - T_1. \qquad (2.101)$$

(continued)

(f) Write the equations that express the constraints.

Answer: The DNA is stationary, so the
acceleration of all beads is zero:

$$a_1 = 0.$$ (2.102)

and

$$a_n = 0.$$ (2.103)

(g) Determine how T_n depends on n.

Answer: Equations 2.100, 2.101, 2.102, and 2.103
together imply that

$$F + T_{n-1} - T_n = 0$$ (2.104)

and

$$F - T_1 = 0.$$ (2.105)

We immediately have

$$T_1 = F,$$ (2.106)

and we can rewrite Eq. 2.104 as:

$$T_n = T_{n-1} + F.$$ (2.107)

Let us look at T_2:

$$T_2 = T_1 + F.$$ (2.108)

But $T_1 = F$, so

$$T_2 = F + F = 2F.$$ (2.109)

We can now see that

$$T_n = nF.$$ (2.110)

(continued)

From this result we infer that the tension in the DNA increases linearly with n, that is, linearly with distance along the DNA from the free end.

2.9.2 Example: Sphere Moving Through a Viscous Fluid

In this example, we examine the motion of an object moving under the influence of a constant external force, such as gravity, through a stationary viscous fluid. The frictional force ($\mathbf{F}$) in this case is given by

$$\mathbf{F} = -\zeta \mathbf{v}, \tag{2.111}$$

where $\mathbf{v}$ is the velocity of the object, and ζ (Greek zeta) is the bead's friction coefficient.

An important feature of an object moving through a viscous fluid under the influence of a constant force is that the object eventually reaches a "terminal velocity" corresponding to the velocity at which the frictional force on the object is exactly equal in magnitude and opposite in direction to the external force(s) on the object. As a result, the object's velocity is constant at the terminal velocity. In this example, our goal is to determine the terminal velocity of a spherical bead falling under the influence of gravity in a viscous fluid in terms of given quantities.

A movie of a number of plastic beads, or more precisely polyoxymethylene[15] beads, falling one after another through canola oil,[16] is shown at

http://www.youtube.com/v/G_nCBT9OOEI.[17]

By examining the movie, you can see that after each bead enters the canola oil, its velocity rapidly decreases to a value that subsequently appears to be constant thereafter. Figure 2.18 shows a screenshot from the movie together with the tracked position of a bead as a function of time. At the start of the track, the velocity is larger but, subsequently, throughout the region highlighted in blue, the bead's velocity is constant, corresponding to its terminal velocity.

---

[15]http://en.wikipedia.org/wiki/Polyoxymethylene

[16]http://en.wikipedia.org/wiki/Canola

[17]http://www.youtube.com/v/G_nCBT9OOEI

(continued)

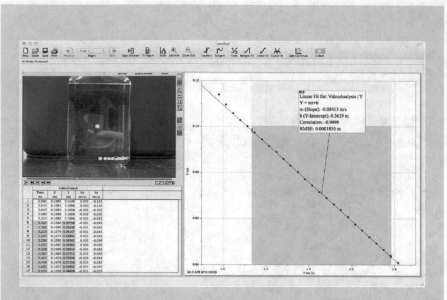

Fig. 2.18 Delrin beads being dropped and then falling through canola oil under the influence of gravity, together with a plot of the position versus time for one bead, determined by tracking the bead's position. Courtesy of Dr. Sidney Cahn

Following the 7-step Strategy from Sect. 2.4:

(a) Sketch the system of interest.

In this case, the movie and Fig. 2.18 provide the required visual representation.

(b) Divide the "system" into parts, each of which can be treated as a point mass.

The bead is the "point mass" in this problem.

(c) Draw a force diagram for each "point mass."

Figure 2.19 shows the force diagram with the three forces acting on the bead: the weight of the bead (mg) downward, the frictional force on the bead (ζv) upwards, and then the buoyant force on the bead $(m_F g)$ acting upwards, where m_F is the mass of the fluid displaced by the

(continued)

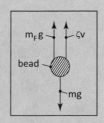

Fig. 2.19 Force diagram for a bead falling under the influence of gravity through a viscous fluid

bead. The buoyant force is un upward force due to the difference in pressure between the fluid just below the bead and the fluid just above the bead, and it is equal to the weight of the mass of fluid displaced by the object $(m_F g)$. We will talk more about this force in Chap. 9. *Fluid Mechanics: Laminar Flow, Blushing and Murray's Law* for now we want to give an intuitive understanding for this force.
To see why the buoyant force is equal to the weight of the fluid displaced by the bead $(m_F g)$, consider the fluid that is to-be-displaced by the bead, before the bead is there. There is no total force on that fluid, or else the fluid would accelerate. But the weight of the fluid is acting down on it. Therefore, there must be an equal and opposite force acting upwards, exerted by the surrounding fluid, that renders the total force on to-be-displaced fluid equal to zero. Now switch out the fluid for the bead. The weight is switched, but the force exerted by the rest of the fluid will be the same. This is the buoyant force, equal and opposite to the weight of the displaced fluid.

(d) Introduce a coordinate system.

We pick the positive x direction to be pointing vertically downwards, parallel to the motion of the bead. Therefore, the acceleration due to gravity is parallel to the positive x direction,

(continued)

and the velocity of the particle is along the positive x direction.

(e) Write the equations of motion.

We have

$$Ma = (m - m_F)g - \zeta v, \qquad (2.112)$$

where a is the bead's acceleration along x, M is the bead's effective inertial mass (that means the mass that it actual has when moving through the fluid, we will say more about this mass at the end of the next example), and v is the bead's velocity along x.

(f) Write the equations that express any constraints.

There is no specific constraint in this case. However, it will be convenient to represent the acceleration as the rate of change of velocity:

$$a = \frac{dv}{dt} = v'. \qquad (2.113)$$

(g) Solve the problem.

Eqs. 2.112 and 2.113 together imply that

$$M\frac{dv}{dt} = (m - m_F)g - \zeta v. \qquad (2.114)$$

To discuss the consequences of Eq. 2.114, for simplicity, we suppose that the bead is released from rest, so that initially the velocity of the bead is zero. It follows from Eq. 2.114, that the bead's acceleration is positive (assuming $m > m_F$). As a result, the bead's velocity becomes positive. This decreases the

(continued)

acceleration, but for a sufficiently small
velocity the acceleration remains positive, so
that the bead's velocity continues to increase.
Eventually, however, the velocity increases to
the point that the right-hand side of Eq. 2.114
becomes zero. At this point, the acceleration
also becomes zero, and, thereafter, the velocity
is constant. The value at which the velocity
is constant is the " terminal velocity." We can
find the terminal velocity (u) by setting the
left-hand side of Eq. 2.114 to zero, and setting
$v = u$. Thus, we find:

$$0 = (m - m_F)g - \zeta u. \tag{2.115}$$

It follows that the terminal velocity is

$$u = \frac{(m - m_F)g}{\zeta}, \tag{2.116}$$

which is the desired result.

Although we have considered a bead moving in a viscous fluid, in fact,
the physics of gel electrophoresis[18] of proteins[19] and DNA,[20] which are
workhorse techniques in biochemical and molecular biological research, may
be described similarly.

---

[18]http://en.wikipedia.org/wiki/Agarose_gel_electrophoresis

[19]http://en.wikipedia.org/wiki/SDS_PAGE

[20]http://en.wikipedia.org/wiki/DNA_electrophoresis

2.9.3 Example: Relaxation Time

Previously, we found the terminal velocity of a bead moving under the
influence of a constant force in a viscous fluid, which corresponds to the
behavior of the bead at long-times, after its acceleration has become zero. In
this example, our goal is to figure out what happens at short times. Therefore,
we need to find a solution to Eq. 2.114 that is valid for arbitrary times.

(continued)

To this end, it is first convenient to re-write Eq. 2.114 in terms of the terminal velocity, u, and the "relaxation time," $\tau = \frac{M}{\zeta}$:

$$\frac{dv}{dt} = \frac{u}{\tau} - \frac{v}{\tau}.$$ (2.117)

(a) Does the "relaxation time," τ, have the dimensions of time?

Answer: The dimensions of ζ are the dimensions of force over velocity, which are the dimensions of mass times acceleration over velocity. Since the dimensions of acceleration over velocity is inverse time. the dimensions of ζ are mass over time. Since τ is mass divided by ζ, we indeed see that it indeed has the dimensions of time.

(b) Use Wolfram Alpha to find the solution to Eq. 2.117, given that the initial velocity at time $t = 0$ is $v(0) = w$.

As we saw in Chap. 1. *Vectors and Kinematics*, Wolfram Alpha helps us to find the solution to Eq. 2.117 by first going to http://www.wolframalpha.com/,[21] then typing

    solve v'(t)=u/tau-v(t)/tau, v(0)=w

and finally hitting "=".[22] Thus, the desired solution is

$$v(t) = u + (w - u)e^{-t/\tau}.$$ (2.118)

Equation 2.118 shows that starting from $v(0) = w$, $v(t)$ approaches u exponentially with the characteristic time scale for the exponential approach, given by the "relaxation time," equal to $\tau = \frac{M}{\zeta}$.

---

[21] http://www.wolframalpha.com

[22] http://www.wolframalpha.com/input/?i=solve+v%27%28t%29%3Du%2Ftau-v%28t%29%2Ftau%2C+v%280%29%3Dw

(continued)

(c) Use Wolfram Alpha to plot your solution for $\tau = 1$, $w = 0$, and $u = 1$ and $\tau = 3$, $w = 0$, and $u = 1$.

```
Answer: The desired plot for τ = 1, w = 0, and u =
1 is at
```

```
    http://www.wolframalpha.com/input/?i=plot++1-
    exp%28-t%29+from+t%3D0+to+9.²³
```

```
For τ = 3, w = 0, and u = 1, the plot is at
```

```
    http://www.wolframalpha.com/input/?i=plot++1-exp%28-
    t%2F3%29+from+t%3D0+to+9.²⁴
```

```
It is apparent from these plots that the
relaxation time quantifies how long the bead
takes to reach its terminal velocity.
```

There is one final "wrinkle," demanded by full disclosure, namely that the value of M on the left-hand side of Eq. 2.112 is not the mass of the bead alone, which is m. Instead, advanced fluid mechanics calculations reveal that $M = m + \frac{1}{2}m_F$, where m_F is the mass of the displaced fluid. The additional mass term,[25] $\frac{1}{2}m_F$, may be thought of as the mass of fluid that is accelerated along with the mass of the bead.

---

[23] http://www.wolframalpha.com/input/?i=plot++1-exp%28-t%29+from+t%3D0+to+9

[24] http://www.wolframalpha.com/input/?i=plot++1-exp%28-t%2F3%29+from+t%3D0+to+9

[25] http://en.wikipedia.org/wiki/Added_mass

2.10 Newton's Second Law in Terms of Total Momentum

So far, we have implicitly assumed point-like masses. To understand the role of force on composite objects, built from a number of point masses held together by internal bonds, it is essential to introduce "momentum." The momentum ($\mathbf{p}$) of a particle of a mass m and velocity $\mathbf{v}$ is:

$$\mathbf{p} = m\mathbf{v}. \tag{2.119}$$

For such a particle, we have

$$\frac{d\mathbf{p}}{dt} = \frac{d}{dt}(m\mathbf{v}) = m\frac{d\mathbf{v}}{dt} = m\mathbf{a} = \mathbf{F}, \tag{2.120}$$

where we used Newton's Second Law in the last step, and we assumed that the particle's mass is constant. It will turn out that $\mathbf{p}$ becomes especially useful and significant when we need to deal with more than one particle. Let us consider a system consisting of N *interacting* particles. It does not matter what the interaction forces are.

Their masses are $m_1, m_2, ... m_N$.

Their position vectors are $\mathbf{r}_1, \mathbf{r}_2, ... \mathbf{r}_N$.

Their momenta are $\mathbf{p}_1, \mathbf{p}_2, ... \mathbf{p}_N$.

The equation of motion of particle n is (Eq. 2.120):

$$\frac{d\mathbf{p}_n}{dt} = \mathbf{F}_n, \tag{2.121}$$

where $\mathbf{F}_n$ is the total force acting on particle n.

Now, $\mathbf{F}_n$ can be separated into two parts: a part that comes from interactions among the particles and a part that comes from forces external to the system:

$$\mathbf{F}_n = \mathbf{F}_n(interactions) + \mathbf{F}_n(external) = \mathbf{F}_n(int) + \mathbf{F}_n(ext) \tag{2.122}$$

in an obvious notation. It follows that

$$\frac{d\mathbf{p}_n}{dt} = \mathbf{F}_n(int) + \mathbf{F}_n(ext). \tag{2.123}$$

Now, let us add together all of these equations, i.e.,

$$\frac{d\mathbf{p}_1}{dt} + \frac{d\mathbf{p}_2}{dt} + ... + \frac{d\mathbf{p}_N}{dt} = \mathbf{F}_1(int) + \mathbf{F}_1(ext) + \mathbf{F}_2(int) + \mathbf{F}_2(ext)$$
$$+ ..... + \mathbf{F}_N(int) + \mathbf{F}_N(ext), \tag{2.124}$$

i.e.,

$$\frac{d}{dt}\Sigma_{n=1}^{N}\mathbf{p}_n = \Sigma_{n=1}^{N}\mathbf{F}_n(int) + \Sigma_{n=1}^{N}\mathbf{F}_n(ext), \tag{2.125}$$

where $\Sigma_{n=1}^{N}$ (Greek capital sigma) means the sum over all terms from $n = 1$ to $n = N$. To decide whether a force is an external force or an internal force, we must first specify precisely what comprises the system. For example, for the "DNA curtains" of Sect. 2.9.1, it is natural to define the system to consist of all of the beads. Then, the external forces acting on the beads comprise the frictional force of each bead and the tension in the connection between the bead, that is attached to the surface, and the surface. All of the other tensions, which act between beads, are forces internal to the system.

Importantly, however, the sum of all the internal forces is zero:

$$\Sigma_{n=1}^{N}\mathbf{F}_n(int) = 0. \tag{2.126}$$

This result follows from Newton's Third Law: Because in Eq. 2.126, we are summing the force from particle n_1 on particle n_2 and the force from particle n_2 on particle n_1. These two forces are equal in magnitude and opposite in direction, according to Newton's Third Law. It follows that their sum is zero, and that is true for all interaction force pairs.

Therefore, introducing the total momentum of the system of particles, namely

$$\mathbf{p}(total) = \Sigma_{n=1}^{N} \mathbf{p}_n = \Sigma_{n=1}^{N} m_n \mathbf{v}_n, \tag{2.127}$$

and the total *external* force, namely

$$\mathbf{F}(ext) = \Sigma_{n=1}^{N} \mathbf{F}_n(ext), \tag{2.128}$$

we have

$$\frac{d\mathbf{p}(total)}{dt} = \mathbf{F}(ext), \tag{2.129}$$

i.e., the rate of change of the total momentum of a system of particles is equal to the total external force on the system of particles.

2.10.1 Conservation of Momentum

If the external force is zero [$\mathbf{F}(ext) = 0$] it follows that $\frac{d\mathbf{p}(total)}{dt} = 0$, and therefore that $\mathbf{p}(total)$ is constant. This is the law of conservation of momentum. Notice that momentum is a vector, each component of which can be conserved, if there is no external force along that direction. For example, consider the collision between blocks on a frictionless table shown in Fig. 2.20. In the collision, the blocks certainly exert horizontal forces on each other, but there are no external forces in the horizontal direction. Because there are no external forces in the horizontal direction, momentum is conserved in this direction. It follows that the initial momentum of the system in the horizontal ($p_i = m_1 v_1 + m_2 v_2$) is equal to the final momentum of the system in the horizontal ($p_f = m_1 w_1 + m_2 w_2$), i.e.,

$$m_1 v_1 + m_2 v_2 = m_1 w_1 + m_2 w_2. \tag{2.130}$$

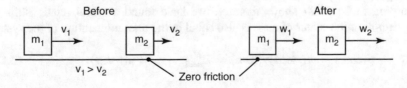

Fig. 2.20 Two blocks before and after their collision on a frictionless table

2.10.2 Center of Mass

The center of mass (COM) is an important concept because it simplifies the description of the motion of complicated systems that are either extended object, i.e., not point like, or a collection of multiple masses. As we will see in this section, the most important result regarding the center of mass is contained in Eq. 2.137 below.

The center of mass of a system is a special position found by the weighted average of the positions of all point masses, or all parts of the system. For instance, for a collection of masses $m_1, m_2, \ldots, m_N$, the center of mass is the position vector:

$$\mathbf{r}(com) = \frac{1}{m(total)} \Sigma_{n=1}^{N} m_n \mathbf{r}_n, \qquad (2.131)$$

where

$$m(total) = \Sigma_{n=1}^{N} m_n \qquad (2.132)$$

is the total mass of the system. This equation defines the center of mass of a collection of masses. To determine the center of mass of an extended object with a continuous mass distribution, we replace the sum by an integral:

$$\mathbf{r}(com) = \frac{1}{m(total)} \int_V \mathbf{r}\rho(\mathbf{r}) dx dy dz, \qquad (2.133)$$

where $\rho(\mathbf{r})$ is the volume mass density of the object at $\mathbf{r}$. In the cases that we will consider, the density will be uniform so that $\rho(\mathbf{r}) = \rho$, a constant.

Using the definition of velocity, we then define the velocity of the center of mass as:

$$\mathbf{v}(com) = \frac{d\mathbf{r}(com)}{dt} = \frac{1}{m(total)} \Sigma_{n=1}^{N} m_n \frac{d\mathbf{r}_n}{dt} = \frac{1}{m(total)} \Sigma_{n=1}^{N} m_n \mathbf{v}_n.$$
$$(2.134)$$

Rearranging this equation by moving the total mass on the left side we have:

$$m(total)\mathbf{v}(com) = \Sigma_{n=1}^{N} m_n \mathbf{v}_n. \qquad (2.135)$$

The right-hand side of this equation is the total momentum of the system $\mathbf{p}(total)$ according to Eq. 2.127 above, therefore we have found a useful relationship, the momentum of the center of mass is also equal to the total momentum of the system:

$$\mathbf{p}(com) = m(total)\mathbf{v}(com) = \Sigma_{n=1}^{N} m_n \mathbf{v}_n = \mathbf{p}(total). \qquad (2.136)$$

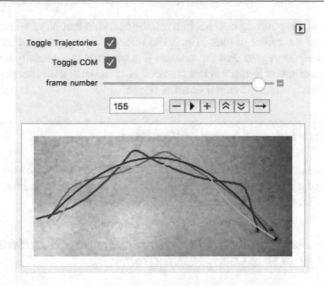

Fig. 2.21 Snapshot of the Wolfram Demonstration at ThreeMasses-Ch02.cdf, showing the trajectories of three masses, that are tied together (orange, cyan, magenta), and the trajectory of their COM (red)

Importantly, Eq. 2.136 informs us that in a frame of reference in which the center of mass of a system of objects is stationary, the total momentum of that system of objects is zero.

What is the significance of the center of mass? If we combine Eqs. 2.129 and 2.136, we find that

$$\mathbf{F}(ext) = m(total)\mathbf{a}(com).\tag{2.137}$$

Equation 2.137 informs us that the center of mass obeys Newton's second Law, where "$\mathbf{F}$" is the *net external force*, "m" is the *total mass*, and "$\mathbf{a}$" is the acceleration of the center of mass. This result provides the justification for treating extended objects, e.g., blocks, as point particles, which we did earlier in this chapter.

The utility of the COM is emphasized in Fig. 2.21, which is a snapshot of the Wolfram Demonstration

```
ThreeMasses-Ch02.cdf,
```

and which shows the individual trajectories of three masses that are tied together, in orange, cyan, and magenta, together with the trajectory of their COM, shown in red. Even though the individual trajectories appear erratic, the trajectory of the COM is clearly a simple parabola, as expected for 2D motion subject to a constant vertical acceleration.

It is also worth pointing out that in a reference frame that has a velocity equal to the velocity of the center of mass, Eq. 2.136 informs us that the total momentum in that frame is equal to zero. For this reason, it is sometimes especially convenient to carry out calculations in a reference frame that is moving with the velocity of the center of mass, namely the center of mass frame.

2.11 Force from a Stream of Particles

We will now encounter representative examples of a class of problems in which a flow/stream of material stopping or reversing direction gives rise to a force, by virtue of Newton's Second Law expressed in the form $d\mathbf{p}/dt = \mathbf{F}$.

2.11.1 General Strategy for Problems Involving a Flow/Stream of Material That Changes Direction

A general strategy for solving these sorts of problems is as follows:

(1) Calculate the mass of material that hits the wall or changes direction in a time Δt.
(2) Calculate the change in velocity of that mass of material. (Change in velocity equals the final velocity minus the initial velocity.)
(3) Then, the change in the momentum of the flow/stream in a time Δt is equal to the product of the mass in (1) multiplied by the change in velocity in (2).
(4) The result obtained in (3) is the change in momentum (Δp) of the stream of material in a time Δt. The external force (F) on the stream is therefore $F = \Delta p / \Delta t$.
(5) Therefore, the force exerted by the stream on whatever is causing the stream to change direction is $-\Delta p / \Delta t$, via Newton's third law.

2.11.2 Example: Force on a Wall from a Stream of Particles

The goal of this example is to determine the mean force on a wall exerted by a stream of particles directed toward the wall, each of mass m, each of velocity v toward the wall, each separated one from another by a distance L, and each of which bounces back from the wall with velocity $-v$. We pick the coordinate system so that the positive x direction is horizontal and to the right, so that the particles' initial velocity is toward the positive x direction. Figure 2.22 shows a sketch of the problem. Evidently, each particle's momentum changes when it bounces off the wall. Therefore, we may expect a time-averaged force on the wall, because of Eq. 2.129.

(continued)

Fig. 2.22 Particles hitting a wall and bouncing back

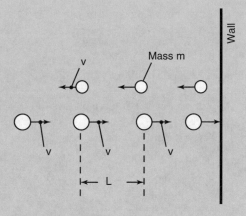

(a) Given that the separation between particles is L, calculate the number of particles that hit the wall in a time Δt.

Answer: In a time Δt, all particles within a distance $v\Delta t$ of the wall and going toward the wall will hit the wall. The number of particles within this length is $\frac{v\Delta t}{L}$, all of which hit the wall in Δt.

(b) What total mass of particles hits the wall in a time Δt?

Each particles has a mass m. Therefore a mass equal to $\frac{mv\Delta t}{L}$ hits the wall in Δt.

(c) What is the change in velocity of each particle caused by hitting the wall.

Answer: Each particle's velocity changes by $-2v$.

(d) What is the total change in momentum of the particle stream in a time Δt (call it Δp)?

Answer: The total change in momentum of the particle stream in Δt is

$$\Delta p = -2v\frac{mv\Delta t}{L} = -\frac{2mv^2}{L}\Delta t. \qquad (2.138)$$

(e) What is the mean force on the stream of particles?

(continued)

Answer: According to Eq. 2.129, the mean force on the stream of particles is therefore

$$F = \frac{\Delta p}{\Delta t} = -\frac{2mv^2}{L}.$$ (2.139)

(f) What is the mean force on the wall?

Answer: The mean force on the wall is therefore:

$$F(wall) = -F = \frac{2mv^2}{L},$$ (2.140)

via Newton's third law.

2.11.3 Pressure of a Gas

We can take the last example a little further by supposing that there is one stream of particles for each area A of the wall. It follows that the force per unit area—the pressure (P)—on the wall is

$$P = \frac{F(wall)}{A} = \frac{2mv^2}{AL}.$$ (2.141)

The quantity AL is the volume occupied by each particle heading toward the wall. It follows that $\frac{1}{AL}$ is the number of particles per unit volume that is headed toward the wall. In a gas, at any point in time, just as many particles are moving away from a wall as are moving toward the wall. Therefore, $n = \frac{2}{AL}$ is the total number of particles per unit volume, including both particles headed toward the wall and particles headed away from the wall. Thus, we see that

$$P = nmv^2.$$ (2.142)

In a gas, not all the particles have the same velocity, but if we write

$$P = nm\left\langle v^2 \right\rangle,$$ (2.143)

where $\left\langle v^2 \right\rangle$ is the mean square x-component of the velocity of gas particles, then we discover an equation for the pressure of a gas that will turn out to be correct.

One of the things that we will discover in Chap. 8. *Statistical Mechanics: Boltzmann Factors, PCR, and Brownian Ratchets* is that

$$\frac{1}{2}m\left\langle v^2\right\rangle = \frac{1}{2}k_B T, \tag{2.144}$$

where k_B is Boltzmann's constant and T is the absolute temperature. Using this result in Eq. 2.143, we discover that

$$P = nk_B T = \frac{N}{V}k_B T \tag{2.145}$$

or

$$PV = Nk_B T = \frac{N}{N_A}(N_A k_B)T = N_m RT, \tag{2.146}$$

where N is the total number of gas molecules in a volume V, $N_m = \frac{N}{N_A}$ is the number of moles, N_A is Avogadro's number, and $R = N_A k_B$ is the gas constant. Equation 2.146 is the ideal gas law, of course. Thus, we now have a mechanical explanation in terms of the momentum change of molecules hitting the container walls. Pretty cool!

2.11.4   Example: Hurricane-Force Winds and Water Jets

The behavior apparent in the video at

```
http://www.youtube.com/watch?v=FxgNuD5M9js[26]
```

demonstrates that the winds associated with hurricanes—can exert considerable forces on a person, comparable to the person's weight. The force exerted by a water jet can be significantly larger because of the much larger density of water compared to air.

This example seeks to calculate the force exerted by a water jet. This calculation is similar to that of the previous section, but here we will treat the water jet as a continuous stream of fluid. Consider (Fig. 2.23) a jet of water of cross-sectional area A and initial velocity v (along x) hits a wall and splashes in all directions in the zy plane in this case. The density of the water is ρ (Greek rho).

(a) In a time Δt, all the water within what distance hits the wall?

Answer: All the water within $v\Delta t$ hits the wall.

---

[26]http://www.youtube.com/watch?v=FxgNuD5M9js

(continued)

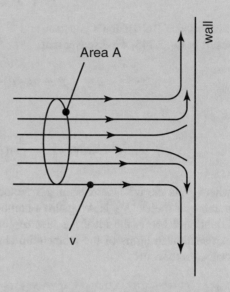

Fig. 2.23 Fluid jet hitting a wall and splashing up and down, but NOT back

Area A

v

wall

(b) What volume of water hits the wall in a time Δt?

    Answer: A volume of water equal to $Av\Delta t$ hits the wall.

(c) What mass of water hits the wall in a time Δt?

    Answer: A mass of waterequal to $\rho Av\Delta t$ hits the wall in a time Δt.

(d) What is the change in the (average) velocity of the water jet upon hitting the wall?

    Answer: The velocity of the water jet along the x-direction before hitting the wall is v. After hitting the wall, the mean velocity along x is zero. The mean velocity along y and z is also zero because the jet shoots out in all directions, so that the (vector) average velocity is zero. Therefore the change in velocity along x is $-v$.

(continued)

(e) What is the change in momentum of the jet in a time Δt (Δp)?

Answer:

$$\Delta p = -v \times \rho A v \Delta t = -\rho A v^2 \Delta t. \tag{2.147}$$

(f) What is the mean force on the jet?

Answer:

$$F = \frac{\Delta p}{\Delta t} = -\rho A v^2. \tag{2.148}$$

(g) What is the mean force on the wall?

Answer:

$$F_{wall} = -F = \rho A v^2. \tag{2.149}$$

Your intuition confirms that the sign here is correct--you know that the wall would feel a force in the same direction as the water velocity. Equation 2.149 shows how wind speed affects the force exerted by hurricane-force winds--there is a quadratic dependence, so, for example, 150 mph winds exert 4 times the force that 75 mph winds exert. It also shows that a water jet exerts much more force than an air jet with the same velocity, because of the larger density of water than air.

2.12 Chapter Outlook

The principles and methods that we have introduced and developed in the present chapter are prerequisite for the material in later chapters that involve forces, especially Chap. 9. *Fluid Mechanics: Laminar Flow, Blushing, and Murray's Law*, Chap. 10. *Oscillations and Resonance*, Chap. 12. *Gauss's Law: Charges and Electric Fields*, and Chap. 17. *Magnetic Fields and Ampere's Law*. In addition, a proper understanding of force and Newton's laws is necessary for a number of the topics appearing in Chap. 3. *Energy, Work, Geckos, and ATP*, Chap. 5. *Random Walks: Brownian Motion and The Tree of Life*, Chap. 8. *Statistical Mechanics: Boltzmann Factors, PCR, and Brownian Ratchets*, and Chap. 11. *Wave Equations: Strings and Wind*. Finally, how to solve Newton's law problems—namely the 7 step approach, described in this chapter—provides a powerful template for how to break up complicated problems into smaller more manageable parts, that can be adapted to many other venues beyond Newton's law problems.

2.13 Problems

Problem 1: Swimming Lessons
Figure 2.24 shows a swimmer.

(a) Explain how the swimmer moves forward?
(b) Explain why swimming is hard work?
(c) Explain why she floats?

Problem 2: Move over, Spiderman
Forget about spiderman, geckos really can climb up vertical walls and across ceilings, as shown in Fig. 2.25 and in the movie at

> https://www.youtube.com/watch?v=pyoE5aj6MYA.[27]

(a) What forces are acting on the gecko shown in Fig. 2.25?
(b) Discuss what aspect of the gecko's anatomy must be special to allow the gecko to walk on walks and ceilings?

Fig. 2.24 US Air Force Senior Airman Megan Stanton swimming, photographed by Tech. Sgt. Samuel Morse. Public domain image from https://www.dvidshub.net/image/982701/perseverance-airmans-commitment-health-triathlon-and-career#.V0XU04TLHvw

---

[27] https://www.youtube.com/watch?v=pyoE5aj6MYA

Fig. 2.25 Gecko (*Uroplatus fimbriatus*) crawling across a vertical glass sheet. Public domain image from https://commons.wikimedia.org/wiki/File:Uroplatus_fimbriatus_(3).jpg

(c) You cannot keep reusing a piece of sticky tape indefinitely. Discuss why not? What about geckos? Perhaps they can they only take a finite number of steps before the stickiness is gone? And is it only young ones that can climb up walls? What do you think?

(d) Discuss whether it will eventually be possible to create a synthetic material that permits a human, appropriately dressed, to actually crawl up or down tall buildings, like spiderman. When do you think such materials will be available?

In Chap. 3. *Energy: Work, Geckos, and ATP*, we will discuss a simplified model of how geckos manage to defy gravity.

Problem 3: Fish Keratocytes
By taking a scale from a fish (even a goldfish) and placing it in the right growth medium, it is possible with a microscope to observe the motions of individual keratocytes, which are the fish's wound healing cells. A movie showing the motion of a keratocyte is at https://www.youtube.com/watch?v=RTjYXBnMcgs.[28]

(a) Why it is important for fish keratocytes to be able to move?

(b) Suggest and explain a list of forces that may play a role in the observed keratocyte motion.

---

[28] https://www.youtube.com/watch?v=RTjYXBnMcgs

Problem 4: Belt and Suspenders

A box sits on a moving conveyor belt without slipping. The belt maintains a constant speed to the right.

(a) Explain how this final situation comes about, starting from the moment when the box is placed (with zero velocity) onto the conveyor belt.
(b) At the moment, immediately after the box is placed on the belt, what are the forces acting on the box? Is there a frictional force? If so in which direction does it point?
(c) When the box is sitting on the moving belt without slipping, explain: what forces are acting on the box? Is there a frictional force? If so in which direction does it point?

Problem 5: Prismatic Tensegrity with Four-Fold Rotational Symmetry

The goal of this problem is to determine the conditions for the tensegrity of Fig. 2.26 to be stable. The separation between the two squares, defined by vertices 1, 2, 3, and 4, and 5, 6, 7, and 8, respectively, is h. The distance from the center of each square to each of its vertices is a. The tension per unit length in the square cables, which connect vertices 1 and 2, 2 and 3, 3 and 4, 4 and 1, 5 and 6, 6 and 7, 7 and 8, and 8 and 5, is t. The tension per unit length in the diagonal cables, which connect vertices 1 and 5, 2 and 6, 3 and 7, and 4 and 8, is d. The tension in the struts, which connect vertices 1 and 6, 2 and 7, 3 and 8, 4 and 5, is s. In order for the cables to be under tension, t and d must be positive. In order for the struts to be in compression, the tension assigned to the struts, s, must in fact be negative.

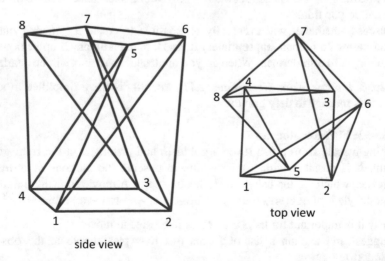

Fig. 2.26 Three-dimensional prismatic tensegrity with four-fold rotational symmetry. This tensegrity consists of four struts that are under compression (blue) and twelve cables that are under tension (red). Left: side view. Right: top view

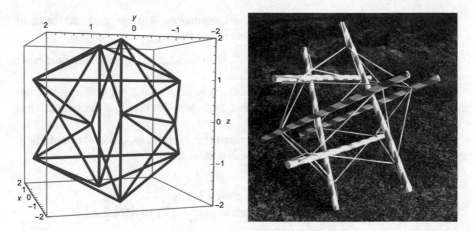

Fig. 2.27 Icosahedral tensegrity. Left: Schematic representation with the axes in units of L. Right: Version constructed with drinking straws, rubber bands, and paper clips

(a) Write an expression for the total force at vertex 1, $\mathbf{F}_1$, in terms of the vectors in which the cables and struts, attached to vertex 1, act ($\mathbf{r}_2 - \mathbf{r}_1, \mathbf{r}_3 - \mathbf{r}_2$, etc.), and the tensions per unit length, t, d, and s.

(b) If the tensegrity is stable, what is the value of the total force at each vertex of the structure? Explain why it is only necessary to consider the force at one vertex. What is the force at vertex 1, namely $\mathbf{F}_1$?

(c) Pick the origin of your coordinate system to be at the center of the bottom square with the z-axis vertically upwards, so that $\hat{\mathbf{k}}$ is the unit vector directed from the center of the lower square toward the center of the upper square. Then, use your answers to (a) and (b) to obtain an equation for d and s by calculating $F_{1z} = \hat{\mathbf{k}} \cdot \mathbf{F}_1$.

(d) Use your answers to (a) and (b) to obtain a second equation for t, d, s, and α by calculating $F_{1x} = \hat{\mathbf{i}} \cdot \mathbf{F}_1$.

(e) Obtain a third equation for t, d, s, and α by calculating $F_{1y} = \hat{\mathbf{j}} \cdot \mathbf{F}_1$.

(f) The answers to (c), (d), and (e) constitute three simultaneous equations for the three unknowns: $\frac{d}{t}$, $\frac{s}{t}$ and α. Solve these equations to find the values of $\frac{d}{t}$, $\frac{s}{t}$ and α in order that the tensegrity is stable (Fig. 2.27).

Problem 6: Icosahedral Tensegrity

Figure 2.27 shows a six-strut icosahedral tensegrity, both a schematic view (left) and a version built out of straws, rubber bands, and paper clips (right). The goal of this problem is to determine the conditions for this tensegrity to be stable. Because of the high symmetry of this structure, the tension in each cable is the same and the compression in each strut is the same. Therefore, it is only necessary to consider the forces acting at the end of one strut. Each strut has a length L, and is separated from the one parallel to it by a perpendicular distance a. It is convenient to pick the coordinate system so that the center of the icosahedron is at the origin and pairs of struts (blue on the left of Fig. 2.27) are parallel to the x-, y-, and z-directions.

(a) The end of one of the struts is at $(\frac{L}{2}, \frac{a}{2}, 0)$. It is attached by cables to four other strut ends. Write their position vectors.
(b) The tension per unit length in the cables is t. The tension per unit length in the strut is s. Write a vector equation that expresses the fact that the tensegrity is stable.
(c) Solve the vector equation to find the conditions for the icosahedral tensegrity to be stable. How is s related to t? How is a related to L?

Problem 7: Tensegrity Subject to an Applied Force
Consider the three-fold rotationally symmetric tensegrity of Sect. 2.7.2.

(a) How are Eqs. 2.50, 2.52, and 2.54 modified when a vertical tensile force, $F\hat{\mathbf{k}}$, is applied to each vertex?
(b) Solve these modified equations using Wolfram Alpha.
(c) What are the values of d, s, and α to linear order in F?

Problem 8: Cubical Tensegrity
A four-fold prismatic tensegrity is shown in Fig. 2.28. The thicker, darker lines are struts, and the thinner, lighter lines are cables. It has 8 vertices given by the following vectors:

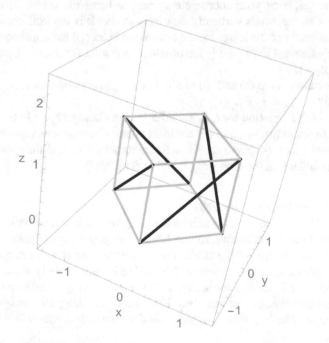

Fig. 2.28 Figure for "cubical tensegrity"

$$\mathbf{r}_1 = (1, 0, 0), \ \mathbf{r}_5 = (1, 0, \sqrt{2}), \ \mathbf{r}_2 = (0, 1, 0), \ \mathbf{r}_6 = (0, 1, \sqrt{2}), \ \mathbf{r}_3 = (-1, 0, 0),$$

$$\mathbf{r}_7 = (-1, 0, \sqrt{2}), \ \mathbf{r}_4 = (0, -1, 0), \ \mathbf{r}_8 = (0, -1, \sqrt{2}), \qquad (2.150)$$

relative to the x, y, z coordinates shown in the figure.

(a) Copy Fig. 2.28 and label each vertex (1–8) at the correct position with respect to the x, y, z coordinate system shown.
(b) What is the length of the cable ℓ_{15} that runs from vertex 1 to vertex 5?
(c) What is the length of the strut ℓ_{16} that runs from vertex 1 to vertex 6?
(d) Given that the tensegrity has the form of a cube, do your answers to (b) and (c) make sense just by using geometrical considerations? Explain.

We call t_1 the tension per unit of length in the cables of the top and bottom area of the tensegrity (which connect the vertices: (bottom) 1 and 2, 2 and 3, 3 and 4, 4 and 1, and (top) 5 and 6, 6 and 7, 7 and 8). We call t_2 the tension per unit of length in the cables that connect the top to the bottom area of the tensegrity (1 and 5, 2 and 6, 3 and 7, 4 and 8). We call s the tension per unit of length in the struts (which connect: 1 and 6, 2 and 7, 3 and 8, 4 and 5).
Furthermore we consider the case in which the tensegrity is subjected to an additional external force $\mathbf{F}$ acting at each vertex as shown in Fig. 2.29. The force $\mathbf{F}$ pulls the vertex outwards along the direction as shown by the arrows in the figure. In particular the force acting on vertex 1 has the vectorial form: $\mathbf{F} =$

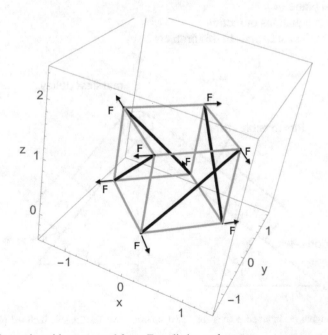

Fig. 2.29 Tensegrity with an external force F applied at each vertex

$F(1, 1, 0)$ and therefore acts outwards along the direction of the line joining vertex 4 to vertex 1.

(e) Write an expression for the total force $\mathbf{F}_1^{tot}$ at vertex 1. Write your answer in terms of the tensions per unit of length t_1, t_2 and s and the appropriate vectors in which the cables and struts (attached to vertex 1) act; also consider the additional external force $\mathbf{F} = F(1, 1, 0)$ pulling outwards as described above.

(f) We want to find the conditions for which the tensegrity in this configuration (with the external force $\mathbf{F}$ applied) is stable. Since all vertices are the same, we can only focus on vertex 1, if the tensegrity is stable, what is the total force at vertex 1, namely what is the value of $\mathbf{F}_1^{tot}$?

(g) Find three equations that express the values of t_1, t_2, and s at stability in terms of the given quantities (note the force $\mathbf{F}$ of magnitude F is a given quantity).

Problem 9: Two Masses, Two Frictionless, Inclined Planes, and a Pulley
A block of mass m_1, on a frictionless inclined plane making an angle θ_1 with the horizontal is connected by an inextensible cord over a small frictionless massless pulley to a second block of mass m_2 on a frictionless plane at an angle θ_2. (See Fig. 2.30) Follow the 7-Step Strategy to solve this problem.

(a) Sketch a force diagram for both masses. Indicate your coordinate system for each mass, i.e., what is your positive x_1 direction? what is your positive x_2 direction? Note that in cases where there is motion along an inclined plane, it is often best (simplest) to pick a coordinate system with an axis parallel to the inclined plane.

(b) Write the equations of motion.

(c) What are the constraints in this problem?

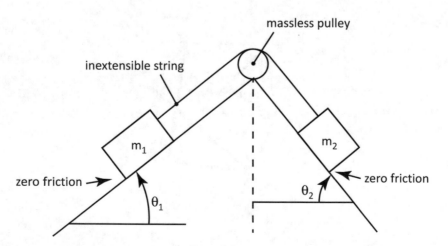

Fig. 2.30 Weights on inclined planes for "Two masses, two frictionless inclined planes, and a pulley"

Fig. 2.31 The five links of "Chains of love"

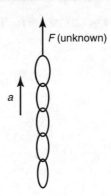

(d) Calculate the acceleration of each block. Do your answers make sense if $m_1 \gg m_2$?

(e) Calculate the tension in the cord.

Problem 10: Chains of Love

A chain consisting of five links, as shown in Fig. 2.31, each with mass m is lifted vertically up with acceleration a. The acceleration due to gravity is g downwards.

(a) Pick a coordinate system for this problem and sketch a force diagram for each link.

(b) What are the given quantities?

(c) What is the total force on each link?

(d) Calculate the force acting between adjacent links.

(e) What is the force (F) exerted on the top link by the agent lifting the chain?

Problem 11: Pulling a Pulley

A vertical force, F, is exerted directly on the axle of the pulley of Fig. 2.32. The pulley and string are massless and the bearing/axle is frictionless. Two objects of masses, m_1 and m_2 are attached as shown to the opposite ends of the string, which passes over the pulley. The object m_2 is in contact with the floor. Follow our procedure to determine:

(a) What is the largest value the force F may have so that m_2 will remain at rest on the floor?

(b) With the force determined in (a), what is the acceleration of m_1?

(c) What happens when $m_1 > m_2$?

Problem 12: Two Blocks, a Rod, and an Inclined Plane

Two objects, with masses m_1 and m_2 are attached to each other by a massless *rod* parallel to the incline on which they both slide. The masses travel down the plane with m_1 trailing m_2. The angle of the incline is θ to the horizontal. The coefficient

Fig. 2.32 Sketch for "Pulling a pulley"

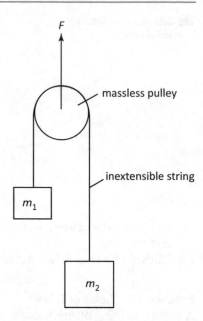

of kinetic friction between m_1 and the incline is μ_1. Between m_2 and the incline, the coefficient of kinetic friction is μ_2. The acceleration due to gravity is g downwards.

(a) Use the 7-Step Strategy to compute the common acceleration of the two objects.
(b) Use the 7-Step Strategy to compute the force transmitted through the rod.
(c) Under what conditions is the rod "in tension"? When is it "in compression." Explain whether your answer makes sense in the limit that $\mu_2 = 0$?

Problem 13: Pulley, Two Masses, and an Inclined Plane
In Fig. 2.33, the weight of object B is $m_F g$ and the weight of object A is $m_A g$. Between object B and the inclined plane the coefficient of static friction is μ_S and the coefficient of kinetic friction is μ_K. The plane is inclined by and angle θ.

(a) Find the frictional force on object B, if it is at rest.
(b) Find its acceleration if object B is moving up the plane.
(c) Find its acceleration if object B is moving down the plane.

Problem 14: Two Blocks Sliding
Two blocks are arranged one on top of each on a table, as shown in Fig. 2.34. The coefficient of kinetic friction between the blocks is μ_K. Provided the two blocks are in relative motion, the magnitude of the frictional force that each block exerts on the other is $\mu_K N$, where N is the normal force that each block exerts on the other. There is no friction between the lower block (mass m_2) and the table. A known force, F, is applied to the upper block (mass m_1) which causes both blocks to accelerate.

Fig. 2.33 Sketch for "Pulley, two masses, and an inclined plane"

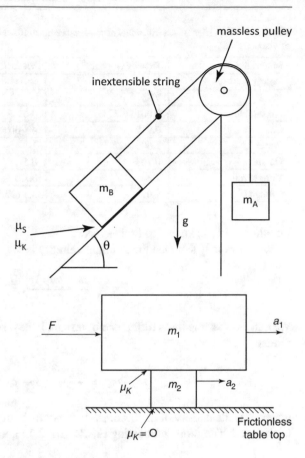

Fig. 2.34 Sketch for "Two blocks sliding"

The goal of this problem is to find the blocks' acceleration, namely a_1 for the upper block and a_2 ($a_2 < a_1$) for the lower block, as shown in the figure, in terms of know quantities: F, m_1, m_2, g, and μ_K.

(a) Sketch the force diagrams, one for each block, indicating all relevant forces.
(b) Write Newton's second law for each mass.
(c) What are the blocks' accelerations in the vertical direction? Give both vertical accelerations.
(d) Using your answers to (b) and (c), determine a_1 and a_2.

Problem 15. Measuring Viscosity
The goal of this problem is to devise a method for measuring fluid "viscosity," which is the material property of a fluid that specifies the frictional forces on objects moving in that fluid.

Advanced texts on fluid mechanics show that the friction coefficient, ζ, for a sphere of radius R in a fluid of viscosity η (Greek eta) is

$$\zeta = 6\pi \eta R. \tag{2.151}$$

Table 2.1 Properties of several relevant materials and fluids. Note that centipoise is the cgs unit of viscosity

Material	Density (g · cm^{-3})	Viscosity at 23 °C (centipoise)
Delrin	1.415	–
Steel	7.85	–
Air bubble	0.001	0.01
Canola oil	0.915	40–60
Water	1.000	1
Octane	0.703	0.5
Karo syrup	1.33	3000
Human blood	1.06	3.5 (37 °C)

Equation 2.151 is known as Stokes' Law.[29]

We previously saw that the terminal velocity is

$$u = \frac{(m - m_F)g}{\zeta}. \tag{2.152}$$

We can express the mass difference in terms of the density difference and the bead radius:

$$m - m_F = \frac{4\pi R^3}{3}(\rho - \rho_F), \tag{2.153}$$

where ρ is the density of the bead, ρ_F is the density of the fluid, and R is the radius of the bead. Therefore, combining Eqs. 2.116, 2.153, and 2.151, we find that

$$u = \frac{2(\rho - \rho_F)g R^2}{9\eta}. \tag{2.154}$$

We see that the terminal velocity is predicted to be proportional to g and to the density difference between the material of the bead and of the fluid, proportional to R^2, and inversely proportional to the fluid viscosity. Densities and viscosities of a number of materials are presented in Table 2.1

(a) What are the SI units for the viscosity, η? Show that you can extract these units by using dimensional analysis of $\mathbf{F} = -\zeta\mathbf{v}$ and $\zeta = 6\pi\eta R$.

(b) The values of the viscosity of various fluids in Table 2.1 are given in centipoise, which is the cgs unit for viscosity. By what factor must you multiply the values of viscosity in the table to convert them to the SI units that you wrote in part (a)?

---

[29] http://en.wikipedia.org/wiki/Stokes_Law

(c) Imagine you are the PI (Principal Investigator) of a lab. You are given some spherical beads of density ρ and a fluid with density ρ_F, but of unknown viscosity, η. What would you tell your students to do, step by step, so that they can put together an experiment to extract the unknown value of the viscosity?

(d) Consider a spherical bead, made of Delrin of diameter 1 mm. Estimate the bead's terminal velocity, u, in blood.

Problem 16: Beer Bubbles

Another version of the spheres-falling-through-a-viscous-fluid experiment is shown in the movie at http://www.youtube.com/v/2UUaNQsYJF4,[30] which shows bubbles rising in a glass of beer.

(a) Explain why the bubbles rise.
(b) Explain the direction of the frictional force on the bubbles.
(c) *Estimate* the initial acceleration of a bubble initially at rest.

Problem 17: Gel Electrophoresis

Gel electrophoresis[31] is a tremendously important analysis method, pervasive in biomedical research, that is used to separate different DNAs or proteins, one from another. Gel electrophoresis is one of the key techniques involved in DNA finger-printing, which enables detailed genetic comparisons between and within species and has therefore revolutionized our understanding of evolutionary relationships among and within species. DNA fingerprinting also has societal impact because it which permits forensic scientists to implicate or exonerate[32] suspects in criminal investigations, and has revolutionized justice systems around the world in the last 20 years, including the justice system.

This problem illustrates the role of Newton's second law and terminal velocity in gel electrophoresis. It also illustrates the role that an understanding of terminal velocity can play in providing a conceptual framework within which to learn something new and interesting about the transport properties of (unfolded) proteins through a gel network. Finally, it will permit you to gain experience in examining, analyzing, and explaining experimental measurements, and in developing hypotheses based on those measurements.

The basis of gel electrophoresis is that there is force on a particle with charge Q in an electric field E equal to QE. In gel electrophoresis, the molecules move through a gel, rather than through a simple fluid. Nevertheless the frictional force still has the form of Eq. 2.111. Therefore, in the case of a charged particle in a

---

[30] http://www.youtube.com/watch?v=2UUaNQsYJF4

[31] http://en.wikipedia.org/wiki/Gel_electrophoresis

[32] http://www.theguardian.com/world/2012/dec/07/dna-testing-frees-man-death-row

gel, given that the gravitational force is negligible compared to QE, the particle's equation of motion becomes

$$M\frac{dv}{dt} = QE - \zeta v, \tag{2.155}$$

where M is the effective inertial mass. It follows that the particle's terminal velocity in this case is

$$u = \frac{QE}{\zeta}. \tag{2.156}$$

(a) Explain why the characteristic, relaxation time to reach the terminal velocity, $\tau = \frac{M}{\zeta}$, is very short for gel electrophoresis, and therefore can be neglected in this problem.

(b) Develop a hypothesis that explains why different lengths of DNA or different proteins have different terminal velocities.

(c) Given that different lengths of DNA or different proteins have different terminal velocities explain how gel electrophoresis separates the different species.

(d) Figure 2.35 shows an sodium-dodecyl-sulfate polyacrylamide gel electrophoresis (SDS PAGE) experiment that separates proteins according to their molecular weight. In SDS solution, each protein is believed to unfold and become decorated with a number of negatively charged dodecyl sulfate moieties. It is also believed that the resultant SDS-protein complex takes on a rodlike configuration in which the length of the rod and the number of SDS molecules is both proportional to the protein molecular weight. In turn, the charge of the SDS-protein complex is closely proportional to the number of SDS molecules involved, and therefore is also proportional to the protein molecular weight. The polyacrylamide gel is a cross-linked polymer material in solution—similar to jello in its physical properties. Given that the SDS PAGE gel of Fig. 2.35 was run for 1 hour, briefly describe how you would experimentally determine how the SDS-protein complex terminal velocity for that gel depends on protein molecular weight.

(e) A measurement of the protein terminal velocity is plotted versus molecular weight in Fig. 2.36. Check whether I got it right, by determining for yourself the terminal velocity of the SDS-protein complex in the "ladder" of molecular weight 25 kDa from Fig. 2.35.

(f) Figure 2.36 shows the terminal velocity of SDS-protein complexes, u, versus molecular weight, W, both plotted on logarithmic axes. Evidently, u decreases with increasing W. The solid line in Fig. 2.36, which well describes the behavior of u versus W, corresponds to

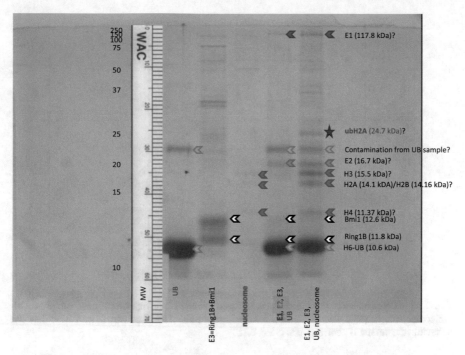

Fig. 2.35 An actual-size image of a sodium-dodecyl-sulfate polyacrylamide gel electrophoresis (SDS PAGE) gel, showing the separation of a protein product of interest, namely ubiquitinated histone H2A (ub-H2A), from several other proteins, including the enzymes that catalyzes ubiquitination (E1, E2, and E3). In this case, the proteins started at the top of the gel—the start location is where the thin darker line running across is apparent—and moved downwards under the influence of an electric field. Included on the left-hand side of the gel is a molecular weight "ladder," consisting of several proteins of different molecular weights. Superimposed at the far left-hand side is a ruler with millimeter divisions. Gel courtesy Catherine McGuiness, Lynne Regan, and Irina Bezsonova

$$u(\mu\text{m s}^{-1}) = \frac{941}{(W(\text{kDa}))^{\frac{3}{2}} \left(1 + \left[\frac{15.25}{w(\text{kDa})}\right]^4\right)^{\frac{3}{8}}}. \tag{2.157}$$

where the $(\mu\text{m s}^{-1})$ and the (kDa) indicate that the terminal velocity and the molecular weight must be measured in micrometers per second and kilodaltons, respectively, in order for this expression to hold true. For large enough W, Eq. 2.157 simples to

$$u(\mu\text{m s}^{-1}) = \frac{941}{(W(\text{kDa}))^{\frac{3}{2}}}, \tag{2.158}$$

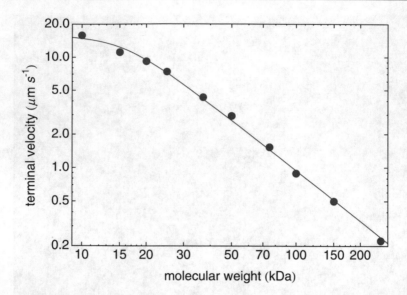

Fig. 2.36 SDS-protein complex terminal velocity (u), plotted on a logarithmic axis, versus protein molecular weight (W), also plotted on a logarithmic axis. The line, which well describes the behavior of u versus W, corresponds to Eq. 2.157

while, for small W, Eq. 2.157 simples to

$$u(\mu m\ s^{-1}) = 15.8. \qquad (2.159)$$

Develop a hypothesis, referring to Fig. 2.37, for the origin of the "crossover" in the curve of u versus W from $W^{-\frac{3}{2}}$-behavior at larger W to W-independent behavior at small W.

(g) In SDS, proteins are believed to become decorated with SDS molecules and for the resultant SDS-protein complex to assume a rodlike configuration for which the length of the rod, L, and the number of SDS molecules is both proportional to the protein molecular weight, W. In turn, the charge of this complex, Q, is proportional to the number of SDS molecules involved, and so is proportional to W. At the same time, for a rod of length L moving through a viscous fluid, the friction coefficient, ζ, is proportional to L, that is, the friction coefficient is proportional to W in a viscous fluid. Explain whether these results/statements make sense in light of the suggestion that the terminal velocity of the SDS-protein complex is independent of W for small W.

(h) Estimate the W-dependence of the friction coefficient of a large protein in a gel.

Problem 18: Conical Pendulum

A mass M hangs by a massless string of length L which rotates at a constant angular velocity, ω (Greek omega). The mass moves in a circular path of constant radius. As

Fig. 2.37 Cartoon of a polyacrylamide gel, showing a large protein, larger than the gel's smallest pore size, and a small protein, smaller than the gel's smallest pore size

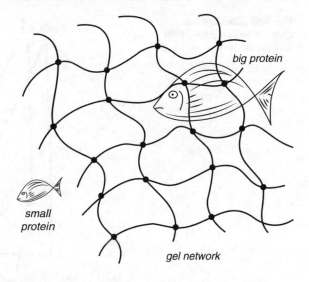

big protein

small protein

gel network

discussed in the context of *Geostationary orbits*, the radial acceleration of any mass moving on a circular path of radius r at angular velocity ω is $a = \omega^2 r$.

(a) Find α, the angle that the string makes with the vertical.
(b) Sketch α as a function of ω.
(c) Explain what happens in the limit of large ω.
(d) What is the minimum possible value of ω?

Problem 19: Talk the Torque

For an extended object, Newton's Second Law states that the product of the object's total mass and the acceleration of its center of mass is equal to the total external force, applied to the object. This result relies on the fact that all internal forces, that characterize interactions between different parts of the object, obey Newton's Third Law, causing these internal forces to cancel out of the equation of motion for the center of mass. For an extended object, attached to a fixed axis of rotation, it is possible to determine an equation that is similar in spirit—often called the Rotational Second Law—that informs us that the product of the object's "moment of inertia" about the axis of rotation and its angular acceleration about the axis of rotation is equal to the "torque" about the axis of rotation. This problem shows how the Rotational Second Law comes from Newton's laws and illustrates torque in an example involving two masses connected together by a string.

Two masses, inner mass m_1 at radius r_1 and outer mass m_2 at radius r_2, are attached to a pivot point and to each other via two inextensible strings as shown in Fig. 2.38. They are both caused to rotate about the pivot point by application of an always-tangential force applied to mass m_2. In fact, the tangential force experienced by mass 1 is the result of the tension in the string connecting the two masses. The fact that the tensional force experienced by mass 2 is equal and opposite the tensional force experienced by mass 1 is a consequence of Newton's third law.

Fig. 2.38 Sketch, force
diagram, and coordinates for
Talk the torque

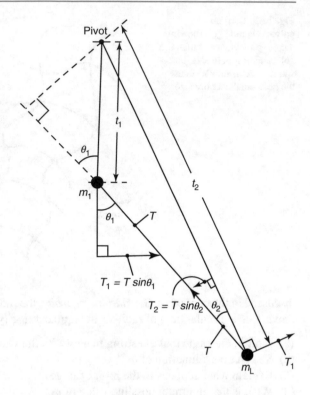

(a) Write the expressions for the tangential accelerations of each mass.
(b) Determine $\sin\theta_2$ in terms of r_1, r_2, and $\sin\theta_1$.
(c) Supposing that both masses possess the same angular acceleration, α, express a_1 and a_2 in terms of α.
(d) Determine α in terms of r_1, r_2, and F only.
(e) Instead of two masses, now suppose that there are N masses, $\{m_i\}$ at radii $\{r_i\}$. How is your result for α modified in this case?

Problem 20: Superman's Super *Pectoralis Major*
It is well known that bullets and other missiles fired at Superman simply bounce off his chest. Suppose that a gangster sprays Superman's chest with bullets of mass m at the rate of n. The speed of each bullet is v and the angle of each bullet with respect to the plane of superman's chest is θ. Suppose too that the bullets bounce off his chest *specularly* with no change in speed. Find the average force exerted by the stream of bullets on Superman's chest.

Problem 21: On the Railroad
An empty freight car of mass M starts at $t = 0$ from rest under a constant external horizontal force, F. Ignore friction, etc. At the same time, grain begins to flow from a grain hopper into the freight car at a rate α. The horizontal velocity of the grain is

zero when it enters the freight car. Your goal in this problem is to find the speed of the freight car plus its grain when a total mass m of grain has been transferred.

(a) What is the net horizontal force on the system consisting of the freight car plus all the grain?
(b) Determine the momentum of the freight car plus the grain at time t, given that at $t = 0$ their velocities are both zero.
(c) What is the mass of the freight car plus the grain carried by the freight car at time t, $m(t)$?
(d) What then is the velocity, $v(t)$, of the freight car and the grain carried by the freight car at time t in terms of given quantities?

Problem 22: Another Water Jet
A water jet of cross-sectional area A, in which the water's velocity is initially $v\mathbf{i}$, hits a stationary, flat surface that is inclined at an angle θ to the jet. The water bounces off as shown in the figure. Its velocity after bouncing off the surface has the same magnitude as before, but its direction is changed, as shown. The (mass) density of water is ρ. (Ignore gravity in this problem.)

(a) What volume of water hits the surface in a time Δt?
(b) What mass of water hits the surface in a time Δt?
(c) Given the geometry shown in the figure, determine the water jet's velocity vector, $\mathbf{w}$ after the jet has bounced off the wall.
(d) What is the change in the velocity vector of the water jet as a result of hitting the surface.
(e) What is the change in momentum of the water jet (call it $\Delta\mathbf{p}$) in a time Δt?
(f) What is the force ($\mathbf{F}_S$) *on the surface*? Sketch this force on the figure.
(g) What is the magnitude of this force? Comment on the θ dependence of the force.

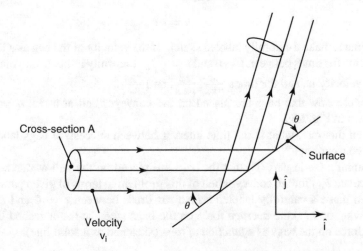

Fig. 2.39 Bags are placed on
the belt at rate n

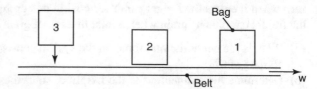

Problem 23: Emotional Baggage

At time $t = 0$, the baggage handlers at JFK place a bag of mass m on a horizontal conveyor belt, that is moving with velocity w in the positive x direction. The bag is initially at rest, but the frictional force between the bag and the belt causes the bag to accelerate. After a short period of time, the velocity of the bag becomes the same as the velocity of the conveyor belt. The coefficient of kinetic friction between the bag and the belt is μ_K. The acceleration due to gravity is g downwards.

(a) What is the momentum of the bag in the x direction (p) at large times, long after the bag is placed on the conveyor belt?
(b) Sketch a force diagram for the bag, applicable immediately after it lands on the conveyor belt, including all forces on the bag and including a clear coordinate system.
(c) Using Newton's 2nd law, write two equations, both applicable immediately after it lands on the conveyor belt: one for the horizontal acceleration (call it a) and one for the vertical acceleration (call it b) of the bag.
(d) Apply any constraints and the usual relationship between kinetic friction and normal force to express the horizontal acceleration, a, in terms of given quantities only.
(e) Given that the velocity (v) of an object, that is initially stationary and then subject to a constant acceleration a, is

$$v = at \qquad\qquad (2.160)$$

at time t, make a carefully labeled sketch of the velocity of the bag as a function of time for times between $t = 0$ and $t = \frac{3w}{\mu_K g}$. Carefully indicate on your sketch the velocity w, and the times $\frac{w}{\mu_K g}$, $\frac{2w}{\mu_K g}$, and $\frac{3w}{\mu_K g}$.
(f) Suppose now that bags are placed on the conveyor belt at a rate $n = \frac{\mu g}{3w}$ as shown in Fig. 2.39
Given this rate, what is the time interval between successive bags landing on the conveyor belt?
(g) Assuming (as in part (f)) that the bags are placed on the belt at rate $n = \frac{\mu g}{3w}$, calculate F_{av} for the conveyor belt of this problem in terms of given parameters. Then make a carefully labeled sketch for times between $t = 0$ and $t = \frac{6w}{\mu g}$ showing both: $i)$ the *average* force on the bags (sketched as a dashed line), $ii)$ the force on the bags as a *function of time* (sketched as a solid line).

Problem 24: Horizontal Pulley

Consider the situation shown in Fig. 2.40. A mass m is sitting on top of a second mass M, which is itself supported on a table, while the two masses are attached

Fig. 2.40 Schematic of the masses in "HorizontalPulley'

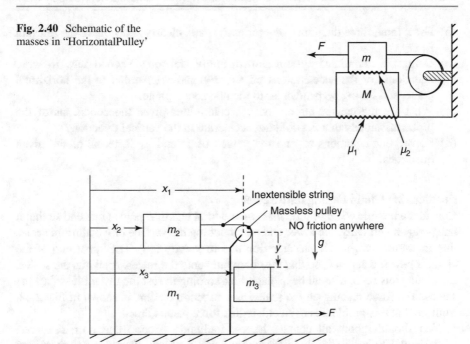

Fig. 2.41 Schematic of the masses in *"Atwood's most complex contraption"*

together by a massless, inextensible string and a horizontal pulley, attached to a wall, as shown. The coefficient of friction between mass M and the table is μ_1 and the coefficient of friction between mass m and mass M is μ_2. The goal of this problem is to find the tension in the string and the accelerations of the masses.

(a) Draw a force diagram for each mass.
(b) Using your force diagrams, write the equations of motion of the two masses. Be sure to indicate your coordinate system for each mass.
(c) What are the constraints in this case?
(d) Using these constraints, solve the equations of motion to find the tension in the string and the acceleration of each mass.

Problem 25: Atwood's Wackiest Contraption

Consider the situation shown in Fig. 2.41. Three given masses, m_1, m_2, and m_3 are arranged as shown. Masses m_2 and m_3 are connected together via a massless, inextensible string that runs over a massless pulley that is rigidly attached to m_1 as shown. There is no friction anywhere. An unknown force F parallel to the surface is applied to m_1 only. The acceleration due to gravity is g in the vertical direction. The force F is just such that mass m_3 does not accelerate in the vertical direction. The goal of this problem is to determine the value of F and the corresponding value of a_1, the acceleration of mass m_1, which is also unknown.

(a) Draw three force diagrams, one for each mass, clearly showing all the relevant forces.
(b) Using the coordinate system shown, apply Newton's Second Law to write equations for the accelerations of m_1, m_2, and m_3 parallel to the horizontal plane, and for m_3 perpendicular to the horizontal plane.
(c) What are the constraints on these accelerations given the conditions of the problem, namely that m_3 does not accelerate in the vertical direction?
(d) Solve your equations to find the values of F and a_1 in terms of the given quantities.

Problem 26: Chain Falling onto Scales

Consider a flexible chain of total mass m and total length L held at one end so that it hangs down vertically, the lower end just touching the surface of a bathroom scales that measures weight. The acceleration due to gravity is g. The upper end of the chain is released at $t = 0$, so that the chain falls onto the scales. As it hits the scales, the chain coils up in a small heap, each link coming to rest the instant it strikes the scales. The force reading on the scales as a function of time is shown in Fig. 2.42. Your goal in this problem is to calculate this force versus time.

You should ignore all contact forces, including friction and normal forces, between the links of the chain throughout this problem. To ensure a consistent notation, suppose that when the top of the chain has fallen a distance z, the velocity of the moving portion of the chain is v and the time is t.

(a) Draw a sketch of the chain-scales situation at time t, including a clearly defined coordinate system.
(b) What is the relationship between v and t and between z and t?

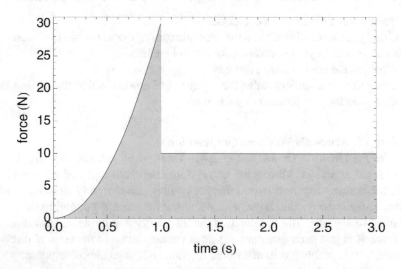

Fig. 2.42 Left: Schematic of falling chain. Right: Force versus time for a falling chain

(c) In a small interval of time Δt, what length of chain hits the scales, given that the velocity of the top of the chain is v?
(d) What is the mass of the length of chain that hits the scales in time Δt?
(e) What is the change in the momentum of the chain in a time Δt?
(f) When the distance that the top of the chain has fallen is z, what is the weight of the chain on the scales?
(g) What is the reading on the scales (F) as a function of time, t, in terms of t (of course) and m, g, and L. Compare your calculation to Fig. 2.42. What value of t separates the two regimes evident in the figure?

Problem 27: Returning to JFK

Two luggage bags, one (of mass m_2) directly on top of the other (of mass m_1), are placed (initial horizontal velocity of zero) onto a conveyor belt that is moving at a constant velocity v to the right. The coefficient of kinetic friction between the top bag and the bottom bag is μ_2. The coefficient of kinetic friction between the bottom bag and the belt is μ_1. The acceleration due to gravity is g (Fig. 2.43).

Assume throughout this problem: (1) that both bags are moving to the right; (2) that the two bags have a relative velocity with respect to each other; (3) that the lower bag has a relative velocity with respect to the conveyor belt. Call a_1 the acceleration of the bag of mass m_1 and a_2 the acceleration of the bag of mass m_2.

(a) Draw a force diagram for each mass.
(b) Write down four equations corresponding to the application of Newton's 2nd Law to the two masses. Clearly show the coordinate system that you have chosen to write your equations.
(c) Use our model for kinetic frictional forces, namely that the magnitude of the frictional force between two objects in contact in relative motion is μN, where μ is the coefficient of kinetic friction of the contact surface and N is the normal force at that surface, and the constraints of this situation to simplify your four equations from part (b).
(d) Solve your equations from part (c) to find the accelerations, a_1 and a_2 of m_1 and m_2, respectively.
(e) Discuss your solution to part (d) in the case that $\mu_1 < \mu_2$. Is it sensible?

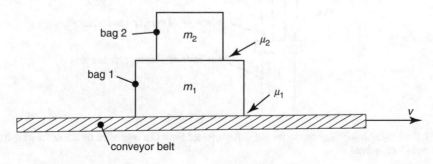

Fig. 2.43 Schematic of two bags, one on top of the other, placed on a conveyor belt at JFK

Problem 28: Summer in the City

It is a hot day in the summer of 1967. Lovin' Spoonful music is playing on the radio, and someone has opened a fire hydrant, which shoots a jet of water, vertically upwards, as shown in the figure. The velocity of the water jet as it comes out of the hydrant ($z = 0$) is v. Furthermore, as also shown in the figure, someone has also placed an upside-down garbage can of mass m on top of the water jet, which is therefore supported at a distance h above the hydrant, i.e., it is supported at $z = h$. The acceleration due to gravity is $-g$. The goal of this problem is to determine h in terms of the given quantities, namely v, m, g, the density of water ρ and the cross-sectional area of the jet as it leaves the fire hydrant, A (Fig. 2.44).

(a) In terms of v, what length (along the z direction) of the water jet shoots out of the hydrant in a small interval of time Δt?

(b) Given that the density of water is ρ, and that the area of the jet coming out of the fire hydrant is A, what mass of water leaves the hydrant in the time interval Δt?

(c) All of the water jet hits the garbage can and therefore your answer to (b) is also the mass of water that hits the garbage can in time Δt. At height h, however, the velocity of the jet is decreased, because of its acceleration due to gravity. We will call the velocity of the jet immediately before it hits the trash can u. You

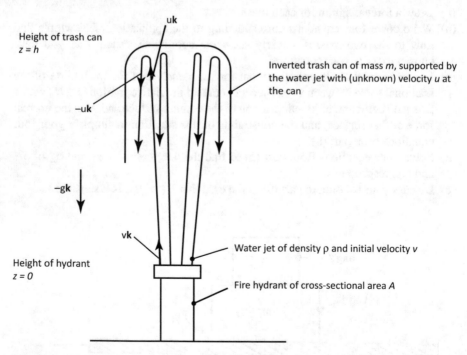

Fig. 2.44 Sketch for "Summer in the city": An inverted trash can supported by the water jet from an open fire hydrant

will determine u below, but for now include u in your answers. The water jet is reflected directly back (downwards) and immediately after it hits the trash can, the water jet's velocity is $-u$. In terms of u, what is the change in the velocity of the water jet as a result of hitting the garbage can?

(d) What is the corresponding change in momentum of the water jet in the time Δt that results from hitting the garbage can—call it $\Delta \mathbf{p}$—in terms of v, u, ρ, A, and Δt only?

(e) Calculate the force (call it $\mathbf{F}_{jet}$) exerted on the garbage can by the water jet in terms of v, u, ρ, A only? Is it directed up or down?

(f) Given that the garbage can is stationary, what is the total force (call it $\mathbf{F}_{tot}$) on the garbage can? Thus calculate u in terms of m, g, v, ρ, and A only.

(g) Using the equations for one-dimensional motion subject to a constant acceleration, applied to a unit mass of the water jet, calculate u in terms of v, g, and h only.

(h) Combine your answers to (f) and (g) to calculate h in terms of given quantities only, namely v, g, m, A, and ρ.

Problem 29: Another system of pulleys and masses
Two masses m_1 and m_2 are connected together via a system of massless pulleys, P1 and P2, and massless, inextensible strings, as shown in Fig. 2.45. One end of the string around pulley P1 is attached to mass m_1 and the other end supports pulley P2, as shown. One end of the string around pulley P2 is attach to mass m_2 and the other end is attached to the floor, maintaining the tension in the string. Your goal in this problem is to find the acceleration of mass m_1.

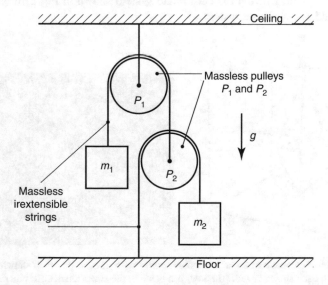

Fig. 2.45 Pulleys for "Another system of pulleys and masses"

(a) Draw three force diagrams, one for each mass and one for pulley P2, clearly showing all the relevant forces. Clearly define any forces that you introduce and a coordinate system for each mass and pulley P2. (You only need to do this for the vertical direction, because nothing moves in the horizontal direction).

(b) Apply Newton's Second Law to write *three* equations that relate the acceleration of each mass to the forces on that mass, and relate the forces on P2 to each other. (You should only do this for forces and acceleration in the vertical direction, because nothing moves in the horizontal direction).

(c) How are the accelerations of the two masses in the vertical direction related to each other, given your coordinate system? That is, what are the constraints in this problem? (Hint: Carefully consider how far and in which direction m_2 is displaced if m_1 is displaced Δx. The ratio of these displacements equals the ratio of the corresponding accelerations.)

(d) Solve your equations to find the acceleration of mass m_1 in terms of the given quantities, namely m_1, m_2, and g.

Problem 30: Octahedral Tensegrity

Figure 2.46 shows a schematic of a three-strut octahedral tensegrity. The goal of this problem is to determine the conditions for this tensegrity to be stable. Because of the high symmetry of this structure, the tension in each cable (red) is the same and the compression in each strut is the same. Therefore, it is only necessary to consider the forces acting at the end of one strut. It is convenient to pick the coordinate system so that the center of the octahedral tensegrity is at the origin, and the struts (blue) lie along x-, y-, and z-axes.

(a) Briefly explain whether the coordinate system shown in Fig 2.46 is a right- or left-handed coordinate system.

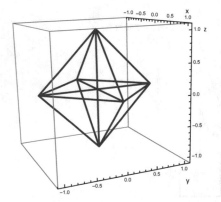
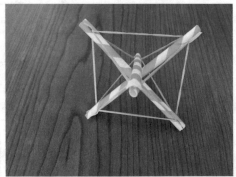

Fig. 2.46 Left: Schematic of an octahedral tensegrity with three struts, each shown in blue, and twelve cables, each shown in red. Although it looks like the struts touch each other at the origin, you should assume that they pass each other without touching. Right: Straw-and-rubber band build of the octahedral tensegrity

(b) The end of one of the struts is at $\mathbf{r}_1 = (1, 0, 0)$. It is attached by cables to four other strut ends. Include in your solution a copy of Fig. 2.46, label the positions of these strut ends, and write the corresponding position vectors. Also, label the other end of the strut in question and write its position vector.

(c) The tension per unit length in the cables is t. The tension per unit length in the struts is s. Write a vector equation that expresses the fact that the tensegrity is stable.

(d) Using your answer to (b), find the condition(s) for the octahedral tensegrity to be stable. How is s related to t?

Problem 31: Luggage on Ramp

Two luggage bags, one (of mass m_1) directly on top of the other (of mass m_2), are placed on a stationary ramp, inclined at an angle θ to the horizontal, where they both slide down toward the baggage carousel. The coefficient of kinetic friction between the top bag and the bottom bag is μ_1. The coefficient of kinetic friction between the bottom bag and the ramp is μ_2. The acceleration due to gravity is g. Call a_1 the acceleration of block m_1 down the ramp, and a_2 the acceleration of block m_2 down the ramp, as shown in Fig. 2.47.

(a) Draw a force diagram for each bag.

(b) Write down four equations corresponding to the application of Newton's 2nd Law to the two masses. Clearly show the coordinate system that you have chosen to write your equations.

(c) Assuming 1) that both bags are moving down the ramp; 2) that the two bags have a non-zero relative velocity with respect to each other; and 3) that the lower bag has a non-zero relative velocity with respect to the ramp, use our model for kinetic frictional forces, namely that the magnitude of the frictional force between two objects in contact in relative motion is μN, where μ is the coefficient of kinetic friction of the contact surface and N is the normal force at that surface, and the constraints of this situation, to achieve four equations in

Fig. 2.47 Schematic of two bags, one on top of the other, placed on an ramp

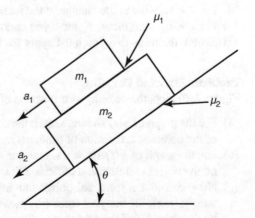

Fig. 2.48 Sketch for "That chain again"

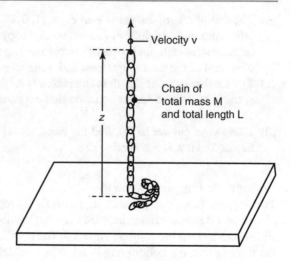

Velocity v

Chain of total mass M and total length L

z

four unknowns, namely the accelerations (a_1 and a_2) of the two bags, and the normal forces at the two contact surfaces (N_1 and N_2).

(d) Solve your equations from part (c) to find the accelerations, a_1 and a_2 of m_1 and m_2, respectively.

Problem 32: That Chain Again

A chain of mass M and total length L is initially coiled on a frictionless horizontal table. At $t = 0$, you grab one end and pull it up vertically at a constant velocity, v. The acceleration due to gravity is $-g$, which is downwards. (We chose that the positive direction is upwards.) The goal of this problem is to determine the force, F, that you must exert as a function of time, t (Fig. 2.48).

(a) At what value of t—call it t_1—does the chain first lose contact with the table?
(b) For $t < t_1$, what length of chain, z, is off the table at time t?
(c) For $t < t_1$, what mass of chain do you set into motion in time Δt?
(d) For $t < t_1$, what is the change in the total momentum of the chain in time Δt?
(e) For $t < t_1$, what force, F, must you exert in terms of M, v, g, L, and t?
(f) Sketch the force that you must apply for $0 < t < 3t_1$.

Problem 33: Salad Dressing

Figure 2.49 plots the velocity as a function of time for a certain object of mass m.

(a) For the graph of $v(t)$ shown, sketch the corresponding graph of the acceleration of the object as a function of time, $a(t)$.
(b) For the graph of $v(t)$ shown, sketch the corresponding graph of the total force on the object as a function of time, $F(t)$.
(c) Now, consider a spherical object moving vertically under the influence of its weight (mg), the buoyant force ($m_F g$, where m_F is the mass of fluid displaced by the object), and fluid friction ($-\zeta v$, where ζ is the object's friction coefficient

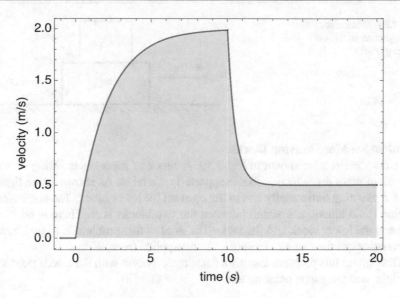

Fig. 2.49 Plot of velocity as a function of time

and v is its velocity) through a beaker containing both canola oil (which has the larger viscosity—larger ζ—and the smaller density) and water (which has the smaller viscosity—smaller ζ—and the larger density). Canola oil and water do not mix. Therefore, because the canola oil is less dense, there is a layer of canola oil on top of a layer of water in the beaker. Given that the magnitude of the terminal velocity, u, of a spherical object in a fluid is given by

$$|u| = \frac{|m - m_F|}{\zeta}, \tag{2.161}$$

and that the characteristic time, τ, for the velocity to approach the terminal velocity is

$$\tau = \frac{m + \frac{1}{2}m_F}{\zeta}, \tag{2.162}$$

and that ζ is proportional to the fluid viscosity, briefly explain whether the velocity-versus-time plot, shown in Fig. 2.49, is (1) appropriate for a bubble rising, or (2) for a bead falling in the beaker, or (3) whether the plot is inconsistent with both of these scenarios, i.e., explain whether it is a bead or a bubble or neither?

Fig. 2.50 Schematic of the configuration in "More moving blocks"

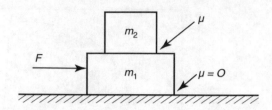

Problem 34: More Moving Blocks

Consider the situation shown in Fig. 2.50. A block of mass m_2 is sitting on top of a block of mass m_1, which is itself supported on a table. As shown in the figure, a force F is acting horizontally and to the right on the lower block. The coefficient of friction (both kinetic and static) between the two blocks is μ. There is no friction between the lower block and the table. The goal of this problem is to find how the accelerations of the two blocks vary with the applied force, F.

Throughout this problem assume a coordinate system with the x-axis pointing to the right, and the y axis pointing up.

(a) For each block, draw a force diagram, clearly labeling each force.
(b) Using your force diagrams, write the equations of motion of the two blocks, assuming that the two blocks are in relative motion. Call the horizontal accelerations a_1 and a_2 for blocks 1 and 2, respectively.
(c) Solve the relevant equations of motion to find the horizontal acceleration of each block in terms of the given quantities,i.e., find a_1 and a_2.
(d) For what range of values of F is your calculation, which assumes that the blocks are in relative motion, valid?
(e) Write the acceleration of the two blocks if they are not in relative motion, but instead are accelerating together.
(f) Using your answers to (c), (d), and (e), sketch a_1 and a_2 versus F from $F = 0$ to $F = (m_1 + m_2)g$. Be sure to make clear in your sketch which curve is a_1 and which curve is a_2.

Problem 35: Going Boldly

During its five-year mission, the starship Enterprise is travelling through an asteroid belt at a constant velocity $-u$ along the x axis, when it encounters a stream of asteroids, each of mass m, each with velocity $+w$ along the x axis, relative to the Enterprise, and each separated one from another by a distance L. When an asteroid hits the Enterprise's deflector shield, it bounces back with a velocity $-w$ along x, relative to the Enterprise. In this problem, assume that (because of its powerful warp drive) the velocity of the Enterprise is unaffected by the collisions with asteroids, and work in the frame of reference of your choice.

(a) During its transit of the asteroid belt, how many asteroids hit the Enterprise in a time Δt?

(b) What is the change in the velocity of each asteroid following its collision with the Enterprise?

(c) What is the change in the x-component of the momentum of the asteroids in a time Δt? Call your answer Δp.

(d) What is the force on the Enterprise as a result of its collisions with asteroids? Call it F.

Problem 36: Regular Icosahedral Tensegrity

Consider the "icosahedral" tensegrity shown in Fig. 2.51. Problem 6 shows that, without any other intervention, for the simple "icosahedral" tensegrity, shown in Fig. 2.51 to be stable, the separation (a) between each pair of parallel struts must equal one-half of the length of each strut (L), i.e., for such a tensegrity to be stable requires that $a = \frac{L}{2}$.

By contrast, in order for the the vertices of this sort of structure to form a regular icosahedron (one of the Platonic solids) requires that the distance between parallel struts must be equal to the length of the cables. The goal of this problem is to determine the force, F, that must be applied at each vertex, to create such a regular icosahedron from the tensegrity shown.

Fig. 2.51 Icosahedral tensegrity constructed with drinking straws (struts), rubber bands (cables), and paper clips

Because of the high symmetry of these structures, the tension in each cable is the same, the compression in each strut is the same, and the length of each cable is the same.

It is convenient to pick the coordinate system so that the center of the icosahedron is at the origin and pairs of struts are parallel to the x-, y-, and z-directions. The end of one of the struts is at $\mathbf{r}_1 = (\frac{L}{2}, \frac{a}{2}, 0)$. The other end of that same strut is at $\mathbf{r}_6 = (-\frac{L}{2}, \frac{a}{2}, 0)$. The end of the strut at $\mathbf{r}_1$ is attached by cables to four other strut ends at

$\mathbf{r}_2 = (\frac{a}{2}, 0, \frac{L}{2})$,
$\mathbf{r}_3 = (\frac{a}{2}, 0, -\frac{L}{2})$,
$\mathbf{r}_4 = (0, \frac{L}{2}, \frac{a}{2})$, and
$\mathbf{r}_5 = (0, \frac{L}{2}, -\frac{a}{2})$.

(a) What is the vector from the vertex at $\mathbf{r}_1$ to the vertex at $\mathbf{r}_2$ in terms of $\mathbf{r}_1$ and $\mathbf{r}_2$?
(b) Calculate the length squared of the cables in terms of a and L. Call it L_C^2.
(c) As noted above, in order for the tensegrity to form a regular icosahedron, it is necessary that $L_C^2 = a^2$. What is the corresponding value of a in terms of L only? Remember that a is a length and so *must* be positive. (You may also find it helpful to know one of the following solutions from the quadratic formula:

For $0 = -a^2 - La + L^2, a = \frac{\pm\sqrt{5}-1}{2}L$;
for $0 = -a^2 + 2La + 2L^2, a = \pm\sqrt{3} + 1L$;
and for $0 = -a^2 + 2La + L^2, a = \pm\sqrt{2} + 1L$.)

(d) To stabilize this regular icosahedral tensegrity, we may apply a force at the end of each strut, directed toward or away from the parallel strut. For the vertex at $\mathbf{r}_1$, this force, $\mathbf{F}$, is parallel to the y-direction:

$$\mathbf{F} = (0, F, 0), \tag{2.163}$$

where F is to-be-determined. If the tension per unit length in the cables is t, and the tension per unit length in the struts is s, write down a *vector equation* that expresses the fact that the regular icosahedral tensegrity is stable, when these forces are applied as described. Give your answer in terms of $\mathbf{r}_1, \mathbf{r}_2, \mathbf{r}_3, \mathbf{r}_4, \mathbf{r}_5$, $\mathbf{r}_6, t, s$, and F.
(e) Re-write your answer to (d) in terms of t, s, L, a, and F.
(f) Using your answers to (c) and (e), find the values of F and s necessary for the regular icosahedral tensegrity to be stable. Give your answers in terms of t and L.

Problem 37: Skid Row
Three equal masses, all of mass m, are placed with zero initial velocity on a frictionless table top, one on top of other. Mass 1 is on top, mass 2 is in the middle,

Fig. 2.52 Three equal
masses, (each of mass m) one
on top of the other, on a table.
A horizontal force, F, is
applied to mass 2

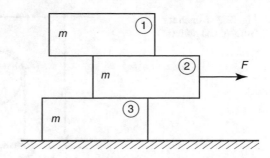

and mass 3 is on the bottom. The coefficient of kinetic friction at both the mass
1-mass 2 contact surface and at the mass 2-mass 3 contact surface is μ. A constant
horizontal force, F, is applied to mass 2, as shown in Fig. 2.52. As a result, the
masses acquire non-zero accelerations and relative velocities. Take that a_1 is the
horizontal acceleration of mass 1, a_2 is the horizontal acceleration of mass 2, and a_3
is the horizontal acceleration of mass 3. The acceleration due to gravity is g.

(a) Draw a force diagram for each mass.
(b) Write six equations corresponding to the application of Newton's 2nd Law to
the three masses. Clearly specify the coordinate system that you have chosen to
write your equations.
(c) What are the constraints in this problem?
(d) Use our model for kinetic frictional forces (namely that the magnitude of the
frictional force between two objects in contact in relative motion is μN, where
μ is the coefficient of kinetic friction of the contact surface and N is the normal
force at that surface) and the constraints from part (c), to solve your equations
from part (b) to find the accelerations, a_1, a_2, and a_3.
(e) In order for your solution to be sensible, F must exceed some minimum value.
Using your solutions for a_1, a_2, and a_3, explain why this is the case, and
determine this minimum value of F.

Problem 38: Masses and Pulleys
Figure 2.53 shows an arrangement of two masses, m_1 and m_2, and two pulleys,
connected together via a massless inextensible string. Your goal in this problem is
to analyze the motion of the two masses. One pulley is attached to the ceiling at a
fixed distance from the ceiling. Mass m_1 is attached to the end of the string. The
other pulley, which is directly attached mass m_2, is supported by the string, so its
position varies as the end of the string moves. A force, F, is applied to the end of
the string, as indicated in the figure. There is no friction in this problem.

(a) Sketch force diagrams, one for each mass, indicating and labeling all relevant
forces.

Fig. 2.53 Sketch for "Masses and pulleys"

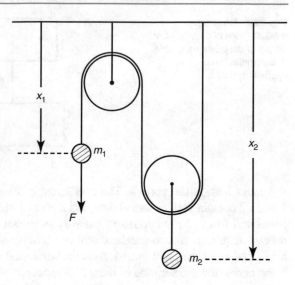

(b) Write two equations that reflect the application of Newton's second law to both masses in vertical direction, Please use the coordinate system indicated in the figure.

(c) How are a_1 and a_2 related to each other? That is, what is the constraint in this problem?

(d) Using your answers to (b) and (c), solve for a_1 and a_2 in terms of given parameters.

(e) Suppose that $m_1 = 0$. In this circumstance, for what range of values of F does mass m_2 increase in height?

References

1. A.J. Engler, S. Sen, H.L. Sweeny, D.E. Discher, Matrix elasticity directs stem cell lineage specification. Cell **126**, 677–689 (2006)
2. N. Wang, J.P. Butler, D.E. Ingber, Mechanotransduction across the cell surface and through the cytoskeleton. Science **260**, 1124–1127 (1993)
3. N. Fletcher, T. Rossing, *The Physics of Musical Instruments* (Springer, Berlin, 1998)
4. E.C. Greene, S. Wind, T. Fazio, J. Gorman, M.L. Visnapuu, DNA curtains for high-throughput single-molecule optical imaging. Methods Enzymol. **472**, 293–315 (2010)

Energy: Work, Geckos, and ATP

> *It is important to realize that in physics today, we have no knowledge of what energy* is.
>
> Richard P. Feynman [1].

Fig. 3.1 Doughnuts. Public domain image from https://commons.wikimedia.org/wiki/File:Krispy_Kreme_Doughnuts.JPG

© The Author(s), under exclusive license to Springer Nature Switzerland AG 2023
S. Mochrie, C. De Grandi, *Introductory Physics for the Life Sciences*,
Undergraduate Texts in Physics, https://doi.org/10.1007/978-3-031-05808-0_3

3.1   Introduction

In this chapter we discuss energy, including the different types of energy that exist, and importantly how energy can be transferred to or removed from a system. We believe that the total energy of the entire universe is fixed, that is, that the total energy of the universe is conserved. For a small part of the universe—a system—what we observe is the transfer of energy in and out of the system or its transformation from one form to another. A system can be simply a moving ball, or the entire Earth and a satellite. Deciding what constitutes the system in question is an important first step to be able to quantify the energy transfers of the system.

Energy can be added to a system in the form of heat. We know that taking an ice cube outside a fridge will cause it to melt. Heat from the environment is a transfer of energy to the ice cube, melting it. In addition, the energy of a system can be increased or decreased by applying a force. For instance, by throwing a ball we give it an energy of motion (called kinetic energy). This energy comes from the work we do applying a pushing force to the ball for a short amount of distance. We will see in this chapter that "work" in this context is actually a very well-defined physical quantity (defined as the integral of force times displacement) and is the amount of energy that can be transferred to an object through an external applied force. Work can also remove energy. For example, the force of friction with the ground will cause a moving object to come to a stop. In this case, the "work" done by the force of friction takes away kinetic energy from the object and transforms it into heat (the surfaces in contact will warm up) and/or sound (coming from the surface scraping past each other). To understand transfers of energy in general, we will discuss a theorem that describes how the work done by forces can increase or decrease the energy of a system, called the *Generalized Work-Energy Theorem*, which will be one of the major results of this chapter. In the second part of the chapter, we will exploit the important connection between the potential energy of a system and the corresponding force between two parts of the system to, for example, model how geckos' feet enable geckos to walk across ceilings, and understand the buckling of beams, drinking straws, or microtubules, when their ends are pushed together. We will also explore a number applications of energy concepts to atomic and molecular bound states. In particular, we will see how an understanding of the binding energies involved in molecules helps us understand their stable configurations.

3.2   Your Learning Goals for This Chapter

By the end of this chapter, you should be able to:

- Apply the law of conservation of energy and the generalized work-energy theorem to analyze processes, taking into account that energy can take the form of kinetic energy, binding energy, gravitational potential energy,

electrostatic potential energy, chemical energy, thermal energy, radiant energy (e.g., light), etc.
- Describe the difference between conservative forces and non-conservative forces, and how conservative forces are related to potential energy.
- Explain the energetics of chemical and nuclear bonding and reactions.
- Calculate the force in a given direction given the functional form of the potential energy versus position.
- Sketch the force versus position given a sketch of the potential energy versus position, and *vice versa*.

3.3   Types of Energy

The law of conservation of energy is, as far as is known, exact and often serves to facilitate analysis of processes in which one form of energy transforms into another. To this end, it is necessary to include all of the different possible forms of energy that may be relevant to the process in question. Many different forms of energy can be identified, including:

- *Kinetic energy.* This is the energy of motion: if a particle of mass m has velocity **v**, its kinetic energy is $\frac{1}{2}mv^2$.
- *Gravitational potential energy.* In systems containing two or more masses, there is a contribution to the total energy of the system that depends on the relative positions of the masses. Any contribution to the energy that depends on coordinates is called "potential energy." Potential energy that depends on relative positions of masses is gravitational potential energy.
- *Electrostatic potential energy.* In systems containing two or more charges, there is a contribution to the total energy of the system that depends on the relative positions of the charges. This energy is the electrostatic potential energy of the system in question.
- *Elastic potential energy.* When we stretch a rubber band, it acquires elastic energy, which depends on the extension of the rubber band. Similarly, when we compress a spring, it acquires elastic energy, which depends on the compression of the spring. Thus, this energy is also a potential energy.
- *Thermal energy*, a.k.a. internal energy. The temperature of an object is an important manifestation of energy. As we will discuss in detail in Chap. 8. *Statistical Mechanics: Boltzmann Factors, PCR, and Brownian Ratchets*, an object's temperature is a measure of the mean kinetic energy of the material's atomic constituents. Specifically, if an object is at a temperature T, each atom within the object undergoes random thermal motions. The corresponding velocities give rise to a mean kinetic energy per atom of approximately $\frac{3}{2}k_BT$,

where k_B is Boltzmann's constant. The corresponding displacements in solid materials give rise to a mean potential energy per atom of approximately $\frac{3}{2}k_B T$.

- *Chemical binding energy.* Chemical reactions, which reconfigure the chemical bonds of the reactants to form the products, can be either exothermic, releasing energy and increasing the temperature of the environment correspondingly, or endothermic, taking in energy and decreasing the temperature of the environment. Different chemical bonds have different "binding" energies. In this chapter we will discuss that chemical binding energies are in fact a manifestation of the kinetic energies and the electrostatic potential energies of the electrons involved in the bonding.
- *Nuclear binding energy.* Just as there are chemical binding energies when atoms are bound close to each other, there are also nuclear binding energies when protons and neutrons are bound together in nuclei. In this case, however, the "strong" force is responsible for the interactions that lead to binding.
- *Radiant energy.* Electromagnetic photons/waves (e.g., radio waves, microwaves, light, x-rays, *etc.*) carry energy. Specifically, the energy (E) of a photon is related to its frequency via $E = h\nu = \hbar\omega$, where h is Planck's constant, $\hbar = \frac{h}{2\pi}$, ν is the photon/wave frequency, and ω is the photon/wave angular frequency.

The SI unit for energy is the Joule:

$$1\,\text{Joule} = 1\,\text{Newton} \times 1\,\text{meter}.$$

The definition of work that we'll introduce in the next section in Eq. 3.5 will allows us to make sense of this unit, more specifically we'll see that work is a form of energy and has units of force times distance. First, in the next Example, we'll get an understanding of how big is a Joule by comparing it with the energy in food (donuts more precisely), which is usually measured in calories or Calories—the capital letter matters—1 Calorie = 1000 calories = 1 kcal). One calorie is 4.2 Joules of energy.

3.3.1   Example: The Most Important Meal of the Day

Before proceeding, it is useful and fun to seek a sense of how different energies compare. This example provides some interesting answers to the question: how much energy is that?

We use gasoline as the fuel in our cars. Evidently, gas is a way to store energy. According to Wikipedia, the energy density of gasoline is 42.4[1]   MJ/kg, this is the energy released per unit of mass.

[1] http://en.wikipedia.org/wiki/Gasoline

(continued)

We use jelly donuts (and similar treats) to fuel our bodies (Fig. 3.1). Therefore, jelly donuts must be another way to store energy. The energy content of a single jelly donut is about 240 Calories, which is equivalent to 240×4200 J $\simeq 1$ MJ, which sounds like a lot! (The conversions are: 1 Calorie $= 1$ kcal. Since 1 cal $= 4.2$ J, 240 Calories $= 240 \times 4200$ J $\simeq 1$ MJ.) Joule is the SI unit for energy, 1 Joule $= 1$ Newton $\times 1$ meter. In comparison, how much energy do we need each day? For many college students we can estimate[2] they will need 2400 Calories per day, which is 10 MJ, equivalent to 10 jelly donuts per day. Yum! (Yuk?) http://en.wikipedia.org/wiki/Food_energy[3] tells us that fats have an energy density of 37 MJ/kg, while carbohydrates have an energy density of about 17 MJ/kg. So for a jelly donut we might guess 25 MJ/kg-ish, somewhat less than gasoline, but comparable.

(a) In Stage 1 of the 2014 Tour de France, riders averaged a mechanical power output of 295 W over the 5 h 33 min/126 mile (203 km) duration of the race. How many croissants (assume 1 croissant $= 1$ jelly donut) should a rider eat to restore his body's energy stores, given that top athletes turn croissants into mechanical energy with an efficiency of 25%?

```
Tour-de-France riders output an average of
20000 s × 295 W ≃     5.9 MJ of mechanical energy,
corresponding to 5.9 croissants. Because they
are 25% efficient, however, they need to eat
about 5.9/0.25 = 24 croissants to replenish.
```

(b) A car gets 25 miles to the gallon. Given that 1 US gallon weighs about 3 kg, how many jelly donuts would be used by the car per mile?

```
1 US gallon is about 3 kg, corresponding to
120 MJ per gallon of gas, containing the energy
equivalent of 120 jelly donuts. Therefore, the
car uses about 5 donuts per mile.
```

(c) Suppose that you ate a jelly donut, but you wish that you had not. So, you decide to go to the gym to "work it off" on the weight machine that we encountered in Chap. 2. *Force and Momentum: Newton's Laws and How to Apply Them*. You load up the weight machine with 50 kg (110 lbs). How many 1 m lifts do you need to do in order to do an amount of work equivalent to one jelly donut, assuming you are 25% efficient?

---

[2]https://en.wikipedia.org/wiki/Basal_metabolic_rate
[3]http://en.wikipedia.org/wiki/Food_energy

(continued)

```
One lift corresponds to a work of 50 kg×10 m
s⁻²×1 m = 500 J. If you are 25% efficient you
expend 2000 J per lift. It follows that you need
to perform 500 repetitions to account for that 1 MJ
jelly donut. If you are in great shape and can do
one repetition in 7 s, then 500 repetitions will
take you one hour! In comparison, sitting for one
hour, you will use about 0.5 MJ.
```

(d) According to http://en.wikipedia.org/wiki/Dynamite,[4] the energy density
    of dynamite is about 5 MJ/kg, five time less than the energy density of a
    jelly donut. Should we be worried about jelly donuts exploding in our
    stomach? Why not?

```
Once ingested, the donuts are converted into
energy via chemical reactions that transform
sugar and oxygen into carbon dioxide and water
with the release of energy. These reactions
happen very slowly, far from looking like an
explosion. In dynamite instead the conversion of
energy happens very quickly.
```

(e) The electrical power output of the Millstone nuclear reactor power station
    in Waterford, Connecticut, USA is 2020 MW.[5] In Millstone, this power
    is generated via the fission reaction of U^{235} with a neutron, producing
    fission products and energy. How many jelly donuts per second of
    electrical power is that?

```
2020 MW corresponds to about 2020 jelly
donuts per second of electrical power. (The
thermal power generated is three times-ish
the electrical power, i.e., Millstone is 30%
efficient.)
```

---

[4]http://en.wikipedia.org/wiki/Dynamite
[5]http://en.wikipedia.org/wiki/Millstone_Nuclear_Power_Plant

3.4 Energy Transfers

Conservation of energy applies to the universe as a whole. However, if we are
focused on a particular "system" that constitutes a small subset of the universe,
we must be aware of the possibility of energy transfers into and out of our system.
In this case, we may express the law of conservation of energy as

$$E_i + E_{in} = E_f + E_{out},$$ (3.1)

which states that the initial energy of the system, E_i, plus the energy transferred into the system by an external agent, E_{in}, is equal to the final energy of the system, E_f, plus the energy transferred out of the system to the remainder of the universe, E_{out}. We may rearrange Eq. 3.1 to read

$$E_f - E_i = E_{in} - E_{out},$$ (3.2)

which recapitulates conservation of energy as the statement that the change in energy of the system (left-hand side of Eq. 3.2) equals the transfer of energy into the system minus the transfer of energy out of the system (right-hand side of Eq. 3.2). For a system, which is isolated from the rest of the universe, E_{in} and E_{out} are both equal to zero, and therefore the total change of energy of the system is zero:

$$\Delta E = E_f - E_i = 0.$$ (3.3)

3.4.1 Examples of Energy Transfers: Heat and Work

There are many different ways to transfer energy into, E_{in}, or out of a system, E_{out}. For example, the Sun transfers energy to the Earth via photons, manifesting radiant energy transfer. Another important type of energy transfer is "heat": if two objects at different temperatures are placed in contact with each other, heat, Q, will flow from the hotter object to the cooler one, leading to a change in temperature, ΔT, given by:

$$Q = C\Delta T,$$ (3.4)

where C is the heat capacity.

Another way to transfer energy into or out of a system is by doing "work" on the system, which is the energy transferred into a system by an external force. Specifically, the work, W, done by a force $\mathbf{F}$ acting between two points $\mathbf{r}_i$ and $\mathbf{r}_f$ is the *line integral* of the force along the path that joins the initial and final points:

$$W = \int_{\mathbf{r}_i}^{\mathbf{r}} \mathbf{F}(\mathbf{r}) \cdot d\mathbf{r}.$$ (3.5)

Importantly, the integrand in Eq. 3.5 is the dot product of two vectors. Because of the properties of the dot product, which we discussed in the first chapter, an essential aspect of work is that solely the component of the force that is parallel to the displacement contributes to the work. In particular, if the force acts along the same direction of the displacement of the object (e.g., a rope pulling a heavy object forward), the work done by that force is *positive* and therefore energy is added to the system (for instance increasing the kinetic energy of the object); if the force acts in the opposite direction of the displacement of the object (e.g., the frictional force with the ground on an object that is pulled forward), the work done by that force is *negative* and therefore energy is removed from the object (for instance kinetic energy is transformed into heat). Therefore according to each situation the work done by a force can add (E_{in}) or remove (E_{out}) energy from the system under consideration. It is important to note that the component of a force perpendicular to the displacement of an object (e.g., a normal force) does not do any work and so does not transfer any energy.

3.4.2   Kinetic Energy and the Work-Energy Theorem

Consider the simplest possible "system," namely a particle, whose dimensions are not important for the description of its motion. Our goal for this section is to introduce the energy for this simple one-particle system – its kinetic energy.

To understand how the energy of a point particle of mass m and velocity $\mathbf{v}$ changes when the particle is subjected to a force—that is, when work is done on the particle—we examine the time rate of change of the quantity, $K = \frac{1}{2}mv^2$, which has dimensions of energy, and which will turn out to be *kinetic energy*—the energy of motion.

$$\frac{d}{dt}\left[\frac{1}{2}mv^2\right] = \frac{d}{dt}\left[\frac{1}{2}m\,\mathbf{v}\cdot\mathbf{v}\right] = \frac{d}{dt}\left[\frac{1}{2}m(v_x^2 + v_y^2 + v_z^2)\right]$$

$$= m\left(\frac{dv_x}{dt}v_x + \frac{dv_y}{dt}v_y + \frac{dv_z}{dt}v_z\right) = m\frac{d\mathbf{v}}{dt}\cdot\mathbf{v} = m\mathbf{a}\cdot\mathbf{v} = \mathbf{F}\cdot\mathbf{v},$$

$$(3.6)$$

where in the last step we used Newton's Second Law: $\mathbf{F} = m\mathbf{a}$, where $\mathbf{F} = \mathbf{F}(\mathbf{r})$ is the force acting on the particle at position $\mathbf{r}$. Multiplying Eq. 3.6 by Δt informs us that

$$\frac{d}{dt}\left[\frac{1}{2}m\,\mathbf{v}\cdot\mathbf{v}\right]\Delta t = \mathbf{F}(\mathbf{r})\cdot\mathbf{v}\Delta t. \qquad (3.7)$$

Assuming Δt to be small, we can write the velocity as $\mathbf{v} = \frac{\Delta \mathbf{r}}{\Delta t}$ and therefore we have that

$$\frac{d}{dt}\left[\frac{1}{2}m\,\mathbf{v}\cdot\mathbf{v}\right]\Delta t = \mathbf{F}(\mathbf{r})\cdot\Delta\mathbf{r}. \qquad (3.8)$$

Equation 3.8 applies to a small element of the particle's trajectory. We can envision the entire trajectory to be composed of many such elements. For element j, the time interval is Δt_j, the particle's displacement is $\Delta\mathbf{r}_j$, and the particle is at $\mathbf{r}_j$. Summing all these elements together, we arrive at:

$$\Sigma_j\frac{d}{dt}\left[\frac{1}{2}m\mathbf{v}_j\cdot\mathbf{v}_j\right]\Delta t_j = \Sigma_j\mathbf{F}(\mathbf{r}_j)\cdot\Delta\mathbf{r}_j. \qquad (3.9)$$

Taking the limit of infinitesimal Δt and $\Delta \mathbf{r}$ transforms the sums of Eq. 3.9 into integrals with the result that

$$\int_{t_i}^{t_f} \frac{d}{dt}\left[\frac{1}{2}m\mathbf{v}\cdot\mathbf{v}\right] dt = \int_{\mathbf{r}_i}^{\mathbf{r}_f} \mathbf{F}(\mathbf{r})\cdot d\mathbf{r}, \tag{3.10}$$

where $\mathbf{r}_i$ is the position at time t_i and $\mathbf{r}_f$ is the position at time t_f. The quantity on the right-hand side of Eq. 3.10 is the work done, W, by the force $\mathbf{F}$. Work is the *line integral* of the force along the path of the particle. The line integral can best be conceptualized via Eq. 3.9 as the sum of all of the tiny displacements that constitute the particle's path, each "dotted"(i.e., taking the dot product) with the force on the particle at the location of each such displacement.

The fundamental theorem of calculus, namely that integration is the inverse of differentiation, then informs us that

$$\frac{1}{2}mv_f^2 - \frac{1}{2}mv_i^2 = \int_{\mathbf{r}_i}^{\mathbf{r}_f} \mathbf{F}(\mathbf{r})\cdot d\mathbf{r}, \tag{3.11}$$

where $\mathbf{v}_i$ is the velocity at time t_i and $\mathbf{v}_f$ is velocity at time t_f. The left-hand side of Eq. 3.11 is the difference between the final and initial values of the quantity $K = \frac{1}{2}mv^2$. The right-hand side is the work done by the force, which is the energy transferred to the system by the force. Thus, for a system comprising of a point particle we have found:

$$K_f - K_i = \int_{\mathbf{r}_i}^{\mathbf{r}_f} \mathbf{F}(\mathbf{r})\cdot d\mathbf{r}. \tag{3.12}$$

This result tells us that the work done by the force acting on the system (right-hand side) is equal to the change of the quantity $K = \frac{1}{2}mv^2$ (left-hand side). But the work done is the change in the energy of the system. It follows that the quantity

$$K = \frac{1}{2}mv^2 \tag{3.13}$$

is the *kinetic energy* of the system. This expression (Eq. 3.13) can be generalized to a system of many particles. $K_{\text{tot}} = K_1 + K_2 + K_3 + \ldots\ldots$ is the total kinetic energy of many particles, equal to a sum of the kinetic energy of each particle in the system.

This result, Eq. 3.12, is the *work-energy theorem*. In general the force $\mathbf{F}$ appearing on the right-hand side of Eq. 3.12 is the total net force acting on the system. The work-energy theorem states that the total work done on a system comprising a point particle is equal to the change in its kinetic energy. The work-energy theorem is a particular case of the *generalized work-energy theorem*, Eq. 3.31 that we will define a little later in the chapter and that is more broadly applicable.

To apply Eq. 3.12, it is necessary to follow a stepwise strategy (similar to the stepwise strategies discussed in the previous chapter):

1. Decide what exactly comprises the system
2. Decide which process you are interested in analyzing, and therefore identify what is the initial time t_i of the process
3. Identify what is the final time t_f of the process of interest
4. Identify all interactions between the system and its environment, and the corresponding total energy transfers into the system, E_{in}, and out of the system, E_{out} (for instance work done by forces on the system).

3.4.3 Example: The Work-Energy Theorem for the Gravitational Force and Escape Velocity

To apply the work-energy theorem, the system should effectively be an isolated point mass, subject to external forces. In this example, we examine what the work-energy theorem tells us about the motion of a projectile of mass m, that is shot upwards from the surface of the Earth (radius R), subject to the the gravitational force of the Earth (mass M), which is the external force in this case. For a center of mass-to-center of mass separation r, the force on the projectile is in the radial direction and is

$$\mathbf{F}(r) = -\frac{GMm}{r^2}\hat{\mathbf{r}}, \qquad (3.14)$$

according to Newton's universal law of gravitation, where $\hat{\mathbf{r}}$ is a unit vector in the radial direction. The minus sign corresponds to the fact that the force is in the direction of decreasing r.

(a) Calculate the work done on the projectile by the Earth's gravitational force when the projectile moves from radius R to radius r. Is the work done positive or negative?

```
Answer: Following the prescription given above:
1. The system is comprised of the projectile,
   treated as a point mass,
2. The initial time corresponds to the
   projectile being at radius R with velocity
   v_i,
3. The final time corresponds to the projectile
   being at radius r with velocity v,
4. The only significant interaction of the
   projectile with its environment is the
```

(continued)

gravitational force between the projectile and the Earth.

Because the force is in the radial direction, only the radial component of the path, $\hat{r}dr$ contributes to the work integral. Specifically, the work done on the projectile by the gravitational force as it moves from radius R to a radius r is

$$-\int_R^r \frac{GMm}{r^2}dr = \frac{GMm}{r}\bigg|_R^r = \frac{GMm}{r} - \frac{GMm}{R}. \qquad (3.15)$$

Because $R < r$, the work done on the projectile by the gravitational force of the Earth is negative, which makes sense since the projectile is moving in the direction opposite to the attractive gravitational force, i.e., gravity acts toward the Earth, but the projectile moves outwards away from the Earth.

(b) What is the projectile's kinetic energy when it is at radius r?

Answer: The work-energy theorem informs us that

$$\frac{1}{2}mv^2 - \frac{1}{2}mv_i^2 = \frac{GMm}{r} - \frac{GMm}{R}, \qquad (3.16)$$

where v is the velocity at the final radius, r, and v_i is the velocity at the initial radius, R. Therefore, rearranging this equation, the kinetic energy at radius r is:

$$\frac{1}{2}mv^2 = \frac{1}{2}mv_i^2 + \frac{GMm}{r} - \frac{GMm}{R}. \qquad (3.17)$$

(c) What is the minimum launch velocity, namely the "escape velocity," for the projectile to completely escape from the gravitational attraction of the Earth?

Answer: To answer this question, we must recognize (1) that to completely escape is to reach $r = \infty$, and (2) that the minimum initial velocity necessary to achieve $r = \infty$ will

(continued)

occur if the final velocity at $r=\infty$ is zero,
corresponding to zero kinetic energy at that
point. So, evaluating Eq. 3.16 for $r=\infty$ and $v=0$,
we find

$$-\frac{1}{2}mv_i^2 = -\frac{GMm}{R},$$ (3.18)

which leads to

$$v_i = \sqrt{\frac{2GM}{R}}.$$ (3.19)

Equation 3.19 is the escape velocity. To escape
the Earth's (Sun's) gravitational attraction,
starting from the surface of the Earth, this
expression yields v_i = 1.12×10^4 m s^{-1} (4.21×10^4 m s^{-1}).

3.4.4 Example: The Work-Energy Theorem for a Sliding Block with Friction

In this example, our goal is to determine the distance, s, that a block of mass, m, moves, when it slides across a horizontal surface, subject to a constant frictional force, f, having started with an initial velocity, v_i.

(a) Use the work-energy theorem to relate the distance the block moves, s, to its initial velocity, v_i, its mass, m, and the frictional force, f.

Answer: Again, following the prescription given
above, in this case, we pick
1. The system is comprised of the (center of
   mass coordinate of the) block,
2. The initial time corresponds to the block
   with initial velocity, v_i, and position $x=0$,
3. The final time corresponds to the block
   having zero velocity, and position $x=s$,
4. The only significant interaction of the
   block with its environment is the frictional
   force, f.

(continued)

The work-energy theorem informs us that

$$0 - \frac{1}{2}mv_i^2 = \int_0^s -f\,dx = -f\int_0^s dx = -fs. \qquad (3.20)$$

This is the desired relation. In this case, the work is negative, corresponding to a transfer of energy out of the system.

(b) Use your answer to (a) to determine s in terms of v_i, m, and f.

Answer: Solving Eq. 3.20 for s yields

$$s = \frac{mv_i^2}{2f}. \qquad (3.21)$$

We see that the distance that the block skids is proportional to v_i^2 and inversely proportional to f.

(c) Discuss what the principle of conservation of energy tells us in this case regarding the energy transferred into the system E_{in} and the energy transferred out of the system E_{out}.

Answer: The initial (kinetic) energy of the block is non-zero, $E_i = \frac{1}{2}mv_i^2$. The final kinetic energy of the block is zero, $E_f = 0$. The work is $-fs$, which is equal to $E_{in} - E_{out}$. Therefore, energy has been transferred out of the system. What has happened to this energy? $-fs$ is the amount of energy transferred to the system by the frictional force. Because $-fs$ is negative, energy has left "the system." But energy is conserved so the energy itself must have gone somewhere. What has happened is that it has become thermal energy that has caused the temperature of the block and the surface to increase. But is not the block part of "the system"? And so does not raising the temperature of the block cause the energy of "the system" to increase? Actually, when we apply the work-energy theorem in this example, we conceive

(continued)

"the system" to be the center of mass coordinate
of the block. The energy of the center of
mass coordinate is only kinetic energy and is
unaffected by the temperature of the block.
By a series of careful experiments on the
heating of water by a paddle wheel James Joule[6]
in the mid-nineteenth century was first able
to show that the loss of kinetic energy was
accompanied by an increase in an equivalent
amount of thermal energy, signaled by an
increase in the temperature of the water. Joule
concluded that in his experiments thermal energy
plus mechanical energy was conserved. In the
case at hand, the block and the surface material
are slightly hotter after the sliding event than
before.
Heat transfers to or from a system at a
different temperature from its surroundings
represent an additional route for energy
transfers. Often, the correct explanation for
an apparent violation of energy conservation is
a previously unaccounted-for energy dissipation
mechanism into thermal energy.

---

[6]http://en.wikipedia.org/wiki/James_Joule

3.5   Potential Energy and Work Done by Conservative Forces

In some cases the work done by a force (Eq. 3.5) does not depend on the path taken
but only on the end points of the integral. In these cases, the force is said to be a
conservative force and it is possible to introduce a potential energy associated with
the force and the two interacting objects that are subject to the force in question.
Examples of conservative forces are:

• The gravitational force ($\hat{\mathbf{r}}$ is a unit vector in the radial direction):

$$\mathbf{F}(\mathbf{r}) = -\frac{GM_1M_2}{r^2}\hat{\mathbf{r}}.$$ (3.22)

• The gravitational force near the Earth's surface:

$$\mathbf{F}(z) = -mg\hat{\mathbf{k}}.$$ (3.23)

- The force of a spring (Hooke's Law):

$$\mathbf{F}(x) = -kx\hat{\mathbf{i}}. \tag{3.24}$$

- The electrostatic force between charges Q_1 and Q_2, separated by a distance r:

$$\mathbf{F}(\mathbf{r}) = \frac{Q_1 Q_2}{4\pi\epsilon_0 r^2}\hat{\mathbf{r}}. \tag{3.25}$$

What all these forces have in common is that they characterize the interaction between two objects (e.g., two masses, two charges, the two ends of a spring). The work integral of a conservative force does not depend on the path. Therefore, when "the system" contains the two interacting objects, we can define a contribution to the total energy of this at-least-two-object system—*potential energy, U*—via:

$$- U(\mathbf{r}) + U(\mathbf{r}_i) = \int_{\mathbf{r}_i}^{\mathbf{r}} \mathbf{F} \cdot d\mathbf{r}. \tag{3.26}$$

The hallmark of any potential energy is that it depends only on coordinates that describe the relative position of the two interacting objects within the system.

To calculate the potential energy corresponding to each of conservative forces given above, we apply Eq. 3.26 along any convenient path for carrying out the work integral. Thus, we find the following potential energies:

- The gravitational potential energy of masses M_1 and M_2 with separation r:

$$U(\mathbf{r}) = -\frac{GM_1 M_2}{r}. \tag{3.27}$$

- Gravitational potential energy at height z near the Earth's surface:

$$U(z) = mgz. \tag{3.28}$$

- The energy of a spring (from Hooke's Law), displaced a distance x from its equilibrium position:

$$U(x) = \frac{1}{2}kx^2. \tag{3.29}$$

- The electrostatic potential energy of charges Q_1 and Q_2, separated by a distance r:

$$U(\mathbf{r}) = \frac{Q_1 Q_2}{4\pi\epsilon_0 r}. \tag{3.30}$$

Forces that correspond to potential energies are called "conservative forces," because they conserve "mechanical" energy, defined as the sum of kinetic energy and potential energies. Forces for which the work integral depends on the path followed and therefore that do not correspond to potential energies are non-conservative. Frictional forces are prime examples of non-conservative forces. When energy is transferred by a non-conservative force, i.e., when a non-conservative force does work, some or all of the transferred energy ends up as thermal energy, making something hotter than before.

3.6   Generalized Work-Energy Theorem

We have so far introduced two types of energy, kinetic energy and potential energy. We also related the change in kinetic energy for a system, comprising a single point mass, to the work done on the system by an external force, namely the work-energy theorem (Eq. 3.12). We are now ready to consider what happens when work is done on a more complicated system, comprising more than a single point mass. Let us define E as the total energy of such a system, including the kinetic energies of its constituents, the potential energy, associated with interactions among these constituents, and any other forms of energy, internal to the system. Then, the *generalized energy-work theorem* states that the change in the total energy of a system subject to an external force, $\mathbf{F}$, equals the work done by the external force on the system:

$$E_f - E_i = \int_{\mathbf{r}_i}^{\mathbf{r}_f} \mathbf{F}(\mathbf{r}) \cdot d\mathbf{r} = W. \qquad (3.31)$$

On the left-hand side of Eq. 3.31 is the change in the total energy of the system, $E_f - E_i$, where E_i is the initial energy of the system, before the external force acted on it, and E_f is the final energy of the system, after the external force acted on it. Therefore $E_f - E_i$, which we can also write as ΔE, is the total change in the energy of the system from the initial time t_i to the final time t_f. The right-hand side of Eq. 3.31 is the work, W, done by the total external force ($\mathbf{F}$), applied to the system between the initial position $\mathbf{r_i}$ and the final position $\mathbf{r}_f$. If there are a number of distinct contributions to the total external force, e.g., gravity and friction, they each give rise to a corresponding contribution to the total work. Please be aware, as we will see in an example below, that, depending on how we define the system of interest, some forces may be external forces for one choice of the system, but the same forces may be internal forces for a different choice of system.

3.6.1 Example: Working Out

You are at the gym and you lift a mass m a distance L. Both the initial and final velocities of the mass are zero.

(a) Considering a system consisting of the mass and the Earth, determine the change in kinetic energy of the system.

Answer: Since both the final and initial kinetic energies are zero, their difference, namely the change in kinetic energy, is also zero.

(b) Considering a system consisting of the mass and the Earth, determine the change in potential energy of the system.

Answer: In this case, it is simplest to represent the gravitational force between the Earth and the mass as mg. Then, the change in potential energy when you raise the mass up a distance L is mgL, which is positive, corresponding to an increase in the potential energy.

(c) What work, W, did you do to raise the mass?

Answer: Here, because the system is more that an isolated point mass, we must apply the generalized work-energy theorem, which informs us that

$$W = E_f - E_i. \qquad (3.32)$$

In this case, the difference in energy is solely the difference in potential energy, namely mgL. Therefore,

$$W = mgL. \qquad (3.33)$$

This is the work that you did to raise up the mass.

(d) Comment on the sign of the work you did.

(continued)

Answer: You did positive work, since the change
in potential energy, mgL, is positive. This sign
is consistent with the fact that the force that
you exert to lift the mass is upwards while the
motion of the mass is also upwards.

3.6.2   Example: The Earth-Projectile System Revisited

In Sect. 3.4.3, we determined the velocity of a projectile shot up from the
surface of the Earth as a function of the distance from the center of the Earth
to the center of mass of the projectile. There, we chose our system to be the
projectile alone and the gravitational force was an external force that did work
on the projectile, thereby changing the projectile's kinetic energy, according
to the work-energy theorem. We now reconsider this situation in light of the
generalized work-energy theorem and considering the energy of the system
as composed of both the kinetic energy and the potential energy. Indeed if
we consider a system which includes both the projectile and the Earth, this
leads us to recognize another form of energy, the potential energy U, which
depends on interactions between the projectile and the Earth.

From this new point of view:

1. The system is comprised of both the projectile and the Earth
2. The initial time corresponds to the projectile being at radius R with
   velocity v_i
3. The final time corresponds to the projectile being at radius r with velocity
   v
4. There are now no external forces acting, because the gravitational interaction between the Earth and the satellite is now internal to the system.
   It follows that no work is done on the system. Thus, according to the
   generalized work-energy theorem Eq. 3.31, $E_f = E_i$.

In this case the total energy is the sum of kinetic and potential energy, in
particular:

$$E_i = \frac{1}{2}mv_i^2 - \frac{GMm}{R}$$

is the initial total energy, and

$$E_f = \frac{1}{2}mv^2 - \frac{GMm}{r}$$

(continued)

is the final total energy. Therefore the generalized work-energy theorem states that:

$$\frac{1}{2}mv^2 - \frac{GMm}{r} = \frac{1}{2}mv_i^2 - \frac{GMm}{R}. \tag{3.34}$$

This expression is equivalent to what we previously found in Eq. 3.16 (after rearranging the terms), indeed such result must remain correct irrespectively of how we chose to describe the system. Equation 3.34 may be written more generally as:

$$K_f + U_f = K_i + U_i, \tag{3.35}$$

expressing that when there is no energy transfer—e.g., no work by an external force to the Earth-satellite system—the final total energy of the system is equal to the initial total energy of the system.

3.7 Force from Potential Energy

In Sect. 3.5, we discussed how the existence of an interaction force between two parts of a system can give rise to potential energy. It is also possible to determine force from potential energy by taking the derivative of Eq. 3.26. For example, for a force directed in the radial direction, such as the gravitational force, Eq. 3.26 becomes

$$\int_{r_i}^{r} F(r)dr = -U(r) + U(r_i). \tag{3.36}$$

Taking the derivative of both sides of this equation with respect to r, we find

$$F(r) = -\frac{dU}{dr}. \tag{3.37}$$

Equation 3.37 is an important result because it allows us to find the force if we know the potential energy. For the gravitational interaction, where

$$U = -\frac{GMm}{r}, \tag{3.38}$$

the corresponding force is

$$F = -\frac{dU}{dr} = -\frac{GMm}{r^2}, \tag{3.39}$$

with the correct negative sign, indicating that the force is in the direction of decreasing r. The force given by Eq. 3.37 is internal to the two-mass system that we are considering. In the gravitational case, it corresponds to the attractive force between the two masses. It is not an external force and consequently is not involved in any work on the system. Instead, the existence of this internal force is manifest as an internal-to-the-system potential energy.

The generalization of Eq. 3.37 to three dimensions is that

$$\mathbf{F} = -\frac{dU}{dx}\hat{\mathbf{i}} - \frac{dU}{dy}\hat{\mathbf{j}} - \frac{dU}{dz}\hat{\mathbf{k}}, \qquad (3.40)$$

which is also an important result. Examination of Eq. 3.40 reveals that the component of the force along the x direction is

$$F_x = -\frac{dU}{dx}, \qquad (3.41)$$

and similarly along y and z, i.e., the force along a given direction is determined by how the potential energy depends on the coordinate along that direction. Although not all forces give rise to potential energy, whenever it is possible to write an energy that depends on coordinates—a potential energy—it is possible to calculate the corresponding (conservative) force.

3.7.1   Example: Force from Potential Energy Graphically

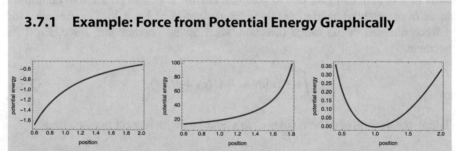

The figure above shows three potential energies as a function of position. Sketch the corresponding force-versus-position curves.

Answer: Force is related to potential energy via $F = -\frac{dU}{dx}$, that is, the force-versus-position curve is the negative of the slope of the potential energy as a function of position. The force-versus-position curves corresponding to the potential energy curves above are sketched below.

(continued)

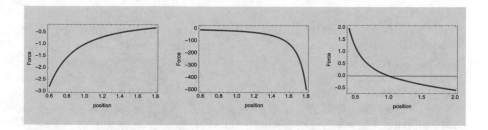

3.8   Potential Energy Models to Predict the Force

Physicists love to build mathematical models of the natural world and then compare those models to experiments. Often it is possible to measure a force, and often the most straightforward route to predict what that force is for a system of interest, is to first calculate the potential energy of the system, and then apply Eq. 3.37.

3.8.1   Example: Energy, Force, and Geckos' Feet

Geckos can walk on ceilings and up walls and, in fact, can easily support their weight on a single toe. This feat is accomplished by virtue of millions of "setae" and "spatulae" on their feet, as shown in Fig. 3.2. The hierarchical branched structures on geckos' foot pads, branching first into setae and then

(continued)

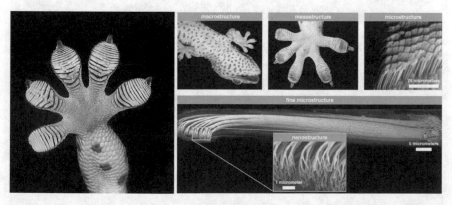

Fig. 3.2 Left: Underside of the foot of a gecko, walking on glass. Public domain image from https://commons.wikimedia.org/wiki/File:Tokay_foot.jpg. Right: Structural hierarchy of gecko toes From Ref. [2] and https://en.wikipedia.org/wiki/File:Micro_and_nano_view_of_gecko%27s_toe.jpg. This work is licensed under the Creative Commons Attribution-ShareAlike 3.0 License

into spatulae, mean that the gecko's foot pads are very soft on a macroscopic scale and therefore can conform to almost any surface profile. As a result, the aggregate area of billions of spatulae is in nanometer scale contact with a surface. Because there is a negative surface binding energy between the spatulae and a surface, the foot pad adheres to the surface. However, there is a seeming paradox: if the gecko's foot pad adhesion can support many times the gecko's weight, how then can the gecko even move its own feet? How can a gecko control the release of foot-surface adhesion, so that it can walk?

The goal of this example is to develop a simple model for the potential energy, U, of adhesion of a gecko's spatula to a surface, and then calculate the corresponding force along the x-direction via $F = -\frac{dU}{dx}$. This approach will allow us to understand how a gecko is able to control the adhesion of its feet to a surface and thus walk across a ceiling.

As we will see, the force required to release the spatulae depends strongly upon the angle at which the force is applied. The gecko controls this angle to control release. Actually, this aspect may not come as a surprise, because it is similar to what happens when you unpeel sticky tape: only a small force is needed if you peel the tape back, but if you try to pull it off the other way, the force required for release is very large.

A simple model for a gecko's spatula is shown in Fig. 3.3. As shown in this figure, a portion of a gecko's spatula is in contact with a horizontal surface. The surface-spatula interaction energy per unit area is σ, which for an attractive, bonding interaction, must be negative. At the same time, a portion of the spatula is at an angle θ to the horizontal.

(a) Consider a fixed-length spatula, and, on the basis of the geometry shown in Fig. 3.3, calculate the displacement along the pulling direction, $x - x_i$,

Fig. 3.3 The geometry for geckos' feet and sticky tape. When the length of the spatula/tape adhered to the surface changes by $L - L_i$, the displacement of the end of the spatula/tape parallel to the unattached spatula/tape changes correspondingly by $x - x_i$

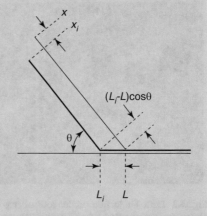

(continued)

that corresponds to a change in the length, $L - L_i$, of the adhered portion of the spatula. Since it specifies the length of the spatula, that is adhered to the surface, L increases towards the left. Then, inspection of the figure shows that when $x - x_i$ is positive, $L - L_i$ is negative.

Answer: It follows directly from the geometry shown that

$$L_i - L = (L_i - L)\cos\theta + x - x_i, \tag{3.42}$$

i.e.,

$$x - x_i = (L - L_i)(\cos\theta - 1). \tag{3.43}$$

(b) Given that the spatula binding energy per unit area is σ and that the width of the spatula is W, calculate the change in potential energy, ΔU, for a given displacement along the pulling direction, $x - x_i$.

Answer: For a displacement $x - x_i$ along the pulling direction, the length of the spatula attached to the surface changes by $L - L_i$, which is a negative quantity. The surface binding energy changes correspondingly:

$$U - U_i = \sigma W(L - L_i), \tag{3.44}$$

where W is the width of the spatula/tape. It follows that

$$U - U_i = \sigma W(L - L_i) = \frac{\sigma W}{\cos\theta - 1}(x - x_i). \tag{3.45}$$

Note that because both σ and $L - L_i$ are negative, $U - U_i$ is positive.

(c) Calculate the corresponding interaction force between the spatula and the surface.

Answer: We may calculate the force along the direction of the released spatula/tape via

$$F = -\frac{dU}{dx} = -\frac{\sigma W}{\cos\theta - 1}. \tag{3.46}$$

(continued)

Since both σ and $\cos\theta$ —   1 are negative, F is
also negative, i.e., F is in the direction of
decreasing x.

(d) What is the sign of the force required to pull the spatula off the surface?

Answer: In light of the answer to (c), to pull
the spatula off the surface the gecko's leg
muscles must provide an equal (or greater)
opposing external force:

$$F_{pull\ off} = \frac{\sigma W}{\cos\theta - 1},\qquad(3.47)$$

which is positive, in the direction of
increasing x.

(e) Explain how a gecko can control the adhesion of its feet.

Answer: The θ-dependence of Eq. 3.47 reveals
the angle dependence of the pull-off force:[7]
For θ = π, corresponding to peeling the
spatula/tape backwards, the pull-off force is
relatively small, equal to $-\frac{1}{2}\sigma W$. However, for
θ progressively smaller, the pull-off force
becomes progressively larger. Therefore, as
shown schematically in Fig. 3.4, if the gecko
places its feet so that its weight effectively
hangs from the spatulae, in that orientation the
pull-off force is very large, and the gecko will
be able to walk up a walk. However, later on,
when the gecko wants to step, if it then moves
its foot, so as to peel off the spatulae, the
pull-off force is small and the gecko can take
the step. You are familiar with this phenomenon
from when you peel off sticky tape.

---

[7]http://www.wolframalpha.com/input/?i=plot+1%2F%281-cos%28x%29%29+for+x%3D0.01+to+pi

Fig. 3.4 Schematic sketch of how the magnitude of the pull-off force depends on the direction of the pull-off force

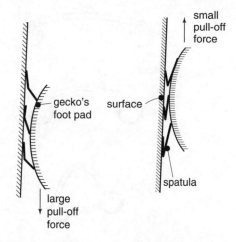

3.8.2   Example: Microtubule Buckling

The eukaryotic cytoskeleton[8] is built from three types of "biological beams," namely actin filaments,[9] intermediate filaments,[10] and microtubules,[11] which all perform structural roles within cells. In particular cases, cells build specific structures by combining individual cytoskeletal filaments. For example, bundles of actin filaments create protrusions—microvilli[12] —on the surface of your intestinal epithelial cells—striated border cells—one of whose key functions is to increase cellular surface area, through which nutrients can be absorbed from the GI tract.

As shown in Fig. 3.5 a microtubule under a compressive load may undergo buckling, as indeed may any beam under a compressive load. Our goal in this example is to determine the conditions under which we may expect microtubule buckling. To accomplish this goal, we will assume that a buckled microtubule takes the shape of a circular arc. Then, its elastic energy may be expressed as

$$U = \frac{1}{2}\frac{\kappa L}{R^2}, \tag{3.48}$$

[8]http://en.wikipedia.org/wiki/Cytoskeleton

[9]https://en.wikipedia.org/wiki/Microfilament

[10]https://en.wikipedia.org/wiki/Intermediate_filament

[11]https://en.wikipedia.org/wiki/Microtubule

[12]https://en.wikipedia.org/wiki/Microvillus

(continued)

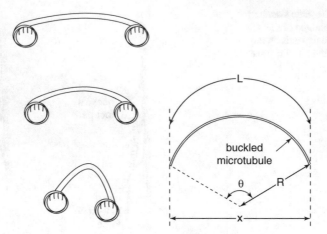

Fig. 3.5 Left: Sketch of a single 20 μm-long microtubule, held between two glass micro-beads at several different micro-bead separations, adapted from Ref. [3]. Using optical tweezers to move the beads together causes the microtubule to progressively buckle as the bead separation is decreased. Right: Schematic of a buckled microtubule, showing the relevant variables

where L is the total length of the microtubule, κ is its "bending modulus," and R is its radius of curvature. Equation 3.48 specifies the bending energy of a beam that is bent into a circular arc and applies to beams of any size, from actin filaments and microtubules, to straws, to our long bones, to beams in buildings.

The bending modulus is the product of the "Young's modulus" of the material that the beam is made of, and a factor that depends on the size and shape of the beam cross-section. Young's modulus is a material property[13] that specifies how stiff the material is (Fig. 2.4).

(a) Find a plastic drinking straw, or something similar. This is your "micro-tubule." Put one end on a table and on the other end press toward the table with your finger. What happens to the shape of the "microtubule" for the smallest forces? What happens for larger forces (but not so large that you put a kink in it)?

---

[13]https://en.wikipedia.org/wiki/List_of_materials_properties

(continued)

For the smallest forces, nothing happens to the shape of the straw. For forces beyond a critical value, the straw buckles and becomes curved. In fact, for any beam placed under end-to-end compression, there is a critical compressive force beyond which the beam buckles, but before which it remains straight.

(b) Consider such a microtubule, held between two micro-beads, separated by a distance x, as shown in Fig. 3.5. Determine the geometrical relationship between θ, R, and L.

Answer: The microtubule approximately conforms to a circular arc of length L. It follows that

$$\theta = \frac{L}{R}. \tag{3.49}$$

(c) Determine the geometrical relationship between θ, R, and x.

Answer:

$$x = 2R \sin \frac{\theta}{2}. \tag{3.50}$$

(d) Using your answers to (b) and (c), eliminate R to find x in terms of θ and L for small values of θ.

Answer: Rearranging Eq. 3.49 implies $R = \frac{L}{\theta}$. Using $R = \frac{L}{\theta}$ in Eq. 3.50 yields

$$x = \frac{2L}{\theta} \sin \frac{\theta}{2} \simeq L - \frac{L\theta^2}{24} + \frac{L\theta^4}{1920}, \tag{3.51}$$

where we used Wolfram Alpha[14] to find the small-θ result. Often, we might drop all terms except the lowest order in θ. Here, we keep the θ^4-term, because it will turn out that it is needed later in the problem.

---

[14]http://www.wolframalpha.com/input/?i=Series+2%2Fh*sin%28h%2F2%29

(continued)

(e) Using Eq. 3.48 and your answer to (b), write the elastic energy of the buckled microtubule in terms of θ.

Answer:

$$U = \frac{1}{2}\frac{\kappa L}{R^2} = \frac{1}{2}\frac{\kappa \theta^2}{L}. \tag{3.52}$$

(f) Calculate the force (F) exerted by the buckled microtubule in the horizontal (x) direction as a function of θ for small θ.

Answer: The buckled microtubule elastic energy depends on the coordinates of the microtubule. Therefore, it is a potential energy whose negative derivative in the x-direction is the force in the x-direction. Using the chain rule of differentiation, we have that

$$F = -\frac{dU}{dx} = -\frac{dU}{d\theta}\frac{d\theta}{dx}. \tag{3.53}$$

Using

$$\frac{dU}{d\theta} = \frac{\kappa\theta}{L}, \tag{3.54}$$

and

$$\frac{dx}{d\theta} = -\frac{L\theta}{12} + \frac{L\theta^3}{480}, \tag{3.55}$$

we find

$$F = \frac{12\kappa}{L^2(1 - \frac{\theta^2}{40})} \simeq \frac{12\kappa}{L^2}(1 + \frac{\theta^2}{40}). \tag{3.56}$$

This equation shows that F is positive, indicating that it is in the direction of increasing x, i.e., outwards.

(g) Briefly explain whether your answer to (f) accounts for the behavior that you observed in (a)?

(continued)

Answer: In order for the buckled microtubule to be stable, the total force on the microtubule must be zero, which implies that the external compressive force that maintains the buckling must be equal and opposite to the force exerted by the microtubule. The result of (f) is that for small values of θ, a buckled microtubule--more generally, any buckled beam--exerts a more-or-less constant force in the direction of increasing x. Only when the external compressive force is equal to or larger than a "critical force," $\frac{12\kappa}{L^2}$ can this external force balance the force exerted by the microtubule. For external compressive forces less than $\frac{12\kappa}{L^2}$ a balance is not possible and it must be that the microtubule does not buckle. This prediction matches, and therefore accounts for, the behavior observed in (a). Pretty cool.

(h) For a buckled microtubule, how does θ depend on the magnitude of the external compressive force, $|G|$, for small values of θ.

Answer: For a stable, mechanical equilibrium, we require that

$$|G| = F = \frac{12\kappa}{L^2(1 - \frac{\theta^2}{40})} \simeq \frac{12\kappa}{L^2}(1 + \frac{\theta^2}{40}), \qquad (3.57)$$

i.e.,

$$\theta = \sqrt{40}\sqrt{\frac{|G|L^2}{12\kappa} - 1}, \qquad (3.58)$$

which applies for $|G| > \frac{12\kappa}{L^2}$. For $|G| < \frac{12\kappa}{L^2}$, there is no buckling and $\theta = 0$.

(i) Briefly explain how the analysis in this example could lead to a method for determining the microtubule bending modulus, κ.

Answer: A measurement of the buckling force as a function of microtubule length would be an

(continued)

elegant way to measure the bending modulus.
Clearly, the optical tweezers-microscopy
experiments, shown in Fig. 3.5 can measure the
length of the microtubule [3]. In fact, optical
tweezers measurements can also measure force and
therefore can determine the critical buckling
force for different length microtubules, as
indeed was done in Ref. [3].
Conversely, if the critical buckling force
is known, then a measurement of the angle,
θ--i.e., a measurement of the shape of the
microtubule--is a measurement of the compressive
force. This observation was used in Ref. [4]
to determine the force-versus-velocity
relationship of growing microtubules. An
analogous measurement for actin filaments
appears in Ref. [5].

3.9   Bound States and Binding Energies

In Sect. 3.5, we discovered that potential energy is a form of energy, that is
internal to a two-body system. However, in chemistry and biochemistry, which
focus on phenomena occurring on the atomic and molecular scale, we generally
talk about binding energies, rather than potential energies. Because what binds to
what is fundamentally important across biology and medicine, bound states and
their binding energies are correspondingly important. Any time a process involves
a change in binding, the application of energy conservation to that process must
incorporate the difference between the final and initial binding energies. In this
chapter, we encounter a number of different binding energies, including the binding
energy of a gecko's-foot spatula to a surface in Sect. 3.8.1, that enables geckos to
climb up vertical walls and walk upside down across ceilings, atomic and molecular
binding energies in this section, and nuclear binding energies in Problem 5. In Chap.
8. *Statistical Mechanics: Boltzmann Factors, PCR, and Brownian Ratchets* a proper
understanding of binding energy will be essential throughout. In this section, we
seek to clarify what binding energy is and how it is related to potential energy.

3.9.1   Atomic and Molecular Binding

The defining feature of any bound state is that the particles involved necessarily
remain in close proximity to each other. A prototypical example of a microscope
bound state is atomic hydrogen, involving an electron and a proton. Because

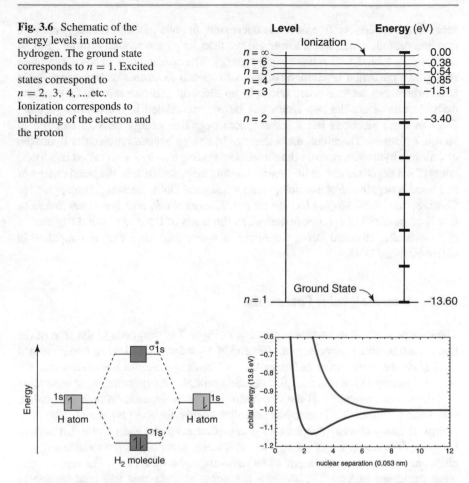

Fig. 3.6 Schematic of the energy levels in atomic hydrogen. The ground state corresponds to $n = 1$. Excited states correspond to $n = 2, 3, 4, ...$ etc. Ionization corresponds to unbinding of the electron and the proton

Fig. 3.7 Left: Schematic of bond formation in molecular hydrogen via the double occupancy of the molecular bonding orbital. from the atomic hydrogen orbitals. Right: Theoretical energy levels of the bonding (red) and antibonding (blue) orbitals as a function of the internuclear separation for two hydrogen atoms. The minimum in the energy of the bonding orbital corresponds to molecular hydrogen. The internuclear separation is specified in units of the Bohr radius (0.53 nm)

quantum mechanical behavior is realized on the atomic scale, only certain quantized values of the binding energy are permitted for atomic hydrogen. The energy levels of atomic hydrogen, each corresponding to a different bound state, are all negative, relative to the energy of an electron and a proton far apart (Fig. 3.6). Therefore, it is necessary to add energy to the system—the hydrogen atom—to separate the electron and the proton. The binding energy of an atomic bound state is always more negative than the energy of the same particles unbound and far apart from each other.

Beyond isolated atoms, we learn in chemistry classes about molecular orbitals and their corresponding energy levels. Figure 3.7 illustrates the results of an approximate quantum mechanical calculation of how the ground-state energy levels of two hydrogen atoms vary as a function of the internuclear separation. As the separation between the two nuclei is decreased, the two atomic orbitals generate a bonding orbital, which has a lower energy than its parent atomic orbitals, and an antibonding orbital, which has a higher energy. The bonding orbital has a minimum energy at a particular separation, which corresponds to molecular hydrogen. Since each orbital can accommodate one spin up electron and one spin down electron, both electrons from the two atoms can be accommodated in the bonding orbital. Each of these electrons has a lower (more negative) energy than an electron in atomic hydrogen. Therefore, the difference in energy between molecular hydrogen and atomic hydrogen, namely the molecular binding energy - also called the "bond energy" - is negative, just as for atomic binding energies. In fact, the bond energy of any bond is negative. For example, when we discuss DNA "melting" (unzipping) in Chap. 8. *Statistical Mechanics: Boltzmann Factors, PCR, and Brownian Ratchets*, it will be essential to appreciate that when the bases of DNA are bound together in pairs in double-stranded DNA, the energy is lower than when they are unpaired in single-stranded DNA.

3.9.2   Intermolecular Potentials

The energy versus internuclear separation in Fig. 3.7 corresponds to the sum of the kinetic and potential energies of an electron in a bonding orbital about two hydrogen nuclei. At the same time, as far as the two hydrogen *atoms* are concerned, the summed energy of the two occupied bonding orbitals versus internuclear separation describes how the energy of the two hydrogen atoms depends on their separation, that is, it describes the interaction potential energy for the two atoms. However, energy is always energy (because of energy conservation) even if the distinctions between different types of energy—in this case, kinetic and potential energy of electrons versus potential energy of atoms—may be a little fuzzy. For reasons that were explained in Sect. 3.7, we take the point of view that any contribution to the energy of a system that depends on spatial coordinates is correctly a type of "potential energy."

The interaction potential energy of two hydrogen atoms (Fig. 3.7) is just one example of an interparticle potential energy. More generally, many molecular and atomic interactions —not only the covalent bonding interactions, relevant to molecular hydrogen—are described by a potential energy qualitatively similar in shape to that of Fig. 3.7, possessing a negligible interaction at large separations, an attraction at intermediate separations, and a repulsion at small separations. This sort of interparticle potential energy is also able to account for the properties of materials composed of collections of atoms and molecules. For example, the repulsion at small distances explains the "normal force," that prevents objects from overlapping each other. And the interparticle attraction at intermediate separations underlies a gecko's ability to climb up walls.

A comparison between the Coulomb potential energy of two charges and the interaction potential energy between two hydrogen atoms nicely illustrates the difference between a "fundamental" interaction, such as the Coulomb interaction (which will be discussed in Chap. 12. *Gauss's Law: Charges and Electric Fields*), which depends on nothing but the charges and their separation, and an interaction that is not fundamental, such as the interaction potential energy between hydrogen atoms, which subsumes the energies and behavior of the underlying electronic states into an interaction between atoms. Because different atoms realize different electronic states, interparticle potentials and bond energies must depend on the electronic states of the specific molecules involved.

3.9.3 Example: Energy from ATP Hydrolysis

An especially important chemical reaction in biology, because the binding energies of the reactants are different from the binding energies of the products, is ATP hydrolysis.[15] How does energy conservation and what we have learned about molecular bonds apply to this reaction? Figure 3.8 shows the chemical species and the chemical reaction involved in ATP hydrolysis. Relative to their atomic constituents, ATP^{4-}, ADP^{3-}, H_2O, and H_2PO^{2-} also all possess negative binding energies. Furthermore, just as for the bond in molecular hydrogen, each bond in ATP^{4-}, including the particular bond in ATP^{4-} that breaks upon ATP^{4-} hydrolysis, has a negative energy. Likewise, each bond in ADP^{3-}, H_2O, and HPO_4^{2-} has a negative energy.

Careful inspection of Fig. 3.8 reveals that there are 6 bonds among the reactants ($ATP^{4-} + H_2O$), highlighted in red in the figure, that are not present in the products, and that there are 5 bonds among the products ($ADP^{3-} + HPO_4^{2-} + H^+$), highlighted in blue in the figure, that are not present in the reactants. Therefore, there is one additional bond on the reactant side of the reaction ($ATP^{4-} + H_2O$), in comparison to the number of bonds on the product side of the reaction ($ADP^{3-} + HPO_4^{2-} + H^+$). Assuming that all other bond energies are the same for both ATP and ADP, Fig. 3.9 shows four proposals for energy level diagrams, relevant to ATP hydrolysis. Which of (A-D) is correct?

All binding/bond energies are negative, ruling out (A) and (B), which propose positive bond energies. The observation that there are 6 bonds on the reactant side and only 5 on the produce side might naively lead us to think that the binding energy of the reactants would be more negative than the binding energy of the products, as proposed in (D) In fact, however, in order for ATP hydrolysis to provide (a positive amount of) energy that can

---

[15]https://en.wikipedia.org/wiki/ATP_hydrolysis

(continued)

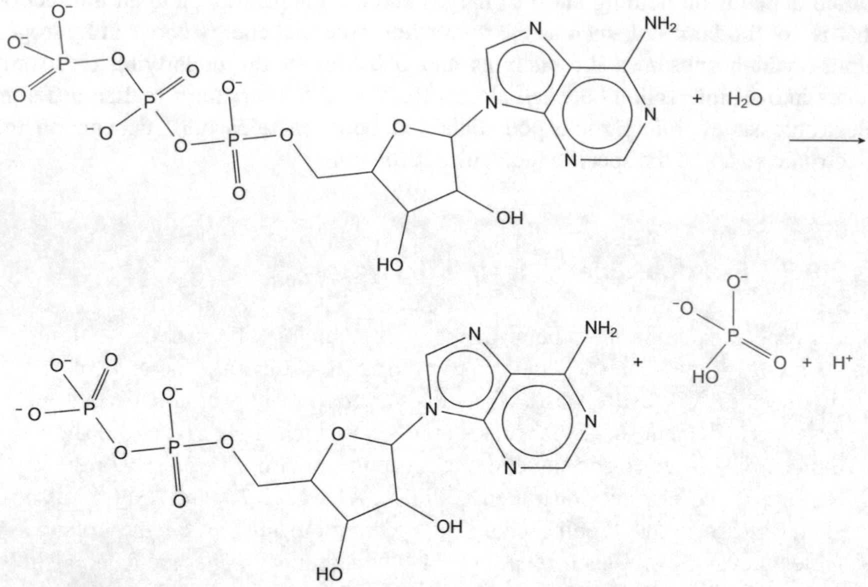

Fig. 3.8 ATP hydrolysis: At physiological pH, the reactants, ATP^{4-} and H_2O, undergo a chemical reaction that yields the products, ADP^{3-}, inorganic phosphate (HPO_4^{2-}), and H^+. (The ionization states depicted correspond to the majority of molecules at physiological pH)

do useful work in the cell, it is necessary that the sum of the binding energies of the products, even though fewer bonds contribute, is more negative than the sum of the binding energies of the reactants. It is in this sense that ATP^{4-} could be said to be a "high-energy" molecule. Because the sum of the binding energies of the products is more negative than the sum of the binding energies of the reactants, energy conservation informs us that energy is released on ATP hydrolysis (that may then be used to do useful work in the cell). It follows that the correct energy level diagram is (C).

Fig. 3.9 Four proposed energy level diagrams for ATP hydrolysis. In each of the four panels (A-D), the energies of the 6 ATP-only bonds in ATP are shown on the left of each panel, while the energies of the 5 ADP-only bonds are shown on the right of each panel

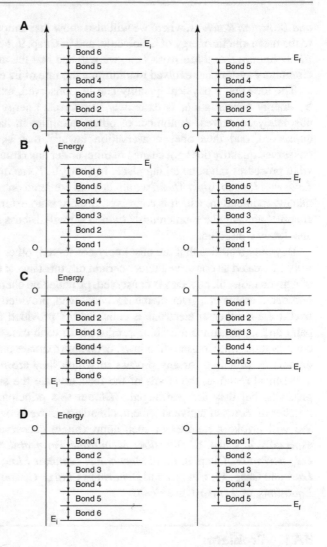

3.10   Chapter Outlook

Few concepts are as important and widely applicable as energy and energy conservation. In this chapter, we introduced a number of key energy-related quantities including work, kinetic energy, potential energy, and binding energy and discussed their connections. We also developed a methodology to create force models, by first writing simpler model potential energies. Finally, we discussed chemical bonds and ATP hydrolysis, and, in particular, that a chemical bond invariably has a negative energy. Looking forward, a proper understanding of binding energy will be important throughout Chap. 8. *Statistical Mechanics: Boltzmann Factors, PCR, and Brownian Ratchets*, where we will also show that temperature is directly related to the mean kinetic energy of a molecule, and in Chap. 9. *Fluid Mechanics: Laminar Flow, Blushing, and Murray's Law*, we will see that the architecture of the human circulatory system has evolved to minimize the rate of its energy consumption.

The idea that a physical quantity can be conserved, exemplified in this chapter by energy conservation, is extremely important. Energy is, as far as we know, absolutely conserved. A number of other quantities in nature are also absolutely conserved, and thus obey conservation laws;[16] that is, the total amount of a conserved quantity does not change in time, under any circumstances, irrespective of what processes might be taking place. In Chap. 2. *Force and Momentum: Newton's Laws and How to Apply Them*, we already encountered one such conserved quantity, namely momentum, which is conserved if there is no external force acting. Electric charge[17] and angular momentum[18] (which we will discuss in later chapters) are also absolutely conserved.

Beyond quantities that are absolutely conserved, other quantities may also usefully be treated as conserved under certain circumstances: the number of molecules of a given molecular species is conserved, provided no chemical reactions occur; the number of atoms of a given element is conserved, provided that no nuclear reactions occur; the number of electrons is conserved,[19] provided that no electron-positron pairs and no neutrinos are being produced. In each case, there is an approximate conservation law. Conservation laws, even approximate ones, provide an important organizing principle for any process in which they are relevant. For example, in a chemical reaction, the atoms of the reactants are the same as the atoms of the products, but they are rearranged. Without this principle of conservation of the number of atoms of a given element, chemistry as we know it would be impossible. We will invoke a number of such approximate conservation laws in the future, especially in Chap. 6. *Diffusion: Membrane Permeability and the Rate of Actin Polymerization*, Chap. 9. *Fluid Mechanics: Laminar Flow, Blushing, and Murray's Law*, and Chap. 14. *Circuits and Dendrites: Charge Conservation, Ohm's Law, Rate Equations, and Other Old Friends*.

3.11   Problems

"Science progresses best when observations force us to alter our preconceptions." Vera Rubin.

Problem 1: Talking About Energy
Danielle, Eric, Jay, Lucie, and Sarah are discussing ATP hydrolysis, shown in Fig. 3.8 [6].

---

[16] http://en.wikipedia.org/wiki/conservation_laws

[17] http://en.wikipedia.org/wiki/Electric_charge

[18] http://en.wikipedia.org/wiki/Angular_momentum

[19] http://en.wikipedia.org/wiki/Lepton_number

Lucie: "I've heard that the O-P bond in ATP, that gets hydrolyzed, is a "high-energy bond", because the energy released on hydrolysis of this bond is large, and is used to do useful work in the body at the molecular and cellular level, such as make a muscle contract, fire a neuron's electrical pulse, etc. "

Jay: "Hey.... idk about ATP, but I *do* know about hydrogen, because I am taking graduate quantum mechanics. We just solved Schrodinger's equation for the energy levels of atomic hydrogen, and they all have a negative energy. It must be that you have to put energy in order to break the electron-proton bond in hydrogen. How can it be that a bond stores energy?"

Eric: "Hmmmm.... You are talking about the energy levels in an atom. Maybe things are different in molecules?"

Danielle: "I don't think so... I remember from my chemistry classes, that when we talked about bonding, we talked about how molecular orbitals emerge from atomic orbitals. As the simplest example, we talked about the formation of molecular hydrogen by bringing together two hydrogen atoms...."

Danielle then sketches the curves in Fig. 3.7 and continues.

Danielle: "What I remember is that there are two orbitals, one from each hydrogen atom. As the two hydrogens come together, a bonding orbital and an antibonding orbital form. The energy of the bonding orbital decreases with decreasing internuclear separation and eventually reaches a minimum. On the other hand, the antibonding orbital has a higher energy. Since each orbital can accommodate two electrons, the two electrons, one from each hydrogen atom, both go into the bonding orbital. As a result the energy of the H_2 molecule is more negative than the energy of two hydrogen atoms. Surely, this implies that breaking molecular bonds also requires us to put energy in? How then can it be that any bond stores energy?"

Lucie: "Look, we just did this whole thing where we discussed that gasoline and jelly donuts are a way to store energy. This energy has to be stored in bonds, and it has to be the same way for ATP. Where else could it be stored?

Eric: "Perhaps, in ATP, jelly donuts, etc. the energy is stored in antibonding orbitals?"

Sarah: "The ATP has four negative charges. That could be one reason that ATP is high energy."

Danielle: "Hmmm.... now I think about it, they also said something in chemistry about "resonance structures" making the inorganic phosphate a low energy molecule."

(a) Why did Danielle say that each orbital can accommodate two electrons?
(b) What line of reasoning lead Danielle to infer that breaking the O-P bond in ATP bond requires energy?
(c) Discuss Sarah's point about charges?
(d) Discuss and explain who is right or wrong, and reconcile the points of view expressed.

Problem 2: Block on an Inclined Plane Revisited

Consider a block sliding down a plane inclined at an angle θ to the horizontal. Using Newton's laws, we can show that the component of the force along the surface, which we pick to be the x-direction, is:

$$F_x = mg(\sin\theta - \mu_K \cos\theta). \qquad (3.59)$$

(a) The block is initially held stationary, but is then released, and starts to move. Using the work-energy theorem, calculate the block's kinetic energy, $\frac{1}{2}mv^2$, after it has moved a distance d along the inclined plane.

(b) Express your answer to (a) in terms of the change in the height of the block, z, instead of d.

(c) Interpret your answer to (b), in light of the law of conservation of energy.

Problem 3: Pulley Force and Work

Figure 3.10 shows an arrangement of (massless) pulleys designed to facilitate the lifting of a heavy load. Assume that friction can be ignored everywhere and that the load has a weight W.

(a) What is the minimum applied force P that can lift the load?

(b) How much work is done by gravity when the weight W is lifted a distance L?

(c) Through what distance must the minimum lifting force be exerted to lift the load a distance L?

(d) How much work must be done by the minimum applied force P to accomplish this task?

(e) What important principle do your answers reflect?

(f) Why do we use pulleys?

Problem 4: Dark Matter

Consider an object of mass m in a circular orbit of radius r and angular velocity ω about a larger object of mass M and radius R, where $M \gg m$.

(a) What is the potential energy of the mass-m object in its circular orbit, assuming that $r > R$?

(b) In Chap. 1. *Vectors and Kinematics*, we showed that the acceleration in the radial direction of an object in a circular orbit of radius r and angular velocity ω is $\omega^2 r$. Use this result, and the result that the velocity of such an object is ωr to calculate the mass-m object's kinetic energy in terms of G, M, m, and r.

(c) Calculate the corresponding velocity of the mass-m object in terms of G, M, m, and r.

(d) What is the total mechanical energy of the mass-m mass-M system?

(e) How do the potential, kinetic, and total energy compare in this gravitational bound state?

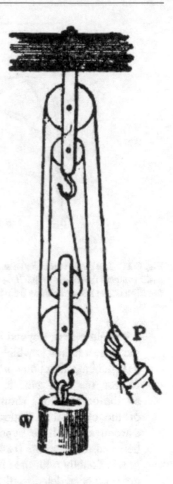

Fig. 3.10 The pulley
arrangement for "Pulley force
and work." Public domain
image from https://commons.
wikimedia.org/wiki/File:
NSRW_Block_and_tackle_arrangements.
jpg

(f) Now, suppose that the mass density of the mass-M object at radius s is $\rho(s)$.
    What is the total mass within radius r?

(g) It turns out that for an object orbiting at radius r, the relevant gravitational mass
    is solely the mass within radius r. How is Newton's Second Law modified in
    this case? What then is the corresponding velocity at radius r?

(h) In the 1970s, Vera Rubin,[20] Kent Ford, Jr., and collaborators measured the
    velocities of stars within galaxies. Figure 3.11 displays several so-called galaxy
    rotation curves from Ref. [7], revealing that the mean velocities of stars in a
    number of galaxies as a function of the stars' distance from the galactic center.
    In each case, the velocity increases from near zero at the galactic center to reach
    a constant, plateau value by about 3 kpc, which is approximately the radius
    which contains most of the observable matter of the galaxy (1 parsec equals

---

[20] http://en.wikipedia.org/wiki/Vera_Rubin

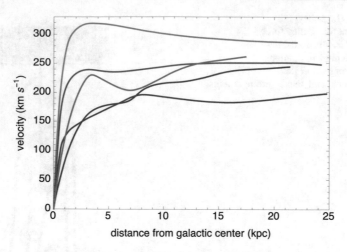

Fig. 3.11 The mean velocities of stars in a number of galaxies—Sa NGC 4378 (orange), Sbc NGC 3145 (magenta), Sbc-Sc NGC 7664 (blue), Sb NGC 1620 (red), and Sb NGC 2590 (brown) – as a function of the stars' distance from the galactic center. Adapted from Ref. [7]

3.26 light years). Beyond 3 kpc, the velocity remains approximately constant to the largest radius at which measurements are possible (25 kpc). What do these data indicate about how $\rho(r)$ depends on r?

In fact, the total galactic mass and the galactic mass distribution, indicated by the observations shown in Fig. 3.11, cannot be understood on the basis of the observable matter comprising the galaxies in question. Therefore, astrophysicists have hypothesized the existence of dark matter,[21] which is believed to constitute 0.845 of the mass of the universe. Dark matter interacts gravitationally with the observable matter of the galaxies, but otherwise it has not (yet) been detected directly. Dark matter is one of science's most important unsolved problems.

Problem 5: Fission Energetics
The basis for power generation by nuclear fission at many nuclear reactors is shown in Fig. 3.12, where the blue dots show the energy per nucleon—each proton and neutron is a nucleon—in keV for all isotopes that have been measured. (1 eV corresponds to 1.6×10^{-19} J.) Highlighted in red is U^{235}, which, if it captures a neutron, undergoes fission often into the two fission fragments of Ba^{141}, highlighted in magenta, and Kr^{92}, highlighted in orange. At the same time, this fission process releases three neutrons, which can go on to cause further U^{235} nuclei to themselves undergo fission, releasing more neutrons and thus propagating a nuclear chain reaction, which we will discuss in detail Chap. 8. *Rates of Change: Drugs, Infections, and Weapons of Mass Destruction.*

---

[21] http://en.wikipedia.org/wiki/Dark_matter

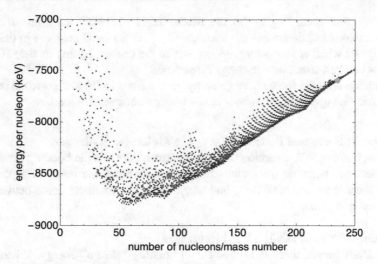

Fig. 3.12 Energy per nucleon (blue dots), including for U^{235} (light blue dot), Ba^{141} (purple dot), and Kr^{92} (orange dot), plotted versus mass number (nucleon number)

(a) Explain on the basis of this plot how you can be sure that energy is released in the fission reaction.
(b) Using this graph or Wolfram Alpha, estimate how much energy is released in each instance of the fission reaction of U^{235} into Ba^{141} and Kr^{92}.
(c) A certain critical mass—52 kg for a sphere of U^{235}—is needed for a nuclear chain reaction to occur, because with less than the critical mass of material, too many neutrons escape the volume of the U^{235} without causing a fission event, and a chain reaction does not occur. The energy released in a nuclear explosion, in which all of the U^{235} nuclei in a 52 kg critical mass contribute, corresponds to how many jelly donuts?

Problem 6: Rubber Band Redux
The potential energy of a rubber band of relaxed length L when it is stretched to a length ℓ may be shown to be

$$U = \frac{1}{6}k\left(\ell^2 - 3L^2 + \frac{2L^3}{\ell}\right),\tag{3.60}$$

where k is a constant.

(a) Calculate the corresponding force (F) as a function of length.
(b) Sketch the force versus length.
(c) Show that the rubber band obeys Hooke's law for small extensions and determine the corresponding rubber band stiffness.

(d) Show that according to (a) the rubber band obeys Hooke's law for large extensions and determine the corresponding rubber band stiffness in this case. Explain whether in reality (and not just in the context of this model) Hooke's law can be correct for very large extensions.

(e) Explain why the force curve given by your answer to (a) and sketched in (b) is not in fact applicable to rubber bands under a compressive load.

Problem 7: Force and Potential Energy in Molecular Hydrogen
On a copy of Fig. 3.7, assuming that two electrons populate the bonding orbital and zero electrons populate the antibonding orbital, indicate the range of separations where there is an attractive force and where there is a repulsive force between the two hydrogen atoms.

Problem 8: Twisted DNA, a.k.a. Plectonemes
When DNA bends, there is an associated bending "elastic" energy. When DNA twists, there is an associated twist elastic energy. On a molecular scale, interatomic bonds are being stretched or compressed away from their minima, increasing the bond energies (Fig. 3.7).

(a) Find yourself a piece of bungy cord, or something equivalent. A section of bungy cord 30 cm long is perfect. This is your model "DNA" in this problem. Twist your "DNA" one complete turn, or a turn and a bit, without pulling it too hard. To accommodate the twist of the ends, instead of twisting, the "DNA" bends into a loop. The "DNA" behaves in this way because the elastic energy of bending is less than the elastic energy of twisting, just as for real DNA.

(b) Describe what happens to the loop size when you pull on your twisted "DNA."

(c) You will now build a simple theory for the force-versus-extension behavior of a total length L of twisted DNA using the expression for the elastic energy of a length a of DNA, bent into a circular arc of radius R:

$$U = \frac{1}{2}\frac{\kappa a}{R^2},\qquad (3.61)$$

where κ is the "bending modulus" of DNA. Approximating the "DNA" loop as a circle, use Eq. 3.65 to determine the elastic energy of a single loop in DNA of radius R.

(d) In a certain "magnetic tweezers"[22] experiment, one end of a segment of DNA is attached to a flat surface, while the other end is attached to a magnetic particle. By subsequently using a magnetic field to rotate the magnetic particle, the DNA is twisted therefore incorporating a loop of radius R, as shown in Fig. 3.13. In this situation, what is the distance (x) from the end of the DNA, attached to the

---

[22] http://en.wikipedia.org/wiki/Magnetic_tweezers

Fig. 3.13 Configuration of
the magnetic tweezers
experiment of "Twisted DNA,
a.k.a. plectonemes"

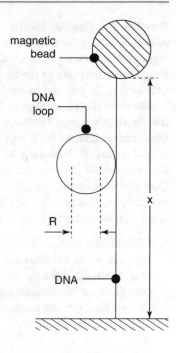

surface, to the end of the DNA, attached to the magnetic particle, given that the
total length of the DNA is L?

(e) What force must be applied to the DNA end to maintain a particular value of R?
How do you predict that this force depends on R? Does your prediction agree
with your experience with bungee cord?

Problem 9: Bones

We showed in this chapter that for a beam with bending stiffness κ, there is a
critical compressive load, $\frac{12\kappa}{L^2}$, beyond which the beam buckles. For a beam with
a circular cross-section of radius R, it is possible to show that $\kappa = \frac{\pi Y R^4}{4}$, where Y
is the Young's modulus[23] of the material that the beam is made of. Therefore, the
critical compressive load for a circular beam is $\frac{3\pi Y R^4}{L^2}$. The composition of bone,
and therefore its Young's modulus, does not vary from bone to bone. Therefore,
the L- and R-dependence of this expression completely specify how the shape of
a cylindrical bone affects whether or not it buckles under compression. Discuss
whether the need to resist buckling explains why the bones of larger animals, such
as elephants, tend to be proportionately thicker than the bones of smaller animals,
such as gazelle.

---

[23] https://en.wikipedia.org/wiki/Young%27s_modulus

Problem 10: Jurassic World

A simple model for a therapod dinosaur, such as *Tyrannosaurus rex*, is a mass m supported on two leg bones, each of length L. In this problem, you will investigate the possible buckling of the dinosaur's leg bones, given that the acceleration due to gravity is g. To this end, consider the sketch shown in Fig. 3.14.

For small dinosaur weights, the dinosaur's leg bones are straight. However, when the dinosaur's weight becomes too large, the leg bones buckle. Suppose that under these circumstances, both legs each take the shape of a circular arc of arc length L and radius R. Referring to the geometry shown in the figure, we know that the relationship between arc length (L), angle (α), and radius (R) is: $L = 2\alpha R$. Correspondingly the "belly" of the dinosaur will also move downwards in the vertical direction of a distance z as in the figure. (Note that z is positive downwards and $z = 0$ when the legs are straight.)

(a) What is the change in gravitational potential energy, U_G, of the dinosaur-Earth system, when the dinosaur's weight is displaced a distance z from its straight-leg position? (Take the zero of the gravitational energy to correspond to when the dinosaur has straight legs.)

(b) When the leg bones are buckled, there is also an elastic potential energy of both legs together defined as:

$$U_E = \frac{\kappa L}{R^2},$$ (3.62)

where κ (Greek kappa) is the leg bones' bending modulus. Show that U_E is proportional to α^2, and find the coefficient of proportionality in terms of given parameters only (i.e., in terms of κ and L, only). In other words show that $U_E = C\alpha^2$ and find what is C.

(c) One can show that for small α (as it is the case here)

$$\alpha^2 = \frac{6z}{L} + \frac{9z^2}{5L^2}.$$ (3.63)

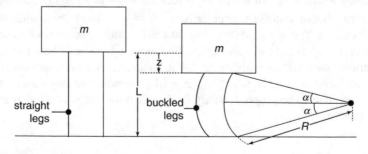

Fig. 3.14 Sketch of two therapod dinosaurs for "Jurassic world," one with straight legs (left) and one with buckled legs (right)

Use this expression to write U_E in terms of z and given parameters.
(d) Using your results to part (a) and (c), calculate the *total* force (F) on the dinosaur, when its leg bones are buckled.
(e) If the dinosaur is at rest with its legs buckled, what is the value of z?
(f) Sketch z as a function of mg for $0 < mg < 3 \times \frac{24\kappa}{L^2}$.

Problem 11: Surface Tension
A simple mechanical model of the cell nucleus hypothesizes that its mechanical properties can be understood similarly to the mechanical properties of a balloon. Specifically, there is a surface energy per unit area (γ)—generally called "surface tension"—that originates from stretching the rubber in the case of the balloon and from stretching the nuclear envelope in the case of the nucleus, together with a pressure difference (P) between the interior and exterior, that results from air pressure in the case of the balloon and from euchromatin in the case of the nucleus. The goal of this problem is to determine the response to force of a spherical nucleus, or balloon, when squeezed between flat plates, i.e., its stiffness.

(a) For a nucleus of radius R, what is the contribution to its total energy from the surface of the nucleus, *i.e.*, from the nuclear envelope (E_S)?
(b) Calculate the force (F_S) exerted by the nuclear envelope.
(c) What is the force per unit area resulting from the nuclear envelope?
(d) Using your answers to (b) and (c), determine the relationship between γ, R, and P.
(e) Now, suppose that the nucleus is squeezed between two parallel plates that apply a force F, which cause the surfaces of the balloon in contact with the plates to become flat, so that opposite surfaces of the balloon are each displaced a distance x from their position in the absence of the forces, exerted by the plates. The corresponding change in volume[24] of each side nucleus may be shown to be $-\pi R x^2$. It may also be shown that the change in surface area under compression is negligible. Given that changing the volume of a balloon at pressure P by a volume ΔV increases the balloon's energy by $P\Delta V$, what is the change in energy of the nucleus (ΔE) when it is compressed by a distance x?
(f) How does the force required, F, depend on x? Does the nucleus obey Hooke's Law? If so, what is the spring constant?

Problem 12: That Chain Again Again
A chain of mass M and total length L is initially coiled on a frictionless horizontal table. At $t = 0$, you grab one end and pull it up vertically at a constant velocity, v. The acceleration due to gravity is $-g$, which is downwards (Fig. 3.15).

---

[24] https://en.wikipedia.org/wiki/Spherical_cap

Fig. 3.15 Sketch for "That
chain again again"

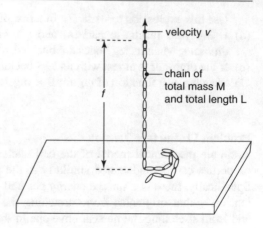

In Chap. 2. *Force and Momentum: Newton's Laws and How to Apply Them*, you
showed that when the top end of the chain is a height z ($z < L$) above the surface
of the table, the force, F, that you must exert in terms is

$$F = \frac{Mv^2}{L} + M\frac{z}{L}g \tag{3.64}$$

for $t < t_1 = \frac{L}{v}$. For $t > t_1$, the force you must exert becomes $F = Mg$. Throughout
the rest of this example, consider only times $t < t_1$.

(a) Considering the chain and the Earth as the system, so that the chain-Earth
gravitational force is internal to the system, what is the total work—call it W—
that has been done by the external forces when the end of the chain is a distance
z above the table top?

(b) What is the kinetic energy of the chain when its end is a height z above the table
top? What is the chain's change in kinetic energy—call it ΔK—between this
kinetic energy and the initial kinetic energy at $t = 0$.

(c) What is the change in potential energy—call it ΔU—of the chain-Earth system
between when the end of the chain is a distance z from the table top and $t = 0$?

(d) Calculate the change in mechanical energy, $\Delta K + \Delta U$.

(e) Under what circumstances is the Law of Conservation of Energy violated?

(f) The generalized work-energy theorem in this case informs us that the energy
transferred into the system by external forces, namely the work, is equal to the
change in energy of the system. Therefore, we should expect the change in the
total energy of the system to be equal to the work done. Compare $\Delta K + \Delta U$
to W in this case. Are they the same? Are they different? If they are different,
propose a hypothesis for the "missing" energy.

Problem 13: Force from Potential

(a) Consider the energy for the bonding and antibonding orbital in Fig. 3.7 (right
side). Explain why these curves can be considered the interaction potential
energy between two hydrogen atoms.

(b) Sketch a graph of the force corresponding to both curves and explain what this force is.

(c) From your graph in (b), what can you say about the behavior of the force corresponding to the bonding orbital in the region where the separation between the two hydrogen nucleii varies from 2 to 4 Bohr radii.

(d) At $t = 0$, the separation between nucleii is 2 Bohr radii and their velocity is zero. Describe the subsequent motion of the nucleii.

(e) Consider the function $U(x) = \frac{1}{2}kx^2$, where k is a constant. Sketch $U(x)$ for $k = 1$.

(f) Sketch the force associated with this potential.

(g) Suggest a physical realization of such a potential energy and force.

Problem 14: DNA Toroids

In spermatozoa, DNA is packaged many times more densely than in somatic cells. This result is achieved by replacing the histones of somatic cells with protamines, which are small positively charged proteins that interact with negatively charged DNA to form "DNA toroids" [8]. Figure 3.16 illustrates the configuration of DNA in an optical tweezers experiment that seeks to investigate the interaction of DNA and protamines. The DNA in the figure forms a "toroid" containing $N = 4$ turns. Each turn is bound to its neighboring turn as a result of protamines between the turns of DNA, as shown in the figure. In this problem, you will build a simple theory for the force-versus-extension of such DNA toroids.

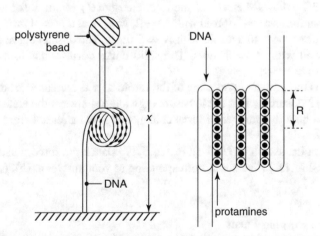

Fig. 3.16 Configuration of DNA in a DNA toroid, being studied via optical tweezers. *Left*: Overall configuration of DNA in the optical tweezers experiment: one end of the DNA is attached to a flat surface, while the other end is attached to a small polystyrene sphere, that is held by the optical tweezers a distance x above the flat surface. Subsequently, protamines are introduced into the solution surrounding the segment of DNA, which then forms an N-turn DNA toroid of radius R, as shown here for the particular case of $N = 4$. *Right*: Side-view of an $N = 4$-turn toroid of radius R, highlighting the protamines between the DNA turns of the toroid, that give rise to turn-turn binding

(a) A certain DNA toroid contains exactly N turns of DNA of radius R. (N is an integer—only integer numbers of turns are assumed in our theory.) What is the total length of DNA contained within this N-turn DNA toroid?

(b) When a length L of DNA is bent into a circular arc of radius R, there is an associated elastic energy:

$$U_E = \frac{1}{2}\frac{\kappa L}{R^2}, \tag{3.65}$$

where κ is the "bending modulus" of DNA. Determine the elastic energy of our N-turn toroid in terms of κ, R, and N. Call it U_E.

(c) The protamine-mediated interaction between turns gives rise to a turn-turn binding energy *per unit length* of ϵ. Briefly explain whether ϵ positive or negative?

(d) By considering the length of the turn-turn contact in an N-turn DNA toroid (see Fig. 3.16), write down the total binding energy of an N turn DNA toroid in terms of ϵ, N, and R. Call it U_B.

(e) What is the total energy of the DNA toroid? Call it U.

(f) In the case of an N-turn toroid, what is the distance (x) from the end of the DNA, attached to the surface, to the end of the DNA, attached to the polystyrene sphere, given that the total length of the segment of DNA is ℓ?

(g) Calculate the force exerted by the DNA toroid in the x direction, either by writing your answer for U in terms of x and ℓ instead of R, or otherwise.

(h) Figure 3.17 shows sketches of the total energy (U) versus x for a segment of DNA that forms a DNA toroid with $N = 2$, $N = 3$, and $N = 4$ turns. The dotted curve corresponds to a toroid with $N = 2$ turns. The dashed curve corresponds to a toroid with $N = 3$ turns. The solid curve corresponds to a toroid with $N = 4$ turns.

Which is the lowest energy state of the toroid as x is increased from $x = 0.2$ to $x = 0.7$? Assuming that the lowest energy state is always the state that occurs, describe how the number of turns changes as x is increased from $x = 0.2$ to $x = 0.7$.

(i) Based on the sketch of U given in Fig. 3.17, sketch the force-versus-extension relationship, i.e., F versus x, corresponding to your answer to (h), from $x = 0.2$ to $x = 0.7$.

Problem 15: Ripping Yarns

Consider, as shown at the top of Fig. 3.18, a long segment of dsDNA (many thousands of base pairs). At one end of the segment, one of the DNA's strands is attached to a microscope cover slip, while the complementary strand is attached to a bead held in an optical tweezers. The external force that must be applied to unzip the DNA is shown in the bottom of Fig. 3.18.

(a) Suppose that the length of a zipped base pair is a. If the end of the DNA strand is displaced a distance x, how many initially zipped base pairs unzip? Call

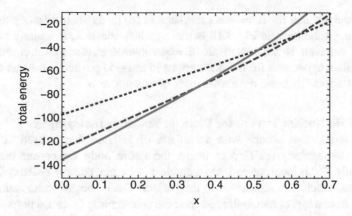

Fig. 3.17 Total energy of DNA toroids with $N = 2$ turns (dotted curve), 3 turns (dashed curve), and 4 turns (solid curve) as a function of x. For these plots, x is measured in units of ℓ, energy is measured in units of $\frac{2\pi^2 \kappa}{\ell}$, and the value of ϵ was chosen to be $|\epsilon| = 200 \times \frac{2\pi^2 \kappa}{\ell}$

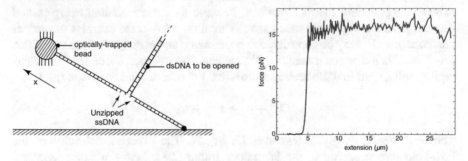

Fig. 3.18 Unzipping a strand of dsDNA from phage λ using optical tweezers to apply the force, adapted from Ref. . Left: Schematic of the experiment Right: Applied force versus displacement (DNA extension). The region of near-zero force and displacements below about 5 μm corresponds to pulling straight the "handles" of the λ DNA, indicated as pTYB1 on the schematic (Top). Once the handles have been straightened, the force increases rapidly, initially with little further displacement. However, at a force of about 15 pN unzipping begins. Each base pair unzipped increases the displacement by about 0.6 nm. Since the change in displacement during the unzipping phase is about 23 μm, we may deduce that $\frac{23000}{0.6} = 38000$ bp were unzipped in this trace. In comparison, λ DNA contains 48000 bp

    this number m, i.e., Write an expression for m. As shown in figure, positive x corresponds to unzipping.

(b) The free energy to unzip one base pair is ΔG. The free energy to unzip m base pairs is $m\Delta G$. Using your answer to (a), write down the free energy of a DNA strand whose end has been displaced a distance x. Call your answer U.

(c) Explain whether your answer to (c), namely U, can be considered a potential energy.

(d) By taking an appropriate derivative or otherwise, determine the force exerted by the end of the DNA.

(e) Explain whether the force you calculated in (d) tends to zip or unzip the DNA.
(f) Although the trace in Fig. 3.18 is more-or-less constant, it actually fluctuates. This behavior is not the result of experimental errors. Develop and briefly explain a hypothesis for the fluctuations in unzipping force, based on what you know about DNA and the earlier parts of this problem.

Problem 16: Nuclear Physics for Positron Emission Tomography

F^{18} is a radioactive isotope with a half-life of 110 min that is used in positron emission tomography (PET)[25] to image the entire body for cancer metastases. Specifically, F^{18} is incorporated into a glucose analogue, namely 2-deoxy-2-fluoro-D-glucose, which is administered to the patient. Then, because metastatic cancers typically have a higher metabolic rate than the surrounding tissues, a preponderance of the F^{18} ends up at any metastases. Thus, it is possible to image the entire body for cancer metastases.

At PET facilities, often a target containing O^{18} is bombarded by protons of energy E_0. Although the O^{18} is actually bound in H_2O and although it is experiencing random thermal motions, because electronic binding energies and thermal kinetic energies are much smaller than the MeV-scale energies of nuclear reactions, the O^{18} may be considered to be unbound and at rest in the context of this problem. When a proton enters the O^{18} nucleus, the resultant unstable F^{19} nucleus rapidly splits apart into the desired radioactive F^{18} nucleus and a neutron (n), i.e.,

$$O^{18} + p \rightarrow F^{18} + n. \tag{3.66}$$

The final kinetic energy is less than E_0 by E_1. The experimental value of the threshold proton energy in the laboratory frame (E_T), below which energy the reaction does not proceed, is 2.57 MeV [9].

(a) Develop a hypothesis to explain the existence of E_1.
(b) By examining this reaction in the center of mass frame, determine the theoretical minimum, threshold value of E_0—call it E_T—for which F^{18} can be produced in terms of E_1? (The COM frame is the frame that moves with the same velocity as that of the COM, so in the COM frame, the COM itself is stationary.)
(c) In fact, according to Wolfram Alpha (http://www.wolframalpha.com/input/?i=O18+binding+energy and http://www.wolframalpha.com/input/?i=F18+binding+energy), the binding energy per nucleon is 7.7670 MeV for O^{18} and 7.6316 MeV for F^{18}. It follows that the difference in binding energy between F^{18} plus a neutron and O^{18} plus a proton is 2.437 MeV. On this basis, what do you predict for the threshold energy in the laboratory frame (E_T)? How does this predicted value compare to the value given above (2.57 MeV).

---

[25] http://en.wikipedia.org/wiki/Positron_emission_tomography

After it is made, F^{18} is quickly incorporated into a glucose analogue, 2-deoxy-2-fluoro-D-glucose (FDG), which is then administered to a patient. Within the patient, cells that have a high metabolic rate, such as brain and metastatic cancer cells, take up the FDG. The presence of fluorine in FDG inhibits normal cell metabolism, until the F^{18} decays to stable O^{18}. The decay of F^{18} thus makes it possible to visualize cancer metastases, using positron emission tomography (PET), which relies of the radioactive decay of F^{18}:

$$F^{18} \rightarrow O^{18} + e^+ + \nu, \qquad (3.67)$$

where e^+ is the positron[26] and ν is a neutrino.[27] The positron travels only a short distance (less than 1 mm or so) before encountering an electron with which it mutually annihilates, creating two 0.511 MeV γ-rays, with energy given by mc^2, where m is the mass of the electron and also of the positron. The two γ-rays exit the patient in opposite directions and are detected in coincidence, allowing a FDG/high metabolism map of the patient to be created, which reveals any metastases.

Problem 16: Creating Radioactive N^{13} for Medical Imaging

N^{13}[28] is a radioactive isotope of nitrogen, whose medical importance derives from its use in positron emission tomography (PET).[29] Specifically, N^{13} can be incorporated into ammonia molecules and administered immediately for PET-based myocardial perfusion imaging[30] [10]. Because its half-life is 9.97 min, N^{13} must be produced immediately before use. Generally, N^{13} production is accomplished by causing protons with velocity v to collide with a target containing stationary O^{16} (such as water), as shown at the right of Fig. 3.19. The proton-O^{16} collision causes the reaction

$$H^1 + O^{16} \rightarrow N^{13} + He^4. \qquad (3.68)$$

The N^{13} so-obtained is then incorporated into ammonia and administered to the patient. Within the patient, the ammonia is taken up by the myocardium (heart muscle). Subsequently, N^{13} undergoes radioactive decay via

$$N^{13} \rightarrow C^{13} + e^+ + \nu, \qquad (3.69)$$

---

[26] http://en.wikipedia.org/wiki/Positron

[27] http://en.wikipedia.org/wiki/Neutrino

[28] http://en.wikipedia.org/wiki/Nitrogen-13

[29] http://en.wikipedia.org/wiki/Positron_emission_tomography

[30] http://en.wikipedia.org/wiki/Myocardial_perfusion_imaging

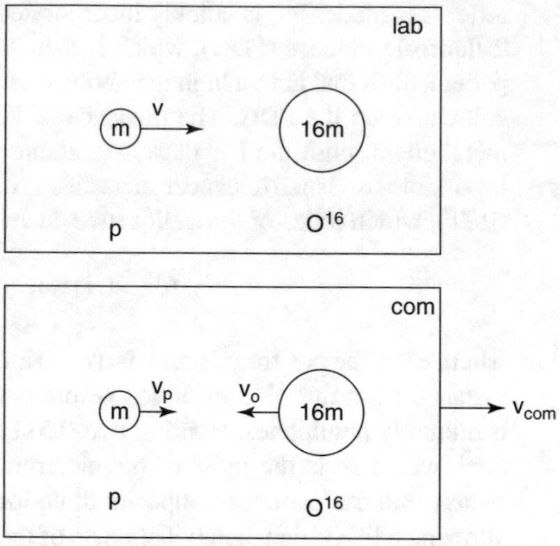

Fig. 3.19 Schematic of the situation before the collision used to produce N^{13}, in the laboratory frame (top) and in the center of mass (COM) frame (bottom)

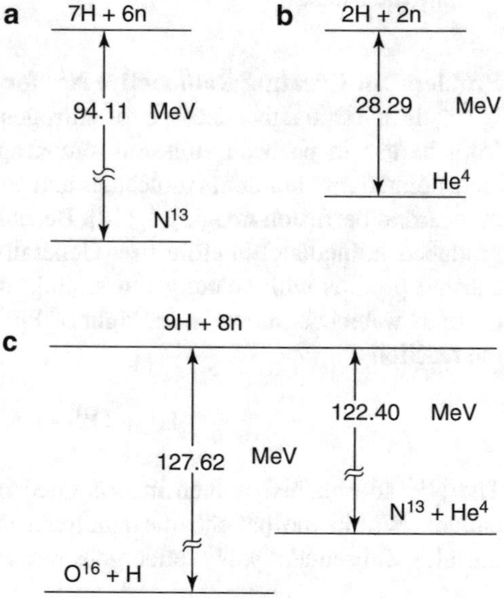

Fig. 3.20 Representation of the binding energies of (**a**) N^{13}, (**b**) He^4, and (**c**) $O^{16}+H$ and $N^{13}+He^4$, relative to their constituent isolated hydrogen atoms and neutrons

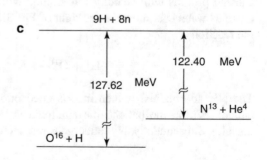

where e^+ is a positron[31] and ν is a neutrino.[32] The positron travels only a short distance (less than 1 mm or so) before stopping and encountering an electron with which it mutually annihilates, creating two γ-rays. These two γ-rays exit the patient in opposite directions and are detected simultaneously, which allows clinicians to obtain a map of the patient's myocardium with a resolution of about 1 mm.

---

[31] http://en.wikipedia.org/wiki/Positron

[32] http://en.wikipedia.org/wiki/Neutrino

Nuclear binding energies are accurately known and readily available. It is therefore straightforward to examine the energetics of nuclear reactions quantitatively. Figure 3.20 shows the binding energies that are relevant for the reaction used to make N^{13} (Eq. 3.68). Relative to their constituent isolated hydrogen atoms and isolated neutrons, the binding energy of N^{13} is -94.11 MeV,[33] the binding energy of He^4 is -28.29 MeV,[34] and the binding energy of O^{16} is -127.62 MeV.[35] The sum of the binding energies of N^{13} and He^4 is -122.40 MeV. Therefore, the binding energy of the reaction products (N^{13} and He^4) is $E_0 = 5.22$ MeV larger than the binding energy of the reactants (O^{16} and H). Our goal in this example is to find the minimum, or threshold, kinetic energy of the proton for the reaction to proceed in terms of E_0, given that protons of mass m and velocity v collide with effectively stationary O^{16} nuclei of mass $16m$, as shown in Fig. 3.19.

(a) Why is it not simply the case that if $\frac{1}{2}mv^2 > E_0 = 5.22$ MeV, then the reaction may proceed?
(b) Use conservation of energy to relate the initial and final kinetic energies to E_0.
(c) However, in the center of mass (COM) frame, the total momentum is zero. Therefore, in the COM frame there can be zero kinetic energy after the collision/reaction, if the reaction products are stationary in the COM frame. Therefore, the threshold kinetic energy in the COM frame is E_0. Because of this observation, it is convenient to analyze the collision in the COM frame. As illustrated in Fig. 3.19, suppose that the initial velocity of the proton in the COM frame be v_P, that the initial velocity of the O^{16} nucleus in the COM frame be $-v_O$, and that the velocity of the COM frame relative to the laboratory frame be v_{COM}. What is the relationship between v_O and v_P?
(d) What is the relationship between v, v_{COM}, and v_P?
(e) What is the relationship between v, v_{COM}, and v_O?
(f) Solve the equations from (c), (d), and (e) to find v_{COM}, v_P, and v_O in terms of v only.
(g) Calculate the kinetic energy in the COM frame.
(h) Calculate the minimum initial kinetic energy in the laboratory frame for the reaction to proceed.

References

1. R.P. Feynman, R.B. Leighton, M. Sands, *The Feynman Lectures on Physics* (Addison Wesley, Boston, 1964)
2. K. Autumn, How gecko toes stick. Am. Sci. **94**, 124–132 (2006)

---

[33] http://www.wolframalpha.com/input/?i=N+13+binding+energy&a=*DPClash.IsotopeP.binding+energy-_*TotalBindingEnergy-
[34] http://www.wolframalpha.com/input/?i=He+4+binding+energy&a=*DPClash.IsotopeP.binding+energy-_*TotalBindingEnergy-
[35] http://www.wolframalpha.com/input/?i=O+16+binding+energy&a=*DPClash.IsotopeP.binding+energy-_*TotalBindingEnergy-

3. M. Kurachi, M. Hoshi, H. Tashiro, Buckling of a single microtubule by optical trapping forces: Direct measurement of microtubule rigidity. Cell Motility Cytoskeleton **30**, 221–228 (1995)
4. M. Dogterom, B. Yurke, Measurement of the force-velocity relation for growing microtubules. Science **278**, 856–860 (1997)
5. D.R. Kovar, T.D. Pollard, Insertional assembly of actin filament barbed ends in association with formins produces picoNewton forces. Proc. Nat. Acad. Sci. U.S.A. **101**, 14725–14730 (2004)
6. J. Svoboda Gouvea, V. Sawtelle, B.D. Geller, C. Turpen, A framework for analyzing interdisciplinary tasks: Implications for student learning and curricular design. CBE Life Sci. Educ. **12**, 187–205 (2013)
7. V.C. Rubin, N. Thonnard, W.K. Ford Jr., Extended rotation curves of high-luminosity spiral galaxies. IV systematic dynamical properties, SA through SC. Astrophys. J. **225**, L107–L111 (1978)
8. N.V. Hud, K.H. Downing, Cryoelectron microscopy of λ phage DNA condensates in vitreous ice: The fine structure of dna toroids. Proc. Nat. Acd. Sci. U.S.A. **98**, 14925–14930 (2001)
9. J.K. Bair, Total neutron yields from the proton bombardment of ^{17,18}O. Phys. Rev. C **8**, 120–3 (1973)
10. H.R. Schelbert, Positron emission tomography measurements of myocardial blood flow: assessing coronary circulatory function and clinical applications. Heart **98**, 592–600 (2012)

Probability Distributions: Mutations, Cancer Rates, and Vision Sensitivity

4

> *Misunderstanding of probability may be the greatest of all impediments to scientific literacy.*
>
> Stephen Jay Gould [1].

Brazil's Pele scores a goal in the 1958 FIFA World Cup Final against Sweden. This Swedish photograph is in the public domain because it is non-artistic and was created before 1969

4.1   Introduction

In this chapter, we will first review the rules of probability. Then, we will introduce the important concept of a random variable and the equally important concept of a probability distribution and express the mean, variance, and standard deviation

© The Author(s), under exclusive license to Springer Nature Switzerland AG 2023
S. Mochrie, C. De Grandi, *Introductory Physics for the Life Sciences*,
Undergraduate Texts in Physics, https://doi.org/10.1007/978-3-031-05808-0_4

of a random variable in terms of its probability distribution. We will then focus on a certain, very-widely applicable probability distribution, namely the Poisson distribution, which specifies the probability of counting a certain number of occurrences or events in a given region or interval under many circumstances. We will then examine several applications of the Poisson distribution, including its application to counting the number of bacterial colonies on a Petri dish and its application to the incidence of retinoblastoma, a childhood eye cancer. An understanding of probability is indispensable for properly designing appropriate experiments to test hypotheses and for evaluating the experimental evidence so-obtained. In particular, whenever an experiment involves counting, the Poisson distribution is more than likely applicable. For example, in this chapter, we will discuss: (1) how to design a virology experiment to ensure that 95% of bacteria are infected by virus; (2) how to interpret measurements that characterize the sensitivity of human vision; and (3) how to determine a bacteria's mutation rate. All of these examples involve an application of the Poisson distribution. We will then introduce another important distribution, namely the binomial distribution, in preparation for its extensive application in the context of random walks in the next chapter. We will then discuss the mean and variance of a composite, "big" random variable, that is formed as the sum of several "small" random variables. Importantly, we will see that the mean of such a "big" random variable is the sum of the means of the "small" random variables, and that its variance is the sum of the variances of the "small" random variables. The chapter will finish by introducing the exponential and Gaussian distributions, and the "central limit theorem," which states that a composite random variable has a Gaussian distribution, provided its component random variables are numerous enough.

4.2   Your Learning Goals for This Chapter

By the end of this chapter, you should be able to

- Calculate probabilities, using the rules of probability, namely that

    1. Probabilities are positive.
    2. Probabilities are normalized.
    3. The sum rule of probability.
    4. The multiplication rule of probability, a.k.a. Bayes theorem.
    5. The addition rule of probability.

- Explain what a random variable is.
- Explain what discrete and continuous probability distributions are.
- Calculate the probability of realizing a particular value or range of values of a random variable of interest given the applicable probability distribution.

(continued)

- Calculate the mean, variance, and standard deviation of a random variable governed by a given discrete or continuous probability distribution.
- Recognize and apply the binomial, Poisson, exponential, and Gaussian distributions.
- Calculate the mean and variance of a random variable that is the sum of a number of independent random variables, whose means and variances are given.
- Know that the distribution of a random variable that is the sum of many independent random variables is well-described by a Gaussian.

4.3 The Rules of Probability

Probability theory[1] is often based on the idealization of an "ensemble"—i.e., a collection of trials—of similarly prepared systems, each one of which yields a particular outcome, or set of outcomes. For example, a collection of M identical coin tosses constitutes an ensemble, containing M members, and the outcome in question could be whether a coin comes up heads or tails. Alternatively, we could consider the ensemble to consist of a sequence of identical trials, carried out one after another, such as a sequence of tosses of a single coin.

Envision, then, compiling a histogram giving the number of occurrences (M_A) of outcome A for an ensemble consisting of M similarly prepared systems. In terms of this histogram, in the limit of large M and large M_A, the *probability* of outcome A is the ratio of M_A to M, that is,

$$P(A) = \frac{M_A}{M}.$$ (4.1)

Two important rules of probability follow from this definition:

- $P(A)$ is non-negative:

$$P(A) \geq 0.$$ (4.2)

- $P(A)$ is normalized:

$$P(A_1) + P(A_2) + P(A_3) + \cdots + P(A_m) = \Sigma_A P(A) = 1,$$ (4.3)

where Σ_A indicates a sum over all possible outcomes, A_1 through A_m.

Instead of focusing on a single outcome, namely outcome A, consider in addition a second possible outcome, namely outcome B, and suppose that, in a given trial,

---

[1] https://www.youtube.com/watch?v=3ER8OkqBdpE

either one or the other or both or neither of these outcomes can be realized. In a total of M trials, suppose that M_{AB} is the number of times that outcome A and outcome B both happen in a single trial, that $M_{A\bar{B}}$ is the number of times that outcome A happens and outcome B does not happen in a single trial, that $M_{\bar{A}B}$ is the number of times that outcome B happens and outcome A does not in a single trial, and that $M_{\bar{A}\bar{B}}$ is the number of times that neither outcome A nor outcome B happens in a single trial. The notation $\bar{A}$ is a standard notation that means "not A." Since every possible outcome may be categorized into one of these four cases, we must have that

$$M_{AB} + M_{A\bar{B}} + M_{\bar{A}B} + M_{\bar{A}\bar{B}} = M. \tag{4.4}$$

In addition, the definition of probability as a ratio of numbers implies that:

- The probability that A and B both occur (a.k.a. the "joint probability" of A and B) is

$$P(A\&B) = \frac{M_{AB}}{M}. \tag{4.5}$$

- The probability that A occurs, irrespective of whether B occurs, is the ratio of the number of A outcomes ($M_{AB} + M_{A\bar{B}}$) to the total number of outcomes (M):

$$P(A) = \frac{M_{AB} + M_{A\bar{B}}}{M}. \tag{4.6}$$

- The probability that B occurs is

$$P(B) = \frac{M_{AB} + M_{\bar{A}B}}{M}. \tag{4.7}$$

- The probability that either A or B or both occur is

$$P(A||B) = \frac{M_{AB} + M_{A\bar{B}} + M_{\bar{A}B}}{M}. \tag{4.8}$$

- The "conditional probability" that A occurs, given that B occurs, is the ratio of the number of A and B outcomes (M_{AB}) to the number of B outcomes ($M_{AB} + M_{\bar{A}B}$):

$$P(A|B) = \frac{M_{AB}}{M_{AB} + M_{\bar{A}B}}. \tag{4.9}$$

- The conditional probability that B occurs, given that A occurs, is

$$P(B|A) = \frac{M_{AB}}{M_{AB} + M_{A\bar{B}}}. \tag{4.10}$$

From these collected results, we can establish three important additional rules of probability, namely **the sum rule**, **the multiplication rule**, and **the addition rule**, as follows.

4.3.1   The Sum Rule

$P(A)$ is obtained by summing over all possibilities for the second outcome, i.e.,

$$P(A) = P(A\&B) + P(A\&\bar{B}). \tag{4.11}$$

More generally, for a collection of second outcomes, labeled B_1 through B_n, we can write

$$P(A) = P(A|B_1) + P(A|B_2) + P(A|B_3) + P(A|B_4) + \cdots + P(A|B_n) = \Sigma_B P(A\&B), \tag{4.12}$$

where the Σ_B represents the sum is over all possible second outcomes. In probability language, we say that we obtain $P(A)$ from $P(A\&B)$ by "marginalizing" outcome B. Sometimes, $P(A)$ is called the marginal probability of A. Similarly, to find $P(B)$, we sum over—marginalize—A:

$$P(B) = \Sigma_A P(A\&B). \tag{4.13}$$

4.3.2   The Multiplication Rule

Equations 4.6, 4.5, and 4.10 imply that the probability that A and B both occur is

$$P(A\&B) = \frac{M_{AB}}{M} = \frac{M_{AB}}{M_{AB} + M_{\bar{A}B}} \times \frac{M_{AB} + M_{\bar{A}B}}{M} = P(A|B)P(B), \tag{4.14}$$

which may be stated as: the probability that outcomes A and B both occur equals the probability that outcome B occurs multiplied by the conditional probability that A occurs given that B occurs. Equation 4.14 is the multiplication rule of probability. Rearranging Eq. 4.14, we find

$$P(A|B) = \frac{P(A\&B)}{P(B)}. \tag{4.15}$$

Equation 4.15 is known as Bayes' theorem.[2]

If the probability of outcome A depends on outcome B, A and B are said to be "dependent" events. However, we are often lead to consider "independent" events, for which the probability of outcome A does not depend on outcome B and

---

[2] https://en.wikipedia.org/wiki/Thomas_Bayes

vice versa. In this case, $P(A|B) = P(A)$. It then follows from Eq. 4.14 that, for independent events,

$$P(A\&B) = P(A)P(B). \tag{4.16}$$

4.3.3 The Addition Rule

Equations 4.6, 4.7, 4.5, and 4.8 imply that the probability of A or B is

$$P(A||B) = \frac{M_{AB} + M_{A\bar{B}} + M_{\bar{A}B}}{M} = \frac{M_{AB} + M_{A\bar{B}}}{M} + \frac{M_{AB} + M_{\bar{A}B}}{M} - \frac{M_{AB}}{M}$$

$$= P(A) + P(B) - P(A\&B). \tag{4.17}$$

For "mutually exclusive" outcomes, such that if outcome A occurs then outcome B cannot occur, it follows that $P(A\&B) = 0$, and, therefore,

$$P(A||B) = P(A) + P(B). \tag{4.18}$$

This is the addition rule of probability for mutually exclusive events. In turn, it follows that

$$P(A) + P(\bar{A}) = 1, \tag{4.19}$$

or

$$P(\bar{A}) = 1 - P(A), \tag{4.20}$$

since A and $\bar{A}$ are mutually exclusive, and together represent all outcomes.

4.3.4 Example: Las Vegas Vacation

Die Hard

What is the probability that "1" or "4" appears on a single roll of a die?

These outcomes are mutually exclusive, and we have $P(1) = 1/6$ and $P(4) = 1/6$, so $P(1||4) = 1/6 + 1/6 = 1/3$ by the addition rule for mutually exclusive events.

A Throw of Dice

You throw two dice. What is the probability of a "1" on the first and a "4" on the second?

(continued)

In this case, we envision that the trial is to throw two dice. The outcomes from each die are independent, so we have $P(1\&4) = 1/6 \times 1/6 = 1/36$ by the multiplication rule for independent events.

Queen Margot

What is the probability that when a card is drawn from each of two decks of cards that one or other or both of these two cards is a queen?

Equation 4.17 informs us that the required result is the probability of a queen from the first deck, namely $\frac{1}{13}$, plus the probability of a queen from the second deck, also $\frac{1}{13}$, minus the probability of a queen from both, which is $\frac{1}{169}$, since the results from each deck are independent, that is $\frac{1}{13} + \frac{1}{13} - \frac{1}{169} = \frac{25}{169}$.

Alternatively, we can conceive the desired probability to be one minus the probability that neither card is a queen. The probability that the card from the first deck is not a queen is $\frac{12}{13}$. The probability that the card from the second deck is also not a queen is also $\frac{12}{13}$. Therefore, the probability that neither card is a queen is $\frac{12}{13} \times \frac{12}{13} = \frac{144}{169}$. It follows that the desired probability is $1 - \frac{144}{169} = \frac{25}{169}$.

Aces High

What is the probability that when two successive cards are drawn from the same decks of cards that the first one is an ace and the second one is also an ace?

In this case, Eq. 4.14 informs us that the required result is the probability of an ace from the first draw, namely $\frac{1}{13}$, multiplied by probability of an ace from the second drawn, given that one of the original four aces is gone, namely $\frac{3}{51}$, that is $\frac{1}{13} \times \frac{3}{51} = \frac{1}{221}$.

4.3.5   Example: Anti-HIV Drug Cocktails

Current HIV treatments involve a triple drug cocktail of reverse transcriptase and protease inhibitors. Our goal in this example is to understand why a triple drug cocktail, rather than either a single drug or a two-drug cocktail? Some relevant facts and numbers concerning HIV and HIV infection are as follows: The HIV genome[3] consists of 10^4 RNA bases, each one of which can be one of 4 different molecules, C, G, A, U. The HIV mutation rate is such that the probability that any one of HIV's bases is incorrectly copied is $p = 3 \times 10^{-5}$. All possible mutations are equally likely. Each day about 10^{10} HIV particles is formed in a patient living with HIV, of which 1% go on to infect a CD4$^+$ T cell. Therefore, there are a total of $M = 10^8$ new T-cell infections per day in a patient living with HIV. Each antiviral drug recognizes a specific amino acid sequence, which depends on the RNA base sequence, and each antiviral drug can be defeated by one specific base mutation of the HIV genome (Fig. 4.1). For example, suppose that a mutation from A to C at base pair 999 is the mutation that defeats antiviral drug X, a mutation from U to G at base pair 4999 is the mutation that defeat antiviral drug Y, and a mutation from C to A at base pair 8999 is the mutation that defeats antiviral drug Z.

(a) What is the probability that a particular HIV genome has a mutation at the correct location to defeat a particular antiviral drug?

   A base is either mutated or not. Therefore, the probability that there is a mutation at the correct base to defeat the particular antiviral is

$$p = 3 \times 10^{-5}. \qquad (4.21)$$

---

[3]http://en.wikipedia.org/wiki/HIV_genome

(continued)

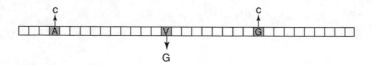

Fig. 4.1 Schematic of the HIV RNA genome, consisting of about 10^4 bases. The three specific mutations that defeat the three antiviral drugs in question are indicated. Viral immunity to the antiviral drug depends of the location of the mutation and the specific identity of the mutated-to base pair

(b) What is the probability (P_1) that an HIV genome has undergone a mutation at the correct location and that the mutation is to the correct base pair to defeat a particular antiviral drug?

We assume that all mutations are equally likely. Therefore, the probability of occurrence of the mutation that defeats a particular antiviral drug, given that the correct base pair is mutated is $\frac{1}{3}$. Equation 4.14 then tells us that the probability that the mutation is at the correct location and that it is to the correct base pair at that location is

$$P_1 = \frac{1}{3} \times p = 10^{-5}. \tag{4.22}$$

(c) What is the probability (P_2 or P_3, respectively) that an HIV genome has *the* specific two or three mutations necessary at the particular locations necessary to defeat two or three antivirals?

Mutations at different locations are independent events. It follows that we can use the multiplication rule for independent outcomes with the results that

$$P_2 = P_1^2 = 10^{-10} \tag{4.23}$$

or

$$P_3 = P_1^3 = 10^{-15}. \tag{4.24}$$

(d) How many HIVs with *the* mutation to defeat one antiviral drug are expected to infect T cells each day in a patient living with HIV?

$$M \times P_1 = 10^3, \tag{4.25}$$

i.e., There are 1000 T cell infected with a virus particle that is resistant to one antiviral drug each day. Verdict on this treatment: ineffective.

(continued)

(e) How many HIVs with *the* two (three) specific mutations to defeat two antiviral drugs are expected to infect T cells each day in a patient living with HIV?

$$M \times P_2 = 10^{-2}, \tag{4.26}$$

```
i.e., There is one T cell infected with a
virus particle that is resistant to a two-drug
cocktail every 100 days (1 semester). Seems like
it could almost work!!
```

$$M \times P_3 = 10^{-7}, \tag{4.27}$$

```
i.e., One infected virus every 10,000,000 days
(10 million days) or 27000 years--essentially
"never" as far as the individual patient is
concerned. This treatment should be highly
effective.
```

(f) Based on your answers, explain why a cocktail of three antiviral drugs, administered simultaneously, is used to treat patients living with HIV.

```
See the comments under (e) and (f).
```

4.4   Discrete and Continuous Random Variables

Often the outcome of a trial may be specified by giving the value of a *random variable*.[4] Two types of random variable can be distinguished: "discrete" and "continuous." An example of a discrete random variable and an example of a continuous random variable both emerge from consideration of the fluorescent signal emitted by a fluorescent fusion protein, consisting of a protein of interest, fused to a fluorescent protein, such as green fluorescent protein (GFP). Within the last two decades, such fusion proteins have revolutionized cell biology.

When subjected to excitation illumination, many species of fluorescent protein undergo stochastic switching events back and forth between a fluorescent, "on" state and a non-fluorescent, "off" state. This phenomenon is known as "blinking." (Eventually, the protein converts to a permanently "off" state, which is "bleaching.") On the left-hand side of Fig. 4.2 is shown a representative measurement of the fluorescence intensity from a single fluorescent fusion protein over a period of 28 s.

---

[4] https://www.youtube.com/watch?v=IYdiKeQ9xEI

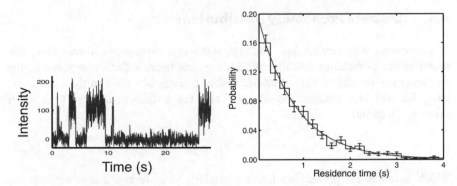

Fig. 4.2 Lifetime of the active state of a single fluorescent protein as an example of a continuous probability distribution. Left: Fluorescence intensity versus time for a single fluorescent protein, called mEos. Under illumination, the protein switches back and forth between a fluorescent-active, "on" state, corresponding to the high intensity in the figure, and dark, "off" states, corresponding to the near-zero intensity in the figure, in a process known as blinking. The duration of each spell in the "on" state represents a measurement of the lifetime of that state. This duration can take on any positive value and therefore is an example of a continuous random variable. Right: Normalized histogram (black line) of the measured on-state lifetimes for a fluorescent protein. The red line corresponds to the best fit to these data by an exponential probability distribution. Courtesy Peter Koo and Matthew Weitzman

Initially, this fluorescent protein was in the off state. It then blinked on for a brief period near 1 s and then off again. It blinked on again near 3 s and off again near 4 s. It then blinked on again near 6 s, remained on for about 3 s, and blinked off near 9 s. It blinked on and off for a short period near 11 s. Finally, it blinked on at 26 s and remained on until the trace ended at 28 s. The duration of each period that the fluorescent protein is in the on state represents a measurement of the lifetime of the on state. Based on many such traces. the right-hand side of Fig. 4.2 shows a histogram (black line) of the measured on-state lifetime distribution. In comparison, the red line is the best fit of the histogram to an exponential function of time, which clearly provides a good description of the experimental histogram.

In the fluorescent protein example, whether or not the fluorescent protein realizes its on state or its off state provides an example of a "discrete random variable." In this case, the random variable takes one of two possible values. More generally, discrete random variables take one of a number of possible values, depending on the character of the random variable. On the other hand, the on-state lifetime can take any value within a continuous range of values and therefore provides an example of a "continuous random variable."

4.5   Discrete Probability Distributions

Associated with any random variable, there is a *probability distribution*, that specifies the probability that the random variable takes a particular value in the discrete case, or takes a value that lies within a range of values in the continuous case. We will first consider the discrete case. For a discrete random variable of interest, the quantity

$$P(k), \tag{4.28}$$

which specifies the probability that the random variable has the value k, is the "probability distribution" of the random variable. For example, if we throw a die and take the value of the random variable k to be the number that comes up, then in this case $P(k) = \frac{1}{6}$ for $1 \leq k \leq 6$ and zero otherwise.

4.6   Mean and Variance

An important quantity for any probability distribution is its *mean*:

$$< k >= \Sigma_k k P(k), \tag{4.29}$$

where Σ_k indicates a sum is over all possible values of k. For our die example, $< k >= \Sigma_{k=1}^6 k \times \frac{1}{6} = 3.5$.

Another important quantity is the probability distribution's *variance*:

$$\sigma_k^2 =< (k- < k >)^2 >= \Sigma_k (k- < k >)^2 P(k). \tag{4.30}$$

Again, the sum is over all possible values of k. We can find a useful alternative expression for σ_k^2 by expanding the parenthesis to find:

$$\sigma_k^2 =< (k- < k >)^2 >=< k^2 - 2k < k > + < k >^2 >=< k^2 >$$
$$- 2 < k >< k > + < k >^2=< k^2 > - < k >^2 . \tag{4.31}$$

For a die, $\sigma^2 = \Sigma_{k=1}^6 k^2 \times \frac{1}{6} - < k >^2= \frac{35}{12}$. The standard derivation, which represents a measure of the width of the distribution, is defined as the square root of the variance.

Consider a random variable A that is proportional to the random variable k, so that

$$A = ak, \tag{4.32}$$

where a is a constant. In this case, we have

$$< A >= \Sigma_k ak P(k) = a\Sigma_k k P(k) = a < k >, \tag{4.33}$$

and

$$\sigma_A^2 = <(ak)^2> -a^2 <k>^2= \Sigma_k (ak)^2 P(k) - a^2 <k>^2= a^2 \Sigma_k k^2 P(k)$$
$$- a^2 <k>^2= a^2 <k^2> -a^2 <k>^2= a^2 \sigma_k^2 \tag{4.34}$$

4.7 Poisson Distribution

An especially important discrete probability distribution, that is widely applicable across the sciences and beyond, is the Poisson distribution (PD).[5] The Poisson distribution for the random variable k is

$$P(k) = \frac{\lambda^k}{k!} e^{-\lambda}. \tag{4.35}$$

It is clear from Eq. 4.35 that the Poisson distribution depends on a single parameter, λ. Notice that λ must be a dimensionless number. It could be the mean number of counts in a time t. It could be the mean number of cancer-causing mutations in a gene that develop in a time t. It could be the mean number of bacterial colonies in an area A. It could be the mean number of virus particles in a bacterium. Whatever it is, it *must* be a *number*.

We can show using Wolfram Alpha that:

- The Poisson distribution is normalized.[6]
- The mean of a Poisson-distributed random variable, k, is

$$\langle k \rangle = \lambda. \tag{4.36}$$

- The mean of k^2 for a Poisson-distributed random variable, k, is

$$\langle k^2 \rangle = \lambda^2 + \lambda, \tag{4.37}$$

so that

$$\sigma_k^2 = \langle k^2 \rangle - \langle k \rangle^2 = \lambda. \tag{4.38}$$

Remarkably, the mean and the variance of a Poisson distribution are equal to each other. This equality is a hallmark of the Poisson distribution (Fig. 4.3).

---

[5] http://www.wolframalpha.com/input/?i=Plot+Poisson+distribution+1.5
[6] http://www.wolframalpha.com/input/?i=sum+from+k%3D0+to+infinity+of++L%5Ek%2Fk%21*exp%28-L%29

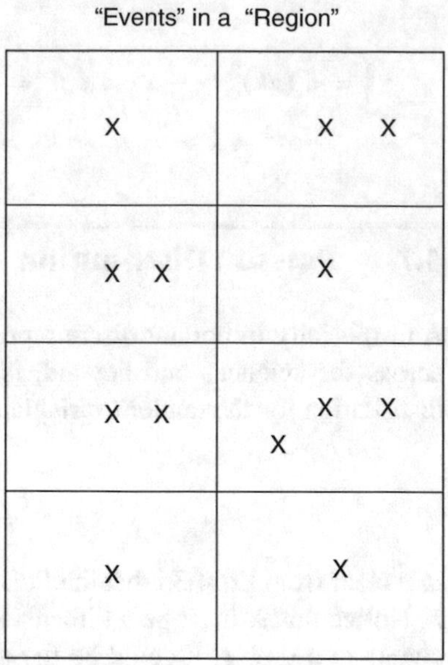

Fig. 4.3 Events and regions. The Poisson distribution gives the probability of realizing k independent "events" in a "region," given that the mean number of independent events in the region is $\langle k \rangle = \lambda$. The events and region can be anything. It is clear that, In this case, there are eight regions and thirteen events. Our best estimate of the mean number of events per region (λ) is therefore $\frac{13}{8}$

"Events" in a "Region"

X	X X
X X	X
X X	X X X
X	X

When is a Poisson distribution appliable? In general, the Poisson distribution applies whenever an experiment or a problem involves counting independent, random events. The following characteristics of the Poisson distribution help identify when the PD is applicable:

- The Poisson distribution gives the probability that k "events" occur in a given "region" (Fig. 4.3).
- The "region" can be a period of time, a three-dimensional volume of space, an area, a bacterium, a population ......
- The "events" can be anything that can be counted: molecules (in a volume or area); photons (detected in a period of time), cancer-causing mutations[7] (in a cell), virus particles (in a cell), disease occurrences (in a population), etc.
- The possible values of k run from 0 to ∞. Even if the possible values of k are limited to $k \leq N$ (for example), but the probabilities for realizing values of k near N are effectively zero ($P(N) \simeq 0$), it is still accurate to apply the Poisson distribution.
- The probability that an event occurs in a region is the same for all regions of equal size.
- The probability that an event occurs in a region is independent of whether or not an event has occurred previously in that region or in a nearby region.

---

[7] http://www.ncbi.nlm.nih.gov/pmc/articles/PMC389051/pdf/pnas00079-0129.pdf

4.7.1 Example: Counting Bacterial Colonies

(a) A Petri dish has an area A. The mean number of bacterial colonies per unit area is ν (Greek "nu"). What is the mean number of bacterial colonies in a Petri dish?

Answer: νA.

(b) What probability distribution determines the probability of observing k bacterial colonies in the Petri dish of part (a)?

Answer: The Poisson distribution is applicable.

(c) What is the probability of observing no bacterial colonies in the Petri dish?

Answer: $P(0) = e^{-\nu A}$.

(d) What is the probability of observing one bacterial colony in the Petri dish?

Answer: $P(1) = \nu A e^{-\nu A}$.

(e) What is the probability of observing two or more bacterial colonies in the Petri dish?

Answer: $1 - P(0) - P(1) = 1 - e^{-\nu A} - \nu A e^{-\nu A}$.

4.7.2 Example: Breast Cancer Susceptibility

It is known from large population studies that for women ages 40–49, 1 in 1000 of such women will develop a new case of breast cancer within their next year of life. In a smaller study of 1000 women ages 40–49, whose mothers had previously contacted breast cancer, 4 of these women were observed to develop breast cancer over a period of a year. Does this observation suggest that there is a genetic component to breast cancer susceptibility or not?

To rationally answer this question, we take the occurrence of breast cancer to follow a Poisson

(continued)

distribution. Then, the "events" in question are
the number of cases of breast cancer in 1000 women
ages 40-49 in a year. The "region" in question is
the next year for 1000 women, ages 40-49. Given
that the mean number of cases of breast cancer in
1000 women ages 40-49 in one year is 1, we have
that λ = 1 for the given "region." Therefore, the
corresponding probability that there are 4 cases
of breast cancer in a population of 1000 women
ages 40-49 in 1 year is: $P(4) = \frac{(1)^4}{4!}e^{-1} = 0.015$, from
Eq. 4.35, assuming there is no genetic component.
Because this probability is small, the observation
of 4 cases of breast cancer in 1000 women ages
40-49, whose mother had breast cancer might lead
us to suspect that there may well be a genetic
component to breast cancer susceptibility.

   A formal way to decide whether or not to reject
the hypothesis under which we calculated this
probability--namely the hypothesis that there
is no genetic component to breast cancer--is to
ask whether the so-called p-value is less than
some value--conventionally chosen to be 0.05--in
which case this so-called null hypothesis is
rejected. The p-value is equal to the probability
of obtaining a result equal to, or more extreme
than, the one actually obtained. Therefore, in
this case the p-value is equal to the probability
of finding 4 or more cases of breast cancer in a
population of 1000 women aged 40-49, whose mothers
had breast cancer. In this case, the p-value is
0.019, which is less than 0.05, thus, according
to convention, leading us to reject the null
hypothesis that breast cancer susceptibility has
no genetic component.

   What accounts for the use of 0.05 to denote
statistical significance? The arbitrary choice
of 0.05 originated with R. A. Fisher[8] and merely
represents convention.

---

[8]http://en.wikipedia.org/wiki/Ronald_Fisher

4.7.3   Example: The Threshold of Human Vision

In a landmark experiment in neuroscience and physiology in 1942, Hecht, Shlaer, and Pirenne investigated the minimum number of light photons that can yield a visual response in human subjects. Their experiment consisted of allowing a known number of photons in a short flash to reach the eye of an experimental subject, who had been previously acclimated to zero light conditions, and asking the subject to say whether he or she saw the flash. At each value of the number of photons in the flash (I), the number M_+ of positive responses in a total of M trials was measured. Thus, Hecht et al. were able to obtain the experimental probability of a positive response via $P_+ = \frac{M_+}{M}$. Hecht et al.'s results for P_+ are plotted as the circles in Fig. 4.4. In this example, we develop a theoretical model for $P_+(I)$, based on the following two hypotheses:

- There is a linear relationship between the number of photons in each flash and the mean number of rod cells producing nerve impulses in each flash. (Rod cells, a.k.a. rods, are the type of photoreceptor cells in our retinas that are most sensitive to light.)
- In order to register a positive result, the number of rods producing a nerve impulse should be greater than or equal to some minimum integer, whose value is to be determined.

(a) Given that you have been lead to count the number of rods producing nerve impulses per flash, what probability distribution is applicable?

(continued)

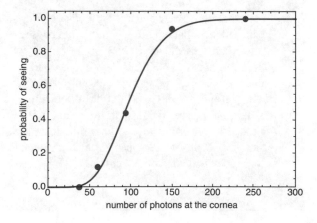

Fig. 4.4 Experimentally determined probability, $P_+(I)$ that a flash of 1 ms duration will be seen (circles) plotted versus the number of visible-light photons in the flash at the cornea (I). The line is the theoretical model described in the example, evaluated with the best fit parameters. Adapted from Ref. [2]

```
Answer: Since we are counting the number of
rods producing nerve impulses per flash, we
may expect the Poisson distribution to be
applicable.
```

(b) Given that I_0 is number of photons necessary for a rod cell to produce a nerve impulse, what then, according to the first hypothesis above, is the mean number of rods that produce a nerve impulse—call it λ—if the number of photons in a flash is I?

```
Answer:
```

$$\lambda = \frac{I}{I_0}. \tag{4.39}$$

(c) If m is the minimum number of rod cells that must produce a nerve impulse in order for the flash to be observed, write down a theoretical expression for probability, P_+, that a flash is observed, in terms of m, I, and I_0. (It is fine to leave your answer in terms of a sum.)

```
Answer:
```

$$P_+ = \Sigma_{k=m}^{\infty} \frac{\lambda^k}{k!} e^{-\lambda} = 1 - \Sigma_{k=0}^{m-1} \frac{\lambda^k}{k!} e^{-\lambda} = 1 - \Sigma_{k=0}^{m-1} \frac{(I/I_0)^k}{k!} e^{-I/I_0}. \tag{4.40}$$

```
Equation 4.40 corresponds to Hecht et al.'s
theoretical model for P+ and to the solid line
in Fig.4.4.
```

Hecht et al. carried out a least-mean-squares fit of their model to their experimental data, with I_0 and m as fitting parameters. The best fit model curve is shown as the line in Fig. 4.4, which provides a very good description of their data. The observed good agreement strongly supports the hypotheses. The best fit values of m and I_0 are 8 and 12, respectively. That is, in order to see a light flash, 8 or more of your rods must initiate a nerve impulse. At the same time, about 12 photons at the cornea are needed for each of these rods to initiate a nerve impulse. That is not to say that 12 photons are needed at the rod. It could be less, because of losses between the cornea and the rod.

4.7.4 Example: Unilateral vs. Bilateral Retinoblastoma

Retinoblastoma is the most common childhood eye cancer. In a landmark study, which eventually lead to the concept of a tumor suppressor gene (TSG), Alfred Knudson Jr. developed the hypothesis that retinoblastoma is a cancer caused by two mutational events – the "two-hit" hypothesis. Part of Knudsen's analysis focused on children with a germline mutation in one of their two RB1 tumor suppressor genes (TSGs). Mutation of the other RB1 TSG in a one of such a child's retinoblasts then leads to a tumor (retinoblastoma). Using the Poisson distribution, Knudson was able to predict the fraction of these patients, who develop bilateral retinoblastoma (one or more tumors in both eyes), and the fraction of these patients, who develop unilateral retinoblastoma (one or more tumors in one eye only).

In this example, we will reproduce Knudson's theoretical analysis, relying on both the Poisson distribution and the rules of probability. We will call the mean number of tumors per child with a germline RB1 mutation, m. Knudsen determined the value of m by comparing his theory to clinical data.

(a) Assuming that the Poisson distribution is applicable, what is the probability that a child with a germline RB1 mutation does *not* develop retinoblastoma?

    Answer: In this case, the events are "tumors," each "region" is a child with a germline RB1 mutation. Therefore, $P(0) = e^{-m}$.

(b) $P(B)$ is the probability that a child with a germline RB1 mutation develops bilateral retinoblastoma, i.e., develops one or more tumors in both eyes. $P(U)$ is the probability that a child aged 0-6 with a germline RB1 mutation develops unilateral retinoblastoma, i.e., develops one or more tumors in one eye only. Write an equation that relates $P(B)$ and $P(U)$ and your answer to (a).

    Answer: A child with a germline RB1 mutation either does not develop retinoblastoma, or develops unilateral retinoblastoma, or develops bilateral retinoblastoma. These are all of the possible outcomes and they are mutually

(continued)

exclusive. Therefore, because the probabilities of mutual exclusive outcomes add and because probabilities are normalized, we have

$$e^{-m} + P(U) + P(B) = 1, \qquad (4.41)$$

i.e.,

$$P(B) = 1 - e^{-m} - P(U). \qquad (4.42)$$

(c) Given that a child with a germline RB1 mutation develops a total of n retinoblastomas, what is the probability that they are all in the left eye OR all in the right eye? That is, that the patient develops unilateral retinoblastoma?

Answer: A tumor is equally likely to occur in the right eye or in the left eye. Therefore, the probability that all n are in the right eye is $\left(\frac{1}{2}\right)^{n} = \frac{1}{2^n}$. The probability that all n are in the left eye is $\frac{1}{2^n}$. These two outcomes are mutually exclusive. Therefore the probability that the tumors are unilateral, given that there are n of them, is $P(U|n) = \frac{2}{2^n} = \frac{1}{2^{n-1}}$.

(d) What is the probability that a child with a germline RB1 mutation develops a total of n retinoblastomas AND they are unilateral?

Answer: We are seeking $P(U\&n)$--often also written $P(U,n)$. The multiplication rule tells us that $P(U\&n) = P(U|n)P(n)$. $P(n)$ is given by the Poisson distribution. The answer to (c) is $P(U|n)$. Therefore,

$$P(U\&n) = \frac{1}{2^{n-1}} \frac{m^n e^{-m}}{n!}. \qquad (4.43)$$

(e) Write a sum that gives $P(U)$. Be sure to specify the lower and upper limits on the sum. Use Wolfram Alpha to do the sum.

(continued)

Answer: Here we are seeking $P(U)$, irrespective of n. Therefore, we should sum $P(U \& n)$ over n (a.k.a. marginalize n), i.e.,

$$P(U) = \Sigma_{n=1}^{\infty} \frac{1}{2^{n-1}} \frac{m^n e^{-m}}{n!} = 2(e^{-\frac{m}{2}} - e^{-m}). \tag{4.44}$$

The sum starts at $n = 1$, because $n = 0$ corresponds to no tumor at all, and goes to infinity

(f) Using your answers to (b) and (e), to determine $P(B)$.

Answer:

$$P(B) = 1 - e^{-m} - P(U) = 1 - e^{-m} - 2(e^{-\frac{m}{2}} - e^{-m}) = 1 - 2e^{-\frac{m}{2}} + e^{-m}. \tag{4.45}$$

Thus, we now know each of $P(0)$, $P(U)$, and $P(B)$ in terms of m.[9] The clinical data showed that bilateral retinoblastoma is about twice as prevalent as unilateral retinoblastoma among children with a germline RB1 mutation, indicating that $m \simeq 3.2$, and $P(0) = 0.04$.

---

[9] https://www.wolframalpha.com/input/?i=Plot+e%5E%7B-x%7D%2C+2%28e%5E%7B-x%2F2%7D-e%5E%7B-x%7D%29%2C+1-2e%5E%7B-x%2F2%7D%2Be%5E%7B-x%7D+for+x%3D0+to+6

4.8   Binomial Distribution

Another important discrete probability distribution is the binomial distribution (BD). The recipe for constructing a random variable (k), distributed according to the binomial distribution,[10] is as follows. Take n "unfair" coins, for each of which the probability of a toss resulting in a "heads" is p, and the probability of the toss resulting in a "tails" is $1 - p$. Toss all n coins together. The value of the random variable, k, is the number of "heads (Fig. 4.5)."

Since the result of each coin toss is independent of every other coin toss, the probability of any particular permutation of k heads and $n - k$ tails is

$$p^k (1 - p)^{n-k}, \tag{4.46}$$

using the product rule.

---

[10] https://www.youtube.com/watch?v=O12yTz_8EOw

Fig. 4.5 Schematic of the binomial distribution, which gives the probability of achieving k "heads" out of a total of n trials, given that there are two possibilities at each trial, "heads" or "tails," and that the probability of achieving "heads" on one trial is p and the probability of "tails" in one trial is $1 - p$. "Heads" and "tails" can represent any two choices

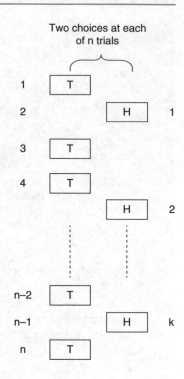

However, the number of permutations of k heads and $n - k$ tails is

$$\frac{n!}{k!(n - k)!}.$$ (4.47)

The notation $n!$ is "n factorial," namely

$$n! = n \times (n - 1) \times (n - 2) \times (n - 3).....3 \times 2 \times 1.$$ (4.48)

This expression—often called "n choose k", more properly a binomial coefficient—is the number of ways of picking k objects out of a total of n objects, without regard to order.

Since different permutations are mutually exclusive outcomes, by the addition rule for mutually exclusive outcomes, the probability of achieving any combination of k heads and $n - k$ tails, irrespective of order, is

$$P(k) = \frac{n!}{k!(n - k)!} p^k (1 - p)^{n-k}.$$ (4.49)

Equation 4.49 is the binomial distribution (BD)[11] for the random variable k in the case that there are n binary choices, each one of which yields 1, a.k.a. "heads," with probability p and 0, a.k.a. "tails," with probability $1 - p$. In general, "heads" and "tails" can represent any two choices. The binomial distribution will be very important when we talk about random walks in Chap. 5. *Random Walks: Brownian Motion and the Tree of Life*. In addition, binomial coefficients will also be very important when we discuss ligand binding in Chap. 8. *Statistical Mechanics: Boltzmann Factors, PCR, and Brownian Ratchets*.

We can show using Wolfram Alpha that:

- $P(k)$ is normalized:

$$\Sigma_{k=0}^{n} P(k) = \Sigma_{k=0}^{n} \frac{n!}{k!(n-k)!} p^k (1-p)^{n-k} = 1. \tag{4.50}$$

- The mean value of k is

$$<k> = np, \tag{4.51}$$

- The mean value of k^2 is

$$<k^2> = np((n-1)p+1) = n^2 p^2 - np^2 + np = np(1-p) + n^2 p^2, \tag{4.52}$$

so that the variance of k is

$$\sigma_k^2 = <k^2> - <k>^2 = np(1-p). \tag{4.53}$$

4.9 Sums of Independent Random Variables

As noted above, if outcome A and outcome B are independent, then the probability of both outcomes is the product of the probability of outcome A and the probability of outcome B:

$$P(A\&B) = P(A)P(B). \tag{4.54}$$

Suppose that outcome A is that a particular random variable takes the value k_1 and that outcome B is that another given random variable takes the value k_2, and that k_1 and k_2 correspond to independent random variables. Then, Eq. 4.54 implies that the probability that the value of the first random variable is k_1 and the probability that the value of the second random variable is k_2 is

---

[11] http://demonstrations.wolfram.com/BinomialDistribution/

$$P(k_1 \& k_2) = P(k_1) P(k_2), \tag{4.55}$$

where $P(k_1)$ and $P(k_2)$ are the probability distributions of the first and second random variables, respectively. These two probability distributions do not need to be identical, but their individual means and variances, we take to be known.

We will often be interested in the sum of a number of independent random variables. The simplest such case is a random variable, s, that is the sum k_1 and k_2:

$$s = k_1 + k_2, \tag{4.56}$$

and its mean:

$$\langle s \rangle = \Sigma_{k_1} \Sigma_{k_2} (k_1 + k_2) P(k_1) P(k_2) = \Sigma_{k_1} \Sigma_{k_2} k_1 P(k_1) P(k_2) + \Sigma_{k_1} \Sigma_{k_2} k_2 P(k_1) P(k_2), \tag{4.57}$$

where Σ_{k_1} indicates a sum over all possible values of k_1. Rearranging we find

$$\langle s \rangle = [\Sigma_{k_1} k_1 P(k_1)][\Sigma_{k_2} P(k_2)] + [\Sigma_{k_1} P(k_1)][\Sigma_{k_2} k_2 P(k_2)] = \langle k_1 \rangle + \langle k_2 \rangle . \tag{4.58}$$

Thus, for the sum of two independent random variables, the mean of the sum equals the sum of the means.

Similarly, we can show that

$$< s^2 > = \left\langle (k_1 + k_2)^2 \right\rangle = \left\langle k_1^2 \right\rangle + 2 \langle k_1 \rangle \langle k_2 \rangle + \left\langle k_2^2 \right\rangle . \tag{4.59}$$

Therefore, the variance of s is

$$\sigma_s^2 = \left\langle s^2 \right\rangle - \langle s \rangle^2 = \left\langle k_1^2 \right\rangle + 2 \langle k_1 \rangle \langle k_2 \rangle + \left\langle k_2^2 \right\rangle - (\langle k_1 \rangle + \langle k_2 \rangle)^2$$

$$= \left\langle k_1^2 \right\rangle - \langle k_1 \rangle^2 + \left\langle k_2^2 \right\rangle - \langle k_2 \rangle^2 = \sigma_{k_1}^2 + \sigma_{k_2}^2 . \tag{4.60}$$

Thus, for the sum of two independent random variables, the variance of the sum equals the sum of the variances.

In fact, Eqs. 4.58 and 4.60 generalize from the sum of two independent random variables to the sum of n independent random variables, that is, for

$$s = \Sigma_{i=1}^n k_i . \tag{4.61}$$

In this case,

$$\langle s \rangle = \Sigma_{i=1}^n \langle k_i \rangle . \tag{4.62}$$

The mean of the sum equals the sum of the means. In addition,

$$\sigma_s^2 = \Sigma_{i=1}^n \sigma_{k_i}^2 . \tag{4.63}$$

The variance of the sum equals the sum of the variances. In the special case that all of the k_is have the same probability distribution, we have

$$\langle s \rangle = n \langle k_i \rangle \tag{4.64}$$

and

$$\sigma_s^2 = n\sigma_{k_i}^2. \tag{4.65}$$

The results for the binomial distribution are consistent with Eqs. 4.64 and 4.65: for a single unfair coin toss, the mean and variance are p and $p(1 - p)$, respectively, while for n unfair coin tosses, the mean and variance are np and $np(1 - p)$.

4.10   Continuous Probability Distributions

Returning now to Fig. 4.2, which shows the intensity emitted by a fluorescent protein (fluorescent protein) as it blinks back and forth between its on and off states. The time that the fluorescent protein spends in the on state corresponds to measurement of the fluorescent protein's on state lifetime, which is an example of a continuous random variable.

Actually, Fig. 4.2 is a normalized histogram, constructed by counting the number of times a fluorescent protein shows an "on"-state lifetime within each 0.2 s time increment, and then dividing by the total number of counts. This plot makes it clear that for random variables that take a value from a continuous range of possible values, it is natural to consider the probability that the random variable falls within some range of values, especially when examining experimental data. One widely used quantity in the context of continuous probability distributions is the cumulative probability that a random variable X realizes a value less than x:

$$P_X(x). \tag{4.66}$$

The cumulative probability can be interpreted as the number of times a value less than x is found, divided by the total number of trials. In the context of Fig. 4.2, we identify the random variable, X, as the lifetime. Then, $P_X(x)$ would be the sum of all bars of the histogram for which the lifetime has a value less than x.

It is also useful to consider the probability that the random variable, X, has value between x and $x + dx$. This quantity is the difference between cumulative distributions:

$$P_X(x + dx) - P_X(x) \simeq \frac{dP_X}{dx}dx = p_X(x)dx, \tag{4.67}$$

where we have now introduced the probability density, a.k.a. the probability distribution, $p_X(x) = \frac{dP_X}{dx}$. We can interpret $p_X(x)$ in terms of the probability that the random variable X has a value between x and $x + dx$:

$$p_X(x)dx = \frac{\text{number of systems with } x < X < x + dx}{M}. \tag{4.68}$$

Thus, for Fig. 4.2, $p_X(x)dx$ is the height of the bar of the histogram located at a lifetime of x and the value of $p_X(x)$ at x is that height, divided by $dx = 0.2$ s. For brevity, we will often drop the X subscript, i.e., we will write $p_X(x) = p(x)$. Note that $p_X(x)dx$ is a probability and therefore is dimensionless. It follows that $p_X(x)$ has dimensions of $1/x$.

Several key properties of $p_X(x)$ are:

1. It is positive, i.e.,

$$p_X(x) \geq 0; \tag{4.69}$$

2. It is normalized, i.e.,

$$\int_{-\infty}^{\infty} p_X(x)dx = 1; \tag{4.70}$$

3. The probability that X has a value less than x, is the cumulative probability (written with a capital P):

$$P_X(x) = \int_{-\infty}^{x} p_X(u)du. \tag{4.71}$$

Note that $P_X(x)$ is a probability and therefore is dimensionless.
4. The probability density may be obtained from the cumulative probability via differentiation:

$$p_X(x) = \frac{dP_X(x)}{dx}. \tag{4.72}$$

5. The mean value of X is

$$\langle X \rangle = \int_{-\infty}^{\infty} xp_X(x)dx. \tag{4.73}$$

6. The variance of X is

$$\sigma_X^2 = \left\langle (X - \langle X \rangle)^2 \right\rangle = \left\langle X^2 \right\rangle - \langle X \rangle^2 = \int_{-\infty}^{\infty} x^2 p_X(x)dx - \langle X \rangle^2. \tag{4.74}$$

4.10.1 Example: A Certain Continuous Probability Distribution

A certain random variable X has a probability density, $p(x)$, given by

$$p(x) = Cx(1 - x), \text{ for } 0 < x < 1$$
$$p(x) = 0, \text{ otherwise,} \tag{4.75}$$

where C is a constant.

(a) Sketch $p(x)$.

```
Answer:   http://www.wolframalpha.com/input/?i=Plot+6*x
          %281-x%29+for+x%3D0+to+1¹²
```

(b) Determine the value of C.

```
Answer: Any probability density is required to
be normalized. Thus, we require
```

$$\int_{-\infty}^{\infty} p(x)dx = \int_{0}^{1} Cx(1 - x)dx = 1, \tag{4.76}$$

```
i.e.,
```

$$\int_{0}^{1} Cx(1 - x)dx = \int_{0}^{1} C(x - x^2)dx = C[\frac{1}{2}x^2 - \frac{1}{3}x^3]_{0}^{1} = \frac{C}{6}, \tag{4.77}$$

```
i.e.,
```

$$C = 6 \tag{4.78}$$

(c) Determine the mean value of x, $\langle x \rangle$.

```
Answer:
```

$$\langle x \rangle = \int_{-\infty}^{\infty} xp(x) \, dx = \int_{0}^{1} 6(x^2 - x^3)dx = 6\left(\frac{1}{3} - \frac{1}{4}\right) = \frac{1}{2}, \tag{4.79}$$

```
which makes perfect sense, of course.
```

(d) Determine the mean square value of x, $\langle x^2 \rangle$.

---

[12]http://www.wolframalpha.com/input/?i=Plot+6*x%281-x%29+for+x%3D0+to+1

(continued)

Answer:

$$\langle x^2 \rangle = \int_{-\infty}^{\infty} x^2 p(x) \ dx = \int_0^1 6(x^3 - x^4) \ dx = 6\left(\frac{1}{4} - \frac{1}{5}\right) = \frac{3}{10}.$$
$$(4.80)$$

(e) Determine the variance of x, σ_x^2, and its standard deviation, σ_x.

Answer:

$$\sigma_x^2 = \langle x^2 \rangle - \langle x \rangle^2 = \frac{3}{10} - \left(\frac{1}{2}\right)^2 = \frac{1}{20} \tag{4.81}$$

and

$$\sigma_x = \frac{1}{\sqrt{20}}. \tag{4.82}$$

(f) Determine the probability that X has a value between 0 and W.

Answer:

$$\int_0^W 6x(1 - x)dx = \int_0^W 6(x - x^2) \ dx = W^2(3 - 2W). \tag{4.83}$$

4.11   Exponential Distributions: Fluorescence Lifetimes, Radioactive Decay, and Drug Elimination from the Body

A widely applicable continuous probability distribution is the exponential distribution. The probability that a drug molecule in the body, or a radioactive nucleus, or a fluorescent protein under laser illumination survives a time t decreases exponentially with t.

The reason that an exponential lifetime distribution is so widely applicable is that it follows from the widely applicable hypothesis that a drug molecule in the body, (or a radioactive nucleus, or a fluorescent protein under laser illumination) has a constant elimination rate, (or decay rate or switching rate). For a rate Γ, the probability that the drug molecule is eliminated in an infinitesimal time period Δt is

$$\Gamma \Delta t. \tag{4.84}$$

We may work out the consequences of this hypothesis as follows. Since the probability that the molecule is eliminated in Δt is $\Gamma \Delta t$, the probability that the molecule is not eliminated in Δt is

$$1 - \Gamma \Delta t, \tag{4.85}$$

using the addition rule for mutually exclusive events. Then, using the multiplication rule for independent events, the probability that it is not eliminated in a total time $t = n\Delta t$ is

$$(1 - \Gamma \Delta t)^n = \left(1 - \frac{\Gamma t}{n}\right)^n. \tag{4.86}$$

If we now invoke the mathematical result, valid in the limit of large n for any λ, that

$$\lim_{n \to \infty} \left(1 - \frac{\lambda}{n}\right)^n = e^{-\lambda}. \tag{4.87}$$

Therefore, the probability that a molecule is not eliminated in a time t is

$$e^{-\Gamma t}. \tag{4.88}$$

It follows that the probability that it is eliminated between t and $t + dt$, namely $p(t)dt$, is equal to the probability that it is not eliminated between 0 and t multiplied by the probability it is eliminated in dt, i.e.,

$$p(t)dt = e^{-\Gamma t} \times \Gamma dt. \tag{4.89}$$

Therefore, the probability density of the lifetime is an exponential function:

$$p(t) = e^{-\Gamma t} \Gamma. \tag{4.90}$$

It is properly normalized as may be shown by integrating t from 0 to ∞.

The mean lifetime is

$$\langle t \rangle = \int_0^\infty t e^{-\Gamma t} \Gamma dt = \frac{1}{\Gamma} \int_0^\infty s e^{-s} ds = \frac{1}{\Gamma}. \tag{4.91}$$

The longevity of radioactive nuclei[13] and of drugs in the body[14] are often specified by giving their half-life,[15] $t_{\frac{1}{2}}$, defined by $e^{-\Gamma t_{\frac{1}{2}}} = \frac{1}{2}$, or equivalently $t_{\frac{1}{2}} = \frac{\log 2}{\Gamma}$.

4.12 Gaussian Distributions

Another very important continuous probability density is the Gaussian probability density:

---

[13] http://en.wikipedia.org/wiki/List_of_isotopes_by_half-life
[14] http://en.wikipedia.org/wiki/Biological_half-life
[15] http://en.wikipedia.org/wiki/Halflife

$$p(x) = \frac{1}{\sqrt{2\pi\sigma^2}}e^{-(x-x_0)^2/(2\sigma^2)}. \qquad (4.92)$$

For the Gaussian distribution of Eq. 4.92, the mean and variance are $<x> = x_0$ and $\sigma_x^2 = \sigma^2$, respectively.

The corresponding cumulative probability is

$$P(x) = \frac{1}{\sqrt{2\pi\sigma^2}}\int_{-\infty}^{x} dx' e^{-(x'-x_0)^2/(2\sigma^2)} = \frac{1}{2}[1 + \mathrm{erf}(x - x_0/\sqrt{2}\sigma)], \qquad (4.93)$$

where $\mathrm{erf}(x)$ is the "error function."

4.13   The Central Limit Theorem

One of the reasons that the Gaussian distribution is so important is the "central limit theorem," a consequence of which is that the probability distribution of a random variable, s, that is the sum of n independent random variables, k_i, is well-represented by a Gaussian distribution with mean $\langle s \rangle = \Sigma_{i=1}^{n} \langle k_i \rangle$ and variance $\sigma_s^2 = \Sigma_{i=1}^{n}\sigma_{k_i}^2$, that is,

$$P(s) = \frac{1}{\sqrt{2\pi\sigma_s^2}}\exp\left(-\frac{(s - \langle s \rangle)^2}{2\sigma_s^2}\right). \qquad (4.94)$$

Equation 4.94 is valid for sufficiently large n. How large does n have to be to achieve a good approximation to a Gaussian distribution? Often, quite small.

To see an example of the central limit theorem in practice, we can consider a random variable, s, that is the sum of n random variables—call them k_i—that each take the value 1 or 0 with probability p and $1 - p$, respectively. It is straightforward to show that $\langle k_i \rangle = p$ and $\sigma_{k_i}^2 = p - p^2 = p(1 - p)$. According to the central limit theorem, we should expect the distribution of $s = \Sigma_{i=1}^{n}k_i$ to be a Gaussian with mean $\langle s \rangle = np$ and variance $\sigma_s^2 = np(1 - p)$. On the other hand, we know that the distribution of s is actually given by the binomial distribution, $P(s) = \frac{n!}{s!(n-s)!}p^s(1 - p)^{n-s}$, for which $\langle s \rangle = np$ and $\sigma_s^2 = np(1 - p)$. Evidently, the mean and variance are given correctly by the central limit theorem. Beyond these averages, Fig. 4.6 shows that the binomial distribution for $n = 200$ and $p = \frac{1}{2}$ is very, very similar to a Gaussian distribution with the same mean and variance, just as the central limit theorem tells us it must be. The central limit theorem is so powerful, because it holds for any distribution of k_i, including when k_i has a Poisson distribution, an exponential distribution, or a Gaussian distribution.

Interestingly, when he was an undergraduate at Cambridge University in 1934, Alan Turing (mathematician, computer scientist, code breaker, and theoretical biologist) wrote a dissertation, proving the central limit theorem, unaware that the

Fig. 4.6 Comparison between the binomial distribution (points) for $n = 200$ and $p = \frac{1}{2}$ ($\langle s \rangle = np = 100$, $\sigma_s^2 = np(p - 1) = 50$) and the Gaussian distribution (line) with the same mean and variance

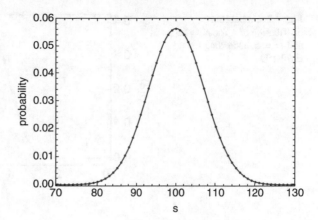

central limit had been proven twelve years previously by Jarl Lindeberg. Quite the senior project though!

4.14 Parameter Estimation, Experimental Errors, and Counting Statistics

In this section, we inquire more carefully how to determine the parameters of a probability distribution, given experimental measurements and, in particular, what are the errors in those parameters. Specifically, we suppose that in a photon counting experiment, we have acquired the following counts: $k_1, k_2, k_3, \ldots k_N$ in each one of N sequential one second periods. Since we are counting, we can expect the Poisson distribution to be applicable. Furthermore, we can expect that the correct way to estimate the Poisson parameter λ is to calculate the mean number of counts, i.e., $\lambda = \frac{1}{N} \Sigma_{n=1}^{N} k_n = \frac{K}{N}$, where K is the total counts. However, because we are dealing with a random process, we can expect different experiments to yield different sets of counts, and, therefore different estimates of λ. How, then, can we estimate not only the value of λ, but also the error in the value of λ without repeating the experiment multiple times? We will answer this question in this section.

Assuming the Poisson distribution is applicable, the probability of counting k_n photons in one second, given that the Poisson parameter is λ, is:

$$P(k_n|\lambda) = \frac{\lambda^{k_n} e^{-\lambda}}{k_n!}. \tag{4.95}$$

Because the counts in different one second periods are independent of each other, the probability of counting $k_1, k_2, k_3, \ldots k_N$ photons in N one second periods is a product:

Fig. 4.7 Probability distribution of λ for $K = 25$ and $N = 5$, according to Eq. 4.100

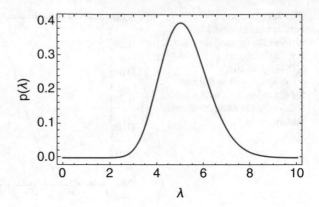

$$P(k_1, k_2, k_3, ...k_N|\lambda) = P(k_1|\lambda)P(k_2|\lambda)P(k_3|\lambda).....P(k_N|\lambda)$$

$$= \frac{\lambda^{k_1}e^{-\lambda}}{k_1!}\frac{\lambda^{k_2}e^{-\lambda}}{k_2!}\frac{\lambda^{k_3}e^{-\lambda}}{k_3!}....\frac{\lambda^{k_N}e^{-\lambda}}{k_N!} = P(\text{data}|\text{parameters}),$$

(4.96)

where the last equality introduces a more schematic notation. However, after an experiment the usual situation is that the data are known, but the model parameters are unknown. Therefore, if we now envision (our measurement of) λ to be a random variable, we would like to know $P(\text{parameters}|\text{data}) = P(\lambda|k_1, k_2, k_3, ...k_N)$. Then, the mean of this distribution would be the experimental estimate of λ and the standard deviation of that distribution would be the experimental error in that estimate (Fig. 4.7).

Bayes' theorem (see Eq. 4.15 found earlier in this chapter) tells us that

$$p(\text{parameters}|\text{data}) = \frac{P(\text{data}|\text{parameters})\,p(\text{parameters})}{P(\text{data})}.$$

(4.97)

On the right-hand side of this equation, we know $P(\text{data}|\text{parameters})$, while $P(\text{data})$ is the independent of λ and so cannot affect the shape of the distribution of λ. $p(\text{parameter}) = p(\lambda)$ is the so-called prior probability of λ—the probability that λ has a certain value before any measurements—and in principle can depend on λ. Before any measurements, however, it is reasonable and simplest to suppose that all values of λ are equally likely, i.e., that $p(\lambda)$ is also independent of λ, corresponding to a uniform probability density.

Assuming a uniform prior probability, we have that

$$p(\lambda|k_1, k_2, k_3, ...k_N) = C \times P(\text{data}|\text{parameters}) = C'\lambda^{k_1+k_2+k_3+.....+k_N}e^{-N\lambda}$$

$$= C'\lambda^K e^{-N\lambda}$$

(4.98)

where C' ensures that $p(\lambda|k_1, k_2, k_3, ...k_N)$ is normalized and is given by

$$\frac{1}{C'} = \int_0^\infty \lambda^{k_1+k_2+k_3+\cdots+k_N} e^{-N\lambda} d\lambda = \int_0^\infty \lambda^K e^{-N\lambda} d\lambda. = \frac{K!}{N^{K+1}}. \qquad (4.99)$$

It follows that the probability density of λ is

$$p(\lambda|k_1, k_2, k_3, \ldots k_N) = \frac{N^{K+1}\lambda^K e^{-N\lambda}}{K!}. \qquad (4.100)$$

Using Eq. 4.100, we can calculate the mean value of lambda, $\langle\lambda\rangle$, and mean square value of λ, $\langle\lambda^2\rangle$:

$$\langle\lambda\rangle = \int_0^\infty \lambda \frac{N^{K+1}\lambda^K e^{-N\lambda}}{K!} d\lambda = \frac{K+1}{N} \qquad (4.101)$$

and

$$\langle\lambda^2\rangle = \int_0^\infty \lambda^2 \frac{N^{K+1}\lambda^K e^{-N\lambda}}{K!} d\lambda = \frac{(K+1)(K+2)}{N^2}. \qquad (4.102)$$

Then, the variance of λ is

$$\sigma_\lambda^2 = \langle\lambda^2\rangle - \langle\lambda\rangle^2 = \frac{(K+1)(K+2)}{N^2} - \frac{(K+1)^2}{N^2} = \frac{K+1}{N^2}. \qquad (4.103)$$

We see that the mean value of λ is close to the expected value, namely $\frac{K}{N}$ and that the expected experimental error is $\sigma \simeq \frac{\sqrt{K}}{N}$. Therefore, the relative error is

$$\frac{\sigma_\lambda}{\langle\lambda\rangle} = \frac{1}{\sqrt{K}}, \qquad (4.104)$$

equal to the square root of the total number of counts. This result, namely that the relative error decreases as the square root of the total number of counts, is colloquially known as "counting statistics" and is as broadly applicable as the Poisson distribution itself.

4.15 Chapter Outlook

Probabilistic phenomena are central to myriad biologically relevant processes, including all of the phenomena that we will discuss in Chap. 5. *Random Walks: Brownian Motion and the Tree of Life*, Chap. 6. *Diffusion: Membrane Permeability and the Rate of Actin Polymerization*, and Chap. 8. *Statistical Mechanics: Boltzmann Factors, PCR, and Brownian Ratchets*. For example, in Chap. 5. *Random Walks: Brownian Motion and the Tree of Life*, we will show how a simple model of random "genetic drift," governed by probabilistic ideas, leads to the prediction that the genetic divergence between two organisms serves as a molecular clock, irrespective of the organisms' population sizes over history, and, thus, reveals when the last common ancestor of the two organisms lived. The material of the current chapter is prerequisite for all of these later topics.

4.16   Problems

"Sometimes it is the people no one can imagine anything of who do the things no one can imagine." Alan Turing

Problem 1: Three Dice Throw
You throw 3 dice.

(a) What is the probability that you throw 3 "sixes"?
(b) What is the probability that you throw either a 6 or a 1 each time?
(c) What is the probability that you throw zero "sixes"?
(d) What is the probability that you throw one or more "sixes"?

Problem 2: Three-Coin Toss
The recipe for constructing a certain random variable (k) is as follows. Take 3 (unfair) coins and toss them. For each "heads," add 1. For each "tails," add zero.

There are 8 ways the coins can land. The corresponding sequence of heads and tails, the value of k, and the value of $P(k)$ are as follows:

Sequence	Value of k	$P(k)$
000	0	1/64
Any of 100, 010, 001	1	9/64
Any of 110, 101, 011	2	27/64
111	3	27/64

(a) Verify that $P(k)$ is positive or zero for all k.
(b) Verify that $P(k)$ is normalized.
(c) What is the mean value of the random variable, k? That is, what is $\langle k \rangle$?
(d) What is the variance of k?
(e) What is the standard deviation of k?

Problem 3: Strong Poisson
For the following situations explain whether or not you would expect each case to be described by a Poisson distribution.

(a) The number of yeast colonies on a Petri dish.
(b) The number of x-rays detected per second at a particular diffractometer setting, during a protein crystallography experiment.
(c) The number of proteins in a given volume of solution.
(d) The number of messages that arrives in your inbox in 10 min.

(e) The number of messages that arrives in my inbox in 10 min.
(f) The number of snowy days during a week in winter.
(g) The height of people in a college classroom.

Problem 4: Virology Experimental Design
You have a culture containing 10^7 bacteria that you would like to infect with lambda phage. How many phage do you need to add to be confident that 95% of the bacteria are infected (with at least one phage)? What about a 99% infection rate? (Assume that every phage added finds its way into a bacterium.)

Problem 5: Restriction Enzymes
Restriction enzymes are routinely used in molecular biology to cut DNA at specified locations. A large number of different restriction enzymes are available commercially, each of which recognizes a different DNA sequence and cuts the DNA at the location of the sequence that they recognize.

The DNA of lambda phage (a virus that infects *E. coli*) consists of 48000 base pairs. The particular restriction enzyme HindIII recognizes the six base sequence AAGCTT. (DNA is built out of 4 bases: A, T, C, and G.)

(a) Assuming all bases are equally likely, how many restriction sites do you expect there to be in lambda DNA?
(b) In fact, HindIII makes six cuts in Lambda DNA at base pair numbers 23130, 25157, 27479, 36895, 37459, 37584, and 44141. The p-value is the theoretical probability of achieving a result as extreme or more extreme than the result actually achieved, according to the "null" hypothesis, which in this case, is the hypothesis is that the sequence of lambda DNA is random. It follows that in this case the p-value is the theoretical probability of observing 6 or fewer cuts $[P(\leq 6)]$. What is the numerical value of the p-value in this case? Can we reject the hypothesis that the sequence of lambda DNA is random according to the (somewhat-arbitrary) criterion that if the p-value is less than 0.05, we should reject the null hypothesis?
(c) Another restriction enzyme, EcoRI, recognizes a different 6 base sequence, but EcoRI only cuts lambda DNA in 5 places. In light of this result, comment on the assumption that lambda DNA is a random sequence?

Problem 6: Counting bacteria in a Microscope
(a) In a study involving *E. coli* bacteria, a certain microscope slide has v (Greek "nu") bacteria per unit area. In a microscope equipped with a ×20 objective lens, the microscope field of view has an area A. What is the mean number of bacteria in the microscope field of view?
(b) What is the probability of observing zero bacteria in this microscope field of view?
(c) What is the probability of observing three bacteria in this microscope field of view?

(d) The ×20 microscope objective lens is now switched out for a ×60 objective lens. What is the probability of observing one bacterium in the microscope field of view now?

Problem 7: Bacteria and Phage Revisited
A culture dish contains N bacteria and n phage. Assume that every phage is in a bacterium.

(a) What is the mean number of phage per bacterium?
(b) What probability distribution determines the probability that the number of phage that infect a given bacterium is k?
(c) What is the probability that a given bacterium is not infected?
(d) What is the probability that a given bacterium is infected?
(e) Calculate n in order to ensure that a fraction f of the bacteria are infected in terms of N and f.

Problem 8: Fluorescence Correlation Spectroscopy
In the technique known as fluorescence correlation spectroscopy (FCS), a laser beam is focused to a micrometer-sized volume (call it the laser focal volume). When fluorescent proteins are in that volume, they are caused to emit fluorescent light. Therefore, the intensity of fluorescent light, which is what is measured experimentally, is proportional to the number of protein molecules in the laser focal volume. In this problem, you will explore how FCS can be used to measure the number of proteins in the laser focal volume and hence the concentration of those proteins [3].

(a) Suppose that the measured intensity of fluorescent light is I_1 if there is one protein in the laser focal volume. What then is the measured intensity of fluorescent light if there are k proteins in the laser focal volume?
(b) As the number of proteins, k, in the laser focal volume changes (fluctuates), the measured intensity of fluorescent light changes (fluctuates) correspondingly, but the mean intensity averaged over a long time of data collection is well defined. Likewise, the variance of the fluorescent intensity averaged over a long time is also well defined. What is the mean intensity ($\langle I \rangle$) of fluorescent light in terms of $\lambda = \langle k \rangle$ and I_1?
(c) What is the variance of the intensity (σ_I^2) of fluorescent light in terms of λ and I_1?
(d) Calculate the square of the mean intensity in terms of λ and I_1.
(e) Calculate the square of the mean intensity divided by the variance of the intensity in terms of the given quantities (λ and I_1).
(f) Explain how FCS permits you to determine the mean number of proteins in the laser focal volume (i.e., λ) without needing to know I_1, which may be difficult to determine experimentally.

(g) Given the value of λ and the value of the laser focal volume V, determine the mean protein concentration (number of proteins per unit volume), n, at the location of the laser focal volume.

Alternative to the method worked out in this problem, we could imagine calibrating the intensity and using the measured intensity ($I = kI_1$) to determine k and λ. However, calibrating to determine I_1 can be difficult and the calibration may change, depending on the details of the experimental setup and the sample under study. On the other hand, for the method worked out in this problem, it is not necessary to know I_1. Thus, it does not suffer from setup-to-setup variations and is very "robust" and reliable.

Problem 9: Measuring Mutation Rate

Ordinarily, *E. coli* is susceptible to infection by phage T1, a bacterial virus. T1 binds to specific receptors on *E. coli*'s surface, and is then able to enter the bacterium and kill it. Occasionally, however, during *E. coli* DNA replication, a mutation occurs in the T1 receptor, which prevents binding. These so-called Ton^R mutants are therefore immune to T1 infection. Here, your goal is to experimentally determine the mutation rate, α, which is the mean number of Ton^R mutations per *E. coli* DNA replication.

A total of M test tubes are prepared with growth media and each inoculated with n bacteria. In the initial growth phase, these bacteria are then permitted to grow by cell division until there are a total of N bacteria in each test tube, where $N \gg n$.

Next, every test tube is inoculated with T1 phage, soon killing all susceptible bacteria. However, after an additional period of time to permit further bacterial growth, it is observed that while there are m test tubes with no bacteria, there are also $M - m$ test tubes containing living bacteria. In these $M - m$ test tubes, during the initial growth phase, one or more times, there must have occurred a mutation of the T1 receptor, giving rise to a Ton^R *E. coli*, which subsequently reproduced to measurable quantities by the end of the experiment.

(a) If there are N bacteria in each test tube by the end of the initial growth phase, how many DNA replications had occurred in each test tube by the end of the initial growth phase? You should assume $N \gg n$ and therefore that n can be neglected compared to N.

(b) Given that α is the number of mutations per DNA replication, what is the expected mean number of mutation events per test tube in the initial growth phase?

(c) What is the probability that a test tube has zero bacteria with mutations in the initial growth phase in terms of α and N?

(d) What is the measured probability that a test tube has zero bacteria with mutations in the initial growth phase in terms of m and M?

(e) Determine α in terms of the experimental parameters, N, M, and m?

In fact, this method for determining the mutation rate, using the Poisson distribution, was an essential component of one of the most important experiments in genetics, described in Ref. [4], which earned Salvadore Luria and Max Delbrück the 1969

Nobel Prize for Physiology or Medicine. Although the distribution of mutation events follows a Poisson distribution, the number of bacteria with mutations does not, because once a bacterium has undergone a mutation event, it divides and both progeny possess the mutation and so on down the generations. Thus, a single mutation event can lead to many mutant bacteria and will lead to very many mutant bacteria, if the mutation event occurs early on in the initial growth phase.

Problem 10: Measuring the Number of Bacteria in a Test Tube
In "Measuring mutation rate," we treated the final number of bacteria in a test tube N, as given. However, if you actually were seeking to measure the mutation rate, you would need to measure N. Suppose, then, that you are given a test tube containing a large, but unknown number of bacteria (N). Devise a procedure for measuring N, including a discussion of the accuracy of your method.

Problem 11: The 2014 FIFA World Cup
It has been suggested that the number of goals in World Cup games is given by a Poisson distribution. A total of 171 goals were scored at the 2014 World Cup[16] in a total of 64 games, implying a mean number of goals per game per team of 1.336. Figure 4.8 shows the observed probability distribution for the number of goals scored per team per game in the 2014 World Cup in blue, determined on the basis of goals per game per team actually scored, in comparison to a Poisson distribution in red corresponding to the actual mean number of goals per team per game ($\lambda = 1.336$) in the 2014 World Cup. Evidently, the observed distribution is quite similar to the Poisson distribution, which leads to the hypothesis that the number of goals that a team scored per game is a Poisson-distributed random variable with, in this case, $\lambda = 1.336$.

(a) But what about Germany's 7 goals in their semi-final game against Brazil? According to the hypothesis that the number of goals that a team scores per game is a Poisson-distributed random variable with mean $\lambda = 1.336$, what is the probability that a team scores seven goals in a match?
(b) The p-value is the probability of achieving a result as extreme or more extreme than the result actually achieved. Therefore in this case the p-value is the probability of a team scoring seven or more goals in a match. What is the numerical value of the p-value in this case according to our hypothesis?
(c) One criterion for rejecting a hypothesis is that the p-value is less than 0.05. According to this criterion, should we reject the proposal that the number of 2014 World Cup goals is a Poisson-distributed random variable with mean 1.336?
(d) Would you expect a Poisson distribution to be applicable?

---

[16] http://www.fifa.com/worldcup/matches/index.html

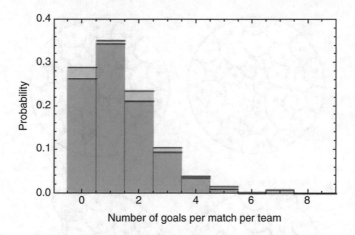

Fig. 4.8 Probability distribution of the number of goals scored per team per game (including extra time where appropriate, and excluding penalty shoot-out goals) in the 2014 World Cup, shown in blue, compared to a Poisson distribution corresponding to the mean number of goals per team per match ($\lambda = 1.336$) in the 2014 World Cup, shown in red

Problem 12: Therapy-Resistant Cancers

Cancer therapies are sometimes unsuccessful because cancer cells can undergo a mutation that confers resistance to a chemotherapy drug that ordinarily kills these cells. For example, treatment of chronic myelogenous leukemia (CML)[17] with imatinib[18] can fail as a result of a single point mutation in the abnormal BCR-ABL fusion protein, which drives CML. Consider then the growth of a cancer, initiated from a single cancer cell, to the point that the cancer is detected, when there are n cancer cells in total. The rate at which mutant cancer cells, resistant to therapy, are generated is k per DNA replication.

(a) What is the probability that the cancer will be resistant to treatment? Assume that all cancer cells do not die and that the fitness of mutated cancer cells is that same as that of non-mutated cancer cells.
(b) Discuss what your answer to (a) says about the value of early cancer detection.

Problem 13: Bilateral Retinoblastoma

Up until age 6 years, during the development of a child's retina, there is a more-or-less constant number of retinoblasts—retina stem cells—that each regularly undergo cell division in a child's growing retina. Each retinoblast cell division produces a differentiated retina cell and a replacement retinoblast. By age 6 years, the retina is fully developed and all remaining retinoblasts become differentiated retina cells.

   Retinoblastoma is the most common childhood eye cancer. Approximately one-half of all retinoblastoma patients suffer from a germ line mutation in one of each

---

[17] http://en.wikipedia.org/wiki/Chronic_myelogenous_leukemia
[18] http://en.wikipedia.org/wiki/Imatinib

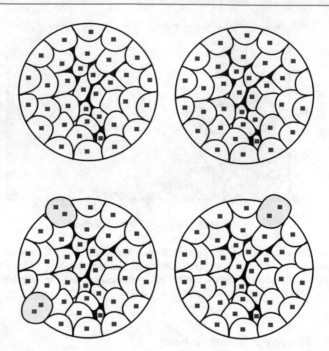

Fig. 4.9 Schematic of the retinas of retinoblastoma patients with RB1 germ line mutations. The two retinas in the top row correspond to the two retinas of a patient at birth with an RB1 germ line mutation, resulting in mutated RB1 gene in every retinoblast at birth, depicted as the blue dots. The bottom row shows the progression with increasing age. Each red dot indicates a second mutated RB1 gene that appears at a replication event during the development of the patient's retina during the first several years of life. Two dots in a single retinoplast represent two RB1 mutations in that retinoblast, which subsequently leads to its uncontrolled growth and division, that is, a retinoblastoma. In the lower pair of retinas, a case of bilateral retinoblastoma is indicated, with two retinoblastomas in the left eye and one in the right eye

cell's two RB1 tumor suppressor genes (TSGs). Following retinoblast cell division, a mutation in the progeny retinoblast's second RB1 gene causes the cell in question to develop improperly into a retinoblastoma, as illustrated schematically in Fig. 4.9, which shows the eyes of a patient, who eventually develops so-called bilateral retinoblastoma (bottom pair of eyes) with one or more cancers in both eyes. The occurrence of bilateral retinoblastoma indicates that the patient has the germ line mutation with very high probability.

You are involved in a medical research effort to determine the mutation rate of the second RB1 gene. The medical data that is available to you is the fraction of patients, who eventually show bilateral retinoblastoma, who do not present retinoblastoma until age t. Your goal in this problem is to develop a theoretical expression for the conditional probability that retinoblastoma does not present until age t, given that eventually the patient manifests bilateral retinoblastoma, which is directly comparable to this data.

(a) For a population of age-t children with the germ line mutation ($t < 6$ years), the accumulated number of retinoblast cell divisions is M per eye. If the mutation rate (mutations per retinoblast cell division) is α, what is the mean number of retinoblasts per eye in this population of children.

(b) Given that the Poisson distribution is applicable, what is the probability, $P(A)$, that an age-t child with the germ line mutation has zero retinoblastomas in both the left eye and the right eye?

(c) When retina development is complete, which occurs at an age of about 6 years, each eye will have experienced an accumulated total of N retinoblast cell divisions ($N > M$). What is the probability, $P(B)$, that a child with the germ line mutation eventually shows one or more retinoblastomas in the left eye and one or more retinoblastomas in the right eye, when the child is 6 years or older?

(d) What is the expected mean number of left eye retinoblastomas that develop between age t (when a patient has experienced a total of M retinoblast cell divisions per eye) and age 6 or more years (when the patient has experienced a total of N retinoblast cell divisions per eye).

(e) What is the probability, $P(C)$, that a child with the germ line mutation develops one or more retinoblastomas in their left eye after age t and one or more retinoblastomas in their right eye after age t?

(f) What is the joint probability, $P(A\&C)$, that a child with the germ line mutation develops zero left eye retinoblastomas and zero right eye retinoblastomas up until age t (outcome A) AND develops one or more left eye retinoblastomas after age t and one or more right eye retinoblastomas after age t (outcome C)?

(g) How is the probability you calculated in (f) related to the joint probability, $P(A\&B)$, that a patient shows zero left eye retinoblastomas and zero right eye retinoblastomas up until age t AND eventually shows one or more left eye retinoblastomas and one or more right eye retinoblastomas?

(h) Calculate the conditional probability, $P(A|B)$, that the patient's left and right eyes are retinoblastoma-free at age t, given that the patient eventually develops one or more left eye retinoblastomas and one or more right eye retinoblastomas, using your answers to (g) and (h) and Bayes' theorem, namely that

$$P(A|B) = \frac{P(A\&B)}{P(B)}. \tag{4.105}$$

(i) Sketch your answer to (h) versus t for $0 < t < 6$, given that $\alpha N = 1$ and $\alpha M = \frac{t}{6}$.

Problem 14: Measuring Cell Numbers with a Hemocytometer
Wikipedia tells us that a "hemocytometer is a device originally designed and usually used for counting blood cells.... and consists of a thick glass microscope slide with a rectangular indentation that creates a chamber. This chamber is engraved with a laser-etched grid of perpendicular lines. The device is carefully crafted so that the area bounded by the lines is known, and the depth of the chamber is also known. By observing a defined area of the grid, it is therefore possible to count

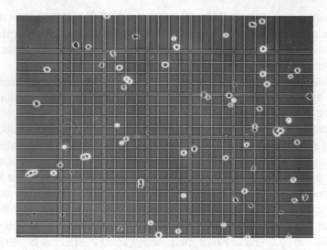

Fig. 4.10 Microscopy image showing cells in a hemocytometer. together with the hemocytometer's grid squares. Public domain image from https://upload.wikimedia.org/wikipedia/commons/b/bf/Neubauer_improved_with_cells.jpg

the number of cells or particles in a specific volume of fluid, and thereby calculate the concentration of cells in the fluid overall." Now, hemocytometers are widely used in biomedical research to count cells of all types.

For your own research project, you need to measure the total number of cells, N, in a suspension of cells of total volume, V. The first step in your experiment is to draw a sample of the suspension and fill a hemocytometer, separated into square grid of M sub-areas. Each of the M grid squares corresponds to a volume, W. Figure 4.10 shows yeast cells in a hemocytometer. In this case, nine grid squares are shown.

(a) What is the predicted mean number of cells (λ) in each grid square of the hemocytometer in terms of the given variables?

(b) What is the predicted probability—call it p—that one particular grid square contains zero cells?

(c) In fact, out of a total of M grid squares, you count that m of those grid squares contain zero cells. What then is the experimental probability—call it p again— that there are zero cells in a grid square in your experiment?

(d) Combine your answers to (b) and (c) to determine N in terms of V, W, M, and m.

(e) There is experimental error associated with your determination of N and, for your research, it is important to know what that error is.

You can understand why there is experimental error by noting that if you were to repeat the experiment several times, each time you would/could count a different number of grid squares with zero cells, i.e., each time you would/could find a different value of m. This reasoning shows that m is a random variable. Your answer to (d) immediately informs us that corresponding the value of N, which you find, is also a random variable. You can identify the experimental error in your determination of N with the standard deviation of N, which we can calculate theoretically without repeating any measurements.

The first step in calculating the standard deviation of N (σ_N) is to calculate the standard deviation or the variance of m (σ_m or σ_m^2). To this end, notice each of the M grid squares either has zero cells with probability p, or it has more than zero cells with probability $1 - p$. It follows that the binomial distribution determines the distribution of the number of empty grid squares, m. Given that the binomial distribution is applicable, what is the variance of m in terms of M and m only?

(f) In physics laboratory classes, we learn that, if N is a function of m, then the standard deviation of N and the standard deviation of m are related via

$$\sigma_N = \left| \frac{dN}{dm} \right| \sigma_m, \tag{4.106}$$

where the two vertical lines denote the absolute value. Thus, calculate σ_N in terms of σ_m, V, W, M, and m.

(g) Combine your answers to (e) and (f) to calculate $\frac{\sigma_N}{N}$ in terms of m, and M only.

Problem 15: Patch Clamp

Figure 4.11 shows the electrical current measured through a small area of cell membrane, that contains a single ion channel, obtained by the so-called the patch clamp technique.[19] What you can see from these data is that, roughly speaking, at different times, the current has two possible values: low and high. The higher value corresponds to the ion channel being closed, so that ions and the electrical current they carry do not flow across the cell membrane; the lower value corresponds to the ion channel being open, so that ions and their electrical current can flow. In each of these two states, the measured signal is not perfectly well-defined as a result of experimental errors, but, in this case, it is straightforward to see that for some of the recording time, the ion channel is open, and for some of the time it is closed.

(a) Explain how to experimentally determine the probability that an ion channel is in the open and closed states on the basis of patch clamp data, such as that shown in Fig. 4.11.

(b) In a particular patch clamp experiment the membrane under study contains n ($n > 1$) ion channels.[20] The fraction of time that any one channel is open, allowing ions to traverse the membrane is p. Assuming that the channels are independent of each other, what is the probability that k of these n channels will be open at any given instant of time?

(c) If the current through a single open channel is i (not the unit vector here), and the current through a closed channel is zero, what is the mean current through the membrane? What is the variance of this current?

---

[19] http://en.wikipedia.org/wiki/Patch_clamp

[20] http://en.wikipedia.org/wiki/Potassium_channels

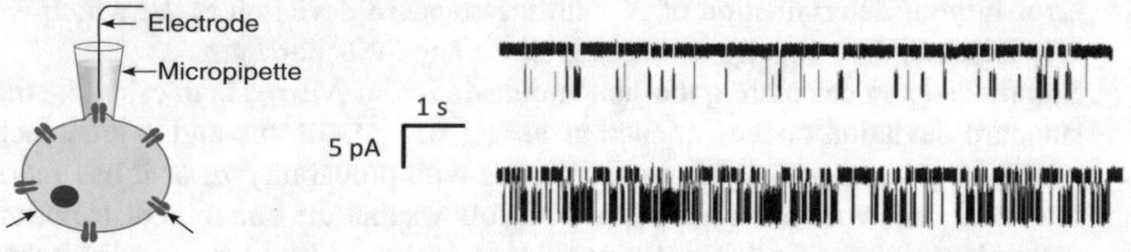

Fig. 4.11 Current through a single ion channel as an example of a discrete probability distribution. Left: Schematic of a patch clamp experiment. Right: Measurements of the current through an individual ion channel at two different conditions. The current, plotted on the vertical axis, versus time on the horizontal axis, measured through a single ion channel in a cell membrane. From https://commons.wikimedia.org/wiki/File:SingleMoleculeData_IonChannel_TwoConcentrations.png. This file is licensed under the Creative Commons Attribution-Share Alike 3.0 Unported license

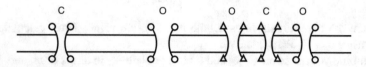

Fig. 4.12 Ion channels in a cell membrane can be either open or closed

Problem 16: More Ion Channels

A certain biological membrane system is prepared with two types of ion channels: There are n Na^+ channels and m Cl^- channels (with little circles and triangles on them, respectively, as suggested in Fig. 4.12). Both types of channel can be either open (O) or closed (C). Under the conditions of this problem, the probability that a Na^+ channel is open is p and the probability that a Cl^- channel is open is q. When a single Na^+ channel is open a current i flows through the membrane. When a single Cl^- channel is open, a current $-i$ flows through the membrane. When either type of channel is closed, there is zero current flow. Whether any given channel is open or not is independent of all other channels.

(a) If there are n Na^+ channels altogether, and p is the probability that a given Na^+ channel is open, what is the mean number of open Na^+ channels?
(b) What is the variance in the number of open Na^+ channels?
(c) If there are m Cl^- channels altogether, and q is the probability that a given Cl^- channel is open, what is the mean number of open Cl^- channels?
(d) What is the variance in the number of open Cl^- channels?
(e) When a single Na^+ channel is open a current i flows through the membrane. When a single Cl^- channel is open, a current $-i$ flows through the membrane (Fig. 4.12). What is the mean total current through the membrane?
(f) What is the variance of this total current?

Problem 17: Cool Way to Measure π
(a) Consider the square in the xy-plane, whose corners occur at the coordinates $((0,0), (1,0), (1,1), (0,1)$. What is its area?
(b) Now, consider a circle of unit radius. What is the area of the quadrant of this circle that lies within the square of part (a) in terms of (the irrational number) π?
(c) What is the area, still inside the square of (a), but outside the circle quadrant of (b) in terms of π?
(d) Now suppose that you have a computer program that generates random numbers lying between 0 and 1. In this way you can create pairs of random numbers, for which both members of a pair lie between 0 and 1. Each such pair of random numbers may be interpreted as the (x,y) coordinate of a "random point," which by construction, lies somewhere within the square of part (a). What then is the probability that such a random point lies within the circle quadrant of part (b)?
(e) What is the probability that such a random point lies outside the quadrant?
(f) Now suppose that you create N such points, and create a random variable by adding one for each random point that lies within the quadrant and adding zero for each random point that is outside the quadrant. What is the mean value of this random variable?
(g) How good do you expect the estimate to be for $N = 64$ and $N = 4096$? That is, what is the expected error in π, measured this way? And why?

Problem 18: Exponential Height Distribution
(a) The probability that a molecule in the atmosphere is at a height between z and $z+dz$ above the surface of the Earth is proportional to $e^{-z/\Lambda}\, dz$. Determine the constant of proportionality. Sketch this distribution.
(b) What is the mean molecular height, i.e., the mean height of a molecule?
(c) What is the variance in the molecular height?
(d) What is the standard deviation in the molecular height?

Problem 19: Another Continuous Random Variable
A certain random variable X has a probability density, $p(x)$, defined so that the probability that X has a value between x and $x + dx$ is $p(x)dx$. In fact, for this particular random variable

$$p(x) = Cx^2 \text{ for } 0 < x < 1$$

$$p(x) = 0 \text{ otherwise,} \qquad (4.107)$$

where C is a constant that you must determine.

(a) Sketch $p(x)$.
(b) Determine the value of C.
(c) Determine the mean value of x, namely $\langle x \rangle$.
(d) Determine the mean square value of x, namely $\langle x^2 \rangle$.
(e) Determine the variance of x, namely $\sigma_X^2 = \langle (X - \langle X \rangle)^2 \rangle = \langle X^2 \rangle - \langle X \rangle^2$.

(f) Determine the probability that X has a value between 0 and $\frac{1}{2}$.
(g) Another random variable Y has the exact same probability density as X, what is the variance and standard deviation of $X + Y$?

Problem 20: Yet Another Continuous Probability Distribution
Another random variable Y has a probability density, $p(y)$, defined so that the probability that Y has a value between y and $y + dy$ is $p(y)\,dy$. In this case,

$$p(y) = Cy \text{ for } 0 < y < 1$$

$$p(y) = 0 \text{ otherwise} \tag{4.108}$$

where C is a constant to be determined.

(a) Determine the value of C.
(b) Determine the mean value of y, $\langle y \rangle$.
(c) Determine the mean square value of y, $\langle y^2 \rangle$.
(d) Determine the variance of y, σ_y^2, and its standard deviation, σ_y.
(e) Determine the probability that Y has a value between V and 1.

Problem 21: Genetic Mapping—the Old-Fashioned Way
Fruit flies' genetic material is encoded on four chromosomes. Usually, we think that each chromosome of a baby fruit fly originates either from the father fruit fly or the mother fruit fly. If this is the case, when we consider two genetic traits (A and B) that correspond to genes on the same chromosome, then the baby fruit fly's A and B traits will both correspond to the traits of the father or both to the traits of the mother.

However, sometimes there occurs a "crossover," where the part of the chromosome in question ahead of the crossover comes from the mother and the part behind the crossover comes from the father (or *vice versa*). The probability of there being k crossovers between baby fruit fly's gene A and baby fruit fly's gene B is given by a Poisson distribution.

Even with today's technologies, it would be a tremendous effort to experimentally determine the number of crossovers between gene A and gene B. What can be measured, however, is whether the same parent originated gene A and gene B, corresponding to an even number of crossovers between gene A and gene B, or whether different parents originated gene A and gene B, corresponding to an odd number of crossovers between gene A and gene B.

(a) Given that the crossover rate is γ crossovers per base pair and the number of base pairs between gene A and gene B is L, what is the mean number of crossovers in that portion of the baby fruit fly's genome between gene A and gene B?
(b) What is the probability that there is one crossover between gene A and gene B?

(c) What is the probability that there are two crossovers between gene A and gene B?

(d) What is the probability—call it P_D—that gene A and gene B are observed to originate from different parent fruit flies? Hint: To answer this question, you may find it useful to know one of the following results:

$$\Sigma_{k=0,2,4,6...}^{\infty} \frac{\lambda^k e^{-\lambda}}{k!} = \frac{1}{2}(1 + e^{-2\lambda}); \qquad (4.109)$$

$$\Sigma_{k=1,3,5,7...}^{\infty} \frac{\lambda^k e^{-\lambda}}{k!} = \frac{1}{2}(1 - e^{-2\lambda}); \qquad (4.110)$$

$$\Sigma_{k=1,2,3,4,5,6...}^{\infty} \frac{\lambda^k e^{-\lambda}}{k!} = 1 - e^{-\lambda}. \qquad (4.111)$$

(e) In 1913, while he was an undergraduate student working in the laboratory of Thomas Hunt Morgan at Columbia University, Alfred Sturtevant constructed the first ever genetic map by applying your expression for P_D to many pairs of fruit fly genes. Sketch P_D as a function of the genome separation, L, between gene A and gene B. Be sure to show the correct behavior at small and large L. Explain whether the behavior of P_D for small and large L is sensible.

Problem 22: Genome Distance

Researchers have created a library of yeast strains, each of which contains two fluorescently labeled gene loci. In each case, one locus (A in Fig. 4.13) is at a known genomic location; the second (B in Fig. 4.13) is at an unknown genomic location. However, in order to carry out their research program, the researchers must determine the genomic separation between the two fluorescent loci in each strain. They could sequence the genomes, but for a large number of strains sequencing would be expensive. In fact, the researchers can determine the separation of the two

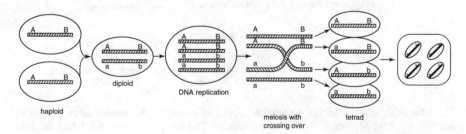

Fig. 4.13 Schematic of yeast mating, meiosis, and tetrad formation, in the case that there is one crossing over event. In this case, the tetrad phenotype is $T = \{AB, Ab, aB, aa\}$, where one spore (AB) contains two fluorescent loci, two spores (Ab and aB) contain one fluorescent locus each, and one spore (ab) contains zero fluorescent loci

loci by relying on the fact that the probability of crossing over between the two loci during meiosis, following mating with a non-fluorescent strain (with corresponding non-fluorescent alleles a and b), depends on the genomic separation of the two loci. In this problem, you explore how to determine the genomic separation by applying the Poisson distribution.

Meiosis in yeast—illustrated in Fig. 4.13—is especially straightforward to study, because all four meiotic products are packaged together as four haploid spores in a tetrad that can be dissected apart. Then, each spore can be grown-up into a colony, whose genotype can then be analyzed. In general, multiple crossovers are possible. However, the number of crossovers is not itself directly observable. Instead, there are just three possible, distinct, observable tetrad phenotypes. These phenotypes are conventionally called $PD = \{AB, AB, ab, ab\}$ (parental ditype), $T = \{AB, Ab, aB, ab\}$ (tetra ditype), and $NPD = \{Ab, Ab, aB, aB\}$ (non-parental ditype).

Considering the effect of increasing the number of crossovers from k to $k + 1$ leads to the conclusion that all of the PD phenotypes with k crossovers become T phenotypes with $k + 1$ crossovers, all of the NPD phenotypes with k crossovers become T phenotypes with $k+1$ crossovers, and, as illustrated in Fig. 4.14, one-half of the T phenotypes with k crossovers remain T phenotypes with $k + 1$ crossovers. These three routes to phenotype T given $k + 1$ crossovers are mutually exclusive. It follows that the probability of realizing phenotype T given $k + 1$ crossovers is the sum of the probabilities of each of these routes. Therefore, if the probability of realizing a T or PD or NPD tetrad, given that there are k crossovers is $P(T|k)$ or $P(PD|k)$ or $P(NPD|k)$, respectively, the rules of probability lead us to write

$$P(T|k + 1) = P(PD|k) + P(NPD|k) + \frac{1}{2}P(T|k). \qquad (4.112)$$

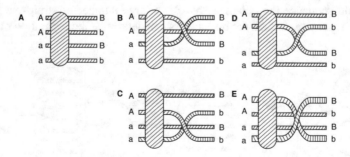

Fig. 4.14 Schematic illustrating the possible effects of increasing the number of crossovers by one to $k + 1$ from phenotype T with k crossovers. (**a**) The original k crossovers, which result in phenotype T, are represented as the hatched blob. In (**b**) through (**e**), the original k crossovers are also represented as the hatched blob, while the additional crossover is depicted explicitly. The two crossovers shown in (**b**) and (**c**) take phenotype T with k crossovers to phenotype T with $k + 1$ crossovers. The crossover shown in (**d**) takes phenotype T with k crossovers to phenotype PD with $k + 1$ crossovers. The crossover shown in (**c**) takes phenotype T with k crossovers to phenotype NPD with $k + 1$ crossovers

(a) Show that increasing the number of crossovers by from k to $k + 1$ takes both phenotype PD and phenotype NPD to phenotype T with probability 1 via analogous sketches to those shown in Fig. 4.14.

(b) Use Eq. 4.112 and the requirement that probabilities are normalized to obtain a recursion relation for $P(T|k)$.

(c) Solve your recursion relation for $P(T|k)$, using Wolfram Alpha or otherwise.

(d) Using your answer to (c), find $P(PD|k)$ and $P(NPD|k)$.

(e) The number of crossovers k in a genomic distance g is given by a Poisson distribution:

$$P(k) = \frac{(\alpha g)^k}{k!} e^{-\alpha g}, \qquad (4.113)$$

where α is the number of crossovers per unit genomic distance. Find the probability of realizing phenotype T in a genomic distance g. Call it $P(T)$. You may find it useful to know that

$$\Sigma_{k=0}^{\infty} \frac{\lambda^k}{k!} = e^{\lambda}. \qquad (4.114)$$

(f) Find the probability of realizing phenotype PD and NPD in a genomic distance g. Call them $P(ND)$ and $P(NPD)$.

(g) Sketch $P(T)$, $P(PD)$, and $P(NPD)$ versus genomic distance.

Problem 23: Translational Bursting

Translation—the process in which a messenger RNA (mRNA), which codes for a protein, is acted on by ribosomes to synthesize copies of the protein in question—is a stochastic process. Recent experiments have shown there is a "burst" of proteins synthesized by a particular mRNA molecule [5, 6]. The number of proteins in each burst from one mRNA is a random variable, governed by a discrete probability distribution. The goal of this problem is to determine this probability distribution, using the rules of probability.

(a) Suppose that the rate at which translation occurs from an mRNA transcript is α. What then is the mean number of proteins, synthesized in a time t?

(b) What probability distribution do you expect to be applicable for the number of proteins synthesized in a time t?

(c) Given your answers to (a) and (b) what is the probability that k proteins are synthesized from one mRNA transcript in a time t?

(d) If the rate of mRNA degradation is β, what is the probability that an mRNA realizes a lifetime that lies between t and $t + dt$? That is, what is the appropriate probability density for mRNA lifetime?

(e) In fact, your answer to (c) gives the probability that k proteins are synthesized from one mRNA transcript, given that the lifetime of the mRNA transcript in question is t, namely $P(k|t)$. Use your answers to (c) and (d) and the multiplication rule of probability to write down the joint probability, $P(k, t)dt$, that an mRNA leads to the synthesis of k proteins AND has a lifetime between t and $t + dt$.

(f) Use your answer to (d) and the sum rule of probability to determine $P(k)$, the probability that an mRNA gives rise to k proteins. (Use Wolfram Alpha to calculate the integral that you need.)

Problem 24: Genome Distance Revisited

(a) Following on from "Genome distance," suppose that in an experiment, the intrepid researchers find N_T tetrads of T phenotype, N_{PD} tetrads of PD phenotype and N_{NPD} tetrads of NPD phenotype. What is the probability— which in this context is generally called the likelihood, L—of making such a set of observations in terms of $P(T)$, $P(PD)$, $P(NPD)$, N_T, N_{PD}, and N_{NPD}?

(b) Explain how you would determine the maximum likelihood value of the genomic distance, g, given measurements of N_T, N_{PD}, and N_{NPD}. (Assume that α is known.)

Problem 25: Cell Division

In the bacterium, *E. coli* the number of certain RNAs can be quite small. When an *E. coli* undergoes cell division, each RNAs is equally likely to end up either of the two (equally sized) daughter cells, namely daughter 1 and daughter 2.

(a) What probability distribution gives the probability that k of them end up in daughter 1 and $n - k$ end up in daughter 2, given that there are a total of n RNAs. Call this probability $P(k|n)$.

(b) Suppose that the probability distribution, $P(n)$ of the number of RNAs in the parent cell, n, is given by a Poisson distribution with mean λ, i.e.,

$$P(n, \lambda) = \frac{\lambda^n e^{-\lambda}}{n!}. \tag{4.115}$$

What is the probability that the parent cell contains n RNAs and daughter 1 contains k RNAs. Call this probability $P(n\&k)$.

(c) Compare your answer to (b) to the product of two Poisson distributions, one for each daughter *E. coli*.

Problem 26: Relating Base Pair Differences to Time Since Most Recent Common Ancestor

A certain protein corresponds to a DNA sequence containing M base pairs. When the DNA sequence of that protein is examined in humans and in rhesus monkeys, it is found that there are m base pair differences between these two species. It is believed and, in this problem, we assume it to be true, that such changes in protein DNA sequence correspond to "neutral mutations" and do not affect species fitness. Suppose that α is the rate at which a base pair in the protein in question switches identity. (We assume that all types of base pair switch are equally likely.) In Chap. 5. *Random Walks: Brownian Motion and the Tree of Life* we will show that α is independent of population size, and that, therefore, the number of base pair differences serves as a molecular clock, that tells the time t since the most recent common ancestor (MRCA) of rhesus and humans lived. Your goal in this problem is to actually relate the observed number of base pair differences to t, assuming that we know α.

(a) What is the mean number switches between humans and rhesus that a particular base pair experiences in the time, t, which is the elapsed time, since the most recent common ancestor (MRCA) of humans and rhesus was living? Call your answer λ.

(b) Using your answer to (a) and assuming that the number of base pair switches follows a Poisson distribution, write the probability that a particular base pair in the protein has not experienced a base pair switch between humans and rhesus?

(c) While the number of base pair switches follows a Poisson distribution, analysis of the connection between observed base pair differences and t is complicated by the fact that after two or more switches a base pair can end up back at its original identity, e.g., in the case of two switches, a base pair could start as AT, then switch to CG, then switch back to AT. Therefore, your answer to (b) is correct for the probability of zero switches, but it underestimates the probability that a base pair appears unchanged. In fact, the conditional probability that a base pair returns to its original identity (i.e., that it appears unchanged), given that it has undergone k switches, is

$$Q(U|k) = \frac{1}{4}\left(1 + 3\left(-\frac{1}{3}\right)^k\right). \tag{4.116}$$

Using the rules of probability, write the probability that a base pair has undergone k switches and appears unchanged from its original identity. Call this probability, $Q(U, k)$.

(d) Given that outcomes with different values of k are mutually exclusive, write a sum that gives the probability that a base pair is unchanged, $Q(U) = \Sigma_{\cdots}......$, irrespective of the the number of base pair switches. Do not evaluate the sum.

(e) It is possible to carry out the summation that you wrote down in (d) with the result that the theoretical probability that a base pair appears unchanged in time t is

$$Q(U) = \frac{1}{4} + \frac{3}{4}e^{-\frac{4\lambda}{3}}. \tag{4.117}$$

What is the probability, $Q(C)$, that a base pair is changed from its initial identity in time t?

(f) What is the experimentally measured probability that a base pair in the protein in question appears changed between human and rhesus?

(g) Using your answers to (e) and (f), determine t in terms of the experimental parameters, α, M, and m.

References

1. S.J. Gould, *Dinosaur in a Haystack: Reflections in Natural History* (Harmony Books, 1995)
2. S. Hecht, S. Shlaer, M. H. Pirenne, Energy, quanta, and vision. J. Gener. Physiol. **25**, 819–840 (1942)
3. P. Cluzel, M. Surette, S. Leibler, An ultrasensitive bacterial motor revealed by monitoring signaling proteins in single cells. Science **287**, 1652–1655 (2000)
4. S.E. Luria, M. Delbrück, Mutations of bacteria from virus sensitivity to virus resistance. Genetics **28**, 491–511 (1943)
5. L. Cai, N. Friedman, X.S. Xie, Stochastic protein expression in individual cells at the single molecule level. Nature **440**, 358–362 (2006)
6. J. Yu, J. Xiao, X. Ren, K. Lao, X.S. Xie, Probing gene expression in live cells, one protein molecule at a time. Science **311**, 1600–1603 (2006)

Random Walks: Brownian Motion and the Tree of Life

5

Seeing is believing.

Recorded by John Clarke in Parœmiologia Anglo-Latina (1639).

A journey of a thousand miles begins with a single step.

Laozi, sixth century BCE

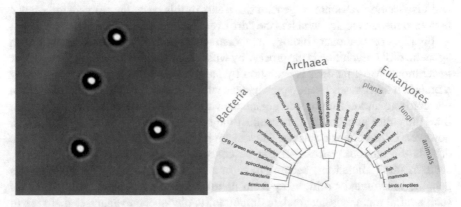

Left: A screenshot of micron-sized beads undergoing Brownian motion. Right: Public domain (CC0) image from https://commons.wikimedia.org/wiki/File:Simplified_tree.png

Electronic Supplementary Material The online version of this article (https://doi.org/10.1007/978-3-031-05808-0_5) contains supplementary material, which is available to authorized users.

5.1   Introduction

The concept of a "random walk" is extraordinarily important across the sciences with applications from evolutionary biology[1] to enzyme kinetics[2] to magnetic resonance imaging.[3] This chapter introduces random walks and explores their application to three biologically relevant phenomena. The first is Brownian motion, which all microscopic particles are subject to, including those in biological situations. Albert Einstein's theory of Brownian motion and its subsequent confirmation by Jean Perrin were pivotal in finally convincing skeptics at the outset of the twentieth century that atoms and molecules really exist, rather than being a convenient bookkeeping device. A key quantity that will emerge from a discussion of a particle's Brownian motion is the particle's diffusion coefficient or diffusivity, D.

Actin polymerization is our second application of the random walk concept. The growth of an actin filament via the association of further actin monomers to the tip of the filament, namely actin polymerization, is an important example of biological dynamics, and underlies many instances of eukaryotic cell motility. In this case, the connection to a random walk may be visualized as follows: when an actin filament adds a monomer, the tip takes a step in the positive direction, and when an actin filament loses a monomer, the tip takes a step in the negative direction. Therefore, since adding and losing monomers are probabilistic events, we expect the location of the tip of the actin filament to perform a random walk. Because the association and dissociation rates are in general different, in this case, the tip generally has a non-zero mean velocity, which is the "drift velocity" of the random walk.

Finally, in evolutionary biology, a random walk has a central role in evolution via "genetic drift." Evolution is the process by which biological populations change over generations. Most familiar is evolution by natural selection: organisms possessing a mutation that causes phenotypic changes which confer a selective advantage will out-compete and out-reproduce organisms without the beneficial mutation, and in time the entire population will manifest the beneficial mutation. Many mutations, however, offer neither a selective advantage nor a selective disadvantage. Genetic drift is the random process—it can be conceived as a random walk—by which such "neutral" mutations, become fixed in a population, leading the population's genome to evolve, or drift, away from its initial state, over many generations, as such neutral mutations accumulate. Importantly, phylogenetic analyses that seek to establish evolutionary relationships among organisms and among populations of a single organism critically rely on the properties of evolution by genetic drift.

---

[1] https://en.wikipedia.org/wiki/Genetic_drift

[2] https://en.wikipedia.org/wiki/Diffusion_limited_enzyme

[3] https://en.wikipedia.org/wiki/Motional_narrowing

This chapter builds directly on Chap. 4. *Probability Distributions: Mutations, Cancer Rates, and Vision Sensitivity*, because we will describe a random walk that steps randomly forwards and backwards using the binomial distribution. In the case of more general random walks that can take more than two types of step, we will rely on the result from Chap. 4. *Probability Distributions: Mutations, Cancer Rates, and Vision Sensitivity* that the mean of a sum of random variables is equal to the sum of the means of those random variables and that the variance of a sum of random variables is equal to the sum of their variances. In addition, we will rely on the rules of probability to explain a key feature of evolution by genetic drift and the theoretical basis for phylogenetic trees, conceived as evolutionary histories. This module also builds on Chap. 2. *Force and Momentum: Newton's Laws and How to Apply Them* by relating the friction coefficient, ζ, encountered in that chapter in the context of a particle falling though a viscous fluid, to the diffusion coefficient, D of the same particle in the same viscous fluid. The connection between ζ and D also exploits kinematics from Chap. 1. *Vectors and Kinematics*.

5.2   Your Learning Goals for This Chapter

At the end of this chapter, you should be able to:

- Describe and explain the phenomenon of Brownian motion and its basic statistical properties.
- Describe and explain what a random walk is, what its basic statistical properties are, and the connection between Brownian motion and random walks.
- Explain the Einstein relation and the reasoning that lead to the general acceptance of the atomic hypothesis by the scientific community in the early twentieth century.
- Appreciate that actin polymerization and chemical reactions more generally may be conceived as random walks.
- Calculate the mean displacement and the variance in the displacement, and the corresponding drift velocity and diffusion coefficient, for any one-dimensional random walk, but in particular for actin polymerization, given the rates of stepping forward and backward.
- Understand a simple mathematical model of evolution via "genetic drift" (namely the Moran model) within a population of N alleles, and how this model is equivalent to a certain random walk.
- Understand that the probability that a single neutral mutation becomes fixed within a population of N alleles is $1/N$, and be able to explain how this result follows from the addition and multiplication rules of probability, and

how it leads to a "molecular clock"[4] and provides the theoretical basis for the construction of phylogenetic trees, in which the branch lengths represent time estimates, including in particular the human phylogenetic tree with "mitochondrial Eve"[5] as the matrilineal most recent common ancestor of all living humans, who lived in Africa between 140,000 and 200,000 years ago.

---

[4]http://en.wikipedia.org/wiki/Molecular_clock
[5]http://en.wikipedia.org/wiki/Mitochondrial_Eve

5.3   Brownian Motion

The video at

https://www.youtube.com/watch?v=mTwGL19W20I[6]

shows five $1.2\,\mu$m-diameter polystyrene particles (beads), which are tethered via a $0.66\,\mu$m-long segment of double-stranded DNA (dsDNA) to a microscope coverslide. The beads were visualized using a $\times 100$-magnification microscope objective lens. Images were captured at a frame rate of 13.7 frames per second, and are played back in real-time. Strikingly, each bead undergoes a more-or-less random, "jiggling" motion. This is Brownian motion, named after botanist Robert Brown,[7] who first described organelles in pollen grains undergoing Brownian motion [1]. Figure 5.1 shows one frame of this video.

5.3.1   Brownian Motion and the Atomic Hypothesis

The hypothesis that matter is composed of atoms, which combine together into molecules via the formation of chemical bonds, gained increasing acceptance throughout the nineteenth century. However, even near the turn of the twentieth century, it was still possible for leading scientists to say:

  I don't believe that atoms exist!

as physicist and philosopher Ernst Mach[8] reportedly said, following an 1897 lecture by Ludwig Boltzmann.[9] Another leading scientist who was unconvinced of the

---

[6] https://www.youtube.com/watch?v=mTwGL19W20I
[7] https://en.wikipedia.org/wiki/Robert_Brown_(botanist,_born_1773)
[8] http://en.wikipedia.org/wiki/Ernst_Mach
[9] http://en.wikipedia.org/wiki/Ludwig_Boltzmann

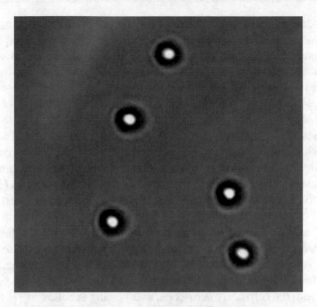

Fig. 5.1 One frame of the video at https://www.youtube.com/watch?v=mTwGL19W20I, obtained using a ×100-magnification microscope, showing five 1.2 μm-diameter polystyrene particles, which are tethered via a 0.66 μm-long segment of double-stranded DNA (dsDNA) to a microscope coverslide. It was acquired at a frame rate of 13.7 frames per second and there are 11.75 pixels per μm. The jiggling motion of the beads in the movie is Brownian motion. The video is courtesy of Peter Koo

existence of atoms at this time was chemist Wilhelm Oswald, whose contributions included introducing the "mole,"[10] and who was awarded the 1909 Nobel Prize for chemistry, "in recognition for his investigations into the fundamental principals governing chemical equilibria and rates of reaction." However, by 1909, Oswald wrote in his book, "Grundriss der allgemeinen Chemie":

I am now convinced that we have recently become possessed of experimental evidence of the discrete or grained nature of matter, which the atomic hypothesis sought in vain for hundreds and thousands of years. The isolation and counting of gaseous ions, on the one hand, which have crowned with success the long and brilliant researches of J. J. Thomson,[11] and, on the other, agreement of the Brownian movement with the requirements of the kinetic hypothesis, established by many investigators and most conclusively by J. Perrin,[12] justify the most cautious scientist in now speaking of the experimental proof of the atomic nature of matter. The atomic hypothesis is thus raised to the position of a scientifically well-founded theory, and can claim a place in a text-book intended for use as an introduction to the present state of our knowledge of General Chemistry.

---

[10] http://en.wikipedia.org/wiki/Mole_(unit)

[11] http://en.wikipedia.org/wiki/J_J_Thomson

[12] http://en.wikipedia.org/wiki/Jean_Perrin

Oswald, and everyone else, became convinced of the reality of atoms and molecules as a result of the agreement between quantitative experiments on Brownian motion, performed by J. Perrin[13] in 1908, for which he was awarded the 1926 Nobel Prize in Physics, and the corresponding theory of Brownian motion, worked out by Albert Einstein[14] in 1905.

What causes Brownian motion? Our hypothesis—the same as Albert Einstein's in 1905—is that Brownian motion is the result of collisions between each bead and molecules of the water in which it is suspended. But is this reasonable? After all, the water molecules are tiny (10^{-10} m and 10^{-27} kg). How can collisions with such tiny masses affect the motion of the bead? In addition, since the bead undergoes many, many collisions with water molecules each second (10^{12} s^{-1}), is not that much too fast to be relevant for motions that we see via video microscopy at 13.7 Hz?

We will see that the resolution of these objections is that the bead performs a *random walk*, composed of many, many tiny steps, which give rise to humanly observable displacements on humanly observable time scales, and when we observe Brownian motion, we are seeing the effects of molecular collisions with our own eyes, albeit with the help of a microscope, and "seeing is believing."

5.3.2   Statistical Properties of Brownian Motion

A statistical analysis of two beads' Brownian motion is summarized in the Mathematica demonstration TetheredBead.cdf at TetheredBead-Ch05.cdf. A movie of the two beads is reproduced in the top panel of the demonstration with each bead's previous 32 positions superimposed on each frame as the red and blue tracks. Each line segment of a track corresponds to the displacement of the bead between neighboring frames. Hit the "play" button to run the movie.

The two beads are identical and are attached to the coverslide via identical DNA. Therefore, we expect both beads' Brownian motion in both the x and y directions to show identical statistical properties. This expectation is born out by the similar appearance of all four traces in the top left subpanel of the demonstration's bottom panel, which are the x (red line) and y (magenta line) coordinate of bead 1 (upper bead) and the x (blue line) and y (cyan line) coordinate of bead 2 (lower bead). A detailed analysis (not shown) confirms that these traces show identical-within-error statistical properties. It follows that we can average together measurements along x and y and for both beads.

In the lower panel of the demonstration (and on the right-hand side of Fig. 5.2), the lower four subpanels show histograms of the beads' displacements between frames separated by time intervals of 1 (middle left), 2 (middle right), 3 (bottom left), and 4 frames (bottom right). The solid lines are Gaussian functions, each with

[13] http://en.wikipedia.org/wiki/Jean_Perrin

[14] http://en.wikipedia.org/wiki/Annus_Mirabilis_papers

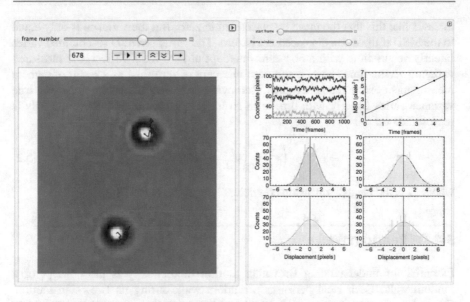

Fig. 5.2 Screenshots from TetheredBead-Ch05.cdf. Left panel, which reproduces the top panel of the demonstration: the tracks of each bead through the previous thirty-two images, shown red and blue, plotted on top of the current image. Right panel, which reproduces the bottom panel of the demonstration—top left: x (red line) and y (magenta line) coordinate of bead 1 and x (blue line) and y (cyan line) coordinate of bead 2. Right panel—middle and bottom: histograms of the beads' displacements between frames separated by time intervals of 1 (middle left), 2 (middle right), 3 (bottom left), and 4 frames (bottom right), compared to zero mean Gaussian functions, shown as solid lines, each with a variance equal to that of the corresponding experimental distribution. Right panel—top right: variance of the displacement for times separated by 1, 2, 3, and 4 frames, determined using 4 sets of displacements (2 beads each with x and y displacements), and a total of 512 frames. It is conventional to refer the variance of the displacement as the mean square displacement (MSD) in cases like this that the mean displacement is zero. But the variance is only equal to the MSD if the mean displacement is zero

a mean equal to zero and a variance equal to the variance of the corresponding experimental displacements:

$$\frac{N \times (0.2 \text{ pixels})}{\sqrt{2\pi\sigma^2}} e^{-\frac{x^2}{2\sigma^2}}, \tag{5.1}$$

where σ^2 is the variance of the displacements for the time delay in question and 0.2 pixels is the histogram bin size. The number of frames that contributes to the histograms (N) and which frames contribute can be controlled by the sliders. Clearly, when a sufficiently large number of frames are included in the analysis, the experimental histograms become well-described by these Gaussian functions.

The top right subpanel plots the variance of the displacement for times separated by 1, 2, 3, and 4 frames, determined using all four sets of displacements (2 beads each with x and y displacements), and a total of 512 frames. It is conventional to refer to the variance of the displacement as the mean square displacement (MSD) in cases like this that the mean displacement is zero. But the variance is only equal to the MSD if the mean displacement is zero. The variance/MSD so-obtained varies linearly versus time with a non-zero intercept at zero time. In fact, the measured variance-versus-time shows a non-zero intercept at zero time as a result of camera measurement errors (camera noise). Removing the effect of camera noise, the true variance varies linearly with time with an intercept of zero, and conventionally is represented as:

$$\langle x^2 \rangle = \langle y^2 \rangle = 2Dt, \tag{5.2}$$

where D is the beads' "diffusion coefficient" or "diffusivity" [2].

5.4   Random Walks

Essential for understanding Brownian motion theoretically is the concept of a random walk. Each bead performs a random walk during its Brownian motion. Examples of one-, two-, and three-dimensional random walks are depicted in Figs. 5.3, 5.4, and 5.5, respectively, which show screenshots from Wolfram demonstrations that permit you to manipulate one-, two-, and three-dimensional random walks.

Fig. 5.3 Screenshot from RandomWalk1D-Ch05.cdf showing three one-dimensional random walks, each consisting of 1024 "time" steps, randomly up or down with equal probability. The vertical coordinate shows the net displacement from the starting position after a given number of "time" steps along the horizontal coordinate

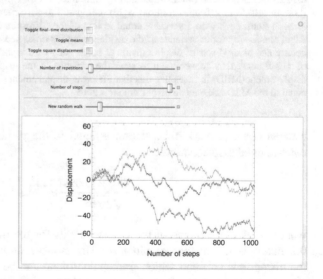

Fig. 5.4 Screenshot of a 2D
random walk from http://
demonstrations.wolfram.com/
LatticeRandomWalkIn2D/

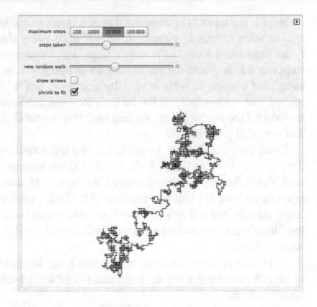

Fig. 5.5 Screenshot of a 3D
random walk from http://
demonstrations.wolfram.com/
LatticeRandomWalkIn3D/

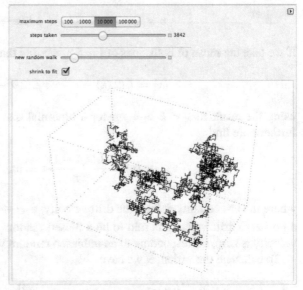

5.4.1 One-Dimensional Brownian Motion, Modeled as a 1D Random Walk

We will now introduce a simplified model of 1D Brownian motion, modeled as a
1D random walk. Since Brownian motion in the y and z directions is just the same
as in the x direction, given the behavior in x, we will be able to build 2D and 3D
random walks.

Here is our model [3]: Suppose that as a result of collisions with water molecules a bead steps randomly either $+L$ or $-L$ once every time step τ. Then, in a time t, the bead takes a total of $n = t/\tau$ steps. For Brownian motion the probability of stepping $+L$ is $\frac{1}{2}$ and the probability of stepping $-L$ is also $\frac{1}{2}$. This is certainly a simplified version of what is actually going on, but it will turn out to provide the insight that we are looking for. In fact, for later applications, it will be convenient to consider the slightly more general case that the bead steps $+L$ with probability p and $-L$ with probability $1 - p$.

Then, the total distance travelled (x) is itself a random variable with a binomial distribution (BD). In Chap. 4. *Probability Distributions: Mutations, Cancer Rates, and Vision Sensitivity*, we discussed the binomial distribution (BD) for a "big" random variable (k) that was the sum of n "little" random variables each of which could take the value 0 or 1 (for heads or tails, respectively). The wrinkle now is that the "little" random variables take the values $-L$ or $+L$, not 0 and 1. We have to account for this.

But this is possible when we realize that k can be re-interpreted to be the number of steps forward and $n - k$ to be the number of steps backwards. Therefore, x and k are related via:

$$x = Lk - L(n - k) = L(2k - n). \tag{5.3}$$

If we take the mean of both sides of Eq. 5.3, we find that

$$\langle x \rangle = L(2 \langle k \rangle - n) = nL(2p - 1), \tag{5.4}$$

using the result that $< k >= np$ for a binomial distribution. Using $n = \frac{t}{\tau}$, we furthermore find

$$\langle x \rangle = \frac{L(2p - 1)}{\tau} t = vt, \tag{5.5}$$

where the last equality defines the drift velocity, $v = \frac{L(2p-1)}{\tau}$. A random walk with a non-zero drift velocity is said to be a biased random walk. For $p = \frac{1}{2}$, the drift velocity is zero, corresponding to an unbiased random walk.

To calculate the variance, we have

$$\sigma_x^2 = \left\langle (x - \langle x \rangle)^2 \right\rangle = \left\langle [L(2k - n) - L(2 \langle k \rangle - n)]^2 \right\rangle = 4L^2 \left\langle [k - \langle k \rangle]^2 \right\rangle = 4L^2 \sigma_k^2. \tag{5.6}$$

Thus, the variance of x is proportional to the variance of k.

$$\sigma_x^2 = \left\langle (x - \langle x \rangle)^2 \right\rangle = 4L^2 \sigma_k^2 = 4L^2 np(1 - p) = \frac{4L^2 p(1 - p)}{\tau} t = 2Dt, \tag{5.7}$$

where we used the result for a binomial distribution that $\sigma_k^2 = np(1 - p)$ and where the last equality defines the diffusion coefficient for this model:

$$D = \frac{2L^2 p(1-p)}{\tau}. \qquad (5.8)$$

Specializing to the case, relevant to Brownian motion, namely $p = \frac{1}{2}$, we find

$$D = \frac{L^2}{2\tau}. \qquad (5.9)$$

When $p = \frac{1}{2}$ and the mean displacement is zero, Eq. 5.7 informs us that the variance of the bead's displacement, a.k.a. its mean square displacement, if the mean displacement is zero, varies linearly with t, as found experimentally (Fig. 5.2). The mean square displacement informs us about the mean distance that the bead moves, as becomes clear by considering the square root of the mean square displacement, namely the root mean square displacement (RMSD), which we is the mean distance that the bead travels from its starting point. For a one-dimensional random walk with zero mean displacement the root mean square displacement is $\sqrt{2Dt}$, which evidently grows with time, and eventually becomes large enough for us to see it as Brownian motion. What this means is that if we wait long enough, eventually the distance that the bead moves will inevitably become large enough to observe.

Based on a picture of collisions between the particle of interest and tiny molecules, we have theoretically reproduced the behavior observed experimentally as Brownian motion. On this basis, we may conclude that the tiny molecules indeed exist, i.e., atoms really exist. We are directly seeing the effects of molecules when we see Brownian motion. This calculation, carried out by Einstein in 1905,[15] was important in convincing scientists of the day once and for all that atoms really do exist.

5.4.2 Random Walks in the Continuum Limit

The central limit theorem (discussed in Sect. 4.13) informs us that a random variable that is itself the sum of many statistically independent random variables has a Gaussian probability distribution with a mean that is equal to the sum of the means of the random variables in the sum and with a variance that is the sum of the of the variances of the random variables in the sum. Therefore, since x is a random variable that is the sum of statistically independent steps, we may expect the probability that the particle's position is x $[P(x)]$ is well-described as a Gaussian function with the same mean and variance as the exact distribution:

$$P(x) = \frac{L}{\sqrt{4\pi Dt}} e^{-\frac{x^2}{4Dt}}, \qquad (5.10)$$

---

[15] http://en.wikipedia.org/wiki/Albert_Einstein#1905_-_Annus_Mirabilis_papers

where we used $< x >= 0$, appropriate for an unbiased random walk and Brownian motion, and $\sigma_x^2 = 2Dt$. The factor of L in the numerator ensures that this expression is dimensionless, as a probability must be. Numerical comparison shows that indeed a binomial distribution is well approximated by a Gaussian distribution of the same mean and variance.

If we envision the limit that $L \to 0$, while D remains constant, then $p(x) = P(x)/L$ is the probability density for x:

$$p(x) = \frac{1}{\sqrt{4\pi Dt}} e^{-\frac{x^2}{4Dt}}. \tag{5.11}$$

Equation 5.11 corresponds to the probability density for the position, x, of an unbiased one-dimensional random walk at a time t, that starts at $x = 0$ at $t = 0$. Although we have so far focused on a random walk that must either step forward or step back at each time step, it is important to appreciate that the key results, $< x >= vt$ and $\sigma^2 = 2Dt$, hold independently of the microscopic stepping details.

5.4.3   Two- and Three-Dimensional Random Walks

For unbiased two- and three-dimensional (2D and 3D) random walks, the steps in x and y and in x, y, and z, respectively, are statistically independent random variables (Fig. 5.2). Therefore, in 2D, the mean square displacement is the sum of the mean-squared displacements along x and y:

$$\left\langle r^2 \right\rangle = \left\langle x^2 \right\rangle + \left\langle y^2 \right\rangle = 2 \times 2Dt = 4Dt. \tag{5.12}$$

In addition, the joint probability density for x and y is a product:

$$p(x, y) = p(x)p(y) = \frac{1}{(4\pi Dt)} e^{-\frac{x^2+y^2}{4Dt}} = \frac{1}{(4\pi Dt)} e^{-\frac{r^2}{4Dt}}, \tag{5.13}$$

where $r^2 = x^2 + y^2$ in 2D. Notice that this function depends only on r, the distance from where the particle originates at $t = 0$, as we should expect for a 2D random walk. Similarly, in 3D, the mean square displacement is the sum of the mean-squared displacements along x and y and z:

$$\left\langle r^2 \right\rangle = \left\langle x^2 \right\rangle + \left\langle y^2 \right\rangle + \left\langle z^2 \right\rangle = 3 \times 2Dt = 6Dt. \tag{5.14}$$

In addition, the joint probability density for x, y, and z is a product:

$$p(x, y, z) = p(x)p(y)p(z) = \frac{1}{(4\pi Dt)^{3/2}} e^{-\frac{x^2+y^2+z^2}{4Dt}} = \frac{1}{(4\pi Dt)^{3/2}} e^{-\frac{r^2}{4Dt}}, \tag{5.15}$$

where $r^2 = x^2 + y^2 + z^2$. This function depends only on the 3D expression for r, the distance from where the particle originates at $t = 0$, as we should expect for a 3D random walk.

5.5   The Einstein Relation[16] and How We Know There Are Atoms

An additional key element of Einstein's theory of Brownian motion is a remarkable connection between diffusion and fluid friction, called the Einstein relation. As we discussed in Chap. 2. *Force and Momentum: Newton's Laws and How to Apply Them*, a bead, moving with a non-zero velocity (v) through a fluid, experiences macroscopic fluid friction ($-\zeta v$). However, fluid friction is result of the collisions with fluid molecules, that give rise to Brownian motion. From this point of view, it is plausible that there could be a connection between the bead's diffusion coefficient (D) and its friction coefficient (ζ). In fact, there is a precise relationship, namely the Einstein relation:

$$D = \frac{k_B T}{\zeta},$$ (5.16)

where k_B is Boltzmann's constant and T is the (absolute) temperature. His paper on Brownian motion and the Einstein relation was one of Einstein's four great papers of 1905,[17] along with papers on special relativity, the photoelectric effect, and $E = mc^2$.

To elucidate the connection between friction and diffusion, we will employ essentially the same simplified one-dimensional model that we used previously. Thus, there is one collision every time step τ. In a small elaboration, we'll suppose that upon each collision, the bead acquires an initial velocity $u = \pm L/\tau$, independent of its velocity before the collision. Then, the bead propagates freely at constant velocity for a time τ, until the next collision. Subsequent to each collision, the two velocity choices, $\pm L/\tau$, are equally likely. In this way, the particle indeed steps $\pm L$ every time step of τ, just as before.

But what happens when we add a constant force, F? Suppose that at time t, a bead is at $x(t)$ and undergoes a collision so that its velocity becomes $u = \pm L/\tau$. Kinematics for motion subject to a constant acceleration then informs us that at a time $t + \tau$, we have

$$x(t + \tau) = x(t) + u\tau + \frac{F}{2m}\tau^2,$$ (5.17)

---

[16] http://en.wikipedia.org/wiki/Einstein_relation_(kinetic_theory)

[17] http://en.wikipedia.org/wiki/Alber_Einstein#1905_-_Annus_Mirabilis_papers

where m is the mass of the bead. Thus, the change in the bead's position in the time interval τ is:

$$x(t+\tau) - x(t) = u\tau + \frac{F}{2m}\tau^2. \tag{5.18}$$

The mean value of the change in position, averaged over many collisions, is then

$$\langle x(t+\tau) - x(t)\rangle = \langle u\rangle \tau + \frac{F}{2m}\tau^2 = \frac{F}{2m}\tau^2, \tag{5.19}$$

where we used $< u >= 0$, since u is equally likely to be $\pm L/\tau$ and its mean is therefore zero. It follows that in the presence of a force, F, the bead acquires a mean velocity, a.k.a. drift velocity, (v) given by

$$v = \frac{\langle x(t+\tau) - x(t)\rangle}{\tau} = \frac{\tau}{2m}F, \tag{5.20}$$

which is linearly proportional to the applied force. This behavior is what we observed for objects falling in a viscous fluid in Chap. 2. *Force and Momentum: Newton's Laws and How to Apply Them*, where we saw that a bead falling under the influence of a force, F, achieves a "terminal velocity," given by

$$v = \frac{1}{\zeta}F. \tag{5.21}$$

Comparing Eqs. 5.20 and 5.21, it is apparent that we have derived a microscopic expression for the friction coefficient:

$$\zeta = \frac{2m}{\tau}. \tag{5.22}$$

To relate this friction coefficient to the diffusion coefficient of our model, we collect together a number of results from the model, namely:

$$D = \frac{L^2}{2\tau}, \tag{5.23}$$

$$\zeta = \frac{2m}{\tau}, \tag{5.24}$$

and

$$\langle u^2\rangle = \frac{L^2}{\tau^2}, \tag{5.25}$$

where $\langle u^2 \rangle$ is the mean square velocity along the x-direction. Multiplying Eqs. 5.23 and 5.24 and then using Eq. 5.25, we find

$$D\zeta = \frac{L^2}{2\tau}\frac{2m}{\tau} = m\frac{L^2}{\tau^2} = m < u^2 > . \quad (5.26)$$

In Chap. 8. *Statistical Mechanics: Boltzmann Factors, PCR, and Brownian Ratchets,* we will discover that

$$\frac{1}{2}m\langle u^2 \rangle = \frac{1}{2}k_B T. \quad (5.27)$$

Combining Eq. 5.27 in Eq. 5.26, we find

$$D = \frac{k_B T}{\zeta}, \quad (5.28)$$

which is the Einstein relation. Although we have used a simplified collisional model to derive the Einstein relation, more sophisticated treatments give the same result. Importantly, the Einstein relation relates two quantities, D and ζ that can be independently measured: D can be determined from microscopy measurements that characterize Brownian motion, while ζ can be determined from measurements of the drift velocity under the influence of a known force, such as gravity. Scientists at the outset of the twentieth century knew about the absolute temperature, T. Therefore, the Einstein relation permitted an experimental determination of Boltzmann's constant, $k_B = 1.38 \times 10^{-23}$ JK^{-1} for the first time. Because they also knew about the ideal gas law, and the value of the gas constant. $R = N_A k_B = 8.31\,J\,K^{-1}\,mol^{-1}$, where N_A is Avogadro's number. they could determine a value for Avogadro's number for the first time: $N_A = \frac{R}{k_B} = 6.022 \times 10^{23}\ mol^{-1}$, and, because they knew that N_A molecules of a gas occupy a volume equal to 22.4 liters at standard temperature and pressure, they could determine the mean volume occupied by one molecule for the first time. Since the size of a molecule must be equal to or smaller than the mean volume it occupies, the Einstein relation lead to the conclusion that atoms and molecules have linear dimensions less than 20 nm. That Einstein's theory both leads to molecular-size estimates that were considered reasonable and can explain Brownian motion convinced the skeptics of the 1900s that atoms and molecules are real.

5.6 Actin Polymerization

The concept of a random walk has broad applicability, far beyond the Brownian motion of a particle in a fluid. For example, it is possible to conceive of the progression of a chemical reaction as a random walk with the transition from reactant to product as a step in the positive direction and a transition from product to reactant as a step in the negative direction. In this section, we will explore a

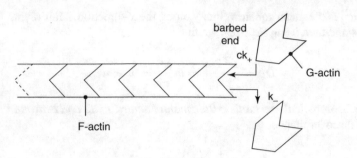

Fig. 5.6 Cartoon showing actin polymerization at the barbed end. The barbed-end tip undergoes a random walk, stepping in the positive direction when a G-actin monomer is added, and stepping in the negative direction when an actin monomer dissociates

particular chemical reaction from this point of view, namely the growth of an actin filament,[18] a.k.a. actin polymerization. In this case, the connection to a random walk is readily visualized: when an actin filament adds a monomer, the tip takes a step in the positive direction, and when an actin filament loses a monomer, the tip takes a step in the negative direction. Therefore, since adding and losing monomers are probabilistic events, we can indeed expect that the location of the tip of the actin filament performs some sort of random walk.

Actin is a key cytoskeletal and muscle protein that can exist in the cell either as an individual protein monomer (G actin), or as a component building block of actin filaments (F-actin), which give mechanical support to cells and permit certain cells to move (cell motility). As shown in the movie at http://www.pnas.org/content/suppl/2001/12/10/211556398.DC1/5563Movie_1.avi,[19] from Ref. [4] and schematically in Fig. 5.6, actin filaments grow (or polymerize) by adding actin monomers to their ends. In turn, this permits the filaments to push the cell forward, i.e., to exert a force. On a molecular scale F-actin is relatively stiff. Therefore, pushing with actin can be conceived as pushing with a stiff rod or stick on subcellular length scales. Each actin filament has a definite directionality with distinct "barbed" (+) and "pointed" (−) ends. The barbed ends generally grow faster than the pointed ends.[20] In fact, because the ends are different, it is possible, because of the involvement of ATP hydrolysis, for one end to grow while the other end shrinks. This behavior shown at

http://www.youtube.com/watch?v=xbswna2lIbk[21]

is called "treadmilling".

---

[18] http://www.youtube.com/watch?v=ks2C9FJ9pFY&feature=related

[19] http://www.pnas.org/content/suppl/2001/12/10/211556398.DC1/5563Movie_1.avi

[20] http://www.ncbi.nlm.nih.gov/pmc/articles/PMC2114620/pdf/jc10362747.pdf

[21] http://www.youtube.com/watch?v=xbswna2lIbk

Phagocytosis of bacteria by macrophages depends on actin polymerization, as shown in this famous movie, made by David Rogers of Vanderbilt University in the 1950s, which shows a neutrophil chasing and engulfing the offending *Staphylococcus aureus* bacterium:

http://www.youtube.com/watch?v=JnlULOjUhSQ&feature=related,[22]

Actin polymerization can be nefariously exploited by bacteria, such as listeria, as shown in

https://www.youtube.com/watch?v=sF4BeU60yT8.[23]

Actin polymerization enables fish skin cells (keratocytes) to move, as shown in

https://www.youtube.com/watch?v=RTjYXBnMcgs,[24]

permitting wounds to heal. Actin polymerization is also an essential component of how neurons connect to other neurons in our developing brains. Sarcoma[25] cells, namely cancer cells arising from connective tissues, are particularly adept at amoeboid movement,[26] which relies on actin polymerization, and which leads to the high rate of sarcoma metastasis.

5.6.1 Actin Polymerization as a Random Walk

To analyze actin polymerization, we will exploit what we know about how to combine the means and variances of statistically independent random variables. Specifically, we will exploit the facts that for a "big" random variable (x), that is the sum of n statistically independent "little" random variables (w_i), the mean of x is

$$\langle x \rangle = \Sigma_{i=1}^{n} \langle w_i \rangle = n \langle w \rangle, \qquad (5.29)$$

and the variance of x is

$$\sigma_x^2 = \Sigma_{i=1}^{n} \sigma_{w_i}^2 = n\sigma_w^2, \qquad (5.30)$$

---

[22] http://www.youtube.com/watch?v=JnlULOjUhSQ&feature=related
[23] https://www.youtube.com/watch?v=sF4BeU60yT8
[24] https://www.youtube.com/watch?v=RTjYXBnMcgs
[25] http://en.wikipedia.org/wiki/Sarcoma
[26] http://en.wikipedia.org/wiki/Amoeboid_movement

where the last equalities in Eqs. 5.29 and 5.30 hold only if the "small" random variables all have the same mean, $< w >$, and variance, σ_w^2. Furthermore, the central limit theorem informs us that x has a Gaussian probability density, for large enough n. If the "little' random variables all have the same distribution, the distribution of x is

$$p(x) = \frac{1}{\sqrt{2\pi n \sigma_w^2}} e^{-\frac{(x-n\langle w \rangle)^2}{2n\sigma_w^2}}. \tag{5.31}$$

We will now calculate the drift velocity and diffusion coefficient of the barbed end of an actin filament. Chemistry classes inform us that the rate at which actin monomers associate with the tip of an actin filament is ck_+, where c is the concentration of monomeric actin and k_+ is the association rate constant. Why this form? Because the rate of association reactions is linearly proportional to the number of actin monomers per unit time that come close enough to the tip to react, and that number must be proportional to the concentration of actin monomers. In fact, we will discuss this form in more detail in the next chapter, Chap. 6. *Diffusion: Membrane Permeability and the Rate of Actin Polymerization.* Given this association rate, the probability that the filament adds an actin monomer at the barbed end in a time period Δt is

$$p_{+1} = ck_+\Delta t, \tag{5.32}$$

corresponding to an increase in the length of the actin filament by a. Similarly, the rate at which an actin monomer dissociates from the barbed-end tip is k_-, so that the probability that an actin monomer dissociates from the barbed-end tip in a time period Δt is

$$p_{-1} = k_-\Delta t, \tag{5.33}$$

corresponding to a change in filament length of $-a$. We pick Δt to be sufficiently small that the only non-negligible possibilities in a time period Δt are for the tip to gain a monomer, or for the tip to lose a monomer, or for nothing to happen. Therefore, the probability that nothing happens in Δt is

$$p_0 = 1 - ck_+\Delta t - k_-\Delta t. \tag{5.34}$$

With these probabilities, we can now calculate the mean displacement of the tip in Δt:

$$\langle w \rangle = (+a)p_{+1} + (0)p_0 + (-a)p_{-1} = a(ck_+ - k_-)\Delta t. \tag{5.35}$$

We can also calculate the mean square displacement in time period Δt:

$$\left\langle w^2 \right\rangle = (+a)^2 p_{+1} + 0^2 p_0 + (-a)^2 p_{-1} = a^2(ck_+ + k_-)\Delta t. \tag{5.36}$$

Notice that there are not any minus signs in the final answer in this case. It follows that the variance in the time period Δt is

$$\sigma_w^2 = a^2(ck_+ + k_-)\Delta t - a^2(ck_+ - k_-)^2(\Delta t)^2. \tag{5.37}$$

As indicated above, in order to determine the mean and variance of the tip displacement after a time t, we will use what we know about how to combine the means and variances of statistically independent random variables, namely that the mean of the sum equals the sum of the means and that the variance of the sum equals the sum of the variances. In this case, we envision $n = \frac{t}{\Delta t}$ trials, all of which have the same mean and variance, which we have just calculated. Therefore, the mean displacement of the tip in a time t is

$$\langle x \rangle = n \langle w \rangle = \frac{t}{\Delta t} \times a(ck_+ - k_-)\Delta t = a(ck_+ - k_-)t. \tag{5.38}$$

The predicted linear dependence of tip displacement on time is consistent with the measurements shown in Fig. 5.7 [5]. The corresponding drift velocity of the tip is

$$v = \frac{\langle x \rangle}{t} = a(ck_+ - k_-). \tag{5.39}$$

There is a "critical" concentration, c_C, at which the actin tip has zero drift velocity, which we can calculate by setting $a(c_C k_+ - k_-) = 0$, thus finding $c_C = \frac{k_-}{k_+}$. For concentrations greater than c_C the tip proceeds forward on average with a mean velocity (drift velocity) that increases linearly with actin monomer concentration. For concentrations less than c_C the tip recedes backwards on average. This predicted behavior is consistent with the measurements presented in Fig. 5.8 [5], which shows that $c_C = 2 \; \mu M$ for the barbed end and $c_C = 5 \; \mu M$ for the pointed end. For actin monomer concentrations between 2 and 5 μM, an actin filament grows at the barbed end and recedes at the pointed end in a "treadmilling" behavior.

In the case of the variance of the tip position, we have

$$\sigma_x^2 = n\sigma_w^2 = \frac{t}{\Delta t} \times \left[a^2(ck_+ + k_-)\Delta t - a^2(ck_+ - k_-)^2(\Delta t)^2 \right]$$

$$= a^2 \left[(ck_+ + k_-) - (ck_+ - k_-)^2(\Delta t) \right] t. \tag{5.40}$$

However, in the limit that $\Delta t \to 0$, we may drop the term in Eq. 5.41 proportional to Δt, because it is so much smaller than the other term. As a result, the variance in the tip position simplifies to read:

$$\sigma_x^2 = a^2(ck_+ + k_-)t. \tag{5.41}$$

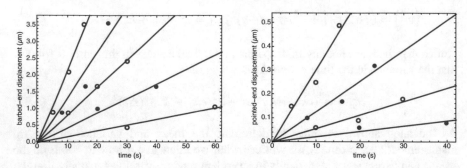

Fig. 5.7 Growth of actin filaments from the barbed and pointed ends versus time. Left: Barbed-end displacement *vs.* time (open and closed circles) for actin monomer concentrations of 2.5, 5. 7.5, 10, and 15 μM from bottom to top. Right: Pointed-end displacement *vs.* time (open and closed circles) for actin monomer concentrations of 5. 7.5, 10, and 15 μM from bottom to top. Adapted from Ref. [5]

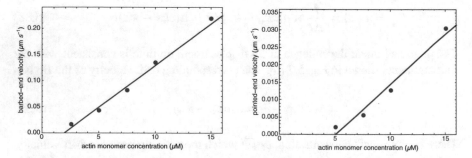

Fig. 5.8 Actin-monomer-concentration dependence of the growth velocity at the barbed and pointed ends of actin filaments. Left: Barbed-end velocity *vs.* concentration (closed circles), determined from the data of Fig. 5.7. Bottom right: Pointed-end velocity *vs.* concentration (closed circles), determined from the data of Fig. 5.7. The lines are linear fits to the data, revealing that the critical concentration, below which the tip velocity is negative, is 2 μM for the barbed end, and 5 μM for the pointed end

For a 1D random walk with diffusivity D, we expect the variance in the position to be $2Dt$. Therefore, here the diffusion coefficient of the tip is $D = \frac{1}{2}a^2(ck_+ + k_-)$. Since the tip of an actin filament has a drift velocity, like a biased random walk ($p \neq \frac{1}{2}$), and it has a diffusion constant, like a random walk, then by the "quacks-like-a-duck" rule, it must be that the tip of an actin filament indeed performs a random walk. In Chap. 8. *Statistical Mechanics: Boltzmann Factors, PCR, and Brownian Ratchets* we will return to the idea that the actin filament tip performs a random walk, when we discuss how actin filaments exert a force.

5.6.2   Example: Molecular Motors as Random Walkers

The molecular motor myosin V walks along a biofilament track, namely actin, taking steps of size $a \simeq 74$ nm, as shown in Fig. 5.9. Myosin's forward stepping rate is k steps per second. Therefore, you should take the probability, that myosin steps forward in a time Δt, to be $k\Delta t$. Myosin does not step backwards. Therefore, stepping forward or remaining stationary are the only two possibilities for myosin.

(a) What is the probability that myosin remains stationary in a time Δt?

   Answer:

$$1 - k\Delta t. \tag{5.42}$$

(b) Calculate the mean distance that myosin moves in a time Δt.

   Answer:

$$< w >= a \times k\Delta t + 0 \times (1 - k\Delta t) = ak\Delta t. \tag{5.43}$$

(c) Calculate the mean square distance that myosin moves in a time Δt.

   Answer:

$$< w^2 >= a^2 \times k\Delta t + 0^2 \times (1 - k\Delta t) = a^2 k\Delta t. \tag{5.44}$$

(continued)

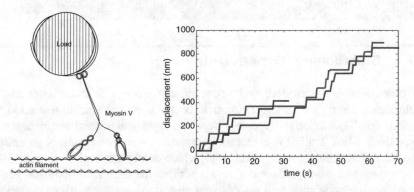

Fig. 5.9 Left: Schematic of myosin V walking on an actin filament. Right: Displacement versus time for three myosin V molecules. Adapted from Ref. [6]

(d) Given your answers to (a) and (b), calculate the variance in distance that myosin moves in a time Δt. Take Δt to be small. Therefore, in your answer, you should set terms that are proportional to $(\Delta t)^2$ equal to zero, but keep terms proportional to Δt.

Answer:

$$\sigma_w^2 = <w^2> - <w>^2 = a^2 k \Delta t - (ak\Delta t)^2 = a^2 k \Delta t, \qquad (5.45)$$

indeed neglecting $(\Delta t)^2$-terms.

(e) Calculate the mean distance that myosin moves in $n = t/(\Delta t)$ time steps, i.e., the mean distance the motor moves in a time t.

Answer:

$$<x> = n<w> = \frac{t}{\Delta t} \times ak\Delta t = akt. \qquad (5.46)$$

(f) Calculate the variance in the distance that myosin moves in $n = t/(\Delta t)$ time steps, i.e., the variance in distance that the motor moves in a time t.

Answer:

$$\sigma_x^2 = n\sigma_w^2 = \frac{t}{\Delta t} \times a^2 k \Delta t = a^2 kt. \qquad (5.47)$$

Interestingly, even though this random walker (myosin) can only step forward, and never steps backwards, it nevertheless shows a diffusion constant: $D = \frac{1}{2}a^2 k$.

5.7   Evolutionary "Genetic Drift"

As noted above, the random walk concept can emerge in many contexts. In evolutionary biology, a certain random walk has a central role in evolution via "genetic drift." Evolution is the process by which biological populations change over generations. Most familiar is evolution by natural selection: organisms possessing a mutation that causes phenotypic changes which confer a selective advantage will out-compete and out-reproduce organisms without the beneficial mutation, and in time the entire population will manifest the beneficial mutation. Many mutations, however, offer neither a selective advantage nor a selective disadvantage. Genetic

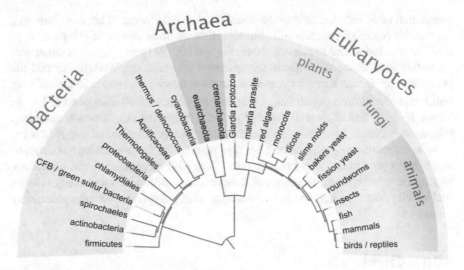

Fig. 5.10 Public domain (CC0) image from https://commons.wikimedia.org/wiki/File: Simplified_tree.png, created by Madeleine Price Ball

drift is the random process—it can be conceived as a random walk—by which such neutral mutations, become fixed in a population, leading the population's genome to evolve, or drift, away from its initial state, over many generations, as such neutral mutations accumulate. (Note that "genetic drift" is not a "drift velocity" of any kind.) Importantly, phylogenetic analyses that seek to establish evolutionary relationships among organisms and among populations of a single organism critically rely on the properties of evolution by genetic drift.

An example "phylogenetic tree" for all life on Earth—the tree of life—is shown in Fig. 5.10. Such phylogenetic trees specify the extent that the DNAs of different entries differ from each other. For the tree shown, branch points toward the outside correspond to relatively small differences in sequence, while branch points near the center correspond to large differences in sequence. In fact, mutations are incorporated into genetically isolated sub-populations at a constant rate, so that the number of mutations serves as a "molecular clock." From this point of view, the location of each branch point in Fig. 5.10 also specifies how long ago the most recent common ancestor (MRCA) of the two branching lineages lived.

A particularly important and interesting phylogenetic analysis appeared in a 1987 *Nature* paper [7], which analyzed the sequence divergence of 133 distinct examples of human mitochondrial DNA (mtDNA), which is inherited maternally. On this basis, Cann et al. [7] determined that all living humans are descended from a single maternal MRCA, who lived in Africa approximately 200,000 years ago, whom the media named "Eve." It is not immediately clear, however, why mutations should be a molecular clock. The rate at which mutations arise in an individual—called it k—must be independent of the size of the population from which that individual originates. Therefore, the rate at which mutations arise in a genetically isolated sub-population of N individuals must be equal to $N \times k$. How can it be, therefore, that the number of mtDNA mutations in humans, that we observe now, is independent of how the population size has changed over the last 200,000 years, which is unknown?

In this section, we will answer this question theoretically. We will see that the probability is $\frac{1}{N}$ that a neutral mutation survives many generations to overtake an entire reproducing sub-population, while the probability that it disappears is $1 - \frac{1}{N}$, where N is the size of the sub-population. As a result, the rate at which mutations take over—become *fixed* in—genetically isolated sub-populations of size N is $Nk \times \frac{1}{N} = k$, independent of population size. As a result, how the population has changed through the generations is not expected to affect phylogenetic analyses. Therefore, phylogenetic trees, based on genetic differences, may indeed be interpreted as clocks that specify how long ago species, or lineages within a species, became genetically separated.

5.7.1   Gambler's Ruin

Before embarking on a discussion of a model of genetic drift, it is instructive to consider a mathematically similar, but somewhat simpler, problem, namely "Gambler's ruin." "Gambler's ruin" refers to a process in which two gamblers repeatedly bet against each other—heads or tails—until one of them has won everything, and the other has gone bust. We will focus on one of the gamblers, and calculate the probability that they end up with \$$N$, where \$$N$ is the total initial stake of both players, given that they start with an initial stake of \$$i$ and bets \$1 each time.

The Gambler's ruin probability calculation is illustrated schematically in Fig. 5.11. We envision the amount of money that our gambler has at a given point in time to be the location of the gambler on a ladder with N rungs. Each time the gambler wins or loses a bet, they take either a step up or a step down on the ladder, respectively. The gambler performs a 1D random walk up and down the ladder, and we are interested in the probability that they make it to the top.

Fig. 5.11 Schematic calculation of the probability of a gambler eventually winning \$$N$

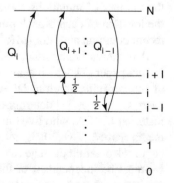

First, we introduce the probability that the gambler will eventually win everything, starting on rung i, Q_i. This outcome can occur via one of two mutually exclusive routes: either the gambler wins the next bet and then wins everything starting from rung $i + 1$; or the gambler loses the next bet and then wins everything starting from rung $i - 1$. These two possibilities are mutually exclusive, so their probabilities obey the addition rule. Therefore, the probability that the gambler wins starting from rung i (Q_i) is equal to the probability that the gambler wins the next bet ($\frac{1}{2}$) multiplied by the probability that the gambler wins starting from rung $i + 1$ (Q_{i+1}) plus the probability that the gambler loses the next bet ($\frac{1}{2}$) multiplied by the probability that the gambler wins starting from rung $i - 1$ (Q_{i-1}), that is,

$$Q_i = \frac{1}{2} \times Q_{i+1} + \frac{1}{2} \times Q_{i-1} = \frac{1}{2}(Q_{i-1} + Q_{i+1}). \tag{5.48}$$

This sort of equation, which relates the probabilities at one step of the process or one time to the probabilities at another step or another time, is sometimes called a "master equation." We can find the solution of Eq. 5.48 using Wolfram Alpha,[27] with the result that

$$Q_i = c_2 i + c_1, \tag{5.49}$$

where c_1 and c_2 are independent of i. Because the probability of winning is zero, if the gambler has already lost, and the probability of winning is unity, if the gambler has already won, we must also have that

$$Q_0 = 0 \tag{5.50}$$

and

$$Q_N = 1. \tag{5.51}$$

These two equations determine the values $c_1 = 0$ and $c_2 = \frac{1}{N}$. It follows that

$$Q_i = \frac{i}{N} \tag{5.52}$$

solves Eqs. 5.48, 5.50, and 5.51. Thus, we see that the probability that the gambler wins everything is equal to the gambler's fraction of the total initial stake. Not surprisingly, the less he starts with, the worse his chances of winning are. In particular, if he starts with \$1 and his opponent starts with \$$(N - 1)$, the probability

---

[27] http://www.wolframalpha.com/input/?i=Solve+Q%28j%29+%3D+1%2F2*Q%28j%2B1%29+%2B+1%2F2*Q%28j-1%29

that he ends up with everything is $\frac{1}{N}$. Since casinos generally have (a lot) more money than their clients, the reason for the name "Gambler's ruin" now becomes clear.

5.7.2 Example: Stacked Odds

A certain casino decides to use unfair coins in their game of gamblers' ruin.

(a) Modify Eq. 5.48 to incorporate an unfair coin into Gamber's ruin. Take the probability of winning the coin toss to be p.

    Answer: In this case, the probabilities of stepping up and down the ladder should be modified to be p and $1 - p$, respectively. Then, Eq. 5.48 is modified to read:

$$Q_i = p Q_{i+1} + (1 - p) Q_{i-1}. \tag{5.53}$$

(b) Solve your equation from (a) for Q_i.
    Wolfram Alpha can find the solution:

   http://www.wolframalpha.com/input/?i=Solve+Q%28j%29+%3D+p*Q%28j%2B1%29+%2B+%281-p%29*Q%28j-1%29 .

    Incorporating the requirements that $Q_0 = 0$ and $Q_N = 1$, we find that the solution for Q_i is

$$Q_i = \frac{1 - (\frac{1}{p} - 1)^i}{1 - (\frac{1}{p} - 1)^N}. \tag{5.54}$$

5.7.3 The Moran Model

What does Gambler's ruin have to do with genetic drift? The answer is that in a population of N individuals, in which i individuals possess allele[28] A of a certain gene and $N - i$ individuals possess allele B of that gene, the number of A alleles in the next generation is determined probabilistically. It follows that how the number of A alleles varies in time is a random walk. Therefore, the number of A alleles in the population is analogous to the number of dollars our gambler has, with the number of A alleles corresponding to the gamber's position on the hypothetical ladder.

---

[28] http://en.wikipedia.org/wiki/Allele

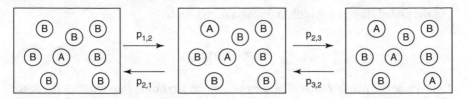

Fig. 5.12 Schematic showing transitions and the corresponding transition probabilities in the Moran model

What makes genetic drift mathematically different from Gamber's ruin is that the probability of increasing or decreasing the number of A alleles in the population at each generational step is not simply $\frac{1}{2}$.

The Moran model is a simple model of evolution via genetic drift that describes how the genetic makeup of a population subject to neutral mutations evolves in time. Specifically, the Moran model

(a) considers a gene for which there are two alleles A and B, *neither of which confers any evolutionary advantage compared to the other,*
(b) considers a population of N genes (which could correspond to a population of $N/2$ diploid organisms, i.e., organisms with two copies of each gene, such as humans).
(c) supposes that, at a certain point in time, there are i individuals with A alleles and $N - i$ individuals with B alleles.
(d) assumes that the population size remains constant in time. That is, for every "birth," there is a corresponding "death."
(e) randomly picks who will die at the next time step with every allele having equal probability to die (Fig. 5.12).
(f) randomly picks who will reproduce at the next time step with every allele having equal probability to reproduce. (The same allele can reproduce and die.) (Fig. 5.12).

The Moran model is certainly a simplified description. Nevertheless, it captures essential features of evolutionary processes. Given (c) and (e) above, the probability of an A death at the next time step is

$$d_A = \frac{i}{N} \tag{5.55}$$

and the probability of a B death at the next time step is

$$d_B = 1 - \frac{i}{N}. \tag{5.56}$$

Given (c) and (f) above, the probability of an A birth at the next time step is

$$b_A = \frac{i}{N} \tag{5.57}$$

and the probability of a B birth at the next time step is

$$b_B = 1 - \frac{i}{N}. \tag{5.58}$$

To increase the number of A alleles by one at the next time step requires the death of a B allele and the birth of an A allele. Thus, the probability that the number of A alleles increases by one is

$$p_{i,i+1} = b_A \times d_B = \frac{i}{N}(1 - \frac{i}{N}). \tag{5.59}$$

To decrease the number of A alleles by one at the next time step requires the death of a A allele and the birth of an B allele. Thus, the probability that the number of A alleles decreases by one is

$$p_{i,i-1} = b_B \times d_A = \frac{i}{N}(1 - \frac{i}{N}). \tag{5.60}$$

The only three possibilities at each time step are for the number of A alleles to increase by one, or decrease by one, or stay the same. Therefore, the probability that the number of A alleles stays the same is

$$p_{i,i} = 1 - p_{i,i+1} - p_{i,i-1} = 1 - 2\frac{i}{N}(1 - \frac{i}{N}). \tag{5.61}$$

5.7.4 Probability That a Mutation Becomes Fixed in the Population and Phylogenetic Trees

Now we will calculate the probability, Q_i, that subsequently the entire population of N genes corresponds to allele A, given that there are now i copies of allele A in the population. This calculation is illustrated schematically in Fig. 5.13, and follows

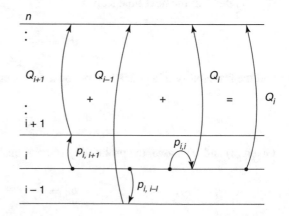

Fig. 5.13 Schematic calculation of the probability, Q_i, that subsequently the entire population of N genes corresponds to allele A, given that there are now i copies of allele A in the population

similar reasoning to our earlier Gambler's ruin probability calculation. In this case, if our genetic random walk is currently located at i, there are three ways to proceed: either the next step is up, with probability $p_{i,i+1}$, or it is down, with probability $p_{i,i-1}$, or it is a step of zero, with probability $p_{i,i}$. If the next step is a step of zero, the probability that subsequently the entire population of N genes corresponds to allele A remains equal to Q_i. If the next step is a step up, the probability that subsequently the entire population of N genes corresponds to allele A is Q_{i+1}. If the next step is a step down, the probability that subsequently the entire population of N genes corresponds to allele A is Q_{i-1}. Therefore, according to the rules of probability, the probability to be at i and eventually reach N is equal to the sum of the probability to step up multiplied by the probability to be at $i + 1$ and eventually reach N, plus the probability to step zero multiplied by the probability to be at i and eventually reach N, plus the probability to step down multiplied the probability to be at $i - 1$ and eventually reach N. Thus,

$$Q_i = p_{i,i+1} Q_{i+1} + p_{i,i} Q_i + p_{i,i-1} Q_{i-1}. \tag{5.62}$$

It follows that

$$(1 - p_{i,i}) Q_i = p_{i,i+1} Q_{i+1} + p_{i,i-1} Q_{i-1}. \tag{5.63}$$

However, normalization requires that $p_{i,i+1} + p_{i,i} + p_{i,i-1} = 1$. Therefore, $1 - p_{i,i} = p_{i,i+1} + p_{i,i-1}$, and so

$$Q_i = \frac{p_{i,i+1}}{p_{i,i+1} + p_{i,i-1}} Q_{i+1} + \frac{p_{i,i-1}}{p_{i,i+1} + p_{i,i-1}} Q_{i-1}. \tag{5.64}$$

We can now use the Moran-model results (Eqs. 5.60 and 5.59), namely $p_{i,i+1} = p_{i,i-1} = \frac{i}{N}(1 - \frac{i}{N})$, with the result that

$$Q_i = \frac{1}{2}(Q_{i+1} + Q_{i-1}), \tag{5.65}$$

which coincidentally, but very conveniently, is identical to Eq. 5.48, and which therefore implies that

$$Q_i = \frac{i}{N} \tag{5.66}$$

in this case too!

If the entire population eventually becomes all allele A, what happens to the number of A alleles at later times? Since according to the Moran model, $p_{N,N-1} = 0$, as soon as the entire population corresponds to allele A, there is no longer the possibility for B alleles to be born. Therefore, when allele A takes over the entire population, the entire population will remain allele A (until there is a mutation that sends A to C or back to B).

Similarly, when the entire population corresponds to allele B, $i = 0$ and we have that $p_{0,1} = 0$ and $Q_0 = 0$, and the population remains entirely allele B. How then does a single allele A arrive into the population, when before there was none? The answer is that it arrives only by mutation of B into A. Now suppose that $i = 1$, as if a mutation had just occurred, turning one of the previously B alleles into an A allele. What is the probability that this mutation, allele A, will eventually take over the entire population, i.e., what is the probability that this mutation will become fixed in the population? The answer is $Q_1 = \frac{1}{N}$.

Given that the mutation rate per gene (i.e., the mean number of mutations per generation per gene) is k, and that there are N genes, we may expect Nk mutations to appear per generation. However, the probability of a given mutation surviving to become fixed in the population is $\frac{1}{N}$. Therefore, the rate at which mutations become fixed in the population is equal to the rate that mutations appear, namely kN, multiplied by the probability that any given mutation becomes fixed in the population, namely $\frac{1}{N}$. Remarkably, the rate at which mutations become fixed in a population is $kN \times \frac{1}{N} = k$ per generation, independent of the population size, N, as promised.

Consider then two species with a most recent common ancestor (MRCA) M generations ago. Since the time at which that MRCA lived, each species will have been (on average) subjected to kM mutations that became fixed. Therefore, they will have on average $2kM$ genes/alleles that are different from each other. Therefore, if we measure their genetic discrepancy ($2kM$) and if we know k, we can determine M, namely how many generations ago their two lineages split apart from each other. And indeed we possess a molecular clock. This result is the basis on which phylogenetic trees can be interpreted as giving the time at which species and lineages within species diverged from each other. Thus, our understanding of the evolution of life on Earth—the tree of life—depends in an essential fashion on probabilistic ideas, and, in particular, on a certain random walk in allele space.

It is, of course, possible to simulate Moran process random walks. A demonstration that does just this (Fig. 5.14) is at

GeneticDrift-Ch05.cdf.

This demonstration also implements the effect of a selective advantage/disadvantage, which can be controlled using the slider label "Fitness."

5.8   Chapter Outlook

Looking forward, we will see in Chap. 6. *Diffusion: Membrane Permeability and the Rate of Actin Polymerization* that diffusion from a region of high concentration to a region of low concentration occurs as a result of individual particles' Brownian motion/random walks, and in Chap. 8. *Statistical Mechanics: Boltzmann Factors, PCR, and Brownian Ratchets*, we will exploit the idea that actin polymerization, and chemical reactions more generally, can be treated as random walks to better understand how the actin polymerization process can exert a force, and how helicase

Genetic drift

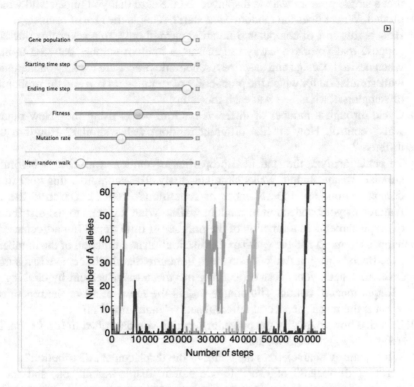

Fig. 5.14 Screenshot from the Mathematica Demonstration implementing the Moran model at GeneticDrift-Ch05.cdf, showing a simulation of several Moran process random walks, each colored differently. Each random walk starts with a mutation that introduces a solitary, new allele into the otherwise homogeneous allele population, which then evolves as described above

motor proteins can unzip dsDNA, for example. It is wonderous that biology works in spite of Brownian motion. What is truly amazing is that, in these examples, biology works *because* of Brownian motion.

5.9   Problems

Problem 1: Random Walk Simulations and Brownian Motion
A Mathematica simulation of a one-dimensional (1D) random walk is available at

RandomWalk1D-Ch05.cdf

(a) Fire up the demonstration and adjust the "Number of repetitions" control so that a single random walk is displayed. Make sure that you understand what is plotted. Where does the random walk start? What is the size of each step?
(b) How is this sort of computer simulation carried out? To answer this question, suppose that you have a way to generate a random number between 0 and 1 when desired. Design and describe a schematic procedure to carry out a random walk simulation for which the probability of stepping up is p and the probability of stepping down is $1 - p$ at each time step.
(c) Cycle through a number of different random walks using the "New random walk" control. How are the different random walks similar? How are they different?
(d) To better analyze the statistical properties of random walks, it is helpful to simulate many random walks simultaneously. To accomplish this goal in the demonstration, use the "Number of repetitions" control. Describe the key features of your collection of random walks. What happens to the distribution of displacements as a function of the number of time steps? In particular, what happens to mean and the width of the distribution as a function of the number of time steps? Can you think of processes in nature that might behave in a similar fashion? Check your ideas concerning the mean displacement by clicking the "Toggle means" button. The orange line is the mean of all of the repetitions. What is the mean for several different sets of random walks?
(e) Describe how to modify the procedure, which you described in (b), to simulate many ($nreps$) random walks.
(f) What quantity characterizes the width of the displacement distribution?
(g) Set "Toggle means" off, set "Toggle square displacement" on, and set the number of repetitions equal to one. Cycle through a number of different random walks using the 'New random walk' control. How are the different random walks similar? How are they different?
(h) To better analyze the statistical properties of the square displacements, it is again helpful to look at many random walks simultaneously, using the "Number of repetitions" control. "Toggle means" on and describe the behavior of the mean square displacement (MSD), shown as the orange line, for many random walks. How does the MSD change as the number of random walks analyzed changes? What function *quantitatively* describes the variation of the MSD as a function of the number of time steps?
(i) What insight does your analysis of a one-dimensional (1D) random walk provide into Brownian motion?

Problem 2: Two- and Three-Dimensional Random Walk Simulations
(a) Run the Wolfram Demonstration simulation of an unbiased two-dimensional random walk

http://demonstrations.wolfram.com/LatticeRandomWalkIn2D/.[29]

---

[29] http://demonstrations.wolfram.com/LatticeRandomWalkIn2D/

(b) Run the Wolfram Demonstration simulation of an unbiased three-dimensional random walk

http://demonstrations.wolfram.com/LatticeRandomWalkIn3D/.[30]

In each case, carry out the simulations for a range of values of the number of steps in the random walk. Notice that, without looking carefully at the axes, it is often difficult to tell how big of a simulation you are running.

Problem 3: A Discrete Random Variable

A certain random variable, k, can take the values -2, 0, or 2, only.

These values of k, and the corresponding values of $P(k)$ are as follows:

Value of k	$P(k)$
-2	$(1-p)^2$
0	2p(1-p)
2	See part (a)

(a) Determine the value of $P(2)$ in terms of p.
(b) Calculate the mean value of the random variable, k, i.e., what is $< k >$?
(c) Calculate the mean square value of k, i.e., what is $< k^2 >$?
(d) Using your answers to (b) and (c), calculate the variance and standard deviation of k?
(e) In fact, we are interested in a random variable—call it S—that is the sum of N copies of the random variable k—call them k_1 through k_N. What is the mean of S, i.e., $< S >$.
(f) What is the variance and standard deviation of S? How do these quantities depend on N?

Problem 4: Ministry of Silly Walks[31]

A certain random walker takes steps of size $-2a$, 0, or a, only.

These step sizes, and the probabilities that each occur in a time period Δt are as follows:

(a) What is the probability that the random walker takes a step of size 0 in a time Δt?
(b) Calculate the mean displacement in time Δt.
(c) Calculate the mean square displacement in time Δt.

---

[30] http://demonstrations.wolfram.com/LatticeRandomWalkIn3D/
[31] https://www.youtube.com/watch?v=eCLp7zodUiI

Step size	Probability
$-2a$	$k \Delta t$
0	See part (a)
a	$3k \Delta t$

(d) Calculate the variance of the displacement in time Δt.
(e) What is the mean displacement in a time t, given that the time t corresponds to $t/(\Delta t)$ steps of the random walker?
(f) Similarly, what is the variance of the displacement in time t? Take $\Delta t \to 0$.
(g) What is the drift velocity of this random walker?
(h) What is its diffusion coefficient?

Problem 5: Biased Random Walks
Consider a molecular motor performing a biased one-dimensional random walk with a positive drift velocity. The molecular motor starts moving from $x = 0$ at time $t = 0$.

Consider three later times: t_1, t_2, and t_3 with $0 < t_1 < t_2 < t_3$. Assume that each of these times is much longer than the time of a single step of the molecular motor.

Make a single plot that shows a sketch of the probability distribution of the molecular motor's position along the x-axis at each of t_1, t_2, and t_3.

Problem 6: DNA Zipping and Unzipping and Polymerase Chain Reaction (PCR)
In this problem, you will explore DNA zipping and unzipping, which are essential processes in DNA multiplication by polymerase chain reaction (PCR).[32] Kary Mullis won the 1993 Nobel Prize in Chemistry for inventing PCR, which has revolutionized not only biotechnology, but also justice systems around the world[33] by enabling DNA testing[34] from minute initial amounts of DNA. The first step in PCR is to raise the temperature above the DNA "melting temperature," T_M. At this elevated temperature, each dsDNA strand unzips to become two ssDNA strands. In the second step, when the temperature is subsequently reduced in the presence of oligonucleotide primers, nucleotides, and DNA polymerase, each previously unzipped ssDNA strand templates its own conversion to dsDNA. These steps double the original number of dsDNA strands, because a new dsDNA strand is created for each ssDNA. PCR involves repeating this temperature cycling process multiple (N) times, with the result that the initial number of dsDNA molecules is multiplied by a factor of 2^N. Thus, initially tiny quantities of dsDNA can be greatly amplified, and subsequently sequenced.

---

[32] http://en.wikipedia.org/wiki/Polymerase_chain_reaction

[33] http://topics.nytimes.com/top/reference/timestopics/subjects/d/dna_evidence/index.html?inline=nyt-classifier

[34] http://en.wikipedia.org/wiki/DNA_testing

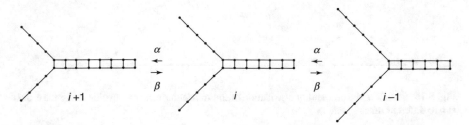

Fig. 5.15 The possible reactions of a DNA strand with i zipped base pairs, either undergoing isomerization to a DNA strand with $i - 1$ base pairs or isomerization to a DNA strand with $i + 1$ base pairs

Consider a strand of DNA that contains a junction between a region to the right where the DNA is double stranded DNA (dsDNA) and a region to the left where the DNA consists of 2 strands of single-stranded DNA (ssDNA), as shown in Fig. 5.15. When the DNA unzips one base pair, the dsDNA-ssDNA junction takes a step in the positive x-direction, and when the DNA zips up one base pair, the dsDNA-ssDNA junction takes one step in the negative x-direction. Therefore, since zipping up and unzipping one base pair are probabilistic events, we can expect that *the junction performs a random walk.* Specifically, suppose that the probability that the DNA zips up one base pair in a time Δt is $\alpha \, \Delta t$, while the probability that the DNA unzips one base pair in a time Δt is $\beta \Delta t$. Thus, α is the zipping rate and β is the unzipping rate. When the junction unzips one base pair, it is displaced a distance $+a$. When the junction zips up one base pair, it is displaced a distance $-a$.

(a) Consider a single time step, Δt. What is the probability that the displacement of the dsDNA-ssDNA junction in a time Δt is zero?
(b) Calculate the mean displacement of the junction in a time Δt.
(c) Calculate the mean squared displacement of the junction in a time Δt.
(d) Calculate the variance of the displacement of the junction in Δt.
(e) What is the mean displacement of the junction in a time t?
(f) What is the variance of the junction displacement in a time t? Consider the limit $\Delta t \to 0$.
(g) What is the corresponding diffusion coefficient of the junction?
(h) What is the drift velocity of the junction?
(i) What condition is there on the junction's drift velocity in order for the DNA to unzip?
(j) What is the corresponding condition on α and β in order for the DNA to unzip?
(k) Briefly explain what this has to do with PCR.

Problem 7: Molecular Motor Sherlock Holmes
A certain molecular motor is performing a random walk. In Fig. 5.16 you are given a sketch of the probability distribution of the molecular motor's position along the x axis at two different times t_A and t_B.

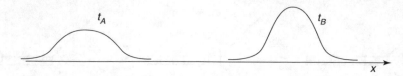

Fig. 5.16 Sketch of the probability distribution of the molecular motor's position along the x axis at two different times t_A and t_B

From the sketch, determine:

(a) Is the molecular motor performing a *biased* random walk or an *unbiased* random walk?
(b) Is $t_A < t_B$ or $t_A > t_B$ or can't we tell?
(c) Is the drift velocity, v_d, of the molecular motor positive, negative, or zero (with respect to the x axis shown in the figure)?
(d) Copy the sketch of Fig. 5.16 and add to it, on the same graph, a sketch of the probability distribution of the molecular motor at a time t_C with $t_C < t_B$ and $t_C < t_A$.

Problem 8: Diffusion Times
Consider a spherical ion of radius 0.1 nm, a spherical protein of radius 5 nm, a spherical virus of radius 100 nm, and a polystyrene bead of radius 1 μm. In each case, estimate how long it takes for the particle in question to diffuse across a eukaryotic cell.

Problem 9: Two Views of a Random Walk
Myosin V walks along a biofilament taking steps of size a, with a forward stepping rate of k steps per second and without stepping backwards. We previous analyzed myosin V's motion as an example in the module.

(a) An alternative point of view concerning myosin V's motion takes note of the fact that it involves a number of steps in a time t. What probability distribution is applicable to counting problems? Is this distribution applicable in this case?
(b) What is the mean number of steps in a time t?
(c) What is the variance in the number of steps in a time t?
(d) How do these answers compare to the corresponding quantities calculated in the module?
(e) Develop a theory of actin polymerization on the basis of two independent Poisson processes: (1) randomly adding monomers to the tip at a rate $k_+ n$, where k_+ is the association rate constant, and n is the concentration of actin monomer, and (2) randomly subtracting monomers from the tip at a rate k_-.

Problem 10: A More Realistic Molecular Motor
Consider a molecular motor that has two (and only two) internal states, state 0 and state 1, with the following properties:

- The probability that the molecular motor is in state 0 is P_0.
- The probability that the molecular motor is in state 1 is P_1.
- Given that the molecular motor is in state 1, the probability that it steps forward (displacement $+a$) in a small time interval, Δt, is $k_+ \Delta t$.
- In state 1, the molecular motor cannot step backward.
- Given that the molecular motor is in state 0, the probability that it steps backward (displacement $-a$) in a small time interval, Δt, is $k_- \Delta t$.
- In state 0, the molecular motor cannot step forwards.

(a) What is the probability that the molecular motor steps forward in time Δt in terms of the given parameters?
(b) What is the probability that the molecular motor steps backward in time Δt?
(c) What is the mean displacement of the molecular motor in time Δt? Call it $\langle w \rangle$.
(d) What is the mean square displacement of the molecular motor in time Δt? Call it $\langle w^2 \rangle$.
(e) What is the mean displacement of the molecular motor in time t? Call it $\langle x \rangle$.
(f) In fact, the biochemists tell us that

$$\frac{P_1}{P_0} = \frac{b[\text{ATP}]}{h}, \tag{5.67}$$

where $[\text{ATP}]$ is the concentration of ATP in the solution about the molecular motor, $b[\text{ATP}]$ is the ATP binding rate, and h is the rate of ATP hydrolysis. In addition, for a certain molecular motor $k_- = 0$. For this molecular motor, eliminate P_0 and P_1 to determine the drift velocity, v, of this molecular motor in terms of a, k_+, $b[\text{ATP}]$, and h only.
(g) Sketch $\frac{1}{v}$ versus $\frac{1}{[\text{ATP}]}$.

Problem 11. Swimming *E. coli* Perform a Random Walk
A bacterium swims at constant speed, v, along trajectories comprising randomly oriented straight runs, separated by fast tumbling motions. The direction of the motion after each such tumble is independent of the direction before each tumble, and from the run length, i.e., run length (L) and run direction (θ) are independent random variables. Ignore the time required for the tumbling. The mean length of a run is $\langle L \rangle$. The mean square run length is $\langle L^2 \rangle = 2 \langle L \rangle^2$.

(a) Given that $\langle \cos^2 \theta \rangle = 1/3$, where the average is taken over all angles, θ, what is the mean square displacement of a run along the x-direction. Call it $\langle (\Delta x)^2 \rangle$. (Hint: express the random variable Δx in terms of the random variables L and θ. Calculate the average of $(\Delta x)^2$, based on the information that you have been given about the random variables, L and θ.)
(b) What is the total mean square displacement $\langle \langle x^2 \rangle \rangle$ along x after n runs and tumbles in terms of $\langle (\Delta x)^2 \rangle$?

(c) Express n in terms of the total time (t), the mean length of each run ($\langle L \rangle$), and the bacterium's speed (v).

(d) Combine your results for (b), (c), and (d), to find the *effective* diffusion coefficient of bacterial motion (D), defined via

$$\langle x^2 \rangle = 2Dt. \qquad (5.68)$$

Problem 12: Gambler's Ruin, Part Deux

A certain unscrupulous casino owner decides to introduce a modified version of Gambler's Ruin, in which two unfair coins at a time are tossed and both must be heads for a gambler to win $1 and both must be tails for the casino to win $1. If there is both a heads and a tails, the coins are tossed again. For both coins, the probability of heads is p and the probability of tails is $1 - p$. To decide whether the game is one to join, a prospective gambler writes down the following equation for the probability that the gambler will beat the house, given that the gambler initially has $i

$$Q(i) = (1 - p)^2 Q(i - 1) + p^2 Q(i + 1) + 2p(1 - p)Q(i). \qquad (5.69)$$

Explain how the gambler arrived at this equation and the origin of each term on its right-hand side.

Problem 13: Life Is not Fair

(a) Explain how would you modify EQ. 5.62 to include the effect of a mutation that does provide a selective advantage?

(b) If, instead of being a neutral mutation, a mutation provides a selective advantage, discuss whether such a mutation is certain to become fixed in the population?

Problem 14: Time in the Moran Model

In this chapter, we discussed the Moran model of evolutionary drift within a population of N alleles. At each elementary step of the evolutionary process, one allele is born and one dies. Which one reproduces and which one dies is determined randomly. Although it seems certain that each elementary birth-death step of the Moran model somehow represents time, it is not immediately clear how to properly relate an elementary step to elapsed time. The goal of this problem is to make the relationship between steps in the Moran model and time more precise. The way that we will do this is that we will calculate the expected lifetime of an allele in terms of the number of alleles in the population. Along the way you will derive a new discrete probability distribution that gives the probability that an allele dies on step m.

(a) Consider a population of N alleles. Given that which allele dies at each step is picked at random, what then is the probability that a given allele dies in a particular birth-death step?

(b) What is the probability that a given allele does NOT die in a particular step?
(c) What is the probability—call it $P(m)$—that a given allele survives for $m - 1$ steps and then dies at step m?
(d) Write a sum that gives the mean number steps at allele death.
(e) What is the value of the sum that you wrote for (d), given that

$$\Sigma_{m=1}^{\infty} x^{m-1} = \frac{1}{1-x}, \tag{5.70}$$

$$\Sigma_{m=1}^{\infty} mx^{m-1} = \frac{1}{(1-x)^2}. \tag{5.71}$$

and

$$\Sigma_{m=1}^{\infty} m^2 x^{m-1} = \frac{1+x}{(1-x)^3}? \tag{5.72}$$

(f) The mean number of steps at allele death corresponds to the mean lifetime of an allele, t_L. What then is the time between successive steps in terms of t_L?
(g) In this chapter, we showed that the probability of a (neutral) mutation becoming fixed in the population is $\frac{1}{N}$. A further analysis of the Moran model reveals that the mean number of steps for such a mutation to become fixed in the population equals N^2. How long is this in terms of t_L?
(h) For the world's current human population of 7 billion, estimate how long it would take for a new mutation to become fixed in this population? In comparison, the age of the Universe is 13.798 ± 0.037 billion years,[35] while the Earth is 4.5 billion years old, and there has been life on Earth for 3.5 billion years. In light of your calculation, discuss whether human evolution is finished.

Problem 15. Zombie Apocalypse
You work for the CDC and have been called in to consult about the appearance in Connecticut of a single zombie. It is known from studies of past zombie epidemics that for this strain of the zombie virus, the probability that a zombie infects one human before the zombie is killed is $P(1) = a$ and the probability that a zombie infects two humans before the zombie is killed is $P(2) = a^2$, and that, for this strain, no-more than two humans can be infected per zombie. Once infected a human quickly turns into a zombie.

(a) What is the probability that a zombie infects zero humans?
(b) Calculate the mean number of people infected by each zombie, $\langle k \rangle$? For what range of values of a is $\langle k \rangle > 1$?
(c) If Q is the probability that the infection from a zombie eventually dies out, we can determine Q by solving:

---
[35] http://en.wikipedia.org/wiki/Age_of_the_universe

$$Q = (1 - a - a^2) + aQ + a^2 Q^2. \tag{5.73}$$

Explain each term in Eq. 5.73.

(d) Find Q and the range of values of a for which the US population as a whole is safe from the zombie apocalypse.

(e) Plot your solution for Q from (d) for $\frac{1}{2} \le a \le \frac{1}{2}(\sqrt{5} - 1)$, and explain the situation for this range of values of a.

Problem 16. Zombie Apocalypse, Part Deux

Having helped save the world from the first zombie epidemic, you now work at the CDC as a team leader, and you have now been called in to consult about the appearance in California of a single zombie who has been infected by a new virulent strain of the zombie virus. For this strain of the zombie virus, the probability that a zombie infects k humans before the zombie is killed is $P(k) = a^k$, where k varies from 1 to ∞. Once infected a human quickly turns into a zombie.

(a) What is the probability that a zombie infects zero humans?

(b) What is the mean number of humans infected, $\langle i \rangle$, and therefore turned into zombies by a single zombie? For what values of a is the mean number of humans infected greater than 1?

(c) If Q is the probability that the infection from a zombie eventually dies out. We can determine Q by solving:

$$Q = 1 - \frac{a}{1-a} + aQ + a^2 Q^2 + a^3 Q^3 + a^4 Q^4 + \ldots = 1 - \frac{a}{1-a} + \frac{aQ}{1-aQ}. \tag{5.74}$$

Explain the form of this equation and find Q.

(d) Use your results for (b) and (c) to determine the range of values of a for which the US population as a whole is and is not safe from the zombie apocalypse. Plot your solution for Q as a function of a.

Problem 17. The Luria-Delbruck Experiment

A basic tenet of evolutionary biology is that evolution occurs as a result of completely random mutations, rather than mutations that occur in response to an environmental challenge to the evolving organism. But how do you/we know this is correct?

The Luria-Delbruck experiment[36] is a seminal 1943 experiment that demonstrated experimentally that genetic mutations occur randomly and spontaneously, and not in response to selective pressure [8]. This experiment contributed greatly to the award of the 1969 Nobel Prize in Physiology or Medicine to Salavatore Luria[37]

---

[36] https://en.wikipedia.org/wiki/Luria-Delbruck_experiment

[37] http://en.wikipedia.org/wiki/Salvador_Luria

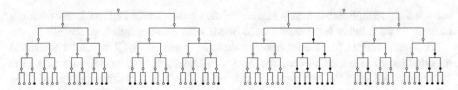

Fig. 5.17 First hypothesis (left): Mutations occur only in the final generation in response to exposure to T1 phage on the plate. Second hypothesis (right): Mutations occur randomly and spontaneously in any generation, leading to a broad distribution of numbers of mutants

and Max Delbruck[38] (together with Alfred Hershey). In an idealized version of the Luria-Delbruck experiment, a large number of test tubes are each inoculated with a single non-mutant *E. coli*. Each *E. coli* grows and multiplies by cell division until there are a total of n bacteria ($n \gg 1$) in each test tube (Fig. 5.17). These n bacteria are then plated onto agar containing T1 bacteriophage, and allowed to grow further. Ordinarily, *E. coli* is susceptible to infection by T1, which proceeds via the virus binding to specific receptors on *E. coli*'s surface, and then being able to enter the bacterium and kill it. Very occasionally, however, a mutation occurs in the T1 receptor, which prevents binding. These so-called TonR mutants are therefore immune to T1 infection. Only TonR mutant *E. coli* can grow on agar containing T1, because bacteria without the mutation are infected and killed. Each TonR mutant *E. coli*, however, grows and forms an observable bacterial colony, which can be counted, yielding the number TonR mutant bacteria, m, out of the original n bacteria placed on each plate. Each plate gives one measurement of m. Many plates give an experimental measurement of the probability that there are m mutants within the total population of n bacteria.

To figure out what such experimental measurements tell us, they must be compared to a hypothesis. The first hypothesis, we consider, is that TonR mutations occur in response to exposure to T1 bacteriophage. In this case, we may expect that each of the n bacteria on a plate will undergo the mutation with some probability α, as a result of exposure to T1. Consequently, in this case, the mean number of TonR mutants is αn and the expected probability distribution is the Poisson distribution with mean αn. Because the experimentally measured distribution could not be described by a Poisson distribution, Luria and Delbruck's measurements refute this hypothesis. The second hypothesis is that TonR mutations occur randomly throughout the growth process. This hypothesis gives rise to a probability distribution for m that is not Poisson. In fact, the so-called Luria-Delbruck distribution shows non-negligible probabilities extending to much larger values of m than does a Poisson distribution, because a mutation event early on in the growth process yields large values of m. The goal of this problem is to find an approximate expression for the Luria-Delbruck distribution—valid for small enough

---

[38] http://en.wikipedia.org/wiki/Max_Delbruck

$n\alpha$—which recapitulates a key feature of the exact Luria-Delbruck distribution, namely that relatively large values of m occur with non-negligible probability.

A simple model for the probability that there are m TonR mutant bacteria in a collection of n bacteria—i.e., for the Luria-Delbruck distribution—assumes the following: All bacteria are descended from a single non-mutant progenitor; the mutation rate is α per cell division; there is no back mutation; TonR mutants have the same fitness, and therefore are equally likely to reproduce as non-mutant bacteria; neither non-mutant nor mutant bacteria die during the growth process; the next bacteria to undergo cell division is picked randomly from the bacterial population. This last assumption is unrealistic in that it allows for the possibility that a cell that has undergone the previous cell division can also undergo the next cell division; it also allows for the the possibility that a cell may never undergo cell division. Nevertheless, with this assumption we need only keep track of n and m.

At each replication event, the number of bacteria increases by one, while the number of bacteria with the TonR mutation either increases by one or remains unchanged in a stochastic fashion. Therefore, we can conceive m as being analogous to the coordinate of a one-dimensional random walk that either steps forward or does not, while n is analogous to the total number of steps in that random walk. (The value of m can take any value from 1 to $n-1$, since we take $m=0$ for $n=1$, i.e., $P((0, 1) = 1.)$ Then, we can ask: What is the probability to arrive at m when there are n bacteria, given that the number of bacteria with the TonR mutation was either $m-1$ or m, when there were $n-1$ bacteria? Writing an answer to this question leads to the following equation for this "Luria-Delbrück" distribution:

$$P(m, n) = \alpha \frac{n-m}{n-1} P(m-1, n-1) + (1-\alpha)\frac{n-1-m}{n-1} P(m, n-1)$$

$$+ \frac{m-1}{n-1} P(m-1, n-1), \tag{5.75}$$

together with

$$P(0, 1) = 1. \tag{5.76}$$

(a) Recently, cancer researchers have become interested in the Luria-Delbrück distribution. Explain why?
(b) Explain Eq. 5.75, i.e., explain the three terms on the right-hand side, the factors in each of these three terms, and why their sum equals the left-hand side.
(c) Equations 5.75 and 5.76 are difficult to solve in general [9]. However, $P(0, n)$ can be calculated straightforwardly by noticing that, for $m = 0$, Eq. 5.75 becomes

$$P(0, n) = (1-\alpha)P(0, n-1). \tag{5.77}$$

Solve this equation and Eq. 5.76 to determine $P(0, n)$, given that $P(0, 1) = 1$.

(d) In addition, given that α is invariably very small, it is reasonable to consider Eq. 5.75 in the limit of small α. Then, we can drop the terms in Eq. 5.75 proportional to α with the result that:

$$P(m, n) \simeq \frac{n - 1 - m}{n - 1} P(m, n - 1) + \frac{m - 1}{n - 1} P(m - 1, n - 1). \qquad (5.78)$$

Show that

$$P(m, n) = \frac{\alpha n}{m(m + 1)} \qquad (5.79)$$

is a solution of Eq. 5.78 by direct substitution of Eq. 5.79 into Eq. 5.78 [10] (Fig. 5.18).

(e) Using the results that

$$\Sigma_{m=1}^{\infty} \frac{1}{m(m + 1)} = 1, \qquad (5.80)$$

and that $P(0, n) \simeq 1 - (n - 1)\alpha \simeq 1 - n\alpha$, for small $n\alpha$, show that our approximate $P(m, n)$ is nevertheless approximately normalized.

Problem 18: A More Realistic Molecular Motor Revisited
A schematic representation of the "more realistic" molecular motor with two internal states (state 0 and state 1), is shown in Fig. 5.19. As shown in this figure, the motor transitions from site n to site $n + 1$ starting from state 1 and ending in state 0. The corresponding transition rate is k_+, and the corresponding displacement is $a - b$, where a is the periodicity of the track and $b < a$. In our model, the motor never steps backwards from site n to site $n - 1$. However, it can transition from state 0 to state 1 at the same site, n, with transition rate, $k_{0 \to 1}$. For this transition, the corresponding displacement is b ($b < a$). It can also transition from state 1 to state 0 at the same site, n, with transition rate, $k_{1 \to 0}$, and displacement, $-b$.

The probability that this molecular motor is in state 1 and at site n at time $t + \Delta t$ can be written

$$p_1(n, t + \Delta t) = k_{0 \to 1} \Delta t p_0(n, t) + (1 - k_+ \Delta t - k_{1 \to 0} \Delta t) p_1(n, t), \qquad (5.81)$$

and the probability that this molecular motor is in state 0 and at location n at time $t + \Delta t$ can be written

$$p_0(n, t + \Delta t) = k_+ \Delta t p_1(n - 1, t) + k_{1 \to 0} \Delta t p_1(n, t) + (1 - k_{0 \to 1} \Delta t) p_0(n, t), \qquad (5.82)$$

where $p_1(n, t)$ is the probability that the molecular motor is in state 1 and at site n at time t, and $p_0(n, t)$ is the probability that the molecular motor is in state 0 and at site n at time t.

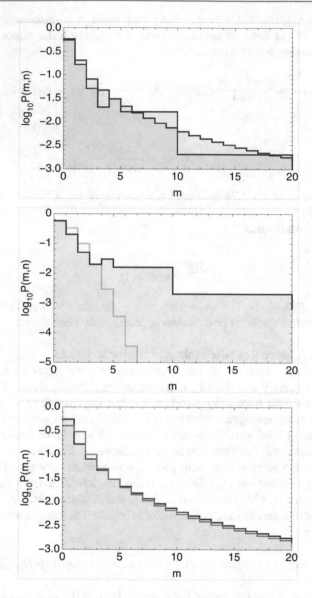

Fig. 5.18 Top: Comparison of the logarithm of the theoretical Luria-Delbruck distribution for $n\alpha = 0.6$ (red) with the logarithm of the experimental distribution (blue) from EXPERIMENT NO. 22 of Ref. [8]. (For the data, there is a single bin each for numbers of mutants between 6 and 10 and numbers of mutants between 11 and 20.) Importantly, both the theoretical and the experimental distribution show measurable probabilities of realizing a relatively large m. Middle: Comparison of the logarithm of the Poisson distribution for $n\alpha = 0.6$ (cyan) with the logarithm of the experimental distribution (blue) from Experiment 22 in Ref. [8]. The Poisson distribution shows much smaller probabilities for realizing a relatively large m than measured experimentally. Bottom: Comparison of the logarithm of the the exact Luria-Delbruck distribution for $n\alpha = 0.6$ (red) with the logarithm of the approximate distribution derived here (orange). Based on their results and analysis, Luria and Delbruck concluded "We consider the above results as proof that in our case the resistance to virus is due to a heritable change of the bacterial cell which occurs independently of the action of the virus"

Fig. 5.19 Schematic of a
two-state motor

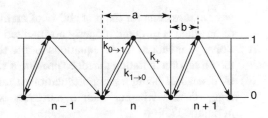

(a) Briefly explain the meaning of each of the three terms on the right-hand side of Eq. 5.82.
(b) In the limit of infinitesimal Δt, EQ, 5.81 becomes

$$\frac{dp_1(n, t)}{dt} = k_{0\to1}p_0(n, t) - (k_+ + k_{1\to0})p_1(n, t), \qquad (5.83)$$

while Eq. 5.82 becomes

$$\frac{dp_0(n, t)}{dt} = k_+p_1(n - 1, t) + k_{1\to0}p_1(n, t) - k_{0\to1}p_0(n, t). \qquad (5.84)$$

Summing these equations over all values of n and introducing the overall probabilities P_0 and P_1—the probability that the motor is in state 0 and the probability that the motor is in state 1, respectively, irrespective of location—leads to

$$\frac{dP_1(t)}{dt} = k_{0\to1}P_0(t) - (k_+ + k_{1\to0})P_1(t) \qquad (5.85)$$

and

$$\frac{dP_0(t)}{dt} = k_+P_1(t) + k_{1\to0}P_1(t) - k_{0\to1}P_0(t). \qquad (5.86)$$

Find the steady-state values of P_0 and P_1 in terms of k_+, $k_{0\to1}$, and $k_{1\to0}$.
(c) Calculate the drift velocity of the molecular motor by calculating the mean displacement—call it w—in a time Δt.
(d) Combine your answers to (b) and (c) to show that the drift velocity is independent of b.

Problem 19: Life Is not Fair, Part Deux
(a) Modify the Moran model to include the effect of a mutation that does provide a selective advantage. Specifically, include a factor f, which increases the relative probability of an allele A birth by a factor of f. Be sure, when f is in place that

your expressions for the probability of an allele A birth and for the probability of an allele B birth are properly normalized.

(b) Solve your Moran-model equations to find the probability of fixation, including the effect of a mutation that does provide a selective advantage or disadvantage.

(c) If, instead of being a neutral mutation, a mutation provides a selective advantage, discuss whether such a mutation is certain to become fixed in the population?

Problem 20: Transcription

Transcription is the process by which RNA polymerase reads a sequence of DNA base pairs and makes the corresponding RNA molecule. The probability that there are k RNA molecules at time $t + \Delta t$, namely $P(k, t + \Delta t)$ may be expressed

$$P(k, t+\Delta t) = \alpha \Delta t P(k-1, t) + (k+1)\beta \Delta t P(k+1, t) + (1 - \alpha \Delta t - k\beta \Delta t) P(k, t),$$
(5.87)

where α is the rate of RNA production by RNA polymerase, β is the rate of RNA degradation, and Δt is a very small increment of time.

(a) Briefly explain each of the three terms on the right-hand side of Eq. 5.87.

(b) Explain why it is important that Δt is small in order for Eq. 5.87 to be appropriate/valid.

References

1. R. Brown, A brief account of microscopical observations made in the months of June, July and August, 1827, on the particles contained in the pollen of plants; and on the general existence of active molecules in organic and inorganic bodies. Philos. Mag. **4**, 161–173 (1828)
2. We may calculate the the numerical value of the diffusion coefficient, D, for the beads, tethered near a surface, given that (according to least-mean-squares fitting) the slope of the line in the plot of MSD versus time in the demonstration is 1.22 ± 0.28 pixels^2 per frame, that 11.75 pixels equals 1 μm, and that 1 frame corresponds to $1/13.7$ s. In terms of the demonstration's units, $D = 0.61\pm0.14$ pixels^2 per frame. Converting to SI units yields $D = 6.1\pm0.14\times10^{-14}$ m^{2}s^{-1}. In comparison, the diffusion coefficient for a free bead of the same size is $D = k_B T/\zeta$, where $\zeta = 6\pi\eta R$ is the bead's friction coefficient, η is the viscosity of water, R is the radius of the bead, T is the absolute temperature, and k_B is Boltzmann's constant, leading to $D = 4 \times 10^{-13}$ m^{2}s^{-1}. We ascribe the discrepancy to the presence of the DNA and the nearby surface, both of which may be expected to increase friction
3. P.C. Nelson, *Biological Physics: Energy, Information, Life* (W. H. Freeman, New York, 2004)
4. K.J. Amann, T.D. Pollard, Direct real-time observation of actin filament branching mediated by arp2/3 complex using total internal reflection fluorescence microscopy. Proc. Nat. Acad. Sci. U.S.A. **98**, 15009–15013 (2001)
5. T.D. Pollard, Rate constants for the reactions of ATP- and ADP-actin with the ends of actin filaments. J. Cell Biol. **103**, 2747–54 (1986)
6. A. Yildiz, J.N. Forkey, S.A. McKinney, T. Ha, Y.E. Goldman, P.R. Selvin, Myosin V walks hand-over-hand: single fluorophore imaging with 1.5 nm localization. Science **300**, 2061–2065 (2003)
7. R.L. Cann, M. Stoneking, A.C. Wilson, Mitochondrial DNA and human evolution. Nature **325**, 31–36 (1987)

8. S.E. Luria, M. Delbrück, Mutations of bacteria from virus sensitivity to virus resistance. Genetics **28**, 491–511 (1943)
9. W.P. Angerer, An explicit representation of the Luria-Delbrück distribution. J. Math. Biol. **42**, 145–174 (2001)
10. D.A. Kessler, H. Levine, Scaling solution in the large population limit of the general asymmetric stochastic Luria-Delbrück evolution process. J. Stat. Phys. **158**, 783–805 (2015)

Diffusion: Membrane Permeability and the Rate of Actin Polymerization

6

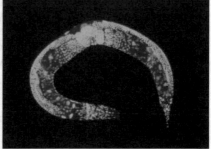

Left: A blue whale, *Balaenoptera musculus*. Public domain image from https://commons.wikimedia.org/wiki/File:Anim1754_-_Flickr_-_NOAA_Photo_Library.jpg. Right: A roundworm, *Caenorhabditis elegans*. Public domain image from https://commons.wikimedia.org/wiki/File:Caenorhabditis_elegans.jpg

6.1   Introduction

The flow of particles by diffusion is tremendously important across the sciences and in medicine. In this chapter, we will see that diffusive flow from a high concentration to a low concentration is an emergent behavior of many particles, which is the result of individual particles' Brownian random walks, and how the consequences of diffusion may be quantified by means of the diffusion equation. This chapter builds directly on Chap. 5. *Random Walks: Brownian Motion and the Tree of Life* because diffusion and the diffusion equation are direct manifestations of the random Brownian motions of individual molecules. The concept of particle number conservation will be especially important, namely that particles cannot

© The Author(s), under exclusive license to Springer Nature Switzerland AG 2023
S. Mochrie, C. De Grandi, *Introductory Physics for the Life Sciences*,
Undergraduate Texts in Physics, https://doi.org/10.1007/978-3-031-05808-0_6

appear or disappear without a good reason (such as that they undergo a chemical reaction). An immediate consequence of particle number conservation is that steady-state particle currents—the number of particles per unit time passing a particular point or a particular radius—are independent of position or radius.

We will study diffusion across membranes and "diffusion to capture" of actin monomers at the tip of an actin filament, involved in actin polymerization. Diffusion across membranes is particularly important in physiology and medicine because the human body can be conceived as consisting of myriad membranes across which oxygen, CO_2, nutrients, waste, therapeutics, etc., all must pass. We will discover that the particle current through a flat membrane is proportional to the difference in concentration across the membrane, providing an example of a "gradient flow." Of course, individual cells also have membranes, across which many of these same quantities must also pass. Actin polymerization is an important example of biological dynamics, underlying many examples of eukaryotic cell motility. Actin polymerization is also a realization of a bimolecular chemical reaction, and our discussion will reveal how essential equations of chemical and biochemical kinetics emerge from particles' Brownian motion and their consequent diffusion, and how diffusion limits the maximum possible rate of a chemical reaction. We will also understand what are the consequences of diffusion for the size of cells, and for the efficiency with which cell surface receptors detect their binding partners.

6.2   Your Learning Goals for This Chapter

By the end of this chapter, you should be able to:

- Explain the diffusion equation, and the concepts of particle flux and particle current and that they are all consequences of individual particles' random walks, together with particle number conservation.
- Explain and apply the concept of "steady-state" diffusion.
- Determine whether a given function is a solution of the one-dimensional, steady-state diffusion equation or of the three-dimensional, spherically symmetric, steady-state diffusion equation, using "direct substitution."
- Apply boundary conditions to determine particular solutions to the "steady-state" diffusion equation in one-dimensional situations and three-dimensional, spherically symmetric situations.
- Use the concepts of particle flux, particle current, particle conservation, and diffusive conductance in one-dimensional situations.
- Calculate the conductance (or permeability) of a planar membrane composed of an arbitrary number of sequential layers "in series."
- Use concepts of flux, current, particle conservation, and diffusive conductance in three-dimensional, spherically symmetric situations.
- Explain the constraints imposed by diffusion on chemical reaction rates.

- Explain how the linear dependence of the association rate of a chemical reaction follows from the diffusion equation and how the association rate constant of a chemical reaction may be conceived as a diffusive conductance.
- Explain the consequences of diffusion for the efficiency with which cell surface receptors detect their binding partners.

6.3 The Diffusion Equation

In our previous discussion of random walks, we focused on a single random walker that starts at $x = 0$ at time $t = 0$, and we found the probability density that it is at x at time t. Now, we turn our attention to many random walkers. If there are many random walkers, they combine to give rise to the phenomenon of *diffusion*. We will find an equation—the diffusion equation—that describes how the average density of many particles behaves if each individual particle is performing a random walk.

Previously, we talked about the probability density $p(x)$ for a particle to be at x. Now, we will talk about the particle density $n(x)$ that is the number of particles per unit volume at x. These quantities are related via $n(x) = Np(x)$, where N is the total number of particles. Actually, sometimes $n(x)$ will be the number of particles per unit length at x. We will be able to decide which it is—per unit length or per unit volume—from the context.

6.3.1 Particle Flux and Fick's Law

To start with, we assume a particle density, $n(x)$, that depends only on x, and we consider the exact same model that we used to discuss Brownian motion, in which each particle randomly steps $\pm L$ at each time step τ. In addition, we consider an area A perpendicular to the x-axis. Then, in a cross-sectional area A, perpendicular to the x-axis, there are $n(x)AL$ particles between $x - L$ and x at a given moment in time, 1/2 of which step L to the left and 1/2 step L to the right in a time interval τ (Fig. 6.1). Similarly, in a cross-sectional area A, there are $n(x + L)AL$ particles between x and $x + L$ at a given moment in time, 1/2 of which step L to the left and

Fig. 6.1 Schematic of our model for diffusion

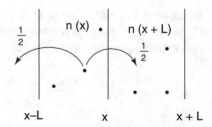

Fig. 6.2 Relationship
between $n(x)$, $n(x + L)$, and
dn/dx

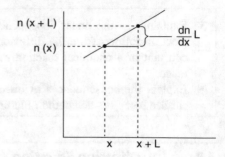

1/2 step L to the right in a time interval τ. It follows that on average the net number
of particles crossing an area A of the plane at x in a period τ is

$$j(x)A\tau = \frac{1}{2}[n(x) - n(x + L)]AL = -\frac{A}{2}\frac{dn}{dx}(L)^2, \qquad (6.1)$$

where we introduced the particle *flux*, $j(x)$, which is the number of particles
crossing unit area of a plane at x per unit time, and we used that

$$n(x + L) - n(x) \simeq \frac{dn}{dx}L, \qquad (6.2)$$

which may be justified on the basis of Fig. 6.2. It is important to recognize that the
sign of j is important: positive j implies that particles are flowing toward larger
values of x; negative j implies particles flow toward smaller values of x. Now we
divide through by A and τ to find:

$$j(x) = -\frac{(L)^2}{2\tau}\frac{dn}{dx}. \qquad (6.3)$$

However, according to our earlier discussion of random walks, the quantity $\frac{L^2}{2\tau}$
is exactly the diffusion coefficient, D, within this hopping-back-and-forth model.
Therefore,

$$j(x) = -D\frac{dn}{dx}. \qquad (6.4)$$

Equation 6.4 is often called Fick's first law or just Fick's law. It is extremely
important in biology, physiology, and medicine. Fick's law relates a gradient in
concentration, namely, $\frac{dn}{dx}$, to a flow of particles, namely, j.

6.3.2   Particle Number Conservation and Fick's Second Law

We can achieve a second equation connecting $j(x)$ and $n(x)$ by examining how the number of particles between $x - L$ and x changes versus time (t). Here, the key idea—particle number conservation—is that the only way for the number of particles between $x - L$ and x to change is by particles hopping (flowing) in from the right or hopping (flowing) out to the left. Particle flux, $j(x)$, specifies the particle flows. It follows that

$$\frac{d}{dt}[n(x)AL] = Aj(x - L) - Aj(x) = -A\frac{dj}{dx}L,   \tag{6.5}$$

where we used that (for small enough L)

$$j(x - L) - j(x) = -\frac{dj}{dx}L.   \tag{6.6}$$

The left hand side of Eq. 6.5 is the change per unit time in the number of particles between x and $x+L$ in a cross-sectional area A. The right hand side is the difference in the number of particles flowing in at $x - L$ per unit time and the number flowing out at x per unit time. Canceling the A and the L, we find

$$\frac{dn}{dt} = -\frac{dj}{dx}.   \tag{6.7}$$

Equation 6.7 is a manifestation of the fact that particles cannot appear from nowhere nor simply vanish, i.e., it is a manifestation of particle number conservation.

Putting Eqs. 6.4 and 6.7 together leads to the equation

$$\frac{dn}{dt} = -\frac{dj}{dx} = -\frac{d}{dx}\left[-D\frac{dn}{dx}\right] = D\frac{d^2n}{dx^2}.   \tag{6.8}$$

This equation,

$$\frac{dn}{dt} = D\frac{d^2n}{dx^2},   \tag{6.9}$$

is the *diffusion equation* (a.k.a. Fick's second law) and likewise is tremendously important in biology and medicine.

In fact, our previous solution for the probability distribution of particles that start at $x = 0$ at $t = 0$, namely,

$$n = \frac{N}{\sqrt{4\pi Dt}}e^{-x^2/(4Dt)},   \tag{6.10}$$

satisfies (is a solution of) the diffusion equation, as can be verified by calculating (using WolframAlpha, for example) the appropriate derivatives (Dd^2/dx^2[1] and d/dt^2) and verifying that they are equal.

6.3.3   The Steady-State, One-Dimensional Diffusion Equation

In many situations, we can learn a great deal by considering the so-called steady-state behavior. In a steady state, the particle density, $n(x)$, does not change in time. Of course, individual particles are still undergoing Brownian motion and diffusing, but the average behavior reflected in $n(x)$ does not change in time. Therefore, in a steady state, and only in a steady state, we have

$$\frac{dn}{dt} = 0. \tag{6.11}$$

Correspondingly, we have

$$0 = D\frac{d^2n}{dx^2}, \tag{6.12}$$

which is the steady-state diffusion equation.

6.3.4   Diffusion Through a Membrane or a Tube

Diffusion through membranes is important biologically, medically, and physiologically. This is because living systems are generally compartmentalized by some sort of "membranes" into subsystems: cells contain organelles, and at organismal level, there are circulatory, respiratory, digestive, lymph, etc., systems. Material is constantly exchanged across the boundaries (i.e., the membranes) of these systems, for example:

- Therapeutics and nutrients from the GI tract into the blood stream.
- Oxygen and nutrients from capillaries into tissues.
- Oxygen in alveoli of lungs moves into capillaries, and CO_2 moves from capillaries into alveoli.

---

[1] http://www.wolframalpha.com/input/?i=d%5E2%2Fdx%5E2%281%2Fsqrt%284*pi*D*t%29+
*exp%28-x%5E2%2F%284*D*t%29%29%29*D

[2] http://www.wolframalpha.com/input/?i=d%2Fdt%281%2Fsqrt%284*pi*D*t%29+*exp%28-x
%5E2%2F%284*D*t%29%29

Diffusion through a tube is also important. One biologically relevant example is the diffusion of ions though an ion channel. In both cases, this transport is generally governed by Fick's law.

Diffusion across a membrane or through a tube can be treated as 1D steady-state diffusion. We will initially consider an over-simplified model of a membrane of thickness d that separates a region $(x < 0)$, where the concentration of a species of interest is n_1, from a region $(x > d)$, where the concentration of the species of interest is n_2. Our goal is to figure out the concentration as a function of x and the particle flux through the membrane. To this end, we need to solve a differential equation, the 1D steady-state diffusion equation, within a region of space, $0 \leq x \leq d$. To carry out such a program, it is useful to adhere to the following steps:

1. Pick the correct equation and its general solution.
   In 1D, in the steady state, the particle density obeys the 1D steady-state diffusion equation:

$$\frac{d^2n}{dx^2} = 0. \tag{6.13}$$

   When confronted with an equation containing derivatives, such as Eq. 6.13, a highly satisfactory method from a mathematical point of view for finding the solution for $n(x)$ is to "guess" the solution and then prove that the guess is indeed correct. This is the solution by direct substitution. In this case, we guess the solution:

$$n(x) = ax + b. \tag{6.14}$$

   We can then verify that Eq. 6.14 is indeed a solution by direct substitution, calculating

$$\frac{d^2n}{dx^2} = \frac{d^2}{dx^2}(ax + b) = \frac{d}{dx}a = 0. \tag{6.15}$$

   Thus, Eq. 6.14 is indeed a solution, irrespective of the values of a and b. We must now find a and b to find the solution that we need in any particular case.
2. Recognize that in order to find the correct solution within a given region, we must be given, or must figure out, the *boundary conditions for that region*, i.e., we must be given the values of $n(x)$ OR $j(x)$ at the boundaries of the region of interest (which here are finite but later in this chapter we'll see can be also at positive or negative infinity). Without knowing the boundary conditions, we cannot find the solution. If we are given the boundary conditions, we can find the solutions within the region.

Here, we have that

$$n(0) = n_1 \qquad (6.16)$$

and

$$n(d) = n_2. \qquad (6.17)$$

These are the boundary conditions in this case.
3. Impose the boundary conditions on the general solution.
   First, we evaluate Eq. 6.14 at $x = 0$ and set it equal
   to Eq. 6.16, with the result that

$$n(0) = n_1 = b. \qquad (6.18)$$

Next, we evaluate Eq. 6.14 at $x = d$ and set it equal
to Eq. 6.17 with the result that

$$n(d) = ad + b = n_2. \qquad (6.19)$$

4. Solve your equations to find the unknown quantities.
   Here, Eq. 6.18 already gives

$$b = n_1. \qquad (6.20)$$

Using Eq. 6.18 in Eq. 6.19, we also find

$$a = (n_2 - n_1)/d. \qquad (6.21)$$

5. Use the values of the unknowns so-obtained to write the desired solution.
   Using these values of a and b, we substitute back
   into Eq. 6.14 with the result that

$$n(x) = (n_2 - n_1)\frac{x}{d} + n_1. \qquad (6.22)$$

This is our solution for the concentration as a function of x within $0 \leq x \leq d$. We see that the concentration varies linearly from one side of the membrane
to the other.
6. Calculate the flux, using Fick's law (Eq. 6.4).
   We can do another interesting thing, namely, to calculate the flux of the species of interest through
   the membrane. The flux at x is given by Fick's law:

$$j(x) = -D\frac{dn}{dx} = -\frac{D}{d}(n_2 - n_1). \qquad (6.23)$$

Notice that if $n_1 > n_2$, the flux is positive, implying that there is a net flow of particles toward positive x, from the larger concentration (n_1) to the lower concentration (n_2), which is what we expect. Similarly, notice that if $n_2 > n_1$, the flux is negative, implying that there is a net flow of particles toward negative x, from the larger concentration (n_2) to the lower concentration (n_1), which is also what we expect.

7. Calculate the particle current.

   Flux is the number of particles per second per unit area, so if we multiply by the area of the membrane, A, we get the particle current, I, namely, the number of particles per second, passing a given point:

$$I = jA = -\frac{DA}{d}(n_2 - n_1). \tag{6.24}$$

Notice that the particle current does not depend on x. It is the same everywhere in the membrane. This is as it must be and corresponds to the same number of molecules passing per second at every value of x within the membrane. Anything different than this result would imply particles either being destroyed or created in the membrane, which would be nonsensical. Notice too that the particle current through the membrane is proportional to the difference in concentration across the membrane, $n_2 - n_1$. This is an example of a so-called gradient flow. Here, the particle current (flow of particles) is linearly proportional to the difference in concentration. Not surprising, the current is inversely proportional to the membrane thickness.

Although we have pursued our analysis of 1D diffusion so far in the context of an over-simplified membrane, the analysis is applicable to all situations in which the density varies along one dimension only and is constant in time, such as a long tube that contacts two particle reservoirs that have different particle concentrations.

6.3.5 Diffusion Through a Membrane: More Realistic Version

Figure 6.3 shows a physicist's impression of a portion of the lipid bilayer membrane of a cell. Its constituent lipid molecules each have a hydrophilic head group that faces the surrounding aqueous solution (a.k.a. water) and two hydrophobic, hydrocarbon (a.k.a. oily) tails. Thus, a physicist, who wants to build a simple model of such a bilayer membrane, might represent the interior of the membrane as oil and the exterior as water. This is what we will do as we consider the diffusive transport of a species of interest across the membrane. We will furthermore suppose that the species of interest is reasonably hydrophilic, i.e., it "likes" to be in water, but it does not "like" to be in oil. In order to represent this tendency, we need to introduce the concept of a "partition coefficient" (B),[3] which you may have

---

[3] http://en.wikipedia.org/wiki/Partition_coefficient

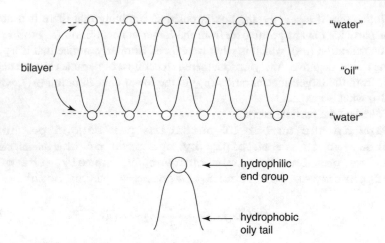

Fig. 6.3 A physicist's rendering of a cell's lipid bilayer membrane (top). A physicist's rendering of a lipid molecule (bottom)

Fig. 6.4 The role of the partition coefficient in establishing the boundary concentrations in a simple model of a lipid bilayer membrane

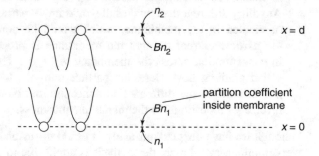

encountered in your chemistry classes. In any case, a partition coefficient is the ratio of concentrations of a species of interest in the two solvent phases of a mixture of two immiscible solvents. For a hydrophilic species, B will be very small. Chapter 8. *Statistical Mechanics: Boltzmann Factors, PCR, and Brownian Ratchets* will further elucidate B.

For our membrane, shown in Fig. 6.4, what this means is that if the concentration of the species of interest in the water just outside the membrane near $x = 0$ is n_1, then the concentration of the species of interest just inside the membrane near $x = 0$ is Bn_1, where B is the partition coefficient. Similarly, if the concentration of the species of interest in the water just outside the membrane near $x = d$ is n_2, then the concentration of the species of interest just inside the membrane near $x = d$ is Bn_2. It turns out that B strongly affects transport across the membrane, so B is a key parameter for any drug/therapeutic whose action occurs within the cell and must diffuse across the cell membrane. As we will discuss in Chap. 13. *Electric Potential, Capacitors, and Dielectrics*, the reason that cell membranes are largely impermeable to ions—and therefore membrane ion channels are needed—is because B is very small for ions in oil/lipid tails.

In comparison to our earlier discussion, the only differences are the boundary conditions at $x = 0$ and $x = d$, namely, that $n(0) = n_1$ and $n(d) = n_2$ are now replaced by $n(0) = Bn_1$ and $n(d) = Bn_2$. It immediately follows that the current through the membrane is now given by

$$I = jA = -\frac{DAB}{d}(n_2 - n_1) \qquad (6.25)$$

Here, the particle current is proportional to B, so is small for hydrophilic species for which B is small inside the oily environment of the bilayer.

6.3.6 Diffusive Conductance and Membrane Permeability

It turns out that it is useful to introduce the diffusive conductance, G, and the membrane permeability, P, via

$$I = -\frac{DAB}{d}(n_2 - n_1) = -G(n_2 - n_1) = -PA(n_2 - n_1). \qquad (6.26)$$

In general, diffusive conductance is the ratio of current to concentration difference, and permeability is conductance per unit area. Sometimes, it is valuable to consider diffusive resistance, which is simply the inverse of diffusive conductance. In Chap. 9. *Fluid Mechanics: Laminar Flow, Blushing, and Murray's Law*, we will encounter that the fluid volume flow rate is linearly proportional to pressure difference. In that case, the constant of proportional is flow conductance. In Chap. 14. *Circuits and Dendrites: Charge Conservation, Ohm's Law, Rate Equations, and Other Old Friends*, we will encounter that the electrical current is proportional to voltage difference (Ohm's law), where the constant of proportionality is the electrical conductance.

6.3.7 Diffusion Through Multilayer Membranes

In Sect. 6.3.5, we considered a single-layer membrane. However, many membranes involve multiple layers, next to each other. For example, in eukaryotic cells, the nuclear envelope consists of three layers, namely, an outer bilayer membrane, a periplasmic space, and an inner bilayer membrane. The membranes of mitochondria and chloroplasts also consist of three such layers.

The concept of particle number conservation, together with Eq. 6.26, permits us to determine the diffusive transport properties of multilayer membranes. Let us consider diffusion across a two-layer membrane for which the individual layer diffusivities, thicknesses, and partition coefficients are D_1, d_1, and B_1 for layer 1 and D_2, d_2, and B_2 for layer 2 (Fig. 6.5). Both layers have the same area, A. As a

Fig. 6.5 Schematic of a two-layer membrane

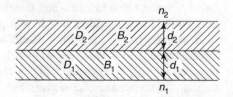

result, the particle currents across layer 1 and layer 2 are

$$I_1 = -\frac{D_1 A B_1}{d_1}(n - n_1) = -G_1(n - n_1) \tag{6.27}$$

and

$$I_2 = -\frac{D_2 A B_2}{d_2}(n_2 - n) = -G_2(n_2 - n), \tag{6.28}$$

respectively, where n_1 is the concentration of the species of interest in the solution neighboring layer 1, n_2 is the concentration of the species of interest in the solution neighboring layer 2, and n is the unknown concentration at the boundary between the layers, and where we have introduced each layer's *diffusive conductance*, $G_1 = \frac{D_1 B_1 A}{d_1}$ and $G_2 = \frac{D_2 B_2 A}{d_2}$ for layer 1 and layer 2, respectively. Particle number conservation demands, however, that the same number of particles per second pass through each membrane. Therefore,

$$I_1 = I_2 = I. \tag{6.29}$$

It follows that

$$\frac{I}{G_1} = n_1 - n \tag{6.30}$$

and

$$\frac{I}{G_2} = n - n_2. \tag{6.31}$$

Adding Eqs. 6.30 and 6.31 yields

$$I\left(\frac{1}{G_1} + \frac{1}{G_2}\right) = n_1 - n_2, \tag{6.32}$$

i.e.,

$$I = -\frac{1}{\frac{1}{G_1} + \frac{1}{G_2}}(n_2 - n_1). \tag{6.33}$$

Now, introducing the diffusive conductance of the two-layer membrane via

$$I = -G_{12}(n_2 - n_1),$$ (6.34)

we see that

$$G_{12} = \frac{1}{\frac{1}{G_1} + \frac{1}{G_2}}$$ (6.35)

or

$$\frac{1}{G_{12}} = \frac{1}{G_1} + \frac{1}{G_2}.$$ (6.36)

For multilayer membranes, in which each layer has the same area, Eq. 6.36 represents a prescription for how to calculate the overall conductance, i.e., the conductance of multiple layers "in series." Incorporating a third layer is straightforward because a third layer is a second layer as far as the two-layer membrane is concerned. Thus,

$$\frac{1}{G_{123}} = \frac{1}{G_{12}} + \frac{1}{G_3} = \frac{1}{G_1} + \frac{1}{G_2} + \frac{1}{G_3}$$ (6.37)

and so on. We will encounter similar relationships in the future in the context of fluid flow and electrical current flow: in Chap. 9. *Fluid Mechanics: Laminar Flow, Blushing, and Murray's Law*, we will see that the fluid conductance of flow channels in series combines in the same fashion as the diffusive conductance for membranes, and in Chap. 14. *Circuits and Dendrites: Charge Conservation, Ohm's Law, Rate Equations, and Other Old Friends*, we will see that the electrical conductance of electrical resistors in series combines in the same fashion as the diffusive conductance of layers in series.

6.3.8 Example: Concentration Between the Layers of a Two-Layer Membrane

What is the concentration, n, between the two layers of a two-layer membrane (Fig. 6.6)?

Answer: To answer this question, we set Eq. 6.27 equal to Eq. 6.33 and then solve for n with the result that

$$n = n_1 + \frac{G_2}{G_1 + G_2}(n_2 - n_1) = \frac{G_1 n_1 + G_2 n_2}{G_1 + G_2}.$$ (6.38)

Fig. 6.6 Abstract representation of diffusive transport through two diffusive conductances in series, represented as the boxes. The solid circles represent the locations, where the concentrations are n_2, $n = \frac{G_2 n_2 + G_1 n_1}{G_1 + G_2}$, and n_1

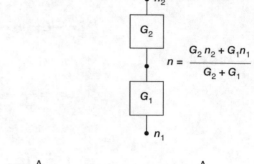

Fig. 6.7 Schematic tube for diffusion along a tube of varying cross-sectional area

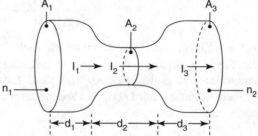

6.3.9 Example: Diffusion Along a Tube of Varying Cross-Sectional Area

Recall that for a region of thickness d and area A, separating locations at which the concentrations are n_2 and n_1, the particle current is $I = -\frac{DBA}{d}(n_2 - n_1) = -PA(n_2 - n_1)$. Now, consider 1D diffusion with diffusion constant D along the tube shown in Fig. 6.7. The tube starts at $x = 0$. Between $x = 0$ and $x = d_1$, the tube's cross-sectional area is A_1. Between $x = d_1$ and $x = d_1 + d_2$, the tube's cross-sectional area is A_2. Between $x = d_1 + d_2$ and $x = d_1 + d_2 + d_3$, the tube's cross-sectional area is A_3. The particle currents in each of these parts of the tube are I_1, I_2, and I_3, respectively.

(a) What is the particle current (I_1) in the first part of the tube in terms of given quantities and the concentration difference across that part of the tube?

    Answer: We take the partition coefficient $B = 1$, assuming that the tube is filled with the same fluid as the reservoirs at its ends. Then

(continued)

$$I_1 = -\frac{DA_1}{d_1}(n - n_1), \qquad (6.39)$$

where n is the concentration at the boundary
between channel 1 and channel 2.

(b) What is the particle current (I_2) in the middle part of the tube in terms of
given quantities and the concentration difference across that part of the
tube?

Answer:

$$I_2 = -\frac{DA_2}{d_2}(m - n), \qquad (6.40)$$

where m is the concentration at the boundary
between channel 2 and channel 3.

(c) What is the particle current (I_3) in the last part of the tube in terms of
given quantities and the concentration difference across that part of the
tube?

Answer:

$$I_3 = -\frac{DA_3}{d_3}(n_2 - m). \qquad (6.41)$$

(d) How are I_1, I_2, and I_3 related? Explain your answer.

Answer:

$$I_1 = I_2 = I_3 = I, \qquad (6.42)$$

because of particle number conservation in the
steady state.

(e) What then is the relationship between the total particle current along the
tube and the concentration difference between one end of the tube and the
other?

Answer: We have

(continued)

$$\left(\frac{d_1}{DA_1} + \frac{d_2}{DA_2} + \frac{d_3}{DA_3}\right) I = -(n - n_1) - (m - n) - (n_2 - m)$$

$$= -(n_2 - n_1), \qquad (6.43)$$

or

$$I = -\frac{1}{\frac{d_1}{DA_1} + \frac{d_2}{DA_2} + \frac{d_3}{DA_3}}(n_2 - n_1). \qquad (6.44)$$

(f) Explain whether your expression makes sense in the limit that $A_2 \ll A_1$ and $A_2 \ll A_3$.

Answer: For $A_2 \ll A_1$ and $A_2 \ll A_3$, the term containing A_2 in the denominator dominates the other two terms in Eq. 6.44, and

$$I \simeq -\frac{DA_2}{d_2}(n_2 - n_1). \qquad (6.45)$$

In this case, the current is limited by the narrow channel 2 and specifically by its small area, which seems sensible.

6.3.10   Example: A Different Process for Transmembrane Transport and Conductances "In Parallel"

In this example, rather than considering diffusive transport through the oily interior of a lipid membrane, we will investigate a different process for transmembrane transport, namely, transport through membrane ion channels. The particle current from one side of a membrane to the other as a result of this mechanism alone—call it I_C—is

$$I_C = -G_C(n_2 - n_1), \qquad (6.46)$$

where G_C is the conductance that we seek to find, n_2 is the concentration on the side of the membrane at larger x, and n_1 is the concentration on the side at lower x.

(continued)

(a) Consider a single open ion channel, which you should think of as a pinhole in the membrane. It has area A_C and length d—the same as the thickness of the membrane itself—since it goes all the way through. The ions' diffusion coefficient is D. What is the particle current through a single such ion channel?

Answer:

$$I_1 = -\frac{D A_C}{d}(n_2 - n_1). \tag{6.47}$$

(b) Suppose that there are σ ion channels per unit area. How many ion channels are there in a membrane area A?

Answer:

$$\sigma A \tag{6.48}$$

(c) Use your results for the last two parts to calculate the total particle current passing through ion channels in an area A of the membrane.

Answer: Because particles can pass through any ion channel to pass through the membrane, the ion channels represent conductances arranged "in parallel" and their conductances add. Therefore, the total conductance is the conductance of one ion channel multiplied by the number of ion channels, namely,

$$I = -\frac{D\sigma A A_C}{d}(n_2 - n_1). \tag{6.49}$$

(d) Determine G_C in this case.

Answer: The ion channel conductance is

$$G_C = \frac{D\sigma A A_C}{d}. \tag{6.50}$$

(e) Of course, the ion of interest can go either through the channels or through the bilayer itself, where the ion's partition coefficient is B. What do you expect for the total conductance of the bilayer, G, in this latter case, in terms of D, d, B, and A?

(continued)

Answer: Because an ion can either go through
the channels or through the bilayer itself,
the total particle current through the membrane
is the sum of the two currents corresponding
to each of these two routes. Since the
concentration difference driving those two
currents is the same, the conductance of the
channels in the membrane and the conductance
of the bilayer itself add together to give the
total conductance. Since the area of the bilayer
is $A - \sigma AA_C$, we find for the total conductance:

$$G = \frac{DBA(1 - \sigma A_C)}{d} + \frac{D\sigma AA_C}{d} = \frac{DA}{d}(B(1 - \sigma A_C) + \sigma A_C) \simeq \frac{DA}{d}(B + \sigma A_C).$$

(6.51)

The channels and the bilayer are said to be
"in parallel." We see here that conductances in
parallel add. This behavior stands in contrast
to conductances in series, for which inverse
conductances add to yield the total inverse
conductance.

(f) Comment on how the relative efficiency of the two transport mechanisms
(through the channels or through the bilayer) depends on B and σ and
A_C.

Answer: If $\sigma A_C > B$, transport through ion
channels will be more efficient than transport
through the bilayer. For ions, B is very
small indeed, and the membrane conductance is
essentially entirely through ion channels, which
is why ion channels must exist.

6.4   Spherical Cows

We now turn to a discussion of the mathematics of diffusion in spherically
symmetric situations. It is important to make clear at the outset that although we will
carry out a number of calculations assuming spherical symmetry, in reality perfect
spherical symmetry rarely exists (Fig. 6.8). Nevertheless, it seems reasonable to treat

Fig. 6.8 A spherical cow

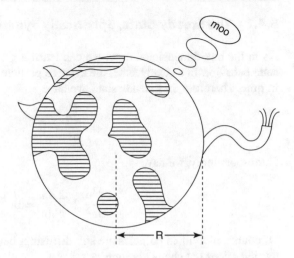

many situations as approximately spherically symmetric, in which case we may expect the results so-obtained to be approximately correct. In fact, consideration of spherically symmetric situations will be very informative: we will gain valuable insight into chemical reaction rates, including the fact that there is a maximum possible chemical reaction rate; we will also learn how cells can take advantage of diffusional transport of attractant and signaling molecules to detect and respond to environmental cues with relatively high efficiency, even with a relatively sparse population of cell-membrane-bound receptors.

In spherically symmetric cases, the diffusion equation depends on the radial coordinate (r), which is the distance from the center of spherical symmetry to a spherical shell at radius r. Although we do not prove it here, it can be shown that the spherically symmetric diffusion equation is

$$\frac{dn}{dt} = D\left(\frac{d^2n}{dr^2} + \frac{2}{r}\frac{dn}{dr}\right),\tag{6.52}$$

where D is the diffusion coefficient and n is the number density (number of particles per unit volume) of diffusing particles. We will use Eq. 6.52 to discuss diffusion in spherically symmetric situations. As may be expected, our earlier expression for the probability density for a three-dimensional random walk:

$$n(r) = \frac{N}{(4\pi Dt)^{3/2}}e^{-r^2/(4Dt)},\tag{6.53}$$

is a solution of Eq. 6.52, which may be verified by direct substitution of Eq. 6.53 into both the left hand side and the right hand side of Eq. 6.52. In spherically symmetric situations, the radial particle flux is

$$j(r) = -D\frac{dn}{dr},\tag{6.54}$$

similar to its one-dimensional counterpart.

6.4.1   The Steady-State, Spherically Symmetric Diffusion Equation

As in the one-dimensional case, we can learn a great deal by considering steady-state behavior. In a steady state, the average particle density, $n(r)$, does not change in time. Therefore, in a steady state, we have

$$\frac{dn}{dt} = 0. \tag{6.55}$$

Correspondingly, we have

$$\frac{d^2n}{dr^2} + \frac{2}{r}\frac{dn}{dr} = 0. \tag{6.56}$$

Of course, individual particles are still diffusing, but the average behavior of many particles does not change in time.

First, we will analyze what happens for a spherically symmetric cell of radius R that needs a nutrient to survive and is in a medium containing nutrient at a concentration n_1 far from the cell. We will furthermore suppose that the cell takes up a fraction of the nutrient molecules that come into contact with the cell radius and that, as a result, the concentration of nutrient is n_2 at the cell's surface ($r = R$). To solve the spherically symmetric diffusion equation, our procedure parallels the procedure that we followed in the one-dimensional case:

1. Draw a sketch.
2. Pick the correct equation and its general solution.

   In the spherically symmetric case, Eq. 6.56 is the correct equation. As usual, we start with a trial general solution:

$$n(r) = \frac{A}{r} + B, \tag{6.57}$$

   where A and B must be picked to satisfy the boundary conditions of the particular problem under study, just as in the one-dimensional case. We may verify that Eq. 6.57 is a solution of Eq. 6.56 for any values of A and B by direct substitution. That is, first, we calculate

$$\frac{dn}{dr} = \frac{d}{dr}\left[\frac{A}{r} + B\right] = -\frac{A}{r^2} \tag{6.58}$$

   and

$$\frac{d^2n}{dr^2} = \frac{d^2}{dr^2}\left[\frac{A}{r} + B\right] = \frac{d}{dr}\left[-\frac{A}{r^2}\right] = \frac{2A}{r^3}. \tag{6.59}$$

Then, by combining Eq. 6.58 (multiplied by $\frac{2}{r}$) and Eq. 6.59, we see that Eq. 6.56 is indeed satisfied by Eq. 6.57.

3. Recognize that in order to find the particular solution within a given region, we must be given, or must figure out, the *boundary conditions for that region*, i.e., we must be given the values of $n(r)$ or $j(r)$ at the boundaries of the region of interest. Without knowing the boundary conditions, we cannot find the solution. If we are given the boundary conditions, we can find the solutions within the region.

Infinitely far from the cell, the concentration of nutrient is

$$n(\infty) = n_1. \tag{6.60}$$

However, because the cell takes up and absorbs a fraction of the molecules that arrive at its surface, at the cell's surface, the concentration is

$$n(R) = n_2, \tag{6.61}$$

where $n_2 < n_1$ if the cell absorbs nutrient. These two conditions are the boundary conditions in this case.

4. Impose each boundary condition on the general solution in turn.

Specifically, we must now impose the boundary conditions to determine a and b. That is, we evaluate Eq. 6.57 at $r = \infty$ and set it equal to Eq. 6.60, with the result that

$$n(\infty) = B = n_1. \tag{6.62}$$

Similarly, we evaluate Eq. 6.57 at $r = R$ and set it equal to Eq. 6.61 with the result that

$$n(R) = \frac{A}{R} + B = n_2. \tag{6.63}$$

5. Solve the resultant equations to find the unknown quantities.

Equation 6.62 gives

$$B = n_1. \tag{6.64}$$

Equation 6.63 together with Eq. 6.64 implies

$$A = R(n_2 - n_1).\qquad(6.65)$$

6. Use the values of the unknowns so-obtained to write the particular solution that applies.

Using these results for A and B in Eq. 6.57, we find

$$n(r) = \frac{R(n_2 - n_1)}{r} + n_1,\qquad(6.66)$$

which is the desired solution for n(r).

7. Calculate the flux, using Fick's law (Eq. 6.54).

The radial flux is, in general,

$$j(r) = -D\frac{dn}{dr}.\qquad(6.67)$$

The radial flux corresponding to Eq. 6.66 is

$$j(r) = -D\frac{d}{dr}\left[\frac{R(n_2 - n_1)}{r} + n_1\right] = \frac{DR(n_2 - n_1)}{r^2}.\qquad(6.68)$$

Equation 6.68 implies that, if $n_2 < n_1$, corresponding to a lower concentration at the cell's surface, then j is negative, and the flow of particles is toward the cell. This result is just what we expect: particles flow from a region of higher concentration (far from the cell) to a region of lower concentration (near the cell). It is important to emphasize that the only motion of individual particles is their random Brownian motion.

8. Calculate the particle current.

The flux, j, is the number of particles per unit time per unit area crossing through a shell at radius r. We may calculate the particle current (I), namely, the number of particles crossing through the spherical shell at radius r, by multiplying j by the area of that shell, which is $4\pi r^2$. It follows that the particle current is

$$I = 4\pi r^2 j = 4\pi DR(n_2 - n_1).\qquad(6.69)$$

Equation 6.69 displays a number of important features. First, the particle current is independent of r. This result reflects conservation of particle number: The only mechanism in our model for particles to disappear is at the surface of the cell, so they should not disappear anywhere else, and indeed they do not.

Second, the particle current is linearly proportional to the radius of the absorbing sphere, R. In particular, the current is not proportional to the surface area of the cell, as we might have naively guessed. Although this result is counterintuitive, it is nevertheless correct and has important consequences.

Finally, inspection of Eq. 6.69 reveals that, just as in the one-dimensional case, the particle current is linearly proportional to concentration difference. Therefore, it is natural in this spherically symmetric situation to also introduce the diffusive conductance, G, via

$$I = 4\pi D R(n_2 - n_1) = G(n_2 - n_1), \tag{6.70}$$

so that the diffusive conductance for a particle starting infinitely far away and ending at the surface of the sphere of radius R is

$$G = 4\pi D R. \tag{6.71}$$

6.4.2 Example: Diffusion Enforces a Fundamental Limit on Cell Size

Blue whales[4] (*Balaenoptera musculus*) are the largest animals ever to have lived on the Earth. Adults are typically 30 m long and weigh 170,000 kg. By contrast, *Caenorhabditis elegans*[5] —a type of roundworm—is about 1 mm long and weighs about 5 μg. That is, the blue whale is about 30 trillion times heavier than *C. elegans*. Each *C. elegans*. has exactly 1031 or 959 cells for the male or hermaphrodite, respectively. Thus, the weight of each cell is about 5 ng, corresponding to a cell radius of approximately 10 μm. Remarkably, the cells of a blue whale are generally also about 10 μm in radius. In fact, irrespective of the size of an animal, its cells are generally about 10 μm in radius. Discuss why is it that animals can differ in size by ten orders of magnitude, and yet it seems that (most of) their cells must be more or less the same size?

---

[4] http://en.wikipedia.org/wiki/Blue_whale

[5] http://en.wikipedia.org/wiki/C._elegans

(continued)

Answer: Equation 6.71 tells us that the maximum possible rate at which a spherical cell can take up oxygen or a nutrient is proportional to the cell's radius, R. On the other hand, it seems reasonable to hypothesize that a spherical cell's metabolic requirements for oxygen or nutrient are proportional to the volume of the cell, namely, proportional to R^3. Therefore, as R gets larger, although the maximum uptake increases with R, the metabolic requirement increases even more rapidly. Eventually, with increasing R, the metabolic requirement would exceed the maximum uptake, and the cell would therefore not be able to live. Thus, diffusion provides a fundamental limit on how big cells are, that even blue whales cannot avoid.

Of course, there are long cylindrical cells, such as your sciatic nerve.[6] But diffusion limits the maximum uptake into cylinders too, similarly giving rise to a diffusion-based limit on the radius of cylindrical cells also (if not on their length).

---

[6]https://en.wikipedia.org/wiki/Sciatic_nerve

6.5   The Rate of Actin Polymerization

In Chap. 5. *Random Walks: Brownian Motion and the Tree of Life*, we conceived of actin polymerization, shown in the movie at http://www.pnas.org/content/suppl/2001/12/10/211556398.DC1/5563Movie_1.avi,[7] from Ref. [1], as a random walk. There, we relied on chemistry classes to tell us (1) that the rate of monomer addition to the filament tip is $k_+ n_1$ (the number of monomers added per unit time), where n_1 is the concentration of actin monomers in solution far from the tip, and k_+ is the association rate constant, and (2) that the dissociation rate from the filament tip is k_- (the number of monomers lost per unit time). As a result, the net rate at which monomers add to the tip is

$$k_+ n_1 - k_-. \tag{6.72}$$

---

[7] http://www.pnas.org/content/suppl/2001/12/10/211556398.DC1/5563Movie_1.avi

In that previous discussion, we did not inquire about the values of k_+ and k_-, or about whether these values might be understandable from a more fundamental point of view. Now, we will apply the results of Sect. 6.4.1 to interpret k_+ as a diffusive conductivity and to see that there exists a maximum possible value for k_+ that depends on the diffusive properties of actin monomers.

6.5.1   Diffusion to Capture and Diffusion-Limited Actin Polymerization

In order for actin polymerization to take place, actin monomers must diffuse to the tip of an actin filament, where they can associate with the tip. We will assume that monomer diffusion to the tip can be described by the spherically symmetric diffusion equation, in the hope, given that the filament is long and thin, and that its presence does not significantly perturb the concentration of actin monomers from the spherically symmetric result (Eq. 6.66), as suggested in Fig. 6.9. It then follows from Eq. 6.70 that the net rate at which actin monomers add to the tip of an actin filament may be expressed as

$$|I| = 4\pi D a (n_1 - n_2), \tag{6.73}$$

where D is the actin monomer diffusion coefficient, n_1 is the concentration of actin monomers far from the tip, and n_2 is the concentration of actin monomers at the "reaction radius," a, which we take to be the size of an actin monomer. In order for association to proceed, the monomer-tip center-to-center separation must reach a. For separations greater than a, monomer-tip association is not possible (Fig. 6.9).

Equations 6.72 and 6.73 purport to describe the same quantity, namely, the net rate of monomer association. To compare these two expressions, we assume that whenever the center of an actin monomer reaches a distance a from the center of the tip, it immediately associates with the tip. Consequently, there are no monomers in solution at $r = a$, so that $n_2 = 0$. We also assume that no monomers dissociate

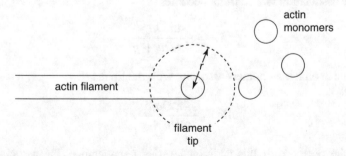

Fig. 6.9 Schematic actin filament tip, treated as a spherical absorber. Actin monomers diffuse to the tip and then associate, leading to growth of the filament

from the tip, so that $k- = 0$. With $n_2 = 0$ and $k_- = 0$, according to Eq. 6.72, the net association rate becomes $k_+ n_1$, and, according to Eq. 6.73, it is $4\pi Da n_1$. Both of these expressions are linearly proportional to n_1. Thus, we see that for actin monomers that immediately associate with the tip upon reaching the reaction radius (a), the association rate constant is given by

$$k_+ = 4\pi Da. \tag{6.74}$$

Because Eq. 6.74 corresponds to every single monomer that reaches radius a, associating with the tip, this equation specifies the maximum possible association rate constant for actin filament growth. Such reactions are said to be "diffusion-limited." Furthermore, comparing Eqs. 6.71 and 6.74, we see in this case that the association rate constant, k_+, that we hear about in chemistry classes is just the diffusive conductance for a molecule starting out at infinity and ending up at radius a. We can now appreciate that diffusion and the diffusion equation provide a fundamental understanding of chemical reaction rates!

6.5.2   Actin Polymerization with Fractional Capture Efficiency

Equation 6.74 corresponds to the situation in which every monomer that reaches the reaction radius a immediately undergoes association with the tip, and there is no dissociation. What happens, if there is still no dissociation, but now not every particle that reaches radius a immediately undergoes association? We may account for this situation by hypothesizing an additional diffusive conductance, g, that represents a further barrier to association, that is in series with the diffusive conductance between infinity and radius a. Then, using the prescription for combining diffusive conductances in series, that we worked out in Sect. 6.3.7, we find

$$\frac{1}{k_+} = \frac{1}{g} + \frac{1}{4\pi Da}, \tag{6.75}$$

so that the association rate constant becomes

$$k_+ = \frac{4\pi Dag}{4\pi Da + g}, \tag{6.76}$$

and the rate at which monomers associate with the tip is

$$|I| = \frac{4\pi Dag}{4\pi Da + g} n_1. \tag{6.77}$$

Even in the presence of this additional series conductance, the association rate remains linearly proportional to concentration, just as we learn in chemistry classes. We may conclude that the linear dependence of the association rate on concentration indeed follows directly from the diffusion equation. Furthermore, Eq. 6.76 shows that when $g \gg 4\pi Da$, $k_+ \simeq 4\pi Da$, corresponding to a diffusion-limited reaction (Eq. 6.74). On the other hand, when $g \ll 4\pi Da$, the reaction is limited by the association step, and $k_+ \simeq g$.

In fact, we can make further contact with chemistry by recalling that in chemistry classes we learn that reaction rates often vary exponentially with inverse temperature. We can account for this observation by hypothesizing that

$$g = g_0 e^{-\frac{\Delta G^\dagger}{k_B T}}, \tag{6.78}$$

where $\Delta G^\dagger$ is the "free energy" of the "transition state," g_0 is the rate constant in the limit of high temperatures, k_B is Boltzmann's constant, and T is the absolute temperature. (We will learn about free energy, Boltzmann's constant, and temperature in Chap. 8. *Statistical Mechanics: Boltzmann Factors, PCR, and Brownian Ratchets.*) For $\Delta G^\dagger \gg k_B T$, Eq. 6.78 tells us that $g \ll 4\pi Da$.

Both 6.74 and 6.76 manifest the result that the chemist's association rate constant is equivalent to a diffusive conductivity. This is an elegant example of how physics is able to bring together ideas that initially appear disparate but in fact are closely connected.

6.5.3 Example: Depletion as a Result of Capture

What is the concentration (n_2) at radius a in the case of fractional capture efficiency?

Answer: Equations 6.73 and 6.77 represent two different expressions for the association rate. Setting these two expressions equal to each other yields

$$4\pi Da(n_1 - n_2) = \frac{4\pi Dag}{4\pi Da + g} n_1. \tag{6.79}$$

Solving for n_2 yields

$$n_2 = \frac{4\pi Da}{4\pi Da + g} n_1. \tag{6.80}$$

For $\frac{4\pi Da}{g} \to 0$, $n_2 \to 0$, as expected.

Fig. 6.10 Inverse of the monomer-filament association rate (k_+^{-1}) versus solution viscosity at the barbed end of the filament, shown in red, and at the pointed end of the filament, shown in blue. The viscosity was varied by adding different amounts of glycerol to the solution. Adapted from Ref. [2]

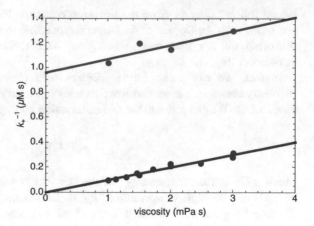

6.5.4 Diffusion Predictions Compared to Experiments

How do these predictions (Eqs. 6.76 or 6.75) compare to experiment? Recall that in Chap. 5. *Random Walks: Brownian Motion and the Tree of Life*, we encountered the Einstein relation, which relates the diffusion (D) and friction (ζ) coefficients via $D = \frac{k_B T}{\zeta}$. In addition, in Chap. 2. *Force and Momentum: Newton's Laws and How to Apply Them*, we encountered Stokes' law: $\zeta = 6\pi \eta a$, for a sphere of radius a in a fluid of viscosity η. It follows that $D = \frac{k_B T}{6\pi a \eta}$. Therefore, one route to varying the kinetics of diffusion is by varying the solution viscosity. Because the inverse of the diffusion coefficient is proportional to the solution viscosity, Eq. 6.75 predicts that k_+^{-1} is linearly proportional to solution viscosity (assuming that g is independent of solution viscosity, which is reasonable). The results of elegant experiments, aimed at testing this prediction, are presented in Fig. 6.10, which shows the viscosity dependence of k_+^{-1}, measured for both the barbed and pointed ends of actin filaments [2]. It is evident from Fig. 6.10 that k_+^{-1} shows behavior consistent with a linear variation versus viscosity for both ends, with similar slopes, as predicted by Eq. 6.75. The zero-viscosity intercept (g^{-1}) is close to zero for the barbed end, indicating that the barbed end shows nearly diffusion-limited growth. By contrast, growth at the pointed end is not diffusion-limited.

6.5.5 Actin Monomer Dissociation

So far, we have focused on monomer-tip association and have largely ignored dissociation. However, in the general case, when dissociation cannot be neglected, we are lead to introduce a monomer concentration at the tip, which we denote K_D, that arises as a result of dissociation from the tip. Then, we may interpret the dissociation rate, k_-, as a current through the same two conductances, $4\pi D a$ and g in series, originating from the concentration K_D. According to this point of view, we have

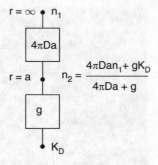

Fig. 6.11 Abstract representation of diffusive transport through the diffusive conductance of the space from $r = \infty$ to $r = a$, in series with diffusive conductance, g, that represents an additional barrier to association and dissociation. These two conductances are represented as the boxes. The solid circles represent the locations where the concentrations are n_1 (namely, $r = \infty$), $n_2 = \frac{4\pi Dan_1 + gK_D}{4\pi Da + g}$ (namely, $r = a$), and K_D

$$k_- = k_+ K_D, \tag{6.81}$$

i.e.,

$$K_D = \frac{k_-}{k_+}. \tag{6.82}$$

Chemists and biochemists call K_D the "dissociation constant." We may express the net rate of monomer-tip association as

$$k_+(n_1 - K_D), \tag{6.83}$$

which makes clear that the filament grows for $n_1 > K_D$ and shrinks for $n_1 < K_D$. We will encounter K_D again in Chap. 8. *Statistical Mechanics: Boltzmann Factors, PCR, and Brownian Ratchets.* An abstract representation of the concentrations and diffusive conductivities in the general case of diffusion to capture, including association and dissociation, is shown in Fig. 6.11.

6.6   Receptors and Biological Signaling Efficiency

Figure 6.12 shows a neutrophil[8]—a type of white blood cell—as it chases after and catches (Yay!!) a bacterium. The neutrophil is able to recognize the presence of the bacterium by virtue of receptors on the neutrophil's surface that sense molecules—attractant molecules, as far as the neutrophil is concerned—that the bacterium emits. Evidently, receptors are important for the body's response to infection.

---

[8] http://en.wikipedia.org/wiki/Neutrophil_granulocyte

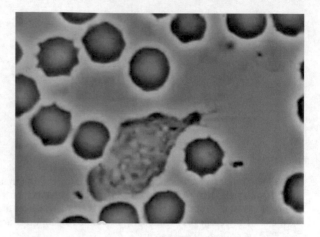

Fig. 6.12 A neutrophil chasing a bacterium from the movie at https://www.youtube.com/watch?v=I_xh-bkiv_c. Taken from a 16-mm movie made in the 1950s by the late David Rogers at Vanderbilt University

An elegant experiment illustrating a similar phenomenon in a more controlled context [3] is shown at

http://www.nature.com/nmeth/journal/v6/n12/extref/nmeth.1400-S2.mov[9]
http://www.nature.com/nmeth/journal/v6/n12/extref/nmeth.1400-S3.mov,[10]

in which the position of a mobile source of N-formylmethionine leucyl-phenylalanine (FMLP), which is a marker for certain bacteria, is manipulated using optical tweezers in order to control the movement of a neutrophil. FMLP interacts with formyl peptide receptors (FPRs) distributed across the neutrophil's surface, which then initiate a cascade of events that gives rise to the cells' response to the attractant. More generally, cells of all sorts use signaling molecules and receptors to communicate and interact. There are many attractants and signaling molecules and correspondingly many different types of receptors, even for a single cell. The medical relevance of receptors includes, for example, breast cancer, where over-expression of HER2,[11] which is an epidermal growth factor receptor (EGFR),[12] occurs in 20–30% of breast cancers and signals an aggressive type of breast cancer. Currently available treatments for HER2+ breast cancers target the HER2 receptor. In the following, we will suppose that the receptors are spread out more or less uniformly over the cell's surface, as suggested in Fig. 6.13. In fact, in some cases, which we will not consider here, a number of receptors are bound together, so that they can interact with each other. For example, HER2 can form dimers either with itself or with other EGFRs.

---

[9] http://www.nature.com/nmeth/journal/v6/n12/extref/nmeth.1400-S2.mov

[10] http://www.nature.com/nmeth/journal/v6/n12/extref/nmeth.1400-S3.mov

[11] http://en.wikipedia.org/wiki/HER2

[12] http://en.wikipedia.org/wiki/Epidermal_growth_factor_receptor

Fig. 6.13 Receptors on the surface of a cell and signaling molecules that react with the receptors, permitting the cell to react to the external signal

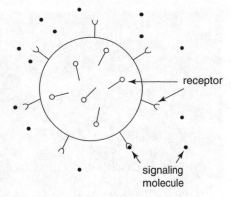

It turns out (Example 6.6) that there is a very interesting feature of the process by which receptors detect their binding partners that follows from the physics of diffusion, namely that it is possible to detect these signaling/attractant molecules with 50% of the maximum possible efficiency, using only 0.02% of the area that would be needed to achieve the maximum possible detection efficiency! This is very beneficial for the cell because it has hundreds of chemicals to detect, and reducing detection efficiency by a factor of 2 from the maximum possible efficiency frees up a lot of cell surface area, permitting many different types of specialized receptors to fit onto the cell's surface.

Example 6.6.1: Receptor Real Estate

Our goal in this example is to figure out how it is that cells can detect signaling molecules with 50% of the maximum possible efficiency, using only 0.02% of the cell surface area that would be needed to achieve the maximum possible detection efficiency. To this end, consider a spherical cell of radius R. The concentration of signaling molecules at $r = \infty$ is $n(\infty) = n_0$. The concentration of signaling molecules at $r = R$ is $n(R) = n_R$, where $n_R < n_0$, because some of the signaling molecules are taken up by the cell. Fitting the general solution to the spherically symmetric diffusion equation to these boundary conditions yields

$$n(r) = \frac{R(n_R - n_0)}{r} + n_0. \qquad (6.84)$$

(a) Calculate the corresponding flux, $j(r)$, and current, I, of attractant/signaling molecules, given that the diffusion coefficient of the signaling molecules is D.

(continued)

Answer: We have

$$j(r) = -D\frac{dn}{dr} = \frac{DR(n_R - n_0)}{r^2} \tag{6.85}$$

and

$$I = 4\pi r^2 j(r) = 4\pi DR(n_R - n_0). \tag{6.86}$$

(b) What is the diffusive conductance, G, between $r = \infty$ and $r = R$?

Answer: We have

$$G_1 = \left|\frac{I}{n_R - n_0}\right| = 4\pi DR. \tag{6.87}$$

(c) In fact, on the cell's surface, there are M receptors for the signaling molecules in question. Each such receptor can be conceived as a diffusive conductance, k, through which the signaling molecules pass before undergoing a chemical reaction, that initiates a signaling cascade within the cell. Thus, if the concentration of signaling molecules at the cell's surface is n_R, the current of signaling molecules into each receptor is

$$kn_R. \tag{6.88}$$

What is the total diffusive conductance, G_2 of M receptors, behaving as M diffusive conductances in parallel?

Answer: For diffusive conductances in parallel, we add conductances. Thus,

$$G_2 = Mk. \tag{6.89}$$

(d) What is the total diffusive conductance, G, for G_1 and G_2 in series, where G_1 is the diffusive conductance between $r = \infty$ and $r = R$ and G_2 is the diffusive conductance of M receptors in parallel?

Answer: For diffusive conductances in series, we have

$$\frac{1}{G} = \frac{1}{G_1} + \frac{1}{G_2}, \tag{6.90}$$

i.e.,

(continued)

$$G = \frac{G_1 G_2}{G_1 + G_2} = \frac{4\pi D R M k}{4\pi D R + M k}. \tag{6.91}$$

(e) In the spirit of the spherical cow, we may approximate the diffusive conductance of each receptor by the diffusive conductance of a sphere of radius a ($a \ll R$), i.e.,

$$k \simeq 4\pi D a. \tag{6.92}$$

Using this approximation, calculate $\frac{G}{G_1}$.

Answer: We have

$$\frac{G}{G_1} = \frac{Ma}{R + Ma}. \tag{6.93}$$

(f) When $\frac{G}{G_1} = \frac{1}{2}$, the current of signaling molecules into the cell has a value equal to one-half of its theoretical maximum possible value. Calculate the ratio of the total surface area of the receptors and the total surface area of the cell for $\frac{G}{G_1} = \frac{1}{2}$. Evaluate your result for $a = 10$ nm and $R = 2\,\mu$m.

A typical neutrophil has a radius of $R = 10\,\mu$m, and a "typical" receptor has a "radius" of $a = 2$ nm. For these values of a and b, we need $M = 5000$ receptors in order to detect a signaling/attractant molecule with 50% of the maximum possible efficiency. Thus, as advertised, to detect 50% of the maximum possible number per second of a certain attractant/signaling molecule, the cell has to devote only $\frac{1}{5000}$ (0.02%) of its surface area to receptors for this purpose. This leaves $\frac{4999}{5000}$ of the cell surface area for other useful stuff: ion channels, ion pumps, receptors for other attractants, passive diffusion, etc. Biology surely benefits from this inevitable consequence of diffusive transport.

6.7 Chapter Outlook

A key result to emerge in this chapter is that a flow of particles—a particle current—is linearly proportional to the difference in concentration. This behavior is an example of a "gradient flow." We will encounter other gradient flows in Chap. 9. *Fluid Mechanics; Laminar Flows, Blushing, and Murray's Law*, where we will see that the fluid volume flow rate is linearly proportional to pressure difference, and in Chap. 14. *Circuits and Dendrites: Charge Conservation, Ohm's Law, Rate*

Equations, and Other Old Friends, where we will see that the electrical current is proportional to voltage difference (Ohm's law).

This chapter also introduced a number of mathematical concepts and techniques that will be critical in later chapters. In this chapter, we employed "solution by direct substitution" for the first time to solve the diffusion equation. We will rely on this method again in Chap. 7 *Rates of Change: Drugs, Infections, and Weapons of Mass Destruction* and in Chap. 9. *Fluid Mechanics: Laminar Flow, Blushing, and Murray's Law*. In this chapter, we also imposed boundary conditions for the first time, which will be essential in Chap. 9. *Fluid Mechanics: Laminar Flow, Blushing, and Murray's Law* and Chap. 11. *Wave Equations: Strings and Wind*.

6.8　Problems

> Mathematical reasoning may be regarded rather schematically as the exercise of a combination of two facilities, which we may call intuition and ingenuity. Alan Turing

Problem 1: Conceptual Problems
(a) **Chemotaxis.** Watch the movie at

       http://www.youtube.com/v/JnlULOjUhSQ,[13]

which shows a neutrophil (the good guy) chasing and capturing (Yay!) a bacterium (the bad guy). Discuss how the neutrophil "sees" the bacterium in order to be able to chase after and eventually capture it?

(b) **New physics in the numbers.** The red line through the data for k_+^{-1} in Fig. 6.10 is the best fit to the barbed end data, yielding $k_+^{-1} = [0.10\eta \text{ (mPa s)}+0.002] \mu M$ s. The blue line is the best fit to the pointed end data, yielding $k_+^{-1} = [0.11\eta \text{ (mPa s)}+0.961] \mu M$ s. The slope in both cases is about $0.1 \mu M \text{ mPa}^{-1} = 10^{-4} M \text{ Pa}^{-1}$. On the other hand, we may estimate a theoretical value for the slope as $\frac{1}{4\pi Da\eta N_A \times 10^3} \simeq \frac{3}{2N_A k_B T \times 10^3} = 1.2 \times 10^{-6} M \text{ Pa}^{-1}$, where N_A is Avogadro's number, and the factor of 10^3 converts units from moles into Molar. Thus, the predicted value for the slope of the viscosity dependence is a factor of 100 smaller than the measured value. This observation implies that the measured, apparently diffusion-limited value of k_+ for barbed ends is 100-fold smaller than predicted based on our simple theory.

Develop a hypothesis to explain this discrepancy. Describe your hypothesis and discuss whether it is really appropriate to describe actin filament growth at the barbed end as "diffusion-limited"?

(c) **Bimolecular reaction rates.** When discussing binary reactions in chemistry classes, the concentrations of the two reactants are specified. In the case of actin

---

[13] http://www.youtube.com/v/JnlULOjUhSQ

polymerization, we have a concentration of actin monomers, which we may write as $n_1 = $ [M] to better conform to chemistry notation. In addition, suppose that there is a concentration of actin filament tips equal to [T]. It follows that the number of monomer-tip association reactions per second per unit volume is

$$k_+[M][T], \tag{6.94}$$

which is now in a form that is familiar from chemistry classes. Therefore, in this chapter, we have seen how this essential equation from chemistry emerges from fundamental ideas concerning random walks and diffusion. How cool is that?

More generally, chemistry and biochemistry classes often discuss binary chemical reactions, such as

$$A+B \underset{k_-}{\overset{k_+}{\rightleftharpoons}} C$$

which specifies the chemical reaction of species A and species B to form species C, all of which are free to undergo three-dimensional diffusion. We learn in those classes that the number of molecules of C formed per unit time per unit volume is equal to

$$k_+[A][B], \tag{6.95}$$

where $[A]$ is the concentration of species A, $[B]$ is the concentration of species B, and k_+ is the association rate constant for this reaction. We also learn in those classes that the number of molecules of C that dissociate into A and B per unit time per unit volume is

$$k_-[C], \tag{6.96}$$

where [C] is the the the concentration of C and k_- is the dissociation rate.

Although we have focused on actin polymerization, our calculations can straightforwardly be applied to bimolecular reactions with a suitable re-interpretation of variables. To apply our expression for k_+ to the reaction of species A and species B, we may re-interpret the actin monomer concentration as [A], and the concentration of actin filament tips as [B]. Then, we must re-interpret D (previously the actin monomer diffusion coefficient) and a (previously the actin monomer-actin tip reaction radius). Now, $a = a_{AB}$ is the species A–species B reaction radius, and $D = D_A + D_B$, where D_A is the diffusion coefficient of A and D_B is the diffusion coefficient of B. Thus, the maximum possible association rate for the reaction of species A and species B is

$$k_+ = 4\pi(D_A + D_B)a_{AB}. \tag{6.97}$$

Explain the form of $D = D_A + D_B$, in terms of the relative motion of species A and species B.

(d) **They do not do random**

> Tom Cronin: "He's making his first mistake."
> Nicki: "It's not a mistake. They don't make mistakes. They don't do random. There's always an objective. Always a target."
> Pamela Landy: "The objectives and targets always came from us. Who's giving them to him now?"
> Nicki: "Scary version? He is."
> The Bourne Supremacy

We have emphasized in this chapter that diffusion is the result of individual particles' probabilistic random walks. However, the predictions that come from the diffusion equation are perfectly deterministic. What has happened? How can it be that the diffusion equation makes perfectly well-defined predictions, if random walks underlie diffusion?

Problem 2: Gram-Negative Bacterial Cell Wall
To a first approximation, as far as their passive transport properties are concerned, the cell walls of gram-negative bacteria[14] may be conceived as consisting of two lipid bilayer membranes, each of thickness d_1 separated by the aqueous periplasmic space of thickness d_2. If the partition coefficient of a certain species of interest in the bilayer membrane is B, and its diffusion coefficient in the membrane is D_1 and in the periplasmic space is D_2, determine the conductance of the Gram-negative bacterial cell wall of area A.

Problem 3: Gradient Flow to Mitochondria: The Spherical Cow Rides Again
To investigate how diffusion might limit the size of eukaryotic cells, consider the situation shown in Fig. 6.14: Far from the cell, there is a concentration of oxygen

Fig. 6.14 Schematic for discussing diffusion to capture at mitochondria

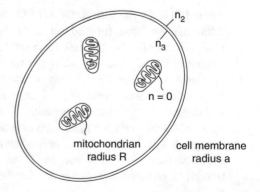

---

[14] http://en.wikipedia.org/wiki/Gram_negative

of n_1; at the exterior surface of the cell membrane, the concentration is n_2; at the interior surface of the cell membrane, the concentration is n_3; and at the surface of each mitochondrion, the concentration is zero under the assumption that every oxygen molecule that reaches the mitochondrion surface is captured.

(a) The oxygen current outside the cell, we approximate as $I_1 = -4\pi Da(n_1 - n_2)$; the oxygen current through the cell membrane, we approximate as $I_2 = -\frac{4\pi a^2 DB}{d}(n_2 - n_3)$; finally, the oxygen current from the interior surface of the cell membrane to all the mitochondria is $I_3 = -4\pi D\frac{Ra}{a-R}Nn_3$, where D is the oxygen's diffusion coefficient, a is the cell's radius, R is the radius of a mitochondrion, B is the partition coefficient of oxygen in the membrane, and N is the number of mitochondria. Briefly, explain where each of these expressions comes from and the important approximations that lead to them.

(b) In a steady state, I_1, I_2, and I_3 are all equal. Using this fact, it is possible to show that

$$I_1 = -\frac{1}{\frac{1}{4\pi Da} + \frac{d}{4\pi Da^2B} + \frac{a-R}{4\pi DNRa}}n_1 = -\frac{4\pi Da}{1 + \frac{d}{aB} + \frac{(a-R)}{NR}}n_1. \qquad (6.98)$$

Briefly, compare the form of this equation to the answer to (e) of Sect. 6.3.9 "Example: Diffusion along a tube of varying cross-sectional area." Discuss whether the fact that I_1 is necessarily smaller in magnitude than $-4\pi Dan_1$ might give insight into the maximum possible size of eukaryotic cells.

Problem 4: Drug Delivery

Biomedical engineers are working to develop a novel, implantable, drug delivery system that consists of a spherical source of drug molecules of radius a, where the concentration of drug molecules is n_a, and that emits L drug molecules per second. Surrounding this source is a spherical shell, composed of a gel-like material, extending from radius a to radius b (call it region 1). Within this gel-like material, the diffusion coefficient of the drug molecules is D_1. For values of the radial coordinate, r, greater than b (call it region 2), the diffusion coefficient of the drug molecule is D_2.

(a) For this drug delivery system, calculate the drug molecule concentration in region 1, $n_1(r)$, in terms of the unknown concentration at $r = b$, n_b, and given quantities.

(b) Calculate the drug molecule concentration in region 2, $n_2(r)$, in terms of n_b, and given quantities.

(c) Calculate the drug molecule current in region 1, I_1, in terms of n_b, and given quantities.

(d) Calculate the drug molecule current in region 2, I_2, in terms of n_b, and given quantities.

(e) How are I_1 and I_2 related to L? Using these relationships, eliminate n_b to determine L in terms of n_a.
(f) The biomedical engineers are interested in adding a second coating layer of inner radius b and outer radius c and diffusion coefficient D_3. Without necessarily doing a detailed calculation, predict the relationship between L and n_a in the case of this modified delivery system. Explain your reasoning.

Problem 5: Calculating Probabilities Using Solutions to the Steady-State Diffusion Equation: Repressor and Operators

A certain DNA binding protein—the repressor (R)—binds to a linear segment of yeast DNA and then undergoes one-dimensional diffusion along the DNA. However, there are two particular sequences of bases on the DNA—operator 1 (O1) and operator 2 (O2)—to which R binds very tightly. If R reaches either O1 or O2, binding is so tight that R may be considered to be captured by O1 or O2.

Your goal in this problem is to determine the probability that R is eventually captured at O1 (call it P_1) and the probability that R is eventually captured at O2 (call it P_2). These two alternatives are the only two possibilities, and they are mutually exclusive.

O1 is located at $x = 0$. R initially binds to the DNA at $x = W$. O2 is located at $x = L$. We have $0 < W < L$, as shown in Fig. 6.15. The diffusion coefficient of R on DNA is D.

The key to finding the desired probabilities is the observation that if we envision a source of R at $x = W$ that fixes the R concentration to be n_0 at $x = W$, then the values of the corresponding R particle currents toward smaller and larger x, respectively, are proportional to the probability that an individual R is captured by O1 or O2, respectively. So if we can calculate those currents, we can calculate the desired probabilities for an individual R. This problem takes you through this calculation.

Importantly, to express that O1 and O2 are sinks from which R does not return, the concentrations at $x = 0$ and $x = L$ are both zero, i.e., $n(0) = 0$ and $n(L) = 0$, where $n(x)$ is the number of Rs per unit length at x.

The solution to the one-dimensional steady-state diffusion equation ($d^2n/dx^2 = 0$) is

$$n(x) = a_1 x + b_1, \qquad (6.99)$$

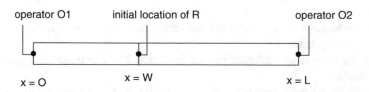

Fig. 6.15 Schematic of the DNA fragment considered in "Repressor and operators." O1 is at $x = 0$. O2 is at $x = L$. R initially lands at $x = W$

Fig. 6.16 Left: Sketch for the tube and membrane "Tube and membrane diffusion"

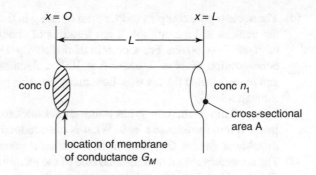

for $0 < x < W$ (Region 1) and

$$n(x) = a_2 x + b_2, \qquad (6.100)$$

for $W < x < L$ (Region 2), where a_1, b_1, a_2, and b_2 are constants that you must determine (Fig. 6.16).

(a) Determine the values of a_1 and b_1, by using Eq. 6.99 together with the boundary conditions that $n(0) = 0$ and $n(W) = n_0$.
(b) Determine the corresponding current of R in Region 1. Call it I_1.
(c) Determine the values of a_2 and b_2, by using Eq. 8.33 together with the boundary conditions that $n(W) = n_0$ and $n(L) = 0$.
(d) Determine the corresponding current of R in Region 2. Call it I_2.
(e) Given that the flux of R toward smaller x is proportional to the probability that an individual R is captured by O1 and that the flux of R toward larger x is proportional to the probability that an individual R is captured by O2, write down an expression for P_1 in terms of I_1 and I_2 only.
(f) Use your answers to (b), (d), and (e) to write down expressions for P_1 and P_2 in terms of known quantities.
(g) What is the value of P_1 in the limit that $L \gg W$? (Hint: Your answer to this part should be very simple.)
(h) Compare P_2 to Q_i in the "Gambler's ruin" example in Chap. 5. *Random Walks: Brownian Motion and the Tree of Life.*

Problem 6: Tube and Membrane Diffusion
(a) As shown in Fig. 6.16, one end (location $x = 0$) of a long thin tube of length L and cross-sectional area A is attached to a fluid reservoir containing no solute. The other end (location $x = L$) is attached to a fluid reservoir containing solute particles of diffusion constant D at concentration n_1. Find the steady-state solute concentration in the tube as a function of x.
(b) Determine the particle current in the tube?
(c) What is the conductivity of the tube? Call it G_T.

(d) The nuclear pore complex (NPC) is an opening in the nuclear envelope between the nucleus and the cytosol. It has a radius of about $R = 50\,\text{nm}$ and a length of about $L = 60\,\text{nm}$. For a protein of diffusivity $D = 3 \times 10^{-11}\,\text{m}^2\text{s}^{-1}$, at a concentration of $n_1 = 1\ \mu\text{M} = 6 \times 10^{20}$ molecules m^{-3} outside the nucleus, and $n_2 = 0$ inside the nucleus, how many proteins per second enter the nucleus through a NPC?

(e) Now, a thin membrane—it has infinitesimal thickness—of conductance G_M is inserted into the tube at $x = 0$. What is the conductivity of the combined tube-membrane system, G_C, in terms of given parameters.

(f) The concentration on the left-hand side of the membrane remains zero, when the filter is added. What are the currents along the tube and through the membrane?

(g) What is the solute concentration immediately to the right-hand side of the membrane, n?

Problem 7: Branched Diffusion

A tube of initial cross-section A_0 and length L_0 splits up into two branches: one tube with cross-section A_1 and another one with cross-section A_2, each of length L, as shown in Fig. 6.17. A number of particles with diffusion coefficient D are diffusing through these tubes because of a concentration difference between the ends: the concentration is n_0 on the left side of the tube and n_2 at the right end of each of the two branches. Assume $n_0 < n_2$.

It is also convenient to introduce the concentration, n_1, at the branch point, as shown in the figure. Importantly, n_1 is *not* a given quantity but depends on the given quantities, namely, n_0, n_2, L_0, L, A_0, A_1, A_2, and D. However, feel free to use n_1 in any part of this problem, unless instructed otherwise.

Throughout this problem, use an x-axis pointing to the right, with the origin ($x = 0$) located at the inlet to the first tube, as shown in Fig. 6.17. Call I_0 the current of particles through the first part of the tube, and I_1 and I_2 the currents through the

Fig. 6.17 Sketch of a tube that splits up into two branches

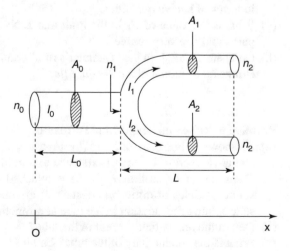

branch with cross-section A_1 and cross-section A_2, respectively, as also shown in the figure.

(a) The generic solution for the steady-state concentration in a one-dimensional diffusive situation: $n(x) = ax + b$. Focusing only on the first part of the tube of length L_0, find the appropriate values of the constants a and b, by imposing the appropriate boundary conditions.

(b) Using your answers from (a), write the solution for the steady-state concentration, $n(x)$, in the first part of the tube, and sketch $n(x)$ for $0 < x < L_0$.

(c) Find the current, I_0, in the first part of the tube, between $x = 0$ and $x = L_0$.

(d) In which direction is the current I_0 flowing? Explain.

(e) Use your answer for $n(x)$ from (b) to calculate the total number of particles, N, that are at any given time between $x = 0$ and $x = L_0$, that is, the total number of particles in the first part of the pipe of length L_0.

(f) What is the conductance of the first part of the tube? Call it G_0.

(g) Using your result for (c) or otherwise, write expressions for I_1 and I_2.

(h) What is the conductance of the branch with current I_1? Call it G_1. What is the conductance of the branch with current I_2? Call it G_2.

(i) Consider the total current flowing between $x = L_0$ and $x = L_0 + L$ through both branches together, $I_B = I_1 + I_2$. How does I_B compare to I_0? Is it bigger, smaller, or the same? Explain.

(j) Call G_B the total conductance of the two branches together. Write an expression for I_B in terms of G_B and the appropriate concentration difference.

(k) Using your answers to (g) and (j), find an expression for G_B in terms of given quantities.

(l) Calculate the overall conductance from $x = 0$ to $x = L + L_0$. Call it G_T.

Problem 8: Mean Passage Time Through a Membrane
The concentration profile *inside* a certain membrane of thickness d $(0 < x < d)$ is

$$n(x) = Bn_1 \left(1 - \frac{x}{d}\right), \tag{6.101}$$

where n_1 is the concentration of the species of interest outside of the membrane for $x < 0$, 0 is the concentration of the species of interest outside of the membrane for $x > d$, and B is a so-called partition coefficient.

(a) Sketch $n(x)$ versus x.

(b) Calculate the particle flux through the membrane.

(c) If the membrane has a cross-sectional area A, what is the net number of particles per second (call it I) passing through the membrane?

(d) The function $n(x)$ gives the number of particles per unit volume. What then is the number of particles in a volume of cross-sectional area, A, and thickness along x of dx in terms of $n(x)$, A, and dx?

(e) Using your answer to (d) write and carry out an integral to calculate the total number of particles in the membrane (call it N).
(f) Use your answers to (c) and (e) to calculate the mean time (call it τ) that each particle spends inside the membrane.

Problem 9: Calculating Probabilities Again

In this problem, you will use solutions of the three-dimensional, spherically symmetric, steady-state diffusion equation to calculate probabilities (Fig. 6.18).

Here is the problem: imagine a particle initially located at a radius a from the the center of a spherical particle absorber (i.e., a particle sink) of radius R ($a > R$). The problem is to calculate (a) the probability that the particle will eventually be absorbed and (b) the probability that the particle will wander away, never to be absorbed. (These are the only two outcomes that are possible, and they are mutually exclusive.)

This initially seems like a difficult and complicated problem involving a time-dependent solution to the diffusion equation in a funky geometry, but there is an easier way that goes as follows.

Instead of a single particle at radius a, envision that the spherical shell at radius a is a particle source, such that the concentration at radius a is c_0. Furthermore, imagine that enough time has gone by since this situation was set up, that by now, a steady state has been achieved. In this steady-state situation, the concentration at radius R is zero because of the absorbing spherical sink, and the concentration at $r = \infty$ is also zero. It follows that there will be steady-state particle currents toward both smaller and larger values of r.

The key to our problem is that the values of the particle currents toward smaller and larger r are proportional to the probability that an *individual* particle is captured or escapes, respectively. So if you can calculate the currents, you can calculate the desired probabilities.

Earlier in this chapter, we showed that steady-state solutions to the spherically symmetric diffusion equation (in regions where there are no particle sources or sinks) are of the form:

Fig. 6.18 Sketch for "Calculating probabilities again." The 100% absorbing cell of radius R is shown hatched. The initial radius of the particle of interest is a. The current for $R < r < a$ is I_1, and the current for $r > a$ is I_2, driven by zero concentration at infinity, concentration c_0 at radius a, and zero concentration at radius R

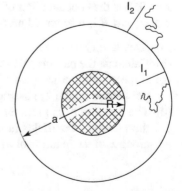

$$c(r) = \frac{A}{r} + B, \tag{6.102}$$

where A and B are to be determined from the boundary conditions.

This equation applies separately for $R < r < a$ (Region 1) and for $a < r < \infty$ (Region 2).

(a) Briefly explain why the values of the particle currents toward smaller and larger r are proportional to the probability that an *individual* particle is captured or escapes, respectively. (Hint: Your answer should make mention of how we originally defined probability, and how we have defined particle fluxes and currents.)

(b) Use Eq. 6.102 in Region 1, and by matching boundary conditions, determine the unknowns A_1 and B_1, which apply in Region 1.

(c) Use Eq. 6.102 in Region 2, and by matching boundary conditions, determine the unknowns A_2 and B_2, which apply in Region 2.

(d) Using your results for (b) and (c) and making use of the fact that probabilities must be normalized, determine the desired probabilities for capture and escape. (Hint: Think carefully (i.e., physically) about what signs to give your particle currents, when you calculate the probabilities.)

Problem 10: Rat Ratchet

Consider a bunch of rats on a staircase as shown in Fig. 6.19. (Possibly they are participating in some psychology experiment?) Each rat performs a random walk on the stair tread that it is on. If a rat encounters the step up, it cannot climb up and so continues its random walk on the same tread. However, if a rat encounters the step down, it falls down to the next tread. (It is a rat ratchet:) In this fashion, the rats proceed downstairs only. Your goal in this problem is to calculate the drift velocity of the rats as they proceed down the stairs.

The rat diffusion coefficient is D. The length of each stair tread is L. This is a 1D problem, and we will assume a rat steady state .

Fig. 6.19 Schematic of the staircase considered in "Rat ratchet," rats and all

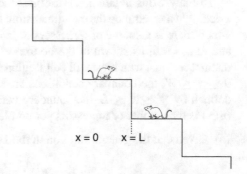

$x = 0$ $x = L$

Consider the step between $x = 0$ and $x = L$ (step 1). The rat density—i.e., the number of rats per unit length at x—on step 1 is $n(x)$ for $0 < x < L$.

(a) To express that $x = L$ corresponds to a "sink" from which a rat does not return, what is the boundary condition for the rat density at $x = L$?

(b) The solution to the one-dimensional steady-state diffusion equation (i.e., $d^2n/dx^2 = 0$) is

$$n(x) = ax + b. \tag{6.103}$$

For the tread between $x = 0$ and $x = L$ ($0 < x < L$), determine one equation relating a and b, using this solution and with the boundary condition that you introduced in (a).

(c) Use your result for (b) to eliminate b from Eq. 6.103.

(d) What is the mean number of rats on step 1, expressed in terms of a and L only? Call it N.

(e) Determine the current of rats at $x = L$. Call it I.

(f) If the mean number of rats on a step is N and I is the current of rats leaving a step, what is the mean time (τ) that each rat spends on a step? Express your answer in terms of the given quantities only. The quantity that you have just calculated is called the "mean first passage time."

(g) Given your answer to (e), what is the mean velocity (v) of the rats?

Problem 11: One-Dimensional Diffusion to Capture on Circular DNA
A certain protein (the repressor) binds a specific DNA sequence (the operator). One hypothesis for how the repressor "finds" the operator is that the repressor lands on DNA and subsequently performs a one-dimensional random walk along the DNA until it encounters the operator to which it then sticks. In this problem, you will analyze this process in the case of bacterial DNA that forms a closed loop of length L. Importantly, in the case of circular DNA, if the repressor is a distance W from the operator in the clockwise direction, it is also a distance $L - W$ from the operator in the counterclockwise direction.

To analyze this situation, it is useful to envision the distance along the DNA as the x-axis and then employ the one-dimensional (1D) steady-state diffusion equation for which there is a source of repressors at $x = W$ and a sink of repressors at $x = 0$ and at $x = L$, as shown at the bottom of Fig. 6.20. Thus, this problem has two distinct regions, which we will call Region 1 and Region 2. In Region 1, defined by $0 \le x \le W$, the boundary conditions are $n(0) = 0$ and $n(W) = n_1$. In Region 2, defined by $W \le x \le L$, the boundary conditions are $n(W) = n_1$ and $n(L) = 0$. ($n(x)$ is the number of repressors per unit length.)

(a) Given that the general solution to the 1D steady-state diffusion equation is

$$n(x) = A_1 x + B_1 \tag{6.104}$$

Fig. 6.20 Top: Schematic of a loop of bacterial DNA, showing the location of the operator and the initial location of the repressor. Bottom: The same DNA now represented to lie along the x-axis. In this representation, there is a sink for repressor (the operator) at both $x = 0$ and $x = L$, i.e., $n(0) = n(L) = 0$. The source of repressor is at $x = W$, i.e., $n(W) = n_1$

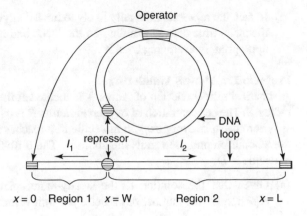

in Region 1, where A_1 and B_1 are constants, determine A_1 and B_1 in terms of the given parameters, namely, n_1, W, and L, and thus write down $n(x)$ in Region 1.

(b) Because $n(x)$ represents the number of repressors per unit length, the relationship between current and $n(x)$ in this problem is

$$I = -D\frac{dn}{dx}, \tag{6.105}$$

where D is the diffusion coefficient of a repressor on DNA. Using this result, calculate the current in Region 1. Call it I_1.

(c) Calculate the number of repressors in Region 1. Call it N_1.

(d) Determine the diffusive conductance of Region 1. Call it G_1.

(e) What is the length of Region 2? By considering how your answer to (d) depends on the length of Region 1, or otherwise, determine the diffusive conductance of Region 2, namely, G_2. Hint: There is no need to do a long calculation.

(f) Region 1 and Region 2 represent two parallel paths for the repressor to reach the operator. Therefore, the total diffusive conductance between $x = W$ and the operator is $G = G_1 + G_2$. Calculate G.

(g) The number of repressors per second that go from $x = W$ to the operator is $|I_1| + |I_2|$. Briefly explain why

$$|I_1| + |I_2| = Gn_1. \tag{6.106}$$

(h) Use Eq. 6.106 and the result that the total number of repressors in both regions together is

$$N = N_1 + N_2 = \int_0^W n(x)dx + \int_W^L n(x)dx = \frac{1}{2}n_1L, \tag{6.107}$$

to determine the mean time, τ, for a repressor at $x = W$ to reach the operator.

(i) In fact, the repressor is equally likely to landed anywhere on the DNA. Calculate the mean time between landing on the DNA and reaching the operator, $\langle \tau \rangle$, by appropriately averaging over W.

Problem 12: Aimless Wandering

A spherical mitochondrion of radius $\frac{R}{2}$ is located at the center of a spherical cell of radius R. The concentration of oxygen at radius R is n_1. The concentration at radius $\frac{R}{2}$ is zero because every oxygen molecule that reaches its surface is taken up by the mitochondrion and used in its metabolism. The diffusion coefficient of an oxygen molecule is D.

(a) Given that the solution to the steady-state spherically symmetric diffusion equation for the number of particles per unit volume is

$$n(r) = \frac{a}{r} + b, \qquad (6.108)$$

apply the given boundary conditions, $n(R) = n_1$ and $n(\frac{R}{2}) = 0$, to find the constants a and b.

(b) Using your result from (a), sketch $n(r)$ versus r for $\frac{R}{2} \leq r \leq R$.

(c) Calculate the flux of oxygen in the region between the mitochondrion and the cell surface ($\frac{R}{2} \leq r \leq R$).

(d) Calculate the current of oxygen molecules, I, in the region between the mitochondrion and the cell surface ($\frac{R}{2} \leq r \leq R$).

(e) Calculate the total number of oxygen molecules, N, in the region between the mitochondrion and the cell surface ($\frac{R}{2} \leq r \leq R$).

(f) Given that N is the number of oxygen molecules between the cell surface and the mitochondrion and that I is the number of oxygen molecules per unit time entering and leaving this region, determine the mean time, τ, that an oxygen molecule spends in this region, in terms of given parameters (D, R, and n_1).

Problem 13: Molecular Motors Revisited

In Chap. 5. *Random Walks: Brownian Motion and the Tree of Life*, we found the mean displacement and the variance of the displacement of a molecular motor by considering the possible outcomes—stepping forward, stepping backward, or remaining still—in a small increment of time, Δt. We can apply a similar approach to derive the so-called master equations for the probability that the motor is at location n at time t.

(a) The probability that a molecular motor is at location n at time $t + \Delta t$ is $P(n, t + \Delta t)$. In the limit of small Δt, explain why

$$P(n, t+\Delta t) = k_+ \Delta t\, P(n-1, t) + k_- \Delta t\, P(n+1, t) + (1 - k_+ \Delta t - k_- \Delta t) P(n, t).$$
$$(6.109)$$

(b) Using Eq. 6.109, we see (in the limit of infinitesimal Δt) that

$$\frac{dP(n,t)}{dt} = \frac{P(n, t+\Delta t) - P(n,t)}{\Delta t} = k_+[P(n-1,t) - P(n,t)]$$
$$+ k_-[P(n+1,t) - P(n,t)]. \tag{6.110}$$

Equations for the time evolution of the probability are generally called "Master equations." Thus, the equation,

$$\frac{dP(n,t)}{dt} = k_+[P(n-1,t) - P(n,t)] + k_-[P(n+1,t) - P(n,t)], \tag{6.111}$$

is a master equation that specifies the probability that a molecular motor is at location n at time t. To elucidate Eq. 6.111, it is helpful to go to a continuum description. To this end, write $P(n,t) = P(x,t)$, $P(n+1,t) = P(x+a,t)$, and $P(n-1,t) = P(x-a,t)$, where $x = na$, and carry out a Taylor series expansion to quadratic order in a. How does the resulting differential equation for $P(x,t)$ compare to the diffusion equation?

(c) Show, using WolframAlpha or otherwise, that

$$P(x,t)\frac{1}{\sqrt{4\pi Dt}}e^{-(x-vt)^2/(4Dt)}, \tag{6.112}$$

corresponding to a biased random walk, is a solution of the equation that you found in (b).

(d) Write a master equation corresponding to a random walk that steps $\pm L$ at a rate $\frac{1}{\tau}$ and thus derive the diffusion equation and the corresponding diffusion coefficient for this model.

Problem 14: Synaptic Diffusion

A synapse between nerve cells (neurons) permits a signal to pass from one nerve cell to another. In a chemical synapse, synaptic vesicles, containing small molecules, called neurotransmitters, are caused to fuse with the cell membrane. As a result, the neurotransmitters are released into the synaptic cleft that separates the pre-synaptic neuron from the post-synaptic neuron. The neurotransmitters diffuse across the synaptic cleft and react with receptors at the surface of the post-synaptic neuron.

In this problem, you will model neurotransmitter transport across the synaptic cleft (Fig. 6.21). The height of the synaptic cleft is L and its area is A. Therefore, you should approximate the synaptic cleft to be a short tube of cross-sectional area A and length L. Take the diffusion coefficient of the neurotransmitters to be D. Assume that the concentration of neurotransmitters has reached a steady state, so that the 1D steady-state diffusion equation is applicable, and the concentration of neurotransmitters at the surface of the pre-synaptic neuron ($x = 0$) is fixed at a value n_1, and the concentration of neurotransmitters at the location of the receptors at the surface of the post-synaptic neuron ($x = L$) is fixed at n_2 ($n_2 < n_1$).

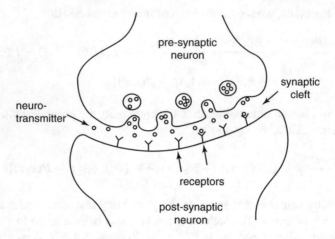

Fig. 6.21 Schematic side view of a synapse

(a) Given that the general solution for the steady-state concentration in a one-dimensional diffusive situation is

$$n(x) = ax + b, \qquad (6.113)$$

where $n(x)$ is the number of particles per unit of volume, determine the concentration of neurotransmitters as a function of distance, x, across the synaptic cleft, in terms of given quantities.

(b) Calculate the current, I, of neurotransmitters across the synaptic cleft. Explain whether the current is positive or negative.

(c) Calculate the diffusive conductance, G_A, of the synaptic cleft.

(d) If every neurotransmitter molecule that reaches the surface of the post-synaptic neuron reacts with the receptors there, and therefore is absorbed, the correct boundary condition is $n(L) = n_2 = 0$. What is the corresponding current in this case—call it I_1—and what is the corresponding rate—call it k_1—at which neurotransmitters arrive at and react with the receptors, i.e., what is the number of neurotransmitters per second that arrive at and react with receptors?

(e) In fact, when a neurotransmitter reaches the post-synaptic neuron, it either reacts with a receptor with probability p or it diffuses away again with probability $1 - p$. In this case, what is the new rate—call it k_2—at which neurotransmitters react with receptors?

(f) Using your answers to (b) and (e), calculate n_2 in terms of n_1 and p.

(g) Briefly explain why the total number of neurotransmitter molecules in the synaptic cleft, N, is given by the following integral:

$$N = A \int_0^L n(x)dx. \qquad (6.114)$$

Evaluating this integral leads to

$$N = \frac{1}{2}AL(n_1 + n_2). \qquad (6.115)$$

(h) Calculate the mean time, τ, that a neurotransmitter spends in the synaptic cleft in terms of p, D, and L.

Problem 15: Travelling by Tube

In a protein crystal growth procedure, protein crystals are separated from a protein solution of concentration, n_0, by a tube of area of length L and area A. Every protein molecule that reaches the crystal surface becomes incorporated into the crystal. To model this situation, consider one-dimensional (1D) diffusion with diffusion constant D along a tube of length L and cross-sectional area A. The tube starts at $x = 0$ and ends at $x = L$. The particle concentration at $x = 0$ is n_0. Because every protein particle that arrives at $x = L$ sticks to the surface of the growing crystal there, the particle concentration at $x = L$ is zero. (Throughout this problem, take the partition coefficient, B, to equal 1.)

(a) Determine the concentration of the diffusing particles, $n(x)$, in the tube, given that the general solution of the 1D steady-state diffusion equation is $n(x) = ax + b$.
(b) Calculate the total number of particles (N_1) between $x = 0$ and $x = \frac{L}{2}$ (region 1 of the tube) and the total number of particles (N_2) between $x = \frac{L}{2}$ and $x = L$ (region 2).
(c) What are the particle currents, I_1 and I_2, in region 1 and region 2, respectively, of the tube in terms of given quantities?
(d) Calculate the mean time that a particle spends in each of region 1 and region 2, namely, τ_1 and τ_2, respectively, in terms of given quantities. Does the average particle spend more time in region 1 of the tube than in region 2 of the tube or less time in region 1 of the tube than in region 2 of the tube or does it spend the same amount of time in each region of the tube?

Problem 16: Droplet Growth by Diffusion

Phase separation is currently a hot topic in cell biology in the context of "non-membrane-bound organelles." In this problem, we consider the growth by diffusion of a spherical droplet of radius R, containing phase-separated protein at high concentration, which is suspended in a dilute solution of the protein in question at concentration n_1. The droplet grows as protein molecules in the solution (solution proteins) diffuse to and are captured by the droplet. Every protein molecule that arrives at the surface of the droplet is captured by the droplet and adds its volume to that of the droplet. The diffusion coefficient of the solution proteins is D.

(a) Given that the solution to the steady-state spherically symmetric diffusion equation for the number of particles per unit volume, $n(r)$, is

$$n(r) = \frac{a}{r} + b, \tag{6.116}$$

apply the given boundary conditions, $n(\infty) = n_1$ and $n(R) = 0$, to find the constants a and b and $n(r)$.

(b) Calculate the flux of solution proteins for $r > R$.

(c) Calculate the current of solution proteins, I, for $r > R$.

(d) When a solution protein is captured by the droplet, the volume of the droplet increases by the volume of the capture protein molecule, v. Write down an expression for the rate of change of the volume of the droplet, $\frac{dV}{dt}$, in terms of v and I.

(e) Use your answers to (c) and (d) and the result that $\frac{dV}{dt} = 4\pi R^2 \frac{dR}{dt}$ (because $V = \frac{4\pi}{3} R^3$) to write an equation for $\frac{dR}{dt}$ in terms of R, D, v, and n_1.

(f) Show by direct substitution that the equation you found in (e) has a solution

$$R = A(t + s)^\alpha, \tag{6.117}$$

and find the values of A and α that are required for the solution. s depends on the initial conditions. (Note that $\frac{d}{dt}(t+s)^\alpha = \alpha(t+s)^{\alpha-1}$.) Evidently, according to this model, the radius of the protein droplet grows approximately as a power-law function of time.

Problem 17: Diffusion near an Absorbing Wall

(a) At $t = 0$, a large number, N, of particles all begin random walks from $x = L$. Sketch the number of particles per unit length, $n(x)$, versus x for a value of t ($t > 0$) for which a significant fraction of the particles, but not most, are at negative values of x. Your sketch represents a solution to the diffusion equation. (Be sure that your sketches throughout this problem cover all relevant values of x.)

(b) Given that the particle flux, j, is given by $j(x) = -D\frac{dn}{dx}$, sketch the flux corresponding to your sketch for $n(x)$ in (a).

(c) Now suppose that an adsorbing wall is located at $x = 0$, which adsorbs all of the particles that reach it. As a result, if we call the number of particles per unit length in the presence of the adsorbing wall $w(x)$, we have $w(0) = 0$. Because the diffusion equation is a linear equation, it obeys the principle of superposition, namely that the sum of two solutions of the diffusion equation is also a solution to the diffusion equation. One way to arrange for $w(0) = 0$ for all values of t, as required, is to subtract from the $n(x)$ of (a) its mirror image. That is, the new expression for the number of particles per unit length in the region $x > 0$ in the presence of the absorbing wall is

$$w(x) = n(x) - n(-x). \tag{6.118}$$

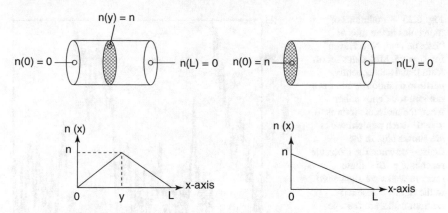

Fig. 6.22 Figure for "Escape from New Haven." The sketch and the plot on the left-hand side apply to (a) and (b). The sketch and the plot of the right-hand side apply to (c) through (g)

Sketch $n(x)$, $n(-x)$, and w(x) versus x all on the same graph.
(d) Sketch the particle flux corresponding to w, i.e., now sketch $j_w(x) = -D\frac{dw}{dx}$.
(e) How does the flux at $x = 0$ in (d) compare to the flux at $x = 0$ in (b)?

Problem 18: Escape from New Haven

Consider a tube that starts at $x = 0$ and ends at $x = L$ and has a cross-sectional area A. At $x = y$, there is a source of particles that maintains the concentration (the number of particles per unit volume) equal to n at that point. Region 1 is between $x = 0$ and $x = y$. Region 2 is between $x = y$ and $x = L$. Solution of the one-dimensional steady-state diffusion equation in Region 1 leads to a concentration in this region of

$$n(x) = n\frac{x}{y}. \tag{6.119}$$

Similarly, in Region 2, the concentration is

$$n(x) = n\frac{L-x}{L-y}. \tag{6.120}$$

This situation and the corresponding $n(x)$ are shown in the left column of Fig. 6.22.

(a) Given that the diffusion coefficient is D, calculate the currents through Region 1 and Region 2, namely, I_1 and I_2, respectively.
(b) An individual molecule is located at $x = y$. Use your answer to (a) to determine the probability—call this probability $P(E|y)$—that a molecular escapes to L given that it is at y.
(c) Now consider a different problem, involving the same tube. But, now, the concentration is n at $x = 0$ and zero at $x = L$. Adapt Eq. 6.120 to write the concentration $n(x)$ in this case.

Fig. 6.23 Simulations of
molecules in the tube of
"Escape from New Haven"
for $L = 25$. Molecules set off
from 0 and subsequently
perform a random walk. Each
random walk ends either
when the molecule returns to
$x = 0$—such random walks
are shown blue in the
figure—or when the molecule
reaches $x = 25$—these
random walks are shown red
in the figure. These red
random walks are "escape
paths"

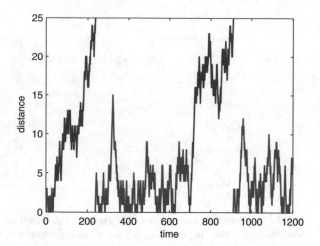

(d) Calculate the total number of particles in the tube, N, in this new situation.

(e) Use your answers to (c) and (d) to write down the probability that a particle in
this new situation is between x and $x + dx$, namely, $p(x)dx$.

(f) In this new situation, molecules start out from $x = 0$, and some of them escape
the tube by reaching $x = L$. However, not all molecules that depart $x = 0$
escape to $x = L$. Some of them return to $x = 0$ after wandering around in
the tube for a while. The goal of this part is to determine the probability that a
molecule in the tube is on an escape path (Fig. 6.23).

Use your answers to (b) and (e) and the product rule of probability to write
down the probability that a particle is between x and $x + dx$ and that it escapes.
Call this probability $p(E, x)dx$.

(g) Use your answer to (f) to calculate the probability that a particle in the tube is
on an escape path, $P(E)$.

References

1. K.J. Amann, T.D. Pollard, Direct real-time observation of actin filament branching mediated by
Arp2/3 complex using total internal reflection fluorescence microscopy. Proc. Nat. Acad. Sci.
USA **98**, 15009–15013 (2001)
2. D. Drenckhahn, T.D. Pollard, Elongation of actin filaments is a diffusion-limited reaction at the
barbed end and is accelerated by inert macromolecules. J. Biol. Chem. **261**, 12754–58 (1986)
3. H. Kress, J.-G. Park, C.O. Mejean, J.D. Forster, J. Park, S.S. Walse, Y. Zhang, D. Wu, O.D.
Weiner, T.M. Fahmy, E.R. Dufresne, Cell stimulation with optically manipulated microsources.
Nat. Methods **6**, 905–909 (2009)

Rates of Change: Drugs, Infections, and Weapons of Mass Destruction

7

HIV is certainly character-building. It's made me see all of the shallow things we cling to, like ego and vanity. Of course, I'd rather have a few more T-cells and a little less character.

Randy Shilts[1]

No-one really thought of fission before its discovery.

Lise Meitner

In some sort of crude sense which no vulgarity, no humor, no overstatement can quite extinguish, the physicists have known sin; and this is a knowledge which they cannot lose.

J. Robert Oppenheimer [1]

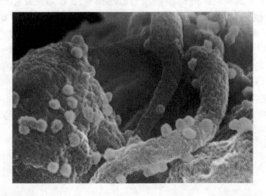

Electron microscopy image of HIV particle (green) budding from an infected cultured lymphocyte. From the Public Health Image Library at https://phil.cdc.gov/Details.aspx?pid=11279, provided by the CDC and C. Goldsmith, P. Feorino, E. L. Palmer, and W. R. McManus

Electronic Supplementary Material The online version of this article (https://doi.org/10.1007/978-3-031-05808-0_7) contains supplementary material, which is available to authorized users.

[1] http://www.nytimes.com/1993/04/22/garden/at-home-with-randy-shilts-writing-against-time-valiantly.html?pagewanted=all&src=pm

S. Mochrie, C. De Grandi, *Introductory Physics for the Life Sciences*, Undergraduate Texts in Physics, https://doi.org/10.1007/978-3-031-05808-0_7

7.1   Introduction

This chapter seeks to develop your ability to use mathematical modeling to describe living systems and to apply quantitative analyses to interpret biological data by focusing on how a number of quantities, relationships, and processes can indeed be modeled mathematically and thus how quantitative predictions can be developed for comparisons with data. In fact, we will encounter examples in which disparate processes are described by similar equations, which therefore have similar solutions. As a result, the understanding gained in one case can be transferred to the other. For example, in this chapter, we will see that the equations that describe the time-course of HIV viral load in a single patient can be re-cast to describe an atomic explosion; the equations that describe the number of a particular species of radioactive nuclei in a medically relevant radioactive decay chain are formally identical to equations that describe the occurrence of retinoblastoma—the most common childhood eye cancer—that result from somatic mutations; these equations are also similar to those that describe the serum concentration of an orally administered drug. Physicists love it when different physical phenomena can be described by similar equations.

   This chapter builds upon material from a number of previous chapters. It builds on Chap. 2. *Force and Momentum: Newton's Laws and How to Apply Them*: we will discuss an equation that describes the serum concentration of a drug during intravenous (IV) infusion, which is *formally* the same—i.e., is of the same mathematical form—as the equation of motion of a bead falling through a viscous fluid, that follows from applying Newton's second law to the bead's motion. It builds on Chap. 6. *Diffusion: Membrane Permeability and the Rate of Actin Polymerization*, because we will encounter processes—the passage of drugs from the GI tract into the bloodstream, for example—and phenomena—the existence of a critical mass of U^{235}, above which an atomic explosion occurs—which rely on diffusion for their conceptual understanding. It depends too on understanding gained in Chap. 3. *Energy, Work, Geckos, and ATP*, where we discussed the energy released when binding energy becomes more negative. Thus, kinetic energy is released when a U^{235} nucleus undergoes fission.

7.2   Your Learning Goals for This Chapter

By the end of this chapter, you should be able to

- Appreciate that it is possible to represent physical quantities, relationships and processes mathematically via "chemical rate equations."
- Recognize linear and non-linear equations.
- Recognize that exponential growth or decay often follows when the rate equations are linear equations or may be approximated as linear equations, and *vice versa*.
- Understand that different physical processes (including biological and medically relevant processes) can be described by similar equations and that similar equations have similar solutions.
- Understand and build simple mathematical models of physical processes, such as the initial progression of an infection, such as HIV, within an individual's body, the progression of a contagious disease, such as measles or the flu or COVID-19, through a susceptible population, and even the detonation of an atomic bomb.
- Solve linear equations that involve two coupled variables using the new concepts of *eigenvalues*, *eigenmodes*, and *eigenvectors*.
- Use direct substitution of a given trial solution to find a general solution, and then impose initial conditions to find the particular solution.
- Exploit the principle of superposition, namely that an arbitrary sum of two independent solutions of a set of linear differential equations is also a solution to that set of equations to find general solutions of coupled linear differential equations.
- Recognize that at long times, rate equations may realize a *steady-state*, in which the quantities of interest are constant in time.
- Read and interpret data plots and comparisons between data and theory.

7.3   Rates of Change

Previously, we saw that an object's acceleration is equal to the rate of change of its velocity, and that its velocity is equal to the rate of change of its position. In this chapter, we discuss rates of change again, but now in the context of "chemical reactions," and "chemical rate equations," which describe how the concentrations of "reactants" and "products" change in time. We have added the quotation marks, because these sorts of "chemical rate equation" actually have relevance far beyond chemistry, including for pharmacists, clinicians, epidemiologists, ecologists, physicists, *etc*. As we will see, this is because it is possible to write "chemical rate equations" that describe, for example, the dynamics of viral infections by the human immunodeficiency virus (HIV), the virus that causes AIDS. Similar equations also describe the progression of infectious diseases through human populations, drug concentrations in the body, population dynamics in populations of predators and prey or of hosts and parasites, and even the nuclear chain reaction that gives rise to the detonation of an atomic bomb! In all cases, the equations in question are differential equations that relate the time derivatives of the quantities of interest to the values of the quantities of interest.

7.4   Administering Therapeutics: Continuous Infusion with Elimination

7.4.1   The Same Equations Have the Same Solutions

In Chap. 2. *Force and Momentum: Newton's Laws and How to Apply Them*, we examined the motion of an object falling under the influence of a constant external force in a viscous fluid. We found that the velocity of a bead that is initially at rest relaxes exponentially with a characteristic time τ from an initial velocity of zero toward a terminal velocity (u). In fact, it turns out that the equation that applies to the motion of an object falling under the influence of a constant external force in a viscous fluid also can describe how the concentration of a chemical species varies as a function of time, if we suitably re-interpret the quantities appearing in the equation.

To appreciate this statement, let us consider a certain drug species (drug P) in the body. Let the number of molecules of drug P in the bloodstream at time t be $N = N(t)$. Intravenous infusion increases the number of molecules of drug P at a rate of α molecules per second. In addition, drug P is eliminated by the machinery of the body, such that the probability that any molecule of drug P is eliminated in one second is β. The total number of drug P molecules in the body that are eliminated per second is therefore βN, i.e., βN is the drug P elimination rate. It follows that the net change in the number of molecules of drug P (ΔN) in a time Δt is

$$\Delta N = \alpha \Delta t - \beta N \Delta t. \tag{7.1}$$

Dividing through by Δt:

$$\frac{\Delta N}{\Delta t} = \alpha - \beta N, \tag{7.2}$$

or in the limit of infinitesimal Δt:

$$\frac{dN}{dt} = \alpha - \beta N. \tag{7.3}$$

In comparison to Eq. 7.3, previously, we had for the velocity of a bead falling in a viscous fluid:

$$\frac{dv}{dt} = \frac{u}{\tau} - \frac{1}{\tau}v. \tag{7.4}$$

Comparing these equations, we see that if we replace $v(t)$ by $N(t)$, and if we replace u/τ by α, and if we replace $1/\tau$ by β, then these two equations (Eqs. 7.3 and 7.4) are formally the same, i.e., they have the same mathematical form. It follows that their solutions too must have the same mathematical form. This means that the chemical rate equation (Eq. 7.3) will show a "steady-state" number of drug P molecules at long times that *mathematically* corresponds to the terminal velocity in the falling bead problem. Both Eq. 7.3 and Eq. 7.4 are *linear* equations. Equation 7.3 is linear in that it involves N and $\frac{dN}{dt}$. The equation would remain linear if terms proportional to $\frac{d^2N}{dt^2}$ or $\frac{d^3N}{dt^3}$ were included. However, adding terms proportional to N^2, N^3, or $N\frac{dN}{dt}$ would yield an equation that is no longer linear.

7.4.2   Solution by Direct Substitution Revisited

When confronted with an equation containing derivatives, such as

$$\frac{dN}{dt} = \alpha - \beta N, \tag{7.5}$$

or equivalently

$$\frac{dN}{dt} + \beta N = \alpha, \tag{7.6}$$

or equivalently

$$N = \frac{\alpha}{\beta} - \frac{1}{\beta}\frac{dN}{dt}, \tag{7.7}$$

a highly satisfactory method from a mathematical point of view for finding the solution for $N(t)$ is to "guess" the solution and then prove that the guess is indeed correct. This is solution by direct substitution. Note that in the context of equations like Eq. 7.5, α and β are known quantities.
Our guess for the solution of Eq. 7.5 is

$$N(t) = A + Be^{\lambda t}, \tag{7.8}$$

where t is the time, and A, B, and λ are to-be-determined constants. Why might we expect this function to work? Eq. 7.7 informs us that $N(t)$ is a function, part of which is a constant, and part of which is proportional to its time-derivative. Exponential functions have the property that they are proportional to their derivatives. Therefore, it is indeed natural to try a function of the form of Eq. 7.8. To verify that this guess is correct and determine the values of A, B, and λ:

(a) First, we calculate

$$\frac{dN}{dt}, \tag{7.9}$$

with the result that

$$\frac{dN}{dt} = \lambda B e^{\lambda t}. \tag{7.10}$$

(b) Next, we calculate

$$\frac{dN}{dt} + \beta N, \tag{7.11}$$

using Eqs. 7.10 and 7.8:

$$\frac{dN}{dt} + \beta N = \lambda B e^{\lambda t} + \beta(A + B e^{\lambda t}) = \beta A + B(\beta + \lambda)e^{\lambda t}. \tag{7.12}$$

(c) Third, we compare Eqs. 7.6–7.12 in order to determine A and λ. For Eq. 7.8 to be the desired solution, we must insist that

$$\beta A + B(\beta + \lambda)e^{\lambda t} = \alpha. \tag{7.13}$$

The only way for Eq. 7.13 to be satisfied for all values of t is to pick

$$\lambda = -\beta \tag{7.14}$$

and

$$A = \frac{\alpha}{\beta}. \tag{7.15}$$

These are the desired values of A and λ.

(d) Then, we substitute these values of A and λ into Eq. 7.8 to find a general solution to Eq. 7.6, i.e., a solution that always satisfies Eq. 7.6:

$$N(t) = \frac{\alpha}{\beta} + B e^{-\beta t}. \tag{7.16}$$

This procedure is solution by direct substitution.

7.4.3   Initial Conditions

Our solution for $N(t)$, namely Eq. 7.16, still involves the to-be-determined constant B. How can we determine B? This requires that we know $N(0)$ or $N'(0)$. For example, suppose that we know that

$$N(0) = 0. \tag{7.17}$$

This is called an "initial condition." Then, we can determine the corresponding value of B and write the particular solution that satisfies both Eq. 7.6 and the initial condition, Eq. 7.17, by equating the two expressions that we now have for $N(0)$, namely Eqs. 7.17 and 7.16, evaluated for $t = 0$, i.e.,

$$N(0) = 0 = \frac{\alpha}{\beta} + B, \tag{7.18}$$

i.e.,

$$B = -\frac{\alpha}{\beta}. \tag{7.19}$$

It follows that

$$N(t) = \frac{\alpha}{\beta} - \frac{\alpha}{\beta}e^{-\beta t} = \frac{\alpha}{\beta}(1 - e^{-\beta t}), \tag{7.20}$$

which is the solution that we want. Wolfram Alpha provides the same solution, as may be seen at

http://www.wolframalpha.com/input/?i=solve+M%27%28t%29%3Da-b*M%28t%29%2C+M%280%29%3D0.[2]

Wolfram Alpha can also plot the solution. For example, the solution for $\alpha = 1$ and $\beta = \frac{1}{2}$ is shown at

http://www.wolframalpha.com/input/?i=plot+2%281-e%5E%28-0.5+t%29%29+from+t%3D0+to+10.[3]

Equation 7.20 shows that the serum number of molecules of drug P, administered by intravenous infusion at a rate α, starts at zero and then increases in a decaying exponential fashion to eventually reach a steady-state value of $\frac{\alpha}{\beta}$, where β is the

---

[2] http://www.wolframalpha.com/input/?i=solve+M%27%28t%29%3Da-b*M%28t%29%2C+M%280%29%3D0

[3] http://www.wolframalpha.com/input/?i=plot+2%281-e%5E%28-0.5+t%29%29+from+t%3D0+to+10

elimination rate of drug P, and corresponds to how quickly the drug achieves its steady-state value: if β is large (small), the steady-state value is achieved in a short (long) time.

7.5   Administering Therapeutics: Oral Dosage with Elimination

Of course, many therapeutics are administered, not by continuous infusion, but orally instead. For example, the solid circles in Fig. 7.1 show the measured serum concentration of sildenafil in healthy male volunteers as a function of time after oral administration. Also shown in this figure as the solid line is a best fit of a simple model to these data. (See Eq. 7.44 later in the chapter for the functional form of this model). Evidently, the model provides an excellent description of the data.

To describe the serum concentration of a medication quantitatively, we must build a mathematical model. To this end, first we note that a drug administered orally enters the bloodstream by diffusing through the gastrointestinal (GI) wall. The GI wall is an elaborate multilayer membrane, but we saw in Chap. 6. *Diffusion: Membrane Permeability and the Rate of Actin Polymerization* that, even for complicated multilayer membranes, the physics of diffusion gives rise to a current of particles across a membrane that is proportional to the concentration difference across the membrane, i.e., $I = -G(C_1 - C_2)$, where G is the conductivity of the membrane, I is the net number of particles per unit time going from the GI tract (concentration C_2) into the bloodstream (concentration C_1). To find the rate of change of serum concentration, corresponding to the particle current across the GI wall, we divide I by the blood volume, V, with the result that this contribution to the rate of change of serum concentration is $-\kappa(C_1 - C_2)$, where $\kappa = \frac{G}{V}$. Furthermore, assuming that the elimination rate of the drug in either the bloodstream or the GI tract is γ, and that the volume of the GI tract equals the blood volume, we arrive at two equations that describe the rate of change of C_1 and the rate of change of C_2:

$$\frac{dC_1}{dt} = -\gamma C_1 - \kappa(C_1 - C_2) = -(\gamma + \kappa)C_1 + \kappa C_2 \qquad (7.21)$$

Fig. 7.1 Serum concentration of sildenafil in healthy male volunteers as a function of time after oral administration

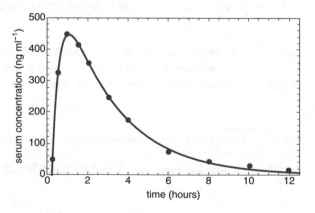

serum concentration (ng ml^{-1})

time (hours)

and

$$\frac{dC_2}{dt} = -\gamma C_2 + \kappa (C_1 - C_2) = -(\gamma + \kappa)C_2 + \kappa C_1, \tag{7.22}$$

which constitute two linear differential equations for the two variables C_1 and C_2, which we must now solve.

7.5.1 Eigenvalues, Eigenvectors, Eigenmodes, and Superposition

To solve Eqs. 7.21 and 7.22, it is necessary to introduce the concepts of "eigenvalues," "eigenvectors," "eigenmodes," and "superposition," which are encountered whenever we are faced with linear equations involving two or more variables, such as Eqs. 7.21 and 7.22. It turns out that these concepts are tremendously important in physics, and we will encounter multiple additional examples later in this chapter and in later chapters. Fortunately, we may specify a procedure for solving such eigenvalue problems:

1. Write the appropriate linear equations.
2. Assume an exponential time dependence with eigenvalue Λ, so that $\frac{dC_1}{dt}$ may be replaced by ΛC_1 and $\frac{dC_2}{dt}$ by ΛC_2.
3. Solve the resultant simultaneous equations for Λ and $\frac{C_1}{C_2}$.
4. For each eigenvalue, find the corresponding eigenvector, namely your choice of $(\frac{C_1}{C_2}, 1)$ or $(1, \frac{C_2}{C_1})$.
5. Construct a general solution as the superposition of all of the eigenmodes.
6. Match the general solution at $t = 0$ to given initial conditions, in order to determine the unknown constants that appear in the general solution to achieve the particular solution appropriate for the initial conditions in question.

We will now solve Eqs. 7.21 and 7.22 using this procedure.

1. Equations
We wrote the appropriate equations in Eqs. 7.21 and 7.22.

2. Exponential Time Dependence
Following our program, we assume that C_1 and C_2 both vary exponentially in time as $e^{\Lambda t}$, that is, we assume

$$C_1(t) = C_{01}e^{\Lambda t} \tag{7.23}$$

and

$$C_2(t) = C_{02}e^{\Lambda t}, \tag{7.24}$$

where C_{01} and C_{02} are constants, independent of time. It follows that

$$\frac{dC_1}{dt} = \Lambda C_{01}e^{\Lambda t} = \Lambda C_1 \tag{7.25}$$

and

$$\frac{dC_2}{dt} = \Lambda C_{02}e^{\Lambda t} = \Lambda C_2. \tag{7.26}$$

Substituting ΛC_1 for $\frac{dC_1}{dt}$ and ΛC_2 for $\frac{C_2}{dt}$ into Eqs. 7.21 and 7.22, these equations become

$$(\gamma + \kappa + \Lambda)C_1 = \kappa C_2 \tag{7.27}$$

and

$$(\gamma + \kappa + \Lambda)C_2 = \kappa C_1. \tag{7.28}$$

3. Eigenvalues

Multiplying the right hand sides and the left hand sides of these equations together, then dividing by $C_1 C_2$, and then taking the square root of both sides, we find:

$$(\gamma + \kappa + \Lambda) = \pm\kappa, \tag{7.29}$$

i.e.,

$$\Lambda = -\gamma - \kappa \pm \kappa. \tag{7.30}$$

Thus, we have the two eigenvalues: $\Lambda_+ = -\gamma$ and $\Lambda_- = -(\gamma + 2\kappa)$.

Why do we use exponential functions only as the trial solutions in this case without the constant term that we used in Eq. 7.16? The answer is that there is not a constant term in either of Eqs. 7.21 and 7.22, but there is in Eq. 7.3.

Corresponding to each of these eigenvalues is a particular value of $\frac{C_2}{C_1}$. Specifically, using Eq. 7.27, for $\Lambda = \Lambda_+ = -\gamma$, we have

$$\frac{C_2}{C_1} = \frac{\gamma + \kappa + \Lambda_+}{\kappa} = 1, \tag{7.31}$$

and for $\Lambda = \Lambda_- = -(\gamma + 2\gamma)$, we have

$$\frac{C_2}{C_1} = \frac{\gamma + \kappa + \Lambda_-}{\kappa} = -1. \tag{7.32}$$

4. Eigenvectors

How can we express the fact that for each eigenvalue, $\frac{C_2}{C_1}$ has a particular value? The answer is to envision a coordinate system in which C_1 is the horizontal axis and C_2 is the vertical axis, as illustrated in Fig. 7.2. Then, a definite value of $\frac{C_2}{C_1}$ corresponds to a definite angle with respect to the horizontal—call it θ_+ for $\frac{C_2}{C_1} = 1$ and θ_- for $\frac{C_2}{C_1} = -1$—for which $\tan \theta = \frac{C_2}{C_1}$. Each of these angles defines a definite direction, which we can therefore represent as a vector. Thus, for each eigenvalue, there is a corresponding eigenvector:

$$\mathbf{e} = (1, \tan \theta) = (1, \frac{C_2}{C_1}).$$
(7.33)

It follows that the eigenvectors in the current problem are

$$\mathbf{e}_+ = (1, 1)$$
(7.34)

and

$$\mathbf{e}_- = (1, -1).$$
(7.35)

These two eigenvectors are illustrated in Fig. 7.2. Similarly to when we were discussing vectors in the xy-plane, the first entry inside the parenthesis is the component of the eigenvector along the C_1-direction and the second entry is the component of the eigenvector along the C_2-direction. An equally valid choice for the eigenvector is

$$(\frac{C_1}{C_2}, 1),$$
(7.36)

which is parallel to $(1, \frac{C_2}{C_1})$, but has a different length and/or sign.

Fig. 7.2 Schematic of the $C_1 C_2$-plane and the two eigenvectors corresponding to our model of oral dosage with elimination

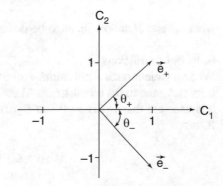

In this example, we first found the eigenvalues and then the eigenvectors. In some cases, the algebra works out so that it is necessary to find the eigenvalues and eigenvectors simultaneously.

5. Superposition
Writing out each eigenvector, multiplied by its corresponding time dependence (specified by the corresponding eigenvalue) we have

$$(C_1, C_2) = A\mathbf{e}_+ e^{\Lambda_+ t} = A(1, 1)e^{-\gamma t}, \tag{7.37}$$

and

$$(C_1, C_2) = B\mathbf{e}_- e^{\Lambda_- t} = B(1, -1)e^{-(\gamma+2\kappa)t}, \tag{7.38}$$

where A and B are constants to-be-determined. Eqs. 7.37 and 7.38 are the "eigen-modes" of Eqs. 7.21 and 7.22. Each eigenmode, namely each of Eqs. 7.37 and 7.38, satisfies both Eqs. 7.21 and 7.22.

An important feature of any set of *linear* equations is that if you have two solutions of the equations, then if you multiply each of those solutions by any constant factor and then sum these results, then that sum is also a solution to the set of linear equations in question. This is called *superposition*: if X and Y are solutions, then $aX + bY$ is also a solution, where a and b are arbitrary scalars. Superposition is applicable whenever we encounter linear equations, which is a lot in physics: the diffusion equation, the wave equation, which describes wave motion, and Maxwell's equations, which describe electromagnetism, are all linear equations, which realize superposition. It is also the case that *any and every* solution to a set of linear equations may be expressed as a linear combination of the eigenmodes. It follows that a completely general solution to Eqs. 7.21 and 7.22 is the superposition of the two eigenmodes, namely

$$(C_1, C_2) = A(1, 1)e^{-\gamma t} + B(1, -1)e^{-(\gamma+2\kappa)t}, \tag{7.39}$$

where A and B are constants to-be-determined.

6. Initial Conditions
When a patient takes a pill initially all of the drug molecules are in the patient's GI tract and none in the bloodstream. Therefore, the appropriate "initial conditions" in this case are that $C_1(0) = 0$ and $C_2(0) = C_0$, which can be expressed in vector form:

$$(C_1(0), C_2(0)) = (0, C_0). \tag{7.40}$$

At an arbitrary time, we have

$$(C_1, C_2) = A(1, 1)e^{-\gamma t} + B(1, -1)e^{-(\gamma+2\kappa)t}. \tag{7.41}$$

Evaluating this expression at $t = 0$ and setting it equal to $(0, C_0)$, we find

$$(0, C_0) = A(1, 1) + B(1, -1). \tag{7.42}$$

Solving for A and B yields $A = \frac{C_0}{2}$ and $B = -A$. Thus, the solution of Eqs. 7.21 and 7.22 for $C_1(t)$ and $C_2(t)$, which satisfies the specified initial conditions, is

$$(C_1, C_2) = \frac{C_0}{2}\left((1, 1)e^{-\gamma t} - (1, -1)e^{-(\gamma+2\kappa)t}\right), \tag{7.43}$$

or, equivalently, writing out the individual components:

$$C_1(t) = \frac{C_0}{2}\left(e^{-\gamma t} - e^{-(\gamma+2\kappa)t}\right) \tag{7.44}$$

and

$$C_2(t) = \frac{C_0}{2}\left(e^{-\gamma t} + e^{-(\gamma+2\kappa)t}\right). \tag{7.45}$$

Figure 7.1 shows the serum concentration of sildenafil in healthy male volunteers as a function of time after oral administration, together with a least mean squares fit of Eq. 7.44[4] to these data. To achieve a good fit, it was necessary to include in the model an additional delay time of 12 minutes after the pill is swallowed before the model form discovered in this section is applicable. We hypothesize that the additional 12 minutes delay is the time required to dissolve the pill in the stomach, which delays the drug being available in the GI tract. The best fit parameter values are $C_0 = 1420$ ng/ml, $\kappa = 1.175$ hr^{-1}, and $\gamma = 0.369$ hr^{-1}.

A key feature of $C_t(t)$ versus t—both data and model—is that the serum concentration increases from zero to a peak and then subsequently decreases eventually to zero. We may determine the time at which C_1 takes its maximum value, using the fact that $C_1(t)$ takes its maximum value when $dC_1/dt = 0$, i.e., for

$$\gamma e^{-\gamma t} = (\gamma + 2\kappa)e^{-(\gamma+2\kappa)t}. \tag{7.46}$$

Solving for this value of t, we find that

$$t = \frac{1}{2\kappa}\log\frac{\gamma + 2\kappa}{\gamma}. \tag{7.47}$$

---

[4] http://www.wolframalpha.com/input/?i=Plot+710.*%28exp%28-0.369*t%29-exp%28-%282*1.175%2B0.369%29*t%29%29+%28t+from+0+to+12%29

For the best fit parameter values, the peak occurs at $t = 0.84$ hours $= 50$ minutes. Adding 12 minutes to dissolve the pills, we find the peak serum concentration to occur at 62 minutes, which we infer is the time after ingestion at which sildenafil is likely to be most effective.

Is it possible to pick some bizarre initial conditions, which cannot be achieved by a linear combination of the two eigenmodes? The answer is "no." Any location (position vector) in the $C_1 C_2$-plane can be reached by an appropriate linear combination of the two eigenmodes.

7.5.2   Example: Tumor Suppressor Genes and Tumorigenesis: The Case of Retinoblastoma

Another situation in which eigenvalues and eigenvectors play a key role is in the context of "two-hit" models of tumorigenesis, for example, retinoblastoma. Retinoblastoma[5] is the most common childhood eye cancer, affecting approximately 300 patients per year in the USA. It occurs in a child's rapidly growing and dividing progenitor retina cells, called retinoblasts. Beyond about 5 years, retina growth is complete, all retinoblasts have differentiated into photoreceptor cells or nerve cells, and the occurrence of new retinoblastoma cases is extremely rare.

Retinoblastoma occurs as the result of mutations in a retinoblast RB1 genes. Ordinarily, each cell has two copies of RB1, and, in order for a retinoblastoma to develop, both copies must be mutated, so that no functional protein is produced by the cell in question. Mutation of both RB1 genes, and the resultant absence in the cell of functional protein, leads to uncontrolled cell growth and division. This is a retinoblastoma. The pivotal role of RB1 means that RB1 is a so-called tumor suppressor gene (TSG).

In fact, retinoblastoma patients fall into two major groups: patients with unilateral retinoblastoma—tumor in one eye—and patients with bilateral retinoblastoma—tumor in both eyes. Figure 7.3 recapitulates from Ref. [2] the number of patients with retinoblastoma plotted versus patient age, for patients with either unilateral retinoblastoma, shown as the stepwise, red curve, or bilateral retinoblastoma, shown as the stepwise, blue curve. Evidently, the two types of retinoblastoma show different dependences on patient age. The reason for the difference is illustrated schematically in Fig. 7.4 and is as follows. For the preponderance of patients with unilateral retinoblastomas, the disease arises as a result of mutations of both RB1 genes within a single somatic retinoblast. For these patients, the distribution in Fig. 7.3, therefore, reflects the fact that two sequential mutations are required for the disease to occur. By contrast, the preponderance of patients with bilateral retinoblastomas have a germline mutation, leading to one

---

[5]http://en.wikipedia.org/wiki/Retinoblastoma

(continued)

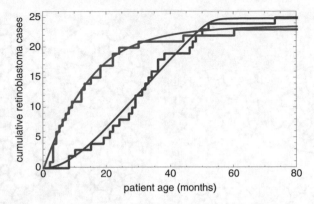

Fig. 7.3 Number of patients (N) with retinoblastoma plotted versus patient age in months for patients with unilateral retinoblastoma, shown as the red stepwise curve, and bilateral retinoblastoma, shown as the blue stepwise curve. The smooth red curve corresponds to the theoretical model, described in the text with $Nu_1u_2 = 0.0138$ per month per month. The smooth blue curve is the best fit exponential function. The data are from Ref. [2]

defective RB1 gene in every cell in the body. For these patients, therefore, the distribution in Fig. 7.3 reflects the fact that only a single mutation is required for disease onset. In fact, these data and the corresponding analysis—shown as the thinner smooth lines in Fig. 7.3—represent an important milestone in our understanding of cancer [2], including leading to the TSG concept [3].

In this section we will focus on patients with unilateral retinoblastoma for whom two mutations in a retinoblast are the cause of retinoblastoma, i.e., "two hits." We will suppose that in the first 5 or so years of life the growing retina contains a number of retinoblasts that grow and divide. For simplicity, we will suppose that when a retinoblast divides, one of the progeny cells differentiates into a photoreceptor cell or a nerve cell, and the other is preserved as a retinoblast to continue growing and dividing. In this way, the number of retinoblasts stays constant in time, while the retina is able to grow. At about age five, the retina becomes fully developed and the remaining retinoblasts stop dividing and differentiate. We will also suppose that the rate at which a retinoblast with two intact RB1 genes undergoes a mutation in one of its RB1 genes is u_1, and the rate at which a retinoblast with one already-mutated RB1 gene undergoes a mutation in its second RB1 gene is u_2. Finally, we will assume that once a retinoblast loses both functional RB1 genes, it very rapidly develops into a retinoblastoma.

With these assumptions, we can write the equations that relate the probabilities that a retinoblast (more precisely a line of successive retinoblasts) possesses two, one, or zero functional TSGs (P_2, P_1, and P_0, respectively) at patient age $t + \Delta t$ to these probabilities at patient age t:

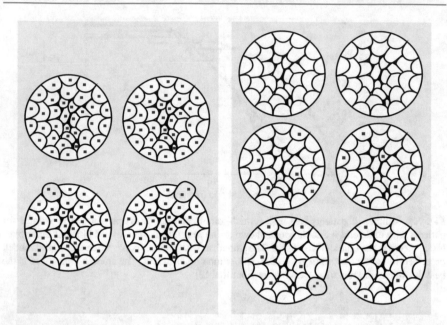

Fig. 7.4 Schematic of the genetic progressions involved in the retinas of retinoblastoma patients with RB1 germ line mutations (left panel) and RB1 somatic mutations (right panel) in their retinoblasts. The two retinas in the left panel of the figure correspond to the two retinas of a patient with an RB1 germ line mutation, resulting in mutated RB1 gene in every retinoblast at birth, depicted as the blue dots. Shown in the right panel is the situation for patients without a germline mutation. For these patients, at birth every cell possesses zero RB1 mutations and therefore two functional RB1 genes. The progression for both types of patients is shown with increasing age from top (birth) to bottom (retinoblastoma). Each red dot indicates a mutated RB1 gene, that appears at a replication event during the development of the patient's retina during the first several years of life. Two dots, either a blue dot and red dot or two red dots, in a single retinoblast indicate two RB1 mutations in that retinoblast, which subsequently leads to its uncontrolled growth and division, that is, retinoblastoma. In patients with a germline RB1 mutation a single subsequent mutation in any retinoblast leads to retinoblastoma, as shown in the lower pair of retinas in the left panel, where a case of bilateral retinoblastoma is indicated. By contrast, in patients without a germline RB1 mutation, two sequential RB1 mutations in the same retinoblast are needed in order to progress to retinoblastoma, as shown in the lowest pair of retinas in the right panel, which shows a case of unilateral retinoblastoma

$$P_2(t + \Delta t) = (1 - u_1 \Delta t) P_2(t), \tag{7.48}$$

$$P_1(t + \Delta t) = u_1 \Delta t P_2(t) + (1 - u_2 \Delta t) P_1(t), \tag{7.49}$$

(continued)

and

$$P_0(t + \Delta t) = P_0(t) + u_2 \Delta t \, P_1(t). \tag{7.50}$$

Equation 7.48 states that the probability that a retinoblast possesses two TSGs at patient age $t + \Delta t$ is the product of the probability that it possesses two TSGs at patient age t, multiplied by the probability that no mutation occurs in Δt, namely $1 - u_1 \Delta t$. Equation 7.49 states that the probability that a retinoblast possesses one TSGs at patient age $t + \Delta t$ is the product of the probability that it possesses one TSGs at patient age t, multiplied by the probability that no mutation occurs in Δt, namely $1 - u_2 \Delta t$, plus the probability that it possesses two TSGs at age t multiplied by the probability that a mutation occurs in Δt, namely $u_2 \Delta t$. Equation 7.50 states that the probability that a retinoblast possesses zero TSGs at patient age $t + \Delta t$ is the product of the probability that it possesses one TSGs at patient age t, multiplied by the probability that a mutation occurs in Δt, namely $u_2 \Delta t$, plus the probability that it possesses zero TSGs at age t. These equations are valid for patients up to about 5 years old. Beyond about age 5, there are no further changes, because the retina is then fully grown.

In the limit of small Δt, it follows that

$$\frac{P_2(t + \Delta t) - P_2(t)}{\Delta t} = \frac{dP_2}{dt} = -u_1 P_2, \tag{7.51}$$

$$\frac{P_1(t + \Delta t) - P_1(t)}{\Delta t} = \frac{dP_1}{dt} = u_1 P_2 - u_2 P_1, \tag{7.52}$$

and

$$\frac{P_0(t + \Delta t) - P_0(t)}{\Delta t} = \frac{dP_0}{dt} = u_2 P_1. \tag{7.53}$$

To solve Eqs. 7.51 and 7.52 for P_2 and P_1 as a function of time, t, we follow the procedure outlined below. Subsequently, we can predict P_0—the probability that a retinoblast has zero TSGs and thus develops into a retinoblastoma—versus patient age by integration of Eq. 7.53. A comparison of this prediction for P_0 to data is shown in Fig. 7.4.

Our procedure is as follows:

1. Write the appropriate linear equations.
   Eqs. 7.51   and 7.52   are the appropriate linear
   equations.
2. Assume an exponential time dependence with eigenvalue Λ.

(continued)

Thus, we assume that $P_2 = P_{20}e^{\Lambda t}$ and $P_1 = P_{10}e^{\Lambda t}$ in order to transform Eqs. 7.51 and 7.52 into algebraic equations, with the result that

$$\Lambda P_2 = -u_1 P_2, \qquad (7.54)$$

and

$$\Lambda P_1 = u_1 P_2 - u_2 P_1, \qquad (7.55)$$

or, equivalently,

$$(\Lambda + u_1)P_2 = 0, \qquad (7.56)$$

and

$$(\Lambda + u_2)P_1 = u_1 P_2. \qquad (7.57)$$

3. Solve the resultant simultaneous equations for Λ and $\frac{P_2}{P_1}$.
   The zero on the right-hand side of Eq. 7.56 means that we need a slightly different approach to finding the eigenvalues and vectors in this case. Inspection of Eq. 7.56 reveals that this equation can be satisfied in two ways. One way is to divide by P_2, revealing that $\Lambda = u_1$ is one of the solutions. Using $\Lambda = \Lambda_+ = -u_1$, we then find

$$\frac{P_2}{P_1} = \frac{u_2 - u_1}{u_1}. \qquad (7.58)$$

Alternatively, Eq. 7.56 is satisfied if $P_2 = 0$, or $\frac{P_2}{P_1} = 0$. Using $P_2 = 0$ in Eq. 7.57, we find the second eigenvalue:

$$\Lambda_- = -u_2. \qquad (7.59)$$

4. For each eigenvalue construct the corresponding eigenvector.
   Because $P_2 = 0$ in this case, it is necessary to choose the eigenvectors to be of the form $(1, \frac{P_2}{P_1})$ or $(\frac{P_2}{P_1}, 1)$, and not $(1, \frac{P_1}{P_2})$ or $(\frac{P_1}{P_2}, 1)$. Therefore, we write

$$\mathbf{e}_+ = (1, \frac{P_2}{P_1}) = (1, \frac{u_2 - u_1}{u_1}) \qquad (7.60)$$

(continued)

and

$$\mathbf{e}_- = (1, \frac{P_2}{P_1}) = (1, 0). \tag{7.61}$$

5. Construct a general solution as the superposition of all of the eigenmodes. The required linear superposition in this case is

$$(P_1(t), P_2(t)) = A(1, 0)e^{-u_2 t} + B(1, \frac{u_2 - u_1}{u_1})e^{-u_1 t}, \tag{7.62}$$

   where A and B are constants to be determined by the boundary conditions.
6. Match the general solution at $t = 0$ to given initial conditions in order to determine the unknown constants that appear in the general solution to achieve the particular solution appropriate for the initial conditions in question.

Answer: We have that at $t = 0$, $P_2(0) = 1$ and $P_1(0) = 0$. Therefore

$$(P_1(0), P_2(0)) = (0, 1) = A(1, 0) + B(1, \frac{u_2 - u_1}{u_1}). \tag{7.63}$$

It follows that

$$A + B = 0, \tag{7.64}$$

and

$$B \frac{u_2 - u_1}{u_1} = 1, \tag{7.65}$$

i.e.,

$$B = \frac{u_1}{u_2 - u_1}, \tag{7.66}$$

and

$$A = -\frac{u_1}{u_2 - u_1}. \tag{7.67}$$

Using these results for a and b, we finally have

(continued)

$$(P_1(t), P_2(t)) = -\frac{u_1}{u_2 - u_1}(1, 0)e^{-u_2 t} + (\frac{u_1}{u_2 - u_1}, 1)e^{-u_1 t}, \qquad (7.68)$$

which is the desired solution. In particular,
looking at the P_1 component, we have

$$P_1(t) = \frac{u_1}{u_2 - u_1}\left(-e^{-u_2 t} + e^{-u_1 t}\right). \qquad (7.69)$$

In fact, in the case of unilateral retinoblastoma, both $u_1 t \ll 1$ and $u_2 t \ll 1$. Therefore, it is pertinent to examine Eq. 7.69 for small $u_1 t$ and small $u_2 t$. Then, we may use Taylor series expansions:

$$e^{-u_2 t} \simeq 1 - u_2 t + \dots \qquad (7.70)$$

and

$$e^{-u_1 t} \simeq 1 - u_1 t + \dots \qquad (7.71)$$

It follows after some algebra that the rate of appearance of retinoblastoma is

$$\frac{dP_0}{dt} = u_2 P_1 \simeq u_1 u_2 t + \dots \qquad (7.72)$$

and, therefore, that the cumulative number of retinoblastoma cases at age t is

$$P_0(t) \simeq \frac{1}{2}u_1 u_2 t^2 + \dots \qquad (7.73)$$

Equation 7.73 corresponds to the thin red line in Fig. 7.3 for ages less than about 41 months, while for ages greater than about 41 months, our model for the cumulative number of retinoblastoma cases crosses over to a constant. This model well describes the observed number of cases versus patient age, demonstrating that the cumulative number of retinoblastoma cases increases quadratically with patient age up to about 41 months. This observation and its subsequent explanation in terms of "two hits," i.e., two sequential mutations, lead to the proposal for the existence of tumor suppressor genes (TSGs).

7.5.3 Example: Gene Expression

In this example, we will build a simple model for unregulated, procaryotic gene expression, that is, we will turn the cartoon shown in Fig. 7.5 into rate equations.

(a) Suppose that RNA polymerases produces messenger RNA (m) at a constant rate α_m, and that mRNA is degraded at a rate $\beta_m m$, that is proportional to m. Write an equation that specifies the rate of change of m.

Answer:

$$\frac{dm}{dt} = \alpha_m - \beta_m m. \tag{7.74}$$

(b) Suppose that ribosomes produce protein (p) at a rate $\alpha_p m$, that is proportional to m, and that they are degraded at a rate, $\beta_p p$ proportional to p. Write an equation that specifies the rate of change of p.

Answer:

$$\frac{dp}{dt} = \alpha_p m - \beta_p p. \tag{7.75}$$

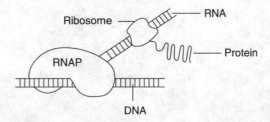

Fig. 7.5 Schematic of (unregulated) procaryotic gene expression. An RNA polymerase transcribes the DNA genetic code into messenger RNA. Subsequently, the messenger RNA is translated by a ribosome into protein. For procaryotic gene expression transcription and translation can essentially coincide, as shown in the figure. However, in the case of eukaryotic gene expression, transcription takes place inside the nucleus, while translation takes place outside the nucleus

(c) The correct answer to (a) is formally the same—i.e., has the same mathematical form—as Eq. 7.10. Therefore, it must have the same solution,
including a steady-state value of m, realized at long times—call it m^*.
Determine m^*.

Answer: The value of m^* is determined by setting
the left-hand side of EQ. 7.74 equal to zero, with
the result that

$$m^* = \frac{\alpha_m}{\beta_m}. \tag{7.76}$$

(d) Is there a steady-state value of p (p^*)? If so determine it.

Answer: There is a non-zero, steady-state value of
p, which we can find by setting the left-hand side
of EQ. 7.75 equal to zero, and then using m^*, with
the result that

$$p^* = \frac{\alpha_p m^*}{\beta_p} = \frac{\alpha_p \alpha_m}{\beta_p \beta_m}. \tag{7.77}$$

We see that the steady-state protein concentration
is proportional to the rate of RNA production
by RNA polymerase. Indeed, "repressors" and
"activators" implement gene regulation by
effectively modulating the rate of RNA production.

(e) There is an important difference between the equation found in (a), and
the two-coupled-variable equations, that we have encountered previously.
In contrast to those equations, Eq. 7.74 includes a constant term, that is
independent of m and p. This behavior stands in contrast to all of the other
terms which are proportional to either m or p. Nevertheless, even with the
addition of a constant term, our equations for gene expression are linear,
so that the principle of superposition is applicable. In practice, what this
means is that the general solution to Eq. 7.74 and Eq. 7.75 is the sum of
the steady-state solution, namely (m^*, p^*) and the solution to the same
equations, but with $\alpha_m = 0$. (This solution is reminiscent of the solution
for the drug concentration during IV infusion in the single-variable case.)
Find the eigenvalues and eigenvectors of the equations, found in (a) and
(b), for $\alpha_m = 0$.

(continued)

Answer: Taking $\alpha_m = 0$ and assuming an exponential time dependence as usual, Eqs. 7.74 and 7.75 become

$$(\Lambda + \beta_m)m = 0 \qquad (7.78)$$

and

$$(\Lambda + \beta_p)p = \alpha_p m. \qquad (7.79)$$

Similar to Eq. 7.56, the zero on the right-hand side of Eq. 7.78 means that this equation can be satisfied in two ways, either

$$\Lambda = \Lambda_+ = -\beta_m, \qquad (7.80)$$

or $m = 0$. In the first case, it follows that

$$\mathbf{e}_+ = (\frac{m}{p}, 1) = (\frac{\beta_p - \beta_m}{\alpha_p}, 1). \qquad (7.81)$$

For the second eigenmode, we have

$$\mathbf{e}_- = (0, 1) \qquad (7.82)$$

and

$$\Lambda_- = -\beta_p. \qquad (7.83)$$

These are the desired results.

(f) What is the general solution to your complete gene expression equations?

Answer: Adding the steady-state solution to our usual expression for the general solution in the absence of a constant term, we arrive at

$$(m(t), p(t)) = (m^*, p^*) + A(\frac{\beta_p - \beta_m}{\alpha_p}, 1)e^{-\beta_m t} + B(0, 1)e^{-\beta_p t}. \qquad (7.84)$$

(continued)

(g) At $t = 0$, gene expression is initiated by exposing the cells in question to a suitable so-called innducer. Find the particular solution in this case.

Answer: If there is no gene expression before $t = 0$, and it is then initiated at $t = 0$, we must take $p(0) = 0$ and $m(0) = 0$. It then follows that

$$(0, 0) = (m^*, p^*) + A(\frac{\beta_p - \beta_m}{\alpha_p}, 1) + B(0, 1). \tag{7.85}$$

Solving for A and B, we find

$$A = -\frac{\alpha_p m^*}{\beta_p - \beta_m} \tag{7.86}$$

and

$$B = \frac{\alpha_p m^*}{\beta_p - \beta_m} - p^*. \tag{7.87}$$

Thus,

$$(m(t), p(t)) = (m^*, p^*) - \frac{\alpha_p m^*}{\beta_p - \beta_m}(\frac{\beta_p - \beta_m}{\alpha_p}, 1)e^{-\beta_m t}$$

$$+ (\frac{\alpha_p m^*}{\beta_p - \beta_m} - p^*)(0, 1)e^{-\beta_p t}$$

$$= (\frac{\alpha_m}{\beta_m}(1 - e^{-\beta_m t}), \frac{\alpha_p \alpha_m (\beta_m(1 - e^{-\beta_p t}) - \beta_p(1 - e^{-\beta_m t})}{\beta_m \beta_p (\beta_m - \beta_p)}.$$

$$\tag{7.88}$$

(h) What is the behavior of $m(t)$ and $p(t)$ at early times?

Answer: Expanding for small t, we find

$$(m(t), p(t)) = (\frac{\alpha_m}{\beta_m}(1 - e^{-\beta_m t}), \frac{\alpha_p \alpha_m (\beta_m(1 - e^{-\beta_p t}) - \beta_p(1 - e^{-\beta_m t})}{\beta_m \beta_p (\beta_m - \beta_p)}$$

$$\simeq (\alpha_m t, \frac{1}{2}\alpha_p \alpha_m t^2).$$

$$\tag{7.89}$$

7.6   Progression of HIV Infection in an Individual Patient

We now turn to a discussion of the progression of human immunodeficiency virus (HIV) infection within an individual patient within the first few days/weeks of infection. HIV,[6] of course, is the virus that eventually causes acquired immunodeficiency syndrome (AIDS). The main target of HIV is a type of white blood cell called a $CD4^+$ T cell[7]—we will call them T cells, for brevity—which are essential for a robust immune response. How the number of uninfected T cells, the number of HIV infected T cells, and the number of free virus particles behave and interact is illustrated schematically in the cartoon "reaction scheme" shown in Fig. 7.6.

If untreated, after a much longer period of time than we consider here, the number of T cells gradually falls, eventually leading the patient to be susceptible to opportunistic infections. This is AIDS. The story of the early years of the AIDS epidemic is brilliantly told in Randy Shilts' 1987 book "And The Band Played On"[8] [4]. However, here we focus on the progression of the HIV infection in the first few days and weeks, well before the infected individual progresses to AIDS.

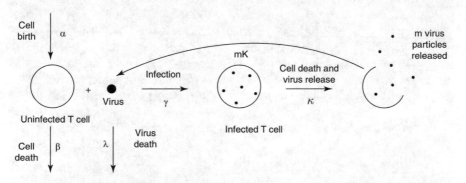

Fig. 7.6 Cartoon of viral infection, following Ref. [5]. In brief, a free virus particle first invades an uninfected T-cell at a rate γ; the virus replicates inside the T-cell, which then dies at a rate κ, thus releasing m copies of the virus; the released virus particles then proceed to infect other uninfected T-cells, and so on

[6] http://en.wikipedia.org/wiki/HIV

[7] http://en.wikipedia.org/wiki/T_helper_cell

[8] http://www.amazon.com/And-Band-Played-On-20th-Anniversary/dp/0312374631/ref=sr_1_1?
ie=UTF8&qid=1347572043&sr=8-1&keywords=And+the+band+played+on

7.6.1   Example: Building a Model of HIV Infection

We will now construct a set of "chemical rate equations" that constitute a mathematical model of the dynamics of HIV[9] infection within an infected individual. Our model involves three variables, namely $U = U(t)$, the number of uninfected T cells in the body, $I = I(t)$, the number of infected T cells in the body, and $V = V(t)$, the number of free virus particles in the body, all of which are functions of the time, t, after infection by V_0 virus particles at $t = 0$. How these quantities behave and interact is illustrated schematically in the cartoon "reaction scheme" of Fig. 7.6. We will proceed by turning this schematic representation of the relevant biologically processes into a corresponding mathematical representation as follows:

First, we examine what contributes to the rate of change of the number of uninfected cells, U.

- Uninfected T cells are produced by the body at a rate α.
- In the absence of virus, uninfected T cells die naturally at a rate βU.
- Uninfected T cells turn into infected T cells at a rate $\gamma V U$. The factor of V reflects the fact that the more virus particles there are, the greater the infection rate of uninfected cells. It is important to note that the parameter γ subsumes all of the microscopic details of how a virus particle come into contact with a T cell and then inveigles its way into the doomed T cell.

It follows that the total rate of change of the number of uninfected cells, U, is:

$$\frac{dU}{dt} = \alpha - \beta U - \gamma U V. \tag{7.90}$$

Next, we examine the rate of change of the number of infected cells, I.

- The number of infected cells increases by one when an uninfected T cell become infected. Therefore, there is a contribution to the rate of change of infected cells of $\gamma V U$, equal and opposite to the rate of infection of uninfected T cells.
- Infected T cells die at a rate κI, releasing m virus particles each time an infected T cell dies, and increasing the number of free virus particles correspondingly.

---

[9]http://en.wikipedia.org/wiki/HIV

(continued)

It follows that the total rate of change of I is:

$$\frac{dI}{dt} = \gamma UV - \kappa I. \tag{7.91}$$

Finally, we examine the rate of change of the number of free virus particles, V.

- The number of free virus particles decreases by one when an uninfected T cell become infected. Therefore, there is a contribution to the rate of change of free virus particles of $-\gamma VU$, equal and opposite to the rate of infection of uninfected T cells.
- Infected T cells die at a rate κI, releasing m virus particles each time an infected T cell dies, thus increasing the number of free virus particles at a rate $m\kappa I$.
- Finally, free virus particles are eliminated by the body at a rate λV.

Therefore, the total rate of change of V is:

$$\frac{dV}{dt} = m\kappa I - \lambda V - \gamma VU. \tag{7.92}$$

These three equations—Eqs. 7.90, 7.91 and 7.92—provide a way to transform the schematic understanding of HIV progression, represented by Fig. 7.6, into a quantitative description. One way to think about this sort of mathematical model is that it is a hypothesis, whose predictions are to be tested against experiments, like any hypothesis.

7.6.2 Numerical Solution of HIV Infection Model

Because Eqs. 7.90, 7.91, and 7.92 contain terms consisting of the product $U \times V$, they are "non-linear." By contrast, linear equations would involve only terms linear in U, I, and V. Non-linear equations are often challenging to solve analytically. However, as before, numerical solutions to non-linear equations can be found computationally, using Mathematica, for example. If we suppose that an individual is exposed to virus and becomes infected at $t = 0$, then, we can visualize the predicted progression of the individual's HIV infection by having Mathematica solve Eqs. 7.90, 7.91, and 7.92 for $U(t)$, $I(t)$, and $V(t)$ for given initial conditions. Such a solution is presented in the Wolfram Demonstration at HIVProgression-Ch07.cdf and in Fig. 7.7, using a linear time axis in the horizontal, and a *logarithmic* number axis in the vertical. In both the demonstration and the figure, the red line is the number of uninfected cells, the green line is the number of infected cells, and the orange line is the number of free virus particles.

HIV Progression

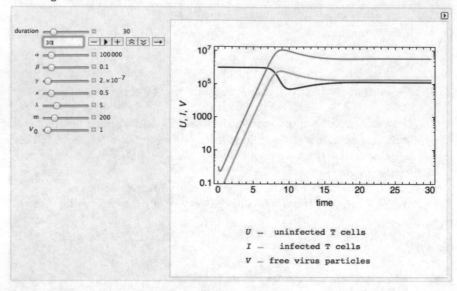

Fig. 7.7 Predicted numbers of uninfected T cells (red), infected T cells (green) and free virus particles (orange), according to the model of Eqs. 7.90, 7.91, and 7.92 as a function of time following infection. The curves shown correspond to the parameters, listed on the left, and which are approximately appropriate for SIV in macaques (Fig. 7.8). The parameters might be different for HIV in humans. In the demonstration such changes can be accounted for with the sliders

For the initial parameters of the simulation, both the number of infected cells (green curve) and the number of free virus particles (orange curve) increase exponentially as a function of time. We can deduce the exponential increase at early times, because at early times both these curves appear as straight lines on the plot which has a linear time axis along the horizontal and a logarithmic vertical U, I, V axis. After a period of exponential growth, both I and V reach a peak in viral load, which corresponds to so-called acute HIV infection. The peak is followed by a decrease to constant steady-state values. Correspondingly, U initially remains essentially constant, then decreases to a minimum, during acute HIV infection, and then increases again to its own constant steady-state value.

7.6.3   Steady-State Solutions: Living with HIV

What we mean by a steady-state is that nothing changes in time. The steady-state values of U, I, and V can be determined by setting the left hand sides of Equations

7.90, 7.91, and 7.92 equal to zero and solving for the corresponding values of U, I, and V. Thus, one can show that[10]

$$U = \frac{\lambda}{\gamma(m-1)},$$ (7.93)

$$I = \frac{\alpha\gamma(m-1) - \beta\lambda}{\kappa\gamma(m-1)},$$ (7.94)

and

$$V = \frac{\alpha\gamma(m-1) - \beta\lambda}{\gamma\lambda}.$$ (7.95)

These values correspond to the patient "living with HIV."

7.6.4   HIV Infection Predictions Compared to Experiments

How do these predictions compare to experiment? Obviously, it is not appropriate to deliberately infect humans with HIV, but the levels of simian immunodeficiency virus-associated (SIV-associated) RNA in twelve macaques[11] following their inoculation with SIV by Nowak et al.[12] [6] are shown in Fig. 7.8. These data, which may be taken as a proxy for the total amount of virus in each macaque, reproduce an exponential growth at short times, a peak in the number of virus at intermediate times, and seem consistent with a constant, steady-state level of virus at long times. Evidently, the model prediction seems to be strongly supported by these experiments! The mathematical description of the SIV levels in a macaque actually works! How cool is that? In humans, the peak in viral load and the dip in the number of uninfected T cells at early times corresponds to "acute" HIV infection[13] and actually occurs 2 to 4 weeks post exposure. The subsequent steady-state is "chronic HIV." AIDS develops on the time scale of years and is outside our current model.

7.7   Nuclear Chain Reactions and Atomic Bombs

Surprisingly, another significant application of equations closely similar to Eq. 7.90 through 7.92 is to an atomic explosion (Fig. 7.9)! A schematic representation of the

---

[10] http://www.wolframalpha.com/input/?i=Solve+A-B*U-G*U*V%3D0%2C+G*U*V-K*J
%3D0%2C+M*K*J-L*V-G*V*U%3D0+for+U%2C+J%2C+V

[11] http://en.wikipedia.org/wiki/Macaques

[12] http://www.ncbi.nlm.nih.gov/pmc/articles/PMC192098/

[13] http://en.wikipedia.org/wiki/AIDS#Acute_infection

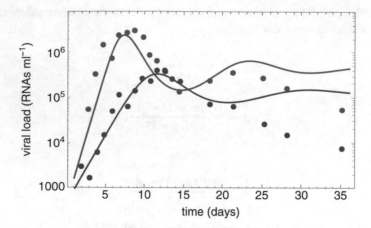

Fig. 7.8 Levels of simian immunodeficiency virus-associated RNA in two macaques following their inoculation with SIV (red and blue circles), adapted from Ref. [6]. The lines are plots of our mathematical model for the number of virus particles versus time, using parameters chosen to approximately describe the data

Fig. 7.9 The "Baker" atomic explosion, carried out by the USA military at Bikini Atoll, Micronesia, July 25 1946. Public domain image from https://en.wikipedia.org/wiki/File: Operation_Crossroads_Baker_(wide).jpg

nuclear chain reaction, which leads to an atomic explosion, is given in Fig. 7.10. Such a chain reaction starts when a neutron is captured by a U^{235} nucleus. The resultant U^{236} nucleus is unstable to nuclear fission[14] and soon splits into fission fragments, releasing several additional neutrons. These neutrons are then captured

---

[14] http://en.wikipedia.org/wiki/Nuclear_fission

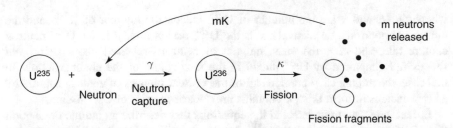

Fig. 7.10 Schematic cartoon representation of chain reaction involved in an atomic explosion: a U^{235} nucleus captures a neutron. The resultant U^{236} nucleus undergoes fission into a number of fission fragments, simultaneously releasing a number of neutrons and a large amount of kinetic energy. The released neutrons are themselves captured by other U^{235} nuclei, leading to further fission reactions, and so on. The most prevalent fission fragments are Ba^{141} and Kr^{92} but U^{235} fission sometimes produces other nucleii

by other U^{235} nuclei and the process repeats, multiplied several fold, and so on. Because each fission event is accompanied by an energy release of about 180 MeV to kinetic energy of the fission fragments, equivalent to 70 TeraJoules/kg of U^{235} (1 TeraJoule = 10^{12} J), the ensuing chain reaction leads to an atomic explosion.[15]

The story of the discovery of nuclear fission in Europe in the late 1930s, of the personalities involved—Lise Meitner[16] and Otto Hahn,[17] in particular—and of the scientific community's response is fascinating from a historical and sociological perspective [7]. The US Government's eventual response to the discovery of nuclear fission, namely the Manhattan Project,[18] profoundly shaped world history, and lead to physicist J. Robert Oppenheimer to make the statement quoted at the beginning of the chapter [8].

Using similar reasoning as for HIV progression, the rate equations, corresponding to the processes illustrated in Fig. 7.10, are as follows:

$$\frac{dQ}{dt} = -\gamma Qn, \tag{7.96}$$

$$\frac{dU}{dt} = \gamma Qn - \kappa U, \tag{7.97}$$

and

$$\frac{dn}{dt} = m\kappa U - \gamma nQ, \tag{7.98}$$

[15] http://www.youtube.com/watch?v=rxIIDmuj8ZE

[16] http://en.wikipedia.org/wiki/Lise_Meitner

[17] http://en.wikipedia.org/wiki/Otto_Hahn

[18] http://en.wikipedia.org/wiki/Manhattan_Project

where Q, U, and n are the number of U^{235} nuclei, the number of U^{236}, and the number of neutrons, respectively, κ is the U^{236} fission rate, γ is the U^{235}-neutron capture rate, and m is the mean number of neutrons released upon fission *and* successfully captured by U^{235} nuclei in the next round of the chain reaction. In this case, the parameter γ contains all of the nuclear physics of neutron capture by a U^{235} nucleus to form U^{236}; κ contains the nuclear physics of U^{236} fission.

In fact, one way to understand the equations that describe an atomic detonation (Eqs. 7.96 through 7.98) is to return to the equations that describe HIV progression (Eqs. 7.90 through 7.92) and re-interpret the number of uninfected T cells as the number of U^{235} nuclei, the number of infected T cells as the number of U^{236} nuclei, and the number of free virus particles as the number of free neutrons, that is, we set $U = Q$, $I = U$, and $V = n$. In addition, because U^{235} nuclei do not spontaneously appear, we may set $\alpha = 0$; because the half-life of the radioactive decay of U^{235} is 700 million years, we can set its decay rate to zero on the time scale of an atomic explosion, which is very fast (microseconds or less), that is, $\beta = 0$; and because the lifetime of the neutron is 11 minutes, which is also much longer than microseconds, we can also set the decay rate of the neutron equal to zero on this time scale, that is, $\lambda = 0$. With these assignments, we achieve Eqs. 7.96 through 7.98. Evidently HIV progression and a nuclear chain reaction—two completely different phenomena—nevertheless are described by very similar equations.

Because HIV progression and a nuclear chain reaction are described by very similar equations, the solutions will necessarily be similar. In the case of HIV infection, we saw that at early times, the number of infected T cells and the number of virus particles grows exponentially, while the number of uninfected T cells remains approximately constant. Analogously, in the nuclear chain reaction case, at early times, we may expect an exponential growth in the number of neutrons and U^{236} nuclei, while the number of U^{235} nuclei remains approximately constant. These features are apparent in the numerical solution of Eqs. 7.96 through 7.98, which is carried out in the Mathematica Demonstration at NuclearChainReaction-Ch07.cdf.

7.7.1   Example: Linear Chain Reaction Equations at Early Times

To analytically investigate the early time, exponential behavior relevant for atomic bombs, we may exploit the fact that the number of U^{235} nuclei, i.e., Q, is effectively constant at early times. Then, the above equations become

$$\frac{dU}{dt} = \gamma Q n - \kappa U, \tag{7.99}$$

and

$$\frac{dn}{dt} = m\kappa U - \gamma Q n, \tag{7.100}$$

(continued)

where $n = n(t)$ and $U = U(t)$ are variable, but now we take Q to be constant. Because Q is now taken to be constant, Eqs. 7.99 and 7.100 are two *linear* equations for U and n, which may therefore be solved analytically.

(a) Determine the eigenvalues of Eqs. 7.99 and 7.100. Call them Λ_+ and Λ_-. Using

$$\frac{dU}{dt} = \Lambda U_0 e^{\Lambda t} = \Lambda U(t) = \Lambda U \qquad (7.101)$$

and

$$\frac{dn}{dt} = \Lambda n_0 e^{\Lambda t} = \Lambda n(t) = \Lambda n, \qquad (7.102)$$

in Eqs. 7.99 and 7.100, we find

$$\Lambda U = \gamma Q n - \kappa U \qquad (7.103)$$

and

$$\Lambda n = m\kappa U - \gamma Q n. \qquad (7.104)$$

To solve Eqs. 7.111 and 7.112 for Λ and $\frac{n}{U}$, it is convenient to rewrite them as

$$(\Lambda + \kappa)U = \gamma Q n \qquad (7.105)$$

and

$$(\Lambda + \gamma Q)n = m\kappa U. \qquad (7.106)$$

Multiplying the left hand side of Eq. 7.105 by the left hand side of Eq. 7.106 and setting this product equal to the right hand side of Eq. 7.105 multiplied by the right hand side of Eq. 7.106, and then dividing by Un on both sides we find a single equation for Λ, in terms of "given quantities" only (γQ, κ, and m), namely

$$(\Lambda + \kappa)(\Lambda + \gamma Q) = \gamma Q m \kappa \qquad (7.107)$$

or

$$\Lambda^2 + (\kappa + \gamma Q)\Lambda - (m - 1)\kappa \gamma Q = 0. \qquad (7.108)$$

(continued)

Eq. 7.108 is a quadratic equation for Λ. Therefore, there are two possible solutions for Λ, which we will call Λ_+ and Λ_-, and which are[19]

$$\Lambda_\pm = \frac{1}{2}\left(\pm\sqrt{(\gamma Q + \kappa)^2 + 4\gamma Q\kappa(m-1)} - (\gamma Q + \kappa)\right). \qquad (7.109)$$

(b) Determine the corresponding values of $\frac{u}{U}$

Corresponding to each eigenvalue is a particular value of $\frac{n}{U}$. Specifically, using Eq. 7.105, we find that for $\Lambda = \Lambda_\pm$,

$$\frac{n}{U} = \frac{\Lambda_\pm + \kappa}{\gamma Q} = \frac{\frac{1}{2}\left(\pm\sqrt{(\gamma Q + \kappa)^2 + 4\gamma Q\kappa(m-1)} + \kappa - \gamma Q\right)}{\gamma Q}.$$

$$(7.110)$$

(c) The results for (a) and (b) are cumbersome. To simplify the algebra, henceforth, suppose, that $m = 3$, which is realistic, and that γQ is equal to 2κ. (The dimensions of γQ and κ are both inverse time.) What are the eigenvalues in this case?

The two equations that we must solve now are

$$\Lambda U = 2\kappa n - \kappa U \qquad (7.111)$$

and

$$\Lambda n = 3\kappa U - 2\kappa n, \qquad (7.112)$$

i.e.,

$$(\Lambda + \kappa)U = 2\kappa n \qquad (7.113)$$

and

$$(\Lambda + 2\kappa)n = 3\kappa U. \qquad (7.114)$$

Thus,

$$(\Lambda + 2\kappa)(\Lambda + \kappa) = 6\kappa^2, \qquad (7.115)$$

which leads to $\Lambda_+ = \kappa$, and $\Lambda_- = -4\kappa$.

---

[19]http://www.wolframalpha.com/input/?i=Solve+%28L%2F%28m*k%29%29%2B1%2Fm%29*%28L%2FG%2B1%29-1+%3D+0+for+L

(continued)

(d) Determine the eigenvectors.

For $\Lambda_+ = \kappa$, we find $\frac{n}{U} = 1$, and, for $\Lambda_- = -4\kappa$, we find $\frac{n}{U} = -\frac{3}{2}$. The corresponding eigenvectors are

$$\mathbf{e}_+ = (1, 1), \tag{7.116}$$

and

$$\mathbf{e}_- = (1, -\frac{3}{2}). \tag{7.117}$$

(e) Given initial conditions of a single neutron, $n(0) = 1$, and zero U^{236} nuclei, $U(0) = 0$, calculate how n and U subsequently vary in time. The general solution to our linearized atomic bomb equations is the superposition of the two eigenmodes, namely

$$(U, n) = A(1, 1)e^{\kappa t} + B(1, -\frac{3}{2})e^{-4\kappa t}. \tag{7.118}$$

At time zero $(t = 0)$ there are zero U^{236} nuclei and one neutron. Therefore, we have $U(0) = 0$ and $n(0) = 1$ or $(U, n) = (0, 1)$ at $t = 0$. It follows that we can write $(U, n) = (0, 1)$ for the LHS of Eq. 7.118, while on the RHS, we set $t = 0$ to find:

$$(0, 1) = A(1, 1) + B(1, -\frac{3}{2}) = A(1, 1) + B(1, -\frac{3}{2}). \tag{7.119}$$

Equating components, we thus find:

$$0 = A + B \tag{7.120}$$

and

$$1 = A - \frac{3}{2}B. \tag{7.121}$$

Solving these equations, we find

$$A = \frac{2}{5} \tag{7.122}$$

(continued)

and

$$B = -\frac{2}{5}. \tag{7.123}$$

Thus, the particular solution in this case is

$$(U, n) = \frac{2}{5}\left((1, 1)e^{\kappa t} - (1, -\frac{3}{2})e^{-4\kappa t}\right), \tag{7.124}$$

or

$$U(t) = \frac{2}{5}e^{\kappa t} - \frac{2}{5}e^{-4\kappa t} \tag{7.125}$$

and

$$n(t) = \frac{2}{5}e^{\kappa t} + \frac{3}{5}e^{-4\kappa t}. \tag{7.126}$$

We can check that $n(0) = 1$ and $U(0) = 0$. We can also check that $U(t)$ and $n(t)$ are both positive for all values of t, as required for a physically sensible number of U^{236} nuclei and neutrons.

(f) Explain how the solution can represent an atomic explosion.
The eigenmode with the positive eigenvalue corresponds to exponential growth in the number of neutrons and U^{236} nuclei in time. Since each fission event that follows the creation of a U^{236} nucleus releases about 200 MeV in the form of kinetic energy of its fission products, there is an exponentially increasing energy release, which constitutes the atomic explosion.

7.8   Chapter Outlook

Looking forward, this chapter introduced a number of mathematical concepts and techniques that are critical to later chapters. We will invoke the steady-state of chemical rate equations in *8 Statistical Mechanics: Boltzmann factors, PCR, and Brownian ratchets*. We will encounter linear equations, eigenvalues, eigenvectors, eigenmodes, superposition, and initial conditions, in *10 Oscillations and resonance*, *11 Wave equations: strings and wind*, and *16 Biologic: Genetic circuits and feedback*. Exponential decay and exponential growth as a function of time are a particular focus of this chapter. We will focus on exponential decays again in *14 Circuits and dendrites: Charge conservation, Ohm's law, rate equations, and other old friends*. More generally, as made clear by the diverse examples presented in the chapter, the process of building mathematical models of physical and biological processes in order to make a hypothesis precise is pervasive across the sciences from ecology to epidemiology, to cell biology, to chemistry, to physics, to climate science.

7.9 Problems

Problem 1: And the Band Played On
Set aside some time to read Randy Shilts acclaimed and best-selling account of the early years of the AIDs epidemic, *And the band played on: Politics, people, and the AIDS epidemic*.[20] This important book should be read by everyone, but especially by clinicians and medical researchers. Briefly describe what you know about human immunodeficiency virus (HIV). Explain what is acute HIV and what is chronic HIV.

Problem 2: Administering Therapeutics: IV Bolus
The equations that describe the number of molecules of a drug in the body in the case that C molecules of the drug are all injected intravenously at $t = 0$ (IV bolus) are:

$$N(0) = C, \tag{7.127}$$

and

$$\frac{dN}{dt} = -\beta N \tag{7.128}$$

which have the solution:[21]

$$N(t) = Ce^{-\beta t}. \tag{7.129}$$

In this equation, β is the elimination rate of the drug in question, which is related to the characteristic time, τ, that we saw in Chap. 1, via $\beta = 1/\tau$. A commonly used alternative to τ or β is the "half life",[22] $t_{\frac{1}{2}}$, which is defined as the time in which the drug concentration falls by a factor of 2. Thus, by definition we have that

$$e^{-\beta t_{\frac{1}{2}}} = \frac{1}{2} \tag{7.130}$$

---

[20] https://www.amazon.com/And-Band-Played-On-20th-Anniversary/dp/0312374631/ref=sr_1_1?ie=UTF8&qid=1347572043&sr=8-1&keywords=And+the+band+played+on

[21] http://www.wolframalpha.com/input/?i=Solve+M%27%28t%29+%3D+-b*M%28t%29%2C+M%280%29%3DC

[22] http://en.wikipedia.org/wiki/Biological_half-life

or

$$t_{\frac{1}{2}} = \frac{\log_e 2}{\beta},$$ (7.131)

which is the half-life, expressed in terms of β.

(a) Given that the half-life of morphine is 2 to 3 hours, what is its elimination rate?
(b) The elimination rate of intravenously administered epinephrine (adrenaline) is 0.01 s^{-1}. What is its half-life?

Problem 3: Bacterial Log-Phase Growth
For a small population of bacteria, which originates with a single parent bacterium, bacterial divisions are initially more-or-less synchronized. However, in large bacterial populations, the synchronicity disappears, and we may characterize the population growth via the bacterial division rate, k.

(a) In a large bacterial population in vigorously shaken, rich medium, what is the change in the number of bacteria (ΔN) in a time Δt?
(b) What then is the rate of change in the number of bacteria ($\frac{dN}{dt}$)?
(c) Show that a function of the form $N(t) = A + Be^{\lambda t}$ is a solution to your equation for $\frac{dN}{dt}$ and find the values of λ and A.
(d) Given that there was one bacterium at $t = 0$, express the number of bacteria at time t in terms of the bacterial generation time, τ.

Problem 4: Extracting the Growth Rate
A certain population of bacteria grows exponentially in time as

$$N(t) = N_0 e^{\kappa t},$$

where N_0 is the number of initial bacteria at time $t = 0$ and κ is what we call the rate of growth. The dimensions of κ are inverse time (1/time). A sketch of such a population growth is shown in Fig. 7.11. Extract from this graph estimates of

1. The initial number of bacteria N_0.
2. The growth rate κ (in units of 1/hours).

For both (1) and (2), explain how you determined the desired quantities from the graph.

Fig. 7.11 Figure for "Extracting the growth rate"

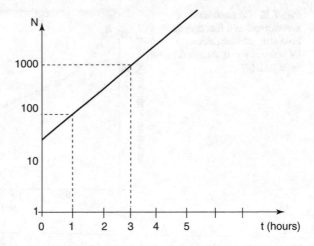

You may find it useful to recall the following properties of the logarithm:

$$\ln e^A = A, \qquad \log_{10} 10^A = A$$

$$\ln(AB) = \ln A + \ln B, \qquad \log_{10}(AB) = \log_{10} A + \log_{10} B$$

where "ln" is the natural logarithm (logarithm base e) and $\log_{10}$ is the logarithm, base 10. To change between logarithms of different bases (for instance, from base d to base b) the rule is:

$$\log_b a = \frac{\log_d a}{\log_d b}.$$

Therefore, if $b = 10$ and $d = e$, the relation between the logarithm base 10 and the natural log ($\log_e = \ln$) is

$$\log_{10} a = \frac{\log_e a}{\log_e 10} = \frac{\log_e a}{2.3} = \frac{\ln a}{2.3}.$$

Problem 5: Serum Antibiotic Concentration

Vancomycin is an antibiotic, recommended for infections, caused by methicillin-resistant *Staphylococcus aureus*, including bloodstream, bone and joint infections, and meningitis. Figure 7.12 shows the serum concentration of vancomycin in a healthy individual as a function of time after administration by IV bolus at $t = 0$ (i.e., the dose is all injected into the bloodstream at once at $t = 0$).

Fig. 7.12 Concentration of vancomycin as a function of time after administration by IV bolus at $t = 0$. Adapted from Ref. [9]

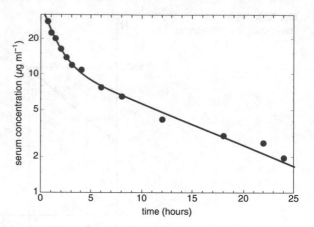

To understand these data, it is valuable to conceive the patient as consisting of two compartments. The first compartment consists of a number of "slow" organs, in particular the brain and lungs, into which vancomycin diffuses slowly. In this compartment, the concentration of vancomycin is $O = O(t)$. The second compartment consists of the patient's bloodstream and the patient's other, "fast" organs. In this compartment, the concentration of vancomycin is $B = B(t)$. The rate at which vancomycin passes from the bloodstream and fast organs to the slow organs is $2\alpha(B - O)$, where α is a constant. In addition, vancomycin is eliminated from the body by the kidneys (which are fast organs) at a rate $3\alpha B$.

(a) Explain why you might expect the rate at which vancomycin passes from the bloodstream to the slow organs to be proportional to $B - O$.

(b) Write the equations that describe the time evolution of $B(t)$ and $O(t)$.

(c) Determine the eigenvalues of these equations.

(d) Determine the corresponding eigenvectors.

(e) Given that the initial conditions are $B(0) = V$ and $O(0) = 0$, determine $B(t)$ and $O(t)$ as a function of time.

(f) Explain what it means that the concentration in Fig. 7.12 appears to show (1) linear behavior, and (2) one linear behavior for times between about 0 and 4 hours and another linear behavior between about 7 and 25 hours with different slopes.

(g) Given that your answer to (d) well describes the bloodstream vancomycin concentration, estimate α from Fig. 7.12.

An interesting aspect of the analysis, carried out in this problem, is that it makes it possible to infer the concentration of vancomycin in the brain, lungs, and other slow-to-reach organs via measurements of vancomycin concentration in the bloodstream. It is clear that this information could be medically important when treating a patient with meningitis, for example.

Problem 6: Michalis-Menten Enzyme Kinetics: The Whole Story
Biochemists are often lead to consider the catalysis by an enzyme (E) of a substrate
(S), yielding product (P). The corresponding reaction scheme is:

$$E+S \underset{k_{-1}}{\overset{k_1}{\rightleftharpoons}} ES \xrightarrow{k_2} E+P$$

 indicating that the reaction occurs via the creation of an enzyme-substrate complex,
ES. Very often, this reaction occurs under conditions of excess substrate, in which
case the concentration of substrate can be considered to be constant. In this case, the
steady-state rate of product production can be described via the Michaelis-Menten
equation:

$$\frac{v_M[S]}{[S] + K_M}, \tag{7.132}$$

where v_M is the maximum rate of the reaction and K_M is the Michaelis constant.

   Under conditions of excess substrate, we can write *linear* equations for the rate
of change of the concentration of enzyme, $E = [E]$, and for the rate of change of the
concentration of the ES complex, $C = [ES]$. To streamline the notation, we will call
$k_1[S] = f$ (f for forward) and $k_{-1} + k_2 = b$ (b for backwards). Then, the chemical
rate equations that describe how E and C change versus time are:

$$\frac{dE}{dt} = -fE + bC, \tag{7.133}$$

and

$$\frac{dC}{dt} = -bC + fE. \tag{7.134}$$

Your goal in this problem is to solve these equations, find their steady-state
solutions, determine the values of v_M and K_M in terms of the given rate constants
and the total enzyme concentration, E_T, and estimate how long the reaction mixture
takes to reach a steady-state for C and E.

(a) Determine the eigenvalues, Λ, of these two equations.
(b) Calculate the corresponding eigenvectors.
(c) Using your answers to (b), write the general solution for $E(t)$ and $C(t)$ in terms
    of $\Lambda_\pm$, E_T, b, and f, and two constants to be determined, A and B.
(d) Find the particular solution in the case that $C(0) = 0$ and $E(0) = E_T$.
(e) Calculate the rate of product production.

(f) What is the steady-state rate of product production? By comparing your expression to the Michaelis-Menten equation, namely Eq. 7.132, determine v_M and K_M.

(g) How long is it necessary to let the reaction proceed in order to achieve a steady-state?

Problem 7: Hemodialysis Revisited

Hemodialysis is one of the most common hospital procedures in the USA, resulting in nearly one million hospital visits in 2011. In hemodialysis, a fraction of the blood flow is directed out of the body to flow through a channel, whose walls consist of semi-permeable membrane, before returning to the patient. Counter-circulating outside the semi-permeable-membrane channel is a large volume of dialyzate. The dialyzate contains glucose, sodium chloride, *etc.* at the same concentrations as in blood, so there is no net flow of these solutes through the semi-permeable membrane. However, unwanted solutes that at present in the blood, but not present in the dialyzate, such as urea and creatinine, diffuse across the semi-permeable membrane from the blood into the dialyzate, thus leaving the fluid volume of the body. Suppose that the volume of fluid in the patient is V_1, that the volume of dialyzate is V_2, and that the diffusive conductivity of the semi-permeable membrane is G.

(a) Calculate the current of urea molecules across the semi-permeable membrane from the dialyzate to the patient, given that the urea concentration in the patient is c_1 and the urea concentration in the dialyzate is c_2.

(b) Determine the rate of change of urea concentration within the patient ($\frac{dc_1}{dt}$) in terms of G, V_1, c_1, and c_2.

(c) Determine the rate of change of urea concentration within the dialyzate ($\frac{dc_2}{dt}$) in terms of G, V_2, c_1, and c_2.

(d) Using a trial solution of the form

$$c_1(t) = Ae^{\Lambda t}, \tag{7.135}$$

and

$$c_2(t) = Be^{\Lambda t}, \tag{7.136}$$

find the eigenvalues, Λ, and eigenvectors, $(1, \frac{A}{B})$, of your equations from (b) and (c).

(e) Using superposition, write the general solution to the rate of change of concentration equations.

(f) What is the particular solution, if, at the start of hemodialysis, $c_1(0) = c$ and $c_2(0) = 0$? What is the corresponding time dependence of the concentration of urea in the patient?

(g) How does the concentration of urea in the patient at long times depend on V_1 and V_2? What is the requirement on V_2 to achieve a low concentration of urea in the patient?

(h) What is the relaxation time, τ, for the approach to this ultimate concentration? How does this time depend on V_1 and V_2?

Problem 8: The Nuclear Physics of Nuclear-Medicine-Based Imaging

About 85% of nuclear-medicine-based diagnostic imaging procedures, such as single-photon emission computed tomography (SPECT),[23] use technetium-99 (Tc^{99m}),[24] as the radioactive tracer, amounting to 20 million procedures each year. Tc^{99m} decays with a half-life of 6 hours via emission of a 141 keV-gamma ray, which is used for imaging. Tc^{99m}'s parent nucleus, namely molybdenum-99 (Mo^{99}), is commonly a byproduct of uranium-235 (U^{235}) fission in a nuclear reactor. Mo^{99} has a half-life of 67 hours, which makes it possible for specialized suppliers to purify and then transport Mo^{99} worldwide to provide Tc^{99m} at the point of care via Tc^{99m} generators, a.k.a. "moly cows." "Milking the cow" for Tc^{99m} at the point of care is accomplished by an elution process enabled by the different chemistry of Tc^{99m} versus Mo^{99}. From the point of view of a Tc^{99m} generator, production of a gamma ray for imaging is a two-step process. First, Mo^{99} undergoes beta decay to Tc^{99m} nucleus:

$$Mo^{99} \rightarrow Tc^{99m} + e^- + \bar{\nu}, \tag{7.137}$$

where e^- is an electron and $\bar{\nu}$ is an anti-neutrino. The corresponding decay rate is α. Subsequently, the metastable (hence the "m") Tc^{90m} nucleus decays with decay rate λ to Tc^{99} via the emission of a gamma ray:

$$Tc^{99m} \rightarrow Tc^{99} + \gamma, \tag{7.138}$$

where γ is a gamma ray with an energy of 141 keV, respectively (Fig. 7.13).

(a) Let M and T be the number of Mo^{99} and Tc^{99m} nuclei, respectively, and write the "chemical rate equations" that describe how M and T vary in time.

(b) Explain whether there is a formal mathematical identity between this and any other problems in this chapter.

(c) Determine the eigenvalues and eigenvectors of your rate equations in (a).

(d) Initially, $M(0) = M_0$ and $T(0) = 0$. Determine T as a function of time.

(e) Calculate the rate at which gamma rays are emitted as a function of time.

---

[23] http://en.wikipedia.org/wiki/SPECT
[24] http://en.wikipedia.org/wiki/Technetium-99m#Medical_uses

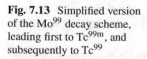

Fig. 7.13 Simplified version of the Mo^{99} decay scheme, leading first to Tc^{99m}, and subsequently to Tc^{99}

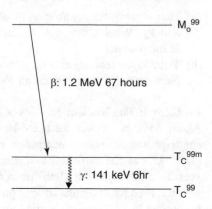

(f) Explain approximately how long should the wait time be between "milkings" of a Tc^{99m} generator, assuming that at the end of each "milking" the amount of Tc^{99m} has been reduced to zero. Explain approximately how long can a Tc^{99m} generator go between Mo^{99} rechargings.

Problem 9: HIV Infection Revisited: The Effect of Protease Inhibitors

In 1995, sequential papers in Nature [10, 11] reported that treatment with a protease inhibitor lead to an exponential decrease in patients' HIV viral loads and a corresponding increase in their CD4 cell counts. These results, and the mathematical modeling that was an integral part of the *Nature* papers, represented a major breakthrough in our understanding of HIV infection. This enhanced understanding soon leads to the idea to treat HIV with a triple drug cocktail of antivirals[25] that now successfully prevents progression of HIV infection to AIDS in many patients.

In this problem, you will further develop our mathematical model of HIV infection to describe what happens when a patient with chronic HIV infection is treated with a protease inhibitor. First, to understand the starting point, prior to treatment, we must understand chronic HIV infection. As shown in Fig. 7.7 and discussed in the chapter, at long times the number of HIV viral particles (V), the number of uninfected T cells (U), and the number of infected T cells (I) all achieve non-zero, steady-state values. Given the realistic assumption that $\beta\lambda \ll \alpha\gamma(m-1)$, these steady-state values become:

$$U = \frac{\lambda}{\gamma(m-1)}, \tag{7.139}$$

$$I = \frac{\alpha}{\kappa}, \tag{7.140}$$

---
[25] http://en.wikipedia.org/wiki/Management_of_HIV/AIDS

and

$$V = \frac{\alpha(m-1)}{\lambda}. \tag{7.141}$$

These values correspond to a patient with (untreated) chronic HIV infection. HIV protease inhibitors[26] prevent infected T-cells from giving rise to infectious HIV particles. Instead, now when an infected T-cell dies, it releases non-infectious particles.

(a) Re-draw the cartoon representation of Fig. 7.6, so that it describes the situation in the presence of protease inhibitor.
(b) Write an equation for $\frac{dN}{dt}$, where N is the number of non-infectious viral particles.
(c) Modify Eqs. 7.91 and 7.92, as needed, to describe the new viral dynamics in the presence of a protease inhibitor.
(d) What are the appropriate initial conditions, if your equations are to describe what happens when an HIV protease inhibitor is administered to a patient at $t = 0$.
(e) The equations that you wrote in (c) for V and I become linear, if we make the approximation that $U \simeq \frac{\lambda}{m-1}$ is essentially constant over the relevant time scales. To further simplify, we also take $m \gg 1$, which is realistic. Then solving the resultant linearized equations for $\frac{dV}{dt}$ and $\frac{dI}{dt}$, we find

$$V = \frac{\alpha m}{\lambda} e^{-\lambda t} \tag{7.142}$$

and

$$I = \frac{\alpha(\kappa e^{-\lambda t} - \lambda e^{-\kappa t})}{\kappa(\kappa - \lambda)}. \tag{7.143}$$

We can then show by direct substitution into your equation for $\frac{dN}{dt}$ that

$$N(t) = \frac{m\alpha}{\kappa - \lambda} \left(\frac{\lambda(e^{-\lambda t} - e^{-\kappa t})}{\kappa - \lambda} + \kappa t e^{-\lambda t} \right). \tag{7.144}$$

Figure 7.14 shows experimental measurements of the total HIV load of three patients versus time, following administration of an HIV protease inhibitor, and compares these measurements to these predictions for the total viral load, namely $V(t)+N(t)$ [12]. Evidently, the best fit provides an excellent description of the data over the seven day period studied for all three patients, with the amount of HIV RNA decreasing exponentially by a factor of about 100 over

---

[26] http://en.wikipedia.org/wiki/Protease_inhibitor_(pharmacology)

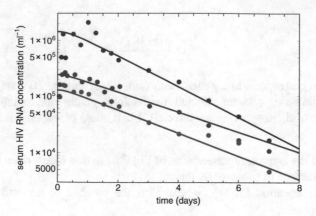

Fig. 7.14 Serum concentrations of HIV RNA, which is a proxy for the total, infectious and non-infectious viral load, for three patients, following the initiation of protease inhibitor treatment at $t = 0$ (red, blue, and magenta circles). The lines are the best fit to the sum of the infectious viral load and the non-infectious viral load, $V(t) + N(t)$, according to the model described in "HIV infection revisited: the effect of protease inhibitors." Adapted from Ref. [12]

Fig. 7.15 Mean rate of new colon cancers for women (red circles) and men (blue circles) in their 40s, 50s, 60s, and 70s versus mean patient age, adapted from http://www.cdc.gov/cancer/colorectal/statistics/age.htm

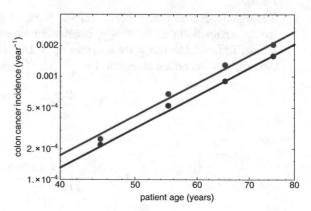

this time period in each case. Given that the infected cell death rate (κ) is significantly smaller than the viral death rate (λ), from these plots, estimate κ for each patient.

Problem 10: Age-Dependence of Cancer Incidence
(a) Other cancers can often show a stronger age-dependence than retinoblastoma.[27] In particular, the mean rate of new colon cancers versus mean patient age for women (red circles) and men (blue circles) is shown in a log-log plot in Fig. 7.15. Explain what it means that there are straight lines in such a log-log plot.

---

[27] http://www.nature.com/nrc/journal/v1/n2/pdf/nrc1101-157a.pdf

(b) As patient age increases from 40 to 80, the rate of colon cancer incidence increases approximately by a factor of sixteen, as shown by the lines in the figure. What does this observation imply about how the rate of colon cancer incidence varies with patient age?

(c) Discuss what the patient-age dependence of the rate of colon cancer incidence might suggest about the mechanisms of colon cancer carcinogenesis.

Problem 11: Reverse Transcriptase Inhibitor (RTI) Treatment of HIV

In the chapter, we discussed that immediately after initial infection with HIV, when the number of uninfected T cells may be considered constant, the number of virus particles (V) and the number of infected T cells (I) in a patient both increase exponentially in time. The equations that describe the time evolution of V and I during this initial phase are

$$\frac{dV}{dt} = -4\kappa V + 3\kappa I \tag{7.145}$$

and

$$\frac{dI}{dt} = -2\kappa I + 5\kappa V. \tag{7.146}$$

Here, all of the parameters of the model have been expressed as multiples of a certain rate κ.

Solving Eqs. 7.145 and 7.146, subject to the initial conditions that $V(0) = 1$ and $I(0) = 0$, leads to

$$V(t) = \frac{3}{8}e^{\kappa t} + \frac{5}{8}e^{-7\kappa t} \tag{7.147}$$

and

$$I(t) = \frac{5}{8}e^{\kappa t} - \frac{5}{8}e^{-7\kappa t}. \tag{7.148}$$

In Problem 9, we discussed treatment of HIV infection, using a protease inhibitor, which prevents infected T-cells from giving rise to infectious HIV particles. A complementary HIV treatment is to administer a reverse transcriptase inhibitor (RTI), which largely prevents new infections of uninfected cells. However, treatment with a certain RTI does not completely prevent new infections. Nevertheless, it reduces the rate of new infections significantly, so that the equations that describe the time evolution of V and I become

$$\frac{dV}{dt} = -4\kappa V + 3\kappa I \tag{7.149}$$

and

$$\frac{dI}{dt} = -2\kappa I + \kappa V. \qquad (7.150)$$

In the presence of this RTI treatment, the term that previously was $5\kappa V$ in Eq. 7.146 is reduced by a factor of five to become κV in Eq. 7.150, corresponding to the now-reduced rate of infection of uninfected T cells by virus particles. The goal of this problem is to determine whether administering this particular RTI is an effective treatment for HIV infection.

(a) By examining Eqs. 7.147 and 7.148, and without doing any calculation, write the eigenvalues of Eqs. 7.145 and 7.146.
(b) By following our usual procedure (or otherwise), determine the eigenvalues of Eqs. 7.149 and 7.150. Call them Λ_+ and Λ_-.
(c) Determine the corresponding eigenvectors of Eqs. 7.149 and 7.150, $\mathbf{e}_+$ and $\mathbf{e}_-$.
(d) Write an expression for the general solution of Eqs. 7.149 and 7.150.
(e) Given that the initial values of V and I at the start of RTI treatment ($t = 0$) are $V(0) = I(0) = W$, determine $V(t)$ and $I(t)$ for $t > 0$.
(f) Using a logarithmic vertical axis and a linear horizontal axis (log-linear plot), sketch both $V(t)$ from Eq. 7.147 and $V(t)$ that you found in part (e) on the same plot. For Eq. 7.147, $t = 0$ corresponds to the time of the initial HIV infection. In this case, do not worry about what happens at very early times. We are interested in the behavior for times significantly greater than $\frac{1}{7\kappa}$. When you plot $I(t)$ from (e), take $t = 0$ to correspond to the time at which the RTI treatment is first administered.
(g) Explain whether you expect treatment by this particular RTI (not RTIs in general) to be effective, given your answers to (e) and (f).

Problem 12: Critical Mass
To enable a nuclear chain reaction and the ensuing atomic explosion, it is necessary for there to be a sufficient quantity of fissile material, such U^{235} or Pu^{239}, in a sufficiently compact volume. The shape with the smallest critical mass is a sphere. This minimum mass of fissile material for an atomic explosion to be possible is called the "critical mass" and is 52 kg for U^{235} and 10 kg for Pu^{235} in the case of bare spheres of these materials. The critical mass is reduced for spheres of fissile material that are surrounded by another material that tends to reflect neutrons back into the fissile material.

(a) Examine Eq. 7.109. For what range of values of m is there exponential growth in the number of neutrons and U^{236} nuclei?
(b) How is the observation that m must lie within a range of values in order to achieve exponential growth in the number of neutrons and U^{236} nuclei, related to the fact that a critical mass of fissile material is needed for an atomic explosion to occur?

(c) Develop and briefly describe a hypothesis that explains the existence of a critical mass.

Problem 13: Deadly Epidemic

It is possible to model mathematically how epidemics proceed through a population of susceptible individuals using the methods of this chapter. In this problem, we will model an always-fatal infection, which has an incubation period during which an individual has been infected but is not yet themselves infectious. Let S be the number of *susceptible* individuals, let E be the number of *exposed* (and infected but not infectious) individuals, and let I be the number of *infectious* individuals. Then, for a particular disease, we can model the progression of the disease through the population via the following equations:

$$\frac{dS}{dt} = \alpha - \beta S - \kappa SI, \tag{7.151}$$

$$\frac{dE}{dt} = \kappa SI - \gamma E, \tag{7.152}$$

and

$$\frac{dI}{dt} = \gamma E - 4\gamma I. \tag{7.153}$$

In the early stages of a possible epidemic, S is very large and can be approximated as a constant, permitting us to treat Eqs. 7.152 and 7.153 as linear equations.

(a) In fact, for the early stages of a particular epidemic, it turns out that $S = \frac{40\gamma}{\kappa}$. Determine the corresponding eigenvalues, Λ_+ and Λ_- in this circumstance.

(b) Determine the corresponding eigenvectors, e_+ and e_-.

(c) Write an expression for the general solution of Eqs. 7.152 and 7.153.

(d) Given that the initial values of E and I at the start of the epidemic ($t = 0$) are $E(0) = 0$ and $I(0) = 1$, determine $E(t)$ and $I(t)$ for $t > 0$.

(e) Briefly interpret/explain each of the terms on the right-hand sides of Eqs. 7.152 and 7.153.

(f) Sketch $I(t)$ versus t using a logarithmic scale for I and a linear scale for t.

(g) More generally, we cannot set $\kappa S = 40\gamma$. Determine the eigenvalues of Eqs. 7.152 and 7.153, in terms of γ and κS in this more general situation.

(h) For what range of values of $\frac{\kappa S}{\gamma}$ is at least one of the eigenvalues positive? What are the implications of this observation for the progression of the epidemic?

Problem 18: Positive Feedback

A genetic network consists of a number of genes, each of which either switches another gene on (activation) or turns another gene off (repression). By creating a network with appropriate interactions among the constituent genes, it is possible to engineer switches, oscillators, and sensors, and to perform logic computations, all realized in bacteria, yeast, or mammalian cells, with applications ranging from biological computation to medical diagnostics and biofuel production.

In this problem, we will examine a simple genetic network, consisting of two genes that each activates the expression of the other. Thus, production of a protein effectively leads to more production of that same protein. This situation is called positive feedback. We will focus on the protein products of these two genes, namely protein P and protein Q, with all the detailed biochemistry (transcription, translation, *etc.*) subsumed into the rate constants. Because of their mutual activation, the rate at which protein P is produced is proportional to Q and the rate at which protein Q is produced is proportional to P. Thus, the rate of change of the number of P proteins—we call this number P—and of the number of Q proteins—we call this number Q—may be written

$$\frac{dP}{dt} = -\beta P + \alpha Q, \tag{7.154}$$

and

$$\frac{dQ}{dt} = \alpha P - \beta Q, \tag{7.155}$$

respectively.

(a) What is the physical meaning of the terms proportional to β in Eqs. 7.154 and 7.155?
(b) Determine the eigenvalues of Eqs. 7.154 and 7.155, namely Λ_+ and Λ_-.
(c) Determine the corresponding eigenvectors, e_+ and e_-.
(d) Write an expression for the general solution of Eqs. 7.154 and 7.155.
(e) At $t = 0$, $P(0) = 0$ and $Q(0) = 100$. Determine the subsequent time evolution of the two proteins in this case.
(f) All on the same graph, sketch $Q(t)$ versus t using a logarithmic scale for Q and a linear scale for t for the following values of α: $\alpha = 0$, $\alpha = \frac{1}{2}\beta$, $\alpha = \beta$, and $\alpha = 2\beta$. Be sure to specify which curve is which.

References

1. J.R. Oppenheimer, *The Open Mind* (Simon and Schuster, New York, 1955)
2. A.G. Knudson Jr., Mutation and cancer: statistical study of retinoblastoma. Proc. Nat. Acad. Sci. USA **68**, 820–823 (1971)
3. D.E. Comings, A general theory of carcinogenesis. Proc. Natl. Acad. Sci. USA **70**, 3324–3328 (1973)
4. R. Shilts, *And the Band Played on: Politics, People, and the AIDS Epidemic* (St. Martin's Press, New York, 1987)
5. M.A. Nowak, R.M. May, *Virus Dynamics: Mathematical Principles of Immunology and Virology* (Oxford University Press, Oxford, 2000)

6. M.A. Nowak, A.L. Lloyd, G.M. Vasquez, T.A. Wiltrout, L.M. Wahl, N. Bischofberger, J. Williams, A. Kinter, A.S. Fauci, V.M. Hirsch, J.D. Lifson, Viral dynamics of primary viremia and antiretroviral therapy in simian immunodeficiency virus infection. J. Virol **71**, 7518–7525 (1997)

7. A fascinating account of the discovery of nuclear fission is presented by Prof. Ruth Lewin Sime. http://www.youtube.com/watch?v=RRDQhBFhuiE

8. R. Muller, *Physics for Future Presidents: The Science Behind the Headlines* (W. W. Norton, New York, 2008)

9. D.J. Krogstad, R.C. Moellering, Jr., D.J. Greenblatt, Single-dose kinetics of intravenous vancomycin. Clin. Pharmacol. **20**, 197–201 (1980)

10. X. Wei, S.K. Ghosh, M.E. Taylor, V.A. Johnson, E.A. Emini, P. Deutsch, J.D. Lifson, S. Bonhoeffer, M.A. Nowak, B.H. Hahn, M.S. Saag, G.M. Shaw, Viral dynamics in human immunodeficiency virus type 1 infection. Nature **373**, 117–22 (1995)

11. D.D. Ho, A.U. Neumann, A.S. Perelson, W. Chen, J.M. Leonard, M. Markowitz, Rapid turnover of plasma virions and CD4 lymphocytes in HIV-1 infection. Nature **373**, 123–126 (1995)

12. A.S. Perelson, A.U. Neumann, M. Markowitz, J.M. Leonard, D.D. Ho, HIV-1 dynamics in vivo: virion clearance rate, infected cell life span, and viral generation time. Science **271**, 1582–1585 (1996)

Statistical Mechanics: Boltzmann Factors, PCR, and Brownian Ratchets

8

The laws of thermodynamics may easily be obtained from the principles of statistical mechanics, of which they are an incomplete expression.

J. W. Gibbs[1] [1]

Electronic Supplementary Material The online version of this article (https://doi.org/10.1007/978-3-031-05808-0_8) contains supplementary material, which is available to authorized users.

[1] http://en.wikipedia.org/wiki/Josiah_Willard_Gibbs

8.1   Introduction

This chapter presents "statistical mechanics." Chemists and biochemists often call similar material "physical chemistry." Importantly, statistical mechanics provides a simple and powerful prescription for calculating probabilities and probability densities, and consequently means and variances. As indicated in the quotation by Gibbs above, statistical mechanics encompasses thermodynamics, including the second law of thermodynamics:

> The law that entropy always increases holds, I think, the supreme position among the laws of Nature. If someone points out to you that your pet theory of the universe is in disagreement with Maxwell's equations, then so much the worse for Maxwell's equations. If it is found to be contradicted by observation, well, these experimentalists do bungle things sometimes. But if your theory is found to be against the second law of thermodynamics I can give you no hope; there is nothing for it but to collapse in deepest humiliation. A. S. Eddington[2] [2]

Several concepts, which we have previously encountered, are brought together in this chapter: our understanding of energy, especially binding energy, obtained in Chap. 3. *Energy: Work, Geckos, and ATP* will be tremendously important in this chapter; statistical mechanics will inform us about the ratio of forward and backward chemical reaction rates that appear in chemical rate equations of the sort discussed in Chap. 6. *Diffusion: Membrane Permeability and the Rate of Actin Polymerization* and Chap. 7. *Rates of Change: Drugs, Infections, and Weapons of Mass Destruction*; we will build on the understanding of random walks gained in Chap. 5. *Random Walks: Brownian Motion and the Tree of Life* to discuss Brownian ratchets, which provide important insights into the second law of thermodynamics. Specific realizations of Brownian ratchets that we will discuss include DNA unzipping by helicase, which is a type of molecular motor, and force generation by actin polymerization, which is the basis of many examples of eukaryotic cell motility, including how the neurons in our brains connect themselves together. Additional topics that we will encounter in this chapter include: membrane ion channels, protein and RNA folding, ligand binding, including the cooperative binding of oxygen by hemoglobin, thermal DNA unzipping (a.k.a. DNA melting) and polymerase chain reaction (PCR), the ideal gas law, and temperature. We all know what hot and cold are, but, in fact, as we will see temperature is related to the mean kinetic energy of molecules moving randomly.

---

[2] http://en.wikipedia.org/wiki/Arthur_Eddington

8.2   Your Learning Goals for This Chapter

At the end of this chapter, you should be able to:

- Explain what a microstate is and be able to enumerate the number/multiplicity of microstates in a variety of cases.
- Apply the Boltzmann factor, properly incorporating the number/multiplicity of microstates, to calculate probabilities for systems held at a temperature T.
- Analyze connections between the Boltzmann factor, reaction rates, the direction of the reaction, the change in free energy, and the second law of thermodynamics in the context of isomerization reactions, such as protein folding/unfolding and DNA zipping/unzipping.
- Explain what a *Brownian ratchet* is, and analyze how Brownian ratchets function in actin polymerization, the DNA unzipping activity of a helicase motor protein, and similar examples.
- Apply statistical mechanics to analyze ligand binding in various cases.
- Explain and analyze how statistical mechanics leads to chemistry's law of mass action.
- Explain the relationship between temperature and the mean kinetic energy of molecules in a gas, liquid, or solid, and how that relationship leads to the ideal gas law: $PV = Nk_BT$.

8.3   The Boltzmann Factor

This fundamental law is the summit of statistical mechanics, and the entire subject is either the slide-down from this summit, as the principle is applied to various cases, or the climb up to where the fundamental law is derived and the concepts of thermal equilibrium and temperature clarified. R. P. Feynman,[3] writing about the Boltzmann factor [3].

Our starting point for a discussion of statistical mechanics is to introduce the Boltzmann factor, named after Ludwig Boltzmann,[4] and then, following Feynman, slide down from that summit [4, 5]. The Boltzmann factor gives the probability of realizing a particular "microstate." Essentially, each microstate is each possible microscopic configuration of a system of interest. The Boltzmann factor specifies how the environment that is in contact with this system affects the probability that a particular microstate of the system is realized: the probability of realizing microstate i with energy ϵ_i, when the system is in contact with an environment at a temperature T, is

---

[3] http://en.wikipedia.org/wiki/Richard_Feynman

[4] http://en.wikipedia.org/wiki/Ludwig_Boltzmann

$$p_i = \frac{e^{-\epsilon_i/(k_B T)}}{Z}, \tag{8.1}$$

where T is the absolute temperature in Kelvin, $k_B = 1.38 \times 10^{-23}$ JK^{-1} is Boltzmann's constant, and Z is the factor required to ensure that the probabilities are correctly normalized. Z is called the partition function, which therefore must be given by

$$Z = \Sigma_i e^{-\epsilon_i/(k_B T)}, \tag{8.2}$$

where the sum runs over all possible microstates i.

8.3.1　Example: Ion Channels

Consider a simple model for a membrane ion channel[5] with 2 possible microstates: open or closed. The energy of the closed state (state 1) is 0. The energy of the open state (state 2) is ϵ.

(a) What is the probability that the ion channel is closed (state 1) in terms of Z and given variables? According to the Boltzmann

factor, the probability that the ion channel is closed (state 1) is

$$p_1 = \frac{1}{Z}. \tag{8.3}$$

(b) What is the probability that the ion channel is open (state 0) in terms of Z and given variables?

According to the Boltzmann factor, the probability that it is open (state 2) is

$$p_2 = \frac{e^{-\epsilon/(k_B T)}}{Z}. \tag{8.4}$$

(c) Determine Z in terms of given variables.

We can determine Z by using the requirement that probabilities are normalized. In this case, there are only two microstates. Therefore,

$$p_1 + p_2 = 1. \tag{8.5}$$

---

[5]http://en.wikipedia.org/wiki/Voltage_gated_ion_channel

(continued)

It follows that

$$\frac{1}{Z} + \frac{e^{-\epsilon/(k_B T)}}{Z} = 1. \tag{8.6}$$

i.e.

$$Z = 1 + e^{-\epsilon/(k_B T)}. \tag{8.7}$$

(d) Using your answer to (c) for Z, express p_1 and p_2 in terms of given quantities only.

Using this expression for Z, our final results for p_1 and p_2 are

$$p_1 = \frac{1}{1 + e^{-\epsilon/(k_B T)}} \tag{8.8}$$

and

$$p_2 = \frac{e^{-\epsilon/(k_B T)}}{1 + e^{-\epsilon/(k_B T)}}. \tag{8.9}$$

(e) Figure 8.1 shows measurements of the current versus time through individual sodium ion channels for several voltages applied across the membrane, from Ref. [6]. For clarity, traces have been shifted relative to each other along the current axis. Clearly, in each trace, the measured current takes one of two values, lower or higher, corresponding to the ion channel being either open or closed, respectively. A more negative current indicates that the ion channel is open. Explain how you could use data like these to determine p_1 and p_2.

The probabilities are equal to the fraction of time in each state, so sum up the length of time in one state in a given total length time and then ratio these two lengths of time.

Inspection of these data makes it clear that the probability that the channel is open depends strongly on the voltage across the membrane, with the closed state favored at more negative voltages and the open state favored at less negative voltages. Figure 8.2 from Ref. [7] shows that the probability of realizing the open state may be fairly well described using Eq. 8.9, assuming $\epsilon = Q(V_0 - V)$ with $Q = 3.16e$, where e is the electronic charge, and

(continued)

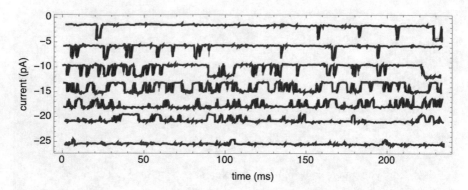

Fig. 8.1 Currents (red) through individual sodium channels, purified from rat brain and reconstituted in a lipid bilayer, for transmembrane voltages, from top to bottom, of −55, −75, −85, −95, −105, −115, and −125 mV. Adapted from Ref. [6]. The black lines show a categorization of these signals into open (more negative current) and closed (less negative current) states. As the transmembrane voltage becomes more negative, a channel is progressively more likely to be open

Fig. 8.2 Probability that a wild-type (red) or mutant (blue) *Shaker* potassium ion channel is open as a function of voltage across the membrane. The solid curves are fit to the model described in the text. Adapted from Ref. [7]

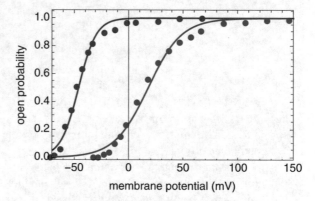

$V_0 = -46.5$ mV for the unmodified, wild-type channel, shown in red, and $Q = 1.67e$ and $V_0 = 17.2$ mV for a channel with a particular mutation, shown in blue, suggesting that in both cases the energy of the open state depends approximately linearly on the voltage across the membrane in which the channel is embedded.

8.3.2   Example: Molecular Height Distribution

The Boltzmann factor also applies to continuous random variables. For example, the Boltzmann distribution gives the distribution of molecular heights in the atmosphere (or in an ultracentrifuge), assumed to have a constant temperature. To discuss how the energy of a molecule in the Earth's atmosphere depends on the height, z, of the molecule, we pick the system to consist of the Earth and the molecule, and therefore, the potential energy of the Earth–molecule interaction is included in the energy of this system, namely fz, where $f = mg$ is the molecule's weight with m its mass, g the acceleration due to gravity, and z (lower case z—not the partition function) is the height of the molecule above the Earth's surface. In this case, a particular microstate corresponds to the location of a particle between z and $z + dz$.

(a) The Boltzmann factor informs us that the probability density for z is

$$p(z) = \frac{1}{Z}e^{-fz/(k_B T)}, \tag{8.10}$$

where Z is the partition function, not to be confused with the coordinate z. Assuming throughout this problem that molecules are confined to the half space $z > 0$, determine Z.

Answer: The value of Z is determined by the requirement that $p(z)$ is normalized:

$$\int_0^\infty p(z)dz = 1. \tag{8.11}$$

It follows that $Z = \frac{k_B T}{f}$ and

$$p(z) = \frac{f}{k_B T}e^{-fz/(k_B T)}. \tag{8.12}$$

(b) Calculate the mean height of a molecule.

Answer: We have

$$\langle z \rangle = \int_0^\infty dz z \frac{f}{k_B T}e^{-fz/(k_B T)} = \frac{k_B T}{f}. \tag{8.13}$$

(c) What is the probability (P) that a certain molecule's height is greater than a?

(continued)

Answer:

$$P = \int_a^\infty dz \frac{f}{k_B T} e^{-fz/(k_B T)} = e^{-fa/(k_B T)}. \tag{8.14}$$

This expression for P will be important later in the Chapter.

8.3.3 Protein and Nucleic Acid Folding–Unfolding and Microstate Multiplicity

In this example, we discuss protein or RNA folding and unfolding, in particular, and isomerization reactions more generally. In an isomerization reaction, one chemical species alternates between different molecular configurations. Protein and nuclei acid folding/unfolding transitions are examples of an isomerization reaction:

$$F \underset{k_f}{\overset{k_u}{\rightleftharpoons}} U$$

k_u is the unfolding rate. k_f is the folding rate.

An important aspect of this example is the concept of multiplicity. Different states—here "folded" and "unfolded"—can realize different multiplicities with important consequences. In the context of protein or nucleic acid folding/unfolding, the folded state (F) generally corresponds to a single microstate because the folded state manifests a single atomically well-defined structure and therefore has multiplicity 1. By contrast, the unfolded state (U) corresponds to many (g) microstates, as illustrated in Fig. 8.3. There are many unfolded microstates because the spatial configuration of a polymer—which is how we conceive the unfolded protein, RNA or DNA—is analogous to the path of a random walk with the number of steps on the path, equal to the number of monomers in the polymer. Therefore, the number of microstates g—the multiplicity g—associated with the unfolded state of an N amino acid protein is approximately equal to the number of N step random walks in 3D. The exact value of g is not important for our discussion (except that $g > 1$).

Introducing ϵ as the energy of each of the g-possible unfolded microstates, relative to the energy of the single folded microstate (ϵ is positive), using the Boltzmann factor (Eq. 8.1), we may write the probability that a protein is folded,

$$p_f = \frac{1}{Z} = \frac{1}{1 + \Sigma_{i=1}^g e^{-\epsilon/(k_B T)}} = \frac{1}{1 + g e^{-\epsilon/(k_B T)}} \tag{8.15}$$

Fig. 8.3 Protein microstates.
There is a single folded
microstate, but there are many
(*g*) unfolded microstates

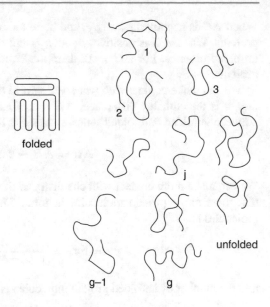

folded

unfolded

and the probability that a protein is unfolded,

$$p_u = \frac{\Sigma_{i=1}^{g} e^{-\epsilon/(k_B T)}}{Z} = \frac{g e^{-\epsilon/(k_B T)}}{1 + g e^{-\epsilon/(k_B T)}}. \tag{8.16}$$

Equations 8.15 and 8.16 involve sums over the *g* unfolded microstates, which simply yield factors of *g* ($\Sigma_{i=1}^{g} = g$) because the energy of each of the unfolded states is the same for all unfolded microstates, namely ϵ. (Different unfolded microstates are mutually exclusive outcomes because if a protein realizes one particular unfolded microstate, it does not realize another unfolded microstate.)

Now, introducing

$$\Delta G = \epsilon - k_B T \ln g, \tag{8.17}$$

which is is the difference in *free energy* between the unfolded state and the folded state, we may re-write Eqs. 8.15 and 8.16 as

$$p_f = \frac{1}{1 + e^{-\Delta G/(k_B T)}}, \tag{8.18}$$

and

$$p_u = \frac{e^{-\Delta G/(k_B T)}}{1 + e^{-\Delta G/(k_B T)}}, \tag{8.19}$$

When ΔG is negative, $p_u > p_f$, and there are more unfolded proteins than folded proteins. When ΔG is positive, $p_u < p_f$, and there are more folded proteins than unfolded proteins. For $\Delta G = 0$, there are equal numbers of folded and unfolded proteins.

We can make contact with what we learn in chemistry classes by noting that, in fact, ϵ is the enthalpy difference, ΔH, and $k_B \ln g$ is the entropy difference, ΔS, between unfolded and folded states, so that Eq. 8.17 may be re-written as

$$\Delta G = \Delta H - T\Delta S. \tag{8.20}$$

We can make more contact with chemistry in the following fashion. Let us suppose that there are n protein molecules in total. Then, the number of folded protein molecules is

$$n_f = np_f = \frac{n}{1 + e^{-\Delta G/(k_B T)}}, \tag{8.21}$$

and the number of unfolded protein molecules is

$$n_u = np_u = \frac{ne^{-\Delta G/(k_B T)}}{1 + e^{-\Delta G/(k_B T)}}, \tag{8.22}$$

while their ratio is

$$\frac{n_u}{n_f} = e^{-\Delta G/(k_B T)}. \tag{8.23}$$

On the other hand, given the protein's folding and unfolding rates, we can write an equation for the time dependence of the number of unfolded proteins:

$$\frac{dn_u}{dt} = -k_f n_u + k_u n_f. \tag{8.24}$$

Eventually, we can expect that n_u will achieve a steady-state value, independent of time, for which

$$0 = -k_f n_u + k_u n_f, \tag{8.25}$$

i.e.,

$$\frac{n_u}{n_f} = \frac{k_u}{k_f}. \tag{8.26}$$

But the left side of this equation is given by Eq. 8.23. Thus, we arrive at

$$\frac{k_u}{k_f} = e^{-\Delta G/(k_B T)}. \tag{8.27}$$

Remarkably, statistical mechanics relates the ratio of forward and backward rates to the free energy of the isomerization reaction.

These ideas have been beautifully realized in optical tweezers based experiments, which unfold RNA [8] and DNA [9] hairpins or proteins [10,11] using force. In these examples, the free energy difference between the unfolded state and the folded state was varied by the application of force/tension (f). Specifically, the difference in free energy between the unfolded state and the folded state is

$$\Delta G(f) = \Delta G - f\ell, \qquad (8.28)$$

where ΔG is the free energy of the unfolded state, relative to the folded state at zero force, and ℓ is the change in length of the protein in the unfolded state relative to the folded state (Fig. 8.5). At zero force, ΔG is positive. However, for large enough f, $\Delta G(f)$ eventually becomes negative, and the nucleic acid or protein unfolds. For forces such that $f\ell$ is comparable to ΔG, the probabilities of being in the folded and unfolded states are comparable, and the protein or nucleic acid hops back and forth between folded and unfolded states, as shown in Fig. 8.4 [11].

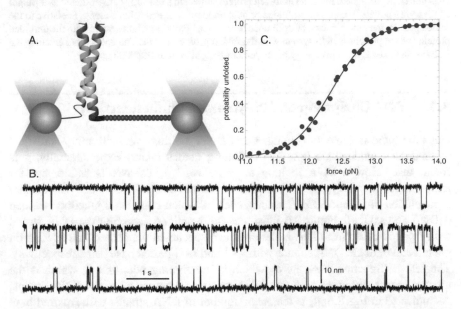

Fig. 8.4 (**a**) Schematic of an optical trapping experiment to characterize protein folding and unfolding in which a coiled coil protein, shown folded, is tethered between two optically trapped beads. Moving the traps apart applies force (tension) to the tether and thus to the protein. (**b**) Representative extension versus time traces, corresponding to changes in the length of the tether between the two trapped beads at 11.8, 12.8, and 13.8 pN from bottom to top. When the extension takes the larger value, the protein is unfolded. When the extension takes the smaller value, the protein is folded. C. Probability that the protein is unfolded versus force. Points are experimental results from data similar to those shown in B. The solid line corresponds to theory, based on the Boltzmann factor [11]. Figure courtesy of Yongli Zhang

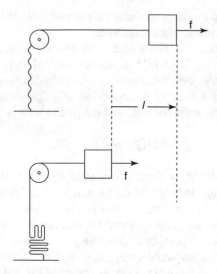

Fig. 8.5 Left: schematic of a folded protein subject to a tension, f, represented as a weight applied via a pulley. Right: schematic of an unfolded protein subject to a tension, f, represented as a weight applied via a pulley. Because the unfolded protein undergoes an extension ℓ along f relative to the folded protein, the potential energy difference between the protein-weight system with the unfolded protein and the protein-weight system with the folded protein is $-f\ell$. The energy that enters in the Boltzmann factor is the total energy of the protein-weight system (Eq. 8.28)

8.4   DNA Unzipping and Polymerase Chain Reaction

Our next topic is DNA unzipping, a.k.a. DNA melting. We will apply statistical mechanics to understand how DNA melting comes about. Experimentally, it is found that below a DNA melting temperature, T_M, dsDNA is stable, but for temperatures above T_M, dsDNA unzips to form two ssDNA strands. We consider a population of identical DNA strands each of which contains a junction between dsDNA and ssDNA. Figure 8.6 illustrates the reactions involving the DNA strand with i paired base pairs. This is the chemical species in the center. The species on the left and right are DNA strands with $i+1$ and $i-1$ paired base pairs, respectively. The relevant reaction rates are α, which is the zipping rate, and β, which is the unzipping rate. When $\alpha > \beta$, the DNA zips up. When $\alpha < \beta$, the DNA unzips. As indicated in Fig. 8.6, n_i is the *mean* number of DNA strands with i paired base pairs, *etc.*

With the help of Fig. 8.6, we can write an equation for the rate of change of n_i in terms of α, β, n_i, n_{i-1}, and n_{i+1}:

$$\frac{dn_i}{dt} = -\alpha n_i - \beta n_i + \alpha n_{i-1} + \beta n_{i+1}. \qquad (8.29)$$

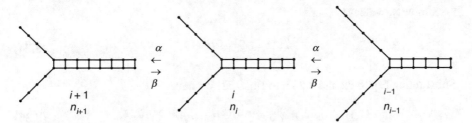

Fig. 8.6 Chemical reaction scheme for DNA zipping and unzipping. The possible reactions of a DNA strand with i zipped base pairs are illustrated, undergoing isomerization to a DNA strand with $i - 1$ or $i + 1$ base pairs. n_{i+1}, n_i, and n_{i-1} are the numbers of DNA molecules with $i + 1$, i, and $i - 1$ zipped-up base pairs, respectively

After a sufficient time, a steady-state situation will have been achieved, for which $dn_i/dt = 0$. Thus,

$$0 = -\alpha n_i - \beta n_i + \alpha n_{i-1} + \beta n_{i+1}. \tag{8.30}$$

It is useful to divide Eq. by n_i to find

$$0 = -\alpha - \beta + \alpha \frac{n_{i-1}}{n_i} + \beta \frac{n_{i+1}}{n_i}. \tag{8.31}$$

The factor n_{i-1}/n_i, which is the ratio of the mean number of DNA strands with $i - 1$ zipped base pairs to the mean number with i zipped base pairs, is equal to the ratio of the probability that a particular DNA strand has $i - 1$ zipped base pairs to the probability that it has i zipped base pairs. Thus, this factor is given by a Boltzmann factor. Specifically, if g_2 is the number of microstates associated with a base pair of dsDNA, and g_1 is the number of microstates associated with a pair of ssDNA bases, then

$$\frac{n_{i-1}}{n_i} = g e^{-\epsilon/(k_B T)} = e^{-(\epsilon - k_B T \ln g)/(k_B T)} = e^{-\Delta G/(k_B T)} \tag{8.32}$$

with $g = \frac{g_1}{g_2}$, and we have again introduced

$$\Delta G = \epsilon - k_B T \ln g, \tag{8.33}$$

where ϵ is the energy of the microstate with one additional unzipped base pair relative to the microstate in which that base pair is zipped up (so ϵ is positive). Because there are two strands of ssDNA but only one strand of dsDNA and because ssDNA is more flexible than dsDNA, we can be confident that $g = \frac{g_1}{g_2}$ is larger than unity.

Similarly, we have

$$\frac{n_{i+1}}{n_i} = e^{+\Delta G/(k_B T)}. \tag{8.34}$$

Substituting Eqs. 8.32 and 8.34 into Eq. 8.31, we have

$$0 = -\alpha - \beta + \alpha e^{-\Delta G/(k_B T)} + \beta e^{+\Delta G/(k_B T)}. \tag{8.35}$$

We can now do some algebra to determine α/β in terms of $e^{\Delta G/(k_B T)}$:

$$\alpha - \alpha e^{-\Delta G/(k_B T)} = \beta e^{+\Delta G/(k_B T)} - \beta, \tag{8.36}$$

i.e.,

$$\frac{\alpha}{\beta} = \frac{e^{+\Delta G/(k_B T)} - 1}{1 - e^{-\Delta G/(k_B T)}}, \tag{8.37}$$

i.e.,

$$\frac{\alpha}{\beta} = e^{+\Delta G/(k_B T)} \frac{1 - e^{-\Delta G/(k_B T)}}{1 - e^{-\Delta G/(k_B T)}}, \tag{8.38}$$

i.e.,

$$\frac{\beta}{\alpha} = e^{-\Delta G/(k_B T)}. \tag{8.39}$$

Equation 8.39 informs us that the DNA unzips, i.e., $\beta > \alpha$, only if $\Delta G < 0$, i.e., only if $\epsilon - k_B T \ln g < 0$. In order for this condition ($\epsilon - k_B T \ln g < 0$) to be satisfied, we require that $T > \epsilon/(k_B \ln g)$. If we define the DNA "melting temperature" to be $T_M = \epsilon/(k_B \ln g)$, we see that the DNA unzips for $T > T_M$, while it zips up for $T < T_M$, exactly as observed experimentally. How cool is that?

DNA melting/unzipping is the basis for DNA multiplication by polymerase chain reaction (PCR), for which Kary Mullis won the 1993 Nobel Prize in Chemistry. The first step in PCR is to raise the temperature above T_M, so that each dsDNA strand unzips to become two ssDNA strands. When the temperature is subsequently reduced in the presence of oligonucleotide primers,[6] nucleotides, and DNA polymerase, each previously unzipped, now-single-stranded ssDNA strand templates its own conversion to dsDNA. This process doubles the original number of dsDNA strands because there is a new dsDNA strand for each of the two ssDNA strands that result from each dsDNA unzipping reaction. PCR involves repeating this temperature cycling process multiple (N, say) times, with the result that the

---

[6] http://en.wikipedia.org/wiki/Primer_%28molecular_biology%29

initial number of DNA molecules is multiplied by a factor of 2^N. Thus, initially tiny amounts of DNA can be hugely amplified and subsequently sequenced. Thus, DNA unzipping has had a major societal impact via polymerase chain reaction (PCR),[7] which relies on the DNA unzipping/melting transition. PCR has revolutionized justice systems around the world via DNA testing,[8] for example.

8.4.1   DNA Zipping and Unzipping as a Random Walk

As a prelude to understanding helicase unzipping of DNA, we will recapitulate our discussion of DNA zipping/unzipping, but from a random walk point of view. Later, we will add helicase translocation on ssDNA into the mix. The key point is that both of these processes can be thought of as a random walk with a non-zero drift velocity (Fig. 8.7).
Consider a dsDNA–ssDNA junction. The probability of zipping up one base pair in a time Δt is $\alpha \Delta t$, and the probability of unzipping one base pair in a time Δt is $\beta \Delta t$, where α and β are exactly the same reaction rates that we talked about in the discussion of thermal zipping and unzipping of DNA. For small enough Δt, it is reasonable to assume that the only three possibilities are: (1) to zip up one base pair, (2) to unzip one base pair, or (3) to not do anything. Therefore, since probabilities sum to unity, we must have that the probability to do nothing is $1 - \alpha \Delta t - \beta \Delta t$.

Given these probabilities, and the length of a base pair, b, we may calculate the mean displacement of the ssDNA–dsDNA junction in a time Δt:

$$\Delta x_J = (-b)(\alpha \Delta t) + (0)(1 - \alpha \Delta t - \beta \Delta t) + (+b)(\beta \Delta t) = b(\beta - \alpha)\Delta t. \quad (8.40)$$

Fig. 8.7 Schematic of helicase translocation on ssDNA and DNA zipping, showing the relevant parameters. In the context of this figure, the coordinate system used in our discussion takes the x-direction to increase toward the right, so that α and k_- correspond to motion in the negative x-direction, and β and k_+ correspond to motion in the positive x-direction

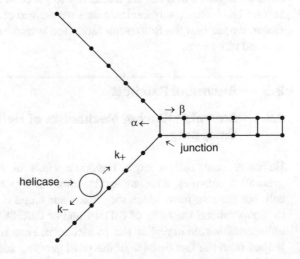

[7] http://en.wikipedia.org/wiki/PCR
[8] http://en.wikipedia.org/wiki/DNA_testing

Since the mean of the sum of n identically distributed, statistically independent random variables is n times the mean of one of them, then in a time $t = n\Delta t$ the mean displacement of the ssDNA–dsDNA junction is

$$x_J = n\Delta x_J = \frac{t}{\Delta t}\Delta x = \frac{t}{\Delta t}b(\beta - \alpha)\Delta t = b(\beta - \alpha)t. \tag{8.41}$$

The corresponding drift velocity of the ssDNA–dsDNA junction is

$$v_J = b(\beta - \alpha). \tag{8.42}$$

This is the drift velocity of a dsDNA–ssDNA junction in terms of the zipping up rate (α) and the unzipping rate (β).

8.4.2   Connection to the Second Law of Thermodynamics

As defined in Eq. 8.33, ΔG is the change in free energy that occurs when one additional base pair is unzipped. Thus, as far as this expression/definition for ΔG is concerned, the final "product" state is the unzipped state, and the initial "reactant" state is the zipped state. Thus, unzipping corresponds to the forward direction of the reaction. This discussion makes contact with what you learned in chemistry classes, namely that a reaction proceeds forward if ΔG is negative. Specifically, we see from Eq. 8.39 that the unzipping reaction proceeds forward, which requires $\beta > \alpha$, only for $\Delta G < 0$, exactly as we are told in chemistry classes. Here, though, this result was derived on the basis of the Boltzmann factor. In fact, the statement that "ΔG must be negative in order for a reaction to proceed forwards" is equivalent to the second law of thermodynamics. Since we arrived at this result via the Boltzmann factor, we see that the Boltzmann factor too is another manifestation of the second law and *vice versa*.

8.5   Brownian Ratchets

8.5.1   Brownian Ratchet Mechanism of Helicase-Catalyzed Unzipping

Helicases constitute a large, important class of motor proteins (a.k.a. force-generating enzymes, a.k.a. molecular motors), which perform myriad tasks in the cell, but the role from which they take their name is the ability of many helicases to unzip/unwind the helix of dsDNA and/or dsRNA [12]. Thus, helicase plays an indispensable role in replication, in particular. How important helicases are may be judged from the fact that 4% of the yeast genome codes for some kind of helicase, which is a lot. Werner syndrome in humans,[9] which gives rise to accelerated aging,

---

[9] http://en.wikipedia.org/wiki/Werner_syndrome

is caused by a mutation in the WRN gene that codes for a certain helicase [13]. Recently, helicases have been proposed as targets for novel anti-viral drugs[10] [14].

We have just seen that for temperatures below the DNA melting temperature, ΔG is negative for DNA zipping. Since helicases unzip DNA, corresponding to a positive ΔG, it initially seems that helicases might permit violation of the second law of thermodynamics. But it is not so: helicases (and other motor proteins) actually catalyze ATP-to-ADP hydrolysis, which gives rise to a corresponding decrease in free energy. As we will see in detail, the net change in free energy, including the contributions both from unzipping the base pair and from ATP-to-ADP hydrolysis, must be negative in order for the helicase to successfully unzip dsDNA, and consequently, the second law is not violated. Phew!

Many animations are available depicting the role of helicase at the replication fork. Wonderful as they are, a misleading aspect of these movies is that they suggest that everything proceeds as regularly as it might on a car production line. In fact, all of the component processes occurring in the helicase movies should be viewed as (biased) random walks. We will see in detail how this plays out in the context of a helicase unzipping dsDNA.

The proposed Brownian ratchet mechanism for how helicase unzips DNA is as follows [15]. The helicase steps unimpeded (via a biased random walk , i.e., a random walk with a drift velocity) on ssDNA toward the replication fork, until it encounters the replication fork, which then prevents its further progress. However, at the replication fork, according to the Boltzmann factor, there is a non-zero probability per unit time for the fork to unzip one base pair. There is then the possibility for the helicase to step into the just-unzipped position. If the helicase does this, it prevents the DNA from zipping back up again. In this way, the replication fork is moved one step toward unzipping. Repeating this process many times leads to the complete unzipping of the DNA. Because this mechanism relies on thermal fluctuations to unzip the DNA, it is said to be a *Brownian ratchet* mechanism. (A ratchet is a device that permits motion in only one direction.)

8.5.2 Helicase Translocation as a Random Walk

It will be useful to consider translocation of a helicase on ssDNA, conceived as a random walk. The probability of the helicase stepping one base pair toward the junction $(+b)$ in a time Δt is $k_+\Delta t$, and the probability stepping one base pair away from the junction $(-b)$ in a time Δt is $k_-\Delta t$, where k_+ and k_- are the rate of stepping toward the junction and the rate of stepping away from the junction, respectively. Since probabilities sum to unity, and we assume that the only three possibilities in a small time Δt are to step toward the junction one base pair or to step away from the junction one base pair or to not do anything, we must have that the probability to do nothing is $1 - k_+\Delta t - k_-\Delta t$.

---

[10] http://www.nature.com/nrd/journal/v4/n10/full/nrd1853.html

Given these probabilities, and the length of a base pair, b, we may calculate the mean displacement of the helicase in a time Δt:

$$\Delta x_H = (+b)(k_+ \Delta t) + (0)(1 - k_+ \Delta t - k_- \Delta t) + (-b)(k_- \Delta t) = b(k_+ - k_-)\Delta t. \tag{8.43}$$

Since the mean of the sum of n identically distributed, statistically independent random variables is n times the mean of one of them, then in a time $t = n\Delta t$ the mean displacement of the helicase is

$$x_H = n\Delta x_H = \frac{t}{\Delta t}\Delta x = \frac{t}{\Delta t}b(k_+ - k_-)\Delta t = b(k_+ - k_-)t. \tag{8.44}$$

The corresponding drift velocity of the helicase is

$$v_H = b(k_+ - k_-). \tag{8.45}$$

This is the drift velocity of a helicase in terms of the stepping toward the junction rate (k_+) and the stepping away from the junction rate (k_-). Just like Eq. 8.42, Eq. 8.45 is appropriate *only* when the helicase is far from the junction.

There is one more important point to make concerning helicase translocation on ssDNA: as we have seen in the examples so far, the ratio of forward and backward rates is given by a change in free energy. Thus, for helicase stepping, we must expect, in analogy with Eq. 8.39, that the ratio of stepping rates is given by

$$\frac{k_+}{k_-} = e^{-\Delta G'/(k_B T)}, \tag{8.46}$$

where $\Delta G'$ is a free energy change. But what free energy change? The answer can be gleaned from the observation that helicases can be thought of as enzymes, which catalyze ATP-to-ADP hydrolysis, which is coupled to the helicase's translocation toward the ssDNA–dsDNA junction. It follows that $\Delta G'$ in Eq. 8.46 corresponds to the free energy difference between ADP-plus-phosphate and ATP. Because the binding energy involved in ADP-plus-phosphate is more negative than the binding energy involved in ATP, $\Delta G'$ is negative. At the same time, according to Eq. 8.46, a negative $\Delta G'$ ensures that $k_+ > k_-$ and the helicase indeed translocates on ssDNA preferentially toward the ssDNA-to-dsDNA junction.

8.5.3 Clash of the Titans

So far, we have considered the dsDNA–ssDNA junction and the helicase in isolation, far apart from each other. However, to determine how helicase unzips dsDNA, we need to figure out what happens when the junction and the helicase come into close proximity, given that they cannot cross each other. They cannot cross because the helicase only translocates on ssDNA.

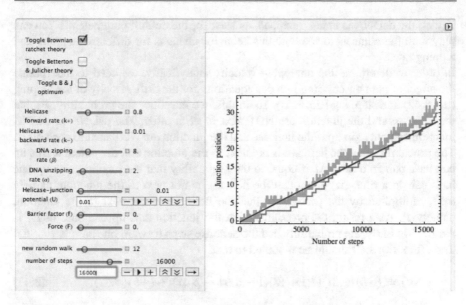

Fig. 8.8 A screenshot of the demonstration at http://demonstrations.wolfram.com/ UnzippingDoubleStrandedDNAByHelicase/, showing two non-crossing random walks with non-zero drift velocities of opposite sign, each random walk consists of 16,000 steps, each one of which is supposed to correspond to a time $\Delta t = 0.01$ s. The orange random walk represents helicase translocation on ssDNA (left) for $k_+ \Delta t = 0.008$ and $k_- \Delta t = 0.0001$. The green random walk represents the position of a ssDNA-to-dsDNA junction for $\alpha \Delta t = 0.02$ and $\beta \Delta t = 0.08$. When the two random walks are in close proximity, evidently, the helicase and the junction track together. Importantly, the helicase moves in the positive direction, which is the same direction as in the absence of the junction. By contrast, the junction moves with the helicase, which is in the reverse direction compared to without the helicase, and corresponds to unzipping dsDNA. Thus, the helicase indeed unzips dsDNA. The magenta line shows our theoretical expression for the unzipping velocity (Eq. 8.58)

First, let us look at the DNA unzipping by helicase Wolfram Demonstration simulation at

http://demonstrations.wolfram.com/UnzippingDoubleStrandedDNAByHelicase/.[11]

A screenshot of the simulation is shown in Fig. 8.8. This simulation shows the location of the helicase as the orange trace and the location of the ss-DNA-to-dsDNA junction as the green trace, plotted on the y-axis versus the number of steps on the x-axis. Both traces are simulated as random walks, but the new wrinkle is that they are not permitted to cross each other, because the helicase does not translocate on dsDNA. In the simulation, with the default parameters, evidently, both random walks track upward together. This implies that the DNA zipping is reversed and,

---

[11] http://demonstrations.wolfram.com/UnzippingDoubleStrandedDNAByHelicase/

in fact, the dsDNA is being unzipped, as least for the default parameters!! You can play with the controls to see how this behavior changes for different stepping and zipping rates.

In order to determine the unzipping velocity analytically, we need to incorporate the effect of the non-crossing into our equations for the drift velocity of the zipping (Eq. 8.42) and stepping (Eq. 8.45). To do this, we introduce the probability, P, that the helicase and the junction are NOT next to each other. The dsDNA-to-ssDNA junction can only zip up if the helicase and the junction are NOT next to each other. The probability that the helicase is not next to the junction and the junction zips up one base pair in a time Δt is equal to the probability that the junction zips up one base pair in a time Δt, given that the helicase is not next to the junction, namely $\alpha \Delta t$, multiplied by the probability that the helicase is not next to the junction, namely P. As a result, the probability that the junction zips up one base in Δt is $P \alpha \Delta t$. Similarly, the probability that the helicase steps forward one base is $P k_+ \Delta t$. Therefore, Eq. 8.40 should be modified to read

$$\Delta x_J = (-b)(\alpha \Delta t)(P) + (0)(1 - \alpha \Delta t - \beta \Delta t) + (+b)(\beta \Delta t), \qquad (8.47)$$

i.e.,

$$\Delta x_J = b(\beta - \alpha P)\Delta t. \qquad (8.48)$$

Similarly, Eq. 8.43 should be modified to read

$$\Delta x_H = b(k_+ P - k_-)\Delta t. \qquad (8.49)$$

It follows that the drift velocities are modified to read

$$v_J = b(\beta - \alpha P), \qquad (8.50)$$

and

$$v_H = b(k_+ P - k_-). \qquad (8.51)$$

However, what we can clearly see from the simulation is that for a large enough number of steps, the junction and the helicase indeed track each other for all times. This is because the helicase steps to, but not beyond the junction, while the junction zips up to, but not beyond the helicase, so that the position of the helicase and the position of the junction necessarily track each other. It follows therefore that while the helicase is unzipping DNA, the helicase and the junction must have the same drift velocity, i.e.,

$$v_H = v_J \qquad (8.52)$$

or

$$b(k_+ P - k_-) = b(\beta - \alpha P). \tag{8.53}$$

We can solve this equation to determine P:

$$k_+ P - k_- = \beta - \alpha P, \tag{8.54}$$

i.e.,

$$\beta + k_- = \alpha P + k_+ P, \tag{8.55}$$

i.e.,

$$P = \frac{k_- + \beta}{k_+ + \alpha}. \tag{8.56}$$

Furthermore, we can use this expression for P (Eq. 8.56) to determine the drift velocity at which the helicase unzips the dsDNA by substituting Eq. 8.56 into Eq. 8.51. Setting $v_J = v_H = v$, we find

$$v = b\left(k_+ \frac{k_- + \beta}{k_+ + \alpha} - k_-\right) = b\frac{k_+ (k_- + \beta) - k_-(k_+ + \alpha)}{k_+ + \alpha} = b\frac{k_+\beta - k_-\alpha}{k_+ + \alpha}. \tag{8.57}$$

It is instructive to divide the numerator and the denominator by αk_+

$$v = b\left(\frac{\frac{\beta}{\alpha} - \frac{k_-}{k_+}}{\frac{1}{\alpha} + \frac{1}{k_+}}\right). \tag{8.58}$$

Notice that the numerator in Eq. 8.58 is the difference of two rate ratios. Using Eqs. 8.39 and 8.46 in Eq. 8.58, we have

$$v = b\left(\frac{e^{-\Delta G/(k_B T)} - e^{\Delta G'/(k_B T)}}{\frac{1}{k_+} + \frac{1}{\alpha}}\right) = be^{-\Delta G/(k_B T)}\left(\frac{1 - e^{(\Delta G + \Delta G')/(k_B T)}}{\frac{1}{k_+} + \frac{1}{\alpha}}\right). \tag{8.59}$$

Remarkably, Eq. 8.59 informs us that whether or not helicase unzips dsDNA depends solely on whether $\Delta G + \Delta G' < 0$ or not, irrespective of the individual values of α, β, k_+, and k_-. For $\Delta G + \Delta G' < 0$, the drift velocity of the helicase plus junction is positive, corresponding to the helicase unzipping the dsDNA.

On the other hand, the net change in free energy for one step forward of both junction and helicase equals $\Delta G + \Delta G'$. According to chemistry classes, this reaction can only proceed forward, i.e., helicase can only unzip dsDNA, if the total change in free energy is negative, that is, if $\Delta G + \Delta G' < 0$. This condition is exactly the same condition that we discovered in the last paragraph from Eq. 8.59.

Once again, we see that the statistical mechanics of thermal ratchets is in agreement with the second law of thermodynamics and with chemistry classes.

In fact, for one base pair, we have $\Delta G \simeq 3k_B T$, while for the hydrolysis of one ATP molecule, we have $\Delta G' \simeq -16k_B T$, so indeed the helicase has a plenty of free energy to do its work. Indeed, from a free-energetic point of view, it seems that one ATP hydrolysis cycle could unzip up to about 4 or 5 base pairs.

Beautiful, single-helicase experiments [16] suggest that the simple Brownian ratchet mechanism of helicase DNA unzipping activity, presented here, should be refined by incorporating both a softer repulsive potential between the helicase and the ds-to-ss junction than the hard-wall potential implicit in our discussion, and suitable free energy barriers between different microstates of the helicase and junction [15]. Our Mathematica Demonstration implements one example of a such soft potential and free energy barrier, and using sliders within the Mathematica Demonstration, it is possible to explore the effects of varying the potential and the barrier, as well as the effects of varying α, β, k_+, and k_-.

8.5.4   Example: Force Generation by Actin Polymerization: Another Brownian Ratchet

We now turn to force generation by actin polymerization, which is the physical basis of eukaryotic cell motility in many cases. Examples of the sort of phenomena that we seek to understand are shown in the movies of Fig. 8.9. The mechanism by which actin or tubulin polymerization exerts a force may also be conceived as a Brownian ratchet [19–21] and is schematically illustrated in Fig. 8.10. In the case of a load, f, applied to a cell membrane, the cell membrane is in turn pushed against the tip of an actin filament, which usually prevents the addition of an additional actin monomer (G-actin) of length a to the tip of the actin filament (F-actin). However, with probability specified by a Boltzmann factor, the membrane's position relative to the tip, z, occasionally fluctuates far enough away from the filament tip ($z > a$) to allow a monomer to fit into the gap. If a monomer does indeed insert and add to the end of the filament, the result is that the filament and therefore the membrane move one step forward, doing work against the load force. Repeating this many times for many such filaments gives rise to cell motility against viscous forces (Figs. 8.11 and 8.12).

(continued)

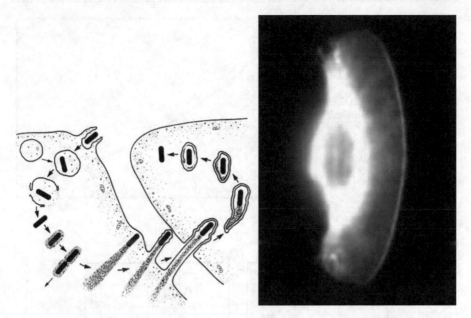

Fig. 8.9 Actin polymerization in action. Left: *Listeria* invade host cells, escape their vacuole, and grow an actin comet tail, which propel each bacterial cell through the cytoplasm and can produce enough force to push into adjacent cells, adapted from Ref. [17] https://commons. wikimedia.org/wiki/File:J_Cell_Biol_2002_Aug_158(3)_409-14,_Figure_1.png. Licensed under the Creative Commons Attribution-Share Alike 3.0 United States license. Right: a portion of a movie frame from Ref. [18], showing a crawling keratocyte, which is a type of skin cell. File licensed under the Creative Commons Attribution 2.5 Generic license

Fig. 8.10 Cartoon illustrating how actin polymerization can do work against a load (f), applied to a membrane against which the polymerizing actin filament (F-actin) abuts. Only if the gap, z, between the tip of the actin filament and the membrane exceeds the length, a, of a G-actin monomer, is it possible for the filament to grow?

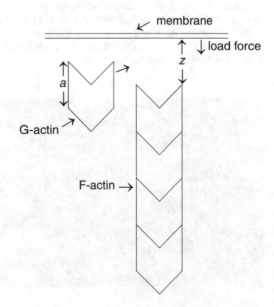

Fig. 8.11 Velocity versus force for microtubule growth from Ref. [20]. Solid squares are experimental measurements, based on the shape and length of microtubules. The solid line is a best fit to the theoretical prediction (Eq. 8.67)

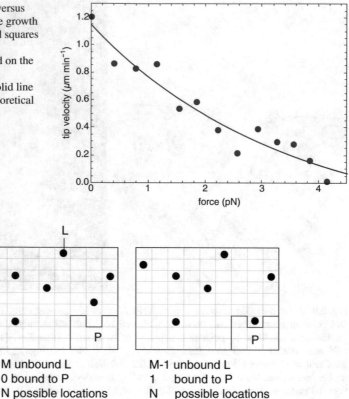

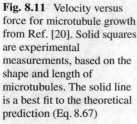

M unbound L
0 bound to P
N possible locations

M-1 unbound L
1 bound to P
N possible locations

Fig. 8.12 Schematic illustration of our simple model of binding. The solution is divided into N cells each of volume v. Left: Protein P does not bind a ligand, so that a total of M ligands are distributed among the N cells. In this case, there are a total of g_1 microstates. Right: Protein P does bind a ligand, so that now a total of $M - 1$ ligands are distributed among the N cells. In this case, there are a total of g_2 microstates

(a) What is the probability that the tip steps forward in a time Δt in the absence of a nearby membrane?

Answer:

$$k_+ n \Delta t, \tag{8.60}$$

where n is the concentration of actin monomers, k_+ is the monomer-tip association rate constant,

(continued)

and k_- is the monomer dissociation rate from the tip.

(b) What is the mean displacement of the filament tip in a time Δt in the absence of a nearby membrane? What is the corresponding tip drift velocity?

Similarly to Eqs. 8.40 and 8.43, we have

$$\Delta x = a(k_+ n - k_-)\Delta t, \tag{8.61}$$

where n is the concentration of actin monomers, a is the size of an actin monomer, k_+ is the actin association rate constant, and k_- is the dissociation rate. The corresponding tip drift velocity is

$$v = a(k_+ n - k_-). \tag{8.62}$$

(c) If the membrane is nearby, however, it is only possible to add an actin monomer, if the tip-membrane distance is greater than a. Letting P be the probability that the filament-membrane distance is greater than a, write the probability that both the filament-membrane distance is a or greater and the tip steps forward in time Δt.

Answer: The probability that both the size of the tip-membrane gap is a or greater and the tip adds a monomer in Δt equals the probability that the tip adds a monomer in Δt, given that the gap is a or greater, namely $k_+ nt$, multiplied by the probability that gap is a or greater, namely P. Thus, the desired probability is

$$P k_+ n\Delta t. \tag{8.63}$$

(d) What is the mean distance that the tip moves in Δt, if the membrane is nearby? What is the corresponding tip drift velocity?

Answer: Equation 8.67 is modified to read

$$\Delta x = a(P k_+ n - k_-)\Delta t. \tag{8.64}$$

(continued)

The corresponding tip drift velocity is

$$v = a(Pk_+n - k_-).\tag{8.65}$$

(e) Suppose that there is a load, f, on the membrane, pushing the membrane toward the tip. What then is P? What is the corresponding tip drift velocity?

Answer: Equation 8.14 informs us that the probability that the membrane-tip gap is greater than a is

$$P = e^{-fa/(k_BT)}.\tag{8.66}$$

It follows that the tip drift velocity is

$$v = a(e^{-fa/(k_BT)}k_+n - k_-).\tag{8.67}$$

This form is compared to experiment in Fig. 8.11.

(f) What is the force exerted by the tip?

Answer: Although a load, f, is applied to the membrane, because the tip's velocity is constant and the membrane moves together with the tip, the membrane's velocity is also constant. Therefore, according to Newton's second law, there can be no net force on the membrane. Therefore, we may deduce that the load is balanced by an equal and opposite force, generated by the polymerization ratchet, exerted by the tip on the membrane.

(g) Calculate the "stall force," namely the force at which the velocity of the tip goes to zero.

Answer: Using Eq. 8.67 with $v = 0$, and then solving for the corresponding value of f, we find that the stall force is given by

(continued)

$$f = \frac{k_B T}{a} \log \frac{k_+ n}{k_-}. \tag{8.68}$$

Equation 8.67 is the force–velocity relationship for a tubulin [20] or actin [21] filament. A beautiful experiment that demonstrates force generation by polymerization of a *single* actin filament is shown in 05902Movie21.mov, which is available as Supplementary Information in Ref. [21]. In this movie, you can see that as the filament polymerizes and elongates, because it is attached at both ends to the microscope cover slide, it necessarily bends. Therefore, the polymerization of the filament provides sufficient force to bend the filament, and, in fact, this observation is the basis for an elegant assay to measure the polymerization force, described in Refs. [20] and [21].

Creating an understanding of initially apparently disparate phenomena, namely helicase unzipping of dsDNA and force generation by actin polymerization, using a common set of basic concepts—in this case random walks and the Boltzmann factor, leading to the concept of a Brownian ratchet—is quintessentially "physics."

8.6   Binding and Reactions

8.6.1   Langmuir Binding Curve

A prominent application of statistical mechanics is to molecular binding, which is tremendously important in biochemistry and molecular biology. In this example, we will consider a single protein, P, that binds ligand, L with binding energy ϵ, which must be negative, of course, to achieve binding, and we will calculate how the probability that the protein binds ligand depends on ligand concentration. An important step in this calculation is to enumerate the number of microstates of ligands in solution. To this end, we employ a simple model for the ligand solution, illustrated in Fig. 8.12, which is as follows: we envision M ligand molecules, each of volume v, in a total volume of fluid V. We will therefore suppose that there are $N = V/v$ possible locations in the total volume V where a ligand can be situated. Each one of these locations is either occupied by a ligand, or it is unoccupied by a ligand.

(a) If there are N possible locations to put a ligand in the solution, how many ways are there to arrange j ligands?

    Answer: The answer to this question is the same
    as the answer to the question: How many ways are

(continued)

there to realize j heads when you toss a coin N
times? The answer is the binomial coefficient "N
choose j," i.e.,

$$\frac{N!}{j!(N-j)!}. \tag{8.69}$$

(b) What is the number of microstates of the system, when no ligand binds
the protein (g_1)?

Answer: if a ligand is not bound to a protein,
then there are M ligands in solution, and g_1 is
given by Eq. 8.69 with $j = M$, that is,

$$g_1 = \frac{N!}{M!(N-M)!}. \tag{8.70}$$

(c) What is the number of microstates of the system, when a ligand is bound
to the protein (g_2)?

Answer: When the protein binds a ligand, the
number of ligand molecules in solution is
decreased from M to $M-1$. Therefore, g_2 is given
by Eq. 8.69 with $j = M-1$, that is,

$$g_2 = \frac{N!}{(M-1)!(N-M+1)!}. \tag{8.71}$$

(d) Apply the Boltzmann factor to calculate the probability (Q) that protein
P binds a ligand in terms of ϵ, $k_B T$, g_1, and g_2.

Answer:

$$Q = \frac{g_2 e^{-\epsilon/(k_B T)}}{g_1 + g_2 e^{-\epsilon/(k_B T)}}. \tag{8.72}$$

(e) Using the answers to (b) and (c), express Q in terms of M and N, instead
of g_1 and g_2.

Answer: It is now convenient to divide both the
numerator and the denominator by g_1 so that

(continued)

$$Q = \frac{\frac{g_2}{g_1} e^{-\epsilon/(k_B T)}}{1 + \frac{g_2}{g_1} e^{-\epsilon/(k_B T)}}. \tag{8.73}$$

Now, using our expressions for g_1 and g_2, we find

$$\frac{g_2}{g_1} = \frac{\frac{N!}{(M-1)!(N-M+1)!}}{\frac{N!}{(M)!(N-M)!}} = \frac{M}{N-M+1}. \tag{8.74}$$

Using this result in Eq. 8.73, we find

$$Q = \frac{\frac{M}{N-M+1} e^{-\epsilon/(k_B T)}}{1 + \frac{M}{N-M+1} e^{-\epsilon/(k_B T)}}. \tag{8.75}$$

(f) Show that for realistic ligand concentrations, for which $N \gg M$ and $M \gg 1$, we may express Q in the form

$$Q = \frac{[L]}{[L] + K_D}, \tag{8.76}$$

where $[L]$ is the ligand concentration and K_D is the ligand–protein dissociation constant, and find K_D in terms of given parameters.

Answer: Using $N \gg M$ and $M \gg 1$, Eq. 8.75 can be further simplified to read

$$Q = \frac{\frac{M}{N} e^{-\epsilon/(k_B T)}}{1 + \frac{M}{N} e^{-\epsilon/(k_B T)}}. \tag{8.77}$$

Furthermore, we may express $[L]$ in terms of N, M, and v: $[L] = M/V = M/(Nv)$, so that $M/N = v[L]$. It then follows that

$$Q \simeq \frac{\frac{M}{N} e^{-\epsilon/(k_B T)}}{1 + \frac{M}{N} e^{-\epsilon/(k_B T)}} = \frac{v[L] e^{-\epsilon/(k_B T)}}{1 + v[L] e^{-\epsilon/(k_B T)}} = \frac{[L]}{[L] + \frac{e^{+\epsilon/(k_B T)}}{v}}. \tag{8.78}$$

This is the typical non-cooperative binding curve, familiar in biochemistry classes, provided we identify the dissociation constant for the binding reaction as

(continued)

$$K_D = \frac{e^{+\epsilon/(k_B T)}}{v}. \tag{8.79}$$

Equation 8.79 represents a microscopic model for the dissociation constant, which is pretty cool. Notice that for strong binding, ϵ is large and negative, so that K is small, which is the correct behavior.

8.6.2 Chemical Equilibrium and the Law of Mass Action

Another key application of statistical mechanics is to chemical reaction equilibria, which are tremendously important in chemistry. To understand what statistical mechanics tells us about chemical reactions, we consider a binary chemical reaction in which species P and species L react to form the complex PL:

$$\text{P+L} \; \underset{k_-}{\overset{k_+}{\rightleftharpoons}} \; \text{PL}$$

where $\frac{k_-}{k_+} = K_D$ is the dissociation constant of this reaction.

In chemistry classes, we learn that

$$\frac{[\text{P}][\text{L}]}{[\text{PL}]} = K_D, \tag{8.80}$$

where [P] is the concentration of species P, [L] is the concentration of species L, and [PL] is the concentration of the complex PL. This is the law of mass action for a binary chemical reaction, which is possibly the most important equation in all of chemistry?

As we learned in Chap. 4. *Probability Distributions: Mutations, Cancer Rates, and Vision Sensitivity*, a probability is the ratio of numbers. Therefore, we may readily calculate the probability that P binds L, which we previously called Q, in terms of these concentrations:

$$Q = \frac{[PL]}{[PL] + [P]}. \tag{8.81}$$

If we multiply top and bottom by K_D and divide top and bottom by $[P]$, we find

$$Q = \frac{\frac{K_D[PL]}{[P]}}{\frac{K_D[PL]}{[P]} + K_D}. \tag{8.82}$$

On the other hand, according to Eq. 8.76, the same quantity is

$$Q = \frac{[L]}{[L] + K_D}. \tag{8.83}$$

By comparing these two expressions for Q, we see that

$$\frac{K_D[PL]}{[P]} = [L] \tag{8.84}$$

or

$$\frac{[L][P]}{[PL]} = K_D, \tag{8.85}$$

which is the law of mass action for a binary chemical reaction. Therefore, we see that the law of mass action—the most important formula in chemistry?—is a result of the Boltzmann factor. How cool is that?

Is there any result in statistical mechanics/physical chemistry that can *not* be derived using the Boltzmann factor? The answer is "no." As Feynman said, the Boltzmann factor is the pinnacle and we are just sliding down.

8.7 Entropy, Temperature, and the Ideal Gas Law

8.7.1 Entropy

In traditional statements of the second law of thermodynamics, the entropy is central but has hardly appeared in our discussions so far. How come? Actually, it has been with us all along. This is because the entropy, S, is k_B times the natural logarithm of the number of microstates, g. Thus, for example, the entropy of the unfolded state of a protein is $S = k_B \log g$. In our DNA melting example, $g = \frac{g_1}{g_2}$ was the ratio of the number microstates for one unzipped base pair to the number of microstates for one zipped-up base pair. In this case, the entropy difference, ΔS, between one unzipped and one zipped-up base pair, is

$$\Delta S = S_1 - S_2 = k_B(\log g_1 - \log g_2) = k_B \log \frac{g_1}{g_2} = k_B \log g, \tag{8.86}$$

where S_1 is the entropy of ssDNA and S_2 is the entropy of dsDNA. Finally, recall that we wrote for the free energy

$$\Delta G = \epsilon - k_B T \log g, \tag{8.87}$$

which, we now see, may be written

$$\Delta G = \epsilon - T\Delta S, \tag{8.88}$$

which expresses the relationship between ΔG and ΔS.

8.7.2   Temperature

The absolute temperature, T, is an essential ingredient when we write the Boltzmann factor, and although we all have a sense of what is temperature—hot versus cold—it is possible to give a more precise expression, which will define temperature as far as we are concerned. Specifically, by applying the Boltzmann factor to determine the distribution of velocities in a gas or a liquid, and the mean square velocity, we can specify precisely what temperature is.

The exponent in the Boltzmann factor is the energy, irrespective of the type of energy. Therefore, it applies to kinetic energy. It immediately follows that the probability density that a molecule in a fluid or a gas or even a solid at a temperature T has velocity $\mathbf{v} = v_x\mathbf{i} + v_y\mathbf{j} + v_z\mathbf{k}$ is

$$p(\mathbf{v}) = p(v_x, v_y, v_z) = \frac{e^{-\frac{mv^2}{2k_BT}}}{Z} = \frac{e^{-\frac{m(v_x^2+v_y^2+v_z^2)}{2k_BT}}}{Z}, \tag{8.89}$$

where Z is the partition function (for normalization), m is the mass of the molecule, and $v^2 = v_x^2 + v_y^2 + v_z^2$. In order to calculate Z in this case, it is convenient to work in terms of v_x, v_y, and v_z with the result that

$$Z = \int_{-\infty}^{\infty} dv_x \int_{-\infty}^{\infty} dv_y \int_{-\infty}^{\infty} dv_z e^{-\frac{m(v_x^2+v_y^2+v_z^2)}{2k_BT}}$$

$$= \int_{-\infty}^{\infty} dv_x e^{-\frac{mv_x^2}{2k_BT}} \int_{-\infty}^{\infty} dv_y e^{-\frac{mv_y^2}{2k_BT}} \int_{-\infty}^{\infty} dv_z e^{-\frac{mv_z^2}{2k_BT}}. \tag{8.90}$$

Notice that this expression for Z is actually the product of three identical integrals because v_x, v_y, and v_z are just integration variables, and therefore, we have that

$$Z = \left(\int_{-\infty}^{\infty} du\, e^{-\frac{mu^2}{2k_BT}} \right)^3, \tag{8.91}$$

where here we are writing the integration variable as u, to emphasize that what we call it does not matter at all. We may evaluate this integral with the result that

$$Z = \left(\frac{2\pi k_BT}{m} \right)^{\frac{3}{2}}. \tag{8.92}$$

Therefore, we have that

$$p(\mathbf{v}) = p(v_x, v_y, v_z) = \left(\frac{m}{2\pi k_B T}\right)^{\frac{3}{2}} e^{-\frac{m(v_x^2 + v_y^2 + v_z^2)}{2k_B T}}$$

$$= \left(\frac{m}{2\pi k_B T}\right)^{\frac{3}{2}} e^{-\frac{mv_x^2}{2k_B T}} e^{-\frac{mv_y^2}{2k_B T}} e^{-\frac{mv_z^2}{2k_B T}}. \tag{8.93}$$

Evidently, Eq. 8.93 is the product of three factors, one for each direction, i.e.,

$$p(v_x, p_y, p_z) = p(v_x)p(v_y)p(v_z). \tag{8.94}$$

Therefore, we may immediately write down the probability density for v_x, for example:

$$p(v_x) = \left(\frac{m}{2\pi k_B T}\right)^{\frac{1}{2}} e^{-\frac{mv_x^2}{2k_B T}}, \tag{8.95}$$

with identical expressions for v_y and v_z (except with v_y and v_z replacing v_x, of course). Equation 8.95 corresponds to a Gaussian distribution of v_x with a mean square value given by

$$\langle v_x^2 \rangle = \frac{k_B T}{m} \tag{8.96}$$

or, equivalently,

$$\frac{1}{2}m\langle v_x^2 \rangle = \frac{1}{2}k_B T. \tag{8.97}$$

It follows that the mean kinetic energy of a molecule is

$$\frac{1}{2}m\langle \mathbf{v}^2 \rangle = \frac{1}{2}m(\langle v_x^2 \rangle + \langle v_y^2 \rangle + \langle v_z^2 \rangle) = \frac{3}{2}k_B T, \tag{8.98}$$

where $\langle \mathbf{v}^2 \rangle$ is the mean square velocity. (Because the velocity of a molecule is equally likely to be directed in any direction, the mean velocity is zero: $\langle \mathbf{v} \rangle = \mathbf{0}$, but the mean square velocity is evidently non-zero.) Importantly, Eq. 8.98 shows that we may interpret temperature in terms of the mean kinetic energy of a molecule. Specifically, the mean kinetic energy of a molecule is $\frac{3}{2}k_B T$. Thus, we see that temperature is directly related to the mean kinetic energy of a molecule.

8.7.3   The Ideal Gas Equation of State

Another important result that follows, in part, from our discussion of temperature concerns the pressure of a gas. Recall from our earlier discussions that the pressure on a wall that results from a number density n of particles, each of mass m, one-half of which are heading toward a wall with a velocity component toward a wall of v_x and then bounce back from the wall with velocity component $-v_x$ was

$$P = nmv_x^2. \tag{8.99}$$

The calculation goes as follows: the change in x-momentum of each molecule when it hits the wall is $2mv_x$. The volume that hits an area A of the wall in a time Δt is $Av_x \Delta t$. The number density of molecules in that volume headed toward the wall is $\frac{1}{2}n$, so that the total change in momentum in Δt is $\Delta p = 2mv_x \times \frac{1}{2}nAv_x \Delta t$. But $\frac{\Delta p}{\Delta t} = F = nmv_x^2 A$. Equation 8.99 then follows from $P = F/A$.

For particles with a distribution of v_xs, we may therefore expect

$$P = nm \left\langle v_x^2 \right\rangle. \tag{8.100}$$

Substituting Eq. 8.97 into Eq. 8.100, we find

$$P = nk_B T = \frac{Nk_B T}{V} = \frac{(N/N_A)(N_A k_B)T}{V} = \frac{N_M RT}{V} \tag{8.101}$$

or

$$PV = N_M RT, \tag{8.102}$$

where N is the number of molecules, N_A is Avogadro's number, $N_M = N/N_A$ is the number of moles, and $R = N_A k_B$ is the gas constant. Equation 8.101 or 8.102 is the ideal gas law!

8.8   Chapter Outlook

We will encounter a number of statistical mechanical arguments in later chapters, but statistical mechanics represents an essential element of the physicist's toolkit for understanding, describing, and predicting nature and is an important destination in its own right. Statistical mechanics elucidates many, many topics across the sciences, including many topics in biology. In this chapter, we have covered just a small selection of the topics about which "stat. mech." has something important to say.

8.9 Problems

Problem 1: Oxygen Binding by Hemoglobin

Figure 8.13 shows a number of curves corresponding to oxygen binding as a function of oxygen partial pressure by canine hemoglobin (solid curves) at various CO_2 partial pressures and equine hemoglobin at a CO_2 partial pressure of 5 mm of Hg. Discuss what features of these curves make hemoglobin an appropriate oxygen delivery vehicle in our bodies.

Problem 2: Chemistry Classes

In chemistry classes, we learn about chemical reactions. Explain what determines whether a chemical reaction proceeds forward or not.

Problem 3: Mean and Variance of the Energy

A certain microscopic system has a total of 5 microstates. It has 3 microstates of zero energy and 2 microstates of energy ϵ.

(a) What is the probability that the system in question has zero energy? You may find it useful to write Z as an intermediate step in your calculation.

(b) What is the probability that the system in question has energy ϵ?

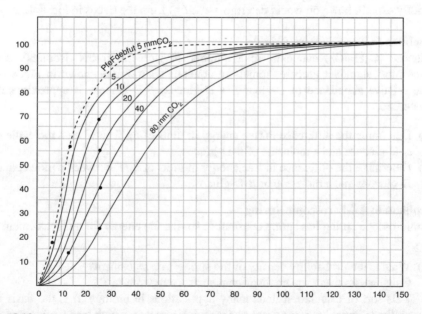

Fig. 8.13 Oxygen binding by dog blood (solid lines) and horse blood (dashed line), expressed as a percent of the maximum possible on the vertical axis, plotted as a function of oxygen partial pressure on the horizontal axis, measured in mm of Hg, at different partial pressures of carbon dioxide, also measured in mm of Hg, reproduced from Ref. [22]

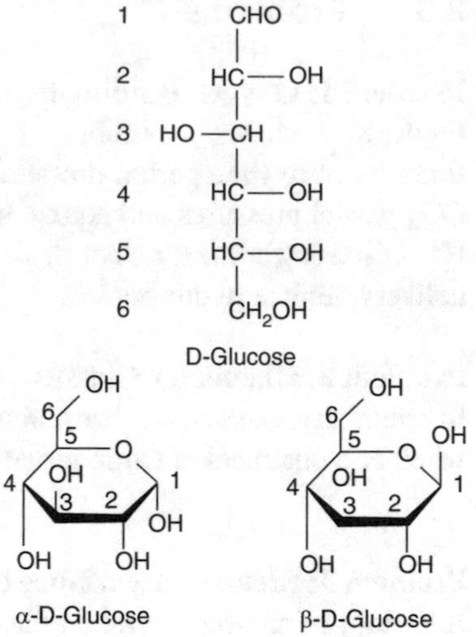

Fig. 8.14 The three most abundant isomers of glucose in aqueous solution

D-Glucose

α-D-Glucose   β-D-Glucose

(c) What is the mean energy, $\langle E \rangle$, of this system?

(d) What is the mean square energy, $\langle E^2 \rangle$, of this system?

Problem 4: Free Energy Differences from Hopping Data

Briefly explain how you would determine $\Delta G(f)$ from the traces in Fig. 8.4.

Problem 5: The Sweetest Thing[12]

Glucose ($C_6H_{12}O_6$) can exist in a number of isomeric forms (isomers). The three most abundant glucose isomers in aqueous solution are shown in Fig. 8.14. The relative abundance of D-glucose:α-D-glucose:β-D-glucose is approximately 1:144:256.

(a) Determine the difference in free energy in units of $k_B T$ between a molecule of α-D-glucose and a molecule of β-D-glucose.

(b) Determine the difference in free energy in units of $k_B T$ between a molecule of D-glucose and a molecule of β-D-glucose.

Problem 6: A Microscopic Spring

A microscope spring has spring constant k, so that the energy for a displacement x is $\frac{1}{2}kx^2$.

(a) What is the probability density of the spring's displacement?

(b) Calculate its mean square displacement at a temperature T.

This sort of calculation finds many applications in biology. It is the basis of methods to calibration optical tweezers instruments, which trap a micrometer-sized bead in the focus of a laser bead, which creates a Hooke's Law-type force

---

[12] https://www.youtube.com/watch?v=5WybiA263bw&list=RD5WybiA263bw#t=58

on the trapped bead. It is also the basis of measurements of the mechanical properties of biological materials, for example, the cytoskeleton. The cytoskeletal network acts like a spring whose spring constant can be measured by monitoring via microscopy the fluctuations in the position of a bead attached to the cytoskeleton.

Problem 7: Forced Unzipping of an RNA Hairpin

An RNA hairpin is attached via DNA "handles" between a microscope cover slide and a polystyrene bead. The bead is held in an optical tweezers—the focus of a powerful laser beam. As the cover slide is moved away from the laser focus, a force f is exerted on the RNA hairpin. At zero applied force, the hairpin is "closed," and the difference in free energy between the open state and the closed state of the RNA hairpin is ΔG, which is positive. When a large enough force is applied, the hairpin transitions to the "open" state. Open and closed are the only possible states of the hairpin (Fig. 8.15).

(a) Explain why the potential energy of the RNA hairpin is $2fL$ greater in the closed state than in the open state, where L is the length of the hairpin.
(b) What is the difference in free energy between the open and closed states? i.e., open state free energy minus closed state free energy.

Fig. 8.15 Schematic of closed (left) and open (right) hairpin states involved in "Forced unzipping of an RNA hairpin"

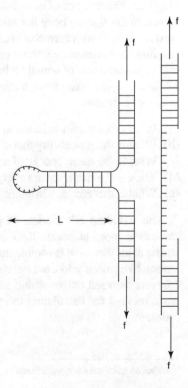

(c) What is the probability that the hairpin is open?
(d) Calculate the force at which the probability is $\frac{1}{2}$, that the hairpin is open.
(e) Calculate the mean free energy of the hairpin at force f, relative to the free energy of the closed state.
(f) Plot $\langle \Delta G(f) \rangle$ and $\Delta G(f)$ versus $\frac{2fL}{k_B T}$ for $0 < \frac{2fL}{k_B T} < 12$ and $\frac{\Delta G}{k_B T} = 6$, and explain your plots.

Problem 8: Partition Coefficient Revisited
Interpret the partition coefficient (B) that we discussed in the context of diffusion through a membrane in light of statistical mechanics, and briefly explain your interpretation.

Problem 9: A Simple Model for Protein Folding
How do I love thee? Let me count the ways. Elizabeth Barrett Browning, Sonnets from the Portuguese.[13]

In order to gain insights into the more complicated real system, theoretical physicists often study simplified models. In this problem, you will study a simplified model of protein folding.

Consider a collection of N two-dimensional self-avoiding "5-mers" consisting of 5 nodes that are located on the vertices of a two-dimensional square lattice, which are held at at a temperature T. Every possible microstate of a 5-mer is shown in Fig. 8.16. The number of non-chain nearest neighbor contacts for each microstate is shown in the figure above the microstate in question. The energy of each 5-mer is zero for 5-mers with zero non-chain nearest neighbor contacts and ϵ for 5-mers with one non-chain nearest neighbor contact. In fact, there are a total of 100 different 5-mer configurations, of which 68 have zero non-chain nearest neighbor contacts, and 32 have one non-chain nearest neighbor contact. 0 have 2 or more non-chain nearest neighbor contacts.

(a) What is the partition function (Z) of a 5-mer?
(b) What is the probability that a 5-mer has 1 non-chain nearest neighbor?
(c) What is the mean energy of a 5-mer?
(d) What is the mean square energy of a 5-mer?
(e) What is the variance in energy of a 5-mer?

The so-called HP models of proteins that theorists have actually studied are somewhat more elaborate than our model, in that they incorporate two types of amino acids that have favorable interactions for non-chain nearest neighbor contacts among like amino acids, but unfavorable interactions for non-chain nearest neighbor contacts between unlike amino acids, and they include more amino acids. But the basic method for calculating the probability that the protein is folded is as in this problem.

---

[13] http://en.wikipedia.org/wiki/Sonnets_From_the_Portuguese

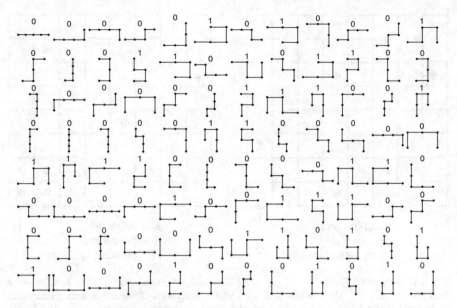

Fig. 8.16 Illustration of every possible distinct microstate for a self-avoiding 5-mer in two dimensions. The numbers above each microstate give the number of non-chain nearest neighbor contacts

Problem 10: Cooperative Binding of Oxygen by Hemoglobin

The goal of this problem is treat the cooperative binding of oxygen by hemoglobin to determine the probability that a hemoglobin molecule binds (four) oxygen molecules (Fig. 8.17).

Follow the same steps as in the chapter, except that, in this case, suppose that either zero or four oxygen molecules can be bound (total binding energy 4ϵ), but that one, two, or three oxygen molecules cannot be bound. This is an example of the so-called cooperative binding. Then, by calculating the number of microstates of the system when four oxygen molecules are bound, you will be able to show that the probability that hemoglobin binds (four) oxygen molecules as a function of the oxygen concentration, $[O_2]$, can be written:

$$P = \frac{[O_2]^4}{[O_2]^4 + K}, \tag{8.103}$$

where K is a quantity—the equilibrium constant—that is independent of $[O_2]$, which you will determine. You should assume that $N \gg M \gg 1$.

(a) How many microstates does this system possess if O_2 is not bound to the hemoglobin?
(b) When hemoglobin binds O_2, the number of O_2 molecules in solution in volume V decreases from M to $M - 4$. How many microstates are there in this case?

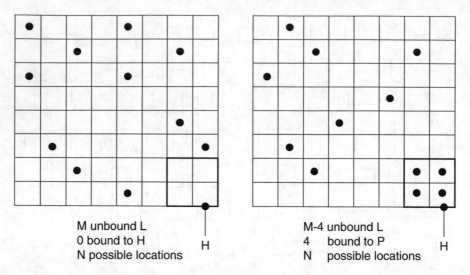

M unbound L
0 bound to H   H
N possible locations

M-4 unbound L
4   bound to P   H
N   possible locations

Fig. 8.17 Schematic illustration of our simple model of cooperative oxygen binding by a hemoglobin molecule. The solution is divided into N cells each of volume v. Left: hemoglobin H does not bind a ligand, so that a total of M ligands are distributed among the N cells. In this case, there are a total of g_1 microstates. Right: hemoglobin binds exactly four ligands; in this simple model, so that now a total of $M - 4$ ligands are distributed among the N cells. In this case, there are a total of g_2 microstates

(c) Apply the Boltzmann factor, and your calculations of the number of microstates in (a) and (b) to calculate the probability (P_{4O_2}) that hemoglobin binds oxygen. Plot your result for $K = 1$ for $0 < [O_2] < 2$.

(d) Oxygen binding by hemoglobin is the canonical example of cooperative binding and gives rise to the characteristically shaped binding curve[14] of Eq. 8.103, which you may have encountered previously in biochemistry or physiology classes? Here, you derived Eq. 8.103 on the basis of the Boltzmann factor. The shape of this curve is, of course, extremely important for our respiration. Briefly explain why.

Problem 11: Gene Expression

A certain system of interest consists of two DNA-binding proteins, call them POL and REPRESSOR, respectively, and a loop of DNA (a plasmid). Both POL and REPRESSOR are always attached to the DNA. There is one copy each of POL and REPRESSOR on the plasmid. Both POL and REPRESSOR can be bound to the DNA at one particularly favorable site for binding, call it the operator, where POL's binding energy is ϵ_P (which is negative), and REPRESSOR's binding energy is ϵ_R (also negative). Or they can each bind at any one of N other sites, where the binding energy is (defined to be) equal to zero. They cannot bind at the same site. Therefore,

---

[14] http://en.wikipedia.org/wiki/Oxygen-hemoglobin_dissociation_curve

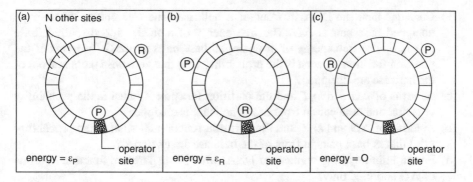

Fig. 8.18 (**a**) POL is bound to the operator, and REPRESSOR is not, (**b**) REPRESSOR is bound to the operator, and POL is not, and (**c**) neither POL nor REPRESSOR is bound to the operator

three configurations of this system may be distinguished, as shown in Fig. 8.18: (a) POL is bound to the operator, and REPRESSOR is not; (b) REPRESSOR is bound to the operator, and POL is not; and (c) neither POL nor REPRESSOR is bound to the operator.

(a) How many microstates are there in the case that the POL (P) is bound to the operator and REPRESSOR (R) is not ?

(b) How many microstates are there in the case that REPRESSOR is bound to the operator and POL is not?

(c) How many microstates are there in the case that neither POL nor REPRESSOR is bound to the operator?

(d) Write the partition function, Z, of this system, i.e., the appropriate normalization factor for the Boltzmann factor.

(e) What is the probability (P_P) that POL binds the operator?

(f) In fact, the rate of expression of a particular gene is often proportional to the probability that POL (a.k.a. RNA polymerase) binds the relevant operator. Briefly explain what happens to P_P and transcription when $e^{-\epsilon_R/(k_B T)} \gg N$ and $e^{-\epsilon_R/(k_B T)} \gg e^{-\epsilon_P/(k_B T)}$.

Problem 12: Unzipping Ratchet Revisited

In this problem, we will develop an alternative model for how helicase unzips DNA. As we discussed, helicase is a motor protein that steps unimpeded, via a biased random walk, on ssDNA toward the ssDNA–dsDNA junction, until it encounters the junction, which then prevents its further progress. However, at the junction, according to the Boltzmann factor, there is a non-zero probability that the junction is unzipped one or more base pairs beyond the helicase. There is then a non-zero probability for the helicase to step into the just-unzipped region. If the helicase does this, it prevents the DNA from zipping back up again. In this way, the junction is moved one step in the unzipping direction. Repeating this process many times leads to the complete unzipping of the DNA.

(a) Consider then the DNA in front of a helicase. The free energy of a single unzipped base pair is ΔG. The free energy of a single zipped (paired) base pair is 0. The free energy of m unzipped base pairs is $m\Delta G$. In terms of the partition function Z, what is the probability (P_0) that zero base pairs in front of the helicase are unzipped?

(b) In terms of ΔG and $k_B T$ and the partition function Z, what is the probability (P_1) that one base pair in front of the helicase is unzipped?

(c) In terms of ΔG and $k_B T$ and the partition function Z, what is the probability (P_m) that m base pairs in front of the helicase are unzipped?

(d) For an infinitely long segment of DNA, what is the partition function in terms of ΔG and $k_B T$ only?

To answer this question, you may find it useful to know that

$$\Sigma_{m=0}^{\infty} e^{-m\Delta G/(k_B T)} = \frac{1}{1 - e^{-\Delta G/(k_B T)}}, \tag{8.104}$$

or that

$$\Sigma_{m=0}^{\infty} m e^{-m\Delta G/(k_B T)} = \frac{e^{-\Delta G/(k_B T)}}{(1 - e^{-\Delta G/(k_B T)})^2}, \tag{8.105}$$

or that

$$\Sigma_{m=1}^{\infty} e^{-m\Delta G/(k_B T)} = \frac{e^{-\Delta G/(k_B T)}}{1 - e^{-\Delta G/(k_B T)}}, \tag{8.106}$$

where $\Sigma_{m=0}^{\infty}$ indicates the sum over all values of m from 0 to ∞.

(e) What is the probability (Q) that one or more base pairs are unzipped in terms of ΔG and $k_B T$ only?

(f) The helicase succeeds in taking a step forward, only if one or more base pairs in front of it are unzipped. What is the probability that the helicase succeeds in taking a step forward in a time Δt, given that its forward stepping rate is k_+?

(g) If the step size is b and the probability that a helicase steps backward in a time Δt is $k_- \Delta t$, what is the mean displacement of the helicase ($\langle w \rangle$) in Δt?

(h) What is the corresponding mean velocity of the helicase ($\langle v \rangle$) in terms of b, $\Delta G, k_B T, k_+, k_-$?

Problem 13: Protein Translocation into Organelles via a Brownian Ratchet Mechanism

There are many instances in the cell where proteins are synthesized in the cytosol and must subsequently pass through a membrane to reach their final location. For example, mitochondrial and endoplasmic reticular proteins must pass through membranes into their respective organelles. Versions of the chaperone protein Hsp70 inside the organelle play a key role in this process via a Brownian ratchet mechanism: inside the organelle, Hsp70 binds to the translocating protein (which is unfolded), preventing it from diffusing back out of the organelle, but nothing

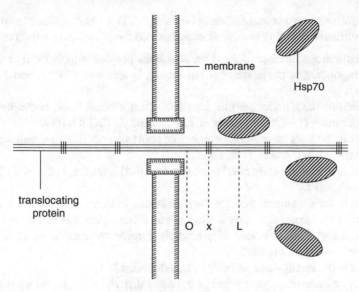

Fig. 8.19 Schematic of the Brownian ratchet mechanism of protein translocation across membranes

prevents diffusion into the organelle. To analyze translocation into organelles, as illustrated in Fig. 8.19, we will envision a translocating protein to be composed of a number of binding sites for Hsp70, each of length L, and we will suppose that the probability that a given Hsp-binding site, exposed within the organelle, is bound by Hsp70 is P.

Your goal in this problem is to determine the mean translocation velocity of proteins into the organelle, which we will do by focusing on the short vertical lines, shown in the figure, that delineate Hsp70 binding sites on the translocating protein, and we consider a population of many translocating proteins so that we can sensibly talk about currents and concentrations of short vertical lines.

If P were equal to unity, this problem would be formally the same as the "Rat ratchet" problem from Chap. 6. *Diffusion: Membrane Permeability and the Rate of Actin Polymerization*, and in a steady state, there would be a current of protein into the organelle, which we will call I. However, P is not unity, and a fraction $1 - P$ of the time, Hsp70, is not bound. As a result, there is a steady-state current $-(1 - P)I$, corresponding to protein motion outward from the organelle. The total current is the sum of these two individual currents.

To determine the number per unit length of these small vertical lines, given an inward current of I and an outward current of $-(1 - P)I$, recall from Chap. 6. *Diffusion: Membrane Permeability and the Rate of Actin Polymerization* that the steady-state current may be correctly conceived as the sum of an inward current, due to a non-zero concentration at boundary 1 ($x = 0$) and zero concentration at boundary 2 ($x = L$), and an outward current due to a non-zero concentration at boundary 2 and a zero concentration at boundary 1. This result follows from the fact that the diffusion equation is a linear equation and the principal of superposition.

(a) Determine the number of short vertical lines per unit length for $0 < x < L$, corresponding to the inward current I only, in terms of I, P, x, and L. Call it $n_1(x)$.

(b) Determine the number per unit length of short vertical lines, corresponding to the current $-(1 - P)I$ in terms of I, P, x, and L. Call it $n_2(x)$.

(c) What is the total steady-state number of short vertical lines per unit length for $0 < x < L$, in terms of I, P, and L?

(d) What is the steady-state number of short vertical lines for $0 < x < L$, in terms of I, P, and L?

(e) What is the net current through the membrane, in terms of I, P, and L?

(f) What is the mean time that a short vertical line spends at locations between $x = 0$ and $x = L$ before fully emerging inside the membrane to reveal the entire Hsp70 binding site?

(g) What is the steady-state velocity of translocation?

(h) Briefly comment on your result in the limit that $P \to 1$, and on what happens if the concentration of Hsp70 is zero?

Problem 14: Ripping Yarns

The goal of this problem is to explore DNA unzipping in the presence of a force that tends to unzip the DNA. Consider a segment of dsDNA. At one end of this segment, one of the DNA's strands is attached to the ceiling. At the same end, the complementary DNA strand is attached to a weight, f, so that the weight tends to unzip the DNA and would unzip the DNA if it were to fall down.

(a) What is the change in the height of the weight if the DNA unzips one base pair of length a?

(b) How then is $\Delta G = \epsilon - k_B T \log g$ modified when a non-zero force, f, is applied?

(c) How is the ratio of the unzipping rate (β) to the zipping rate (α), namely $\frac{\alpha}{\beta}$, modified by the force f?

(d) Describe what happens for $T < T_M$ as the force is increased from zero. Briefly compare the behavior that you predict to what happens when a horizontal force is applied to a mass on a horizontal surface with a non-zero coefficient of static friction.

(e) Given that DNA unzips at a force of 15 pN, that $k_B T = 4.1$ pN nm at room temperature, and that the change in extension on unzipping one base pair is 0.6 nm, determine the ratio of the zipping rate to the unzipping rate at room temperature and zero force, i.e., determine $\frac{\alpha}{\beta}$. What is the corresponding value of ΔG for unzipping one base pair of dsDNA at zero force?

(f) In fact, the trace in Fig. 8.20 shows that the unzipping force is not constant but fluctuates. This behavior is not the result of experimental errors. Develop and briefly explain a hypothesis for the fluctuations in unzipping force, based on what you know about DNA.

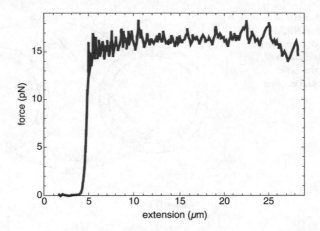

Fig. 8.20 Force versus displacement for unzipping a strand of dsDNA from phage λ using optical tweezers to apply the force, instead of a weight, adapted from Ref. [23]. The region of near-zero force and displacements below about 5 μm corresponds to pulling straight the "handles" of the λ DNA. Once the handles have been straightened, the force increases rapidly, initially with little further displacement. However, at a force of about 15 pN, unzipping begins. Each base pair unzipped increases the displacement by about 0.6 nm. Since the change in displacement during the unzipping phase is about 23 μm, we may deduce that $\frac{23000}{0.6} = 38000$ bp was unzipped in this trace. In comparison, λ DNA contains 48000 bp

Problem 15: Gene Expression Revisited
A certain system of interest, shown in the figure, consists of a DNA-binding protein, call it RNAP, and a loop of DNA (a plasmid). The copy number of RNAP is *two* in this case, i.e., there are two copies of RNAP. RNAP can be bound to the DNA at one particularly favorable site for binding, call it the operator, where the binding energy is ϵ (which is negative), or at one of N other sites, where the binding energy may be chosen to be equal to zero. Within this model, both RNAPs are always attached to the DNA; there is negligible probability that either one is in solution (Fig. 8.21).

(a) How many microstates are there in the case that *one* RNAP is bound to any one of the N other sites?
(b) How many microstates are there in the case that one RNAP is bound to any one of the N other sites and one RNAP is bound to the operator site? Call this number g_1.
(c) How many microstates are there in the case that both RNAPs are each bound to one of the N other sites? Call this number g_2.
(d) Write the partition function, Z, of this system, i.e., the appropriate normalization factor for a Boltzmann factor.
(e) What is the probability, P, that RNAP binds the operator?
(f) What is the variance in the number of RNAPs bound to the operator site, σ_n^2?

Fig. 8.21 Schematic of a
plasmid with two RNAPs.
Left: one RNAP bound to the
operator site and one to N
other sites. Right: both
RNAPs are bound to other
sites

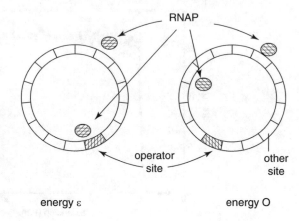

Problem 16: Osmotic Pressure

Osmosis—the movement of solvent molecules through a semi-permeable membrane from a region of lower solute concentration to a region of higher solute concentration—is tremendously important in medicine and biology. The goal of this problem is to understand osmotic pressure.

Consider, then, the steady-state situation shown at the right of Fig. 8.22, which illustrates a very large solvent reservoir that is separated by a semi-permeable membrane from a solution of volume $V = Ah$ containing M solute molecules. The solution is in a tube of cross-sectional area A that is open at the top. The top of the solution is a height $z = h$ above the semi-permeable membrane, which is at the same height ($z = 0$) as the top surface of the solvent reservoir. Solvent molecules pass freely back and forth across the semi-permeable membrane via a large number of tiny, cylindrical channels in the membrane, but solute molecules are too large to pass through these channels, rendering the membrane impermeable to solute particles. Ignore atmospheric pressure throughout this problem.

(a) What is the gravitational potential energy of the solution, U, given that the acceleration due to gravity is g, that the density of the solution is ρ, and that the height of the solvent reservoir is at $z = 0$? (Note the solution density should be taken to be independent of h.)

(b) To determine the free energy of the solution, in addition to its gravitational potential energy, we also need to calculate the number of possible microstates, w, corresponding to a volume $V = Ah$ containing M solute particles. To do this calculation, we use the same model for a solution that we used previously, namely that a solution of M solute particles in a volume V may be conceived as $N = \frac{V}{v}$ possible locations, of which M are occupied by solute particles, where v corresponds to volume of a solute particle. Then,

$$w = \frac{N!}{M!(N-M)!}, \tag{8.107}$$

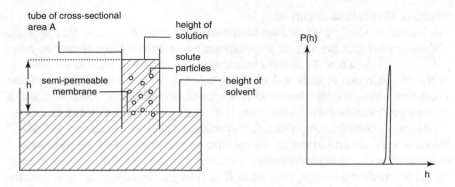

Fig. 8.22 Left: schematic of an osmotic steady state in which a solution is raised above a solvent reservoir. Right: schematic of the probability density of h, namely $p(h)$, vs. h

where $N = \frac{V}{v}$. Using Eq. 8.107 and the facts that, in this case, $V = Ah$ and $N \gg M$, it is possible to show that

$$k_B T \log w \simeq M k_B T \left(\log h + \log \frac{A}{v} - \log M \right). \qquad (8.108)$$

Using your answer to (a) and Eq. 8.108, write the free energy, ΔG, of the solution.

(c) Statistical mechanics, via the Boltzmann factor expression for the probability density of h, namely

$$p(h) = \frac{e^{-\frac{\Delta G}{k_B T}}}{Z}, \qquad (8.109)$$

informs us that the most probable value of h is the value of h that minimizes ΔG because $e^{-\frac{\Delta G}{k_B T}}$ is largest for the smallest possible value of ΔG, i.e., the minimum value of ΔG. (Z is the partition function and is independent of h.)
In fact, $p(h)$ is a very sharp function of h (as schematically illustrated in the right-hand side of Fig. 8.22), so that the value of h that minimizes ΔG is equal to the mean of h and the standard deviation of h from this value is very small. In other words, the most probable value of h is the value of h that we see and measure.
Using your answer to (b), calculate the value of h that minimizes ΔG.

(d) The hydrostatic pressure at a depth d below the surface of a fluid is $P = \rho g d$. Calculate the hydrostatic pressure at the bottom of the solution ($z = 0$) in terms of M, V, and $k_B T$ only. Briefly comment on your result and whether it is similar to other equations that you recognize.

Problem 17: Facilitated Diffusion

We learned in Chap. 6. *Diffusion: Membrane Permeability and the Rate of Actin Polymerization* that the flux of a certain molecule across a membrane is $j_1 = -\frac{BD}{d}(\ell_2 - \ell_1)$, where ℓ_1 is the concentration at surface 1 of a membrane, ℓ_2 is the concentration at surface 2 of the membrane, B is the molecule's partition coefficient within the membrane, D is the molecule's diffusion coefficient, and d is the membrane thickness. Therefore, if B is tiny, the flux of ligand across the membrane is correspondingly tiny. This problem discusses "facilitated diffusion," which is a mechanism to enable a large flux by introducing into the membrane a concentration of substrate molecules that bind a ligand and that effectively carry ligand across the membrane, even when $B \simeq 0$ for the ligand within the membrane (Fig. 8.23).

(a) As we learned in the chapter, when a ligand, L, can bind to a substrate, S, the probability that a particular substrate molecule binds ligand is given by

$$Q = \frac{\ell}{\ell + K}, \tag{8.110}$$

where ℓ is the concentration of ligand and K is the dissociation constant of the binding reaction. We consider then a membrane-confined substrate with a total concentration within the membrane of s_0. In this case, what is the concentration of substrate binding ligand at surface 1 of the membrane, where substrate is in contact with free ligand at concentration ℓ_1? Call it s_1.

(b) What is the concentration of substrate binding ligand at surface 2 of the membrane? Call it s_2.

(c) Assuming that $B = 0$, so that the concentration of free ligand in the membrane is zero, determine the flux of substrate binding ligand across the membrane, given that D_s is the diffusion coefficient of substrate (with or without ligand bound). Call it j_2.

(d) What is the corresponding flux of ligand across the membrane? Call it j_3. Comment on your answer.

(e) What is the flux of substrate that is not binding ligand, across the membrane? Call it j_4. Comment on your answer.

Fig. 8.23 Schematic of the membrane and concentrations for facilitated diffusion

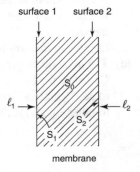

Things are a little more complicated for non-zero B because then the concentrations of ligand, substrate not binding ligand, and substrate binding ligand must satisfy the law of mass action at every position within the membrane (assuming binding is in equilibrium everywhere).

Problem 18: A Protein with Two Binding Sites

In this problem, you will examine ligand–protein binding in the case that the protein has two independent binding sites and therefore can bind zero, one, or two ligands. As in Module 9, we consider a single protein surrounded by a solution of ligand molecules. There are a total of M ligand molecules either in solution or bound to the protein. When one ligand is bound to the protein, the ligand–protein binding energy is ϵ. When two ligands are bound to the protein, the ligand–protein binding energy is 2ϵ. Our model for the solution surrounding the protein is the same as that discussed in the chapter the solution is divided into N cells each of volume v, so that the total volume of the solution is $V = Nv$.

(a) How many microstates does this system possess if ligand is not bound to the protein? Call the answer g_0.
(b) How many microstates are there when one ligand molecule binds the protein? Be sure to account for the fact that the protein has two ligand binding sites. Call the answer g_1.
(c) How many microstates are there when two ligand molecules bind the protein? Call the answer g_2.
(d) Calculate the probabilities (P_0, P_1, and P_2) that the protein binds zero, one, and two ligand molecules, respectively, in terms of g_0, g_1, and g_2.
(e) Given that $\frac{g_1}{g_0} \simeq \frac{2M}{N}$ and $\frac{g_2}{g_0} \simeq \frac{M^2}{N^2}$, calculate the mean number of bound ligand molecules. Call it $\langle n \rangle$.

Problem 19: A Translocation Ratchet

Consider the situation shown in Fig. 8.24, in which an unfolded protein is in the process of translocating through a membrane channel from the cytosol outside a mitochondrion into the mitochondrion. Inside the mitochondrion, "chaperone" proteins may bind to sites on the translocating protein with binding energy ϵ, which is negative. In the absence of chaperones, the rate at which the translocating protein steps into or out of the mitochondrion is α for both stepping in and stepping out. The figure illustrates a situation in which there are M chaperone binding sites, each of length b, extending into the mitochondrion, of which n are actually bound by chaperones. (Outside the mitochondrion, the number of chaperones is negligible.) The goal of this problem is to show that chaperone binding leads the translocating protein to acquire a non-zero drift velocity into the mitochondrion and to calculate that drift velocity.

(a) Ignoring for now the possibility of chaperone binding, what is the probability that the translocating protein takes a step of length b into the mitochondrian (which is in the positive x-direction) in a time Δt?

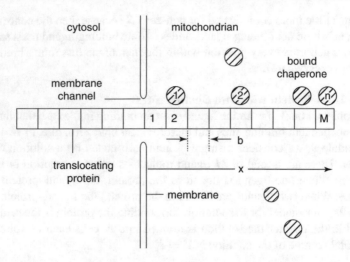

Fig. 8.24 Schematic of a protein—illustrated as a sequence of square binding sites—in the process of translocating through a membrane channel into a mitochondrion. Inside the mitochondrion, the translocating protein may be bound by a number of chaperones, which are illustrated as shaded circles

(b) Still ignoring the possibility of chaperone binding, what is the probability that the translocating protein takes a step of length b outward from the mitochondrian in a time Δt.

(c) Still ignoring the possibility of chaperone binding, what is the translocating protein's drift velocity? Show the steps to your answer.

(d) Still ignoring the possibility of chaperone binding, what is the translocating protein's mean square displacement in a time Δt, namely $\langle w^2 \rangle$?

(e) Still ignoring the possibility of chaperone binding, what is the translocating protein's mean square displacement in a time t, namely $\langle x^2 \rangle$, and its corresponding diffusion coefficient, D?

From now on, we include the possibility that the portion of the translocating protein that is inside the mitochondrion can bind chaperones. We consider the "state" (consisting of many microstates), in which M possible binding sites extend into the mitochondrion, of which n bind chaperones. An example of one of these microstates is shown in Fig. 8.24. The multiplicity of this state, in which M binding sites extend into the mitochondrion, of which n are bound by chaperones, is $g = \frac{M!}{n!(M-n)!}$ and its energy is $n\epsilon$. It is possible to show that $\log g$ may be approximated by

$$\log g = M \log M - n \log n - (M - n) \log(M - n). \qquad (8.111)$$

As a result, the difference in free energy (ΔG) between the state in which M binding sites protrude into the mitochondrion, of which n bind chaperones, and the state with zero bound chaperones is

$$\Delta G = n\epsilon - k_B T \log g = n\epsilon - k_B T \left(M \log M - n \log n - (M - n) \log(M - n) \right).$$
$$(8.112)$$

(f) What is the relationship between M, which is the number of binding sites that extend into the mitochondrion, and x, which is the distance that the translocating protein extends into the mitochondrion?

(g) Using your answer to (f), express ΔG in terms of x, b, n, ϵ, and $k_B T$ only.

(h) Given that your answer to (g) is an energy that depends on the coordinate x and, therefore, can be conceived of as a potential energy—i.e., take $U(x) = \Delta G$—calculate the force, F, in the x-direction exerted on the translocating protein, by chaperone binding.

(i) In addition to the force F, which tends to drive the translocating protein into the mitochondrian, there is a frictional force, $f = -\zeta v$, that also acts on the translocating protein, when it has acquired a velocity v. Because the Einstein relation relates the friction coefficient (ζ) to the diffusion coefficient (D), namely $\zeta = \frac{k_B T}{D}$, this frictional force may be expressed as

$$f = -\frac{k_B T}{D} v. \qquad (8.113)$$

Using Eq. 8.113 and your answers to (e) and (h), use Newton's second law to calculate the "terminal velocity" of the translocating protein through the membrane channel in terms of b, α, n, and M, only.

(j) Briefly explain how you expect $\frac{n}{M}$ to vary as the chaperone-translocating protein binding energy becomes more negative. Figure 8.25 shows a plot of the terminal velocity (v) of a protein translocating through a membrane channel versus $\frac{n}{M}$.

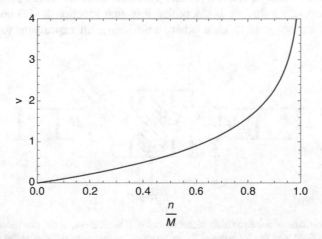

Fig. 8.25 Plot of the terminal velocity a.k.a. drift velocity (v) of a protein translocating through a membrane channel versus $\frac{n}{M}$

Explain how you expect v to vary as the chaperone-translocating protein binding energy becomes more negative.

Problem 20: Thermal Ratchet Model of RNA Polymerase

RNA polymerase (RNAP) is the motor protein that transcribes DNA genes into RNA. In contrast to DNA polymerase, RNA polymerase can unzip double-stranded DNA (dsDNA) into two strands of single-stranded DNA (ssDNA) on its own. On the left of Fig. 8.26 is a representation of RNAP during transcription, which shows dsDNA, opening as it enters the RNAP to form a "bubble" of ssDNA. One side of the bubble is transcribed into RNA, before the two ssDNA strands of the "bubble" close back into dsDNA, as the DNA leaves the RNAP.

In this problem, we consider a thermal ratchet model for RNAP. We suppose that RNAP can only translocate on ssDNA, and so its motion is impeded until the leading dsDNA–ssDNA junction undergoes a thermally induced fluctuation that unzips the dsDNA. There is then a non-zero probability for the RNAP to step forward into the just-unzipped region, preventing the DNA from zipping back up again. In this way, the ds–ss junction in front of the RNAP moves one step in the unzipping direction. In addition, the pair of ssDNA strands that follow the RNAP comes back together after the RNAP has moved forward. From this point of view, there is a second dsDNA–ssDNA junction that follows behind the RNAP (junction 2) in addition to the first dsDNA–ssDNA junction in front of the RNAP (junction 1). The length of a base pair is s. Take the positive x-direction to be directed toward the right.

(a) Supposing that P_1 is the probability that junction 1 (J1) and the polymerase are not next to each other, write down an expression for the mean velocity of junction 1, v_{J1}, in terms of the ssDNA–dsDNA zipping and unzipping rates (α and β, respectively), s, and P_1. Thus, calculate P_1 in terms of v_{J1}, s, α, and β.

(b) Next, supposing that P_2 is the probability that junction 2 (J2) and the polymerase are not next to each other, write down an expression for the mean

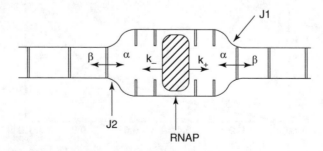

Fig. 8.26 Schematic of our Brownian ratchet model of RNAP, showing the two junctions between dsDNA and ssDNA and the RNAP, and the corresponding stepping rates. α is the DNA zipping rate. β is the DNA unzipping rate. k_+ is the RNAP stepping forward rate on ssDNA. k_- is the RNAP stepping backward rate on ssDNA [phase (right)]

velocity of junction 2, v_{J2}, in terms of the ssDNA–dsDNA zipping and unzipping rates, s, and P_2. Thus, calculate P_2 in terms of v_{J2}, s, α, and β.

(c) Now, write an expression for the mean velocity of the polymerase, v_P, in terms of the polymerase forward and backward translocation rates on ssDNA, k_+ and k_-, respectively, s, and P_1 and P_2.

(d) How are v_{J1}, v_{J2}, and v_P related during RNAP translocation along dsDNA?

(e) Use your answers to (a) for P_1, to (b) for P_2, and to (c) and (d), in order to calculate v_{J1}, v_{J2}, and v_P in terms of α, β, k_+, k_-, and s only.

(f) Assuming that $\beta \gg k_+ + k_-$, how does v_P depend on free energy of unzipping, namely $\Delta G = -k_B T \log \frac{\beta}{\alpha}$? Does the sign of v_P depend on $\Delta G + \Delta G'$, where $\Delta G'$ is the free energy of RNAP translocation? Explain why or why not.

Problem 21: Active Unzipping

In our previous discussion of helicase-catalyzed DNA unzipping, the sole interaction between the helicase and the junction was that they could not cross. In this problem, we investigate what happens to the unzipping velocity when there is a repulsive interaction between the helicase and the junction, when they are adjacent. Roughly speaking, we could say that in this case the helicase is forcing the junction open.

Specifically, as illustrated in Fig. 8.27:

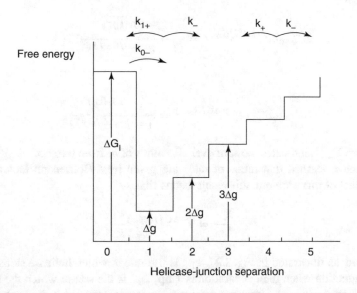

Fig. 8.27 Free energy as a function of the helicase–junction separation. Also shown are the helicase translocation rates, which in this case, depend on the helicase–junction separation, as discussed below

- When the helicase–junction separation is 0 base pairs, the free energy is ΔG_1. In this problem, we envision that ΔG_1 is positive and represents the repulsive helicase–junction interaction. (In the chapter, we took $\Delta G_1 = 0$. That is not the case here.)
- When the helicase–junction separation is 1 unzipped base pair, the free energy is ΔG.
- When the helicase–junction separation is m unzipped base pairs, the free energy is $m\Delta G$.

(a) In terms of ΔG_1 and $k_B T$ and the partition function, Z, what is the probability (P_0) that the helicase–junction separation is 0 bp?

(b) In terms of ΔG and $k_B T$ and the partition function Z, what is the probability (P_1) that the helicase–junction separation is 1 bp?

(c) In terms of ΔG and $k_B T$ and the partition function Z, what is the probability (P_m) that the helicase–junction separation is m bp?

(d) For an infinitely long segment of DNA, what is the partition function in terms of ΔG_1, ΔG, and $k_B T$ only?

To answer this question, you may find it useful to know that

$$\Sigma_{m=0}^{\infty} e^{-m\Delta G/(k_B T)} = \frac{1}{1 - e^{-\Delta G/(k_B T)}}, \tag{8.114}$$

or that

$$\Sigma_{m=0}^{\infty} m e^{-m\Delta G/(k_B T)} = \frac{e^{-\Delta G/(k_B T)}}{(1 - e^{-\Delta G/(k_B T)})^2}, \tag{8.115}$$

or that

$$\Sigma_{m=1}^{\infty} e^{-m\Delta G/(k_B T)} = \frac{e^{-\Delta G/(k_B T)}}{1 - e^{-\Delta G/(k_B T)}}, \tag{8.116}$$

where $\Sigma_{m=0}^{\infty}$ indicates the sum over all values of m from 0 to ∞.

(e) We have learned that ratios of rates are given by a Boltzmann factor. In the context of this problem, this result means that

$$\frac{k_{1+}}{k_{0-}} = e^{-\Delta G_1/(k_B T)} \frac{k_+}{k_-}, \tag{8.117}$$

where, as illustrated in Fig. 8.27, k_{1+} is the rate at which helicase steps toward the junction when their separation is 1 bp, k_{0-} is the rate at which the helicase steps away from the junction when their separation is 0 bp, k_+ is the rate at which helicase steps toward the junction when their separation is greater than 1 bp, and k_- is the rate at which the helicase steps away from the junction when their separation is greater than 0 bp. In this problem, we will assume that

$$k_{1+} = k_+ \tag{8.118}$$

and

$$k_{0-} = e^{\Delta G_1/(k_B T)} k_-, \qquad (8.119)$$

which satisfy Eq. 8.117. Corresponding to these rates, when the helicase–junction separation is m, the helicase velocity is

$$v_m = b(k_+ - k_-), \qquad (8.120)$$

for $m > 0$, and

$$v_0 = -bk_- e^{\Delta G_1/(k_B T)}, \qquad (8.121)$$

for $m = 0$.
Write a general expression for the mean velocity, $\langle v \rangle$, as a sum from $m = 0$ to $m = \infty$ in terms of v_m and P_m.
(f) Use Eqs. 8.120 and 8.121 and your answer to (e), calculate $\langle v \rangle$ in terms of b, k_+, k_-, P_0, and $e^{\Delta G_1/(k_B T)}$, only.
(g) Assuming that $e^{-\Delta G_1/(k_B T)} \ll e^{-\Delta G/(k_B T)}$ and $e^{-\Delta G_1/(k_B T)} \ll 1$, use your answers to (d) and (f), to express $\langle v \rangle$ in terms of b, k_+, k_-, and $e^{\Delta G/(k_B T)}$, only.
(h) Comment on your result here in comparison to the result obtained when the sole interaction between the helicase and the junction is that they cannot cross.

Problem 22: Frayed Ends
In this problem, you consider the end of a strand of double-stranded DNA (dsDNA). Because of thermal fluctuations, a number of based pairs at the end of the strand can unzip, even though the free energy per base pair between unzipped and zipped base pairs, namely ΔG, is positive. Correspondingly, the free energy of m unzipped base pairs in $m\Delta G$, measured relative to the completely zipped microstate.

To answer this question, you may find it useful to know that

$$\Sigma_{m=0}^{\infty} e^{-m\Delta G/(k_B T)} = \frac{1}{1 - e^{-\Delta G/(k_B T)}}, \qquad (8.122)$$

$$\Sigma_{m=0}^{\infty} m e^{-m\Delta G/(k_B T)} = \frac{e^{-\Delta G/(k_B T)}}{(1 - e^{-\Delta G/(k_B T)})^2}, \qquad (8.123)$$

$$\Sigma_{m=1}^{\infty} e^{-m\Delta G/(k_B T)} = \frac{e^{-\Delta G/(k_B T)}}{1 - e^{-\Delta G/(k_B T)}}, \qquad (8.124)$$

and

$$\Sigma_{m=0}^{\infty} m^2 e^{-m\Delta G/(k_B T)} = \frac{e^{-2\Delta G/(k_B T)} + e^{-\Delta G/(k_B T)}}{(1 - e^{-\Delta G/(k_B T)})^3}, \tag{8.125}$$

where $\Sigma_{m=0}^{\infty}$ indicates the sum over all values of m from 0 to ∞.

(a) In terms of the partition function Z, what is the probability (P_0) that zero base pairs are unzipped?

(b) In terms of ΔG and $k_B T$ and the partition function Z, what is the probability (P_1) that one base pair is unzipped?

(c) In terms of ΔG and $k_B T$ and the partition function Z, what is the probability (P_m) that m base pairs are unzipped?

(d) For a long segment of DNA, for which we can take m to run from 0 to ∞, what is the partition function in terms of ΔG and $k_B T$ only?

(e) What is the mean number of unzipped base pairs, $\langle m \rangle$ in terms of ΔG and $k_B T$?

(f) Calculate the variance in the number of unzipped base pairs.

(g) Near the DNA melting temperature, $\Delta G/(k_B T)$ is small. For small $\Delta G/(k_B T)$, we may make the approximation

$$e^{-\Delta G/(k_B T)} \simeq 1 - \frac{\Delta G}{k_B T}. \tag{8.126}$$

Using this approximation, calculate an approximate expression for $\langle m \rangle$, valid for small $\Delta G/(k_B T)$, and then sketch your answer versus $\Delta G/(k_B T)$.

Problem 23: Molecular Motor Redux

In this problem, you will bring together ideas from several previous chapters to show via a different route that a molecular motor can be expected to show "Michalis–Menten kinetics." Consider, then, a certain molecular motor that has two, and only two, internal states, state 0 and state 1. In state 1, the molecular motor binds ATP. In state 0, it does not. Suppose that the probability that the molecular motor is in state 1 is P_1 and that the probability that the motor steps forward in a small time interval, Δt, is $k_+\Delta t$, given that the molecular motor is in state 1. The probability that the motor steps backward is negligible in both states.

(a) What is the probability that the molecular motor steps forward in time Δt?

(b) What is the mean displacement of the molecular motor in time Δt? Call it $\langle w \rangle$, given that the motor's step size is s.

(c) What is the mean velocity, v, of the molecular motor?

(d) In fact, when the molecular motor steps forward, it inevitably transitions from state 1 to state 0, corresponding to simultaneous ATP hydrolysis and unbinding. Thus the rate at which the motor transitions from state 1 to state 0 is k_+, equal to its step-forward rate. The motor transitions from state 0 to state 1, corresponding to ATP binding, at a rate k_-, and it remains at the same location when it does so. Importantly, in this problem, k_- is NOT the rate of backward stepping.

Consider a population containing a total of $N_0 + N_1$ molecular motors with N_0 in state 0 and N_1 in state 1. Then, as discussed in this chapter in the context of protein folding and unfolding, for example, we can expect that the rate equation that describes the rate of change of N_1 to be

$$\frac{dN_1}{dt} = -k_+ N_1 + k_- N_0. \tag{8.127}$$

Using Eq. 8.127, determine what is the steady-state value of P_1 in terms of k_+ and k_- only?

(e) Use your answers to (c) and (d) to express v in terms of s, k_+, and k_- only.

(f) It is possible to derive an expression for the ATP binding rate, k_-, on the basis of the diffusion equation, assuming that the molecular motor is a center of spherical symmetry and that molecular motor-ATP binding is a diffusion-limited reaction. To this end, we take the ATP-molecular motor reaction radius to be R and the ATP diffusion coefficient to be D, and we suppose that the protein diffusion coefficient can be neglected compared to D. The general solution to the spherically symmetric steady-state diffusion equation is given by

$$n(r) = \frac{a}{r} + b, \tag{8.128}$$

for any values of a and b. In Eq. 8.128, $n(r)$ is the number density of ATP molecules at a radial distance r from the molecular motor. Given that the concentration of ATP at $r = \infty$ is $c = [ATP]$, determine b.

(g) Assuming that every ATP molecule that encounters the molecular motor binds to the molecular motor—it is a diffusion-limited reaction—then, the appropriate boundary condition at $r = R$ is $n(R) = 0$. Impose this boundary condition to determine a, and thus write down the complete solution for $n(r)$.

(h) What is the corresponding flux of ATP molecules as a function of the radial distance from the protein?

(i) Using your answer to (h), determine the rate at which ATP molecules react with the protein, i.e., determine k_-.

(j) Write an equation that shows how v depends on [ATP]?

References

1. J.W. Gibbs, *Elementary Principles in Statistical Mechanics* (Charles Scribner's Sons, New York, 1902)
2. A.S. Eddington, *The Nature of the Physical World* (The MacMillan Company, New York, 1929)
3. R.P. Feynman, *Statistical Mechanics: A Set of Lectures* (Addison Wesley, Boston, 1981)
4. D.J. Phillies, *Elementary Lectures in Statistical Mechanics* (Springer, New York, 2000)
5. H.G. Garcia, J. Kondev, N. Orme, J.A. Theriot, R. Phillips, A first exposure to statistical mechanics for life scientists: applications to binding. Unpublished (2007). http://www.rpgroup.caltech.edu/publications/Garcia2007b.pdf

6. B.U. Keller, R.P. Hartshorne, J.A. Talvenheimo, W.A. Catterall, M. Montal, Sodium channels in planar lipid bilayers. Channel gating kinetics of purified sodium channels modified by batrachotoxin. J. Gen. Physiol. **88**, 1–23 (1986)
7. N.E. Schoppa, F.J. Sigworth, Activation of *shaker* potassium channels II. Kinetics of the V2 mutant channel. J. Gen. Physiol. **111**, 295–311 (1998)
8. J. Liphardt, B. Onoa, S.B. Smith, I. Tinoco, C. Bustamante, Reversible unfolding of single RNA molecules by mechanical force. Science **292**, 733 (2001)
9. M.T. Woodside, P.C. Anthony, W.M. Behnke-Parks, L. Larizadeh, D. Hershlag, S. Block, Direct measurement of the full, sequence-dependent folding landscape of a nucleic acid. Science **314**, 1001–1004 (2006)
10. Y. Gao, G. Sirinakis, Y. Zhang, Highly anisotropic stability and folding kinetics of a single coiled coil protein under mechanical tension. J. Am. Chem. Soc. **133**, 12749–12757 (2011)
11. Z.Q. Xi, Y. Gao, G. Sirinakis, H.L. Guo, Y.L. Zhang, Direct observation of helix staggering, sliding, and coiled coil misfolding. Proc. Natl. Acad. Sci. USA **109**, 5711–5716 (2012)
12. A.M. Pyle, Translocation and unwinding mechanisms of RNA and DNA helicases. Annu. Rev. Biophys. **37**, 317–336 (2008)
13. M.D. Gray, J.C. Shen, A.S. Kamath-Loeb, A. Blank, B.L. Sopher, G.M. Martin, J. Oshima, L.A. Loeb, The Werner syndrome protein is a DNA helicase. Nat. Genet. **17**, 100–103 (1997)
14. A.D. Kwong, B.G. Rao, K.-T. Jeang, Viral and cellular RNA helicases as antiviral targets. Nat. Rev. Drug Discov. **4**, 845–853 (2005)
15. M.D. Betterton, F. Jülicher, A motor that makes its own track: helicase unwinding of DNA. Phys. Rev. Lett. **91**(25), 258103 (2003)
16. D.S. Johnson, L. Bai, B.Y. Smith, S.S. Patel, M.D. Wang, Single-molecule studies reveal dynamics of DNA unwinding by the ring-shaped T7 helicase. Cell **129**, 1299–1309 (2007)
17. D.A. Portnoy, V. Auerbuch, I.J. Glomski, The cell biology of *listeria monocytogenes* infection: the intersection of bacterial pathogenesis and cell-mediated immunity. J. Cell Biol. **158**, 409–414 (2002)
18. C.I. Lacayo, Z. Pincus, M.M. Van Duijn, C.A. Wilson, D.A. Fletcher, F.B. Gertler, A. Mogilner, J.A. Theriot, Emergence of large-scale cell morphology and movement from local actin filament growth dynamics. PLoS Biol. **5**, e233 (2007)
19. C.S. Peskin, G.M. Odell, G.F. Oster, Cellular motions and thermal fluctuations: the Brownian ratchet. Biophys. J. **65**, 316–324 (1993)
20. M. Dogterom, B. Yurke, Measurement of the force-velocity relation for growing microtubules. Science **278**, 856–860 (1997)
21. D.R. Kovar, T.D. Pollard, Insertional assembly of actin filament barbed ends in association with formins produces picoNewton forces. Proc. Nat. Acad. Sci. USA **101**, 14725–14730 (2004). https://www.pnas.org/doi/10.1073/pnas.0405902101
22. C. Bohr, K. Hasselbalch, A. Krogh, Uber einen in biologischer Beziehung wichtigen Einfluss, den die Kohlensaurespannung des Blutes auf dessen Sauerstoffbindung ubt. Skand. Arch. Physiol. **16**, 401–421 (1904)
23. U. Bockelmann, P. Thomen, B. Essevaz-Roulet, V. Viasnoff, F. Heslot, Unzipping DNA with optical tweezers: high sequence sensitivity and force flips. Biophys. J. **82**, 1537–1553 (2002)

Fluid Mechanics: Laminar Flow, Blushing, and Murray's Law

9

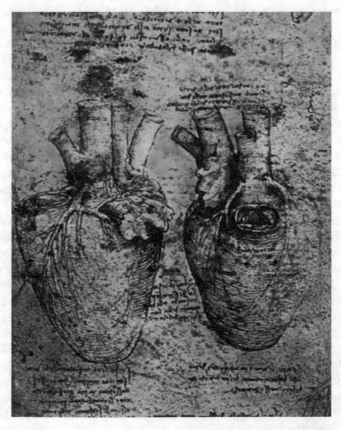

Anatomical sketch by Leonardo da Vinci of the heart and its blood vessels. Public domain image from https://commons.wikimedia.org/wiki/File:Heart_and_blood_vessels_by_da_Vinci.jpg

© The Author(s), under exclusive license to Springer Nature Switzerland AG 2023
S. Mochrie, C. De Grandi, *Introductory Physics for the Life Sciences*,
Undergraduate Texts in Physics, https://doi.org/10.1007/978-3-031-05808-0_9

9.1   Introduction

"Fluid Mechanics" has to do with how fluids (liquids or gasses) respond to applied forces and pressures. This chapter focusses on describing the fluid flow in the so-called laminar regime which means that the fluid flows smoothly without any vortices. This type of fluid flow continues to become increasingly important in biomedical research [1], and is especially relevant to the flow of blood through your circulatory system, consisting of arteries, arterioles, capillaries, venules, veins, and your heart, which is the pump that generates the pressure that drives blood flow. We will conclude this chapter with a discussion of the principles that underlie the physiology of your circulatory system (Murray's Law), which demonstrates how the physics of viscous liquids plausibly has determined human physiology.

Among the key concepts that we explore in this chapter are fluid viscosity and the corresponding viscous forces within fluids. An important consequence of these forces is that the flow rate through a channel is linearly proportional to the difference in pressure across the channel (which we'll refer to as "Ohm's law for fluids"). The coefficient of proportionality is the flow conductance. Another essential concept in the case of liquids, such as water and blood, which are the focus of this chapter, is liquid incompressibility, namely that the volume of a given mass of liquid is essentially independent of the pressure experienced by that liquid. This behavior stands in contrast to the behavior of gasses, which are highly compressible. (To be precise, liquids compress slightly with increasing pressure, but the effect is small, and we will neglect it in this chapter.) A consequence of liquid incompressibility is that the volume flow rate into a particular location in a flow channel or flow circuit must equal the volume flow rate away from that location. This observation will permit us to work out rules for determining the fluid flow through arbitrarily complex flow circuits, containing composite channels built from sequential channels of different flow conductance, and branched channels.

This chapter builds on Chap. 2. *Force and Momentum: Newton's Laws and How to Apply Them* and Chap. 6. *Diffusion: Membrane Permeability and the Rate of Actin Polymerization* in order to understand how frictional, viscous forces act and to introduce viscosity, which is the material property of the fluid in question, that quantifies these forces. Furthermore, it applies Newton's laws to a small volume of fluid within a fluid flow to calculate how the velocity profiles and the volume flow rates of viscous fluids through flow channels depend on the channel dimensions, the fluid viscosity, and the pressure difference between the ends of the channel. Finally, it is worth nothing that this chapter also relies on Chap. 6. *Diffusion: Membrane Permeability and the Rate of Actin Polymerization* for introducing a number of mathematical techniques that we will use in this chapter. Specifically, we will use "solution by direct substitution" to find general solutions for fluid velocity profiles, and we will impose boundary conditions to find the solutions that are relevant in particular situations.

9.2   Your Learning Goals for This Chapter

By the end of this chapter you should be able to:

1. Calculate the force on a surface that is subject to pressure.
2. Appreciate that there are two characteristic types of fluid flow—laminar and turbulent—and be able to describe each one qualitatively.
3. Understand how viscous forces act and how they are quantified by fluid viscosity.
4. Discuss the role of the Reynolds number and whether laminar or turbulent flow may be expected in a given flow situation.
5. Apply Newton's laws to appropriate fluid elements to determine equations for the velocity profiles within flow channels.
6. Calculate the volume flow rate through a laminar flow channel, given the velocity profile in the channel.
7. Apply "Ohm's law for fluids," namely that for flow through a laminar flow channel, there is a linear relationship between volume flow rate (Q) and pressure difference ($P_1 - P_2$). The coefficient of proportionality is the flow conductance (G), i.e., $Q = G(P_1 - P_2)$.
8. Explain the significance of the dependence of the flow conductance of a cylindrical tube on the tube dimensions for phenomena associated with the human circulatory system, such as atherosclerosis and blushing.
9. Apply the requirement for liquids that the flow rate into any location within a flow circuit must equal the flow rate away from that location, which is a consequence of liquid incompressibility.
10. Apply the resulting rules for combining flow conductances of flow channels "in series" and "in parallel."
11. Analyze elaborate laminar flow circuits to determine the overall flow conductance of the entire circuit in terms of the flow conductance of the component channels, and find the pressure at arbitrary points within the flow circuit in terms of the component flow conductances and the inlet and outlet pressures.
12. Calculate the power dissipated in an arbitrary flow circuit, or equivalently the power that must be supplied by an external agent to maintain flow in the circuit.
13. Appreciate that Murray's law, which relates the radii of parent and child blood vessels, before and after branch points in circulatory systems, is the result of minimizing the total energy cost (mechanical and metabolic energy cost) of the circulatory system.

9.3   Pressure and Hydrostatic Pressure

To discuss fluid mechanics, it is necessary to understand pressure. The pressure (P) of a gas or fluid is defined as the force per unit of area that the gas/fluid exerts on an area (A):

$$P = \frac{F}{A}.$$ (9.1)

Pressure is a scalar quantity and is described by the SI unit "Pascal":

$$\text{Pa} = \frac{1N}{1\text{m}^2}.$$

Pressure can be defined at any point within a gas/fluid. By knowing the pressure in a fluid, we can determine the amount of force the fluid exerts perpendicularly to any given area A via:

$$F = PA.$$

Pressure is due to internal collisions between the molecules in the gas/fluid, as discussed, for instance, in Sect. 2.11.3 *Pressure of a gas* through the understanding of the momentum carried by each molecule. Pressure in a fluid is also due to gravity, which pulls downward the molecules, so that the deeper into a fluid the higher the pressure. We call this hydrostatic pressure.

To gain insight into hydrostatic pressure, we will create a model for how the pressure under water increases the deeper down you go. To this end, consider a small volume, ΔV, of water at a depth z under the surface, of thickness Δz and area A (see Fig. 9.1). (Here, we pick z to be positive going down.)

What are the forces on this element of water volume? First, of course, is its weight:

$$(\rho \Delta V)g\hat{k} = \rho A \Delta z g \hat{k},$$ (9.2)

where ρ is the mass density of the liquid.

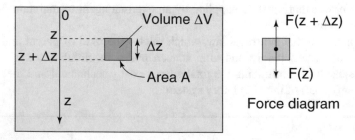

Fig. 9.1 Small volume ΔV under water and corresponding force diagram (right side)

Since our liquid element is not accelerating, there must be an equal and opposite buoyancy force:

$$- (\rho \Delta V) g \hat{k}. \tag{9.3}$$

The buoyancy force is exerted by the rest of the liquid and is independent of the material that ΔV is made of. We introduced this force previously when we discussed the terminal velocity of a particle in a viscous liquid in Chap. 2. *Force and Momentum: Newton's Laws and How to Apply Them.*

In fact, the buoyancy force originates in the pressure within the surrounding liquid. Specifically, there is a force on our volume element from the pressure of the liquid on the large z side (coordinate $z + \Delta z$):

$$F(z + \Delta z) = -AP(z + \Delta z)\hat{k} \tag{9.4}$$

and from the pressure of the liquid on the small z side (coordinate z):

$$F(z) = +AP(z)\hat{k}. \tag{9.5}$$

The negative sign in Eq. 9.4 implies this force is up. The positive sign in Eq. 9.5 implies this force is down. Since the volume element is not accelerating, the sum of these three forces must be zero:

$$\rho A \Delta z g \hat{k} - AP(z + \Delta z)\hat{k} + AP(z)\hat{k} = 0. \tag{9.6}$$

Using

$$P(z) - P(z + \Delta z) = -\frac{dP}{dz} \Delta z, \tag{9.7}$$

valid for infinitesimal Δz, Eq. 9.6 becomes:

$$\rho A \Delta z g \hat{k} - A \frac{dP}{dz} \Delta z \hat{k} = 0, \tag{9.8}$$

and canceling $A \Delta z \hat{k}$ on both sides:

$$\rho g - \frac{dP}{dz} = 0. \tag{9.9}$$

Rearranging, we find:

$$\frac{dP}{dz} = \rho g. \tag{9.10}$$

This equation tells us that the rate of change in pressure (P) with depth (z) is directly proportional to gravity and the density of the fluid. By direct substitution, the solution to Eq. 9.10 may be seen to be:

$$P = P_0 + \rho g z, \tag{9.11}$$

where P_0 is the pressure at $z = 0$ (surface level). This is called hydrostatic pressure, and we see it increases linearly with depth. In fact, the force on an area A at depth z is $P_0 A + \rho A z g$. The quantity $\rho A z g$ is simply the weight of the liquid on top of area A, which is what we might have expected without doing the calculation.

9.4   Two Types of Fluid Behavior: Laminar Fluid Flow and Turbulent Fluid Flow

It is important to know that two characteristic types of fluid flow may be distinguished: laminar flow and turbulent flow. Figure 9.2 shows a sketch of laminar flow, where the flows appear smooth and without vortices. By contrast, Fig. 9.3 shows a number of turbulent flows, that exhibit vortices and time-dependent, often irregular, behavior. Here, we will focus exclusively on laminar flows because they are relevant for most of the human circulatory system and are much simpler to describe mathematically than turbulent flows (which instead often require computer simulations).

9.5   Viscous Forces and Viscosity

You may know from your own experience that once you mix milk with coffee (or milk with your favorite beverage), it is not possible to unmix them. By contrast, as shown in the movie at

https://www.youtube.com/watch?v=X4zd4Qpsbs8,[1]

Fig. 9.2 Sketch of laminar flow around a cylinder. Imagine that flow is moving right to left and is visualized using alternating regions at of dyed fluid (dark regions) and not dyed fluid (blank regions)

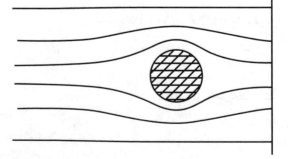

---

[1] https://www.youtube.com/watch?v=X4zd4Qpsbs8

Fig. 9.3 Turbulent fluid flow. Top: sketch of turbulent fluid flow around a cylinder, showing vortices. Imagine that flow is moving right to left and is visualized using alternating regions at of dyed fluid (dark regions) and not dyed fluid (blank regions). Bottom: NASA Landsat 7 image https://commons.wikimedia.org/wiki/File:Vortex-street-1.jpg of clouds around the Juan Fernandez Islands off the Chilean coast revealing a repeating pattern of vortices, arising as a result of the unsteady separation of air flow about the islands. Public domain image from https://commons.wikimedia.org/wiki/File:Vortex-street-1.jpg

it is possible to unmix corn syrup. The key difference between coffee and corn syrup is that the viscosity of corn syrup is much larger than that of coffee. Therefore, we may hypothesize that laminar flows (flows that are smooth and where any mixing is reversible) tend to occur in liquids with higher viscosity, like corn syrup, and turbulent flows (that have vortices and that cannot be unmixed) tend to occur in liquids with lower viscosity, like coffee. But to make this statement precise, we need to know: what exactly is viscosity? Qualitatively, a liquid with a large viscosity is difficult to stir. It flows slowly. It is "thick" and "goopy"—think maple syrup, or honey, which is very viscous. However, to both quantitatively define viscosity and elucidate the role of viscous forces in laminar fluid flows, we will explore the "kinetic theory" model, illustrated in Figs. 9.4 and 9.5.

Specifically, we consider laminar flow in a viscous fluid, confined between two parallel, flat plates, that are separated by a distance d along the x-direction (Fig. 9.4). One of the plates, located at $x = 0$, is stationary. The other plate, located at $x = d$, is moving with velocity v, along the z axis. We assume that the fluid experiences laminar flow, and that as a result of the relative motion of the plates, the fluid has a non-zero velocity along the z-direction that varies depending on x, i.e., we take

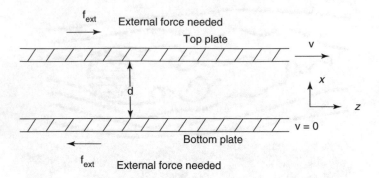

External force needed

Top plate

v

d

x

z

$v = 0$

Bottom plate

f_{ext} External force needed

Fig. 9.4 Two plates in relative motion separated by a layer of viscous fluid

Fig. 9.5 Hopping model for viscosity

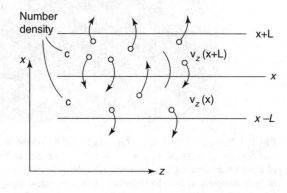

$$v_z = v_z(x) \qquad (9.12)$$

to be non-zero. To maintain the relative velocity of the plates, it is necessary to apply a force, f_{ext}, along the z-direction, which is equal and opposite to the liquid viscous forces that resist the plate's motion, and that we are interested in. (This situation is similar to a so-called Couette cell,[2] which consists of concentric cylinders with viscous fluid between. It that case too, it is necessary to apply force to the handle of the Couette cell to maintain the angular velocity of the inner cylinder, relative to the outer cylinder.)

To describe the liquid between the plate, we invoke a version of our favorite "hopping" model from Chap. 6. *Diffusion: Membrane Permeability and the Rate of Actin Polymerization* in which we assume that at each tick of a τ clock, every molecule in the liquid hops $\pm L$ in x-direction with equal probability (Fig. 9.5). In this case, we take the number density of particles (c) to be constant. However, as discussed previously, we take the mean particle velocity along the z direction to be a function of x: $v_z = v_z(x)$. We consider the plane at x, between the plates, and seek to calculate the net flow of z-momentum across an area A of this plane. The

---

[2] http://www.youtube.com/watch?v=X4zd4Qpsbs8&NR=1

number of molecules in cross-sectional area A of the two "slices" either side of x is cAL. Therefore, the net flow of z-momentum across an area A of the plane at x in a time τ is

$$j_{p_z} A\tau = \frac{1}{2}cALmv_z(x) - \frac{1}{2}cALmv_z(x+L) = -\frac{1}{2}cAL^2m\frac{dv_z}{dx}, \qquad (9.13)$$

where m is the mass of a molecular of the fluid and where we have introduced the flux of z momentum, j_{p_z}, which is the net z-momentum crossing unit area in the x-direction per unit time.

We know, however, that momentum per unit time is force, as discussed in Chap. 2. *Force and Momentum: Newton's Laws and How to Apply Them*. Therefore, dividing Eq. 9.13 by τ, we find

$$f = j_{p_z} A = -\frac{L^2}{2\tau}mcA\frac{dv_z}{dx} \qquad (9.14)$$

is the force along the z-direction that an area A of the fluid below x exerts on an area A of the fluid above x. The viscosity (η) is defined via this equation:

$$f = -\frac{L^2}{2\tau}mcA\frac{dv_z}{dx} = -A\eta\frac{dv_z}{dx}, \qquad (9.15)$$

i.e.,

$$\eta = mc\frac{L^2}{2\tau}. \qquad (9.16)$$

What are the SI units for viscosity? If we use dimensional analysis we know: $[m] = kg$, $[c] = \frac{1}{m^3}$, $[L] = m$, $[\tau] = s$ and we find:

$$[\eta] = kg\frac{1}{m^3}\frac{m^2}{s} = \frac{kg}{m\,s} \quad \text{or} \quad [\eta] = Pa\,s.$$

The Pascal-second (Pa s) is the SI unit for viscosity; another common unit used is the Poise (P) which is the equivalent of Pascal-second in the centimeter-gram-second system of units (cgs), it turns out that 1 Pa s = 10 Poise. The most common units used for viscosity are milli Pascal-second mPa s $= 10^{-3}$ Pa and the centipoise, 1cP $= 10^{-2}$P, therefore 1mPa s $=$ 1cP. In Table 9.1 are some typical viscosity values for common fluids.

Equation 9.16 represents a microscopic model for the viscosity, analogous to the microscopic model for the diffusion constant D that we found in Chap. 5. *Random Walks: Brownian Motion and the Tree of Life* and used again in Chap. 6. *Diffusion: Membrane Permeability and the Rate of Actin Polymerization*. The difference is that D characterizes the transport of particles, while η characterizes the transport of momentum. In fact, $mc = \rho$, the mass density of the fluid, and, according to our

Table 9.1 Viscosity of some fluids

Fluid	Viscosity at 23 °C (mPa/cP)
Octane	0.5
Water	1
Human blood	3.5 (37 °C)
Vegetable oil	40–70
Honey/corn syrup	2000–3000

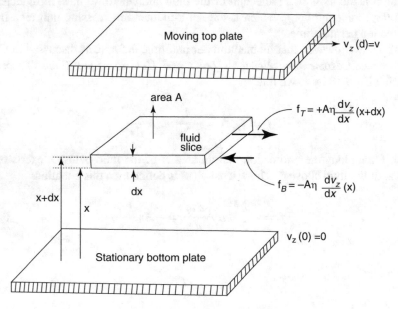

Fig. 9.6 Newton's Second Law applied to the slice of liquid between x and $x + dx$ of area A between two plates, one stationary at $x = 0$ and the other moving along the z direction with velocity v at $x = d$

hopping model, $\frac{L^2}{2\tau} = D$, the diffusion coefficient of the liquid. Thus, we predict $\eta = \rho D$. While this relation is accurate for a gas, it is not accurate for a liquid. Therefore, our hopping model is too simple to properly describe liquid viscosity, and we will consider viscosity and diffusivity to be independent material parameters for any liquid of interest. Nevertheless, Eq. 9.15 correctly describes the relationship between velocity and force, which is what we are interested in.

To determine the complete velocity profile within the fluid between the stationary plate and the moving plate and the viscous forces acting on the plates, we consider an area A of a thin slice of fluid between x and $x + dx$, and apply Newton's second law to this slice, as illustrated in Fig. 9.6. We just worked out that the horizontal force on the bottom of this slice at x is

$$f_B = -A\eta \frac{d}{dx} v_z(x). \tag{9.17}$$

Therefore, Newton's third law informs us that the force on the top of the slice at $x + dx$ from the fluid above is

$$f_T = +A\eta \frac{d}{dx} v_z(x + dx). \tag{9.18}$$

These are the only forces we need consider in this situation. Therefore, since the acceleration of the slice is zero in a steady state, by Newton's second law, we have that

$$0 = f_B + f_T = -A\eta \frac{d}{dx} v_z(x) + A\eta \frac{d}{dx} v_z(x + dx) = A\eta \frac{d^2 v_z}{dx^2} dx, \tag{9.19}$$

where the final equality holds for small-enough dx. Dividing through by A, dx, and η, we find

$$\frac{d^2 v_z}{dx^2} = 0. \tag{9.20}$$

Equation 9.20 describes the fluid velocity between the two plates, and is formally the same as the steady-state one-dimensional diffusion equation, whose solutions we saw in Chap. 6. *Diffusion: Membrane Permeability and the Rate of Actin Polymerization* are of the general form:

$$v_z(x) = Ax + B, \tag{9.21}$$

where A and B are constants to-be-determined by the boundary conditions.

For our two parallel plates, one with velocity v at $x = d$ and one with velocity 0 at $x = 0$, we assert that the velocity of the fluid immediately next to each plate is equal to the velocity of the plate. Therefore, we must have the following boundary conditions on the fluid flow:

$$v_z(0) = 0 \tag{9.22}$$

and

$$v_z(d) = v. \tag{9.23}$$

These boundary conditions yield

$$v_z(0) = A \times 0 + B = 0, \tag{9.24}$$

which implies that $B = 0$, and

$$v_z(d) = Ad + B = Ad = v, \tag{9.25}$$

which implies $A = v/d$. Thus, we have

$$v_z(x) = v\frac{x}{d} \tag{9.26}$$

for the velocity profile, the velocity increases linearly with the distance from the stationary plate.

With this velocity profile, we can calculate the force via Eq. 9.17 that the fluid below exerts on the neighboring fluid above, or the force that the fluid below exerts on the top plate (obviously, fluid molecules cannot jump into the plate, but they can jump on and off the plate, and this process yields the same forces):

$$f = -\eta A\frac{dv_z}{dx} = -\eta A\frac{d}{dx}\left(v\frac{x}{d}\right) = -\frac{\eta A v}{d}. \tag{9.27}$$

The negative sign implies that the force on the plate is in the opposite direction to the velocity.

To maintain the constant velocity of the top plate, an external agent must apply an equal and opposite force:

$$f_{ext} = +\frac{\eta A v}{d}. \tag{9.28}$$

This is essentially the same force that must be exerted in order to keep a Couette cell turning. It is thus clear that viscous forces are frictional forces, albeit with different properties to the kinetic and static frictional forces between solid bodies in contact, that we encountered in Chap. 2. *Force and Momentum: Newton's Laws and How to Apply Them.* We may also evaluate the power that must be supplied by the external agent in order to keep the plate moving:

$$\text{Power} = f_{ext}v = \frac{\eta A v^2}{d}. \tag{9.29}$$

Since there is no change in the fluid's velocity or potential energy as a result of this power input, we may deduce that all of this power ends up as heat and heats up the fluid. Thus, we see that the larger the value of η, the larger are the frictional losses.

9.6   Reynolds Number

Now that we know what viscosity is, we can sensibly address the conditions for laminar versus turbulent flow. As we have seen, we can unmix corn syrup and dyed corn syrup, but we cannot unmix coffee and milk. The key difference between these two cases is that corn syrup has a large viscosity and coffee/milk/water has a small viscosity. But what does it really mean to say that a liquid has a large viscosity or a small viscosity? As we have seen above, viscosity is a quantity that has dimensions of mass per length per time ($[\eta] = \frac{\text{kg}}{\text{m s}}$), viscosity alone cannot predict the behavior of a fluid, because the fluid flow also depends on the geometry of the fluid situation (e.g., size of a moving object in a fluid, or size of a channel inside which a fluid moves), and how fast the fluid is moving. For instance, the behavior of water around a tiny plankton is very different than the behavior of water around a huge whale. In order to be able to predict if a fluid behaves laminarly or turbulently we need to introduce a new quantity that combines together: viscosity, density, velocity, and geometry to achieve a dimensionless quantity, this is the Reynolds number—N_{Re}— that specifies whether the flow is lamellar or turbulent. The Reynolds number is defined as

$$N_{Re} = \frac{\rho v d}{\eta}, \tag{9.30}$$

where ρ is the mass density of the fluid, η is the fluid viscosity, v is a characteristic velocity, associated with the fluid flow, and d is a characteristic dimension, associated with the fluid flow. For flow in a cylindrical channel, v is the average velocity and d is the tube diameter. For flow past a sphere, v is the fluid velocity before it passes the sphere and d is the diameter of the sphere. For flow past a perpendicular cylinder, v is the liquid velocity before it passes the cylinder and d is the cylinder diameter. For small-enough values of N_{Re}, flows are laminar. For large enough values of N_{Re}, flows are turbulent. The value of N_{Re} at which laminar flow disappears—called the critical Reynolds number—depends on the exact flow geometry in question. For flow in a cylindrical tube, such as a blood vessel, flow is laminar for $N_{Re} \leq 2300$. Above this value, flow vortices start to appear and with further increase proliferate. For values of N_{Re} greater than about 5000, the flow is said to be fully turbulent. Problem 4 at the end of this chapter asks you to calculate the Reynolds number in your coronary artery. For flow past a sphere, the flow is laminar for $N_{Re} \leq 10$, above which value vortices start to appear. For flow past a cylinder, the flow is laminar for $N_{Re} \leq 5$. Irrespective of the flow geometry, if the Reynolds number is small enough, the flow will be laminar.

Notice that N_{Re} is proportional to d, v, and ρ, as well as depending inversely on η. This means that, although we saw laminar flow on centimeter length scales with velocities in the centimeter per second range in corn syrup for which η is very large, we also may expect to see laminar flow in water (or blood), which has a viscosity that is many times smaller than that of corn syrup, provided we examine phenomena at sufficiently smaller length and velocity scales. For example, the flow in small vascular systems are laminar, including, for example, the blood flow in a chick embryo's circulatory system [2]. In fact, most normal flows in the human body and all flows in molecular- and cellular-scale biology are laminar, and so in this chapter we will examine laminar flows and their properties in some detail, and turbulent flows not at all, beyond to note their existence at large Reynolds number. Fortunately, laminar flows are also relatively easy to describe mathematically, while turbulent flows are not. This is because the mathematical equations describing turbulent flow cannot be solved analytically but require computer simulations.

9.7   How to Describe Laminar Fluid Flow

9.7.1   Velocity Profile in a Thin Channel

We would like to understand the steady-state fluid flow through a cylindrical tube as a model for an artery. However, we first treat the simpler problem of laminar fluid flow between stationary parallel plates. One realization of this example is a microfluidic channel, such as those in the movie at http://www.youtube.com/watch?v=5QVwljd04Kw.[3]

1. We will derive an appropriate differential equation that describes the steady-state fluid flow between the plates by applying Newton's laws to the fluid between the plates.
2. We will write a general solution to our differential equation and verify that it is a solution.
3. We will impose the boundary conditions onto our general solution to find the particular solution we want.

Notice that the later steps in this procedure mirror how we solved the steady-state diffusion equation discussed in Chap. 6. *Diffusion: Membrane Permeability and the Rate of Actin Polymerization*, so that the techniques that you learned there, we will use again here.

Let us consider a thin channel of thickness d, which is relatively small, length L, which is relatively large, and width, W, which is also relatively large. This is just the situation depicted in Fig. 9.7. The x coordinate is along the vertical direction, with $x = 0$ at the bottom of the channel and $x = d$ at the top of the channel. Now consider a thin slice of fluid that lies between x and $x + dx$. The fluid pressures on the two sides of the thin slice are P_1 (on the left side) and P_2 (on the right side). We will apply Newton's laws to this slice of fluid.

The appropriate force diagram is shown in Fig. 9.8. Using this force diagram, we can write Newton's Second Law for the slice:

$$P_1 W dx - P_2 W dx + W L \eta \frac{dv_z(x + dx)}{dx} - W L \eta \frac{dv_z(x)}{dx} = 0, \qquad (9.31)$$

since the acceleration of this fluid element is zero. As we learned in Chap. 2. *Force and Momentum: Newton's Laws and How to Apply Them*, the sign of the viscous force terms in Eq. 9.31 must be chosen so that this force is directed opposite to the velocity of the fluid slice relative to the adjacent fluid. By writing

$$\frac{dv_z(x + dx)}{dx} - \frac{dv_z(x)}{dx} \simeq \frac{d^2 v_z}{dx^2} dx. \qquad (9.32)$$

---

[3] http://www.youtube.com/watch?v=5QVwljd04Kw

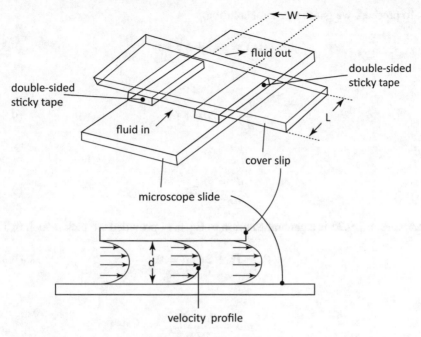

Fig. 9.7 Top: A microscope cover slip separated from a microscope slide by double sided sticky-tape creates a microfluidic channel in which the flow is laminar with a parabolic flow profile. A setup like this is often used in life-science laboratories to examine biochemistry and/or cell biology happening in real-time under a fluorescence or DIC (differential interference contrast) microscope. Bottom: Representation of the velocity profile in the channel

Fig. 9.8 Force diagram for the fluid slice between x and $x + dx$

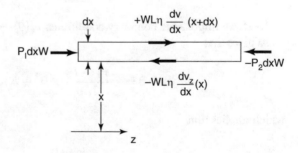

Eq. 9.31 becomes

$$P_1 W dx - P_2 W dx + W L \eta \frac{d^2 v_z}{dx^2} dx = 0, \qquad (9.33)$$

or dividing through by $W dx$,

$$P_1 - P_2 + L \eta \frac{d^2 v_z}{dx^2} = 0. \qquad (9.34)$$

This is the differential equation that governs the fluid flow in the channel.

To proceed, we guess a general solution:

$$v_z(x) = ax^2 + bx + c. \tag{9.35}$$

Then,

$$\frac{dv_z}{dx} = 2ax + b, \tag{9.36}$$

and

$$\frac{d^2v_z}{dx^2} = 2a. \tag{9.37}$$

Therefore, Eq. 9.35 is a general solution to Eq. 9.34 provided we pick a such that

$$P_1 - P_2 + 2L\eta a = 0, \tag{9.38}$$

or

$$a = -\frac{P_1 - P_2}{2L\eta}, \tag{9.39}$$

so that

$$v_z(x) = -\frac{P_1 - P_2}{2L\eta}x^2 + bx + c. \tag{9.40}$$

Next, we impose the boundary conditions: $v_z(0) = 0$ and $v_z(d) = 0$. The first of these gives that

$$v_z(0) = 0 = -\frac{P_1 - P_2}{2L\eta} \times 0^2 + b \times 0 + c = c, \tag{9.41}$$

which implies that

$$c = 0. \tag{9.42}$$

The second gives that

$$v_z(d) = 0 = -\frac{P_1 - P_2}{2L\eta}d^2 + bd, \tag{9.43}$$

i.e.,

$$b = \frac{P_1 - P_2}{2L\eta}d. \tag{9.44}$$

Putting everything together, we find

$$v_z(x) = -\frac{P_1 - P_2}{2L\eta}x^2 + d\frac{P_1 - P_2}{2L\eta}x = -\frac{P_1 - P_2}{2L\eta}x(x - d). \tag{9.45}$$

This corresponds to a parabolic velocity profile, as sketched at the bottom of Fig. 9.7.

9.7.2 Flow Rate

In addition to the velocity profile, another way to describe a fluid flow, is to calculate its (volume) flow rate through the channel, i.e., how fast an amount of fluid flows through a given section of a channel. The volume flow rate Q is defined as volume per unit of time:

$$Q = \frac{\Delta V}{\Delta t} \tag{9.46}$$

therefore it is has dimensions of m^3/s. How to accomplish such a calculation for a channel of arbitrary cross-section may be understood with the help of Fig. 9.9: the volume of fluid passing through an element of the channel of cross-sectional area dA in a time Δt is

$$dA\, v\Delta t, \tag{9.47}$$

where v is the velocity of the fluid in that portion of the channel. Equation 9.47 is similar to expressions that we encountered previously in Chap. 2. *Force and Momentum: Newton's Laws and How to Apply Them* in the context of water jets. Here, however, because v varies across the channel cross-section, it is not correct to say that the flow rate is vA. Instead, to find the total volume of flow in Δt, it is necessary to sum up $v\, dA$ over all the elements of cross-sectional area, i.e., it is necessary to integrate $v\, dA$ over the cross-sectional area of the channel:

$$\int_A v\, dA\ \Delta t. \tag{9.48}$$

Fig. 9.9 Quantities for calculating the flow rate and power dissipation in lamellar flow through a channel

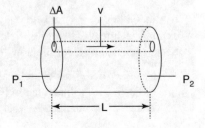

Dividing by Δt, we find the flow rate:

$$Q = \int_A v \, dA. \tag{9.49}$$

The reason that carrying out this integral is necessary is that the velocity v depends on location in the channel cross section.

We can now specialize Eq. 9.49 to the case of the flow through a thin channel of rectangular cross-section (Fig. 9.7) of height d, width W, and length L. For the slice between x and $x + dx$, we have $dA = W \, dx$ and the volume of fluid that passes a plane perpendicular to the z-axis per unit time is

$$W \, dx \, v_z(x). \tag{9.50}$$

To determine the volume flow through the entire channel we must add up the contributions from all such slices, namely we must carry out the integral:

$$Q = W \int_0^d v_z(x) \, dx = -\frac{W(P_1 - P_2)}{2L\eta} \int_0^d x(x - d) \, dx$$

$$= -\frac{W(P_1 - P_2)}{2L\eta} \int_0^d x(x - d) \, dx = \frac{W(P_1 - P_2)d^3}{12L\eta} \tag{9.51}$$

Notice that the flow rate is *not* proportional to the channel area (Wd), as we might have (incorrectly) expected. Notice too that there is a linear relationship between the flow rate Q and the pressure difference $P_1 - P_2$.

9.8   Laminar Fluid Flow in a Cylindrical Tube

In this section we extend the results found in the prevision section for a microfluidic squared channel to the case of a cylindrical channel; this, of course, is relevant to describe fluid flow through blood vessels.

9.8.1   Example: Velocity Profile in a Cylindrical Tube

In this example, we seek to calculate the laminar velocity profile within a cylindrical artery. Therefore, we must carry out a similar calculation to that of Sect. 9.7.1, but now for the flow in a cylinder of radius R and length L.

(a) List the forces on a cylindrical shell of fluid that lies between radius r and radius $r + dr$, illustrated in Fig. 9.10.

(continued)

Fig. 9.10 Schematic of the geometry for fluid flow in a cylindrical geometry. To derive the appropriate differential equation in this case, it is necessary to write down Newton's Second Law for the cylindrical shell shown

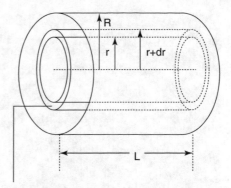

Apply Newton's second law to this cylindrical shell

Answer: The forces on this cylindrical shell are as follows:

- The force from the pressure at one end:

$$P_1 2\pi r dr. \tag{9.52}$$

- The force from the pressure at the other end:

$$- P_2 2\pi r dr. \tag{9.53}$$

Note that $2\pi r dr$ is the area of the end of the cylindrical shell.
- The viscous force from the fluid at smaller radii acting on the shell:

$$- \eta 2\pi r L \frac{dv_z(r)}{dr}. \tag{9.54}$$

Note that $2\pi r L$ is the area of the inside of the cylindrical shell.
- The viscous force from the fluid at larger radii acting on the shell:

$$+ \eta 2\pi (r + dr) L \frac{dv_z(r + dr)}{dr}. \tag{9.55}$$

Note that $2\pi (r + dr)L$ is the area of the outside of the cylindrical shell.

(continued)

(b) In a steady state, where there is no acceleration of the fluid, the sum of these forces must be zero, according to Newton's Second Law. Use Newton's Second Law to obtain an equation that relates $\frac{dv_z}{dr}$ and $\frac{d^2v_z}{dr^2}$ to $P_1 - P_2$.

Answer: Newton's Second Law informs us that the sum of the forces on the cylindrical shell is zero in a steady state, namely that

$$(P_1 - P_2)2\pi r dr - \eta 2\pi r L \frac{dv_z(r)}{dr} + \eta 2\pi (r + dr)L \frac{dv_z(r + dr)}{dr} = 0.$$
(9.56)

In the limit that dr is very small (infinitesimal), this becomes:

$$(P_1 - P_2)2\pi r dr - \eta 2\pi r L \frac{dv_z(r)}{dr} + \eta 2\pi (r + dr)L \left(\frac{dv_z(r)}{dr} + \frac{d^2v_z}{dr^2}dr \right)$$

$$\simeq (P_1 - P_2)2\pi r dr + \eta 2\pi L \left(dr \frac{dv_z}{dr} + r \frac{d^2v_z}{dr^2}dr \right) = 0,$$
(9.57)

where we dropped terms proportional to $(dr)^2$. Next, dividing through by $2\pi r dr L \eta$, we arrive at:

$$\frac{P_1 - P_2}{L\eta} + \frac{1}{r}\frac{dv_z}{dr} + \frac{d^2v_z}{dr^2} = 0.$$
(9.58)

Equation 9.58 is the desired equation.

(c) The solution to the equation found in (b) permits determination of the velocity profile within the tube. Either using WolframAlpha or by direct substitution, show that the solution to this equation that is both finite at $r = 0$ and satisfies the boundary condition that $v_z(R) = 0$ is

$$v_z(r) = \frac{R^2 - r^2}{4L\eta}(P_1 - P_2).$$
(9.59)

Answer: Using WolframAlpha, we have

(continued)

```
http://www.wolframalpha.com/input/?i=Solve+v%27%27%28s
%29%2Bv%27%28s%29%2Fs+%2BP%3D0%2C+v%28S%29%3D0,⁴
```

where we used $P = \frac{P_1 - P_2}{4L\eta}$, and s instead of r. We must set the coefficient of the logarithm equal to zero to ensure that the solution does not exhibit infinite velocity at $r = 0$.

---

[4]http://www.wolframalpha.com/input/?i=Solve+v%27%27%28s%29%2Bv%27%28s%29%29%2Fs+%2BP%3D0%2C+v%28S%29%3D0

9.8.2 Poiseuille's Law, Blushing, and Atherosclerosis

Let us describe the flow rate for a blood vessel and discuss its implications.

Calculate the volume flow rate (Q) through a blood vessel of radius R.

Answer: The required calculation is analogous to the calculation of Sect. 9.7.2. First, we must calculate the flow rate through our cylindrical shell. Then, we must add up the flow rates for all such shells at different radii, which we do by integrating from $r = 0$ to $r = R$. The calculation goes as follows:

The volume flow rate in a cylindrical shell that flows past a plane perpendicular to the axis of the channel is the fluid velocity in the shell multiplied by the cross-sectional area of the cylindrical shell, namely

$$v_z(r) \times 2\pi r dr. \tag{9.60}$$

Therefore, the volume rate through the entire cylinder is

$$Q = \int_0^R v_z(r) 2\pi r dr. \tag{9.61}$$

(continued)

Substituting, Eq. 9.59 into Eq. 9.61 and performing
the integral, we find:

$$Q = \frac{\pi R^4}{8L\eta}(P_1 - P_2). \tag{9.62}$$

Equation 9.62 is called Poiseuille's law and it is important to describe how the fluid flow in a vessel is affected by the radius of the vessel R, the length of the vessel L, the viscosity of the fluid η, and the pressure difference $(P_1 - P_2)$ at the ends of the vessel. In particular, Poiseuille's law tells us that the fluid flow depends strongly (R^4) on the radius, R. This means that a relatively small change in radius produces a large change in volume flow rate. Specifically, if the radius decreases by 20%, the volume flow rate is multiplied by a factor of $(0.8)^4 = 0.41$, so nearly 2.5 times less fluid per second. Evidently, a relatively small change in blood vessel radius can have a dramatic effect on the volume of blood flow. Poiseuille's law helps us understand the consequences of atherosclerosis, which is when the radius of blood vessels is reduced as a result of the accumulation of cholesterol and other substances on the vessel walls. Atherosclerosis limits oxygen supply to tissues and can give rise to a variety of symptoms (e.g., angina, claudication).

Equation 9.62 also expresses the physics of how you blush—being embarrassed causes a relatively modest increase in the radius of the blood vessels in your face, in turn this causes a large increase in the blood volume flow rate, and that is a blush. As to why you blush—that is outside our scope. Groan.

9.8.3   Steady-State Approximation for the Human Circulatory System

The calculations of the previous section were carried out for steady-state flows, such as would be realized if pressure differences were constant in time. However, this assumption is not strictly correct for vertebrate circulatory systems, in which the heart applies pressure in a more-or-less periodic fashion, so that the blood experiences periodic acceleration and deceleration. Insight into how reasonable a steady-state approximation may be, is provided by Fig. 9.11, which displays: the pressure inside the human heart (ventricular pressure), the pressure of the blood when it has just exited the heart in the aorta (aortic pressure), and the pressure of the blood just before it goes back to the heart (atrial pressure), plotted versus time over a couple of heartbeats. From this figure, we see that: the pressure in the heart varies a lot (the ventricular pressure goes from zero to 120 mm Hg), the atrial pressure is always close to zero, and the aortic pressure varies from a minimum of 80 mm Hg to a maximum of 120 mm Hg with a mean of about 100 mm Hg. For the point of view of describing the blood fluid flow, what matters is the pressure 'ahead' and 'behind' of a given vessel, in this case the pressure ahead (P_2) is the atrial pressure, and the pressure behind (P_1) is the aortic pressure. Thus, even for the human circulatory

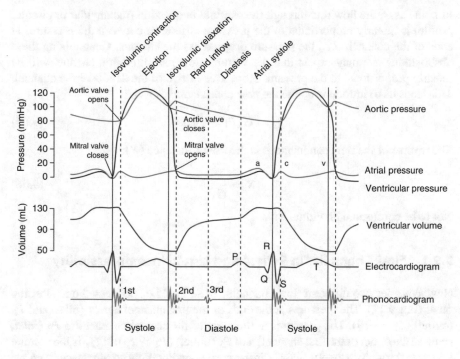

Fig. 9.11 An example Wiggers diagram from http://en.wikipedia.org/wiki/Wiggers_diagram, which includes plots of the atrial and aortic pressure on the vertical axis as a function of time on the horizontal axis, over the duration of two heartbeats

system, it seems reasonable to take $P_1 - P_2 \simeq 100$ mm Hg, and to use the steady-state results, which we have worked out, to calculate the average blood flow rate. In any case, for small Reynolds number flows, such as the blood flow through smaller blood vessels, the ma-term in $F = ma$ is small and therefore negligible compared to F. It follows that for such flows, the steady-state approximation is accurate.

9.9   Flow Conductance and Resistance

In a previous section in this chapter, we investigated the volume flow rate (Q) of laminar flow through a rectangular channel of height d, width W, and length L in the case that $d \ll W$. We found

$$Q = \frac{Wd^3}{12\eta L}(P_1 - P_2).$$ (9.63)

Similarly, for the flow rate for laminar flow through a cylindrical tube of radius R and length L (it is actually called Poiseuille flow), we showed that:

$$Q = \frac{\pi R^4}{8\eta L}(P_1 - P_2).$$ (9.64)

In both cases, the flow rate through the channel in question (rectangular or circular profile) is linearly proportional to the pressure difference between the pressures at ends of the channel, *i.e.*, the pressure drop across the channel. Generalizing these observations, we may expect that the volume flow rate through a channel will be linearly proportional to the pressure drop, irrespective of the details of the channel. This leads us to introduce/define the *flow conductance*, G via:

$$Q = G(P_1 - P_2).\tag{9.65}$$

The inverse of the flow conductance is the flow resistance (R):

$$R = \frac{1}{G},\tag{9.66}$$

Not to be confused with radius R.

9.9.1   Flow Channels "In Series" and Liquid Incompressibility

Now, consider two different flow channels "in series," i.e., connected one after the other (Fig. 9.12). The pressures at the ends of the first channel are P_1 (inlet) and P_2 (outlet) ($P_1 > P_2$). The pressures at the ends of the second channel are P_2 (inlet, same as the outlet of the first channel) and P_3 (outlet) ($P_2 > P_3$). If Q_1 is the volume flow rate through channel 1 and G_1 is the corresponding flow conductance, then we have:

$$Q_1 = G_1(P_1 - P_2).\tag{9.67}$$

Similarly, if Q_2 is the volume flow rate through channel 2 and G_2 is the corresponding flow conductance, then we have:

$$Q_2 = G_2(P_2 - P_3).\tag{9.68}$$

To determine the flow conductance of these two channels arranged "in series," we need an extra ingredient, namely that liquids are (to a good approximation)

Fig. 9.12 Two flow channels "in series"

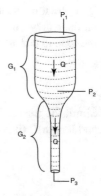

incompressible. This fact requires that (to a good approximation) the volume flow rate in the two channels are the same:

$$Q_1 = Q_2 = Q.$$

Otherwise, liquid would have to accumulate somewhere or miraculously appear from somewhere. Therefore, we have

$$Q = G_1(P_1 - P_2) \tag{9.69}$$

and

$$Q = G_2(P_2 - P_3). \tag{9.70}$$

Dividing Eq. 9.69 by G_1 and Eq. 9.70 by G_2, and then adding the two resultant equations together, we discover that

$$Q\left(\frac{1}{G_1} + \frac{1}{G_2}\right) = (P_1 - P_2) + (P_2 - P_3) = (P_1 - P_3), \tag{9.71}$$

i.e.,

$$Q = \frac{1}{\left(\frac{1}{G_1} + \frac{1}{G_2}\right)}(P_1 - P_3). \tag{9.72}$$

Equation 9.72 shows that for two channels in series the volume flow rate is also linearly proportional to the total pressure drop across the two channels. Thus, we can identify the flow conductance of the channels in series, G_{12}, as

$$G_{12} = \frac{1}{\left(\frac{1}{G_1} + \frac{1}{G_2}\right)}, \tag{9.73}$$

i.e.,

$$\frac{1}{G_{12}} = \frac{1}{G_1} + \frac{1}{G_2}. \tag{9.74}$$

This is a general prescription for combining the flow conductance for flow channels arranged in series.

It is worthwhile pointing out that if we suppose that channel 2 is itself composed of two channels in series—channel 2A and channel 2B, say—then applying Eq. 9.74, we immediately have that

$$\frac{1}{G_2} = \frac{1}{G_{2A}} + \frac{1}{G_{2B}}, \tag{9.75}$$

and

$$\frac{1}{G_{12}} = \frac{1}{G_1} + \frac{1}{G_{2A}} + \frac{1}{G_{2B}}. \tag{9.76}$$

The generalization to N channels in series is clear:

$$\frac{1}{G_{SERIES}} = \Sigma_{i=1}^{N} \frac{1}{G_i}. \tag{9.77}$$

Notice that Eq. 9.74 (as well as Eqs. 9.75, 9.76, and 9.77) is *formally* the same as the equation that tells you how to combine diffusive conductivities for membranes (for example) arranged "in series," i.e., arranged one after the other. In the context of this observation, we may say that flow conductance is *analogous* to diffusive conductance, volume flow rate is analogous to diffusion current, and pressure difference across the flow channel(s) is analogous to concentration difference across the membrane(s). Physicists love this sort of analogy, where we see the same equations applying to different physical phenomena. In Chap. 14. *Circuits and Dendrites: Charge Conservation, Ohm's Law, Rate Equations, and Other Old Friends*, we will see that equations similar to Eq. 9.74 describe how to combine electrical conductances in series. In that case, we often talk about resistance, which is the reciprocal of conductance, and so we have for combining resistances in series:

$$R_{12} = R_1 + R_2, \tag{9.78}$$

which applies equally to electrical resistance and flow resistance. The benefit of emphasizing this sort of analogy is that once you have intuition for one case and understand how to solve problems in one case, then you have intuition and can solve problems for *all* of the analogous cases. In the flow conductance/diffusive conductance/electrical conductance cases, this is a three-fold benefit.

9.9.2   Flow Channels "In Parallel"

How do we determine what is the flow conductance of two channels arranged "in parallel," as in Fig. 9.13? Channels in parallel start from the same place and arrive at the same place. The pressure at the inlet of both channels is P_1. The pressure at the outlet of both channels is P_2. Therefore,

$$Q_1 = G_1(P_1 - P_2) \tag{9.79}$$

and

$$Q_2 = G_2(P_1 - P_2). \tag{9.80}$$

Fig. 9.13 Two flow channels
"in parallel"

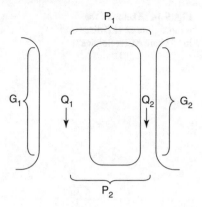

The total (combined) flow rate through both channels is $Q = Q_1 + Q_2$. Thus, we find

$$Q = G_1(P_1-P_2)+G_2(P_1-P_2) = (G_1+G_2)(P_1-P_2) = G_{12}(P_1-P_2), \quad (9.81)$$

where, in this equation,

$$G = G_1 + G_2$$

is the flow conductance of two channels in parallel. The generalization to multiple (N) channels in parallel is clear:

$$G_{PARALLEL} = \Sigma_{i=1}^{N} G_i. \quad (9.82)$$

This is another equation worth remembering. Using Eqs. 9.77 and 9.82, the flow conductance of an arbitrary microfluidic circuit can be determined.

Similarly as we have mentioned for the channels in series case, the result in Eq. 9.82 about conductances, can be equivalently written in terms of resistances as:

$$\frac{1}{R_{PARALLEL}} = \Sigma_{i=1}^{N} \frac{1}{R_i}. \quad (9.83)$$

9.9.3   Example: Square Microfluidics

A biotechnology start-up company has introduced a range of microfluidic devices that are patterned on a square grid. One of their devices is shown in Fig. 9.14. Each straight portion of the channels in this device has a length

(continued)

Fig. 9.14 Sketch of the
microfluidic device discussed
in "Square Microfluidics")

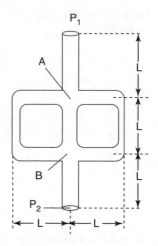

L and a flow conductance G. As can be seen in the figure, the channel that
connects the inlet (where the pressure is P_1) to point A has length L; the three
channels that connect point A to point B have lengths $3L$, L, and $3L$ for
the left, center and right channels, respectively; and the channel that connects
point B to the outlet (where the pressure is P_2) has length L.

(a) What is the flow conductance of the leftmost channel that connects point
    A and point B in terms of G?

    Answer: To calculate the flow conductance of
    a complicated flow circuit, it is generally
    necessary to start with the "innermost" flow
    channels and work "out." This problem does just
    that, by starting with the three conductances in
    series of the leftmost (or rightmost) channel.
    Their combined series conductance is

$$\frac{G}{3}. \tag{9.84}$$

(b) What is the total flow conductance between point A and point B ?

    Answer: The left, middle, and right channels are
    in parallel, so the total flow conductance is

$$\frac{G}{3} + G + \frac{G}{3} = \frac{5G}{3}. \tag{9.85}$$

(continued)

(c) What is the total flow rate Q between the inlet and the outlet?

Answer: First, we calculate the total flow
conductance:

$$\frac{1}{\frac{1}{G} + \frac{1}{5G/3} + \frac{1}{G}} = \frac{5G}{13}. \tag{9.86}$$

It immediately follows that

$$Q = \frac{5G}{13}(P_1 - P_2). \tag{9.87}$$

(d) What are the volume flow rates through the left, middle, and right-hand channels between A and B?

Answer: At a branch point, where a single flow
channel splits into multiple flow channels,
because of fluid incompressibility, the volume
flow rate through the single channel, upstream
of the branch point, is the same as the total
volume flow rate through all of the channels
together, after the branch point, i.e.,

$$Q = Q_1 + Q_2 + Q_3, \tag{9.88}$$

where Q is the total volume flow rate, Q_1 is
the volume flow rate through the left channel,
Q_2 is the volume flow rate through the middle
channel, and Q_3 is the volume flow rate through
the right channel. In this case, because all
three channels connect the same points A and
B, the pressure difference across the three
channels is necessarily the same. Therefore,

$$\frac{Q_1}{G_1} = \frac{Q_2}{G_2} = \frac{Q_3}{G_3}, \tag{9.89}$$

where $G_1 = \frac{G}{3}$, $G_2 = G$, and $G_3 = \frac{G}{3}$. Equations 9.88
and 9.89 constitute three equations for the
three unknowns Q_1, Q_2, and Q_3. Solving these
equations, we find

(continued)

$$Q_1 = \frac{G_1}{G_1 + G_2 + G_3} Q = \frac{Q}{5}, \tag{9.90}$$

$$Q_2 = \frac{G_2}{G_1 + G_2 + G_3} Q = \frac{3}{5} Q \tag{9.91}$$

and

$$Q_3 = \frac{G_3}{G_1 + G_2 + G_3} Q_1 = \frac{Q}{5}. \tag{9.92}$$

It is generally true that for channels in parallel the volume rates are proportional to the conductances.

(e) What is the pressure at point A? (Call it P_A.)

Answer: The flow rate from the inlet to point A equals the total flow rate. Therefore,

$$G(P_1 - P_A) = \frac{5G}{13}(P_1 - P_2). \tag{9.93}$$

Solving for P_A gives

$$P_A = P_1 - \frac{5}{13}(P_1 - P_2) = \frac{8}{13}P_1 + \frac{5}{13}P_2. \tag{9.94}$$

9.10   Power Dissipation in a Viscous Fluid Flow

It is interesting to calculate the power required to maintain a fluid flow Q. We will carry out this calculation, based on the result that work is force times distance. Consider the configuration shown in Fig. 9.9 and an element of fluid of cross-sectional area dA, as shown, within the fluid flow. In a time Δt, this element of fluid moves a distance $v\Delta t$, where v is the velocity of the element in question. The force on the element of fluid as a result of the pressure difference across the flow channel is $dA(P_1 - P_2)$, where dA is a tiny element of cross-sectional area. Therefore, the work done by the agent that causes the external pressure on the element of fluid in a time Δt is:

$$dA(P_1 - P_2)v\Delta t, \tag{9.95}$$

and the corresponding power (work per unit time) that must be supplied by this external agent is

$$dA(P_1 - P_2)v. \tag{9.96}$$

This is the power required to send fluid through the element of cross-sectional area, ΔA. To find the total power required, we must again integrate over the entire cross-sectional area. Thus, we find that the power required to maintain a volume flow rate Q though a channel, across which the pressure difference is $P_1 - P_2$, is

$$(P_1 - P_2) \int_A v \, dA = (P_1 - P_2)Q, \tag{9.97}$$

using Eq. 9.49.

Equation 9.97 can be written, using $Q = G(P_1 - P_2)$, in two equivalent ways. It can be written either in terms of the pressure difference as

$$G(P_1 - P_2)^2, \tag{9.98}$$

or in terms of the flow rate as

$$\frac{1}{G}Q^2. \tag{9.99}$$

One or other of these expressions may be useful in different situations.

Why is it necessary to supply power to keep the velocity profile constant? The answer is that it is necessary to supply energy to make up for frictional losses. The power calculated in Eq. 9.102 equals the rate of energy dissipation as a result of the fluid's viscosity.

9.11   Murray's Law and the "Engineering" of the Human Circulatory System

There is an interesting feature of human circulatory systems that we will now discuss, namely "Murray's Law," which relates the radius of a blood vessel upstream of a branch point (parent blood vessel) to the radius of the smaller blood vessels downstream of the branch point (child blood vessels) [3]. It is that

$$R_P^3 = R_{C_1}^3 + R_{C_2}^3 + \dots + R_{C_N}^3 = \Sigma_{i=1}^N R_{C_i}^3, \tag{9.100}$$

i.e.,

$$R_P = (\Sigma_{i=1}^N R_{C_i}^3)^{1/3}, \tag{9.101}$$

where R_P is the radius of the larger, parent blood vessel, upstream of the branch point, and R_{C_i} is the radius of one of the smaller, branched, child blood vessel,

Fig. 9.15 Schematic of a branch point in the circulatory system, where a "parent" blood vessel of radius R_P branches into two "child" blood vessels of radius R_{C_1} and R_{C_2}, respectively. These are the quantities that appear in Murray's Law (Eq. 9.101)

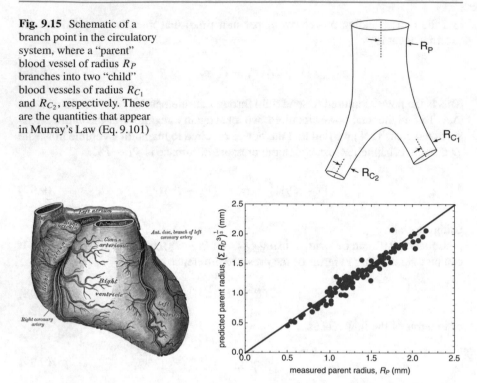

Fig. 9.16 Left: illustration from Gray's anatomy [4] of the blood vessels that serve the heart. From https://commons.wikimedia.org/wiki/File:Gray492.png. Right: scatter plot for branch points in the vasculature of the heart of the predicted radius of the parent blood vessel, (Eq. 9.101), determined from the observed branched child blood vessels' radii, versus the observed parent blood vessel's radius. Adapted from Ref. [5]

downstream of the branch point (Fig. 9.15). N is the number of branches after the branch point—a small number, usually two.

Evidence for the existence of this relationship is presented in Fig. 9.16, taken from Ref. [5].[5] On the left of this figure is shown an illustration from Gray's Anatomy [4] of the blood vessels that serve the (human) heart. On the right of this figure shows a scatter plot for each branch point in the vasculature of the heart of the predicted radius of the parent blood vessel, i.e., the right hand side of Eq. 9.101, determined from the observed branched child blood vessels' radii versus the observed parent blood vessel's radius.

Murray's law is widely applicable to many fluid circulatory systems in zoology [6]. It also holds for water transport in plants [7]. But what is the origin of Murray's Law?

---

[5] http://circres.ahajournals.org/content/38/6/572.full.pdf

We hypothesize (following Murray [3]) that physiological vascular systems have evolved to be optimally efficient. That is, blood vessels have evolved to have a radius R so that the total power dissipation in a blood vessel is as small as it can be. Our goal now is to determine the consequences of this hypothesis.

Our first step is to calculate the power required to maintain a flow of blood. Equation 9.99 informs us that the power dissipation is Q^2/G. For a cylindrical blood vessel, specifically, we have

$$\frac{Q^2}{G} = \frac{8\eta L Q^2}{\pi R^4},$$
(9.102)

since

$$G = \frac{\pi R^4}{8\eta L}$$
(9.103)

for a cylindrical channel. Equation 9.102 corresponds to the power dissipation for a fixed flow rate, Q. Notice that for fixed Q, Eq. 9.102 is proportional to R^{-4}, so frictional losses are significantly smaller for blood vessels of larger radius.

Second, we will inquire what is the metabolic energy cost of a blood vessel of radius R and length L? Blood vessels are filled with... blood, and blood has a metabolic cost that surely is proportional to the volume of blood. Therefore, we may expect the metabolic cost of a blood vessel to be proportional to the volume of the blood vessel in question (V), i.e., equal to

$$KV = K\pi R^2 L,$$
(9.104)

where K is the metabolic energy per unit time per unit volume of blood. Notice that Eq. 9.104 increases with increasing radius. Therefore, the total power required to maintain a fluid flow Q is the sum of the two terms, discussed above, i.e.,

$$P_T = \frac{8\eta L Q^2}{\pi R^4} + \pi K L R^2,$$
(9.105)

shown in Fig. 9.17.

Our hypothesis is that physiological vascular systems have evolved to be optimally efficient. That is, that the radius of blood vessels has evolved so that the total power dissipation in a blood vessel is as small as it can be, subject to constant values of Q, K, η, etc. Following this hypothesis concerning the role of evolution and natural selection, we pick R to minimize Eq. 9.105. This minimum value of R occurs for

$$\frac{dP_T}{dR} = 0,$$
(9.106)

as is clear from the plot of Eq. 9.105,[6] i.e., it occurs when

---

[6] http://www.wolframalpha.com/input/?i=plot+r%5E2%2B1%2Fr%5E4+from+r%3D0.5+to+4.0

Fig. 9.17 Frictional/viscous power loss in a cylindrical blood vessel, the metabolic power cost of keeping the blood vessel and the blood it contains alive, and the sum of these two contributions sketched versus vessel radius

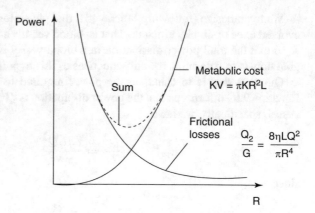

$$\frac{dP_T}{dR} = -\frac{32\eta L Q^2}{\pi R^5} + 2\pi K L R = 0, \tag{9.107}$$

i.e.,

$$R^6 = \frac{16\eta Q^2}{\pi^2 K}, \tag{9.108}$$

or

$$R^3 = \sqrt{\frac{16\eta}{\pi^2 K}}\, Q. \tag{9.109}$$

Equation 9.109 specifies the energetically most efficient value of the blood vessel radius for a given flow rate. Experimental support for this relationship between Q and R is presented in Fig. 9.18 from Ref. [8], which shows a plot of flow rate versus blood vessel diameter for blood vessels within the cremaster muscle of living rats. These data are consistent with a vessel flow rate that varies as the cube of the vessel radius, as predicted.

Now consider a branch point, as shown in Fig. 9.15. Because fluids are incompressible, we must have

$$Q_P = Q_{C_1} + Q_{C_2}. \tag{9.110}$$

Equations 9.110 and 9.109 together then imply that

$$R_P^3 = R_{C_1}^3 + R_{C_2}^3, \tag{9.111}$$

i.e., Murray's Law. More generally, we would have

$$Q_P = \Sigma_{i=1}^N Q_{C_i}, \tag{9.112}$$

Fig. 9.18 Measurements of the flow rate versus blood vessel radius (circles) for blood vessels within the cremaster muscle of living rats. The line corresponds to the theoretical prediction, Eq. 9.109. Adapted from Ref. [8]

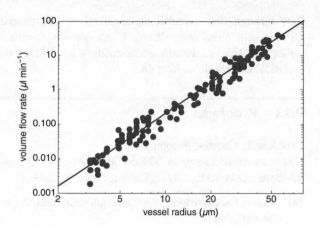

leading to Eq. 9.101, i.e., Murray's Law again .

This result is cool and elegant: we see that the "engineering" of biological vascular systems, including that of our own bodies, is simply and directly a consequence minimizing the total energy cost of the vascular system in question. The process of evolution and natural selection has sought out the most-efficient physiology. We also see that the minimization calculus depends fundamentally on the frictional losses encountered in laminar fluid flows. It is a great example of just how relevant physics is to human physiology.

9.12   Chapter Outlook

In this chapter, we have mathematically modeled the flows of laminar fluids. A direct application of the results is to the understanding of the human circulatory system. We hope you have appreciated how the appropriate relationships between pressure, velocity, and geometry of a fluid channel can help understand a variety of aspects related to blood, for instance: how much blood flow through an artery at a given time, why we blush, why our vessels have a the particular radii that they have. We also showed that the observed physiology of the circulatory system plausibly follows from minimizing the energy cost of circulating blood. Looking forward, a number of the results derived here in the context of fluid flow have analogies in other systems and so turn out to be universal behaviors. For instance, the resultant linear relationship between flow rate and pressure difference that we discussed in this chapter ($Q = G(P_1 - P_2)$) is another example of a "gradient flow," analogous to the linear relationship between the particle current through a membrane and the difference in concentration across the membrane, which we encountered in Chap. 6. *Diffusion: Membrane Permeability and the Rate of Actin Polymerization.* In addition, because analogous rules exist for determining the current flow through electrical circuits to those that we will encounter here for determining the volume flow through flow circuits, the material of this chapter foreshadows Chap. 14. *Circuits and Dendrites: Charge Conservation, Ohm's Law, Rate Equations, and Other Old Friends.* Hence, understanding fluid flows will help you to understand electrical current flows later on.

9.13   Problems

Problem 1: Cardiac Surgery
You are a cardiac surgeon. The coronary artery of one of your patients suffers from atherosclerosis, so that its radius is reduced by 20%.

(a) What is the corresponding fractional reduction in the cross-sectional area of the coronary artery?
(b) How much do you think the blood flow through the artery will increase, after you perform a procedure to restore the artery to its original radius?

Problem 2: Reynolds Number
(a) Given that the frictional force on a spherical bead of radius R moving with velocity v through a fluid of viscosity η is $6\pi\eta Rv$, calculate the Reynolds number for a bead of radius R and density ρ_1, falling at its terminal velocity through a liquid of viscosity η and density ρ_2.
(b) Estimate the Reynolds number for 3 mm-diameter Delrin (polyoxymethylene) beads ($\rho_1 \simeq 1.4$ g cm^{-3}), falling at their terminal velocity in canola oil ($\rho_2 \simeq 0.9$ g cm^{-3}, $\eta \simeq 60$ mPa s). Do you expect laminar or turbulent flow in this case?

Problem 3: Average Flow Speed
It is possible to define an "average" flow velocity, $< v >$, for blood flow rate, Q, through a blood vessel of cross-sectional area A via $Q =< v > A$. The average speed of blood in the aorta is 0.3 m s^{-1} and the radius of the aorta is 1 cm. There are about 2×10^9 capillaries with an average radius of 6 μm. What is the average speed of blood flow in a capillary?

Problem 4: Reynolds Number in Your Coronary Artery
(a) The inside diameter of a healthy adult human coronary artery is about 3 mm. In many cases, however, with advancing age, this diameter is reduced by atherosclerosis.[7] This sort of coronary artery disease is implicated as the leading cause of death worldwide. Calculate the Reynolds number for blood flow in your coronary artery, given that the average flow velocity there is 15 cm s^{-1}, and that the viscosity of blood is 3 times that of water. Do you expect laminar or turbulent flow in your coronary artery?

---

[7] http://en.wikipedia.org/wiki/File:RCA_atherosclerosis.jpg

(b) If an artery is subject to a stenosis that decreases its cross-sectional area locally, discuss whether turbulent flow in the artery is more or less likely near the site of the stenosis than away from it.

(c) A bruit is an abnormal noise that blood makes when it rushes past an obstruction in an artery. Discuss whether your answer to (b) yields insight into the possible origins of this noise.

Problem 5: Emergency Room
A patient requires emergency infusion of intravenous saline as quickly as possible. Is it preferable to use a standard triple lumen central line (length 20 cm, radius of each lumen 0.84 mm) or a standard 16 gauge peripheral IV (length 4 cm, radius 1.2 mm)?

Problem 6: JFK Viscosity
In their continuing quest to help students learn physics, staff at JFK have set up two parallel conveyor belts as shown in Fig. 9.19. One is driven at a speed v_1. The other runs without friction and is undriven. There are bags and baggage handlers on both belts and on the stationary floor next to the undriven conveyor belt. Each time step τ, the baggage handlers on the unpowered belt place $\frac{1}{2}N$ bags, which are each of mass m, onto the powered belt, and $\frac{1}{2}N$ bags onto the floor. At each time step τ, the baggage handlers on the driven belt place $\frac{1}{2}N$ bags (each of mass m and each with initial velocity v_1) onto the unpowered belt. Similarly, at each time step τ, the baggage handlers on the floor place $\frac{1}{2}N$ bags (each of mass m and each with initial velocity 0) onto the unpowered belt. After a while a steady state is achieved in which the undriven, frictionless belt has a constant velocity v.

(a) Calculate the force on the unpowered belt as result of the bags from the belt moving with velocity v_1.

(b) Calculate the force on the unpowered belt as result of the bags from the floor (velocity 0).

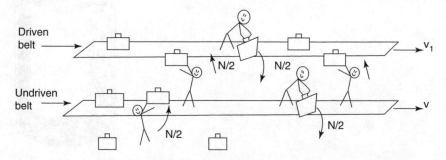

Fig. 9.19 Configuration of conveyor belts and bags for the beginning of "JFK viscosity"

(c) What is the total force on the unpowered belt? Hence determine v in terms of given parameters.

(d) Explain the analogy between this example and how fluid frictional forces cause the fluid between two moving plates to move.

Problem 7: Fish Circulatory System

As shown in Fig. 9.20, a fish heart consists of two chambers: the ventricle and the atrium. The role of the heart, of course, is to generate the pressure difference required to cause blood flow around the fish's circulatory system. The pressure at the exit of the ventricle is P_1. The pressure immediately before the entrance to the atrium is P_2. In order for the flow to occur in the direction indicated, it is necessary that $P_1 > P_2$. In fact, the fish circulatory system can be conceived as consisting of three components in series. The flow conductance of the gills is G_G. The flow conductance of the aorta is G_A. The flow conductance of the systemic capillaries and veins is G_C.

(a) Suppose that the pressure at the point between the gills and the aorta is P_3. If the volume flow rate through the gills is Q, what is the pressure difference across the gills ($P_1 - P_3$) in terms of Q and G_G?

(b) How is the volume flow rate through the aorta (Q_A) related to that through the gills (Q)?

(c) Take the pressure at the point between the aorta and the capillaries to be P_4. Write an equation relating Q_A, G_A, and $P_3 - P_4$.

(d) Combine your answers to (a), (b), and (c) to calculate $P_1 - P_4$ in terms of Q, G_G, and G_A.

(e) Your result to (d) recapitulates how to combine two flow conductances in series. Extend this result to three flow conductances in series to determine $P_1 - P_2$ in terms of G_G, G_A, and G_C and Q.

Fig. 9.20 Schematic of the fish circulatory system

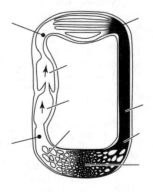

Problem 8

Consider the flow channel shown in Fig. 9.21. Each of its constituent channels has the flow conductance shown. Do NOT worry about any additional effects beyond these flow conductances.

(a) Determine the unknown flow rate, Q, in terms of the given quantities P_1, P_2 G_1, G_2, G_3, and G_4.
(b) What is the power dissipation of this flow circuit in terms of given quantities?
(c) What are the pressures at point A and at point B?

Problem 9: Self-Similar Flow Circuit

Consider the flow channel shown in Fig. 9.22. Each of its constituent channels has the flow conductance shown.

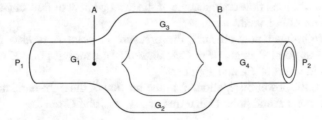

Fig. 9.21 Series and parallel flow circuit

Fig. 9.22 Flow circuit in which the flow conductances of the individual channels are labeled

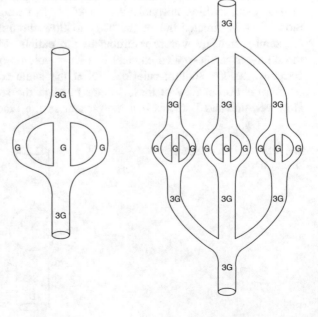

(a) Determine the flow conductance of the flow circuit shown on the left-hand side of the figure (call it G_L) in terms of G.
(b) If the pressure at the top of the top of the left-hand flow circuit is P and the pressure at the bottom is 0, what is the pressure at the top branch point (where one channel divides into three)
(c) Determine the flow conductance of the flow circuit shown on the right-hand side of the figure (call it G_R) in terms of G.

Problem 10: Funky Flows
(a) Calculate the total flow conductance, G_T, of the arrangement of flow channels shown at the left of Fig. 9.23.
(b) In the left-hand flow channel, the pressure at point A is P_A and the pressure at point B is P_B. Calculate the pressure at point C (P_C) in this channel in terms of P_A and P_B only.
(c) Calculate the pressure at point D (P_D) in the left flow channel in terms of P_A and P_B only.
(d) Calculate the total flow conductance of the arrangement of flow channels shown at the right of Fig. 9.23.
(e) In the right-hand flow channel, the pressure at point A is also P_A and the pressure at point B is also P_B. Calculate the pressure at point E (P_E) in this channel in terms of P_A and P_B only.
(f) Calculate the power dissipation, S, in the portion of the right-hand flow channel between point E and point D in terms of P_A, P_B, and G only.

Problem 11: Hemodialysis
Hemodialysis is one of the most common hospital procedures in the US, resulting in nearly one million hospital visits in 2011. In hemodialysis, a fraction of the blood flow is directed out of the body to flow through a channel, whose walls are semi-permeable, before returning to the patient. Counter-circulating outside the semi-permeable-walled channel is a large volume of dialyzate. The dialyzate contains glucose, sodium chloride, etc. at the same concentrations as in blood, so there is no net flow of these solutes through the semi-permeable membrane. However, unwanted solutes that are present in the blood, but not present in the

Fig. 9.23 Flow channels for "Funky flows"

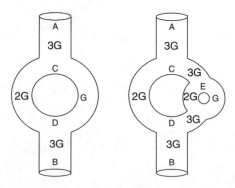

Fig. 9.24 Sketch of a fistula
and the relevant flow
conductances. The flow
circuit Inside the dashed
circles is completed with the
flow conductance indicated

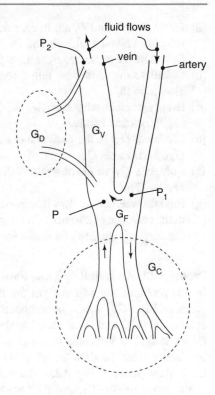

dialyzate, such as urea and creatinine, diffuse across the semi-permeable membrane
from the blood into the dialyzate, thus leaving the fluid volume of the body.

To facilitate long-term hemodialysis treatment, the preferred procedure involves
the surgical creation of an arteriovenous fistula, in which an artery and a vein are
joined together in the dialysis patient's non-dominant arm, as shown schematically
in Fig. 9.24. For hemodialysis, two needles are inserted in the vein, immediately
downstream of the fistula: An "upstream" needle is inserted into the "arterial"
location, close to the vein-artery connection, diverting blood out of the body for
dialysis, while a "downstream" needle is inserted into the "venous" location, further
downstream from the fistula, in order to return cleansed blood to the body.

Your goal in this problem is to explain the rationale for introduction of the fistula.
To this end, suppose, as shown in Fig. 9.24, that the arterial pressure immediately
upstream of the fistula is P_1 and the venous pressure immediately downstream of the
"downstream" needle is P_2. From the artery, blood flows either through capillaries
(fluid conductance, G_C) or through the fistula (fluid conductance, G_F) to the inlet
to the vein. At the inlet to the vein, blood then either flows through the vein itself
(fluid conductance, G_V), or through the dialysis apparatus, before returning to the
vein at the "venous" location.

(a) Given that P_1 and P_2 are fixed, calculate the total volume flow rate, Q between the points labeled with P_1 and P_2.
(b) Calculate the pressure, P, at a the point shown in the figure.
(c) Calculate the fluid flow rate through the dialysis apparatus, Q_D, using your answer to (b).
(d) Briefly explain why $G_F \gg G_C$ in view of the different dimensions of the capillaries and fistula.
(e) Calculate Q_D in the presence of the fistula, assuming that $G_F \gg G_C, G_V \gg G_D$, and $G_F \gg G_V$.
(f) Calculate Q_D in the absence of the fistula, assuming that $G_V \gg G_D$ and that $G_V \gg G_C$.
(g) Explain whether the flow rate through the dialysis apparatus is larger or smaller in the presence or absence of the fistula. Thus, explain the rationale for including the surgical creation of a fistula for dialysis patients.

Problem 12: An Infinitely Long Flow Ladder
In this problem, we will analyze the infinitely long flow ladder shown in the top panel of Fig. 9.25, which is composed of infinitely many flow channels of flow conductance $4G$ and G, arranged as shown. The total flow conductance of the flow ladder is G_T, as indicated in the middle panel of Fig. 9.25. Notice, however, that the ladder after the first "rung" is the same as the entire ladder, as indicated in the bottom panel of Fig. 9.25. The ladder after the second rung is also the same as the entire ladder. These results are a consequence of the ladder being infinitely long. It follows that all three flow circuits shown in Fig. 9.25 have the same flow conductance, equal to the flow conductance of the entire ladder, G_T.

(a) Calculate the flow conductance of the flow circuit, conceived as shown in the bottom panel of Fig. 9.25, in terms of G and G_T.
(b) Given that the total flow conductance of the ladder, namely G_T, must be equal to your answer to (a) write down a single equation that establishes a relation between G_T and G.
(c) Show that your equation is satisfied by picking $G_T = G$. In the remainder of this problem, you should take $G_T = G$.
(d) Calculate the total flow rate, Q, between point 1, where the pressure is P_1, to point 2, where the pressure is P_2.
(e) What is the flow rate through the first rung of the ladder (Q_1)?
(f) What is the pressure difference across the first rung between point A and point C ($P_A - P_C$), in terms of P_1 and P_2 only?
(g) What is the flow rate through the second rung of the ladder (Q_2), given that the flow conductance of the ladder beyond the second rung is G_T?
(h) What is the pressure difference across the second rung between point B and point D ($P_B - P_D$), in terms of P_1 and P_2 only?

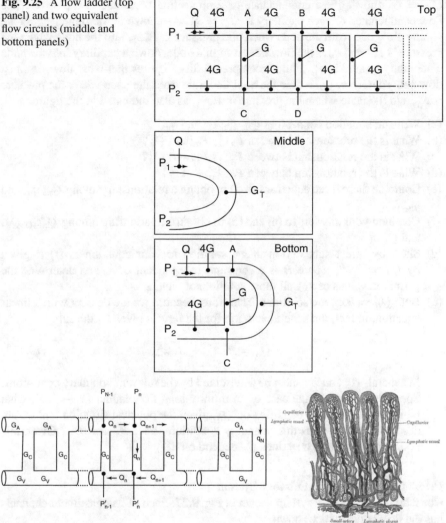

Fig. 9.25 A flow ladder (top panel) and two equivalent flow circuits (middle and bottom panels)

Fig. 9.26 Left: A flow ladder, modeling capillary microcirculation. Right: capillaries in villi of the small intestine, which realize flow ladders similar to those considered in this problem. From https://commons.wikimedia.org/wiki/File:Gray1061.png

Problem 13: Simplified Model of Capillary Microcirculation: Another Flow Ladder

An arteriole and venule are connected by a large number (N) of capillaries as shown in Fig. 9.26, where the top flow channel represents the arteriole and the bottom flow channel represents the venule. Each rung of the ladder represents a capillary. Between successive capillaries, the flow conductance of each section of arteriole is

$G_A = G$, and the flow conductance of each section of venule is $G_V = G$. The flow conductance of each capillary is $G_C = \frac{G}{12}$. As shown in the figure, Q_n is the volume flow rate from the node where the pressure is P_{n-1}, into the node where the pressure is P_n, and q_n is the flow rate from that node, down a capillary into the node where the pressure is P'_n. Fluid incompressibility requires that what flows in must flow out. Therefore, Q_n is also the fluid flow rate from the node where the pressure is P'_n, into the node where the pressure is P'_{n-1}, as also indicated in the figure.

(a) Write an equation connecting Q_n, Q_{n+1}, and q_n.
(b) What is the relationship between P_{n-1}, P_n, and Q_n?
(c) What is the relationship between P'_{n-1}, P'_n, and Q_n?
(d) What is the relationship between P_n, P'_n, and q_n.
(e) Combine these three equations (b-d) to obtain a relationship among Q_n, q_n, and q_{n-1}.
(f) Combine your answers to (a) and (e) to obtain a relationship among Q_{n-1}, Q_n, and Q_{n+1}.
(g) Show by direct substitution that a solution to your equation in (f) is given by $Q_n = Bx^n$, where B is a constant, independent of n, and determine the permitted values of x. Call those solutions x_+ and x_-.
(h) Both $Q_n = Bx_\pm^n$ are solutions. Therefore, because we are dealing with a linear equation, in fact, the general solution for the flow is given by the sum:

$$Q_n = B_+ x_+^n + B_- x_-^n. \qquad (9.113)$$

In general, B_+ and B_- must be determined by the relevant boundary conditions. Specializing now to the case of an infinity-long flow ladder ($n \rightarrow \infty$), what must be the boundary condition on Q_n, given the physical situation that we are describing? How does this boundary condition affect B_+ and B_-? Given this boundary condition, how does Q_n depend on n?

Problem 14: Fractal Circulatory System
Consider the circulatory system shown in Fig. 9.27. Each of its constituent channels has the flow conductance shown.

(a) Determine the flow conductance of the entire system of channels, in terms of G_0, G_1, G_2, and G_3.
(b) Calculate the power dissipation through this circulatory system for a total flow rate Q.
(c) If the pressure at the top is P and the pressure at the bottom is 0, calculate the pressure at the top of the bottom G_0 channel (P_B) and the pressure at the bottom of the top G_0 channel (P_A).
(d) Now, considering only the four G_1 channels, what is the power dissipation in these channels?
(e) Now, considering only the two G_0 channels, what is the power dissipation in these channels?

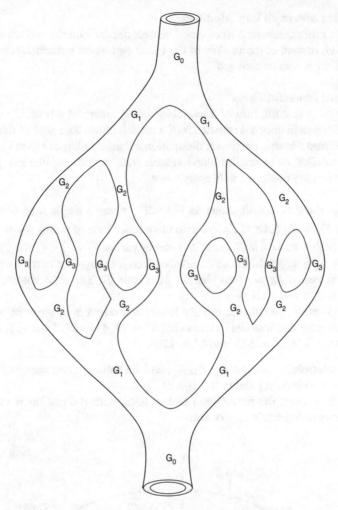

Fig. 9.27 Circulatory system

(f) If this circulatory system follows Murray's Law, i.e., that $R_0^3 = 2R_1^3$, *etc.*, and all the channels are the same length, is more energy dissipated in the R_0-radius channels or in the R_1-radius channels?

(g) Suppose that there are n steps at which each channel splits into 2 smaller channels. What is the ratio of the energy dissipation in the smallest channels to the energy dissipation in the largest?

Problem 15: Murray's Law Modified
Suppose that the metabolic cost of blood vessels depends on the surface area of the blood vessel, instead of the volume of the blood and blood vessels. Determine how Murray's Law would be changed.

Problem 16: Powerful Flows
Murray's law is that the sum of the radii cubed upstream of a branch point equals the sum of the radii cubed downstream of a branch point. The goal of this problem is to understand whether the power dissipation is larger in larger blood vessels (i.e., in your arterioles) or in smaller blood vessels (i.e., in your capillaries), given that human physiology conforms to Murray's law.

Consider the flow circuit shown in Fig. 9.28, where a single tube of length $\frac{1}{2}L$ and radius $R_6 = R$ (Tube 1) divides into three tubes, one of radius $R_3 = \frac{1}{2}R$ (Tube 3), one of radius $R_4 = \frac{2}{3}R$ (Tube 4) and one of radius $R_5 = \frac{5}{6}R$ (Tube 5). Tubes 3, 4, and 5 all have length L. These three tubes then join together into a second single tube also of radius $R_6 = R$ and length $\frac{1}{2}L$ (Tube 2). The total volume flow rate through this flow circuit is Q.

In this problem, you should usually leave your answers in terms of 3^4, 4^4, ... , etc., but you may find it useful to know that $3^3 = 27$, $4^3 = 64$, $5^3 = 125$, $6^3 = 216$, $3^4 = 81$, $4^4 = 256$, $5^4 = 625$, and $6^4 = 1296$.

(a) What relationship among R_3, R_4, R_5, and R_6 follows from applying Murray's Law at each branch point in the figure?
(b) Explain whether the flow circuit of Fig. 9.28, with the particular radii given, conforms to Murray's Law or not.

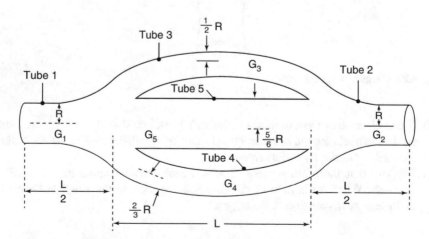

Fig. 9.28 Flow circuit for "Where is the power?" Tube 1 (Tube 2) at the left (right) has radius $R_6 = R$ ($R_6 = R$) and flow conductance $G_1 = 2G$ ($G_2 = 2G$). Tubes 3, 4, and 5 have radii $R_3 = \frac{1}{2}R$, $R_4 = \frac{2}{3}R$, $R_5 = \frac{5}{6}R$, respectively, and flow conductances G_3, G_4, and G_5, respectively

(c) Given that the flow conductance of Tube 1 is $G_1 = 2G$, what are the flow conductances of Tubes 3, 4, and 5—call them G_3, G_4 and G_5, respectively—in terms of G.

(d) If the volume flow rate through Tube 1 is Q, what is the total flow rate through Tubes 3, 4, and 5 in terms of Q only? Call these flow rates Q_3, Q_4, and Q_5, respectively.

(e) What is the power, W_3, dissipated in Tube 3, because of viscous friction, in terms of Q and G?

(f) What is the total power dissipated in Tubes 3, 4, and 5 in terms of Q and G? You may find it helpful to infer from your answer for W_3, the corresponding expressions for W_4 and W_5.

(g) What is the total power dissipated in Tubes 1 and 2 in terms of Q and G?

(h) Which is larger, the power dissipated in Tubes 3 ,4, and 5 ($W_3 + W_4 + W_5$) or the power dissipated in Tubes 1 and 2 ($W_1 + W_2$)? What does this result, and the preceding calculation, tell us about power dissipation within a collection of capillaries, that together carry the same blood volume flow rate as an arteriole, compared to the power dissipation in the arteriole?

Problem 17: Go with the Flow

(a) Calculate the total flow conductance, G_T, of the arrangement of flow channels shown in Fig. 9.29.

(b) For the flow circuit of Figure 9.29, the pressure at point A is P_A and the pressure at point B is P_B. Calculate the pressure at point C (P_C) in this channel in terms of P_A and P_B only.

(c) For the flow circuit of Fig. 9.29, calculate the pressure at point D (P_D) in terms of P_A and P_B only.

(d) The top of Figure 9.30 shows an infinite flow ladder, comprised of an infinite, periodic sequence of flow channels. The flow conductance of this flow ladder is G_T. Briefly explain how the flow conductance of this infinite ladder can be calculated from the flow circuit shown at the bottom of Fig. 9.30. Do not actually calculate G_T in this part.

(e) Using your answer to (a), or otherwise, determine G_T for the flow circuit of Fig. 9.30 in terms of G.

Fig. 9.29 First flow circuit for "Go with the flow"

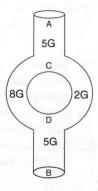

Fig. 9.30 Two
representations of the second
flow circuit for "Go with the
flow"

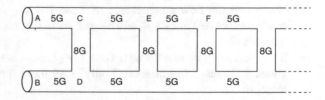

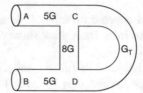

(f) What is the pressure difference between point D and point C in the flow circuit at the top of Fig. 9.30? That is, what is $P_C - P_D$?

(g) If the total flow rate from point A to point C in the flow circuit at the top of Fig. 9.30 is Q, what is the flow rate from point C to point E in terms of Q? Call this flow rate Q_1.

(h) What is the flow rate from point E to point F in terms of Q? Call this flow rate Q_2.

Problem 18: Dialysis Design Project

You are working for a biomedical engineering company on a project to design a next-generation dialysis machine. Your supervisor has asked you to investigate the possibility of using an infinitely long flow ladder in this device. To this end, you draw the flow ladder shown in Fig. 9.31. The patient's blood inlet is at point A, where the pressure is P_A. The outlet is at point B, where the pressure is P_B with $P_B < P_A$. The "rungs" of your flow ladder have conductance G and the "risers" have conductance G_1, where G_1 is to be determined in terms of G in order to achieved the desired flow properties, which include a relatively large conductance and a relatively long length over which the flow decreases along the flow ladder. Because of these considerations, you determine that the total flow conductance of the flow ladder should be $5G$. The goal of this problem is to determine how to pick G_1 and to determine what is the decay length of the flow in the flow ladder.

(a) Draw a flow circuit that is equivalent to the infinite flow ladder of Fig. 9.31, but that replaces the circuit beyond the first rung by a single flow channel of the appropriate flow conductance.

(b) Determine the required value of G_1 by first calculating the total flow conductance of the circuit that you drew in (a) in terms of G and G_1 (or otherwise).

(c) What fraction of the total flow goes through the first rung when G_1 has the value you determined in (b)?

Fig. 9.31 Flow circuit for
"Dialysis design project"

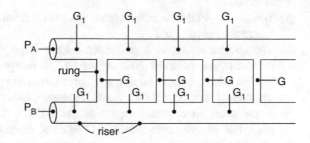

Fig. 9.32 Stuck nucleus

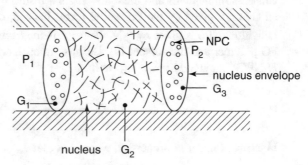

(d) What fraction of the total flow goes through the flow ladder beyond the first
    rung? That is, what fraction of the flow does not go through the first rung?
(e) What fraction of the total flow goes through the flow ladder beyond the second
    rung?
(f) What fraction of the total flow goes through the flow ladder beyond the n-th
    rung?
(g) We can define the decay "length," N, of the flow ladder– it is actually the
    number of rungs over which the flow decays—by asserting that the answer to (f)
    is equal to $e^{-n/N}$. Using this result and your answer to (f), calculate N. Next,
    use the result that for small x, $\log(1 + x) \simeq x$, to simplify your expression
    for N.
(h) Given that N is given by your simplified answer to (g), compare the total
    conductance of N conductances in parallel, each one of conductance G, to the
    total fluid conductance of the flow ladder, namely $5G$.

Problem 19: Nucleus in a Pinch
Consider a cell nucleus that is pinched and stuck in a narrow passage as shown in
Fig. 9.32. In the cytosol either side of the nucleus there are different pressures: P_1 on
the left and P_2 on the right. Because of this pressure difference (assuming $P_1 > P_2$
for clarity), we may expect fluid flow into the nucleus through the nuclear envelope
on the left (flow conductance G_1), then across the nucleus (flow conductance G_2),
and finally out of the nucleus through the nuclear envelope on the right (flow
conductance $G_3 = G_1$).

(a) What is the total flow conductance (call it G) from where the pressure is P_1 to where the pressure is P_2?

(b) The nuclear envelope is decorated with a large number (N at each end) of nuclear pore complexes (NCPs).[8] An NCP is an opening in the nuclear envelope that permits molecules (water, ions, proteins, ...) to pass back and forth between the nuclear volume and the cytosol. However, rather than the NPC being simply a tiny tube, an essential aspect of the NPC is that intrinsically, disordered phenylalanine-glycine repeats (F-G repeats) extend into the NPC channel that connects the nucleus and the cytosol. As a result, the flow conductance of the NPC is *not* given by the Poiseuille form. Instead, the flow conductance of an individual NPS is given by $\frac{\pi a^2}{df}$, where f is the friction coefficient between the FG polymer and the solvent [9]. Because at each end of the nucleus, there are N NPCs, arranged in parallel, we have

$$G_1 = \frac{N\pi a^2}{fd}. \tag{9.114}$$

In terms of material properties, we expect that

$$f \simeq \frac{\eta}{\zeta^2}, \tag{9.115}$$

where η is the solvent viscosity and ζ is the gel's mesh size [9]. Using this form for the friction coefficient we can write

$$G_1 = \frac{N\pi a^2 \xi^2}{\eta d}, \tag{9.116}$$

which may be compared to the corresponding Poiseuille form for a channel of the same dimensions but without a gel:

$$G_{Poise} = \frac{N\pi a^4}{8\eta d}. \tag{9.117}$$

By comparing Eqs. 9.116 and 9.117, we see that, up to a constant factor, we can intuitively consider the mesh to correspond to $\frac{a^2}{\xi^2}$ channels each of radius ξ.

The interior of the nucleus contains chromatin, lamins, *etc.*, and we will suppose that it too may be modeled as a gel with the same friction coefficient. According to this model, what is G_2?

(c) Write the total flow conductance, G.

(d) Discuss whether, within the context of this model, the volume flow rate through the stuck nucleus is limited by the NPCs or by the nuclear volume.

---

[8] https://en.wikipedia.org/wiki/Nuclear_pore

(e) Criticize this model.

(f) Estimate the volume flow rate for a pressure difference driven by a 1 mM concentration difference.

References

1. G.M. Whitesides, The origins and the future of microfluidics. Nature **442**, 368–373 (2006)
2. T.N. Ford, K.K. Chu, J. Mertz, Phase-gradient microscopy in thick tissue with oblique back-illumination. Nat. Methods **9**, 1195–1197 (2012)
3. C.D. Murray, The physiological principle of minimum work: I. The vascular system and the cost of blood volume. Proc. Natl. Acad. Sci. USA **12**, 207–214 (1926)
4. H. Gray, *Anatomy of the Human Body* (Lea and Febiger, Philadephia, 1918)
5. G.M. Hutchins, M.M. Miner, J.K. Boitnott, Vessel caliber and branch-angle of human coronary artery branch points. Circ. Res. **38**, 572–576 (1976)
6. M. LaBarbera, Principle of design of fluid transport systems in zoology. Science **249**, 992–1000 (1990)
7. K.A. McCulloh, J.S. Sperry, F.R. Adler, Water transport in plants obeys Murray's law. Nature **421**, 939–942 (2003)
8. H.N. Mayrovitz, J. Roy, Microvascular blood flow: evidence indicating a cubic dependence on arteriolar diameter. Am. J. Physiol. Heart Circul. Physiol. **245**, H1031–H1038 (1983)
9. M. Tokita, T. Tanaka, Friction coefficient of polymer networks of gels. J. Chem. Phys. **95**, 4613–4619 (1991)

Oscillations and Resonance

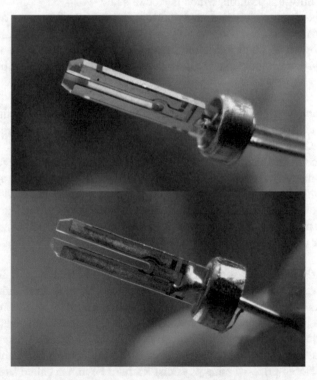

Fig. 10.1 Quartz crystal resonator https://en.wikipedia.org/wiki/Crystal_oscillator, which resonates at 32768 Hz and keeps time in quartz watches. Public domain image from https://en.wikipedia.org/wiki/Crystal_oscillator#/media/File:Inside_QuartzCrystal-Tuningfork.jpg

© The Author(s), under exclusive license to Springer Nature Switzerland AG 2023   499
S. Mochrie, C. De Grandi, *Introductory Physics for the Life Sciences*,
Undergraduate Texts in Physics, https://doi.org/10.1007/978-3-031-05808-0_10

10.1   Introduction

In this chapter, we will examine the motion of a mass or masses attached to a spring or springs, which undergo oscillatory motions, when the friction is small. This motion is called simple harmonic motion. These equations of motion are formally the same as equations that describe many other physical phenomena. Why is it that the equations for simple harmonic motion appear in so many different situations? Or, equivalently, why are Hooke's law forces so prevalent? The answer is that sufficiently near any potential energy minimum, where a system is stable, almost all energy landscapes may be approximated as a quadratic function of displacement from the location of their minimum. Since (as we have seen in Chap. 3. *Energy: Work, Geckos, and ATP*) the force along a particular direction is the negative derivative of the potential energy with respect to the coordinate along that direction, it follows that there is a restoring force linearly proportional to the displacement from the equilibrium position, which is Hooke's law.

Next, this chapter discusses an important, related phenomenon, namely resonance. Resonance corresponds to a system of interest responding strongly when it is driven by an oscillatory external force with a frequency near the frequency at which the system naturally oscillates. Examples of resonance include the following:

- Inside your watch,[1] there is a quartz crystal in the shape of a tuning fork (see Fig. 10.1), which has a resonance at 32768 Hz, which enables your watch to keep time.
- Resonance in an electrical circuit is the basis for how a radio tuner functions.
- Nuclear magnetic resonance, which permits magnetic resonance imaging (MRI), corresponds to the adsorption of radio-frequency (hundreds of MHz) power, which provides the external oscillatory force, by a system of nuclear spins.
- The pattern of strong absorption resonances in the infrared (IR) is a signature of particular drug molecules.

This chapter's final topic is a discussion of coupled oscillators, as, for example, when two masses on springs are coupled together by an additional spring. Because the equations that describe coupled harmonic oscillators are linear equations, involving more than one variable, their solutions involve eigenvalues and eigenvectors and obey the principle of superposition, just as in Chap. 7. *Rates of Change: Drugs, Infections, and Weapons of Mass Destruction*. In the context of simple harmonic motion, the Nobel-prize-winning physicist Richard Feynman called superposition

...probably the most general and wonderful principle of mathematical physics [1].

---

[1] http://en.wikipedia.org/wiki/Quartz_clock

He went on to explain that

> .... a linear system need not be moving in a purely sinusoidal fashion, i.e., at a definite single
> frequency, but no matter how it does move, this motion can be represented as a superposition
> of pure sinusoidal motions. The frequency of each of these motions is a characteristic of the
> system, and the pattern or waveform of each motion is also a characteristic of the system.
> The general motion in any such system can be characterized by giving the strength and the
> phase of each of these modes, and adding them all together. Another way of saying this is
> that any linear vibrating system is equivalent to a set of independent harmonic oscillators,
> with the natural frequencies corresponding to the modes [1].

Feynman's characteristic frequencies are the eigenvalues, his "pattern or waveform
of each motion" is its eigenvector, and his modes are eigenmodes.

10.2   Your Learning Goals for This Chapter

By the end of this chapter, you should be able to:

- Show that the equation of motion corresponding to undamped simple
  harmonic motion has sinusoidal/cosinusoidal, or equivalently imaginary
  exponential, solutions.
- Show that the equation of motion corresponding to damped simple harmonic
  motion has general solutions that are complex exponential functions of time.
- Impose the requirement that any physically sensible solution for a harmonic
  oscillator's motion must be a real function.
- Determine the position and velocity of an undriven damped simple harmonic
  oscillator for all subsequent times given any two initial conditions.
- Calculate the "steady-state" motion of a damped simple harmonic oscillator,
  driven at a constant angular velocity, and be able to explain the phenomenon
  of resonance.
- Apply the principle of superposition to express the motion of a driven,
  damped, simple harmonic oscillator at any time, as a superposition of the
  "steady-state" driven solution and the general undriven solution.
- Determine the eigenvalues and eigenvectors of two coupled oscillators.
- Apply the principle of superposition to express an arbitrary motion of two
  coupled oscillators as a superposition of eigenmodes.
- Determine the amplitude of each eigenmode in any particular case using
  initial conditions.

10.3   Simple Harmonic Motion

Recall the force exerted by a spring, as we discussed in Chap. 2. *Force and Momentum: Newton's Laws and How to Apply Them*. This force depends on the stiffness of the spring. We defined the spring constant, k, as a constant that characterizes how stiff is a spring: the higher the value of k the stiffer is the spring. Consider a mass, m, attached to a spring of spring constant, k, that is displaced a distance x from its unstretched/uncompressed state (see Fig. 10.2). Initially, we will suppose that there is no friction. Hooke's law tells us that the force on the mass is

$$F = -kx. \tag{10.1}$$

In a particular case, the mass is displaced to $x = c$ and then, at $t = 0$, is released with zero velocity. Our goal is to determine the mass's position and velocity as a function of time.

Newton's second law informs us that

$$ma = -kx. \tag{10.2}$$

Writing

$$a = \frac{d^2x}{dt^2}, \tag{10.3}$$

Fig. 10.2 A mass attached at the end of a spring is subjected to a restoring force $F = -kx$ according to Hooke's law

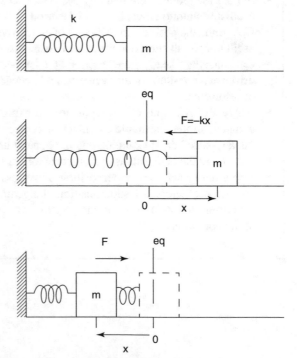

we have

$$\frac{d^2x}{dt^2} = -\frac{k}{m}x. \tag{10.4}$$

Equation 10.4 is a differential equation that has a solution that may be written in a number of apparently different, but actually equivalent, ways.

10.3.1 General Solution That Is the Sum of a Sine and a Cosine

First, following our usual strategy, we will show that a function is a solution of Eq. 10.4 by direct substitution of the function into Eq. 10.4. Our trial solution is

$$x = x(t) = A \cos \omega_0 t + B \sin \omega_0 t, \tag{10.5}$$

where ω_0 and A and B are constants to be determined. We can immediately calculate

$$\frac{dx}{dt} = -\omega_0 A \sin \omega_0 t + \omega_0 B \cos \omega_0 t, \tag{10.6}$$

and then

$$\frac{d^2x}{dt^2} = -\omega_0^2 A \cos \omega_0 t - B\omega_0^2 \sin \omega_0 t = -\omega_0^2 (A \cos \omega_0 t + B \sin \omega_0 t) = -\omega_0^2 x. \tag{10.7}$$

Equation 10.7 leads us to conclude that Eq. 10.4 is indeed satisfied by Eq. 10.5, provided

$$\omega_0^2 = \frac{k}{m}. \tag{10.8}$$

This conclusion is valid for any values of A and B. But what determines A and B? As we saw in Chap. 7. *Rates of Change: Drugs, Infections, and Weapons of Mass Destruction*, these values are determined by the initial conditions. In this case, however, two initial conditions are required since we are dealing with an equation that involves $\frac{d^2x}{dt^2}$. (In the case of equations that involve $\frac{dx}{dt}$ and not $\frac{d^2x}{dt^2}$, we needed one initial condition.) In this example, we are told the two required initial conditions, namely the initial displacement is

$$x(0) = c, \tag{10.9}$$

and the initial velocity is

$$\frac{dx}{dt}(0) = 0. \tag{10.10}$$

Fig. 10.3 Position as a
function of time, $x(t)$, of a
harmonic oscillator, as
described by the cosine form
in Eq. 10.14

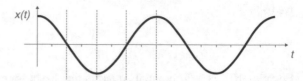

The first of these initial conditions implies that

$$x(0) = c = A\cos(0) + B\sin(0) = A. \tag{10.11}$$

The second implies that

$$\frac{dx}{dt}(0) = 0 = -\omega_0 A\sin(0) + \omega_0 B\cos(0) = \omega_0 B, \tag{10.12}$$

i.e.,

$$B = 0. \tag{10.13}$$

Putting everything together, we have

$$x(t) = c\cos\omega_0 t. \tag{10.14}$$

This is the particular solution that satisfies the initial conditions in this case. A plot
of this kind of sinusoidal time dependence is shown in Fig. 10.3. c represents the
amplitude of this oscillation.

10.3.2 General Solution That Is a Cosine with an Amplitude and a Phase

As an alternative to Eq. 10.5, we may write the general solution to Eq. 10.4 as

$$x = C\cos(\omega_0 t + \phi). \tag{10.15}$$

In this case, C is the amplitude of the oscillation and ϕ is its phase. The fact
that Eq. 10.15 is equivalent to Eq. 10.5 may be seen by recalling the trigonometric
identity:[2]

$$\cos(\delta + \epsilon) = \cos\delta\cos\epsilon - \sin\delta\sin\epsilon, \tag{10.16}$$

which permits us to re-write Eq. 10.15:

$$C\cos(\omega_0 t + \phi) = C\cos\phi\cos\omega_0 t - C\sin\phi\sin\omega_0 t. \tag{10.17}$$

---

[2] http://en.wikipedia.org/wiki/Trigonometric_identity

It follows that Eqs. 10.15 and 10.5 are equal for $A = C \cos \phi$ and $B = -C \sin \phi$, i.e., $\phi = -\arctan(B/A)$ and $C = \sqrt{A^2 + B^2}$.

10.3.3 Energy of an Undamped Simple Harmonic Oscillator

The total mechanical energy of an undamped harmonic oscillator is the sum of its kinetic energy and its potential energy $E_{tot} = K + U$. The kinetic energy is given by $K = \frac{1}{2}mv^2$, where m is the mass of the oscillator and v its velocity. The potential energy associated with a Hooke's law-type force ($F = -kx$) is $U = \frac{1}{2}kx^2$ (as we discussed in Chap. 3. *Energy, Work, Geckos, and ATP*), where k is the spring constant and x is the displacement of the oscillator relative to its resting, equilibrium position. Therefore,

$$E_{tot} = \frac{1}{2}mv^2 + \frac{1}{2}kx^2. \tag{10.18}$$

Each term in the equation above depends on time since we know that in general $x = x(t)$ varies sinusoidally versus time. For example, in the case that $x(t)$ is described by Eq. 10.14, the potential energy at time t is given by

$$U(t) = \frac{1}{2}k(x(t))^2 = \frac{1}{2}k(c \cos \omega_0 t)^2 = \frac{1}{2}kc^2 \cos^2 \omega_0 t. \tag{10.19}$$

The velocity also depends on time. Again assuming $x(t)$ is described by Eq. 10.14, the velocity can be found as the derivative:

$$v(t) = \frac{dx(t)}{dt} = \frac{d(c \cos \omega_0 t)}{dt} = c\omega_0 \sin \omega_0 t. \tag{10.20}$$

It follows that the kinetic energy at time t is

$$K(t) = \frac{1}{2}m(v(t))^2 = \frac{1}{2}m(c\omega_0 \sin \omega_0 t)^2 = \frac{1}{2}m(c\omega_0)^2 \sin^2 \omega_0 t = \frac{1}{2}kc^2 \sin^2 \omega_0 t, \tag{10.21}$$

where in the last step we used the result of Eq. 10.8 substituting for the frequency: $\omega_0^2 = \frac{k}{m}$. Combining Eqs. 10.19 and 10.21, we find an expression for the total mechanical energy:

$$E_{tot} = \frac{1}{2}kc^2 \cos^2 \omega_0 t + \frac{1}{2}kc^2 \sin^2 \omega_0 t = \frac{1}{2}kc^2, \tag{10.22}$$

where we used the trigonometric identity: $\sin^2 y + \cos^2 y = 1$. From this expression, we see that, despite the fact that the kinetic and potential energy themselves vary in

Fig. 10.4 Energy of a harmonic oscillator as a function of time: the potential energy U and kinetic energy K oscillate in time, while the total mechanical energy is constant

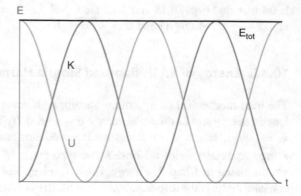

time, their sum is constant, as expected for the total energy of an isolated system. A plot of the behavior of: $K(t)$, $U(t)$, and E_{tot} as a function of time is shown in Fig. 10.4.

10.4 Primer on the Algebra of Complex Numbers

To facilitate discussion of the third alternative solution to Eq. 10.4, which represents this solution in terms of imaginary exponential functions, we now briefly review complex numbers, i.e., numbers that contain $i = \sqrt{-1}$, where i is the imaginary unit satisfying $i^2 = -1$. In addition to this chapter, we will encounter complex numbers again in Chap. 11. *Wave Equations: Strings and Wind*, Chap. 16. *Biologic: Genetic Circuits and Feedback*, and Chap. 18. *Faraday's Law and Electromagnetic Induction*.

Any complex number, call it z, can be written as the sum of a real part ($\Re(z) = x$) and an imaginary part ($\Im(z) = y$) in the following way:

$$z = x + iy, \tag{10.23}$$

where x and y are real numbers. Any complex number can be visualized as a point on a two-dimensional plane that has the real part $\Re(z)$ on the horizontal axis and the imaginary part $\Im(z)$ on the vertical one, as shown in Fig. 10.5.

Operations with Complex Numbers
Imaginary numbers follow the usual rules of algebra, including:

- Addition:

$$z_1 + z_2 = x_1 + iy_1 + x_2 + iy_2 = (x_1 + x_2) + i(y_1 + y_2). \tag{10.24}$$

- Subtraction:

Fig. 10.5 Representation of
a complex number as a point
on a two-dimensional plane

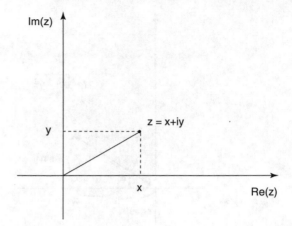

$$z_1 - z_2 = x_1 + iy_1 - (x_2 + iy_2) = (x_1 - x_2) + i(y_1 - y_2). \tag{10.25}$$

- Multiplication:

$$z_1 z_2 = (x_1 + iy_1)(x_2 + iy_2) = x_1 x_2 + i^2 y_1 y_2 + i x_1 y_2 + i x_2 y_1$$
$$= (x_1 x_2 - y_1 y_2) + i(x_1 y_2 + x_2 y_1), \tag{10.26}$$

where in the last step we have used the result $i^2 = -1$, which follows from the
definition of $i = \sqrt{-1}$.

Complex Conjugate

The complex conjugate of $z = x + iy$ is denoted by the star:

$$z^* = x - iy. \tag{10.27}$$

Taking the complex conjugate involves switching the sign of the imaginary part of
z. Note that

$$z + z^* = x + iy + x - iy = 2x \tag{10.28}$$

is real. Similarly,

$$z - z^* = x + iy - (x - iy) = 2iy \tag{10.29}$$

is imaginary. We can show that

$$(z_1 z_2)^* = z_1^* z_2^*. \tag{10.30}$$

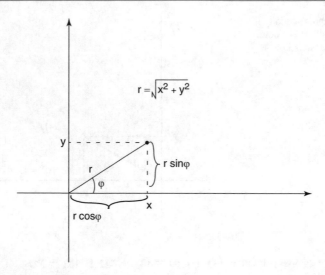

Fig. 10.6 A complex number can also be identified by the radius r—its magnitude—and the angle φ in the complex plane—its phase

Magnitude and Phase of a Complex Number

Referring to the graphical representation in Fig. 10.5, sometimes, instead of identifying the complex number z with the coordinates (x, y), it is useful to represent it in *polar form*, *i.e.*, by providing the radius r, representing the distance of the point z from the origin of the two-dimensional complex plane, and the angle φ that it makes with the real axis, as shown in Fig. 10.6. Then, Pythagoras informs us that

$$r = \sqrt{x^2 + y^2}, \tag{10.31}$$

while trigonometry informs us that

$$x = r\cos(\varphi), \qquad y = r\sin(\varphi). \tag{10.32}$$

r is the magnitude of z; φ is the phase of z. Therefore, in polar form, we can write

$$z = r\cos(\varphi) + ir\sin(\varphi) = r(\cos(\varphi) + i\sin(\varphi)). \tag{10.33}$$

Euler's theorem states that for any angle θ:

$$e^{i\theta} = \cos\theta + i\sin\theta. \tag{10.34}$$

Therefore, using Euler's result in Eq. 10.33, we find the compact polar form for a complex number to be

$$z = re^{i\varphi}. \tag{10.35}$$

This formula will be tremendously important to us throughout this and later chapters. For an elegant discussion of Euler's theorem, and algebra more generally, you are encouraged to read Chap. 22 of the Feynman Lectures on Physics, Volume 1 [1].

10.4.1 General Solution That Is the Sum of Imaginary Exponentials

The third alternative solution of Eq. 10.4, which expresses this solution in terms of imaginary exponential functions, is

$$x = \alpha e^{-i\omega_0 t} + \alpha^* e^{i\omega_0 t} = \alpha e^{-i\omega_0 t} + cc, \tag{10.36}$$

where α is a to-be-determined constant complex number and cc denotes the complex conjugate of the first term. The fact that we constructed Eq. 10.36 as the sum of two terms that are complex conjugates of each other ensures that x is real, as required for a physically meaningful displacement.

To verify that Eq. 10.36 is indeed a solution to Eq. 10.4, first we have

$$\frac{dx}{dt} = -i\omega_0 \alpha e^{-i\omega_0 t} + i\omega_0 \alpha^* e^{i\omega_0 t} \tag{10.37}$$

and then

$$\frac{d^2 x}{dt^2} = (-i)^2 \omega_0^2 \alpha e^{-i\omega_0 t} + i^2 \omega_0^2 \alpha^* e^{i\omega_0 t} = -\omega_0^2 (\alpha e^{-i\omega_0 t} + \alpha^* e^{i\omega_0 t}) = -\omega_0^2 x. \tag{10.38}$$

We again see that Eq. 10.36 is a solution to Eq. 10.4 for $\omega_0^2 = \frac{k}{m}$, as before.

10.5   Damped Simple Harmonic Motion

We may incorporate friction via a term of the same form as used to describe motion through a viscous fluid in Chap. 2. *Force and Momentum: Newton's Laws and How to Apply Them*. This may apply to the case of having a spring oscillating instead of in air in a viscous fluid such as a water or maple syrup (see Fig. 10.7). In this case, Newton's second law becomes

$$ma = -kx - \zeta v, \tag{10.39}$$

where v is the velocity and ζ is the friction coefficient. Equation 10.39 can be written as

$$m\frac{d^2 x}{dt^2} + \zeta\frac{dx}{dt} + kx = 0. \tag{10.40}$$

Equation 10.40 is the equation for damped simple harmonic motion. If the friction is not too big, we can show (see the derivation below in Sects. 10.5.1 and 10.5.2) that a general solution to Eq. 10.40 is

$$x = e^{-\gamma t} \left[A \cos \omega_0 t + B \sin \omega_0 t \right], \tag{10.41}$$

where

$$\gamma = \frac{\zeta}{2m} \tag{10.42}$$

and

$$\omega_0 = \sqrt{\frac{k}{m} - \left(\frac{\zeta}{2m} \right)^2}, \tag{10.43}$$

under the assumption that friction is small: $\left(\frac{\zeta}{2m} \right)^2 \ll \frac{k}{m}$. It follows that $\omega_0 \simeq \sqrt{\frac{k}{m}}$ is a real number. A and B are constants to be determined by the boundary conditions. A plot of such a solution for $x(t)$ (Eq. 10.41 in the case that $B = 0$) is shown in Fig. 10.8. The graph shows behavior typical of a damped oscillator: the amplitude of the sinusoidal function decreases exponentially in time as a result of the $e^{-\gamma t}$ term. The characteristic time $\tau = 1/\gamma$ controls how fast the amplitude goes to zero. The greater the friction (γ), the smaller the characteristic time (τ). Of the three examples of damped harmonic motion, illustrated in Fig. 10.7, which one do you think has the smallest τ?

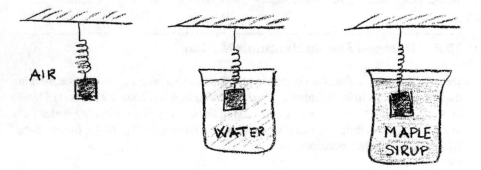

Fig. 10.7 If the spring is oscillating in a viscous medium, it will be subjected to an additional damping force $F = -\zeta v$ that opposes its velocity. This force is the biggest the greater is the viscosity of the medium. For instance, it will be negligible in air, while more significant in water, and much more in maple syrup

Fig. 10.8 Position as a function of time for a damped harmonic oscillator. This graph corresponds to Eq. 10.41 with $B = 0$

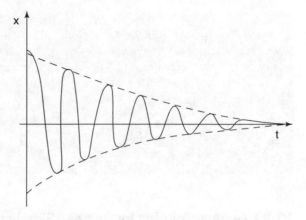

10.5.1 Example: General Solution for Damped Simple Harmonic Motion: Sum of Complex Exponential Functions

Show that

$$x = \alpha e^{st} + \alpha^* e^{s^* t} \tag{10.44}$$

is a general solution of Eq. 10.40 for the appropriate value of s and find that value of s. Both α and s are complex constants. The value of α is determined by the initial conditions in a particular case.

To proceed, we substitute

$$x = \alpha e^{st} \tag{10.45}$$

into Eq. 10.40. This is one-half of what we need in the end, but the other half is achieved simply by adding the complex conjugate. Therefore, using

$$\frac{dx}{dt} = \frac{d}{dt} \alpha e^{st} = s \alpha e^{st} = sx \tag{10.46}$$

and

$$\frac{d^2 x}{dt^2} = \frac{d^2}{dt^2} \alpha e^{st} = \frac{d}{dt} s \alpha e^{st} = s^2 \alpha e^{st} = s^2 x, \tag{10.47}$$

we find

$$ms^2 x + \zeta s x + kx = 0, \tag{10.48}$$

(continued)

which is a quadratic equation for s. Notice that by using a complex exponential function, we have been able to very quickly turn a differential equation into an algebraic equation. Solving Eq. 10.48 for s, we find

$$s = \frac{-\zeta \pm \sqrt{\zeta^2 - 4mk}}{2m} = -\frac{\zeta}{2m} \pm \sqrt{\left(\frac{\zeta}{2m}\right)^2 - \frac{k}{m}}. \tag{10.49}$$

Equation 10.49 is the desired value of s.

10.5.2 General Solution for Damped Simple Harmonic Motion: Exponentially Decaying (Co)sinusoidal Solutions

It turns out that the complex exponential functions that we discovered were the solutions to Eq. 10.40 in Sect. 10.5.1 and may also be written as the product of a decaying exponential function of time and a (co)sinusoidal function of time, if the friction is small. Supposing that the friction is small, so that $\left(\frac{\zeta}{2m}\right)^2 \ll \frac{k}{m}$, it then follows that the factor inside the square root is negative, and the square root is imaginary. In this case, it is convenient to write

$$s = -\frac{\zeta}{2m} \pm i\sqrt{\frac{k}{m} - \left(\frac{\zeta}{2m}\right)^2} = -\gamma \pm i\omega_0, \tag{10.50}$$

where we have introduced $\gamma = \frac{\zeta}{2m}$ and $\omega_0 = \sqrt{\frac{k}{m} - \left(\frac{\zeta}{2m}\right)^2}$, both of which are real. We pick the $-$ sign to be s and the $+$ sign to be s^*. Re-writing Eq. 10.44 in terms of γ and ω_0 yields

$$x = \alpha e^{-\gamma t - i\omega_0 t} + \alpha^* e^{-\gamma t + i\omega_0 t}. \tag{10.51}$$

Then, using Euler's theorem, Eq. 10.34, we find

$$x = \alpha e^{-\gamma t}(\cos \omega_0 t - i \sin \omega_0 t) + \alpha^* e^{-\gamma t}(\cos \omega_0 t + i \sin \omega_0 t)$$
$$= e^{-\gamma t}\left[(\alpha + \alpha^*)\cos \omega_0 t - i(\alpha - \alpha^*)\sin \omega_0 t\right]. \tag{10.52}$$

We learned in Sec. 10.4 that the sum of a complex number and its complex conjugate is real and that the difference between a complex number and its complex conjugate is imaginary. Therefore, both $A = \alpha + \alpha^*$ and $B = -i(\alpha - \alpha^*)$ are real numbers, and we may write

$$x = e^{-\gamma t}\left[A \cos \omega_0 t + B \sin \omega_0 t\right], \tag{10.53}$$

which is the same as Eq. 10.41 as we wanted to prove. We prefer to use Eq. 10.44 in comparison to Eq. 10.53 because it is easier to take derivatives of Eq. 10.44 than Eq. 10.53. However, both Eqs. 10.44 and 10.53 are the general solutions for a damped oscillator for any values of α or A and B, respectively. The values of α or A and B are determined from the boundary conditions.

10.5.3 Using WolframAlpha

The discussion so far has been algebra-intensive. Fortunately, WolframAlpha does algebra. For example, suppose that we want to determine the behavior of a damped harmonic oscillator in the case that the initial velocity of the mass is zero, and its initial displacement from its resting position is c. Then, we must solve Eq. 10.40 subject to the initial conditions: $x(0) = c$ and $\frac{dx}{dt}(0) = 0$. To this end, we may type:[3]

```
Solve m* x''(t)+z*x'(t)+k*x(t)=0, x'(0)=0, x(0)=c
```

and click "=," which produces the desired solution.

As we discussed previously, x' is a shorthand for the derivative of x with respect to its argument, t, and therefore is equivalent to $\frac{dx}{dt}$. Similarly, x'' is a shorthand for $\frac{d^2x}{dt^2}$.

To plot the solution for particular values, we can tell WolframAlpha the values of interest. For example, for $m = 1$, $k = 1$, $z = 0.1$, and $c = 10$, type:[4]

```
Solve x''(t)+0.1*x'(t)+x(t)=0, x'(0)=0, x(0)=10
```

and click "=" to behold decaying oscillations, which are characteristics of damped simple harmonic motion.

10.5.4 Example: Energy Dissipation of a Damped Harmonic Oscillator

In the presence of friction, the total mechanical energy of an oscillator is no longer conserved because the frictional force represents an external force that does (negative) work and thus transfers energy out of the system.

(a) Write an expression for the rate at which the frictional force does work on the oscillator system.

(continued)

---

[3] http://www.wolframalpha.com/input/?i=Solve+m*+x%27%27%28t%29%2Bz*x%27%28t%29%2Bk*x%28t%29%3D0%2C++x%27%280%29%3D0%2C+x%280%29%3Dc

[4] http://www.wolframalpha.com/input/?i=Solve++x%27%27%28t%29%2B0.1*x%27%28t%29%2Bx%28t%29%3D0%2C++x%27%280%29%3D0%2C+x%280%29%3D10

Recalling the rate of work from Chap. 3. *Energy, Work, Geckos, and ATP*

$$\frac{dW}{dt} = \mathbf{F}_f \cdot \mathbf{v} = -\gamma v^2. \tag{10.54}$$

(b) Explain how you can be certain that your answer to (a) corresponds to energy dissipation, *i.e.*, energy loss.

Since v^2 is necessarily positive, we can be sure that $-\gamma v^2$ is negative and hence is a negative change in energy per unit time.

(c) For the oscillator above with $m = 1$, $k = 1$, $z = 0.1$, and an initial position and velocity of $x(0) = 10$ and $v(0) = 0$, respectively, use WolframAlpha to find the total mechanical energy versus t.

WolframAlpha informs us that the solution in this case is

$$x(t) = e^{-0.05t}(0.500\sin(0.999t) + 10\cos(0.999t)). \tag{10.55}$$

Then using WolframAlpha to calculate $\frac{1}{2}mv^2 + \frac{1}{2}kx^2$, we find that the sum of the kinetic and potential energies[5] is

$$50.251e^{-0.1t}\sin^2(0.999t) + 50.e^{-0.1t}\cos^2(0.999t)$$

$$+ 5.006e^{-0.1t}\sin(0.999t)\cos(0.999t) \simeq 50e^{-0.1t}. \tag{10.56}$$

We see that the energy in the oscillator decays approximately as $e^{-2\gamma t}$.

---

[5]http://www.wolframalpha.com/input/?i=1%2F2*(d%2Fdt(e%5E(-0.05+t)+(0.500626+
sin(0.998749+t)+%2B+10+cos(0.998749+t))))%5E2%2B1%2F2*(e%5E(-0.05+t)+(0.
500626+sin(0.998749+t)+%2B+10+cos(0.998749+t)))%5E2

10.6   Forced Damped Simple Harmonic Motion and Resonance

Here, we also consider a damped harmonic oscillator but with the important addition of an applied cosinusoidal external force. This is forced, damped, simple harmonic motion. Now, Newton's second law reads

$$ma = -kx - \zeta v + F\cos(\omega t), \tag{10.57}$$

where m is the mass, a is the acceleration, k is the spring constant, x is the position, ζ is the friction coefficient, v is the velocity, F is the amplitude of the applied cosinusoidal force, and ω is its angular frequency, i.e.,

$$m\frac{d^2x}{dt^2} + \zeta\frac{dx}{dt} + kx = F\cos(\omega t) \tag{10.58}$$

or

$$m\frac{d^2x}{dt^2} + \zeta\frac{dx}{dt} + kx = \frac{1}{2}(Fe^{i\omega t} + Fe^{-i\omega t}), \tag{10.59}$$

using Euler's theorem.

10.6.1 Steady-State Solution at Late Times

As we can see from this WolframAlpha example,[6] after an initial period of time involving transient motions, eventually the forced/driven damped harmonic oscillator shows motion that is periodic at an angular frequency ω, equal to the driving frequency. It is this so-called steady-state, late-time motion that we are interested in now. It will lead us to the important idea of resonance. In fact, we will start off considering

$$m\frac{d^2x}{dt^2} + \zeta\frac{dx}{dt} + kx = \frac{F}{2}e^{-i\omega t} \tag{10.60}$$

with

$$x(t) = \alpha e^{-i\omega t}. \tag{10.61}$$

At the end, to achieve a real solution, we will add the complex conjugate. Our first goal is to determine the behavior of α as a function of the driving frequency ω. Using this expression for α, we will then deduce the frequency dependence of the power dissipation, which is actually what is measured in MRI, etc.
Equation 10.61 implies that

$$v(t) = \frac{dx}{dt} = -i\omega\alpha e^{-i\omega t} \tag{10.62}$$

and

$$a(t) = \frac{d^2x}{dt^2} = -\omega^2\alpha e^{-i\omega t}. \tag{10.63}$$

Substituting Eqs. 10.61, 10.62, and 10.63 into Eq. 10.60 leads to

$$\alpha\left[-m\omega^2 e^{-i\omega t} - i\zeta\omega e^{-i\omega t} + ke^{-i\omega t}\right] = \frac{1}{2}Fe^{-i\omega t}. \tag{10.64}$$

---

[6] https://www.wolframalpha.com/input/?i=Solve++x''(t)%2B0.1*x'(t)%2Bx(t)%3D0.
1*Cos(t),++x'(0)%3D0,+x(0)%3D10

In turn, this yields

$$\alpha = \frac{\frac{1}{2}F}{k - m\omega^2 - i\zeta\omega}. \tag{10.65}$$

To relocate the i so that it appears in the numerator, we multiply top and bottom by $k - m\omega^2 + i\zeta\omega$ with the result that

$$\alpha = \frac{\frac{1}{2}F(k - m\omega^2 + i\zeta\omega)}{(k - m\omega^2 - i\zeta\omega)(k - m\omega^2 + i\zeta\omega)}$$

$$= \frac{\frac{1}{2}F(k - m\omega^2)}{(k - m\omega^2)^2 + \zeta^2\omega^2} + i\frac{\frac{1}{2}F\zeta\omega}{(k - m\omega^2)^2 + \zeta^2\omega^2}. \tag{10.66}$$

This is the desired result for the frequency dependence of α. We can now write the solution to Eq. 10.58 as Eq. 10.61 plus its complex conjugate:

$$x(t) = \alpha e^{-i\omega t} + \alpha^* e^{i\omega t} = (\alpha + \alpha^*)\cos\omega t - i(\alpha - \alpha^*)\sin\omega t$$

$$= 2\Re(\alpha)\cos\omega t - i(2i\Im(\alpha))\sin\omega t$$

$$= \frac{F(k - m\omega^2)}{(k - m\omega^2)^2 + \zeta^2\omega^2}\cos\omega t + \frac{F\zeta\omega}{(k - m\omega^2)^2 + \zeta^2\omega^2}\sin\omega t$$

$$= \frac{F\left[(k - m\omega^2)\cos\omega t + \zeta\omega\sin\omega t\right]}{(k - m\omega^2)^2 + \zeta^2\omega^2}$$

$$= \frac{\frac{F}{m}\left[(\omega_0^2 - \omega^2)\cos\omega t + \frac{\zeta}{m}\omega\sin\omega t\right]}{(\omega_0^2 - \omega^2)^2 + \frac{\zeta^2}{m^2}\omega^2}, \tag{10.67}$$

where in the second step on the first line we used the fact that

$$\cos(\omega t) = \frac{e^{i\omega t} + e^{-i\omega t}}{2} \tag{10.68}$$

$$\sin(\omega t) = \frac{e^{i\omega t} - e^{-i\omega t}}{2i}, \tag{10.69}$$

which follow from Euler's theorem, and where, in the last step, we introduced the resonant frequency,

$$\omega_0 = \sqrt{\frac{k}{m}}. \tag{10.70}$$

Equation 10.67 is the desired solution in the case of a cosinusoidal driving force.

We note that even though the right-hand side of Eq. 10.70 differs from the right-hand side pf Eq. 10.43, we follow the convention and use ω_0 for both frequencies. In fact, Eq. 10.43 reduces to Eq. 10.70 in the limit of small damping.

10.6.2 Resonance and Power Dissipation

Using the results of Sect. 10.3.2 and WolframAlpha,[7] we can show that the solution for a driven harmonic oscillator in Eq. 10.67 can be re-written as $x(t) = C \cos(\omega t + \phi)$, where the amplitude, C, and phase, ϕ, are

$$C = \frac{\frac{F}{m}}{\sqrt{(\omega_0^2 - \omega^2)^2 + \frac{\zeta^2}{m^2}\omega^2}} \tag{10.71}$$

and

$$\phi = -2 \arctan \frac{\omega \frac{\zeta}{m}}{\sqrt{(\omega_0^2 - \omega^2)^2 + \frac{\zeta^2}{m^2}\omega^2} + \omega_0^2 - \omega^2}, \tag{10.72}$$

respectively. Importantly, this amplitude is ω-dependent and displays a peak when the drive frequency, ω, equals ω_0.

We know from Chap. 3. *Energy, Work, Geckos, and ATP* that the power transferred into a system by an external force, $F \cos(\omega t)$, is the product of the force and the velocity of what the force is being applied to. In a steady state, the average power transferred into the system must equal the average power transferred out in order that the total average energy remains constant (steady). Therefore, the average power dissipation, $\langle PD \rangle$, in the case of driven, damped simple harmonic motion equals the energy transfer by the external force and is

$$\langle PD \rangle = \langle F \cos(\omega t) \times v(t) \rangle . \tag{10.73}$$

To calculate the average power dissipation, we need to calculate $v(t) = \frac{dx}{dt}$:

$$v(t) = \frac{F\left[-\omega(k - m\omega^2) \sin \omega t + \zeta \omega^2 \cos \omega t\right]}{(k - m\omega^2)^2 + \zeta^2 \omega^2} . \tag{10.74}$$

It follows that

$$\langle PD \rangle = \left\langle F \cos \omega t \times \frac{F\left[-\omega(k - m\omega^2) \sin \omega t + \zeta \omega^2 \cos \omega t\right]}{(k - m\omega^2)^2 + \zeta^2 \omega^2} \right\rangle . \tag{10.75}$$

---

[7] https://www.wolframalpha.com/input/?i=Solve+-C*sin%28phi%29%3DF%2Fm*z%2Fm*w%2F%28%28w0%5E+-+w%5E2%29%5E2+%2B+z%5E2%2Fm%5E*w%5E2%29%2C+C*Cos%28phi%29%3DF%2Fm*%28w0%5E+-+w%5E2%29%2F%28%28w0%5E+-+w%5E2%29%5E2+%2B+z%5E2%2Fm%5E*w%5E2%29+for+C+and+phi

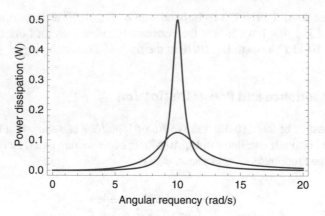

Fig. 10.9 Equation. 10.76 plotted for $F = 1$ N, $m = 1$ kg, $k = 100$ N/m, and $\zeta = 1$ N m/s (red line) and $F = 1$ N, $m = 1$ kg, $k = 100$ N/m, and $\zeta = 4$ N m/s (blue line)

The sort of average that we are talking about here is an average over time, over an integer number of periods of the motion. We denote this average via <>. To calculate the average, we invoke the trigonometric identities: $\sin(\omega t)\cos(\omega t) = \frac{1}{2}\sin(2\omega t)$ and $\cos^2(\omega t) = \frac{1}{2}[1 + \cos(2\omega t)]$ and the results that $\langle\cos(2\omega t)\rangle = 0$ and $\langle\sin(2\omega t)\rangle = 0$, when averaged over an integer number of periods. Applying these results, we arrive at

$$\langle PD \rangle = \frac{\frac{1}{2}F^2\zeta\omega^2}{(k - m\omega^2)^2 + (\zeta\omega)^2}. \tag{10.76}$$

Equation 10.76 is plotted for two sets of parameters in Fig. 10.9. The striking aspect of the power dissipation spectra, shown in Fig. 10.9, is the existence of peaks at a frequency of 10 rad/s. Each of these peaks is an example of a resonance.[8] The peak frequency is the resonance or resonant frequency and occurs at $\omega = \omega_0 = \sqrt{k/m}$. At resonance, i.e., $\omega = \omega_0$, the power dissipation is $\frac{F^2}{2\zeta}$. Counterintuitively, the power dissipation increases as the friction coefficient decreases.

10.7   Superposition Revisited

In all of the equations that we have been solving, x appears linearly, that is, there are no x^2-terms, $x\frac{dx}{dt}$-terms, or any higher power terms in x than linear. As we discussed previously in Chap. 7. *Rates of Change: Drugs, Infections, and Weapons of Mass Destruction*, importantly, the solutions to linear equations obey the principle of superposition, namely that if we have two different solutions of a linear equation,

---

[8] http://en.wikipedia.org/wiki/Resonance

then a linear combination of those two solutions is also a solution of the linear equation in question. This property is a fundamental aspect of linear equations and one that we will repeatedly rely on. In fact, we already used this property when we added together complex conjugates to achieve a real solution.

One cute thing that superposition permits us to do is to write the general solution for forced, damped simple harmonic motion for arbitrary times without doing any additional calculations. Here is how: in our discussion of damped simple harmonic motion, we first focused on the case $F = 0$ and found that a general solution in that case. Call it $x_0(t)$. We then focused on the "steady-state" behavior for non-zero driving force, $F \cos \omega t$. In this case, we found a solution, valid at long times only. Call it $x_F(t)$. Now, notice that if we try a solution

$$x(t) = x_0(t) + x_F(t), \tag{10.77}$$

then

$$m \frac{d^2}{dt^2}(x_0 + x_F) + \zeta \frac{d}{dt}(x_0 + x_F) + k(x_0 + x_F) = m \frac{d^2 x_0}{dt^2} + \zeta \frac{dx_0}{dt} + kx_0$$

$$+ m \frac{d^2 x_F}{dt^2} + \zeta \frac{dx_F}{dt} + kx_F = 0 + F \cos(\omega t) = F \cos(\omega t). \tag{10.78}$$

That is, the superposition, Eq. 10.77, is indeed a solution of Eq. 10.58. Therefore, we now have the general solution for the behavior of a forced damped simple harmonic oscillator at all times, not just at long times, because we can use the two arbitrary constants in x_0 to satisfy the initial conditions on $x(0)$ and $v(0)$.

10.8 Examples of Resonance

Although our discussion of resonance has focused on a mass on a spring, in fact, resonance occurs in many diverse systems:

- Figure 10.10 shows the frequency response of a quartz tuning fork oscillator of the sort pictured at the beginning of this chapter and used in electronic watches.
- Figure 10.11 shows measurements of the amplitude of basilar membrane displacements versus frequency for a Parnell's mustached bat, *Pteronotus parnellii*, from Ref. [3]. This species echolocates its insect prey by emitting monochromatic sound waves of a frequency, which is slightly away from the resonance. However, when this sound is reflected from a moving insect, the sound frequency can be shifted (via the Doppler effect) to the resonant frequency, where the bat's hearing is correspondingly highly sensitive.
- Wine glasses show a resonance at about 500 Hz, which can be excited by sound waves of that same frequency. Singers who can sing sufficiently strongly at the resonant frequency can cause the glass's oscillation amplitude to become so large that the glass shatters.

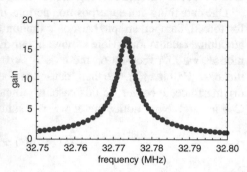

Fig. 10.10 Left: quartz crystal resonator, https://en.wikipedia.org/wiki/Crystal_oscillator which resonates at 32768 Hz, and keeps time in quartz watches. Public domain image from https://en.wikipedia.org/wiki/Crystal_oscillator#/media/File:Inside_QuartzCrystal-Tuningfork. jpg. Right: frequency response of a quartz tuning fork, adapted from Ref. [2], showing a resonance at 32773.3 Hz with a full-width-at-half-maximum (FWHM) of 7.664 Hz [2]. Quartz is "piezoelectric." Therefore, its mechanical response also gives rise to an electrical response, which is what is measured and shown in this figure. In comparison to the resonant frequency, $2^{15} = 32768$. Electronic watches use a quartz tuning fork resonator and count 2^{15} cycles (periods) to time 1 second

- Musical instruments, such as the flute, oboe, violin, etc., all display several resonances, as discussed in more detail in Chap. 11. *Wave Equations: Strings and Wind.*
- To listen to your favorite radio station, you have to tune your radio so that its tuner circuit is at resonance with the broadcast radio waves at your favorite station's frequency.
- In magnetic resonance imaging (MRI), nuclear spins in a magnetic field show resonance.

10.9   Coupled Harmonic Oscillators

We now turn our attention to coupled harmonic oscillators, such as the system of three springs and two masses shown in Fig. 10.12. For simplicity, in our coupled oscillator calculation, we will usually consider the situation that there is zero damping, but the basic ideas of eigenmodes and superposition are equally valid in the case of non-zero damping.

To determine the motion of coupled oscillator systems, we adapt the procedure for solving eigenvalue problems that we introduced in Chap. 7. *Rates of Change:*

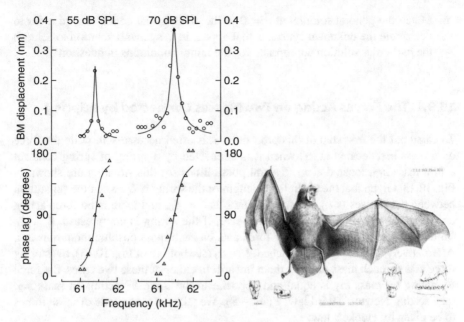

Fig. 10.11 Left: frequency response of the displacement of a portion of the basilar membrane with the ear of the mustached bat, *Pteronotus parnellii*, revealing a resonance at 61.5 kHz with a full-width-at-half-maximum (FWHM) of 200 Hz [3]. Right: drawing of Parnell's mustached bat from https://en.wikipedia.org/wiki/Parnell%27s_mustached_bat#/media/File:ChilonycterisOsburniFord.jpg

Fig. 10.12 Weakly coupled oscillators

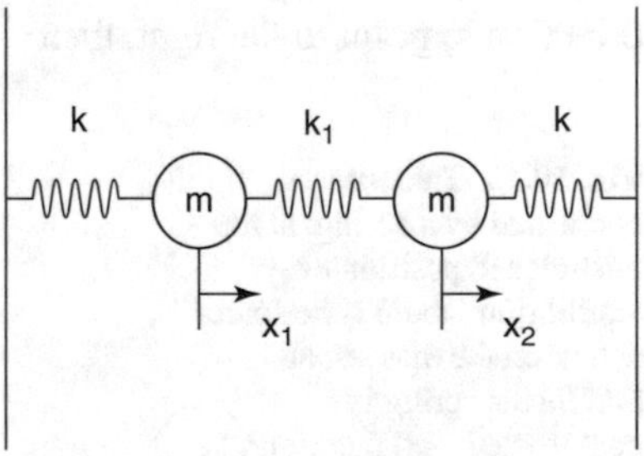

Drugs, Infections, and Weapons of Mass Destruction. For coupled oscillators, the procedure reads:

1. Write appropriate linear equations.
2. Assume a sinusoidal time dependence with eigenvalue/eigenfrequency ω, so that $\frac{d^2x_1}{dt^2}$ may be replaced by $-\omega^2 x_1$ and $\frac{d^2x_2}{dt^2}$ by $-\omega^2 x_2$.
3. Solve the resultant simultaneous equations for ω^2 and $\frac{x_2}{x_1}$.
4. For each eigenvalue, construct the corresponding eigenvector, $(1, \frac{x_2}{x_1})$.
5. Construct a general solution as the superposition (linear sum) of all the eigenmodes.

6. Match the general solution at $t = 0$ to the given initial conditions, in order to determine the unknown constants that appear in the general solution to achieve the particular solution appropriate for the initial conditions in question.

10.9.1 The Forces Acting on Two Masses Connected by a Spring

To carry out the first step of this procedure, it is often necessary to write the force on a mass m_1, located at x_1, when it is connected by a spring of spring constant k to a mass m_2, located at x_2. Several possibilities for this situation are shown in Fig. 10.13. Given that the unstretched length of the spring is ℓ_0, when the separation between the masses is ℓ_0, the spring length is $\ell = \ell_0$, and there is no force acting on the masses (top of Fig. 10.13). However, if the spring is compressed, $\ell < \ell_0$ (middle of Fig. 10.13), a restoring force acts on each mass pushing them outward. Alternatively, if the spring is stretched, $\ell > \ell_0$ (see bottom of Fig. 10.13), a restoring force acts on each mass pushing them inward. In either of these two cases, the force $\mathbf{F}_1$ acting on mass m_1 is equal and opposite to the force $\mathbf{F}_2$ acting on mass m_2, manifesting Newton's third law: $\mathbf{F}_1 = -\mathbf{F}_2$. We take the magnitude of these forces to be given by Hooke's law:

$$|\mathbf{F}_1| = |\mathbf{F}_2| = k(\ell - \ell_0). \tag{10.79}$$

The sign of the force needs to be properly incorporated into Newton's second law, which depends on our choice of coordinate system. If we choose the positive direction to point to the right, then

Fig. 10.13 Two masses connected by a spring: at *top* unstretched position or equilibrium, there is no force acting on the masses; at *middle* the spring is compressed, and therefore, the masses each experience an outward force; at *bottom*, the spring is stretched, and therefore, the masses each experience an inward force

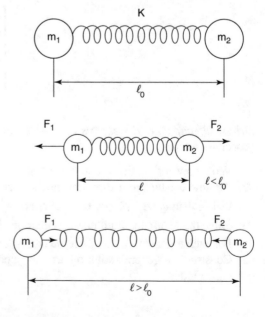

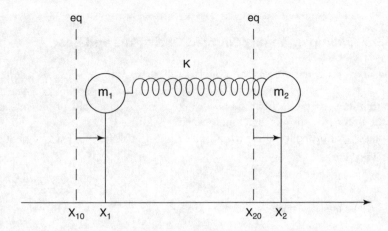

Fig. 10.14 Two masses connected by a spring. The displacement of mass m_1 from its equilibrium position is $X_1 - X_{10}$. The displacement of mass m_2 from its equilibrium position is $X_2 - X_{20}$. In terms of these coordinates, the extension of the spring is $X_2 - X_{20} - (X_1 - X_{10})$

$$\mathbf{F}_1 = k(\ell - \ell_0) \tag{10.80}$$

$$\mathbf{F}_2 = -k(\ell - \ell_0). \tag{10.81}$$

A close look at these two expressions will convince you that they give the correct sign for the forces in both cases: $\ell < \ell_0$ and $\ell > \ell_0$.

We now seek to extend this description to the situation depicted in Fig. 10.14 in which the position of m_1 (m_2) is X_1 (X_2). At equilibrium, when the spring is unstretched, mass m_1 is at position X_{10} and mass m_2 is at position X_{20}. Therefore, according to the x-axis shown in the figure: $\ell_0 = X_{20} - X_{10}$. More generally, when m_1 and m_2 are at arbitrary positions X_1 and X_2, respectively, the length of the spring is given by $\ell = X_2 - X_1$. In terms of these coordinates, the forces $\mathbf{F}_1$ and $\mathbf{F}_2$ acting on each mass can be expressed as

$$\mathbf{F}_1 = k(\ell - \ell_0) = k[(X_2 - X_1) - (X_{20} - X_{10})]$$

$$= k[(X_2 - X_{20}) - (X_1 - X_{10})] \tag{10.82}$$

$$\mathbf{F}_2 = -k(\ell - \ell_0) = -k[(X_2 - X_1) - (X_{20} - X_{10})]$$

$$= -k[(X_2 - X_{20}) - (X_1 - X_{10})]. \tag{10.83}$$

Now introducing $x_1 = X_1 - X_{10}$ and $x_2 = X_2 - X_{20}$, which are the displacements from the masses' equilibrium positions, these expressions for the forces simplify to

$$\mathbf{F}_1 = k(x_2 - x_1) = -k(x_1 - x_2), \tag{10.84}$$

$$\mathbf{F}_2 = -k(x_2 - x_1) = k(x_1 - x_2). \tag{10.85}$$

In the following section, we will write the equations of motion in terms of these displacements, x_1 and x_2.

10.9.2 Example: Weakly Coupled Oscillators and Beats

Consider two equal masses, m, coupled together by three springs as shown in Fig. 10.12. Ignore any damping in this problem. The spring constant of the middle spring, attaching the two masses, is k_1, while the spring constant of the other two springs attaching each mass to the wall is k. Let x_1 be the displacement of the left mass toward the right. Let x_2 be the displacement of the right mass toward the right.

(a) Watch the movie at

      http://www.youtube.com/watch?v=RoSYKPTdlxs,[9]

and briefly describe the behavior of the two pendula.

Although the right pendulum is started with a large-amplitude oscillation, after a period of time the right pendulum has a small amplitude and the left pendulum a large amplitude of oscillation. After a further period of time, the right pendulum has recovered its large-amplitude oscillation and the left pendulum once again has a small-amplitude oscillation. At this point in time, the configuration is the same as at the beginning, so we may infer that this behavior will repeat periodically in time, so that large-amplitude motion alternates back and forth between the two pendulums.

(b) Using the coordinate systems shown in Fig. 10.12, write Newton's second law for each mass.

Answer: x_1 is the displacement of mass 1 from its equilibrium position; x_2 is the displacement of mass 2 from its equilibrium position. Both x_1 and x_2 are positive for rightward displacements. Therefore, in terms of these coordinates, Newton's second law for mass 1 is

$$m\frac{d^2 x_1}{dt^2} = -kx_1 - k_1(x_1 - x_2) \tag{10.86}$$

---

[9]http://www.youtube.com/watch?v=RoSYKPTdlxs

(continued)

and for mass 2, it is

$$m\frac{d^2x_2}{dt^2} = -kx_2 - k_1(x_2 - x_1). \tag{10.87}$$

(c) Determine the eigenvalues (eigenfrequencies) of the system.

Answer: Assuming sinusoidal trial solutions with frequency ω, Eqs. 10.86 and 10.87 become

$$- m\omega^2 x_1 = -kx_1 - k_1(x_1 - x_2) \tag{10.88}$$

and

$$- m\omega^2 x_2 = -kx_2 - k_1(x_2 - x_1), \tag{10.89}$$

which lead to

$$(k + k_1 - m\omega^2) = k_1 \frac{x_2}{x_1} \tag{10.90}$$

and

$$(k + k_1 - m\omega^2)\frac{x_2}{x_1} = k_1, \tag{10.91}$$

respectively. Eliminating $\frac{x_2}{x_1}$ yields

$$(k + k_1 - m\omega^2)^2 = k_1^2. \tag{10.92}$$

Therefore, the two desired eigenvalues are

$$\omega_A^2 = \frac{k}{m} \tag{10.93}$$

and

$$\omega_B^2 = \frac{k + 2k_1}{m}. \tag{10.94}$$

(d) Determine the eigenvectors of the system.

Answer: For eigenmode A, the eigenvector is given by

(continued)

$$(k + k_1 - m\omega_A^2) = k_1 = k_1 \frac{x_2}{x_1}, \qquad (10.95)$$

leading to

$$\frac{x_2}{x_1} = 1 \qquad (10.96)$$

and

$$\mathbf{e}_A = (1, 1). \qquad (10.97)$$

For eigenmode B, the eigenvector is given by

$$(k + k_1 - m\omega_B^2) = k_1 = k_1 \frac{x_2}{x_1}, \qquad (10.98)$$

leading to

$$\frac{x_2}{x_1} = -1 \qquad (10.99)$$

and

$$\mathbf{e}_B = (1, -1). \qquad (10.100)$$

What do these results mean physically? For eigenmode A, we have that $x_1 = x_2$, so that the two masses move together, but for eigenmode B, we have that $x_1 = -x_2$, so that the two masses move opposite to each other. The displacement patterns are schematically illustrated in Fig. 10.15.

(e) Write the general solution for $\mathbf{x}(t) = (x_1(t), x_2(t))$.

Answer:

$$\mathbf{x}(t) = C_A \mathbf{e}_A \cos \omega_A t + S_A \mathbf{e}_A \sin \omega_A t + C_B \mathbf{e}_B \cos \omega_B t + S_B \mathbf{e}_B \sin \omega_B t, \qquad (10.101)$$

where C_A, S_A, C_B, and C_B are constants to be determined by the initial conditions.

(continued)

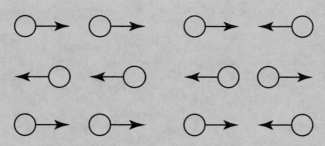

Fig. 10.15 Displacement pattern in eigenmode A (left) and in eigenmode B (right)

(f) Suppose that initially the position of mass 1 is displaced a distance c from its equilibrium position, while mass 2 remains undisplaced. Then, the masses are released from rest at $t = 0$. Determine the subsequent motion.

We are told the initial conditions are $x_1(0) = c$, $v_1(0) = 0$, $x_2(0) = 0$, and $v_2(0) = 0$. Because the equations of motion involve second derivatives, we must have two initial conditions for each mass in order to be able to determine the motion. In the x_1x_2 coordinate system, the initial position and velocity vectors are $\mathbf{x} = (c, 0)$ and $\mathbf{v} = (0, 0)$, respectively. Therefore, at $t = 0$, we have

$$(c, 0) = C_A(1, 1) + C_B(1, -1) \tag{10.102}$$

for the displacement, and

$$(0, 0) = \omega_A S_A(1, 1) + \omega_B S_B(1, -1) \tag{10.103}$$

for the velocity. It follows that $S_A = S_B = 0$ and $C_A = C_B = \frac{c}{2}$. Therefore,

$$\mathbf{x}(t) = \frac{c}{2}\mathbf{e}_A \cos \omega_A t + \frac{c}{2}\mathbf{e}_B \cos \omega_B t, \tag{10.104}$$

i.e.,

$$x_1(t) = \frac{c}{2} \cos \omega_A t + \frac{c}{2} \cos \omega_B t \tag{10.105}$$

and

(continued)

$$x_2(t) = \frac{c}{2} \cos \omega_A t - \frac{c}{2} \cos \omega_B t. \tag{10.106}$$

(g) Your answers in part (d) for $x_1(t)$ and $x_2(t)$ should be the sum of two
cosine terms with different frequencies. Use trigonometric identities to
show that your answers can be written as the product of two cosines or
two sines.

Answer:

$$x_1(t) = \frac{c}{2} \cos \omega_A t + \frac{c}{2} \cos \omega_B t = c \cos \left[\frac{1}{2}(\omega_A + \omega_B)t\right] \cos \left[\frac{1}{2}(\omega_A - \omega_B)t\right] \tag{10.107}$$

and

$$x_2(t) = \frac{c}{2} \cos \omega_A t - \frac{c}{2} \cos \omega_B t = -c \sin \left[\frac{1}{2}(\omega_A + \omega_B)t\right] \sin \left[\frac{1}{2}(\omega_A - \omega_B)t\right]. \tag{10.108}$$

(h) How does your answer to (g) explain the behavior of two weakly coupled
pendulums observed in the movie?

Answer: Equations 10.107 and 10.108 show
that the motion of each mass can be
described as the product of a relatively
high-frequency oscillation at the mean of
the two eigenfrequencies and a relatively
low-frequency oscillation at a frequency given
by the difference in eigenfrequencies. The
low-frequency oscillation is a cosine for one
mass and a sine for the other. Therefore, when
one mass oscillates with a large amplitude,
the other necessarily oscillates with a small
amplitude and vice versa. Thus, this term
indeed accounts for the behavior observed
in the movie in which large-amplitude motion
slowly alternates back and forth between the
two pendulums. The low-frequency oscillations
are called "beats," and $\omega_B - \omega_A$ is the beat
frequency, defined as the frequency at which
large-amplitude motion of one of the masses
occurs. Quantitatively, for $k_1 \ll k$, we may

(continued)

approximate $\omega_B \simeq \sqrt{\frac{k}{m}}(1 + \frac{k_1}{k})$. It follows that the beat frequency is $\frac{k_1}{k}\sqrt{\frac{k}{m}}$, which is small for small $\frac{k_1}{k}$, leading to a beat period, which becomes progressively smaller as the coupling (k_1) becomes weaker.

10.9.3 Example: Three Identical Masses and Four Identical Springs

For any system of N coupled harmonic oscillators, no matter how complicated, there exist exactly N eigenmodes, each of which corresponds to a particular pattern of displacements—the eigenvector of the mode in question—each of which evolves in time with its own definite angular frequency—the eigenvalue or eigenfrequency of the mode in question. The addition of damping does not change these statements, although in that case there may be complex eigenvalues and eigenvectors. Consider three identical masses (Fig. 10.16), each of mass m, in line, connected together by four identical springs, each with spring constant k.

(a) How many eigenmodes are there in this case?

Since there are three masses, m, there must be three eigenmodes.

(continued)

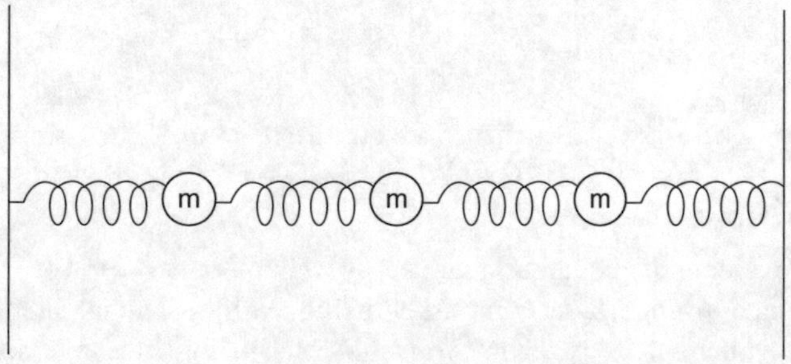

Fig. 10.16 Figure for Example 10.9.3: three identical masses connected by four identical springs

(b) Write appropriate equations of motion.

The required equations of motion are

$$m\frac{d^2x_1}{dt^2} = -kx_1 - k(x_1 - x_2) \tag{10.109}$$

for mass 1,

$$m\frac{d^2x_2}{dt^2} = -k(x_2 - x_3) - k(x_2 - x_1) \tag{10.110}$$

for mass 2, and

$$m\frac{d^2x_3}{dt^2} = -kx_3 - k(x_3 - x_2) \tag{10.111}$$

for mass 3, where x_1, x_2, and x_3 are the displacements of mass 1, mass 2, and mass 3 from their equilibrium positions, respectively.

(c) Assuming sinusoidal solutions, transform your equations of motion into three algebraic equations for three unknowns.

Assuming a sinusoidal trial solution, these equations become

$$(2k - m\omega^2) - k\frac{x_2}{x_1} = 0, \tag{10.112}$$

$$(2k - m\omega^2)\frac{x_2}{x_1} - k\frac{x_3}{x_1} - k = 0, \tag{10.113}$$

and

$$(2k - m\omega^2)\frac{x_3}{x_1} - k\frac{x_2}{x_1} = 0, \tag{10.114}$$

respectively. Equations 10.112, 10.113, and 10.114 are the three desired equations for three unknowns, namely ω^2, $\frac{x_2}{x_1}$ and $\frac{x_3}{x_1}$.

(d) These algebraic equations can be solved by hand. However, solving them is conveniently accomplished using WolframAlpha. Go ahead and solve your equations using WolframAlpha to find the eigenvalues and eigenvectors in this case.

(continued)

Type:[10]

```
Solve (2*k-m*w)-k*r=0, (2*k-m*w)*r-k*s-k=0,
(2*k-m*w)*s-k*r=0 for w and r and s
```

where w stands for ω^2, r stands for $\frac{x_2}{x_1}$, and s stands for $\frac{x_3}{x_1}$, and behold the output,[11] which informs us that the eigenvalues (eigenfrequencies) are

$$\omega_A = \sqrt{(2 - \sqrt{2})k/m},\tag{10.115}$$

$$\omega_B = \sqrt{2k/m},\tag{10.116}$$

and

$$\omega_C = \sqrt{(2 + \sqrt{2})k/m},\tag{10.117}$$

with the corresponding eigenvectors,

$$\mathbf{e}_A = (1, \sqrt{2}, 1),\tag{10.118}$$

$$\mathbf{e}_B = (1, 0, -1),\tag{10.119}$$

and

$$\mathbf{e}_C = (1, -\sqrt{2}, 1),\tag{10.120}$$

respectively.

(e) Sketch the pattern of displacements for each of the eigenmodes.

Answer: See Fig. 10.17.

---

[10]http://www.wolframalpha.com/input/?i=Solve+%282*k-m*w%29-k*r%3D0%2C+%282*k-m*w%29*r-k*s-k%3D0%2C%282*k-m*w%29*s-k*r%3D0+for+w+and+r+and+s

[11]http://www.wolframalpha.com/input/?i=Solve+%282*k-m*w%29-k*r%3D0%2C+%282*k-m*w%29*r-k*s-k%3D0%2C%282*k-m*w%29*s-k*r%3D0+for+w+and+r+and+s

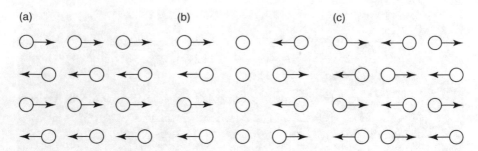

Fig. 10.17 Displacement pattern for (**a**) eigenmode A, (**b**) eigenmode B, and (**c**) eigenmode C

10.9.4 Example: Coupled, Damped, Driven Oscillators

In this chapter, we have examined both (1) the motion of a single, periodically driven, damped oscillator and (2) the motion of undamped coupled oscillators that have been set into motion by imposition of an initial displacement from their equilibrium position. Based on ideas from these two topics, in this section, we seek to understand conceptually the motion of periodically driven, damped, coupled oscillators. Specifically, we consider the system of Fig. 10.12, but, this time, mass 1 experiences a damping force proportional to its velocity, and mass 2 is driven by a cosinusoidal force of angular frequency ω and amplitude F.

(a) How many eigenmodes are there in this case?

   Since there are two masses, m, there must still be two eigenmodes.

(b) Using the coordinate systems shown in Fig. 10.12, write Newton's second law for each mass, including damping.

   Answer: We augment the equations of motion from Sec. 10.8.2 with the addition of damping forces, acting on both masses and the driving force acting on mass 1:

$$m\frac{d^2x_1}{dt^2} = -kx_1 - k_1(x_1 - x_2) - \zeta\frac{dx_1}{dt},\qquad(10.121)$$

   and for mass 2, it is

$$m\frac{d^2x_2}{dt^2} = -kx_2 - k_1(x_2 - x_1) + F\cos\omega t.\qquad(10.122)$$

(continued)

(c) Given that the force is applied to mass 2 and that only mass 1 experiences friction, discuss whether the power dissipation in this system is $\langle F \cos \omega t \times v_1(t) \rangle$ or $\langle F \cos \omega t \times v_2(t) \rangle$.

```
The force is applied to mass 2. Therefore, the
power that must be supplied by that external
force, which is equal to the power dissipation,
is        ⟨F cos ωt × v₂(t)⟩,        even though only mass 1
experiences friction.
```

(d) Without performing a detailed calculation, sketch the steady-state power dissipation as a function of frequency, measured in units of $\sqrt{k/m}$, assuming that the damping is small so that the motion is underdamped.

```
For small friction, from Eqs. 10.93 and 10.94,
we can expect this system will show two
eigenfrequencies, one at ωₐ = √(k/m) and the other
at ω_B = √((k+2k₁)/m). When the drive frequency, ω, is near
ωₐ or ω_B, we can expect the power dissipation to
be similar to EQ. 10.75. Exactly at resonance,
then we expect ⟨PD⟩ ≃ F²/2ζ, independent of ω. In
addition, we can also expect that the power
dissipation falls to one-half of its peak value
when ω ≃ ω_{A,B} ± ζ/2m, i.e., that the width of the
resonance is η/m, which is also independent of
ω.      The power dissipation, plotted versus drive
frequency, consistent with these considerations,
is shown in Fig. 10.18 for m = 1, k = 1, k₁ = 0.125,
and ζ = 0.03. Even though only one of the masses
experiences friction and only one of them
experiences a driving force, both eigenmodes
show damped resonances.
```

10.10 Chapter Outlook

Oscillations are ubiquitous in science. As described in this chapter, an oscillatory behavior is the outcome for any situation in which an object is subject to a restoring force, in other words, a force that wants to keep the object near an equilibrium position, provided the damping is not too large. Oscillations near equilibrium are at the core of the understanding of solid materials, for example, which can be thought of as many coupled oscillators (many atoms connected to each other via

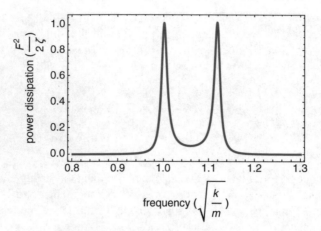

Fig. 10.18 Power dissipation versus frequency for two coupled, damped oscillators. This system exhibits two resonances, one at each of the two eigenfrequencies

interaction forces), similar to the coupled oscillators that we discussed at the end of this chapter. If you let the spring stiffness to relax, these materials can be soft tissues, as well as membranes, and cells. Any single object or complex system has its own natural oscillation frequency, and hence, it will respond best—it will resonate— when stimulated with that specific frequency. This is resonance. The concept of resonance has infinite applications.

We encountered superposition again in this chapter. Superposition is tremen- dously important across physics because many of the key equations that we deal with are linear equations, including the diffusion equation, the wave equation, which describes wave motion (Chap. 11. *Wave Equations: Strings and Wind*), and Maxwell's equations (Chap. 19. *Maxwell's Equations and Then There Was Light!*), which describe electricity and magnetism. In addition, we can often usefully linearize non-linear equations to get an idea of the behavior in certain limits, as we already saw in Chap. 7. *Rates of Change: Drugs, Infections, and Weapons of Mass Destruction* in the context of the early-time behavior of an HIV infection and of an atomic explosion. In Chap. 16. *Biologic: Genetic Circuits and Feedback*, we will usefully linearize chemical rate equations that describe the behavior of certain genetic "circuits."

In the next chapter, Chap. 11. *Wave Equations: Strings and Wind*, we will discuss the importance of oscillations in the context of waves. Waves of any type (sound waves, electromagnetic waves, waves on a string) are always created by something that is oscillating. The very oscillatory motion that we have discussed in this chapter is the source for all possible waves around us, which ultimately allow communication and signaling. Every time you see something oscillating, you should think there is also a wave associated with it. In addition, the discussion of eigenmodes in this chapter foreshadows material in Chap. 11. *Wave Equations: Strings and Wind* concerned with standing waves, which are the eigenmodes of wave motion in finite regions.

10.11 Problems

Problem 1: Undamped Simple Harmonic Motion

A mass, m, on a spring of spring constant, k, initially located at its equilibrium location ($x = 0$) is suddenly set into motion with velocity v at $t = 0$.

(a) Write the general solution for undamped simple harmonic motion.
(b) Specify the appropriate initial conditions in this case.
(c) Are these initial conditions sufficient to determine $x(t)$ for all $t > 0$?
(d) Go ahead and use the appropriate initial conditions to find the particular solution for $x(t)$ that is applicable in this case.

Problem 2: Pendulum Motion

(a) Write Newton's laws in the x- and z-directions for a pendulum composed of a massless rod of length L and mass M at the end of the rod, when the rod is at an angle θ to the vertical. (Introduce any forces that you need. g is the acceleration due to gravity.)
(b) For small enough θ, show that the pendulum undergoes simple harmonic motion and determine the natural oscillation frequency.

Problem 3: Fooling Around with Complex Numbers

This problem provides a little practice in manipulating complex numbers.

(a) What are the real and imaginary parts of $\frac{2+7i}{3+4i}$?
(b) Express $z = 2 + 2i$ in polar form.
(c) Express $z = \exp(-2\pi i/3)$ in cartesian form and calculate z^3.
(d) What is the real part of $re^{2\pi i/5}$? (r is the absolute magnitude.)
(e) Calculate the derivative of $e^{-i\omega_0 t}$ with respect to t.

Problem 4: Why We Love the Polar Form of a Complex Number

(a) If $z = \cos\theta + i\sin\theta$, show that $z^{-1} = \cos\theta - i\sin\theta$ in two ways: (1) by simplifying $\frac{1}{\cos\theta+i\sin\theta}$ and by (2) writing z in polar form, then take the reciprocal.
(b) By examining $(z + z^{-1})^3$, show that $\cos(3\theta) = 4\cos^3\theta - 3\cos\theta$.
Hint: Recall that $(a + b)^3 = a^3 + 3a^2b + 3ab^2 + b^3$.

Problem 5: Conceptual Question on Harmonic Motion

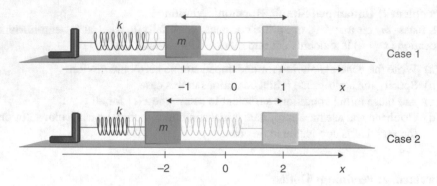

In the two cases shown in the figure, the mass and the spring are identical, but the amplitude of the simple harmonic motion is twice as big in Case 2 as in Case 1.

How are the maximum velocities in the two cases related? Provide *at least* two different explanations for your answer. (Hint: Think about which different concepts you may use: kinematics observations, energy observations, etc.)

Problem 6: Overdamped, Underdamped, and Critically Damped SHM: WolframAlpha to the Rescue

(a) At

> http://www.wolframalpha.com/[12]

enter:

```
solve y''(t)+g*y'(t)+w^2*y(t)=0, y'(0)=v, y(0)=0
```

and click "=".[13]

Here, y' is a shorthand for the derivative of y with respect to its argument, namely t, and y'' is a shorthand for the double derivative of y with respect to its argument, so actually you have just solved the damped harmonic oscillator problem using WolframAlpha in the case that there is an initial velocity (v) and zero initial displacement.

(b) Plot your solution for g=0.1, w=1, and v=1. Briefly describe the solution. This is an example of "underdamped" simple harmonic motion.

(c) What is the correspondence between w and g and k/m and ζ/m, as used in the chapter?

(d) Plot your solution for g=4, w=1, and v=1. Briefly described the solution. This is "overdamped" simple harmonic motion.
(e) Next, in the entry box, type:

```
solve y''(t)+2*w*y'(t)+w^2*y(t)=0,  y'(0)=v,  y(0)=A
```

and click "=".[14] Notice that in this case, the solution involves not only exponential and trigonometric functions but terms that are linear in t. This is a special case, where the damping and the natural frequency have a particular relationship (g equals 2w). What is the corresponding relationship among k, ζ, and m? The motion under these circumstances is said to be "critically-damped" simple harmonic motion. For g greater than 2w, we find overdamped simple harmonic motion. For g less than 2w, we find underdamped simple harmonic motion.
(f) Plot your critically damped solution for w=1, A=0, and v=1.

Problem 7: Radio Tuners
(a) Discuss what will happen if the single cosinusoidal term on the RHS of Eq. 10.58 is replaced by two or more cosinusoidal terms, each with a different drive frequency.
(b) Discuss why the answer to (a) is critical to how an FM tuner works.
(c) Comment on how the power dissipation varies with friction coefficient, ζ, when the driving frequency exactly equals the resonant frequency.

Problem 8: Coupled Pendulums
Consider two pendulums, each composed of a mass m at the end of a string of length L. These two masses are coupled together by a spring of spring constant, k, attached between the masses. Ignoring any damping and making the small angle approximation throughout, it turns out that this problem is mathematically the same as the coupled harmonic oscillator example in the chapter. Make the translation from that example to this problem, and thus write down the eigenfrequencies and eigenvectors of the two modes.

Problem 9: Another Coupled Oscillator Problem
Two masses are arranged on a line, and each is attached to a wall and to each other, as shown in Fig. 10.19. The mass of mass 1 is m. The mass of mass 2 is $2m$. Mass 1 is attached to its wall via a spring of spring constant k. Mass 2 is attached to its wall via a spring of spring constant $2k$. The two masses are also tethered together via another spring of spring constant k. Any friction is completely negligible. The

---

[14] http://www.wolframalpha.com/input/?i=solve+y%27%27%28t%29%2B2*w*y%27%28t%29 %2Bw%5E2*y%28t%29%3D0%2C+y%27%280%29%3Dv%2C+y%280%29%3D0

Fig. 10.19 Masses
connected by springs

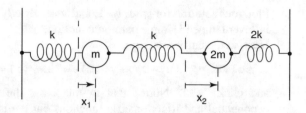

displacement of mass 1 from its equilibrium position is x_1, and the displacement of mass 2 from its equilibrium position is x_2.

(a) Given that there are two masses, how many eigenmodes do you expect for this system?
(b) Write the equations of motion (Newton's second law) for each mass in terms of the coordinate system shown in the figure; x_1 is the displacement of mass m_1 from equilibrium, and it is assumed positive to the right; x_2 is the displacement of mass m_2 from equilibrium, and it is assumed positive to the right.
(c) By assuming that the masses' displacements from their equilibrium positions vary in time sinusoidally with angular frequency ω, find the eigenvalues of these equations.
(d) Determine the corresponding eigenvectors.

Problem 10: Two Masses Hanging from the Ceiling

A mass m hangs from the ceiling via a spring of spring constant k. A second mass m hangs from the first via a second spring of spring constant k. Any friction is completely negligible. x_1 is the displacement of the upper mass from its equilibrium position, and x_2 is the displacement of the lower mass from its equilibrium position.

(a) Write the equations of motion (Newton's second law) for each mass.
(b) By assuming that the masses' displacements from their equilibrium positions vary in time sinusoidally with angular frequency ω^2, find the eigenvalues of these equations.
(c) Determine the corresponding eigenvectors.

Problem 11: Coupled Spring and Pendulum

As shown in Fig. 10.20, mass $m_1 = m$ is supported on a frictionless table and connected to a rigid wall via a spring of spring constant k. At the same time, attached via a massless, inextensible string of length L to mass m_1 is mass $m_2 = m$, i.e., $m_1 = m_2 = m$. The acceleration due to gravity is g.

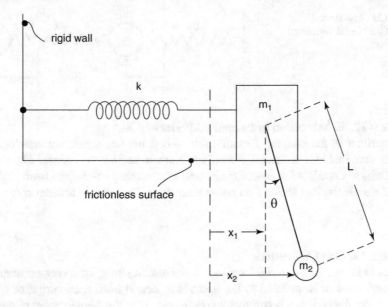

Fig. 10.20 Two masses, $m_1 = m$ and $m_2 = m$, are connected to each other and to a rigid wall by means of a spring and a string

(a) Using the small angle approximation that $\sin\theta \simeq \theta$ and $\cos\theta \simeq 1$, write Newton's second law for each mass in terms of $\frac{d^2 x_1}{dt^2}$ and $\frac{d^2 x_2}{dt^2}$, x_1, x_2, m_1, m_2, g, L, and k.

(b) Calculate the eigenvalues (eigenfrequencies) of these equations in the limit that $g/L \gg k/m$. Specifically, in your final expressions for the eigenvalues, you can neglect k/m compared to g/l, but you cannot neglect k/m on its own. Call them ω_1 and ω_2, with $\omega_1 < \omega_2$.

(c) Calculate the corresponding eigenvectors. Call them $\mathbf{e}_1$ and $\mathbf{e}_2$ for ω_1 and ω_2, respectively.

(d) What is the potential energy (U) of the configuration shown in Fig. 10.20? (Hint: Use the small angle approximation to the second order for the cosine: i.e., $\cos(\theta) \approx 1 - \frac{\theta^2}{2}$).

(e) For certain initial conditions, the solution is

$$(x_1(t), x_2(t)) = \mathbf{e}_1 \cos\omega_1 t. \tag{10.123}$$

What is the potential energy in this case?

(f) What is the kinetic energy (K) in this case?

(g) Calculate the total mechanical energy (E=K+U) in this case. Comment on the time dependence that you find.

Fig. 10.21 Masses and
springs for "Good vibrations"

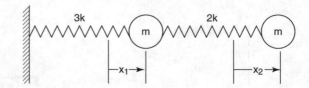

Problem 12: Echolocation in Parnell's Mustached Bat

As described in the chapter, Parnell's mustached bat has a unusual echolocation
system that includes a basilar membrane with a narrow resonance. Develop a
hypothesis to explain what advantage Parnell's mustached bat derives from having a
basilar membrane that shows a narrower resonance, instead of a broader resonance.

Problem 13: Good Vibrations

A mass m is attached to a rigid wall via a horizontal spring of spring constant $3k$.
A second mass m is attached to the first via a second horizontal spring of spring
constant $2k$. Any friction is completely negligible. x_1 is the displacement of the first
mass from its equilibrium position, and x_2 is the displacement of the second mass
from its equilibrium position, as shown in Fig. 10.21.

(a) Write the equations of motion (Newton's second law) for each mass for the
    displacements, x_1 and x_2, of the two masses.
(b) By assuming that the masses' displacements from their equilibrium positions
    vary in time (co)sinusoidally with angular frequency ω^2, find the eigenfrequen-
    cies of your equations. Call the eigenfrequencies ω_A and ω_B.
(c) Determine the corresponding eigenvectors. Call them $\mathbf{e}_A$ and $\mathbf{e}_B$.
(d) Calculate the scalar product of the two eigenvectors.
(e) At $t = 0$, the velocities of both masses are zero, while the upper mass has a
    displacement a and the lower mass has a displacement $2a$. Calculate the masses'
    positions at (a later) time, t.

Problem 14: Approach to Steady State

(a) Figure 10.22 shows the steady-state position (left) and velocity (right) versus
    time for a driven, damped, harmonic oscillator with $\zeta = 1$ N m s^{-1}, $m =
    1$ kg, $\omega_0 = \sqrt{10}$ s$^{-1} \simeq 3.16$ s^{-1}, $\omega = 2$ s^{-1}, $F = 1$ N, and a driving force,
    $F \cos \omega t$. From these figures or otherwise, estimate $x(0)$ and $v(0)$ in this case.
    Briefly explain how these curves would change (if at all) if the driving force
    were $F \sin \omega t$ instead of $F \cos \omega t$.
(b) Consider the same oscillator as in (a) with $\zeta = 1$ N m s^{-1}, $m = 1$ kg, and $\omega_0 =
    \sqrt{k/m} = \sqrt{10}$ s^{-1}, except that now there is no driving force ($F = 0$). Sketch
    the position and velocity of this damped, harmonic oscillator versus time, given
    that its initial position and velocity are given by your answers for $x(0)$ and $v(0)$,

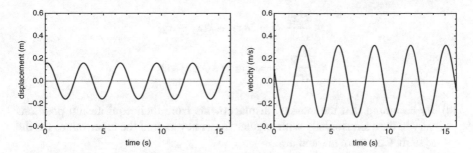

Fig. 10.22 Steady-state position (left) and velocity (right) for a driven, damped, harmonic oscillator with $\zeta = 1\,\text{N m s}^{-1}$, $m = 1\,\text{kg}$, $\omega_0 = \sqrt{10}\,\text{s}^{-1} \simeq 3.16\,\text{s}^{-1}$, $\omega = 2\,\text{s}^{-1}$, and $F = 1\,\text{N}$

respectively, in part (a). Briefly explain how your two curves—$x(t)$ and $v(t)$—are related to each other mathematically.

(c) Considering the same oscillator as in (a) and (b), briefly explain how the functions you sketched in (b) and the functions illustrated in Fig. 10.22 can be used to describe $x(t)$ and $v(t)$ when a driving force $F \cos \omega t$ (with $F = 1\,\text{N}$ and $\omega = 2\,\text{s}^{-1}$) is suddenly applied at $t = 0$ to a damped oscillator at rest (so that $x(0) = 0$ and $v(0) = 0$). Sketch $x(t)$ and $v(t)$ versus t in this case.

Problem 15: Spring Fever

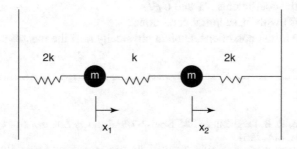

Two masses, both of mass m, are connected to three springs as shown in the figure. The two springs on the sides have spring constant $2k$, and the spring in the middle has spring constant k. We choose the coordinate system shown in the figure: x_1 is the displacement of the mass on the left from equilibrium, and it is assumed positive to the right; x_2 is the displacement of the mass on the right from equilibrium, and it is assumed positive to the right.

By applying Newton's 2nd law to each mass in this system, we find the following equations of motion:

$$m\frac{d^2x_1}{dt^2} = -2kx_1 - k(x_1 - x_2)$$

$$m\frac{d^2x_2}{dt^2} = -2kx_2 - k(x_2 - x_1).$$

(a) By assuming that the masses' displacements from their equilibrium positions vary in time sinusoidally with angular frequency ω, find the eigenvalues for the system. Call them ω_A and ω_B.

(b) Determine the corresponding eigenvectors, and call them $\mathbf{e}_A$ and $\mathbf{e}_B$.

(c) The general solution for the position of the masses is

$$\mathbf{x}(t) = C_A\mathbf{e}_A\cos\omega_A t + S_A\mathbf{e}_A\sin\omega_A t + C_B\mathbf{e}_B\cos\omega_B t + S_B\mathbf{e}_B\sin\omega_B t,$$

where C_A, S_A, C_B, and S_B are constants to be determined by the initial conditions. In the case where the initial velocity of both masses is zero, this expression will simplify. Under the condition $\mathbf{v}(t = 0) = (0, 0)$ or more explicitly, $v_1(0) = 0$ and $v_2(0) = 0$, this expression reduces to

$$\mathbf{x}(t) = C_A\mathbf{e}_A\cos\omega_A t + C_B\mathbf{e}_B\cos\omega_B t.$$

Suppose you are now given the following second initial condition: when $t = 0$, the position of the mass on the left is A, and the position of the mass on the right is $-A$.

What are the coefficients C_A and C_B?

(d) What is $\mathbf{x}(t)$ with these initial conditions?

(e) With these initial conditions, explain physically how the masses are moving.

References

1. R. P. Feynman, R. B. Leighton, and M. Sands. *The Feynman Lectures on Physics*. Addison Wesley, Boston, MA, 1964.

2. Robert D. Grober, Jason Acimovic, Jim Schuck, Dan Hessman, Peter J. Kindlemann, Joao Hespanha, A. Stephen Morse, Khaled Karrai, Ingo Tiemann, and Stephan Manus. Fundamental limits to force detection using quartz tuning forks. *Rev. Sci. Ins.*, 71:2776–2780, 2000.

3. M Kössl and IJ Russell. Basilar membrane resonance in the cochlea of the mustached bat. *Proceedings of the National Academy of Sciences of the United States of America*, 92(1):276–279, January 1995.

Wave Equations: Strings and Wind

11

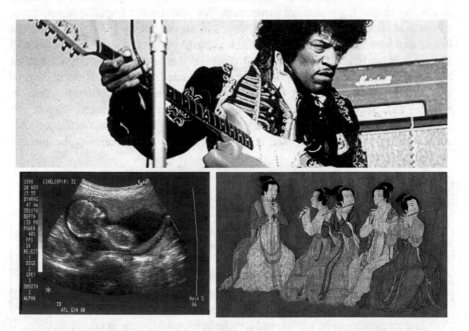

Fig. 11.1 Top: Jimi Hendrix, playing guitar. Public domain image from https://commons.wikimedia.org/wiki/File:Jimi_Hendrix_1967.jpg. Bottom left: Ultrasound image of a fetus. Public domain image (ID# 1666) from the CDC's Public Health Image Library (http://phil.cdc.gov/phil/home.asp). Bottom right: Women playing flutes, a detail from a twelfth-century remake of the tenth-century *Night Revels of Han Xizai*, by Gu Hongzhong

 Electronic Supplementary Material The online version of this article (https://doi.org/10.1007/978-3-031-05808-0_11) contains supplementary material, which is available to authorized users.

11.1   Introduction

In this chapter, we will introduce many of the basic features and properties of wave motion. We will discuss traveling waves, standing waves and the difference between them, and how boundary conditions select particular standing wave wavelengths and frequencies. We will also discuss how boundary conditions at an interface between different media give rise to reflection and transmission of waves at that interface, which is an important topic for all types of waves. For example, on the medical side, reflection of high frequency (1–18 MHz) sound waves at the interfaces between different structures in the body is the basis of medical ultrasonography,[1] which is an essential tool for physicians (Fig. 11.1).

Of course, one of the most important ways that we communicate with each other is via sound waves, which we detect by hearing and which we generate by speaking or singing or playing a musical instrument. We will discuss impedance matching, which enables us to transduce sound waves in air eventually into nerve impulses inside our inner ear. We will also discuss the physics of wind instruments and clarinets, in particular, where we will discover that a clarinet exhibits resonant behavior at frequencies corresponding to the natural frequencies of oscillation of standing sound waves within the instrument. However, as is often the case in physics, even though we specifically consider a particular resonator, namely the clarinet, the essential idea that extended objects exhibit resonances at frequencies corresponding to their natural oscillation frequencies is very broadly applicable and very important.

11.2   Your Learning Goals for This Chapter

By the end of this chapter you should be able to:

- Explain and exploit general solutions to the wave equation, corresponding to both forward and backward propagating waves.
- Explain the difference between transverse waves, such as waves on a string and electromagnetic waves, and longitudinal waves, such as sound waves.
- Represent forward and backward propagating waves in the case of (co)sinusoidal waves, either as (co)sinusoidal functions or as imaginary exponential functions.
- For (co)sinusoidal waves, explain and exploit amplitude, phase, wavelength, wavenumber, frequency, angular frequency, and wave speed, and the relationships among these quantities.

---

[1] http://en.wikipedia.org/wiki/Medical_ultrasonography

- Explain the difference between standing and travelling waves and be able to represent standing waves as the superposition of a forward and a backward propagating wave of equal amplitude.
- Explain that standing waves correspond to the eigenmodes of waves in a finite-length regions and be able to determine the form of the eigenmodes for given boundary conditions.
- Apply the principle of superposition to waves to represent an arbitrary displacement of a string as a sum over standing wave eigenmodes.
- Calculate the reflection and transmission coefficients for travelling waves incident on an interface between media of different impedances.
- Calculate the power carried by a traveling wave.

11.3   The Wave Equation and Transverse Waves on a String

Physics starts with basic principles, e.g., Newton's Laws, to develop a fundamental mathematical description of some aspect of the natural world, namely a model. It then deploys a set of tools (algebra, calculus, etc.) to transform the model and the basic principles into new understanding. In this chapter, this program will result in the wave equation and an understanding of wave motion. We start off by considering waves on a string of total mass m and total length L, held at a tension T and lying along the x-axis. We ignore friction and damping, because they are small for a string in air. Actually, this model is also an approximation for human vocal chords.[2] Wave motion occurs when the string is displaced in a direction perpendicular, to the string, namely the y direction; we call this also the "transverse" direction. To derive the wave equation, which describes small-amplitude wave motions on a tensioned string, we apply Newton's Second Law to the element of the string between x and $x + dx$, first in the x-direction and then in the y-direction. A sketch of the situation is shown in Fig. 11.2. Because the mass per unit length is $\mu = m/L$, the mass of this element of string is $\frac{m}{L}dx = \mu dx$.

Throughout this example, we will take both θ_1 and θ_2 to be small, so that we can reasonably approximate $\cos\theta_1$, $\sin\theta_1$, $\cos\theta_2$ and $\sin\theta_2$ by their Taylor series expansions near $\theta_1 = 0$ and $\theta_2 = 0$. This approximation is usually referred to as the "small-angle approximation" and it is used here as well as in many other contexts in physics where the system under study is well described only by looking at the behavior for small angles. In general the Taylor series expansion of a function $f(x)$ near $x = x_0$ is given by

---

[2] http://en.wikipedia.org/wiki/Vocal_cord

Fig. 11.2 An element of a
string under tension between
x and $x + dx$, while the string
is vibrating

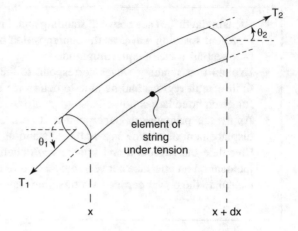

$$f(x) \simeq f(x_0) + f'(x_0)(x - x_0) + \frac{1}{2!} f''(x_0)(x - x_0)^2 + \frac{1}{3!} f'''(x_0)(x - x_0)^3$$

$$+ \frac{1}{4!} f''''(x_0)(x - x_0)^4 + \dots. \tag{11.1}$$

where $f' = \frac{df}{dx}$ is the first derivative respect to x, $f'' = \frac{d^2 f}{dx^2}$ is the second derivative, $f''' = \frac{d^3 f}{dx^3}$ the third, and so forth. Very often, including here when analyzing wave motion, it is enough to keep only the first two terms of this series, i.e., up to the term proportional to $(x - x_0)$, while neglecting the rest, i.e., we keep the linear order terms, while neglecting the quadratic- and higher-order terms. Applying a Taylor series expansion for small θ, we obtain the following "small-angle approximations":

$$\cos \theta \simeq \cos(0) - \sin(0) \times \theta + \dots \simeq 1 + \dots. \tag{11.2}$$

and

$$\sin \theta \simeq \sin(0) + \cos(0) \times \theta + \dots \simeq \theta. \tag{11.3}$$

There is no motion in the x-direction and in particular no acceleration along the x-direction. Therefore, in terms of the variables in Fig. 11.2, applying Newton's second law in the x-direction, we have

$$0 = T_2 \cos \theta_2 - T_1 \cos \theta_1. \tag{11.4}$$

θ_1 and θ_2 are the angles the string makes with the horizontal direction at x and $x + dx$, respectively. Therefore, for small enough θ_1 and small θ_2 (i.e., to linear order), Eqs. 11.4, 11.2, and 11.3 implies

$$T_1 = T_2. \tag{11.5}$$

Henceforth, we will call

$$T_1 = T_2 = T. \tag{11.6}$$

It will also be convenient to write

$$T_2 \cos \theta_2 = T \tag{11.7}$$

and

$$T_1 \cos \theta_1 = T. \tag{11.8}$$

Since $\cos \theta_1 = \cos \theta_2 = 1$ to linear order in θ_1 and θ_2, respectively, this manipulation is technically correct, although admittedly it is tricky. However, these relations will be handy below!

Newton's second law for motion in the y-direction gives

$$(\mu dx)\frac{d^2 y}{dt^2} = T_2 \sin \theta_2 - T_1 \sin \theta_1. \tag{11.9}$$

Dividing by $T = T_2 \cos \theta_2 = T_1 \cos \theta_1$, we have

$$\frac{\mu}{T} dx \frac{d^2 y}{dt^2} = \frac{T_2 \sin \theta_2}{T_2 \cos \theta_2} - \frac{T_1 \sin \theta_1}{T_1 \cos \theta_1} = \tan \theta_2 - \tan \theta_1. \tag{11.10}$$

Next, we use

$$\tan \theta_1 = y'(x), \tag{11.11}$$

where $y' = \frac{dy}{dx} = \tan \theta_1$ is the slope of the string at x, and

$$\tan \theta_2 = y'(x + dx) \tag{11.12}$$

is the slope of the string at $x + dx$. The result is that

$$\frac{\mu}{T} dx \frac{d^2 y}{dt^2} = y'(x + dx) - y'(x). \tag{11.13}$$

Next, carrying out a linear Taylor series expansion of $y'(x + dx)$ about x with the result that $y'(x + dx) = y'(x) + dx \, y''(x)$, we find

$$\frac{\mu}{T} dx \frac{d^2 y}{dt^2} = \frac{d^2 y}{dx^2} dx = y''(x) dx. \tag{11.14}$$

Finally, dividing through by dx and shuffling every term to one side of the equation, we arrive at the wave equation on a string:

$$\frac{d^2 y}{dx^2} - \frac{1}{v^2} \frac{d^2 y}{dt^2} = 0. \tag{11.15}$$

In Eq. 11.15,

$$v = \sqrt{T/\mu} \tag{11.16}$$

will turn out to be the wave speed, which may be seen to increase with increasing tension T and decrease with increasing mass per unit length μ. The latter fact is the main reason why adult male voices are deeper (lower frequency) than adult female voices: for the same length of a vocal cord, adult males have more massive vocal chords, therefore the male sound velocity is smaller, which as a consequence will correspond to a lower frequency. The exact relationship between wave speed and frequency will be discussed a bit later in this chapter in Sect. 11.8 and in particular Eq. 11.61.

Although we have derived the wave equation for waves on a string, Eq. 11.15 is a much more general result and formally identical to the wave equations that are applicable in many other contexts, including the wave equations for sound waves and electromagnetic waves. Importantly, the wave equation is linear in y (i.e., it contains only terms proportional to y or its derivates, but not terms proportional to y^2, y^3, and higher). Therefore, the principle of superposition holds for solutions to the wave equation. In the context of waves, superposition is often called interference, but they are the same thing.

11.3.1 General Solutions to the Wave Equation

The general solution to the wave equation is of the form

$$y(x, t) = f(x - vt) + b(x + vt), \tag{11.17}$$

where f and b are arbitrary functions of a single variable. We often think of a wave as being a sinusoidal disturbance, and indeed such waves are very important. However, solutions to the wave equation—a. k. a. waves—can also take the form of a pulse, as shown in Fig. 11.3, or in fact any other function. The term $f(x - vt)$ describes a "forward" traveling wave, while the term $b(x + vt)$ describes a "backward" traveling wave.

To see that, in fact, $f(x - vt)$ corresponds to a forward traveling wave, consider a wave described by a function $f(x - vt)$ that exhibits a maximum when its argument is zero, as shown in Fig. 11.3. At $t = 0$, the argument is x and therefore the function shows a peak at $x = 0$. At a later time ($t > 0$), however, its argument is $x - vt$, which is zero when $x - vt = 0$ or equivalently $x = vt$. Therefore, at a time $t > 0$, $f(x - vt)$ shows a peak at a positive value of x, given by $x = vt$. Therefore, in this case the pulse is moving in the positive direction with velocity v, that is why we call it a forward traveling wave, because for future times it moves in the positive

Fig. 11.3 An arbitrary function of $x - vt$ (and nothing else) satisfies the wave equation and corresponds to a forward travelling wave. An arbitrary function of $x + vt$ (and nothing else) satisfies the wave equation and corresponds to a backward travelling wave

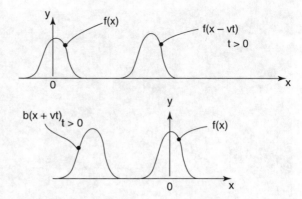

direction. A similar argument establishes that $b(x + vt)$ corresponds to a backwards traveling wave, moving in the negative direction with velocity v.

11.3.2 Example: Only Time Will Tell

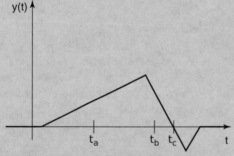

Imagine a wave traveling to the right through a string and passing through point x_0 along the x-axis. The graph above shows the vertical position as a function of time for a point of the string located at x_0.

(a) For the time points labeled t_a, t_b, and t_c, state if the instantaneous velocity of the point on the string is positive negative or zero.

    Answer: $v_a > 0$, $v_b < 0$, and $v_c < 0$. Because this graph represents the position as a function of time the slope of the graph gives the velocity of the point moving up and down. The slope is positive for t_a, and negative for both t_b and t_c.

(continued)

(b) Which of the following graphs could be a snapshot of the wave passing through x_0?

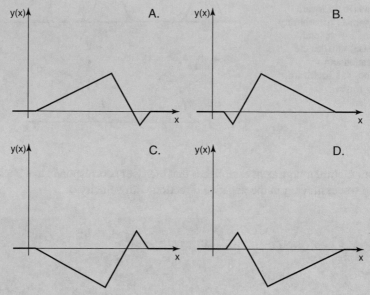

Answer: Graph B must match the wave traveling through the point x_0. The graph of $y(t)$ shows that the point at x_0 moves gradually in the positive y-direction, until it reaches a maximum. Then it drops rapidly toward $y = 0$, continuing to a point where $y < 0$, then turning around and climbing back to $y = 0$. We choose a point where initially $y = 0$ on the right side of the four $y(x)$ graphs and follow it as the shapes travel to the right. For graph A and D, the point initially travels in the negative y-direction. For B and C, the point initially travels in the positive y-direction. For C, the point travels quickly upward, then quickly downward to a large negative y value, then back up to zero. Only graph B would give the right movement, where moving gradually to a maximum positive y-position, quickly down past $y = 0$, then turning around and stopping at $y = 0$.

11.4 Periodicity in Space and Time for Sinusoidal Traveling Waves

It is important to realize that a solution to the wave equation $y(x, t)$, as described by Eq. 11.17, is function of two variables: the position x along the wave, and the time t. Let us first analyze qualitatively the behavior of a sinusoidal (or equivalently co-sinusoidal) wave that means a wave that repeats itself in time and space oscillating as a harmonic wave. We could graph this wave in two possible ways: (1) we choose a position $x = x^*$ along the string and plot the behavior of this point as a function of time, i.e., plotting $y(x^*, t)$ versus t (see left side of Fig. 11.4) (2) we take a snapshot of the wave, effectively choosing a time $t = t^*$ and plot the behavior of the wave at this time for any point x along the wave, i.e., plotting $y(x, t^*)$ as a function of x (see right side of Fig. 11.4). In each representation, we can identify an oscillating behavior that repeats itself in time or in space. For these oscillations we can define two "periodicities": one in time and one in space. We define the period T as the length of time for a full oscillation cycle to be completed; the period is marked on the graph on the left side of Fig. 11.4 as the distance between two consecutive peaks. Similarly, the wavelength λ is the length of space for a full oscillation cycle to be completed and it is marked on the graph on the right side of Fig. 11.4 as the distance between two consecutive peaks. The SI units for period are seconds, while the SI units for wavelength are meters.

Correspondingly, we can define the inverse of these periodicities as two types of "frequencies" (in other words how many cycles happen in a unit of time or length): the angular frequency

$$\omega = \frac{2\pi}{T} \tag{11.18}$$

and the wavenumber

$$k = \frac{2\pi}{\lambda}. \tag{11.19}$$

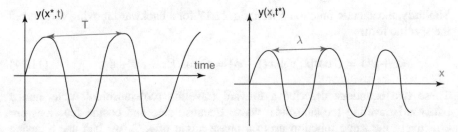

Fig. 11.4 Depiction of a sinusoidal wave. Left side: a point $x = x^*$ along the wave is kept fixed and its motion is graphed as a function of time, the period in time T is marked on the graph. Right side: snapshot of the wave at time $t = t^*$, the periodicity in space λ is marked on the graph

Note that the wavenumber k is not the same as the spring constant k (introduced in *2. Force and momentum: Newton's Laws and how to apply them*) although they are represented by the same symbol. (Physicists tend to use the letter k in too many different contexts. Be sure to pay attention every time, which k is being discussed!). The angular frequency ω is measured in radians per second, while the wavenumber is measured in radians per meter. We also introduce the frequency, ν, which differs from the angular frequency by a factor of 2π,

$$\nu = \frac{1}{T}, \qquad \omega = 2\pi \nu. \tag{11.20}$$

Frequency is measured in Hertz (Hz): $1\,\text{Hz} = \text{s}^{-1}$.

An important relationship that holds true for all waves is the fundamental relationship between the angular frequency, the wavenumber, and the wave speed:

$$\omega = \nu k. \tag{11.21}$$

This relationship comes from the kinematic understanding that the wave covers a length of time λ in a time T, therefore:

$$\nu = \frac{\lambda}{T}. \tag{11.22}$$

Using the definition of ω and k just introduced, we can show that Eqs. 11.21 and 11.22 express the same relationships.

11.5   Sinusoidal Traveling Waves

In the case of a sinusoidal (or co-sinusoidal) traveling wave, the generic function f from Eq. 11.17 for a forward traveling wave takes the specific form

$$f(x - \nu t) = C \cos[k(x - \nu t) - \phi] = C \cos(kx - \omega t - \phi). \tag{11.23}$$

Similarly, the generic function b from Eq. 11.17 for a backward moving wave takes the specific form

$$b(x + \nu t) = C \cos[k(x + \nu t) - \phi] = C \cos(kx + \omega t - \phi). \tag{11.24}$$

These two equations describe a forward traveling (co)sinusoidal wave, and a backward traveling (co)sinusoidal wave. Because sine and cosine functions are effectively the same function up to a phase angle of $\pi/2$, we can use a cosine function without loss of generality, knowing that by varying the value of phase angle ϕ we can also obtain a sine function (e.g., $\sin(x) = \cos(x - \pi/2)$).

The parameters in the equations above are defined as follows:

- C is the amplitude of the wave: it quantifies how large is the displacement of the wave from the equilibrium position (either in the positive $+y$ or negative $-y$ direction)
- v is the wave speed: it quantifies how fast the waves travels (either forward or backward)
- λ is the wavelength: it quantifies the length in space of a full oscillation of the wave (distance in space between two consecutive peaks)
- k is the wavenumber of the wave in question $k = \frac{2\pi}{\lambda}$
- $\omega = vk$ is its angular frequency, measured in radians per second.
- ϕ is the phase of the wave.

11.5.1 Example: Writing the Equation for a Harmonic Wave

Write the $y(x, t)$ function describing a harmonic wave, moving to the left, with amplitude 5m, period 3s, and wavelength 1m.

Answer: The general expression for a harmonic wave traveling to the left is given by

$$y(x, t) = C \sin(kx + \omega t), \qquad (11.25)$$

where C is the amplitude, k is related to the wavelength by $k = \frac{2\pi}{\lambda}$ and $\omega = \frac{2\pi}{T}$ (Note, we have assumed a zero phase shift $\phi = 0$ since we are not given any specific information about the initial conditions of the wave.) Using the given quantities, the expression is

$$y(x, t) = 5\text{m} \sin(\frac{2\pi}{1\text{m}} x + \frac{2\pi}{3\text{s}} t). \qquad (11.26)$$

11.5.2 Example: Making Waves

How can we create a sinusoidal wave? Generally, waves are created by an oscillating source. For example, at the gym, a popular arm exercise is to drive waves on a rope as shown in Fig. 11.5. In this example, we investigate the properties of a wave, created by the application of a transverse force,

Fig. 11.5 A crossfitter (an athlete at a crossfit gym) shakes sinusoidally the end of a rope to create a wave

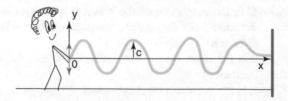

$F \cos \omega t$, to the end of an infinitely long rope of mass per unit length, μ, under tension, T. Specifically, we seek to determine the amplitude and phase of the wave and how much power it carries away from its source. We pick the coordinate system, so that the rope lies along the positive x-axis, and we call the displacement of the rope at time t and position x, $y(x, t)$. The force is applied at end of the rope at $x = 0$.

(a) Apply Newton's second law to the very end of the rope.

Answer: Since the very end of the rope has zero (infinitesimal) mass, Newton's second law reads

$$0 = F \cos \omega t + T \frac{dy}{dx}(0, t), \tag{11.27}$$

where $T\frac{dy}{dx}(0,t)$ is the force in the y direction exerted by the rope on its left-hand end at $x = 0$.

(b) Assuming that in steady state, the force gives rise to a sinusoidal, forward travelling wave, $y(x, t) = C \cos(kx - \omega t - \phi)$, use your answer to (a) to determine the steady-state wave amplitude, C, and phase ϕ.

Answer: Using

$$\frac{d}{dx} C \cos(kx - \omega t - \phi) = -kC \sin(kx - \omega t - \phi), \tag{11.28}$$

now, we have

$$F \cos \omega t = -T \frac{dy}{dx}(0, t) = TkC \sin(\omega t - \phi). \tag{11.29}$$

To satisfy Eq. 11.29 at all times, t, we require that $\phi = -\frac{\pi}{2}$, because $\sin(\omega t + \frac{\pi}{2}) = \cos \omega t$. Then, we have

(continued)

$$F \cos \omega t = -T\frac{dy}{dx}(0, t) = TkC \cos \omega t, \qquad (11.30)$$

allowing us to cancel the cosine on both sides. In turn, it follows that $C = \frac{F}{Tk} = \frac{Fv}{T\omega}$, where we used that the angular frequency (ω), wavenumber (k), and wave speed (v) are related via $\omega = vk$. Thus, the wave, created on the rope, is

$$y(x, t) = -\frac{Fv}{T\omega} \sin(kx - \omega t). \qquad (11.31)$$

(c) Calculate the instantaneous power, P, that must be supplied by the arms of the crossfitter driving the wave.

Answer: Power is the product of force and velocity. The velocity of the end of the rope is $\frac{dy}{dt}(0, t)$. Therefore, the power supplied to the end of the rope is

$$P = F \cos \omega t \times \frac{dy}{dt}(0, t) = \frac{F^2 v}{T} \cos^2 \omega t. \qquad (11.32)$$

(d) What is the mean power, $\langle P \rangle$, supplied by the arms of the crossfitter?

Answer: Since the time-average of $\cos^2 \omega t$ equals $\frac{1}{2}$,

$$\langle P \rangle = \frac{F^2 v}{2T}. \qquad (11.33)$$

(e) Express your answers to (c) and (d) in terms of the amplitude of the wave, C, instead of the applied force.

Answer: Using $C = \frac{Fv}{T\omega}$, we find $F = \frac{\omega T C}{v}$, so that

$$P = \frac{F^2 v}{T} \cos^2 \omega t = \frac{\omega^2 T C^2}{v} \cos^2 \omega t, = \omega^2 \sqrt{\mu T} C^2 \cos^2 \omega t, \qquad (11.34)$$

using $v = \sqrt{T/\mu}$. Similarly,

$$\langle P \rangle = \frac{1}{2}\omega^2 \sqrt{\mu T} C^2. \qquad (11.35)$$

(continued)

(f) What is the power and mean power carried by a sinusoidal wave of amplitude, C, and angular frequency, ω?

Answer: Because of energy conservation, all of the power supplied by the crossfitter's arms is carried away by the wave. Therefore, the answers to (d) and (e) also represent the power carried by the wave. Since all of the parameters in the answers to (c) and (d) correspond to properties of a wave on a string, these expressions apply to any sinusoidal traveling wave on a string, irrespective of how that wave was generated. Therefore, the instantaneous power (at $x = 0$) and mean power carried by such a wave are always given by

$$P = \omega^2 \sqrt{\mu T} C^2 \cos^2 \omega t, \tag{11.36}$$

and

$$\langle P \rangle = \frac{1}{2} \omega^2 \sqrt{\mu T} C^2. \tag{11.37}$$

An important insight from Example 11.5.2 is that a cosinusoidal-in-time source leads to a wave that is sinusoidal in time and space, and that the wave's angular frequency, ω, is determined by the source frequency. Nevertheless, ω specifies the time dependence of the displacement of the wave at a particular point in space, and thus is a key characteristic of the wave. By contrast, the wave speed and the wave impedance (see below) are material properties, dependent solely on the medium in which the wave propagates, and not at all on the source of the wave.

11.6   Sound Waves

11.6.1 Longitudinal Sound Waves

The most fundamental difference between waves on a string and sound waves is that waves on a string are transverse waves, in which the string's displacements are perpendicular to the direction of wave propagation, while sound waves are longitudinal waves, in which the displacements of the medium carrying the waves, are parallel to the direction of wave propagation. In this chapter, we focus on one-dimensional sound waves that propagate along elastic rods or gas-filled tubes, which requires that the diameter of the rod or tube is much less than the sound wavelength.

A gas-filled tube can be, for instance, a flute or clarinet which consists of a tube filled with air to produce music. An elastic rod is, for instance, the rail of a commuter rail train track, you may have noticed you can hear a train coming before you can see it because the sound propagates along the rails and out into the air to your ears.

The displacements in an elastic rod or a column of air during the passage of a sound wave are illustrated in Fig. 11.6. As shown in this figure, we consider a small length located between x and $x + \Delta x$, therefore the Δx is the unstretched length of this element when at rest. Assume a sound wave travels through this element of elastic rod or column of air, a sound wave is a longitudinal propagation and will cause a displacement of the molecules of the medium along the direction of propagation. In particular, the molecules of the medium at x will displace of an amount u, and the molecules at location $x + \Delta x$ will displace of an amount $u + \Delta u$; therefore when the sound wave is coming by the end of the segments are displaced to $x + u$ and $x + \Delta x + u + \Delta u$, respectively. Therefore the full length of the stretched element is $\Delta x + \Delta u$, which implies the extension of the element in question is Δu. Thus, as discussed in 2. *Force and momentum: Newton's Laws and how to apply them*, we can expect Hooke's law to be applicable for small Δu. The appropriate spring constant for an elastic rod or a column of gas is

$$\frac{BA}{\Delta x},$$

where A is the cross-section of the rod or column and B is a material property that measures the elasticity of the medium in which the sound is propagating (e.g., how the medium is reacting to compressions); specifically B is either the Young's modulus in the case of an elastic rod or the gas' bulk modulus in the case of a column of gas. The A-dependence of this spring constant follows because increasing A is analogous to increasing the number of contributing springs in parallel within the element. The Δx-dependence follows because increasing Δx is analogous to increasing the number of contributing springs in series within the

Fig. 11.6 An element of an elastic rod or column of air that lies between x and $x + \Delta x$, when at rest (top figure), and becomes an element that lies between $x + u$ and $x + \Delta x + u + \Delta u$, during passage of a longitudinal sound wave (bottom figure)

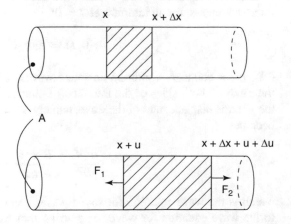

element. Therefore, Newton's Third Law informs us that the force exerted on the element by the rod to the left of the element is

$$F_1 = -\frac{BA}{\Delta x}\Delta u = -BA\frac{\Delta u}{\Delta x} = -BA\frac{du(x,t)}{dx}. \tag{11.38}$$

Because the extension can vary versus position, the force exerted on the element by the rod to the right of the element is

$$F_2 = BA\frac{du(x+\Delta x,t)}{dx} = BA(\frac{du(x,t)}{dx} + \Delta x\frac{d^2u(x,t)}{dx^2}). \tag{11.39}$$

The total force on the element is the sum of these two forces:

$$F_1 + F_2 = -BA\frac{du(x,t)}{dx} + BA(\frac{du(x,t)}{dx} + \Delta x\frac{d^2u(x,t)}{dx^2}) = BA\Delta x\frac{d^2u}{dx^2}. \tag{11.40}$$

The mass of the element is $\rho A\Delta x$, where ρ is the mass density of the medium. Its acceleration is $\frac{d^2u}{dt^2}$. Therefore, Newton's Second Law for the element reads:

$$\rho A\Delta x\frac{d^2u}{dt^2} = BA\Delta x\frac{d^2u}{dx^2}. \tag{11.41}$$

Dividing both sides by Δx we have, i.e.,

$$A\frac{d^2u}{dt^2} = \frac{BA}{\rho}\frac{d^2u}{dx^2}. \tag{11.42}$$

In order to discuss reflection and transmission of sound waves at the boundary between two tubes with different cross-sectional areas, we need to apply the appropriate boundary conditions at the interface, one of which is that the volume of material displaced by the sound wave must be the same on both sides of the interface, that is, for an interface at $x = 0$:

$$A_1u_1(0,t) = A_2u_2(0,t). \tag{11.43}$$

(The other boundary condition on sound waves at interfaces is that the pressure is the same on both sides of the interface.) Equation 11.43 motivates introduction of the volume displacement of the wave, namely $V = A \times u$. In terms of V, Eq. 11.42 becomes

$$\frac{d^2V}{dt^2} = \frac{B}{\rho}\frac{d^2V}{dx^2}, \tag{11.44}$$

which is the wave equation for sound waves. As expected this equation is similar to the wave equation for waves on a string that we found at the beginning of the chapter in Eq. 11.15. In addition, the wave equation in Eq. 11.44 allows us to define the sound wave speed as:

$$v = \sqrt{\frac{B}{\rho}}. \tag{11.45}$$

Comparing this wave speed with the wave speed we found for waves on a string in Eq. 11.16: $v = \sqrt{T/\mu}$ we confirm a general property of wave speeds for any medium: the wave speed increases with the elasticity of the medium (the tension in the string, the Young's modulus, the gas' bulk modulus), and decreases with the weight of the medium displaced.

11.6.2 Example: Loudspeaker Sound Generation and Power Output

Consider a pipe of cross-sectional area A, extending along the positive x-axis. At one end of the pipe ($x = 0$), there is a loudspeaker, also of area A, that oscillates sinusoidally with frequency ω and amplitude C:

$$C \sin \omega t. \tag{11.46}$$

The displacement of air in the tube $x = 0$ in the tube is equal to the displacement of the loudspeaker for all values of t. As a result, there occurs a sinusoidal sound wave in the tube for which the volume displacement is

$$V(x, t) = CA \sin(\omega t - kx) = -CA \sin(kx - \omega t). \tag{11.47}$$

(a) What is the volume flow rate of the gas, $Q(x, t) = \frac{dV(x,t)}{dt}$ at $x = 0$, namely $Q(0, t)$?

Answer:

$$Q(x, t) = \frac{dV(x, t)}{dt} = CA\omega \cos(kx - \omega t). \tag{11.48}$$

Therefore,

$$Q(0, t) = CA\omega \cos(\omega t). \tag{11.49}$$

(b) Calculate the force exerted by the loudspeaker on the air in the tube.

(continued)

Answer: The discussion in Sect.11.6.1 makes
clear that the force exerted by the loudspeaker
on the air in the tube is

$$-BA\frac{du(0,t)}{dx} = -BA\frac{1}{A}\frac{dV(0,t)}{dx} = BAkC\cos\omega t. \tag{11.50}$$

(c) Using your answers to (a) and (b) determine the power, P, supplied by
the loudspeaker.

Answer: The power, P, supplied by the
loudspeaker is the force exerted by the
loudspeaker on the air multiplied by the
velocity of the medium at $x = 0$:

$$P = BAkC\cos\omega t \times C\omega\cos\omega t = \omega kBAC^2\cos^2\omega t$$

$$= \frac{\omega^2 BAC^2}{v}\cos^2\omega t = \omega^2\sqrt{\rho B}AC^2\cos^2\omega t \tag{11.51}$$

Where we have used: the relationship $\omega = vk$ to
re-write $k = \frac{\omega}{v}$; and the expression for the sound
wave speed: $v = \sqrt{\frac{B}{\rho}}$.

(d) What is the average power ($\langle P\rangle$) supplied by the loudspeaker (averaged
over an integer number of periods)?

Answer: Recall that the average over an integer
number of periods of $\sin(\omega t + \phi)^2$ or $\cos(\omega t + \phi)^2$ is
equal to 1/2, i.e., $\langle\sin(\omega t+\phi)^2\rangle = \langle\cos(\omega t+\phi)^2\rangle = \frac{1}{2}$,
where ϕ is any arbitrary time-independent phase.
Therefore,

$$\langle P\rangle = \frac{1}{2}\omega^2\sqrt{\rho B}AC^2. \tag{11.52}$$

(e) Express your answer to (d) in terms of the amplitude of the volume
displacement, V_0.

Answer: We have $V_0 = CA$. Therefore,

$$\langle P\rangle = \frac{1}{2}\omega^2\frac{\sqrt{\rho B}}{A}V_0^2. \tag{11.53}$$

(continued)

Conservation of energy implies that this is the
energy transported away by the sound wave each
second.

11.7 Longitudinal Waves and Transverse Waves: Facts to Get Straight

Here is a quick overview of facts about waves.

- In a transverse wave, the particles of the medium are displaced in a direction perpendicular to the direction of wave propagation.
- The waves on a string under tension are transverse waves.
- In a longitudinal wave, the particles of the medium are displaced in a direction parallel to the direction of wave propagation.
- Sound waves in a fluid are longitudinal waves.
- In solid materials, there are both longitudinal and transverse sound waves.
- Light is a transverse wave. As we will see in *19 Maxwell's equations and then there was light*, light is an electromagnetic wave in which, the electric and magnetic fields both oscillate perpendicularly to the direction of propagation of the wave.
- FM radio waves are transverse. This is because radio waves are electromagnetic waves.
- Ocean surface waves[3] are tricky, because fluid molecules undergo a circular motion in ocean waves, and thus have motion components both parallel and perpendicular to the direction of propagation.

11.8 Standing Waves

So far we have tacitly assumed that there are no boundaries in the region within which our wave propagates. In this section, however, we discuss waves that are confined on a finite length of string with ends at $x = 0$ and $x = L$. Initially, we will suppose that the string is fixed at $x = 0$ and $x = L$ (like a guitar string which is fixed at both sides), so that the boundary conditions on y are: $y(0, t) = 0$ and $y(L, t) = 0$.

We guess that a possible solution of the wave equation on this length of string can be written

$$y(x, t) = A \sin(kx) \sin(\omega t - \phi). \tag{11.54}$$

---

[3] https://www.youtube.com/watch?v=dtVQJCq2cCM

This equation corresponds to a "standing wave",[4] because it does not appear to propagate either to the left or to the right. Initially, it seems that this functional form is not a suitable candidate solution, because it is not obviously of the form $f(x - vt)$ or $b(x + vt)$, that we know solves the wave equation. However, a trigonometric identity implies that

$$y(x, t) = A \sin(kx) \sin(\omega t - \phi) = \frac{A}{2} \left(\cos(kx - \omega t + \phi) - \cos(kx + \omega t - \phi) \right),$$
(11.55)

revealing that this standing wave is actually a superposition of forward and backward travelling waves with equal amplitudes, and therefore is indeed a solution of the wave equation.

Equation 11.54 automatically satisfies the boundary condition $y(0, t) = 0$. In order to satisfy the boundary condition at $x = L$, namely $y(L) = 0$, it then must be that

$$\sin kL = 0. \tag{11.56}$$

Therefore, we require that

$$kL = n\pi, \tag{11.57}$$

or

$$k = \frac{n\pi}{L}, \tag{11.58}$$

where n is an integer. The wavenumber, k, is related to the wavelength, λ, via

$$\lambda = \frac{2\pi}{k}. \tag{11.59}$$

It follows that there is a condition on the wavelength of the standing wave, namely

$$n\frac{\lambda}{2} = L, \tag{11.60}$$

which informs us that, in this case, an integer number of one-half wavelengths must fit onto the string, as shown for the first few modes in Fig. 11.7.

Just as only discrete values of k are allowed, because of the relationship between frequency, wavenumber, and wave speed, similarly only discrete values of the frequency are allowed, namely

$$\omega = \omega_n = kv = \frac{n\pi v}{L}. \tag{11.61}$$

---

[4] http://www.youtube.com/watch?v=no7ZPPqtZEg&feature=endscreen&NR=1

Fig. 11.7 Top: depiction of the first three eigenmodes (standing wave modes) of a string with fixed ends. Middle: depiction of the two eigenmodes of two masses coupled by springs. Bottom: depiction of the three eigenmodes of three masses coupled by springs

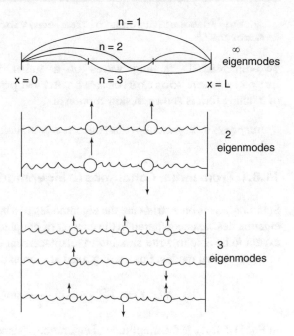

Thus, we see that the frequency (ω) of a standing wave takes one of a set of particular values. In fact, standing waves are the *eigenmodes* for wave motion on the segment of string.

It is important to appreciate that the allowed values of k depend on the details of the boundary conditions. If the boundary conditions are that $y(0, t) = 0$ (as before), but that $\frac{dy(L,t)}{dx} = 0$, instead of $y(L, t) = 0$, it then follows that

$$\cos kL = 0. \tag{11.62}$$

In this case, therefore, the wavenumber, k, is given by

$$k = \frac{(n + \frac{1}{2})\pi}{L}, \tag{11.63}$$

where n is an integer.

It is also important to appreciate that the phenomenon of eigenmodes, a.k.a. standing waves, extends far beyond one-dimensional strings. In two-dimensions, the nodes of the two-dimensional standing waves of a square plate are visualized at

`https://www.youtube.com/watch?v=wvJAgrUBF4w`[5]

and in the Wolfram Demonstration at

---

[5] https://www.youtube.com/watch?v=wvJAgrUBF4w

```
http://demonstrations.wolfram.com/VibrationsOfARectangular
Membrane/.⁶
```

Two-dimensional standing waves of a guitar top plate are visualized at `https://www.youtube.com/watch?v=hpqOT-E7GTE`.[7] An eigenmode of a tuning fork is shown in slow motion at

`https://www.youtube.com/watch?v=pANIvSh2r2A`.[8]

11.8.1 From Initial Conditions to Eigenmode Amplitudes

Standing waves on a string are the eigenmodes of a string segment, analogous to the eigenmodes we encountered in the context of coupled oscillators. Therefore, we can expect to be able to write an arbitrary displacement of the string as a superposition of these normal modes. For a string fixed at both ends, we have

$$y(x, t) = \Sigma_{n=1}^{\infty} A_n \sin \frac{n\pi x}{L} \sin(\omega_n t - \phi). \tag{11.64}$$

In Eq. 11.64, A_n is the amplitude of eigenmode (standing wave) n. The corresponding velocity of each point on the string, moving in the y direction, is

$$\frac{dy(x, t)}{dt} = \Sigma_{n=1}^{\infty} \omega_n A_n \sin \frac{n\pi x}{L} \cos(\omega_n t - \phi). \tag{11.65}$$

In view of what we found in Chap. 10. *Oscillations and Resonance* we should expect the relative amplitudes of different eigenmodes to be determined by the initial conditions. In this case, however, we must specify the initial displacement and velocity all along the string. For example, for a string released from rest with a certain initial displacement, $g(x)$, the initial conditions are

$$y(x, 0) = g(x) \tag{11.66}$$

and

$$\frac{dy(x, 0)}{dt} = 0. \tag{11.67}$$

Equation 11.67 implies that $\cos \phi = 0$, e.g., $\phi = -\frac{\pi}{2}$. Then,

$$y(x, t) = \Sigma_{n=1}^{\infty} A_n \sin \frac{n\pi x}{L} \cos \omega_n t, \tag{11.68}$$

---

[6] http://demonstrations.wolfram.com/VibrationsOfARectangularMembrane/

[7] https://www.youtube.com/watch?v=hpqOT-E7GTE

[8] https://www.youtube.com/watch?v=pANIvSh2r2A

and

$$g(x) = \Sigma_{n=1}^{\infty} A_n \sin \frac{n\pi x}{L}. \tag{11.69}$$

Remarkably, Eq. 11.69 permits us to determine the amplitudes, A_n. To see how, we first consider the integral

$$I = \int_0^L dx \sin \frac{n\pi x}{L} \sin \frac{m\pi x}{L}. \tag{11.70}$$

Using a trigonometric identity, we can re-write I as

$$I = \int_0^L dx \frac{1}{2} \left[\cos \frac{(n-m)\pi x}{L} - \cos \frac{(n+m)\pi x}{L} \right], \tag{11.71}$$

It is then possible to show[9] that

$$I = 0 \tag{11.72}$$

for $m \neq n$ and

$$I = \frac{L}{2} \tag{11.73}$$

for $m = n$.

Motivated by this result, we multiple both sides of Eq. 11.69 by $\sin \frac{m\pi x}{L}$ and integrate from 0 to L with the result that

$$\int_0^L g(x) \sin \frac{m\pi x}{L} dx = \Sigma_{n=1}^{\infty} A_n \int_0^L dx \sin \frac{m\pi x}{L} \sin \frac{n\pi x}{L}. \tag{11.74}$$

Using Eqs. 11.72 and 11.73, the right-hand side of Eq. 11.74 evaluates to

$$\Sigma_{n=1}^{\infty} A_n \int_0^L dx \sin \frac{m\pi x}{L} \sin \frac{n\pi x}{L} = \Sigma_{n=1}^{\infty} A_n I = \frac{L}{2} A_m. \tag{11.75}$$

Thus, we find

$$A_m = \frac{2}{L} \int_0^L g(x) \sin \frac{m\pi x}{L} dx. \tag{11.76}$$

Equation 11.76 demonstrates that we can determine the amplitude of each normal mode (A_m) by carrying out this integral.

---

[9] http://www.wolframalpha.com/input/?i=integrate+cos%28%28n-m%29*x*pi%2FL%29+from+x%3D0+to+x%3DL

11.8.2 Example: Plucking a String

(a) For a string plucked at its center, as shown in Fig. 11.8, for which

$$g(x) = y_0 \frac{2x}{L}, \quad 0 \le x < \frac{L}{2}$$

$$g(x) = y_0 (2 - \frac{2x}{L}), \quad \frac{L}{2} \le x < L, \tag{11.77}$$

calculate the corresponding values of A_n.

Answer: Applying Eq. 11.76, we have

$$A_m = \frac{2y_0}{L} \int_0^{L/2} \frac{2x}{L} \sin \frac{m\pi x}{L} dx + \frac{2y_0}{L} \int_{L/2}^{L} (2 - \frac{2x}{L}) \sin \frac{m\pi x}{L} dx. \tag{11.78}$$

WolframAlpha can evaluate these integrals[10] with the result that

$$A_m = 8y_0 \frac{\sin \frac{m\pi}{2}}{m^2 \pi^2}. \tag{11.79}$$

Thus, $A_1 = \frac{8y_0}{\pi^2}$, $A_3 = -\frac{8y_0}{(3\pi)^2}$, $A_5 = \frac{8y_0}{(5\pi)^2}$, $A_7 = -\frac{8y_0}{(7\pi)^2}$, $A_9 = \frac{8y_0}{(9\pi)^2}$, etc. These amplitudes are illustrated in Fig. 11.9.

---

[10]http://www.wolframalpha.com/input/?i=%28integrate+%282*x%2FL%29*sin%28m*x*pi%2FL%29+from+x%3D0+to+x%3DL%2F2%29%2B%28integrate+%282-2*x%2FL%29*sin%28m*x*pi%2FL%29+from+x%3DL%2F2+to+x%3DL+%29

(continued)

Fig. 11.8 Plucking a string at its center, yielding a triangular displacement

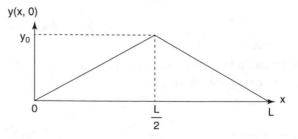

Fig. 11.9 Mode amplitudes versus mode number for standing wave modes on a string, fixed at both ends and plucked at its center

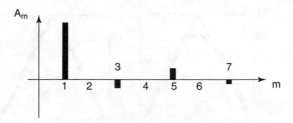

(b) Write the displacement of the string at all times $t > 0$.

Answer:

$$y(x, t) = 8y_0 \sum_{m=1}^{\infty} \frac{\sin \frac{m\pi}{2}}{\pi^2 m^2} \sin \frac{m\pi x}{L} \cos \omega_m t. \tag{11.80}$$

(c) Use WolframAlpha to plot your answer to (b) for $t = 0$ including terms up to $m = 7$.

Answer: Fig. 11.10 shows the sum, including progressively larger numbers of terms. The WolframAlpha output is only physically meaningful for $0 < x < L$.

11.8.3 Standing Sound Waves in a Clarinet

To a good approximation, a clarinet with all stops closed is a tube with one end open and one end closed. For an open-ended tube, the sound wave within the cross-section of the pipe cannot drive the pressure outside the pipe, and therefore the boundary condition at an open end is that the excess pressure of the sound wave is zero. The excess pressure in a sound wave is given by

$$\frac{B}{A} \frac{dV}{dx}, \tag{11.81}$$

where V is the volume displacement, caused by the sound wave. Therefore, the boundary condition for an open tube end is

$$\frac{dV}{dx} = 0. \tag{11.82}$$

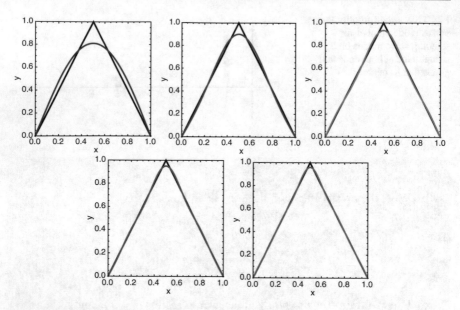

Fig. 11.10 Progressively better approximations to a triangle wave with increasing number of terms: 1 (red), 3 (blue), 5 (orange), 7 (magenta), and 9 (brown)

For a closed tube end, the volume displacement must be zero:

$$V = 0. \tag{11.83}$$

We can ensure that the closed-end boundary condition is satisfied by choosing the coordinate systems so that the closed end is at $x = 0$ and then taking

$$V(x, t) = A \sin(kx)e^{-i\omega t}. \tag{11.84}$$

It follows that

$$\frac{dV}{dx}(x, t) = kA \cos(kx)e^{-i\omega t}. \tag{11.85}$$

Therefore, the boundary condition on an open end at $x = L$ requires

$$\frac{dV}{dx}(L, t) = kA \cos(kL)e^{-i\omega t} = 0, \tag{11.86}$$

i.e.,

$$\cos(kL) = 0. \tag{11.87}$$

Thus, the wavenumber can take discrete values:

$$k = k_n = \frac{\pi}{L}(n - \frac{1}{2}),$$
(11.88)

where $n = 1, 2, 3, 4 \ldots$. Correspondingly, the eigenfrequencies are given by:

$$\omega = \omega_n = ck_n = \frac{\pi c}{L}(n - \frac{1}{2}).$$
(11.89)

11.9 Wave Reflection and Transmission

11.9.1 Reflected and Transmitted Waves

Reflection and transmission of waves at an interface between two different media is an important topic for all types of waves. Reflection of high frequency (1–18 MHz) sound waves is the basis of ultrasound imagining.[11] Later on in this chapter, we will discuss reflection and transmission of sound, but, initially, we focus on reflection and transmission of waves on a string, that for $x < 0$ possesses a mass per unit length of μ_1 (medium 1) and of μ_2 for $x > 0$ (medium 2). Representing waves as complex exponentials greatly facilitates the treatment of the reflection and transmission of waves at interfaces between different media, Of course, the displacements in the actual waves are real numbers. Therefore, when we represent a wave by a complex exponential, what we really mean is that we will represent the wave by the real part of a complex exponential, and that we will take the real part at the end of the calculation.

We envision a steady-state situation (Fig. 11.11), in which in medium 1 there is an incident wave, traveling toward positive x, and a reflected wave, traveling toward negative x, while in medium 2 there is a transmitted wave traveling toward positive x. We represent the incident wave as:

$$e^{i(k_1 x - \omega t)}.$$
(11.90)

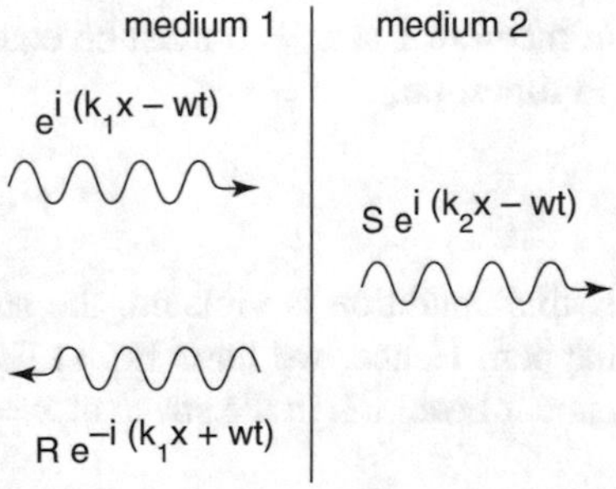

Fig. 11.11 Incident, reflected, and transmitted waves

medium 1 | medium 2

$e^{i(k_1 x - wt)}$

$S e^{i(k_2 x - wt)}$

$R e^{-i(k_1 x + wt)}$

[11] http://en.wikipedia.org/wiki/Medical_ultrasonography

Equation 11.90 is a function of $k_1 x - \omega t = k_1(x - v_1 t)$. Therefore, we can be confident that it corresponds to a forward travelling wave. We chose the incident wave to have an amplitude of 1, because we are interested in the ratio of the transmitted and reflected amplitudes to the incident amplitude. The reflected wave is:

$$Re^{-i(k_1 x + \omega t)}, \tag{11.91}$$

where R is the (amplitude) reflection coefficient, and is one of the goals of our calculations. In general, R is a complex number. Equation 11.91 is a function of $k_1 x + \omega t = k_1(x + v_1 t)$. Therefore, it corresponds to a backwards travelling wave. The time dependence of Eq. 11.91 is the same as that of Eq. 11.90, namely $e^{-i\omega t}$ in both cases. The net displacement in medium 1 is given by the superposition of Eqs. 11.90 and 11.91:

$$e^{i(k_1 x - \omega t)} + Re^{-i(k_1 x + \omega t)}. \tag{11.92}$$

The transmitted wave in medium 2 is

$$Se^{i(k_2 x - \omega t)}, \tag{11.93}$$

where S is the transmission coefficient and is the other quantity, that we seek to calculate. In general, S is complex. The time dependence here is also $e^{-i\omega t}$. The fact that the time dependence is the same for the incident, reflected, and transmitted waves ensures that, when we apply boundary conditions to determine R and S, the results are valid at all times.

11.9.2 Boundary Conditions at the Interface

The next ingredient in quantifying transmission and reflection of waves at a boundary between two different media is the boundary conditions that the wave disturbance must satisfy at that boundary. For waves on a string, composed of regions of different mass per unit length, first, we must have that the displacement in medium 1 at $x = 0$ must be equal to the displacement in medium 2 at $x = 0$ at all times, i.e.,

$$y_1(0, t) = y_2(0, t). \tag{11.94}$$

If this condition is violated, the string would be broken at $x = 0$, which cannot happen. Hence, we have Eq. 11.94. The second boundary condition is that there cannot be a kink in the string at $x = 0$ at any time, i.e.

$$\frac{dy_1(0, t)}{dx} = \frac{dy_2(0, t)}{dx}. \tag{11.95}$$

Physically, we can interpret Eq. 11.95 in light of what we analyzed at the beginning of the chapter when deriving the wave equation on a string (see Eq. 11.9); the net force on an segment of the string is given by the tension pulling from the right side minus the one pulling from the left side:

$$F_{\text{net}} = T_2 \sin\theta_2 - T_1 \sin\theta_1.$$

Under the small angle approximation we can therefore write the net force on the point of the string at $x = 0$ as:

$$F_{\text{net}} = T\frac{dy_2(0, t)}{dx} - T\frac{dy_1(0, t)}{dx}.$$

It needs to be that $\frac{dy_1(0,t)}{dx} = \frac{dy_2(0,t)}{dx}$ (as in Eq. 11.95) otherwise there would be a net non-zero force on zero mass (since the point at $x = 0$ is massless given its zero extension), implying infinite acceleration, which cannot happen.

11.9.3 Calculating Reflection and Transmission Coefficients

Equations 11.94 and 11.95 are the boundary conditions that we need to determine R and S. Equation 11.94 together with Eqs. 11.92 and 11.93 (both evaluated at $x = 0$) imply that

$$e^{-i\omega t} + Re^{-i\omega t} = Se^{-i\omega t}. \tag{11.96}$$

Dividing through by $e^{-i\omega t}$, we find

$$1 + R = S. \tag{11.97}$$

Similarly, Eq. 11.95 together with Eqs. 11.92 and 11.93 imply that

$$ik_1 e^{-i\omega t} - ik_1 Re^{-i\omega t} = ik_2 Se^{-i\omega t}. \tag{11.98}$$

Now, dividing through by $ie^{-i\omega t}$, we find

$$k_1(1 - R) = k_2 S. \tag{11.99}$$

Substituting for S from Eq. 11.97 into Eq. 11.99, we find

$$k_1(1 - R) = k_2(1 + R). \tag{11.100}$$

Solving Eq. 11.100 for R, and then using Eq. 11.97 to solve for S, we find

$$R = \frac{k_1 - k_2}{k_1 + k_2} \tag{11.101}$$

and

$$S = \frac{2k_1}{k_1 + k_2}. \tag{11.102}$$

We may re-write R and S in terms of μ_1 and μ_2 using $k = \frac{\omega}{v} = \omega\sqrt{\frac{\mu}{T}}$ with the result that

$$R = \frac{k_1 - k_2}{k_1 + k_2} = \frac{\sqrt{\mu_1/T} - \sqrt{\mu_2/T}}{\sqrt{\mu_1/T} + \sqrt{\mu_2/T}} \tag{11.103}$$

and

$$S = \frac{2k_1}{k_1 + k_2} = \frac{2\sqrt{\mu_1/T}}{\sqrt{\mu_1/T} + \sqrt{\mu_2/T}}. \tag{11.104}$$

Evidently, R and S are independent of the wave's properties, such as frequency and wavenumber, but depend only on the properties of the strings, namely their masses per unit length. It follows that even though we found R and S for sinusoidal waves, in fact, R and S apply for all waveforms.

11.9.4 Example: Another Boundary Condition, Another Reflected Wave.... or Not

Consider the situation, shown in Fig. 11.12, in which a string under tension is attached at $x = 0$ to a "damper" consisting of a ring, threaded by a fixed rod, that moves in a viscous fluid, giving rise to a frictional force on the ring equal

(continued)

Fig. 11.12 String under tension attached to a massless ring on a rod in a viscous fluid

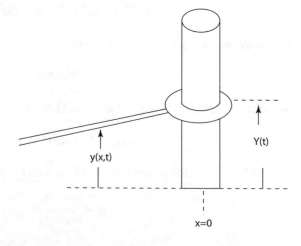

to $-\zeta Y'(t)$, where $Y(t)$ is the displacement of the ring from the x-axis. The string is not subject to any frictional forces.

(a) Assuming complex exponential incident and reflected waves of amplitudes 1 and R, respectively, and that $Y(t) = Y_0 e^{-i\omega t}$, write an equation that quantifies the fact that the ring is always attached to the string and that relates R to Y_0.

Answer:

$$1 + R = Y_0. \tag{11.105}$$

(b) Since the ring is massless the net force on the ring in the y-direction is zero, according to Newton's Second Law. Using this reasoning, find a second independent equation relating R and Y_0.

Answer: There are two forces on the ring along the y-direction: First, the force exerted by the tension in the string. Because the string lies along the negative x-axis and the damper is at x = 0, we see that the force exerted by the string on the damper in this case is given by

$$-T\frac{dy(0,t)}{dx} = -ikT(1-R)e^{-i\omega t}. \tag{11.106}$$

The sign of this force is opposite to the sign that we encountered in Example 11.5.2, because now we are seeking the force exerted by the right-hand end of the string, not the left-hand end of the string. The second force on the damper is the frictional force of the viscous fluid, given by

$$-\zeta Y'(t) = i\omega\zeta Y_0 e^{-i\omega t}. \tag{11.107}$$

These two forces must sum to zero:

$$-ikT(1-R) + i\omega\zeta Y_0 = 0, \tag{11.108}$$

which is the desired second equation relating R and Y_0.

(continued)

(c) Using your answers to (a) and (b), find the reflection coefficient and show that it does not depend on the wave's frequency or wavenumber.

Answer: Combining these two equations, we find

$$-kT(1 - R) + \omega\zeta(1 + R) = 0, \tag{11.109}$$

which may be solved for R with the result that

$$R = \frac{kT - \omega\zeta}{kT + \omega\zeta}. \tag{11.110}$$

Dividing top and bottom by ω and using $\frac{1}{v} = \frac{k}{\omega}$, we may write

$$R = \frac{\frac{T}{v} - \zeta}{\frac{T}{v} + \zeta}. \tag{11.111}$$

We find that, R is independent of the wave's frequency or wavenumber, since v depends on T and μ only.

(d) Explain what happens if $\zeta = \sqrt{\mu T}$.

Answer: Using $v = \sqrt{\frac{T}{\mu}}$, we may re-write R as

$$R = \frac{\sqrt{\mu T} - \zeta}{\sqrt{\mu T} + \zeta}. \tag{11.112}$$

Thus, we see that R is zero, if $\zeta = \sqrt{\mu T}$, i.e., the damper with this particular value of the friction coefficient entirely swallows up the wave without leaving a trace. Yum!

(e) When $\zeta = \sqrt{\mu T}$, what mean power is dissipated in the damper, when the damper is excited by a wave $y(x, t) = C \cos(kx - \omega t)$?

Answer: The power dissipation (PD) in the damper is equal to the product of the y-component of the force on the damper times the y-component of its velocity.

(continued)

$$PD = -T\frac{dy(0, t)}{dx} \times \frac{dy(0, t)}{dt}. \qquad (11.113)$$

To evaluate this expression, we must use the real, physical wave. For an incident wave $y(x, t) = C\cos(kx - \omega t)$, the mean power dissipation (time-averaged over an integer number of periods) is

$$\langle PD \rangle = \left\langle -T\frac{dy(0, t)}{dx} \times \frac{dy(0, t)}{dt} \right\rangle$$

$$= \langle TkC\sin(kx - \omega t) \times \omega C\sin(kx - \omega t) \rangle = \frac{1}{2}\omega^2\sqrt{\mu T}C^2,$$

$$\qquad (11.114)$$

where, in the last step, we used that $k = \frac{\omega}{v}$ and $v = \sqrt{\frac{T}{\mu}}$.

(f) What is the mean power, $\langle P \rangle$, carried by a wave $y(x, t) = C\cos(kx - \omega t)$?

Answer: In (e), we found the power dissipated in the damper, when there is no reflected wave. Conservation of energy then implies that this power was carried to the damper by the wave. But nothing in Eq. 11.114 depends on the damper. Therefore, the mean power transported by a wave under all circumstances is the same expression, namely

$$\langle P \rangle = \frac{1}{2}\omega^2\sqrt{\mu T}C^2. \qquad (11.115)$$

11.9.5 Impedance

Both Examples 11.5.2 and 11.9.4 lead to the same expression for the mean power carried by a wave on a string $y(x, t) = C\cos(kx - \omega t)$, time-averaged over an integer number of periods, namely

$$\langle P \rangle = \frac{1}{2}\omega^2\sqrt{\mu T}C^2. \qquad (11.116)$$

The quantity $\frac{T}{v} = \mu v = \sqrt{\mu T}$ that we encountered in these examples is the so-called impedance, Z, of the string.

For a wave on a string, the impedance is the ratio of the force driving the wave to the resultant velocity of the medium. We can calculate this ratio—i.e., Z— in the case of sinusoidal waves with the result that

$$Z = \frac{-T\frac{dy}{dx}}{\frac{dy}{dt}} = \frac{Tk\cos(kx - \omega t - \phi)}{\omega \cos(kx - \omega t - \phi)} = \frac{T}{v} = \sqrt{\mu T}. \tag{11.117}$$

Z depends only on the properties of the string—its tension and mass per unit length. Although, here, we calculated it in the context of sinusoidal waves, Z is the same for any wave function.

In Example 11.9.4, the quantity, ζ, which is the friction coefficient of the damper, may be identified as the impedance of the damper. Similarly, it depends only on the damper itself. In that example, we discovered that when a string is terminated by a damper with a friction coefficient, ζ, equal to the string's characteristic impedance, Z, then there is no reflected wave.

The time-average power carried by a wave on a string can be expressed in terms of Z as:

$$\langle P \rangle = \frac{1}{2}\omega^2 Z C^2. \tag{11.118}$$

It turns out that, in general, it is possible to determine the mean power carried by a wave, directly from its complex exponential representation. Specifically, for $y(x, t) = Ce^{-i(kx-\omega t)}$, where in general C is complex. the real power carried by this wave is

$$\langle P \rangle = \frac{1}{2}\omega^2 Z C C^*. \tag{11.119}$$

Equation 11.119 is a useful and important result.

Returning to Eqs. 11.103 and 11.104, we may express R and S in terms of the impedances of the two strings, simply by multiplying both the numerator and the denominator by T with the result that

$$R = \frac{Z_1 - Z_2}{Z_1 + Z_2} \tag{11.120}$$

and

$$S = \frac{2Z_1}{Z_1 + Z_2}, \tag{11.121}$$

where $Z_1 = \sqrt{\mu_1 T}$ and $Z_2 = \sqrt{\mu_2 T}$.

11.9.6 Example: Energy Conservation at Wave Reflection

The goal of this example is to investigate how the power carried away from an interface in reflected and transmitted waves is related to the power carried toward the interface by the incident wave.

(a) Calculate the time-averaged power in a wave $e^{i(k_1 x - \omega t)}$ impinging on an interface at $x = 0$ between medium 1 ($x < 0$) and medium 2 ($x > 0$).

Equation 11.119 informs us that the time-averaged incident power is

$$\langle P_i \rangle = \frac{1}{2} \omega^2 Z_1. \tag{11.122}$$

(b) Calculate the time-averaged power in a wave $S e^{i(k_2 x - \omega t)}$ transmitted through an interface at $x = 0$ between medium 1 ($x < 0$) and medium 2 ($x > 0$).

In this case, EQ. 11.119 informs us that the time-averaged transmitted power is

$$\langle P_t \rangle = \frac{1}{2} S S^* \omega^2 Z_2. \tag{11.123}$$

(c) Calculate the time-averaged power in a wave $R e^{-i(k_1 x + \omega t)}$ reflected from an interface at $x = 0$ between medium 1 ($x < 0$) and medium 2 ($x > 0$).

In this case, EQ. 11.119 informs us that the time-averaged reflected power is

$$\langle P_r \rangle = \frac{1}{2} R R^* \omega^2 Z_1. \tag{11.124}$$

(d) Compare the sum of the reflected and transmitted power to the incident power.

We have

$$\langle P_r \rangle + \langle P_t \rangle = \frac{1}{2} \omega^2 (Z_1 R R^* + Z_2 S S^*). \tag{11.125}$$

(continued)

In addition, using Eqs. 11.120 and 11.121, the outbound power becomes

$$\langle P_r \rangle + \langle P_t \rangle = \frac{1}{2}\omega^2 (Z_1 \frac{(Z_1 - Z_2)^2}{(Z_1 + Z_2)^2} + Z_2 \frac{4Z_1^2}{(Z_1 + Z_2)^2})$$

$$= \frac{1}{2}\omega^2 Z_1 (\frac{Z_1^2 - 2Z_1 Z_2 + Z_2^2}{(Z_1 + Z_2)^2} + \frac{4Z_1 Z_2}{(Z_1 + Z_2)^2}) = \frac{1}{2}\omega^2 Z_1,$$

$$(11.126)$$

which equals EQ. 11.122. Thus, we see that the power moving away from the interface equals the power moving toward the interface. This result agrees with our expectation based on the principle of conservation of energy.

11.9.7 Medical Ultrasound Imaging

Comparing Eq. 11.53 to Eq. 11.118 suggests that we should identify $Z = \frac{\sqrt{\rho B}}{A}$ as the impedance of a tube of air. In fact, Eqs. 11.120 and 11.121, which express the reflection and transmission coefficients in terms of impedances, are applicable to waves of all sorts. In particular, they are applicable to sound waves, for which the impedance is $Z = \frac{\sqrt{\rho B}}{A}$, where ρ is the mass density of the medium, $B = -V\frac{dP}{dV}$ is its bulk modulus,[12] and A is the cross-sectional area of the medium in which the sound wave is propagating. For sound waves, impedance is the ratio of excess pressure to volume flow rate:

$$Z = \frac{-\frac{B}{A}\frac{dV}{dx}}{\frac{dV}{dt}} = \frac{B}{Av} = \frac{\sqrt{\rho B}}{A}. \qquad (11.127)$$

Again, the impedance depends only on the properties of the medium in which the wave is travelling. Previously, when we encountered the ratio of a pressure to volume flow rate, it was in Chap. 9. *Fluid Mechanics: Laminar Flow, Blushing, and Murray's Law*, and we called it fluid resistance. Acoustic impedance and fluid resistance have the same units.

The boundary conditions on a sound wave at an interface between medium 1 and medium 2 at $x = 0$ are that the volume displacements are the same in both media,

$$V_1(0, t) = V_2(0, t), \qquad (11.128)$$

---

[12] http://en.wikipedia.org/wiki/Bulk_modulus

and that the pressure is the same in both media,

$$-\frac{B_1}{A_1}\frac{dV_1(0,t)}{dx} = -\frac{B_2}{A_2}\frac{dV_2(0,t)}{dx}. \qquad (11.129)$$

Medical ultrasound images, such as the image of a fetus at the start of this chapter (see Fig. 11.1), depend directly on the reflection of sound pulses at interfaces in the body between regions of different acoustic impedance. There are several different styles of ultrasound probe. The particular probe used to acquire the image of a fetus at the start of the chapter incorporates an array of $\sim$100 piezo-electric transducers, arranged around a circular arc. Thus, each transducer corresponds to a particular azimuthal angle. The location of this arc of transducers corresponds to the upper boundary of the image. Each transducer sequentially generates a sound pulse propagating radially outwards from the arc. The size of the pulse, transverse to its propagation direction, is set by the size of the transducer. When such a pulse encounters an interface between regions with different acoustic impedances, a reflected pulse is generated. That reflected pulse then travels back to the transducer, where it is converted into an electrical signal and detected. The separation in time between when the transducer emits the pulse and when it detects the reflected pulse, t_d-t_e, establishes that the location of the interface giving rise to that reflection is at a depth $\frac{t_d-t_e}{v}$ within the body. In addition to the reflected wave from this first interface, there is a transmitted wave, which propagates beyond the first interface, and which is reflected by any second, deeper interface. The time at which that second reflected wave returns to the transducer establishes the depth of that interface. And so on for additional interfaces. The two-dimensional ultrasound image that we see at the start of the chapter, is constructed by plotting reflected intensity for each transducer, corresponding to azimuthal angle in the image, versus delay time, corresponding to (the change) in radial coordinate in the image.

Table 11.1 Acoustic impedance and sound velocity for several tissues

Tissue	Acoustic impedance per unit area (g $\cdot$ cm^{-2} $\cdot$ s $\times$ 10^{-5})	Velocity (m $\cdot$s^{-1})
Air	0.0004	331
Fat	1.38	1450
Water	1.54	1540
Liver	1.62	1549
Blood	1.61	1570
Muscle	1.70	1585
Bone	7.80	4080

11.10 Clarinets, Flutes, and Ears: Resonance Revisited

One of the most important ways that we communicate with each other is via sound waves, which we detect by hearing and which we generate by speaking or singing or playing a musical instrument. A number of wind instruments, such as the flute and the clarinet, can be conceived as a finite-length tube, with a length much larger than its diameter, within which sound waves propagate effectively as one-dimensional waves. This model is also an approximation to the human ear canal.

At the bottom of Fig. 11.13 are shown the pressure spectra produced by a flute and by a clarinet. These spectra correspond to the pressure within a sound wave output from the clarinet or flute, when subject to a constant-amplitude drive of varying frequency. Evidently, both spectra are comprised of a set of narrow peaks. Each such peak recalls a resonance of the sort we encountered in Chap. 10. *Oscillations and Resonance*, where we found a resonance, when the

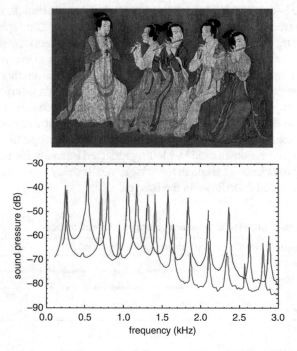

Fig. 11.13 Top: women playing flutes, a detail from a twelfth-century remake of the tenth-century *Night Revels of Han Xizai*, by Gu Hongzhong. Bottom: experimental sound pressure spectra [1] of a clarinet, shown in blue, adapted from http://www.phys.unsw.edu.au/music/clarinet/C4.html, and a flute, shown in red, adapted from https://newt.phys.unsw.edu.au/music/flute/modernC/C4. html) The sound pressure is specified in decibels (dB), which is a logarithmic unit, such that 30 dB corresponds to a factor of 1000, 20 dB corresponds to a factor of 100, 10 dB corresponds to a factor of 10, −10 dB corresponds to a factor of 0.1, −20 dB corresponds to a factor of 0.01, −30 dB corresponds to a factor of 0.001, etc.

frequency of a sinusoidal driving force is equal to an eigenfrequency of the system. (Properly stated, resonance occurs when when the frequency of a sinusoidal driving force is equal to the real part of an eigenvalue of the system.) There are strong resonances at the first few odd multiples of the lowest-frequency resonance (i.e., the fundamental). These measured clarinet resonance frequencies may be compared to the standing wave frequencies given by Eq. 11.89, which predicts standing waves at odd multiples of the fundamental. Nevertheless, there are also noticeable weak peaks at the first few even multiples of the fundamental, and at larger frequencies, the resonances at even and odd multiples of the fundamental are comparable.

For the clarinet and the flute, the eigenmodes are standing sound waves within the instrument, and the eigenfrequencies are the frequencies of those standing sound waves. The existence of standing waves within these musical instruments originates in the large impedance mismatch between inside the tube of the instrument and outside. We saw in Sect. 11.9 that the amplitude of a wave reflected at the boundary between media of different impedances is proportional to the difference in impedance between medium 1 and medium 2, $R = \frac{Z_1 - Z_2}{Z_1 + Z_2}$. In the case of sound waves emerging from an open end of a tube of cross-sectional area A, $Z_1 = \frac{\rho v}{A}$ and $Z_2 = \frac{\pi \rho v}{\lambda^2}$, where λ is the wavelength of the sound wave. For long, thin musical instruments, the length of the tube determines the sound wavelength, λ, to be very much larger than the tube diameter. It follows that $Z_1 \gg Z_2$, so that $R \simeq 1$, the wave is almost fully reflected. Therefore, sound in a long, narrow tube comprises a forward travelling wave and a backward travelling wave of almost equal amplitude (because $R \simeq 1$). The superposition of these two waves is a (near) standing wave. For the clarinet, the two lowest-frequency strong peaks correspond to the frequencies of standing waves with wavelengths equal to 4 and $\frac{4}{3}$ times, respectively, of the length of the clarinet, as expected for a tube open at one end and closed at the other. For the flute, the two lowest-frequency strong peaks correspond to the frequencies of standing waves with wavelengths equal to 2 and 1 times, respectively, of the length of the flute, as expected for a tube open at both ends.

We also saw in Chap. 10. *Oscillations and Resonance*, that the width of a resonance depends on the friction coefficient of the mass, which is how energy is dissipated in that case. For sound waves in a wind instrument, the wave that is transmitted out of the tube ends represents an important route for energy dissipation. The mean power carried away by this wave is $\langle P \rangle = \frac{1}{2} \omega^2 Z_2 C C^*$. (Eq. 11.119.) Since Z_2 is small, the power dissipation is correspondingly small, compared to that of the reflected wave, which remains inside the clarinet. It follows that the resonance is correspondingly narrow.

As suggest by the left of Fig. 11.14, the human ear canal approximates a tube that extends from the external ear—the auricle—to the eardrum. Unsurprisingly, therefore, similar sound wave physics is an important contributor to the sensitivity of human hearing. Like a clarinet, the ear canal is open at the external ear and closed at the eardrum, while its length is about $L = 2.5$ cm. Therefore, we expect that the lowest-frequency standing wave in the ear canal has a wavelength of about

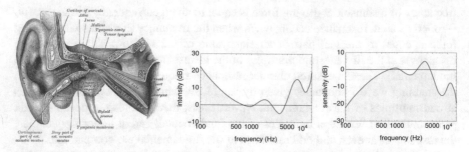

Fig. 11.14 Left: anatomy of the outer ear from Gray's Anatomy, showing the ear canal, a.k.a. the external acoustic meatus, public domain image from https://commons.wikimedia.org/wiki/Gray %27s_Anatomy_plates#/media/File:Gray907.png Center: minimum audible sound intensity versus frequency. Right: sensitivity, defined as the inverse of the minimum audible sound intensity, of human hearing versus frequency. The peak in this plot at about 3500 Hz closely corresponds to the expected frequency of the first harmonic standing wave in the ear canal. The secondary peak in this plot at about 11,000 Hz closely corresponds to the expected frequency of the third harmonic sound standing wave. ear

$\lambda = 4L = 10$ cm and a corresponding frequency $\nu = \frac{v}{\lambda} = \frac{340}{0.1} = 3400$ Hz, while the next lowest frequency would be at three times this frequency, namely at 11,200 Hz.

Shown in center of Fig. 11.14 is a plot versus frequency of the minimum sound wave intensity that is audible to the human ear. The sound intensity is measured in decibels, this is a logarithmic unit that uses the threshold for human hearing as a reference intensity: $I_0 = 1 \times 10^{-12} \frac{\text{W}}{\text{m}^2}$. Then an intensity I measured in decibel is: $\beta = (10 \text{ dB}) \log_{10} \left(\frac{I}{I_0} \right)$. Therefore $\beta = 0$ dB means $I = I_0$, $\beta = 10$ dB means $I = 10 I_0$, $\beta = 20$ dB means $I = 100 I_0$, etc. This plot shows a deep minimum at 3500 Hz informing us that human hearing is most sensitive at 3500 Hz. In fact, the minimum audible intensity is inversely proportional to the ear's sensitivity, which is plotted versus frequency in the right side of Fig. 11.14. It is apparent from this plot that the ear's sensitivity shows peaks at 3500 Hz and 11000 Hz, closely corresponding to the expected frequencies of the first and third harmonic standing wave in the ear canal. Our hearing shows a peak in sensitivity to sound at these frequencies, because the transmission of sound energy through the ear canal shows a maximum at its resonant frequencies, just like wind instruments. The peaks in the frequency dependence of our ears' sensitivity to sound are relatively much broader than for wind instruments. Physically, we can infer that this is because the impedance mismatch at the eardrum is not too large. Biologically, we can infer that this is because we need to hear a broad range of frequencies. Beyond the ear canal, the sensitivity of our hearing also depends upon further impedance matches as the sound energy travels into the inner ear and is transduced into electrical signals in the cochlea. But this is beyond our scope.

11.11 Chapter Outlook

This chapter is an essential stepping stone *en route* to the intellectual high point, presented in Chap. 19. *Maxwell's Equations and Then There Was Light*, of showing that Maxwell's equations of electromagnetism necessarily imply the existence of electromagnetic (EM) waves, including light and x-rays. This chapter is also prerequisite for Chap. 15. *Optics: Refraction, Eyes, Lenses, Microscopes, and Telescopes*, where we will also discuss how refraction of light waves transmitted through an interface between different media can be exploited to form images.

Although we focused on wind instruments in the latter part of this chapter, the general idea that extended objects (and not only the point masses on springs of Chap. 10. *Oscillations and Resonance*) exhibit a resonance at each frequency corresponding to a natural oscillation frequency of the object is very broadly applicable. A classic example of this phenomenon is when a wine glass is driven to its own destruction by a driving sound wave with a frequency equal to one of the wine glass's resonant frequencies, as shown in the movie: `https://www.youtube.com/watch?v=BE827gwnnk4`.[13]

11.12 Problems

Problem 1: Snapshot
The graph below shows a snapshot at $t = 0$ of a wave pulse traveling on a string whose shape and motion are described by the function $f(x + vt)$.

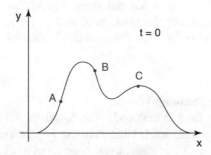

(a) For particles A, B, and C on the string draw an arrow representing the particle's velocity at t=0. Or, if the particle is stationary say so. Explain in a few sentences how you know what the velocity is.
(b) Draw a graph of y vs t for points A, B, and C.
(c) Repeat part (b) if the function was instead $f(x - vt)$.

---

[13] https://www.youtube.com/watch?v=BE827gwnnk4

Problem 2: Vibrating String and Vibrating Membrane Wolfram Demonstrations

Run the mathematica Demonstration at

tp://demonstrations.wolfram.com/TheVibratingString/.[14]

Turn on "display modes" and run for "piecewise linear" and you will see the initial conditions that we considered in the chapter and the time evolution that we calculated. Remarkably, the time-dependent function, which we calculated in the chapter as an infinite sum of trigonometric functions, is the same as the function, built out of time-dependent straight lines, that you see in the Wolfram Demonstration.

Problem 3: Adding Traveling Waves with Phases

(a) Using Euler's theorem, show that $\cos(\alpha \pm \beta) = \cos(\alpha)\cos(\beta) \mp \sin(\alpha)\sin(\beta)$ and $\sin(\alpha \pm \beta) = \sin(\alpha)\cos(\beta) \pm \sin(\beta)\cos(\alpha)$.

(b) Two waves with the same frequency, wavelength, and amplitude are traveling along the x-axis. The two waves can be described by the functions $y_1(x, t) = A\sin(kx - \omega t - \frac{\pi}{6})$ and $y_2(x, t) = A\sin(kx - \omega t + \frac{\pi}{6})$. Draw a graph of $y_1(x, t)$ and $y_2(x, t)$ when $t = 0$. Make sure to accurately indicate on the graph what the value of y is when $x = 0$.

(c) Let $Y(x, t) = y_1(x, t) + y_2(x, t)$. Draw a graph of $Y(x, 0)$ by adding the graphs you drew in part (b).

(d) The sum of the two waves has the form $Y(x, t) = B\sin(kx - \omega t + \phi)$. Determine mathematically the value of B and ϕ. (Hint: use your answers to part a) identifying $\alpha = kx - \omega t$ and $\beta = \pi/6$) Does this match the graph you drew in part c)?

Problem 4: Graphing Standing Waves

A string of length L is fixed at both ends. The length of the string is along the x-direction (i.e., the horizontal) and it vibrates in the y-direction (i.e., the vertical). It vibrates in the third, $n = 3$ harmonic. Draw a graph of $y(x)$ at the instant of time when:

1. the vertical velocity of the string is zero,
2. the vertical velocity of the string is maximum.

For each of your graph: (a) explain why your graph has the shape it does, (b) indicate the regions on the graph where the vertical acceleration is positive, negative, and zero.

---

[14] http://demonstrations.wolfram.com/TheVibratingString/

Problem 5: Plucking a Guitar String

Consider a 1.0 m guitar string. At what distances from the end should you pluck the string if you want to enhance the $n = 4$ mode? Is $n = 4$ the only mode that is excited?

Problem 6: Harmonious Conditions

Consider the case as in the figure below of a string of length L that is free to move up and down along a rod on the left end (at $x = 0$), and fixed at the right end (at $x = L$). The goal of this problem is to write an expression for the second harmonic that can be excited on this string.

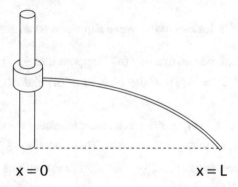

$$x = 0 \qquad\qquad\qquad x = L$$

The boundary conditions for this string are:

$$\frac{dy(0, t)}{dx} = 0,$$

and

$$y(L, t) = 0.$$

(a) Consider the function

$$y(x, t) = A \cos(kx) \cos(\omega t + \phi).$$

By imposing the boundary conditions on this function, find the allowed values of k.

(b) Suppose that the initial condition for this wave is that:

$$y(x, 0) = 0.$$

Calculate ϕ.

(c) Assume the string has tension T and mass density μ. What is the angular frequency ω associated with the second lowest possible value of the wavenumber k in terms of the given quantities.

(d) Write an expression for a standing wave on the string that satisfies the boundary and initial conditions, and that has the second lowest possible value of the wavenumber k.

(e) What is the corresponding wavelength λ for the second harmonic?

(f) For such standing wave, write an expression for the instantaneous velocity of each point on the string (i.e., for each x) for each time.

(g) Draw a graph of $y(x)$ for the second harmonic at the instant of time when the vertical velocity of the string is zero.

Problem 7: General Solutions of the Wave Equation as an Exercise in Using the Chain Rule

Consider a function of the variable u, $f(u)$. Suppose that $\frac{df}{du} = f'(u) = g(u)$, and $\frac{d^2 f}{du^2} = f''(u) = g'(u) = h(u)$, where $g(u) = f'(u)$ and $h(u) = g'(u)$ are also functions of u.

(a) Suppose that $u = x - vt$, where v is a constant. Using the chain rule, calculate $\frac{df}{dt}$ in terms of g, assuming x is constant. Then, calculate $\frac{d^2 f}{dt^2}$ in terms of h.

(b) Repeat (a) but now to calculate $\frac{d^2 f}{dx^2}$, assuming t is constant.

(c) Now calculate $\frac{d^2 f}{dx^2} - \frac{1}{v^2} \frac{d^2 f}{dt^2}$.

(d) What happens when you repeat (a)-(c) for $b(w)$ instead of $f(u)$, where $w = x + vt$, instead of $u = x - vt$?

(e) Is $y(x, t) = f(x - vt) + b(x + vt)$ a solution of the wave equation?

Problem 8: Proof of Superposition Principle for the Wave Equation

A function $y_1(x, t)$ is a solution of the wave equation, that is equivalent to say that it satisfies:

$$\frac{d^2 y_1}{dx^2} - \frac{1}{v^2} \frac{d^2 y_1}{dt^2} = 0.$$

Another function $y_2(x, t)$ is also a solution of the wave equation. Prove that the function $y_3(x, t) = y_1(x, t) + y_2(x, t)$ is also a solution of the wave equation.

Problem 9: Nodes and Antinodes

Consider the 4th harmonic (standing wave with $n = 4$) on a string of length L with fixed ends, mass density μ and tension T.

(a) what is the wavenumber (k_4)? what is the angular frequency (ω_4)? what is the wavelength (λ_4)?

(b) Write the function $y(x, t)$ describing this harmonic assuming at time $t = 0$ the amplitude is zero everywhere, and that the maximum amplitude of the wave is C.

(c) On a standing wave, the nodes are the points that are not moving, and the antinodes the ones that move with the biggest amplitude. How many nodes and antinodes are on the 4th harmonic? Count them and make a graph of the function clearly showing where all the nodes and antinodes are located.

(d) Write an expression for the instantaneous velocity of each point of the string (i.e., for each x) as it moves up and down.

(e) Using your answer to part (d) write the *maximum* velocity v_{max} for each point along the x axis as it moves up and down.

(f) Does your answer to part (e) predict the position of the nodes and antinodes? Explain.

Problem 10: Standing Waves Are Solutions of the Wave Equation
Prove that a standing wave is a solution of the wave equation.

Problem 11: String with One Open End
Consider the case as in the figure below of a string of length L that is fixed at one end, and free to move up and down along a rod on the other end. The boundary conditions in this case are:

$$y(0, t) = 0, \qquad \frac{dy(L, t)}{dx} = 0. \qquad\qquad (11.130)$$

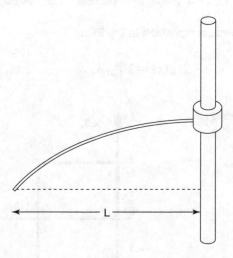

(a) What are the wavenumbers k_n, wavelengths λ_n, and frequencies ω_n allowed in this case? Show your work and the steps to get to your answers.

(b) Sketch the first three harmonics (eigenmodes) on the string, i.e., the standing waves for the three lowest values of k_n. Locate the nodes and antinodes on each of your graph.

Problem 12: Fourier Decomposition of a Square Wave

In Fig. 11.15 is depicted a square standing wave on a string of length L at time $t = 0$. The amplitude of the wave is equal to 1 for half of the length of the string, and -1 for the second half. The goal of this problem is to find the Fourier components of this square wave. This is similar to what is done in Example 11.5.2 in the chapter for a triangular wave. We recommend you first review carefully that example.

(a) Write a function $g(x) = y(x, 0)$ that describes the square wave at $t = 0$ shown in Fig. 11.15 for any x between 0 and L.

(b) Use WolframAlpha to calculate the coefficients A_m as defined in Eq. 11.76:

$$A_m = \frac{2}{L} \int_0^L g(x) \sin \frac{m \pi x}{L} dx, \qquad (11.131)$$

where you will substitute your $g(x)$ from part a). The command to type will look like:

```
integrate (2/L)*g(x)*sin(m*x*pi/L) from x=0 to x=L
```

and you will need to evaluate this for the first six Fourier amplitudes that means for $m = 1, 2, 3, 4, 5, 6$.

(c) Once you have all the first six Fourier amplitudes (i.e., A_m for $m = 1, 2, 3, 4, 5, 6$), make a graph to represent them, analogous to the one in Fig. 11.9.

(d) Using the Fourier decomposition in Eq. 11.69

$$g(x) = \Sigma_{n=1}^{\infty} A_n \sin \frac{n \pi x}{L}, \qquad (11.132)$$

Fig. 11.15 Square standing wave on a string of length L

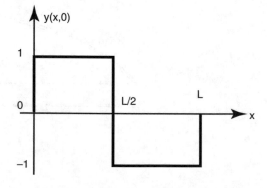

write your $g(x)$ as a sum up to $n = 6$ of sinusoidal functions with the coefficients you found in part (b).

(e) Using WolframAlpha plot your $g(x)$ from part (d) assuming $L = 1$.

(e) How does your function in part (e) compare to the square wave in Fig. 11.15? Are the graphs the same or different? Explain the reason of any discrepancy.

Problem 13: Eigenmodes for Various Boundary Conditions

(a) Find the functions that describe standing waves inside a pipe of length L with both ends open (at $x = 0$ and $x = L$), including any conditions on the wavenumber.

(b) Find the functions that describe standing waves inside a pipe of length L with the end at $x = 0$ closed and the end at $x = L$ open, including any conditions on the wavenumber.

Problem 14: A Handy Shortcut

Consider a sound wave $u(x, t) = \alpha e^{i(kx - \omega t)}$, where $\alpha = A e^{-i\phi}$ is complex, where A and ϕ are real.

(a) Calculate the real wave.

(b) Calculate $\langle (Re[u])^2 \rangle$. Where $\langle \rangle$ denotes the average of an integer number of periods. (Recall that the average over an integer number of periods of $\sin(\omega t + \phi)^2$ or $\cos(\omega t + \phi)^2$ is equal to $1/2$, i.e., $\langle \sin(\omega t + \phi)^2 \rangle = \langle \cos(\omega t + \phi)^2 \rangle = \frac{1}{2}$, where ϕ is any arbitrary time-independent phase).

(c) Calculate $\frac{1}{2} u u^*$ and comment on your answers to (b) and (c).

Problem 15: Sound Wave Energy

In the chapter, we showed that the power carried by a sound wave of frequency ω and volume displacement V_0 propagating in a tube of air of acoustic impedance Z is equal to $\langle P \rangle = \frac{1}{2} \omega^2 Z V_0^2$. Determine the dimensions of the quantities Z and $\frac{1}{2} \omega^2 Z V_0^2$ and comment.

Problem 16: Ultrasound Gel

(a) The speed of sound in tissues is about $v = 1500 \text{ ms}^{-1}$. What is the wavelength of $f = 7$ MHz ultrasound?

(b) Explain why a gel is used between the ultrasound transducer and the patient. Suppose that you are a physician working in a situation, where commercial ultrasound gels are not available. Discuss the criteria required for a suitable ultrasound gel and whether it would be possible, sensible, and safe for your patients to prepare a local substitute.

Problem 17: Violins

Discuss what concepts from this and the previous chapter are relevant to the violin and the guitar and how violins and guitars make music.

Problem 18: Reflection and Transmission of Wave Pulses

(a) Run the Wolfram Demonstration at

   ReflectionAndTransmissionOfWaves-Ch11.cdf

and hit the "run" button to see a simulation of a wave pulse incident on a boundary between two different media.

(b) Comment on the amplitude of the reflected wave as a function of the impedance ratio.

(c) Comment on the length of the transmitted pulse as a function of the velocity ratio.

Problem 19: Reflection and Transmission of Sound Waves

The excess pressure (ΔP) and volume displacement (V) in a sound wave are related via:

$$\Delta P = \frac{B}{A} \frac{dV}{dx}, \tag{11.133}$$

where B is the bulk modulus of the medium in which the sound propagates, and A is its cross-sectional area. The boundary conditions for a sound wave at an interface at $x = 0$ between two different media are that the volume displacement must be the same across the interface

$$V_1(0, t) = V_2(0, t), \tag{11.134}$$

and the excess pressure must be the same across the interface

$$\frac{B_1}{A_1} \frac{dV_1(0, t)}{dx} = \frac{B_2}{A_2} \frac{dV_2(0, t)}{dx}, \tag{11.135}$$

where A_1 and A_2 are the areas of medium 1 and medium 2, respectively, and B_1 and B_2 are the bulk moduli of medium 1 and medium 2, respectively.

Consider a sound wave, travelling toward larger values of x, that is incident on a boundary at $x = 0$ between medium 1 (for $x < 0$) and medium 2 (for $x > 0$).

(a) Impose the appropriate boundary conditions at $x = 0$ to find two equations relating the reflection, R, and transmission, S, coefficients for sound waves at this interface.

(b) Introduce the sound wave impedance, $Z = \frac{B}{vA} = \frac{\sqrt{\rho B}}{A}$, to solve your equations in (a) in terms of Z_1 and Z_2.

(c) How do your results for reflection and transmission of sound compare to the reflection and transmission coefficients for waves on strings.

Problem 20: Sound Through a Wind Instrument

In the chapter, we examined wave reflection and transmission at a single interface between two media with different impedances. We also showed the pressure spectrum of a flute and a clarinet, and presented an accompanying conceptual discussion. In this problem, we carry out a detailed calculation of the sound wave transmission through a tube of length, L, representing a simple model of a wind instrument. The situation we analyze is shown in Fig. 11.16 with a finite-length tube (region 2), representing the musical instrument, sandwiched between two semi-infinite regions (regions 1 and 3). As indicated in the figure, when a sound wave is incident into the tube of interest from the left, there are incident ($e^{i(kx-\omega t)}$) and reflected waves ($Re^{-i(kx+\omega t)}$) in medium 1, there are also transmitted ($se^{i(kx-\omega t)}$) and reflected ($re^{-i(kx+\omega t)}$) waves inside the cavity (region 2), and there is a transmitted wave in region 3 ($Se^{i(kx-\omega t)}$). For these sound waves in air, the wavenumber is the same in regions 1, 2, and 3. The boundary between region 1 and region 2 is at $x = 0$. The boundary between region 2 and region 3 is at $x = L$.

(a) Use Eq. 11.120 to write an expression for $\frac{r}{s}$ in terms of R_{23} and e^{ikL}, where R_{23} is the reflection coefficient going from region 2 to region 3.

(b) The boundary conditions for a sound wave at an interface at $x = 0$ between two media are that the volume displacement must be the same across the interface

$$V_1(0, t) = V_2(0, t), \tag{11.136}$$

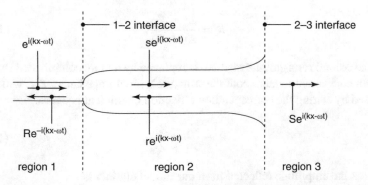

Fig. 11.16 The relevant waves in a clarinet with all holes closed. Region 1 corresponds to $x < 0$. The 1-2 interface is at $x = 0$. The instrument, which corresponds to region 2, has a length L, so that the 2-3 interface is at $x = L$, and region 3 corresponds to $x > L$

and the excess pressure must be the same across the interface

$$\frac{B}{A_1}\frac{dV_1(0,t)}{dx} = \frac{B}{A_2}\frac{dV_2(0,t)}{dx},\tag{11.137}$$

where A_1 and A_2 are the areas of medium 1 and medium 2, respectively, and B is the bulk modulus of air. Impose these boundary conditions at $x = 0$ to find an equation relating r and s to given quantities only.

(c) Combine your answers to (a) and (b) to find r and s.

(d) The volume displacement must be equal on both sides of the 2-3 interface, which relates r and s to S that

$$Se^{ikL} = se^{ikL} + re^{-ikL}.\tag{11.138}$$

Use this equation and your answers in (c) to determine the transmitted wave amplitude, S.

(e) Comment on your answer to (d) for $Z_1 = Z_2$, and, alternatively, for $Z_2 = Z_3$.

(f) For the physicist's clarinet in Fig. 11.16, the diameter of region 1 is much smaller than the diameter of the clarinet (region 2). It follows that $Z_1 \gg Z_2$, so that $R_{12} \simeq 1$. In addition, because the diameter of the clarinet is much less than the wavelength of sound, we also have that $Z_2 \gg Z_3$, so that $R_{23} \simeq 1$. What then are the values of k that permit a large transmitted wave amplitude?

(g) Equation 11.119 informs us that the the fraction of the incident power transmitted through the physicist's clarinet is given by $\frac{\langle P_3 \rangle}{\langle P_1 \rangle} = \frac{Z_3}{Z_1}SS^*$. Use WolframAlpha to plot $\frac{\langle P_3 \rangle}{\langle P_1 \rangle}$ versus $x = kL = \omega L/v$, where v is the sound velocity and ω is the sound frequency, for $R_{12} = 0.95$, and $R_{23} = \frac{1-(\frac{kL}{100})^2}{1+(\frac{kL}{100})^2}$.

Problem 21: Input Impedance of a Clarinet

In the chapter, we saw that the reflection coefficient when a wave goes from a semi-infinite medium of impedance Z_1 to a semi-infinite medium of impedance Z_2 is

$$R = \frac{Z_1 - Z_2}{Z_1 + Z_2}.\tag{11.139}$$

When the second semi-infinite medium is replaced by a more complicated structure, it is often convenient to talk about the structure's input impedance, Z_{in}, which may be defined by setting the corresponding reflection coefficient equal to:

$$R = \frac{Z_1 - Z_{in}}{Z_1 + Z_{in}}.\tag{11.140}$$

Given that the amplitude reflected from our model clarinet is

$$R = \frac{R_{12} + R_{23}e^{2ik_2L}}{1 + R_{12}R_{23}e^{2ik_2L}}, \qquad (11.141)$$

calculate the input impedance of our model clarinet in terms of Z_2, R_{23}, k_2, and L.

Problem 22: Upon Further Reflection
A harmonic wave with wave number k and frequency ω travels along a string in the $+x$-direction approaching the origin, $x = 0$, from the left.

The wave has the form

$$y_i(x, t) = A \cos(kx - \omega t).$$

(a) The end of the string is fixed at the point $x = 0$. This causes a reflected wave which travels in the $-x$-direction. Write a function $y_r(x, t)$ that represents the reflected wave.
(b) Write an equation $Y(x, t)$ that is the superposition of the incident wave and the reflected wave.
(c) The string is fixed at $x = 0$. What is the boundary condition this implies for the function $Y(x, t)$ for any time t?
(d) Use this boundary condition to determine how the amplitudes A and B are related.
(e) The function $Y(x, t)$ has the form

$$Y(x, t) = C \sin(\alpha) \sin(\beta)$$

or

$$Y(x, t) = C \cos(\alpha) \cos(\beta).$$

Which form is it and what are the values of C, α, and β? (Hint: You may need trigonometric identities to write $Y(x, t)$ in this form.)
(f) Is the resulting wave $Y(x, t)$ a wave traveling toward $+x$, a wave traveling toward $-x$ or a standing wave?

Problem 23: Wave Meets Oscillator
Consider the situation, shown in Fig. 11.17, in which a string under tension, T, is attached at $x = 0$ to an oscillator consisting of a mass m, attached to a spring of spring constant κ. The spring gives rise to a restoring force on the mass equal to $-\kappa Y(t)$, where $Y(t)$ is the displacement of the mass from the x-axis. Neither the string nor the mass is subject to any frictional forces.

Fig. 11.17 Snapshot of a string under tension attached to a frictionless oscillator

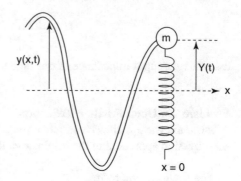

(a) Assuming complex exponential incident and reflected waves, $e^{i(kx-\omega t)}$ and $Re^{-i(kx+\omega t)}$, respectively, and that $Y(t) = Y_0 e^{-i\omega t}$, and that a steady state has been reached, write an equation that quantifies the fact that the mass is always attached to the string and that relates the reflection amplitude, R, to the mass' oscillation amplitude, Y_0.

(b) Use Newton's Second Law to write a second independent equation relating R and Y_0.

(c) Using your answers to (a) and (b), determine the reflection coefficient, R. Express your result for R in the form $R = \frac{A+iB}{C+iD}$ and state your values of A, B, C, and D, where each of A, B, C, and D is a real number, which may, but need not, depend on ω.

(d) What is the oscillation amplitude, Y_0, when the frequency of the wave that drives the oscillator, namely ω, is equal to the resonant frequency of the oscillator, namely $\sqrt{\frac{\kappa}{m}}$? What is the corresponding reflection amplitude, R? That is, what are Y_0 and R when $\omega = \sqrt{\frac{\kappa}{m}}$? Describe how energy flows in this situation.

(e) Express R in polar form, i.e., as the product of an amplitude, $|R|$ and a phase factor, $e^{i\phi}$, i.e., $R = |R|e^{i\phi}$. To this end, you may find it helpful to sketch each of $A + iB$ and $C + iD$ in the complex plane. Then, the angle between $A + iB$ and $C + iD$ in the complex plane is ϕ, while the ratio of their lengths is $|R|$. Describe how energy flows in this general situation.

Reference

1. P. Dickens, R. France, J. Smith, J. Wolfe, Clarinet acoustics: introducing a compendium of impedance and sound spectra. Acoust. Australia **35**, 1–17 (2007)

Gauss's Law: Charges and Electric Fields

12

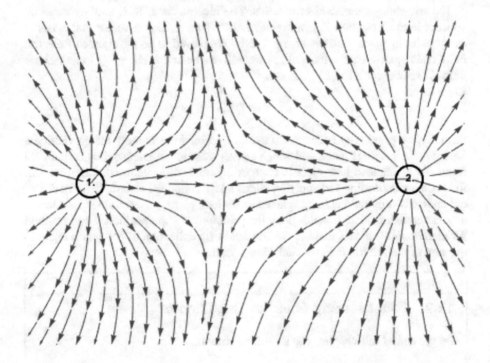

Electronic Supplementary Material The online version of this article (https://doi.org/10.1007/978-3-031-05808-0_12) contains supplementary material, which is available to authorized users.

12.1 Introduction

Our study of electromagnetic phenomena starts in this chapter with electrostatics and Gauss's law. The electric charge, Q, is an essential concept in electrostatics. There are two types of charge, positive and negative. Like charges repel and opposite charges attract. Importantly, charge is conserved: starting from zero charge, if you isolate a charge Q somewhere, there has to be an equal and opposite charge $-Q$ somewhere else. Charge is also quantized. It comes in multiples of the fundamental unit of charge, $\pm e$, where $e = 1.6 \times 10^{-19}$ C; here C indicates the Coulomb, the SI unit for electric charge. The electron has charge $-e$. The proton has charge $+e$. More precisely, quarks, three of which constitute a proton or a neutron, can have charges $\pm \frac{1}{3} e$ or $\pm \frac{2}{3} e$. However, an isolated quark cannot occur. For isolated particles, the charge is always found to be an integer multiple of the electronic charge.

Electric charges create electric fields. The electric field, $\mathbf{E}$, is another essential concept in electrostatics. $\mathbf{E}$ is a "vector field." That is, $\mathbf{E}$ is a vector that exists at every point in space. Positive charge can be conceived to be the source of electric field, and negative charge the sink, as we will see below. A charge Q in an electric field $\mathbf{E}$ experiences a force

$$\mathbf{F} = Q\mathbf{E}.$$

Actually, we previously encountered another vector field, namely the gravitational field, whose source is mass, and which causes mass to experience the gravitational force. To understand how electric charges create electric fields, this chapter will focus on understanding and applying Gauss's law to find the electric field for different charge configurations (e.g., spheres, cylinders, planes). Because Gauss's law is a linear equation, electric fields obey the principle of superposition. Superposition for electric fields means that the electric field of a collection of charges is equal to the sum of the electric fields of each charge in the collection.

12.2 Your Learning Goals for This Chapter

By the end of this chapter, you should be able to:

- Apply Coulomb's law to determine magnitude and direction of the force acting between charges.
- Determine the electric field given a charge distribution using Gauss's law in situations with spherical, cylindrical, or planar symmetry.
- Sketch the electric field lines for an arrangement of charges.
- Apply superposition for electric fields to determine the electric field in more complicated situations.

12.3   Force Between Charges: Coulomb's Law

There are two types of charges, positive and negative. Charges of the same kind repel each other, and charges of the opposite kind attract each other. The strength of the force acting between the charges is quantified by Coulomb's law: given two charges q_1 and q_2 at a separation r, the magnitude of the force acting on each of them—Newton's Third Law requires these forces are equal and opposite—is given by

$$F = \frac{1}{4\pi \epsilon_0} \frac{|q_1||q_2|}{r^2},$$ (12.1)

where the symbol | | means we are taking the absolute value of each charge (since q_1 and q_2 could be either positive or negative), and ϵ_0 is a constant, called the permittivity of free space or the permittivity of the vacuum:

$$\epsilon_0 = 8.85 \times 10^{-12} \frac{C^2}{N\,m^2}.$$

As shown in Fig. 12.1, the direction of the force is along the line joining the two charges, the force is *repulsive* for like charges (*left* and *middle* case of Fig. 12.1), and *attractive* for opposite charges (*right* case of Fig. 12.1). Therefore, in vectorial form, we can write Coulomb's law as

$$\mathbf{F(r)} = \frac{q_1 q_2}{4\pi \epsilon_0 r^2} \hat{\mathbf{r}}.$$ (12.2)

Here, $\hat{\mathbf{r}}$ is the unit vector in the radial direction, which we first encountered in Chap. 1. *Vectors and Kinematics*, and then in Chap. 2. *Force and Momentum: Newton's Laws and How to Apply Them*, and then again in Chap. 6. *Diffusion: Membrane Permeability and the Rate of Actin Polymerization*. When q_1 and q_2 have opposite signs, the force is proportional to $-\hat{\mathbf{r}}$, which means it is the direction of decreasing separation, and thus it is attractive force.

   If there are more than two charges, we need to add the effect of each of the other charges to find the total force acting on a single charge. For instance, given three

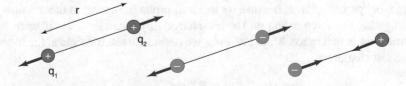

Fig. 12.1 Examples of Coulomb's force acting between two point charges: *left* two positive charges repel each other, and the force is then pointing outward on each of them, pushing them away from each other; *middle* two negative charges repel each other in the same way; *right* positive and negative charges attract each other, and the force is then pointing inward toward the other charge

charges q_1, q_2, and q_3, the force acting on the charge q_1 will be the sum of: (1) the force due to the charge q_2,

$$\mathbf{F}_{12} = \frac{1}{4\pi\epsilon_0} \frac{q_1 q_2}{r_{12}^2} \hat{\mathbf{r}}_{12}, \tag{12.3}$$

where r_{12} is the distance between q_1 and q_2 and $\hat{\mathbf{r}}_{12}$ is a unit vector in the direction from charge 1 to charge 2, and (2) the force due to the charge q_3,

$$\mathbf{F}_{13} = \frac{1}{4\pi\epsilon_0} \frac{q_1 q_3}{r_{13}^2} \hat{\mathbf{r}}_{13}, \tag{12.4}$$

where r_{13} is the distance between q_1 and q_3 and $\hat{\mathbf{r}}_{13}$ is a unit vector in the direction from charge 1 to charge 3. In general, $\hat{\mathbf{r}}_{13}$ and $\hat{\mathbf{r}}_{12}$ are not parallel to each other. Therefore, the total force acting on the charge q_1 will be the vector sum:

$$\mathbf{F}_{tot,1} = \mathbf{F}_{12} + \mathbf{F}_{13}, \tag{12.5}$$

taking into account the direction of each force, as we saw in Chap. 2. *Force and Momentum: Newton's Laws and How to Apply Them.*

12.4   Electric Field

Coulomb's law in Eq. 12.1 describes the force between two point charges and therefore is strictly dependent on the specific two charges under consideration. It turns out that it is very useful to define a more general quantity that describes the effect of a charge on its surrounding space, which can also be applied in more complicated situations than a pair of point charges. This quantity is the *electric field*.

Consider a positive charge Q. This charge—just by being where it is—alters the space around it (attracting or repelling other charges). We capture this idea by saying that the charge Q creates an electric field around itself. The electric field, $\mathbf{E}$, is a vector defined at every point in space as the *force per unit of charge* that the charge Q creates. In other words, see Fig. 12.2, if we bring another charge $q > 0$ near the charge Q, this charge will experience a force. We call the charge q a "test" charge, or "probe" charge, because we use it to probe the space; in other words, by investigating the force acting on the test charge q, we can describe if there is an electric field in that space. More precisely, we define the electric field at the location of the test charge as

$$\mathbf{E} = \frac{\mathbf{F}_{\text{on q}}}{q}. \tag{12.6}$$

From this definition, we know that the units of electric field are N/C, i.e., Newtons over Coulombs.

Fig. 12.2 The charge Q exerts a force $\mathbf{F}_{\text{on q}}$ on the test charge q. The electric field is defined as such force divided by the value of the test charge q

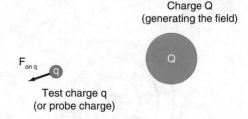

Charge Q
(generating the field)

$F_{\text{on q}}$

Test charge q
(or probe charge)

For any generic configuration of charges, once we know the electric field $\mathbf{E}$ at a given point, then we can always evaluate the force acting on a charge q_1 by multiplying the charge times the electric field:

$$\mathbf{F} = q_1\mathbf{E}. \tag{12.7}$$

We will see how useful this is in many situations. It is important to emphasize though that, while Eq. 12.7 provides the force acting on a charge, the electric field created by a charge, or a configuration of charges, exists irrespectively of whether or not there are charges to experience it.

12.4.1 Electric Field of a Single Point Charge

Combining the definition of electric field in Eq. 12.6 with Coulomb's Law (Eq. 12.1), we have that the force experienced by a test charge q, placed near a charge Q, is

$$\mathbf{F}_{\text{on q}} = \frac{qQ}{4\pi\epsilon_0 r^2}\hat{\mathbf{r}}. \tag{12.8}$$

Dividing by the test charge, we find that the electric field created by the charge Q is

$$\mathbf{E} = \frac{Q}{4\pi\epsilon_0 r^2}\hat{\mathbf{r}}. \tag{12.9}$$

The dependence on q disappears, as expected, because the electric field is solely determined by the charge Q, which we can call the source of the field. The E-field is a property of space and exists for any point at a distance r from its source charge, irrespective of the presence of a test charge, q.

Since r is the only variable appearing in Eq. 12.9, this electric field has spherical symmetry, i.e., it has the same magnitude for all points on a sphere centered around the charge, while the electric field vector points radially outward in the case of a positive charge and radially inward in the case of a negative charge, as shown in Fig. 12.3.

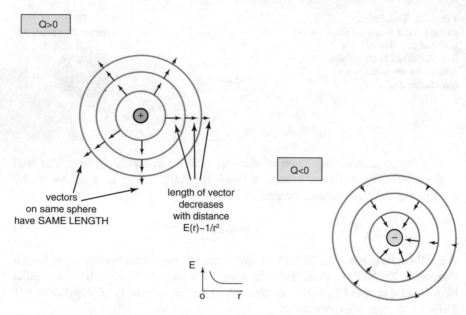

Fig. 12.3 Electric field vectors—**E**-fields—for a positive charge (left) and negative charge (right)

12.4.2 Facts to Know

(a) Is charge quantized? Yes
(b) What is the charge of an electron? -1.6×10^{-19} C
(c) What is the charge of a proton? -1.6×10^{-19} C
(d) What is the charge of a neutron? 0
(e) What is the equation for the force (**F**) on a charge Q in an electric field **E**?
     $\mathbf{F} = Q\mathbf{E}$
(f) Do **E**-fields obey the principle of superposition? Yes

12.4.3 Example: Using Superposition to Find the Electric Field of Two Charges

In this example, we utilize the superposition principle to find the electric field created by two electric charges. Two equal and positive charges are placed on the x-axis. One is placed to the left of the origin and the other one to the right, as shown in Fig. 12.4.

(continued)

Fig. 12.4 Two charges for
Example 12.4.3

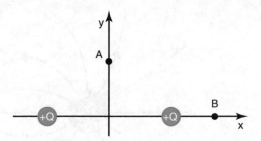

(a) What is the direction of the electric field at point A?
(b) What is the direction of the electric field at point B?
(c) For both A and B, explain your reasoning or justify your answer with an
    appropriate diagram.

12.4.4 Electric Field Lines

In this section, we introduce electric field lines, which provide a very useful tool
for representing and visualizing the electric field vector in space. Electric field lines
respect the following four rules:

1. Electric field lines are continuous curves drawn **tangent** to the electric field
   vectors. Conversely, the electric field vector at any point is tangent to the field
   line at that point.
2. Electric field lines start only from positive charges and end only on negative
   charges, i.e., following the direction of the electric field. (Later on, we will
   see that electric field lines can form closed loops in the presence of changing
   magnetic fields.)
3. Electric field lines never cross.
4. The number of electric field lines per unit area crossing through a surface
   perpendicular to the lines is proportional to the magnitude of the electric field in
   that region. Therefore, the field lines are close together where the electric field
   is strong and far apart where the field is weak.

An example of electric field lines for the case of two equal (i.e., same magnitude)
and opposite charges is shown at the beginning of the chapter and in Fig. 12.5.

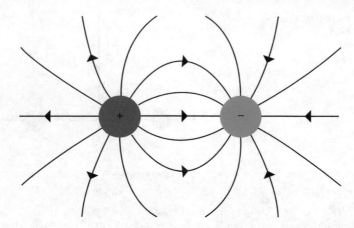

Fig. 12.5 Electric field lines for two equal and opposite charges, namely an electric dipole

12.5   Gauss's Law

Coulomb's law specifies the electric field of a point charge. More generally, the connection between electric field and charge is specified by Gauss's[1] Law:

$$\oint_A \mathbf{E} \cdot d\mathbf{A} = \frac{Q}{\epsilon_0}. \tag{12.10}$$

Gauss's law is one of kMaxwell's[2] (four) equations, about which mathematician Ian Stewart[3] wrote

> Maxwell's equations didn't just change the world. They opened up a new one.

What does Eq. 12.10 mean?

- The left side of Eq. 12.10 is an integral over a closed—the so-called Gaussian—surface, A, of $\mathbf{E} \cdot d\mathbf{A}$. The circle on the integral sign informs us that the integral is carried out over a closed surface, which we will henceforth refer to as a Gaussian surface. Gaussian surfaces are not actual surfaces, but they are conceptual surfaces that we use as a mathematical construct to calculate the integral on the left side of Eq. 12.10. The important thing is that they are closed surfaces (e.g., spheres, cylinders, cubes), so that it is possible to decide what charge is inside the Gaussian surface and which are outside.

---

[1] http://en.wikipedia.org/wiki/Gauss
[2] http://en.wikipedia.org/wiki/James_Clerk_Maxwell
[3] https://en.wikipedia.org/wiki/Ian_Stewart_(mathematician)

- $d\mathbf{A}$ is an element of vector area. This is a vector with magnitude equal to the area under consideration and direction that is perpendicular to the area and directed out of the closed surface.
- $\mathbf{E}$ is the electric field at the location of element $d\mathbf{A}$ of the Gaussian surface.
- The $\cdot$ "dot" sign between $\mathbf{E}$ and $d\mathbf{A}$ denotes the scalar product or dot product. The scalar product, proportional to the cosine of the angle between the two vectors involved, is maximum when two vectors are parallel to each other and zero when they are perpendicular. Therefore, the biggest contributions to the integral on the left side of Eq. 12.10 are for the instances when $\mathbf{E}$ and $d\mathbf{A}$ are parallel to each other, which geometrically is when the electric field is perpendicular to the area, i.e., when the electric field is perpendicular to the Gaussian surface.
- The quantity $\mathbf{E} \cdot d\mathbf{A}$ is said to be an element of "electric flux," while the full integral on the left side of Eq. 12.10 is the "total electric flux," denoted Φ_E, calculated by summing up all the contributions from each element $\mathbf{E} \cdot d\mathbf{A}$. It is important to be aware, however, that it is not a flux in the sense of flux that we encountered in Chap. 6. *Diffusion: Membrane Permeability and the Rate of Actin Polymerization*, where particle flux was the number of particles that cross unit area in unit time; instead, this electric flux is a way to quantify how many electric field lines cross through the given Gaussian surface, as we discuss shortly below. Thus, the right side of Eq. 12.10 is an integral of electric flux over a closed surface.
- The right side of Eq. 12.10 is the total charge enclosed by the Gaussian surface in question. In other words, it is the net sum of all the charges that are inside the Gaussian surface, added taking into account their sign (for instance, if inside the Gaussian surface, there is a charge $+q$ and a charge $-q$, the total charge enclosed is zero, since $Q = +q - q = 0$).
- ϵ_0 is a fundamental constant called the permittivity of free space or the permittivity of the vacuum that we have already encountered in Coulomb's law at the beginning of the chapter (Eq. 12.1).

How can we "translate" Eq. 12.10 into words?

The Total electric flux through any closed surface equals the total net charge inside that closed surface divided by ϵ_0.

Gauss's law initially seems abstract because it deals with quantities—electric field and charge—that are unfamiliar, compared, for example, to mass, acceleration, and force, which are the subjects of Newton's second law. However, the meaning of Gauss's law, and how to use it, will become clearer as we provide alternative explanations and go through examples of its application. Actually, it is reasonable that it will take us a bit to fully understand this major result. After all this law was formulated over the course of half a century, first in 1773 by the Italian–French mathematician and astronomer Joseph-Louis Lagrange, then in 1813 by the German mathematician and physicist Carl Friedrich Gauss. Finally, it will end up being the first of Maxwell's equations, which are very general and powerful equations, formulated by Scottish physicist, James Clerk Maxwell, in 1861, that describe the behavior of any electric and magnetic field configuration. Why the law is called

"Gauss's" instead of "Lagrange-Gauss", or otherwise, we do not know, but it is worth reflecting on the fact that in science the name attached to an important result may often not capture all the contributions to its discovery and dissemination.

12.5.1 Demonstration of Gauss's Law for a Single Point Charge

Gauss's law is very general and applies to any kind of configurations of electric charges and electric fields. Demonstrating the equivalence of Gauss's law and Coulomb's law in general is beyond the scope of this book, and it is helpful to convince ourselves that they are the same in the case of a point charge. Consider a positive point charge Q, and the electric field lines in this case are shown in Fig. 12.6. We choose as Gaussian surface a sphere of radius r centered at the positive charge as shown on the right side of Fig. 12.6.

Our goal is to show that the electric flux through this Gaussian surface (left side of Gauss's law Eq. 12.10) is equal to the charge enclosed in that surface divided by ϵ_0 (right side of Gauss's law Eq. 12.10).

Let us calculate the total electric flux through this Gaussian sphere: $\oint_A \mathbf{E} \cdot d\mathbf{A}$. In this case, we already know from Eq. 12.9 that the electric field is given by

$$\mathbf{E} = \frac{Q}{4\pi\epsilon_0 r^2}\hat{\mathbf{r}}.$$

We can then calculate the flux following these steps:

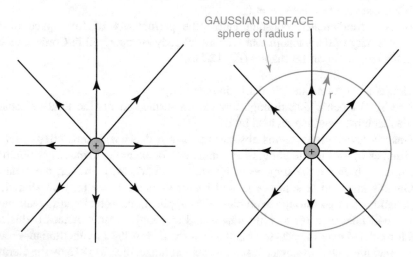

Fig. 12.6 Left side: Electric field lines for a positive point charge. The field lines start on the charge and go out radially outward toward infinity. Right side: A sphere of radius r is chosen as Gaussian surface and marked on the figure to apply Gauss's law to the point charge

$$\Phi_E = \oint_A \mathbf{E} \cdot d\mathbf{A} = \oint_A \frac{Q}{4\pi\epsilon_0 r^2}\hat{\mathbf{r}} \cdot d\mathbf{A} = \oint_A \frac{Q}{4\pi\epsilon_0 r^2}dA$$

$$= \frac{Q}{4\pi\epsilon_0 r^2}\oint_A dA = \frac{Q}{4\pi\epsilon_0 r^2}A_{\text{sphere}}$$

$$= \frac{Q}{4\pi\epsilon_0 r^2}4\pi r^2 = \frac{Q}{\epsilon_0}. \tag{12.11}$$

In the steps above, we have used the following simplifications:

- $\hat{\mathbf{r}}$ and $d\mathbf{A}$ are parallel vectors, since they both point radially outward; therefore, the dot product $\hat{\mathbf{r}} \cdot d\mathbf{A}$ simplifies to $1\,dA$.
- The electric field magnitude $\frac{Q}{4\pi\epsilon_0 r^2}$ is constant when integrating on the sphere of radius r; therefore, it can come out of the integral.
- The surface of a sphere of radius r is $4\pi r^2$.

The steps above prove that indeed the electric flux through a closed sphere is equal to $\frac{Q}{\epsilon_0}$ as Gauss's law would have predicted! We can now say we can prove that Gauss's law is valid for a single point charge.

You may wonder, what if instead of a Gaussian sphere of radius r, we would have chosen a Gaussian sphere of a different radius, for instance $r_1 > r$? By close inspection of Eq. 12.11, you can see that by changing the radius of the Gaussian sphere to be r_1, then the electric field magnitude will also change to $\frac{Q}{4\pi\epsilon_0 r_1^2}$, and accordingly, the surface of the sphere will change to $4\pi r_1^2$. Despite these changes, overall the dependence on r_1 will still cancel out to leave us only with $\frac{Q}{\epsilon_0}$. This is the strength of Gauss's law, and it works with any surface; a larger sphere will cause a smaller electric field, but also a larger surface, so overall the flux remains unchanged, and this is what we expect. No matter which size and shape we chose, the amount of fluxes coming out of a Gaussian surface can be only determined by how much charge is enclosed. This is the main idea of Gauss's law!

12.5.2 Using Gauss's Law to Find the E-Field for a Point Charge

In the previous section, we have shown that starting from the known electric field for a point charge (Eq. 12.9), we can prove Gauss's law. Now we will do the opposite, we will assume that we do not know the electric field for a point charge (and hence that we do not know Coulomb's law, which was used to derive Eq. 12.9) and use Gauss's law to find it.

Consider the same positive point charge Q of the previous section shown in Fig. 12.6. Our goal is to find the electric field of the point charge by applying Gauss's law Eq. 12.10.

- We choose as Gaussian surface a sphere of radius r centered at the positive charge as shown on the right side of Fig. 12.6.
- The *left* side of Gauss's law is the total electric flux through this Gaussian sphere: $\Phi_E = \oint_A \mathbf{E} \cdot d\mathbf{A}$. In this case, $\mathbf{E}$ and $d\mathbf{A}$ are parallel vectors, since they both point radially outward; therefore, the dot product simplifies to $E\, dA$. In addition, since the electric field has radial symmetry, it is constant on the surface of the sphere and only depends on the radial coordinate; therefore, we call it $E(r)$ and can pull it out of the integral. All these steps together give us

$$\Phi_E = \oint_A \mathbf{E} \cdot d\mathbf{A} = \oint_A E\, dA = E(r) \oint_A dA = E(r)\, 4\pi r^2.$$

- The *right* side of Gauss's law is easier to determine. The only charge enclosed in the Gaussian surface is the point charge Q; therefore, the right side of Gauss's law is

$$\frac{Q}{\epsilon_0}.$$

- Now let us set the left side of Gauss's law equal to the right side:

$$E(r)\, 4\pi r^2 = \frac{Q}{\epsilon_0}.$$

If we simplify the equation above and solve for $E(r)$, we find

$$E(r) = \frac{Q}{4\pi \epsilon_0 r^2},$$

which (after adding the appropriate unit vector convention) is identical to the result we previously found in Eq. 12.9.

From this last result, we can also derive Coulomb's law (Eq. 12.2), by using the fact that the force on a charge q is given by $\mathbf{F} = q\mathbf{E}$.

The application of Gauss's law that we showed in this section is the reason why Gauss's law is important in electrostatics: in many situations (that are more complex than simple point charges), we will not know the electric field of a specific charge configuration, and therefore, Gauss's law will be the way we'll go to find it. This will be discussed in the next section.

12.5.3 Using Gauss's Law to Find the E-Field of Highly Symmetric Extended Objects

There are many situations in which we will need to consider charge configurations that are not as simple as a number of point charges, but that are extensive charged objects (for instance, charged planes, spheres, cylinders) and for which we will not know the electric field. In this section, we describe the steps to take to apply Gauss's law to determine the electric field of extended charged objects that have a specific simple symmetry (e.g., planar, spherical, cylindrical symmetry).

We have to note that Gauss's law is general and in principle can be used to find the electric field in any situation, even the non-symmetrical ones; in practice though Gauss's law is convenient to use only in highly symmetrical situations, and those are the ones that we will consider in this chapter. In some cases, it may be convenient to break up a more complicated problem up into a number of parts, each one of which can itself be solved separately. The complete solution can then be obtained using the principle of superposition. After all, as we have seen with the example of the spherical cow in Chap. 6. *Diffusion: Membrane Permeability and the Rate of Actin Polymerization*, approximating objects to spheres, cylinder, and planes may still provide a great deal of valuable insight.

Here is a general procedure to use Gauss's law to find the electric field in highly symmetrical situations:

1. Make a sketch of the problem and assess the symmetry (e.g., determine if it is a planar, spherical, or cylindrical symmetry). Add electric field lines to the sketch, making sure that they respect the symmetry of the problem. In a spherically or cylindrically symmetric situation, **E** will be directed radially. In a planar geometry, **E** will be perpendicular to the plane.
2. Select the appropriate Gaussian surface according to the symmetry of the problem and add it to the sketch. In problems with spherical symmetry, this will be a concentric sphere. In problems with cylindrical symmetry, it will be a coaxial cylinder with flat ends. In problems with planar symmetry, it will be a box—the shape does not matter—with top and bottom, parallel to the symmetry plane.

   If the problem indeed has a simple enough geometry and if you picked the Gaussian surface sensibly, recognize, first, that, on certain portions of your Gaussian surface, **E** is perpendicular to the surface, and hence, it is parallel to $d\mathbf{A}$. On other portions of the Gaussian surface, recognize that **E** is parallel to the surface and hence perpendicular to $d\mathbf{A}$. These latter portions of the surface, therefore, do not contribute to the left side of Gauss's law.
3. Calculate the *left* side of Gauss's law (Eq. 12.10), which means the total electric flux through your Gaussian surface. For the portion(s) of your Gaussian surface for which **E** and $d\mathbf{A}$ are parallel (so that $\mathbf{E} \cdot d\mathbf{A}$ is not zero), recognize (if you picked the surface appropriately) that **E** has a constant magnitude and therefore can be brought outside the integral. For this portion of your Gaussian surface,

$$\int_A \mathbf{E} \cdot d\mathbf{A} = EA, \tag{12.12}$$

where E is the magnitude of the electric field and A is the area of the portion of the Gaussian surface, where $d\mathbf{A}$ and **E** are parallel. For spherical geometries, often you will find $A = 4\pi r^2$, corresponding to a Gaussian surface that is a sphere of radius r. For cylindrical geometries, often you will find $A = 2\pi r L$, corresponding to a Gaussian surface that is a cylinder of radius r and length L. In this case, the end caps do not contribute because, for these surfaces, $\mathbf{E}$ and $d\mathbf{A}$ are perpendicular. For a planar geometry, often A is the cross-sectional area of one or both of the parallel top and bottom surfaces of your Gaussian surface.

4. Calculate the *right* side of Gauss's law; this depends on Q that in Eq. 12.10 is the charge inside the Gaussian surface only. If your surface contains all the charge, then Q is all the charge, but in many examples it will not be all of the charge, and you will have to figure out how much charge is enclosed. Once you have succeeded in determining the enclosed charge, you know the right side of Eq. 12.10.

5. Set your previous two answers equal to each other (i.e., *left* side of Gauss's law = *right* side of Gauss's law) and solve for the electric field E, which is the variable you are looking for.

12.5.4 Example: Gauss's Law Applied to Uniformly Charged Sphere

Consider a sphere of radius R and total charge $Q > 0$ in which the charge is distributed uniformly throughout the volume of the sphere. We will now apply the procedure outlined in the previous section to find the electric field $\mathbf{E}(r)$ for any $r > 0$.

(a) Make a sketch of the problem and assess the symmetry (e.g., determine if it is a planar, spherical, or cylindrical symmetry). Add electric field lines to the sketch, making sure that they respect the symmetry of the problem.

```
Answer: See the sketch at the left of Fig. 12.7.
This is a case of spherical symmetry. The
```

(continued)

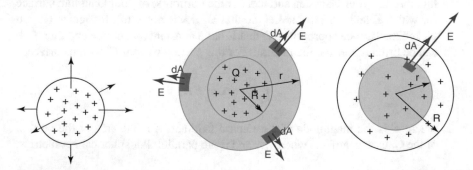

Fig. 12.7 Uniformly charge spheres for Example 12.5.4

electric field lines are coming radially
outward.

(b) First, consider the case $r > R$. What is the appropriate Gaussian surface
in this case? Select the appropriate Gaussian surface according to the
symmetry of the problem and add it to the sketch.

Answer: See the sketch at the center of
Fig. 12.7, where the appropriate Gaussian surface
is marked in blue, and this is a sphere of
radius r   >   R, concentric with the charge
distribution. For this surface, $d\mathbf{A}$ is radially
outward. Because of the spherical symmetry,
everywhere $\mathbf{E}$ points radially from the center
of symmetry.

(c) Calculate the left of Gauss's law for $r > R$.

Answer: In this case, $\mathbf{E} \cdot d\mathbf{A} = E \times dA$, since the
two vectors are parallel, and therefore, the
cosine of the angle between them is equal to 1.
Furthermore, since E is the same everywhere on
the spherical surface (magnitude of electric
field vector is the same everywhere on the
surface), as a result of the spherical symmetry,
then the left side of Gauss's law becomes

$$E \times 4\pi r^2. \tag{12.13}$$

(d) Calculate the enclosed charge for $r > R$.

Answer: For r > R, the enclosed charge is simply
the total charge, $+Q$.

(e) Set your previous two answers equal to each other (i.e., *left* side of
Gauss's law = *right* side of Gauss's law) and solve for the electric field
E, for $r > R$.

Answer: We have

$$4\pi r^2 E = \frac{Q}{\epsilon_0}. \tag{12.14}$$

(continued)

Therefore,

$$E = \frac{Q}{4\pi\epsilon_0 r^2}. \tag{12.15}$$

We can convert Eq. 12.15 into a vector equation by recalling that the E is directed radially:

$$\mathbf{E} = \frac{Q}{4\pi\epsilon_0 r^2}\hat{\mathbf{r}}, \tag{12.16}$$

and Eq. 12.15 is independent of R and therefore applies in the limit that $R \rightarrow 0$, namely for a point charge at the origin. Combining Eq. 12.16 with $\mathbf{F} = Q_1\mathbf{E}$ yields Coulomb's law[4] for the force, $\mathbf{F}$, between charges Q and Q_1 separated by a distance r:

$$\mathbf{F} = \frac{QQ_1}{4\pi\epsilon_0 r^2}\hat{\mathbf{r}}. \tag{12.17}$$

(f) Next, consider the case $r < R$. What is the appropriate Gaussian surface in this case? What is the direction of $d\mathbf{A}$ for this surface?

Answer: See the sketch at the right of Fig. 12.7. The appropriate Gaussian surface is now a spherical Gaussian surface of radius $r < R$, concentric with the charge distribution. For this surface too, $d\mathbf{A}$ is radially outward. Because of the spherical symmetry, everywhere E points radially from the center of symmetry just as before.

(g) Calculate the left side of Gauss's law for $r < R$.

Answer: The steps and calculation are identical to the case for $r > R$. The left side of Gauss's law becomes

$$E \times 4\pi r^2, \tag{12.18}$$

which is the same result as before.

---

[4]http://en.wikipedia.org/wiki/Coulomb%27s_law

(continued)

(h) Calculate the enclosed charge for $r < R$.

Answer: For $r < R$, only a fraction $\frac{r^3}{R^3}$ of the total charge lies inside the Gaussian surface. Therefore, the enclosed charge is

$$+Q\frac{r^3}{R^3}.$$

(i) Set your previous two answers equal to each other (i.e., *left* side of Gauss's law = *right* side of Gauss's law) and solve for the electric field E, for $r < R$.

Answer: Now, Gauss's law informs us that

$$E \times 4\pi r^2 = \frac{Qr^3}{\epsilon_0 R^3}. \tag{12.19}$$

It immediately follows that

$$E = \frac{Qr}{4\pi\epsilon_0 R^3}. \tag{12.20}$$

We see that for $r < R$, inside the uniform charge distribution, the electric field increases linearly with r. Figure 12.8 below provides a sketch of the complete behavior of E versus r.

Fig. 12.8 Magnitude of the electric field of a uniform charge distribution of radius R plotted versus radial coordinate, r

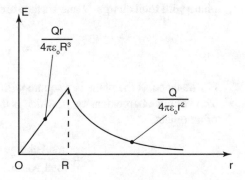

12.5.5 Charge Densities for Uniformly Charged Extended Objects

When discussing extended charged objects such as planes, cylinders, sphere, etc., it is helpful to introduce a charge density, which means a quantity that describes how much charge is distributed per: unit of length (in the case of a 1D object, such as a charged wire), unit of area (in the case of a 2D object, such as a charged plane), unit of volume (in the case of 3D object, such as a charged sphere) of the object. In particular, for objects that are uniformly charged, which means the charge is equally distributed throughout the object, we introduce the following quantities:

- The linear charge density, λ, which is the amount of charges per unit of length of an object

$$\lambda = \frac{\text{total charge}}{\text{total length}} \qquad [\lambda] = \frac{C}{m}. \qquad (12.21)$$

This is most often used to describe charged objects that are mostly 1-dimensional (such as a thin rod or a charged wire). For example, the linear charge density of a charged rod with total charge Q and length L is

$$\lambda = \frac{Q}{L}.$$

- The surface charge density, σ, which is the amount of charges per unit of area of an object

$$\sigma = \frac{\text{total charge}}{\text{total surface area}} \qquad [\sigma] = \frac{C}{m^2}. \qquad (12.22)$$

This is most often used to describe charged objects that are mostly 2-dimensional (such as a charged plane). For example, the surface charge density of a charged plane with total charge Q and surface area A is

$$\sigma = \frac{Q}{A}.$$

For instance, if the plane is a square with side L, then $\sigma = \frac{Q}{L^2}$.
- The volume charge density, ρ, which is the amount of charges per unit of volume of an object

$$\rho = \frac{\text{total charge}}{\text{total volume}} \qquad [\rho] = \frac{C}{m^3}. \qquad (12.23)$$

This is most often used to describe charged objects that are 3-dimensional (such as a charged sphere). For example, the volume charge density of a charged sphere with total charge Q and volume V is

$$\rho = \frac{Q}{V}.$$

In particular, if the radius of the sphere is R, then

$$\rho = \frac{Q}{\frac{4}{3}\pi R^3}.$$

Charge densities are used to find the amount of charges contained in any specific length, area, or volume within the object. We will see this is very helpful in applying Gauss's law, especially when calculating the charge enclosed in a given Gaussian surface (right side of Gauss's law). Here a few examples of how we can use charge densities to determine the amount of charges contained in a specific portion of an extended object:

- A thin charged wire has linear charge density λ. The amount of charges contained in a length dx of the wire is

$$dq = \lambda dx.$$

- A charged plane has surface charge density σ. The amount of charges contained in an area dA of the plane is

$$dq = \sigma dA.$$

- A charged sphere has volume charge density ρ. The amount of charges contained in an volume dV of the sphere is

$$dq = \rho dV.$$

More particularly, considering a uniformly charged sphere of radius R, the amount of charges contained within a sphere of radius $r < R$ is given by

$$q = \rho \frac{4}{3}\pi r^3.$$

12.5.6 Example: Gauss's Law to Find the E-Field of an Infinite Charged Plane

An infinite plane has uniform charge density σ, with $\sigma > 0$ (i.e., σ is the amount of charges per unit of area). The goal of this problem is to find the electric field at a point P at a distance d above the plane.

(a) Make a sketch of the problem and assess the symmetry. Add electric field lines to the sketch, making sure that they respect the symmetry of the problem.

   Answer: A sketch of the problem is shown on
   the left of Fig. 12.9. As shown in the center
   panel of this figure, electric field lines point
   away from positive charges. Because the plane of
   charge is infinite, the field points vertically
   upward above the plane and vertically downward
   below the plane.

(b) Consider the point P at a distance d above the infinite plane as shown on the left of Fig. 12.9. Your goal for this part is to properly use Gauss's law to calculate the electric field at point P.
   1. Choose an appropriate Gaussian surface and draw it on a sketch of the infinite plane. Clearly mark the dimensions of your surface.

(continued)

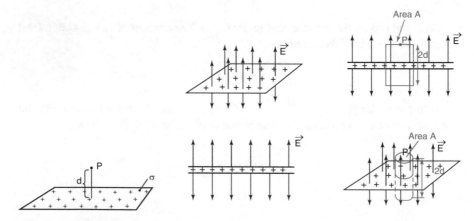

Fig. 12.9 Sketches for Example 12.5.6

Answer: See the right panel of Fig. 12.9. The
Gaussian surface here is a cylinder of height
$2d$ passing through point P and end caps of
area A. (One can also choose a rectangular
box of height $2d$ and end caps of area A.)

2. Calculate the total electric flux through your Gaussian surface (i.e.,
   calculate the left side of Gauss's law).

Answer:

$$\oint \mathbf{E} \cdot d\mathbf{A} = \int_{top} \mathbf{E} \cdot d\mathbf{A} + \int_{side} \mathbf{E} \cdot d\mathbf{A} + \int_{bottom} \mathbf{E} \cdot d\mathbf{A}$$

(12.24)

$$= EA + 0 + EA = 2EA.$$

(12.25)

3. Calculate the total charge enclosed Q_{encl} in your Gaussian surface
   (i.e., calculate the right side of Gauss's law).
4. What is the electric field at point P? Give *both* the magnitude and
   direction. (Give the magnitude in terms of the given quantities and
   indicate the direction with a vector on the sketch and in words).

Answer: Using Gauss's law, $\oint \mathbf{E} \cdot d\mathbf{A} = \frac{Q_{enc}}{\epsilon_0}$,

$$2EA = \frac{\sigma A}{\epsilon_0} \qquad \rightarrow \qquad E = \frac{\sigma}{2\epsilon_0}.$$

(12.26)

At point P, the field points upward, and the
electric field strength is $E = \frac{\sigma}{2\epsilon_0}$.

(c) How would your answer to the previous question change if the point P
    would be at a distance $2d$ instead of d from the plane?

Answer: The answer would not change if the
distance is doubled since the dependence on d
cancels out and does not appear in the final
result for the electric field. The electric
field magnitude will still be equal to

$$E = \frac{\sigma}{2\epsilon_0}.$$

(continued)

(d) Based on your answer to the last two questions, what can you say about the electric field for *any* point above or below an infinite charged plane?

Answer: The procedure above can be generalized to any point above or below the infinite charged plane always leading to the same result. The result will not depend on the location of the point, and it is true for any point; therefore, we can say that the electric field for an infinite charged plane with surface charge density σ has magnitude:

$$E = \frac{\sigma}{2\epsilon_0}. \tag{12.27}$$

The electric field lines are perpendicular to the plane and are coming outward, away from the plane, in the case of a positive charge density, $\sigma > 0$, and would be pointing inward, toward the plane, in the case of a negative charge density, $\sigma < 0$.

(e) Why should we care about infinite planes? Are not real objects always finite?

Answer: While the result for an infinite object may seem too abstract for concrete applications, this does not turn out to be the case. As an example, consider the Euryplatea nanaknihali,[5] the world's smallest fly, measuring only 0.4 mm in size. When a Euryplatea nanaknihali flies very close to a finite charged surface (e.g., the screen of a laptop, the top of car, the leaf of a tree), it is an appropriate approximation to model the plane as infinite since the dimensions of the object, the fly, are three-to-four orders of magnitude smaller than the size of the plane, and so effectively the plane looks infinite to the fly.

---

[5]https://en.wikipedia.org/wiki/Euryplatea_nanaknihali

(continued)

Hence, as spheres can be good approximation
for cows, infinite planes could be a good
approximation if you are interested in analyzing
the effect of the plane on objects that are
close to the plane and small with respect to
the overall size of the plane. Infinite objects,
in general, such as infinite cylinders, infinite
wires, are easier to model mathematically and
can give insight into many situations in which
the finite extent of the object (what we call
"edge effects") is negligible for the purpose of
the question we are asking.

12.5.7 The Electric Field of a Parallel Plate Capacitor

Capacitors are circuit elements that store charge and consequently electrostatic energy. They are used on devices ranging from high-speed computers to heart defibrillators. As an example, the flash on a camera uses energy stored in a capacitor.

A parallel plate capacitor consists of two parallel plates, each of area A, separated by a distance d. In this section, we will determine the electric field between and outside of this capacitor when a charge Q is uniformly distributed on the bottom plate and a charge $-Q$ is uniformly distributed on the top plate.

It is convenient to introduce a coordinate system z pointing upward, as shown in Fig. 12.10, with the positive plane located at $z = 0$ and the negative plane at $z = d$. The surface charge density for each plate of this capacitor has magnitude given by

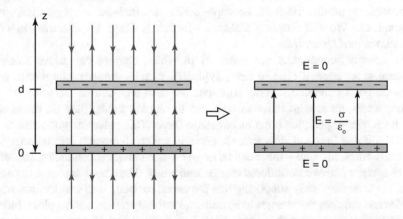

Fig. 12.10 Charges, field lines, and superposition in a parallel plate capacitor

$$\sigma = \frac{|Q|}{A}.$$

The positive plane will have a charge density $+\sigma$, and the negative plane $-\sigma$. Following the result from the previous section (see Eq. 12.27), we know that each plane creates an electric field of magnitude $E = \frac{\sigma}{2\epsilon_0}$. For the positive plane, this field points *away* from the plane; for the negative plane, the field points *toward* the plane, as shown on the left side of Fig. 12.10, where the electric field lines for the positive plane are marked in red, and the electric field lines for the negative plane are marked in blue. It is important to note that we are modeling the two planes as infinite planes, and therefore, the E-field lines of each plane extend to infinite above and below each of them. To find the total electric field of the capacitor, we need to use the principle of superposition and add the two electric fields at every point in space. The result is shown on the right side of the figure. It turns out that the net electric field is zero for $z > d$ and $z < 0$ because in these regions the plates create opposite fields that cancel each other. By contrast, between the plates ($0 < z < d$), the individual fields from each plate add, producing a total electric field pointing upward given by

$$\mathbf{E} = \frac{\sigma}{\epsilon_0}\hat{\mathbf{k}} = \frac{Q}{\epsilon_0 A}\hat{\mathbf{k}}. \tag{12.28}$$

12.6   Chapter Outlook

The examples in this chapter have involved situations in which the distribution of charges was specified. More generally, however, in addition to situations with a specified charge distribution, we will encounter electrostatic problems where the type of material, in which the charges reside, is specified. As far as electrostatics are concerned, two classes of materials can be distinguished: *insulators* and *conductors*.

Insulators, a.k.a. dielectrics, are materials in which charges move only slightly in response to an electric field. Example dielectrics include: most plastics, glass, diamond, etc. We will consider dielectrics further in Chap. 13. *Electric Potential, Capacitors, and Dielectrics.*

By contrast, conductors are materials in which charges can move freely in response to an applied electric field. Typically, metals—copper, aluminum, gold, etc.—are good conductors. Ionic solutions are also conductors. Because charges in conductors are free to move in response to electric fields, they do move until they no longer experience a net electrostatic force. This behavior gives rise to the general rule that in any electrostatic situation, in which all charges are stationary—it is electrostatics, after all—there can be no net **E**-field inside a conductor. In fact, in this chapter, when we considered charge uniformly distributed across a capacitor plate, we were implicitly supposing that the capacitor plate was conducting, which enables and requires the charges to be uniformly distributed across the plate. Indeed, capacitor plates are invariably conducting. The explanation of why some materials are insulators and some materials are conductors is essentially quantum mechanical and is outside our scope.

If an electric field is maintained inside a conductor (by a power supply or battery, for example), charges inside the conductor move in response to that electric field, giving rise to an electrical current, which is a flow of charges. Current in electrical circuits is discussed in Chap. 14. *Circuits and Dendrites: Charge Conservation, Ohm's Law, Rate Equations, and Other Old Friends.*

Looking further forward, Gauss's law is the first of Maxwell's equations. The other three Maxwell's equations will be introduced in Chap. 17. *Magnetic Fields and Ampere's Law*, Chap. 18. *Faraday's Law and Electromagnetic Induction*, and Chap. 19. *Maxwell's Equations and Then There Was Light*. The electric field and electric field lines, which have been the focus of this chapter, are essential preparation for the magnetic field and magnetic field lines, which we will encounter in these later chapters, *en route* to the discovery of electromagnetic waves, which constitutes one of the intellectual triumphs of science.

12.7 Problems

Problem 1: E-Field of Two Charges

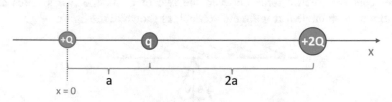

(a) A positive test charge q is released from rest at a distance a away from a fixed charge of $+Q$ and a distance $2a$ away from a fixed charge of $+2Q$, as shown in the figure above. How will the test charge move immediately after being released? Briefly explain your reasoning.

(b) Assume that the charge $+Q$ on the left is located at $x = 0$. Draw a sketch of the magnitude of the electric field as a function of x for all values of x (i.e., from negative to positive infinity).

(c) Run the Wolfram Alpha demonstration
http://demonstrations.wolfram.com/ElectricFieldsFor[6]
ThreePointCharges.[7]
Use the demonstration to plot the field lines for the arrangement of charges above **without the test charge**. Draw a sketch of the electric field for this system.

---

[6] http://demonstrations.wolfram.com/ElectricFieldsForThreePointCharges/
[7] http://demonstrations.wolfram.com/ElectricFieldsForThreePointCharges/

Problem 2: Superposition of Electric Fields from Several Charges
A charge Q is located at cartesian coordinates $(-a, a, 0)$. A charge Q is located at coordinates $(a, -a, 0)$. A charge Q is located at coordinates $(-a, -a, 0)$. Calculate the resultant electric field (**E**-field) at $(a, a, 0)$. Specify the magnitude and the direction of the electric field at $(a, a, 0)$. What happens if the charge Q at $(-a, -a, 0)$ is replaced by a charge $-2Q$?

Problem 3: Electric Field Lines
(a) Run

```
http://demonstrations.wolfram.com/ElectricFieldsForThree⁸
PointCharges/⁹
```

(b) Use this demonstration to plot the field lines of an electric dipole.
(c) Use this demonstration to plot the field lines of an electric quadrupole.

Problem 4: Finding an Unknown Charge Q
Two spheres have the same positive charge Q and the same mass m. They hang vertically from the ceiling, each attached at the end of a string of length L. Because of the repulsive electric force between the two of them, the two spheres are at equilibrium at an angle θ from/to the vertical, as shown in the figure.

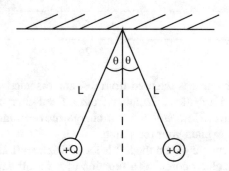

(a) Analyze the forces acting on each sphere and draw a force diagram for each of them.
(b) Find an expression for the charge Q on each sphere as a function of the given quantities L, m, g, and θ only. (Of course, your result will also contain the dielectric constant ϵ_0.)

Problem 5: Electric Flux Through a Wedge

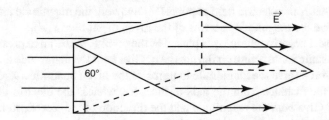

Consider a closed triangular box resting within a horizontal electric field of magnitude E as shown in the figure above. The height of the box is h, the horizontal width is W, and the depth is D into the page. Find an expression for the electric flux through: (a) the vertical rectangular surface, (b) the slanted surface, and (c) the entire surface of the box. Show *each step* of your calculation.

Problem 6: Walk the Line

An infinitely long wire is uniformly charged with a positive linear charge density λ, with $\lambda > 0$ (i.e., λ is the amount of charges per unit of length of the wire).

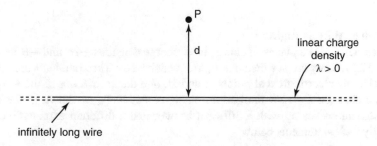

(a) By reflecting on the symmetry of the charge distribution of the system, determine what the E-field lines look like around the wire. Describe the E-field in words and with a simple sketch. In order to give a complete description of the E-field lines, make **two sketches**: (i) one with a side view of the wire, (ii) one with a cross-sectional view of the wire. Make sure to also show the *direction* of the E-field lines.

(b) Consider the point P at a distance d from the wire as shown in the figure above. Your goal for this part is to properly use Gauss's law to calculate the electric field at point P.

    1. Choose an appropriate Gaussian surface and clearly draw it on a sketch of the wire.

    2. For each point of your Gaussian surface, explain if/where the electric field vector is perpendicular, parallel, or at some other angle respect to this surface.

3. Calculate the total electric flux through your Gaussian surface.
4. Calculate the total charge enclosed Q_{encl} in your Gaussian surface.
5. What is the electric field at point P? Give *both* the magnitude and direction (give the magnitude in terms of the given quantities: λ, d).

(c) Suppose a negative charge q is placed at the point P. Write an expression for the electrostatic force acting on the charge q. Give *both* the magnitude and direction of the force (give the magnitude in terms of the given quantities: λ, d, q).

(d) Call r the distance from the axis of the wire, what is the electric field for any $r > 0$? Give *both* the magnitude and the direction of the electric field.

(e) Make a sketch of the magnitude of $E(r)$ for any $r > 0$.

Problem 7: The Infinite Cylinder
Charge is uniformly distributed across the volume of an infinitely long cylinder of radius R. The resultant charge per unit length is λ.

(a) Use Gauss's law to calculate the electric field at a point outside of the cylinder as a function of r, the radial distance from the center of the cylinder ($r > R$).

(b) Use Gauss's law to calculate the electric field at a point inside the cylinder as a function of r, the radial distance from the center of the cylinder ($r < R$).

(c) Make a plot of the electric field magnitude $E(r)$ as a function of r for any $r > 0$.

Problem 8: Who Is Right?
Two parallel, infinite planes of charge have charge densities $+2\sigma$ and $-\sigma$ as shown in Fig. 12.11. They are at a distance L_0 from each other. Three students are asked to determine the electric field at points A and B, at a distance L *above* the top plane, and at a distance L *below* the bottom plane, respectively, as shown in the figure. Each student comes up with a different answer and a different explanation. Read carefully their statements below.

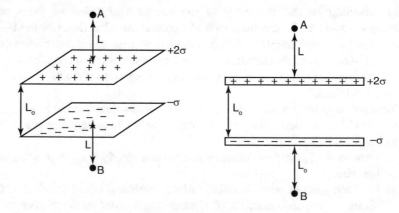

Fig. 12.11   Two equivalent views of the two infinite parallel planes discussed in Problem 8

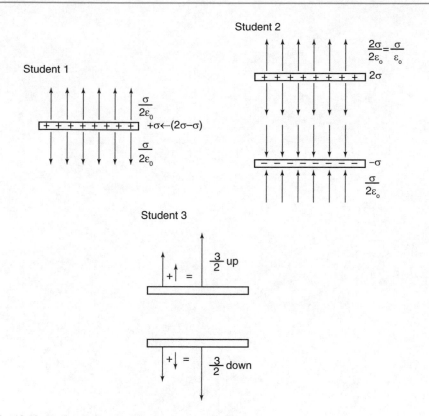

Fig. 12.12 Students' sketches

The students are reminded that the electric field magnitude of an infinite plane of charge with charge density σ is $E = \frac{\sigma}{2\epsilon_0}$.

- Student 1 draws the sketch in Fig. 12.12 [top left] and says: "I think from the outside we can think of the two planes as a single infinite plane with charge density $+2\sigma - \sigma = \sigma$, so then we know the electric field must be $\sigma/(2\epsilon_0)$ pointing *upwards* at point A and also $\sigma/(2\epsilon_0)$ pointing *downwards* at B."
- Student 2 draws the sketch in Fig. 12.12 [top right] and says "I don't agree with you, I think you need to consider the effect of each single plane, so at point A there will be a field *upwards* due to the top plane with magnitude $2\sigma/(2\epsilon_0) = \sigma/\epsilon_0$, and at point B there will be a field *upwards* due to the bottom plane with magnitude $\sigma/(2\epsilon_0)$."
- Student 3 draws the sketch in Fig. 12.12 [bottom] and says "Do we just add the electric fields together? I'm going to add up the field from each plane at both points. So at A I get a field *upward* of magnitude $\sigma/\epsilon_0 + \sigma/(2\epsilon_0) = 3\sigma/(2\epsilon_0)$, and at point B I get the same $3\sigma/(2\epsilon_0)$ but now pointing *downward*."

Evaluate each student's statement. Decide whether or not the statement is correct. If the student explanation contains incorrect reasoning, explain how the argument is flawed.

Electric Potential, Capacitors, and Dielectrics 13

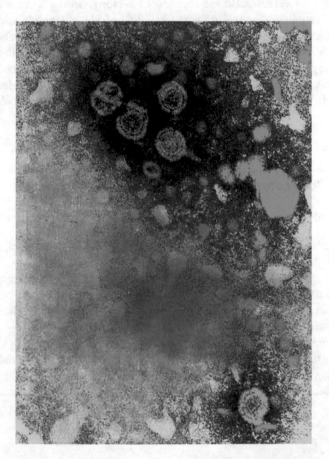

Fig. 13.1 Transmission electron microscopy image of hepatitis B virus particles, highlighted in orange. Public domain image from the Public Health Image Library: https://phil.cdc.gov/details.aspx?pid=10755

© The Author(s), under exclusive license to Springer Nature Switzerland AG 2023   625
S. Mochrie, C. De Grandi, *Introductory Physics for the Life Sciences*,
Undergraduate Texts in Physics, https://doi.org/10.1007/978-3-031-05808-0_13

13.1   Introduction

In this chapter, we continue our study of electrostatics, introducing the concepts of electric potential and capacitance. We will analyze electrical circuits containing capacitors in parallel and in series and learn how energy, electric potential, and electric charge are related in different situations. We will also elucidate electrostatic phenomena inside "dielectric materials," a.k.a. dielectrics and insulators. Both water and oil are dielectrics, and it turns out that the electrostatics of dielectrics explains why ions dissolve much better in water than in oil and therefore why ions do not easily pass through cell membranes in the absence of ion channels. More generally, electrostatics is very important in biology as evidenced by the pH dependence and ion concentration dependence of living processes. However, understanding the character of electrostatic interactions in biology, where ionic solutions are ubiquitous, requires a synthesis to two key sets of ideas: Gauss's law, etc. from "classical electrostatics" and the Boltzmann factor, etc. from statistical mechanics. At the end of the chapter, we will bring these two sets of concepts together and show that for charges in ionic solution the long-ranged Coulomb interaction becomes a short-ranged, screened Coulomb interaction. The *screened* Coulomb interaction is ubiquitous in biology. In particular, we will examine its role in viral capsid assembly (Fig. 13.1).

13.2   Your Learning Goals for This Chapter

By the end of this chapter, you should be able to:

- Calculate the capacitance of a capacitor with spherical, cylindrical, or planar symmetry.
- Calculate the overall capacitance of an arbitrary collection of capacitors in series and/or in parallel.
- Calculate the electrostatic energy of a given arrangement of charges or a given arrangement of electric fields.
- Determine electric field, capacitance, and electrostatic energy in situations where the spaces between charges are filled with dielectric material.
- Explain how electrostatics is modified for charges in ionic solutions.

13.3 Potential Energy Associated with Electrostatic Force

We first introduced potential energy in Chap. 3. *Energy: Work, Geckos, and ATP* (Sect. 3.5). Potential energy is the energy arising from any conservative force between two interacting objects. Because the electrostatic force, described by Coulomb's law, Eq. 12.2, is a conservative force, there is a corresponding electrostatic potential energy:

$$\Delta U = U(\mathbf{r}) - U(\mathbf{r}_i) = -\int_{\mathbf{r}_i}^{\mathbf{r}} \mathbf{F} \cdot d\mathbf{r}. \tag{13.1}$$

Written in vectorial form, the electrostatic force between charges q_1 and q_2, separated by a distance r, is

$$\mathbf{F}(\mathbf{r}) = \frac{q_1 q_2}{4\pi \epsilon_0 r^2} \hat{\mathbf{r}}. \tag{13.2}$$

Therefore, using Eq. 13.1, one can derive the associated electrostatic potential energy:

$$U(r) = \frac{q_1 q_2}{4\pi \epsilon_0 r}. \tag{13.3}$$

The potential energy in Eq. 13.3 resembles very much the gravitational potential energy we have already seen in Sect. 3.5.

13.4 Electrostatic Potential

The potential energy in Eq. 13.3 describes the potential energy of two charges, and therefore it is strictly dependent on which two charges we are considering. However, similarly to what we did in the previous chapter, when we defined the electric field created by a single source charge, it is convenient to also define a more general quantity to describe the electrostatic energy of a source charge, independently of there being a second test charge. Previously, we divided the electrostatic force by the value of test charge to find the electric field of the source charge. Now, we divide the electrostatic potential energy by the test charge to find the *electrostatic (or electric) potential* of the source charge. The electric potential is often denoted with the symbol, V. Why V? Because another name for electrostatic potential difference is "voltage." It is especially common to refer to a potential difference as a voltage in the context of an electrical circuit. Potential is defined as potential energy *per unit of charge*, more specifically from Eq. 13.1, it turns out that a *change* in potential is defined as

$$\Delta V = \frac{\Delta U}{q} = \frac{U(\mathbf{r}) - U(\mathbf{r}_i)}{q} = -\int_{\mathbf{r}_i}^{\mathbf{r}} \frac{\mathbf{F}}{q} \cdot d\mathbf{r} = -\int_{\mathbf{r}_i}^{\mathbf{r}} \mathbf{E} \cdot d\mathbf{r}, \qquad (13.4)$$

where in the last step we have used the definition of electric field, Eq. 12.4. Thus, electrostatic potential has the same relationship to potential energy as electric field has to force. Electric field can be conceived as force per unit charge. Electrostatic potential can be conceived as (electrostatic) potential energy per unit charge. The SI unit of electrostatic potential is Volt = Joule/Coulomb, 1 V = J/C. The name Volt was chosen after Alessandro Volta (1745–1827), an Italian physicist and chemist, who invented the electrical battery.

For any generic configuration of charges, once we know the electrostatic potential V at a given point, then we can always evaluate the potential energy of a charge q at that point by multiplying the charge times the potential:

$$U = qV. \qquad (13.5)$$

This is analogous to what we saw in Chap. 12. *Gauss's Law: Charges and Electric Fields* with the relationship between force and electric field: $\mathbf{F} = q\mathbf{E}$.

Equation 13.4 in a spherically symmetric situation becomes

$$V(r) - V(R) = -\int_R^r E(r')dr'. \qquad (13.6)$$

As is the case for potential energy, we are free to pick the zero of the electrostatic potential in any way we like. However, a common choice is to pick $V(\infty) = 0$, so that

$$V(r) = -\int_\infty^r E(r')dr'. \qquad (13.7)$$

Thus, for a point charge (Q) at the origin, the electrostatic potential created by that charge is

$$V(r) = -\int_\infty^r \frac{Q}{4\pi\epsilon_0 r'^2}dr' = \frac{Q}{4\pi\epsilon_0 r}. \qquad (13.8)$$

In Sect. 3.7 we discussed how it is possible to determine the force given the potential energy by taking the derivative of Eq. 13.1. Specifically, in a spherically symmetric situation, we had

$$F(r) = -\frac{dU}{dr}. \qquad (13.9)$$

Similarly, we can determine the radial component of the electric field from the potential by taking the derivative of Eq. 13.6 to find

$$E(r) = -\frac{dV}{dr}. \qquad (13.10)$$

In one-dimensional situations, the relationship between electrostatic potential and the x-component of the electric field is

$$E(x) = -\frac{dV}{dx} \tag{13.11}$$

and

$$V(x) - V(X) = -\int_X^x E(x')dx'. \tag{13.12}$$

13.4.1 Electrostatic Potential of a Capacitor and Capacitance

Capacitors[1] are important devices. They are the basis of cardiac defibrillators,[2] for example. Defibrillators apply a voltage of several hundreds to 1000 V across a heart in fibrillation, which ends up receiving up to 300 J of energy in the shock. As a result, normal heartbeat can be restored and the patient lives.

For a parallel plate capacitor, with $-Q$ on a flat conducting plate of area A at $x = 0$ and $+Q$ on a flat conducting plate of area A at $x = d$, as shown in Fig. 13.2, we calculated in the previous chapter that $\mathbf{E} = -\frac{Q}{\epsilon_0 A}\hat{\mathbf{i}}$ pointing downward from the positive to the negative plate. Using the definition in Eq. 13.12, we can calculate the potential difference between the positive and the negative plates to be

$$V(d) - V(0) = -\int_0^d -\frac{Q}{A\epsilon_0}dx = \frac{Qd}{\epsilon_0 A}. \tag{13.13}$$

This potential difference, $\Delta V = V(d) - V(0)$, is what we call the *voltage across the capacitor* (shortly, V):

$$V = \frac{Qd}{\epsilon_0 A} \qquad \text{equivalently} \quad V = Ed. \tag{13.14}$$

Fig. 13.2 Parallel plate capacitor

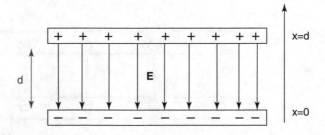

---

[1] http://en.wikipedia.org/wiki/Capacitor
[2] http://en.wikipedia.org/wiki/Defibrillator

We can rewrite this equation by isolating the charge as

$$Q = \frac{\epsilon_0 A}{d} V. \tag{13.15}$$

The capacitance, C, of a capacitor is defined as the ratio of the charge on a capacitor to the voltage across the capacitor:

$$C = \frac{Q}{V}, \tag{13.16}$$

which implies

$$Q = CV. \tag{13.17}$$

Thus, by comparing Eqs. 13.15 and 13.17, we see that for a parallel plate capacitor

$$C = \frac{\epsilon_0 A}{d}. \tag{13.18}$$

Equation 13.18 is an important result which shows that the capacitance only depends on the geometrical properties of the capacitor: the area of plates and the distance between plates. Therefore, the capacitance of a capacitor can be increased by either increasing the area of its plates or decreasing the distance between the plates. Additionally, as we will see a little later in this chapter when discussing dielectrics, the capacitance depends on the medium filling the space between the plates; the capacitance given above in Eq. 13.18 is correct for the case of a capacitor filled by vacuum, if instead the capacitor is filled by a dielectric material, then the constant ϵ_0 will need to be augmented to include the effect of the dielectric. In general, we can say that the capacitance depends on the geometry of the capacitor and the material with which it is filled.

To arrange for the voltage across the plates to be V, we must use a battery or power supply, which generates a voltage V (see Fig. 13.3). We connect the terminals

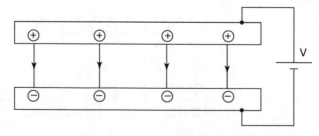

Fig. 13.3 Parallel plate capacitor, charged by a battery. The two unequal line segments, one bigger than the other, that are perpendicular to the conducting wire denote a battery. The larger line segment denotes the positive terminal. The smaller line segment denotes the negative terminal. The voltage, a.k.a. potential difference, created by the battery is V, and as a result there is a charge $+Q$ on the top plate and a charge $-Q$ on the bottom plate, which flows through the conducting wires

of the battery across the capacitor via conducting leads, which then ensures that the voltage across the battery terminals becomes the voltage across the capacitor. On the right side of Fig. 13.3, you can see the circuit symbol that denotes a battery: two unequal line segments, one bigger than the other, that are perpendicular to the conducting wire. The smaller line segment denotes the negative terminal of the battery and the longer one the positive one.

13.4.2 Example: Capacitance of a Spherical Capacitor

Capacitors can be any shape, although the capacitance will be different for different geometries. Figure 13.4 shows a spherical capacitor consisting of two concentric conducting spheres of radius R_1 (inner sphere) and R_2 (outer sphere). We envision that we apply a voltage V to the inner conductor as shown, as a result of which the inner conductor becomes charged with total charge Q, and the outer conductor becomes charged with total charge $-Q$, as shown. The goal of this example is to calculate the capacitance (C) of this spherical capacitor.

(a) Use Gauss's law to determine the electric field outside the outer sphere, which means for any radius $r > R_2$.

Answer: We choose as Gaussian surface a sphere, centered at the center of the capacitor and with radius $r > R_2$.
The total flux through this Gaussian surface (we will call it A) is

$$\Phi = \oint_A \mathbf{E} \cdot d\mathbf{A} = \oint_A E \, dA = E(r) \oint_A dA = E(r) \, 4\pi r^2, \qquad (13.19)$$

where (1) in the first step, we used the fact that the angle θ between the electric field vector $\mathbf{E}$ and the area vector $d\mathbf{A}$ is always zero, since $\mathbf{E}$ is perpendicular to the Gaussian surface at every point, and (2) in the second step, we pulled out of the integral the magnitude E of the electric field, since it is constant over the entire surface (the surface is at a fixed distance r from the center, and the electric field magnitude only depends on the distance from the center). This magnitude $E = E(r)$ is the magnitude at a distance r from the center.

(continued)

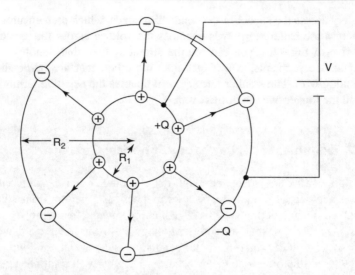

Fig. 13.4 Spherical capacitor, charged by a battery. The two lines one bigger than the other denote
a battery. The big line denotes the positive terminal. The small line denotes the negative terminal.
The voltage, a.k.a. potential difference, created by the battery is V, and as a result, there is a charge
$+Q$ on the inner spherical conductor, which flows there via conducting wire shown, and a charge
$-Q$ on the outer spherical conductors, which flows there through its corresponding conducting
wire

For the charge enclosed in this Gaussian
surface, since $r > R_2$, we know both the charge
on the inner sphere and the charge on the outer
sphere are inside, only and therefore

$$Q_{\text{enclosed}} = +Q - Q = 0. \tag{13.20}$$

Using Gauss's law, we combine these last two
equations to find that the electric field must
be zero:

$$\Phi = \frac{Q_{\text{enclosed}}}{\epsilon_0} = 0 \tag{13.21}$$

$$E(r)4\pi r^2 = 0 \tag{13.22}$$

$$E(r) = 0. \tag{13.23}$$

This electric field is zero everywhere outside
the capacitor.

(continued)

(b) Use Gauss's law to determine the electric field inside the inner sphere, which means for any radius $r < R_1$.

Answer: We proceed similarly to the previous part, and we choose as Gaussian surface a sphere, centered at the center of the capacitor and with radius $r < R_1$.
The total flux through this Gaussian surface (we will call it A) is

$$\Phi = \oint_A \mathbf{E} \cdot d\mathbf{A} = \oint_A E \, dA = E(r) \oint_A dA = E(r) \, 4\pi r^2. \qquad (13.24)$$

For the charge enclosed in this Gaussian surface, since $r < R_1$, and the sphere is a conductor, we know the charge is only on the surface of a conductor, and therefore the charge enclosed in this Gaussian surface must be zero: we know both the charge on the inner sphere and the charge on the outer sphere are inside, only and therefore

$$Q_{\text{enclosed}} = 0. \qquad (13.25)$$

Therefore we find that the electric field is also zero inside the inner shell.

(c) Use Gauss's law to determine the electric field between the two spheres, which means for any radius $R_1 < r < R_2$.

Answer: We proceed similarly to the previous two parts, and we choose as Gaussian surface a sphere, centered at the center of the capacitor and with radius $R_1 < r < R_2$.
The total flux through this Gaussian surface (we will call it A) is

$$\Phi = \oint_A \mathbf{E} \cdot d\mathbf{A} = \oint_A E \, dA = E(r) \oint_A dA = E(r) \, 4\pi r^2, \qquad (13.26)$$

where (1) in the first step, we used the fact that the angle θ between the electric field

(continued)

vector $\mathbf{E}$ and the area vector $d\mathbf{A}$ is always zero,
since $\mathbf{E}$ is perpendicular to the Gaussian surface
at every point, and (2) in the second step, we
pulled out of the integral the magnitude E of
the electric field, since it is constant over
the entire surface (the surface is at a fixed
distance r from the center, and the electric
field magnitude only depends on the distance
from the center). This magnitude $E = E(r)$ is the
magnitude at a distance r from the center.
For the charge enclosed in this Gaussian
surface, since $R_1 \; < \; r \; < \; R_2$, we know only
the charge on the inner sphere is inside and
therefore

$$Q_{\text{enclosed}} = +Q. \tag{13.27}$$

Using Gauss's law, we combine these last two
equations to find the electric field at distance
r:

$$\Phi = \frac{Q_{\text{enclosed}}}{\epsilon_0} \tag{13.28}$$

$$E(r)4\pi r^2 = \frac{Q}{\epsilon_0} \tag{13.29}$$

$$E(r) = \frac{1}{4\pi \epsilon_0} \frac{Q}{r^2}. \tag{13.30}$$

This gives us the magnitude of the electric
field at r, the direction we know already that
is radially outward, we can write the full
vector as

$$\mathbf{E}(r) = \frac{1}{4\pi \epsilon_0} \frac{Q}{r^2} \hat{\mathbf{r}}, \tag{13.31}$$

where $\hat{\mathbf{r}}$ is the unit vector along the radial
direction.

(d) Make a sketch of the magnitude of $E(r)$ for any $r \geq 0$. Clearly show on
your graph the behavior of $E(r)$ for all regions: $0 < r < R_1$, $R_1 < r <
R_2$, and $r > R_2$.

(continued)

Answer: See Fig. 13.5.

(e) What is the potential difference between the inner and outer spheres in terms of Q, R_1, R_2? Hints: Use your result from part (c) and the expression $\Delta V = -\int_{\mathbf{r}_i}^{\mathbf{r}_f} \mathbf{E} \cdot d\mathbf{r}$.

Answer: The electrostatic potential difference between the inner and outer spheres is given by the negative integral of the electric field from the radius of the inner sphere to the radius of the outer sphere:

$$V(R_1) - V(R_2) = -\int_{R_2}^{R_1} \frac{Q}{4\pi\epsilon_0 r^2} dr \qquad (13.32)$$

Evaluating this integral we find

$$V(R_1) - V(R_2) = \frac{Q}{4\pi\epsilon_0}\left(\frac{1}{R_1} - \frac{1}{R_2}\right). \qquad (13.33)$$

(f) How is the potential difference, calculated in the previous part, related to the voltage supplied by the battery?

Answer: They are equal. Since the capacitor is connected to the battery, the assumption is that the system is in electrostatic equilibrium and therefore that means the positive side of the battery and the positive plate of the capacitor are at the same potential and

(continued)

Fig. 13.5 Electric field versus radius for a spherical capacitor

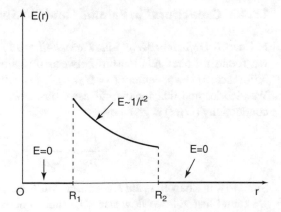

similarly the negative side of the battery and
the negative plate of the capacitor are at the
same potential. As a consequence, the potential
difference between the plates of the capacitor
must be the same as the potential difference
between the terminals of the battery.

(g) What is the capacitance of this spherical capacitor?

Answer: Setting

$$V = \frac{Q}{4\pi\epsilon_0}\left(\frac{1}{R_1} - \frac{1}{R_2}\right) \qquad (13.34)$$

and using

$$Q = CV, \qquad (13.35)$$

it follows that

$$C = \frac{4\pi R_1 R_2 \epsilon_0}{R_2 - R_1}. \qquad (13.36)$$

Similarly to the result previously found for a
parallel plate capacitor (Eq. 13.18), Eq. 13.36
confirms that capacitance only depends on the
geometry of the capacitor (in this case the
inner and outer radius) and the material that
fills the capacitor (ϵ_0).

13.4.3 Capacitors "In Parallel" and "In Series"

In Chap. 6. *Diffusion: Membrane Permeability and the Rate of Actin Polymerization*,
we found a linear relationship between diffusion current, I, and concentration
difference, $n_1 - n_2$, namely $I = G_D(n_1 - n_2)$, where G_D is diffusive conductance.
We also learned that for two diffusive obstacles in series, the combined diffusive
conductivity (G_{DS}) is given by

$$\frac{1}{G_{DS}} = \frac{1}{G_{D1}} + \frac{1}{G_{D2}}. \qquad (13.37)$$

Similarly, in Chap. 9. *Fluid Mechanics: Microfluidics, Blushing, and Murray's Law*,
we found that volume flow rate, Q_F (not to be confused with charge), is linearly

proportional to pressure difference, $P_1 - P_2$, namely that $Q_F = G_F(P_1 - P_2)$, where G_F is the flow conductance. For two flow channels in series, the combined flow conductance (G_{FS}) is given by

$$\frac{1}{G_{FS}} = \frac{1}{G_{F1}} + \frac{1}{G_{F2}}, \tag{13.38}$$

while for two flow channels in parallel, the combined flow conductance (G_{FP}) is given by

$$G_{FP} = G_1 + G_2. \tag{13.39}$$

In this chapter, we have found a linear relationship between charge, Q, and voltage, V: $Q = CV$ or $Q = C\Delta V$ (see Eq. 13.17). Therefore, there is an analogy with the previous examples—specifically that V is analogous to $n_1 - n_2$ and $P_1 - P_2$, respectively, Q is analogous to I and Q_F, respectively, and C is analogous to G_D and G_F, respectively. These analogies are summarized in Table 13.1.

Therefore we may expect that, for two capacitors in series, the combined capacitance is given by

$$\frac{1}{C_S} = \frac{1}{C_1} + \frac{1}{C_2}, \tag{13.40}$$

and, for two capacitors in parallel, the combined capacitance is given by

$$C_P = C_1 + C_2. \tag{13.41}$$

We can confirm these expectations directly. Figure 13.6 shows a number of capacitors of capacitance $C_1, C_2, C_3 ....$ connected (via conducting wires) "in series." The wire at the top is held at a voltage V relative to "ground" (0 V). The symbol at the bottom composed of one longer horizontal line and two shorter horizontal lines indicates a "ground," namely zero voltage. Importantly, two parallel equal-length lines is the symbol used for capacitors in electrical circuit diagrams.

Table 13.1 Linear relationships between quantities in different contexts

Chapter/context	Quantity of interest	Quantity that changes	Proportionality	Relationship
Chapter 6. *Diffusion*	Diffusion current I	Concentration	Diffusive conductance G_D	$I = G_D(n_1 - n_2)$
Chapter 9. *Fluid Mechanics*	Volume flow rate Q_F	Pressure	Flow conductance G_F	$Q_F = G_F(P_1 - P_2)$
Chapter 13. *Electric potential*	Electric charge Q	Electric potential	Capacitance C	$Q = C\Delta V$

Fig. 13.6 Capacitors connected "in series"

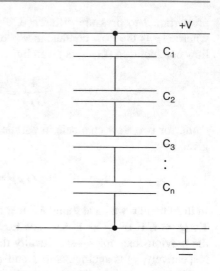

In this case, since the capacitor are connected in series, i.e., one after another, charge conservation requires that each capacitor holds the same charge, Q. In addition, the sum of the voltage drops across all the capacitors must be equal to V, i.e., $V = V_1 + V_2 + V_3 + \ldots$, where $V_1 = Q/C_1$, etc. Therefore

$$V = \frac{Q}{C_1} + \frac{Q}{C_2} + \frac{Q}{C_3} + \ldots = Q\left(\frac{1}{C_1} + \frac{1}{C_2} + \frac{1}{C_3} + \ldots\right) = \frac{Q}{C_S}, \qquad (13.42)$$

where the last equality was used to introduce the total capacitance of capacitors in series, C_S, which is given by

$$\frac{1}{C_S} = \frac{1}{C_1} + \frac{1}{C_2} + \frac{1}{C_3} + \ldots = \Sigma_i \frac{1}{C_i}. \qquad (13.43)$$

Thus, we see that the reciprocal of capacitances in series indeed adds, as expected based on the analogies to diffusion and fluid flow.

Figure 13.7 shows a number of capacitors of capacitance C_1, C_2, $C_3 \ldots$ connected (via conducting wires) "in parallel." The wire at the top is held at a voltage V relative to "ground" (0 V). For capacitors connected in parallel, the voltage drop across all the capacitors is the same, V in this case. Therefore we can write the total charge on all the capacitors as

$$Q = Q_1 + Q_2 + Q_3 + \ldots = C_1 V + C_2 V + C_3 V + \ldots = (C_1 + C_2 + C_3 + \ldots)V = C_P V, \qquad (13.44)$$

where the last equality was used to introduce the total capacitance:

$$C_P = C_1 + C_2 + C_3 + \ldots = \Sigma_i C_i. \qquad (13.45)$$

Thus, we see that capacitances in parallel add, just as expected.

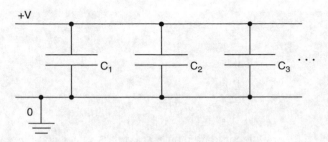

Fig. 13.7 Capacitors arranged "in parallel." Importantly, two parallel equal-length lines is the symbol used for capacitors in electrical circuit diagrams. The symbol (next to the "0") composed of one longer horizontal line and two shorter horizontal lines indicates a "ground," namely zero voltage

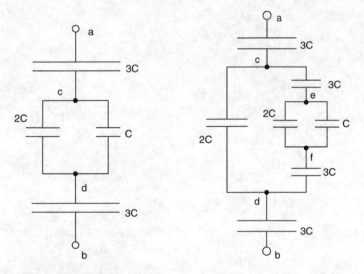

Fig. 13.8 Circuit for "Funky capacitors"

13.4.4 Example: Funky Capacitors

(a) Calculate the total capacitance, C_T, of the arrangement of capacitors shown on the left side of Fig. 13.8.

Answer: We proceed in steps, from the inner to the outer components of the circuit. Here, we first combine together the two capacitors in parallel (2C and C) in the middle part, then we

(continued)

add in series the capacitor at the top (3C) and
at the bottom (3C), and then the desired total
capacitance is given by

$$\frac{1}{C_T} = \frac{1}{3C} \frac{1}{C + 2C} + \frac{1}{3C} = \frac{1}{C}. \tag{13.46}$$

(b) In the left-hand circuit, the voltage at point A is V_A and the voltage at point B is V_B. Calculate the voltage at point C (V_C) in terms of V_A and V_B only.

Answer: We have that the total charge is

$$Q = C(V_A - V_B). \tag{13.47}$$

We also have that

$$Q = 3C(V_A - V_C). \tag{13.48}$$

Equating these two equations for Q, we find

$$V_C = V_A - \frac{1}{3}(V_A - V_B) = \frac{2}{3}V_A + \frac{1}{3}V_B, \tag{13.49}$$

which is the desired answer.

(c) Calculate the voltage at point D (V_D) in terms of V_A and V_B only.

Answer: We have that the total charge is

$$Q = C(V_A - V_B). \tag{13.50}$$

We also have that

$$Q = 3C(V_D - V_B). \tag{13.51}$$

Equating these two equations for Q, we find

$$V_D = V_B + \frac{1}{3}(V_A - V_B) = \frac{1}{3}V_A + \frac{2}{3}V_B, \tag{13.52}$$

which is the desired answer.

(continued)

(d) Calculate the total capacitance of the arrangement of capacitors shown at the right side of the figure at the top of this example.

Answer: In the right-hand side circuit, a single capacitor of capacitance C was replaced by the circuit of (a), which however itself has total capacitance C. Therefore, the capacitance of the right-hand side flow channel is also C.

(e) In the right-hand circuit, the voltage at point A is also V_A and the voltage at point B is also V_B. Calculate the voltage at point E (V_E) in terms of V_A and V_B only.

Answer: In the right-hand circuit, the voltages at point C and point D are V_C and V_D as given in (b) and (c), respectively, while the capacitance between C and D via E and F is C. Therefore, the charge on the capacitors between C and D via E and F--call it q--is

$$q = C(V_C - V_D). \tag{13.53}$$

This charge is also given by

$$q = 3C(V_C - V_E). \tag{13.54}$$

Equating these two expressions for q, we find

$$3C(V_C - V_E) = C(V_C - V_D). \tag{13.55}$$

Solving for V_E, we find

$$V_E = V_C - \frac{1}{3}(V_C - V_D) = \frac{2}{3}V_C + \frac{1}{3}V_D$$
$$= \frac{2}{3}(\frac{2}{3}V_A + \frac{1}{3}V_B) + \frac{1}{3}(\frac{1}{3}V_A + \frac{2}{3}V_B) = \frac{5}{9}V_A + \frac{4}{9}V_B, \tag{13.56}$$

which is the desired answer.

13.5 Energy Stored on a Capacitor

Capacitors are important because they can store electric charge, which can be for instance (as in a defibrillator) discharged as needed. Storing charge means also storing energy, since we know that there is an electrostatic potential energy associated with having charges close to each other. It turns out (as derived in Example 13.5.1) that the electrostatic potential energy for a capacitor with capacitance C and charge Q is given by

$$U = \frac{Q^2}{2C}. \tag{13.57}$$

13.5.1 Example: Electrostatic Energy of a Parallel Plate Capacitor

The goal of this example is to calculate the electrostatic potential energy of a parallel plate capacitor with charge $+Q$ on one plate and $-Q$ on the other.

(a) What is the force exerted by one plate on the other?

Answer: The electric field in the region between the two plates is $-Q/(A\epsilon_0)\hat{\mathbf{k}}$. Only one-half of that field originates from one plate. Therefore, the force exerted by plate 1 on plate 2, which is given by charge times electric field, is

$$Q \times -\frac{Q}{2A\epsilon_0} = -\frac{Q^2}{2A\epsilon_0}. \tag{13.58}$$

The negative sign reflects the fact that the force is in the direction of decreasing plate separation. Interestingly, this force is constant, independent of the separation between the plates.

(b) Calculate the corresponding work done by an external agent.

Answer: In order to separate the plates from a separation of 0 to a separation d, an external force equal and opposite to the force calculated in (a) must be applied. It follows that the work done by the external agent, supplying the force, is

(continued)

$$W = \frac{Q^2 d}{2A\epsilon_0}. \tag{13.59}$$

(c) What is the electrostatic potential energy of the capacitor?

Answer: By energy conservation, the work done by the external agent is equal to the change in electrostatic potential energy of the capacitor. Therefore, the potential energy of the capacitor is

$$U = \frac{Q^2 d}{2A\epsilon_0}. \tag{13.60}$$

We can see that Eq. 13.60 may be re-written in terms of the capacitance and charge alone:

$$U = \frac{Q^2 d}{2A\epsilon_0} = \frac{Q^2}{2C}. \tag{13.61}$$

Although Eq. 13.61 was derived for a parallel plate capacitor, $U = \frac{Q^2}{2C}$ is always the energy of a capacitor of capacitance C, irrespective of the capacitor's geometry. To see this, we imagine two initially uncharged conductors separated by a fixed distance, and we envision taking positive charge away from one plate and depositing it on the other plate. As a result one plate acquires a total charge $+Q$ and the other one is left with a charge $-Q$. Consider a moment in this process at which the charges on the two conductors are $+q$ and $-q$. Consequently, at this moment, the potential difference between the two conductors is

$$V = \frac{q}{C}, \tag{13.62}$$

where C is the capacitance of the arrangement of conductors in question. The change in electrostatic potential energy involved in then increasing the charge separation by a small additional amount, dq, equals

$$dU = V dq = \frac{q}{C} dq. \tag{13.63}$$

The change in potential energy for the entire process, in which the charge on one conductor increases from $q = 0$ to $q = Q$, while that on the other decreases from 0 to $-Q$, is obtained by integrating:

$$U = \int_0^Q \frac{q}{C} dq = \frac{Q^2}{2C}, \tag{13.64}$$

yielding the same expression for U as before, but now in a fashion that does not rely on the parallel plate geometry.

It is also insightful to express Eq. 13.61 in terms of the electric field between the plates: since

$$E = \frac{Q}{A\epsilon_0}, \tag{13.65}$$

we have

$$U = \frac{1}{2}\epsilon_0 E^2 Ad. \tag{13.66}$$

Since Ad is the volume between the plates, which is where there is a non-zero electric field, we see that

$$u = \frac{U}{V} = \frac{1}{2}\epsilon_0 E^2 \tag{13.67}$$

may be conceived as the electrostatic energy density between the plates (i.e., the energy per unit volume). Similarly to Eq. 13.61, Eq. 13.67 is a general result for the electrostatic energy density of any electric field, independent of the parallel plate geometry.

13.6   Dielectric Materials

Dielectrics are materials in which charges are not free to move. In a conductor, where charges are free to move, charges quickly reorganize to reach electrostatic equilibrium, which means a situation in which the electric field inside the conductor is zero and all extra charges are distributed on the surface of the conductor. By contrast, when a dielectric is subject to an applied electric field, the charges within the dielectric material are displaced a relatively small, but usually non-zero, amount from their zero-field locations with positive charges displaced in the opposite direction to negative charges. The dielectric is thus said to become *polarized* in the applied field. Given all positive charges shift a little one way and all negative charges shift the other way, we expect that there is a net charge, induced by the applied electric field, at the surfaces of dielectric materials.

Figures 13.9 and 13.10 illustrate the induced charge and electric in the case of a rectangular dielectric block with its surfaces perpendicular and parallel to the applied field. As a result, an induced surface charge appears on the surfaces perpendicular to the field, which gives rise to a corresponding induced electric field within the material, that is directed in the opposite direction to the applied field.

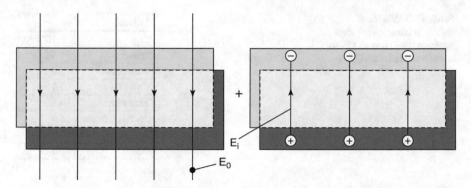

Fig. 13.9 Response of a dielectric material to a downward-directed applied field. When a dielectric material is subjected to a downward electric field (left), the positive charges in the material are slightly displaced downward in response to the electric force they experience, $Q\mathbf{E}$, and the negative charges are slightly displaced upward (right). As shown on the right-hand side of this figure, there is consequently an excess positive charge on the bottom surface and an excess negative charge on the top surface. These excess charges themselves give rise to an induced electric field, which is directed in the opposite direction to the applied field. The total field inside the dielectric is the superposition of the applied external field and the induced field

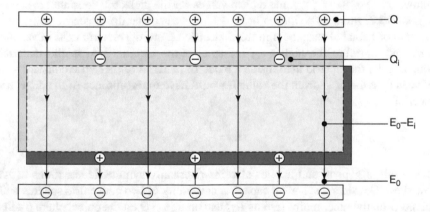

Fig. 13.10 Parallel plate capacitor filled with a dielectric

Therefore, the magnitude of the total electric field within the dielectric block is reduced from its value outside the dielectric. For most materials, the induced field, E_i, is linearly proportional to the applied field E_0, i.e.,

$$E_i = -\alpha E_0. \tag{13.68}$$

It follows that the total field in the material is

$$E = E_0 + E_i = E_0 - \alpha E_0 = (1 - \alpha)E_0 = \frac{E_0}{\epsilon_r}, \tag{13.69}$$

Table 13.2 Values of
dielectric constant for some
common dielectric materials
at room temperature (20°)

Material	ϵ_r
Air	0.1000589
Octane	1.95
Hexadecane	2.05
Olive oil	3.1
Chloroform	4.8
Alcohol (ethanol)	25.3
Glycerol	46.5
Water	80.1

where in the last step we have introduced the dielectric constant,

$$\epsilon_r = \frac{1}{1-\alpha} = \frac{E_0}{E}, \tag{13.70}$$

which is a dimensionless quantity that manifests the ability of a dielectric material
to polarize and therefore to reduce an external electric field. Since $0 \leq \alpha < 1$, it
follows that $1 \leq \epsilon_r < \infty$; some typical values for the dielectric constant are shown
in Table 13.2. From its definition in Eq. 13.70, we see that the dielectric constant is
the ratio of actual electric field in the dielectric (E) to the electric field in vacuum
(E_0), and therefore the higher ϵ_r, the smaller the electric field in the material.
Equivalently, the electric field inside a block of material of dielectric constant ϵ_r is
reduced by a factor $\frac{1}{\epsilon_r}$ from the value it would have in the absence of the dielectric
material:

$$E = \frac{E_0}{\epsilon_r}. \tag{13.71}$$

This result is general and applies also in spherically symmetric situations as we
now show. Consider a spherical capacitor consisting of two concentric spheres with
charge Q on the inner radius (radius R_1) and charge $-Q$ on the outer sphere (radius
R_2). In the absence of dielectric, the field between the spheres is $E_0 = \frac{Q}{4\pi\epsilon_0 r^2}$ (as
we have shown in Example 13.4.2). However, in the case that a dielectric material
fills the space between the two spheres, as in Fig. 13.11, the electric field causes a
small charge separation in this case too. Specifically, the induced electric field is

$$E_i = -\alpha E_0 = -\frac{\alpha Q}{4\pi\epsilon_0 r^2} \tag{13.72}$$

inside the dielectric. Equation 13.72 immediately informs us that the induced
surface charge is $-\alpha Q$. The total field in the dielectric is

$$E = E_0 + E_i = \frac{(1-\alpha)Q}{4\pi\epsilon_0 r^2} = \frac{Q}{4\pi\epsilon_r\epsilon_0 r^2}. \tag{13.73}$$

Fig. 13.11 Spherical capacitor filled with a dielectric

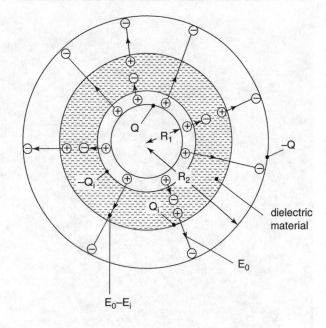

Thus, in this spherically symmetric situation too, it also appears that electrostatics is modified in such a way that ϵ_0 is replaced by $\epsilon_r \epsilon_0$. In fact, this prescription is correct, provided the electric field lines are perpendicular to the surfaces of the dielectrics, which is the case in all situations with planar, spherical, or cylindrical symmetry.

13.6.1 Example: Capacitor Filled with a Dielectric Material

Consider a parallel plate capacitor (as in Fig. 13.10), with $-Q$ on a flat conducting plate of area A at $x = 0$ and $+Q$ on a flat conducting plate of area A at $x = d$. The region between the plates is filled with a material of dielectric constant ϵ_r, so that the electric field between the plates is $\mathbf{E} = -\frac{Q}{A\epsilon_r\epsilon_0}\hat{\mathbf{k}}$.

(a) Calculate the voltage across this capacitor.

```
Answer: Applying Eq. 13.12 and the expression
for the electric field for a parallel plate
capacitor that we found at the end of the last
chapter, the potential difference between the
positive and the negative plates is
```

$$V = -\int_0^d -\frac{Q}{A\epsilon_r\epsilon_0}dx = \frac{Qd}{A\epsilon_r\epsilon_0}. \tag{13.74}$$

(continued)

(b) What is the capacitance, C, of this capacitor?

Answer: By definition, $Q = CV$. Therefore, in this case,

$$C = \frac{A\epsilon_r\epsilon_0}{d}.$$ (13.75)

(c) By considering the change in energy, dU, of this capacitor when its charge is increased from q to dq, calculate the energy, U, of the capacitor when its charge is Q.

Answer: By definition, $Q = CV$. Therefore, in this case,

$$dU = dq \times V = \frac{qdq}{C}.$$ (13.76)

Therefore,

$$U = \int_0^Q \frac{qdq}{C} = \frac{Q^2}{2C} = \frac{Q^2 d}{2A\epsilon_r\epsilon_0}.$$ (13.77)

(d) Express U in terms of the electric field between the plates, $E = \frac{Q}{A\epsilon_r\epsilon_0}$, and the volume between the plates. Hence calculate the energy density between the plates, u, in terms of E.

Answer: We have

$$U = \frac{Q^2 d}{2A\epsilon_r\epsilon_0} = \frac{1}{2}Ad\epsilon_r\epsilon_0\left(\frac{Q}{A\epsilon_r\epsilon_0}\right)^2 = \frac{1}{2}Ad\epsilon_r\epsilon_0 E^2.$$ (13.78)

Since Ad is the volume between the plates, we have that the energy density is

$$u = \frac{1}{2}\epsilon_r\epsilon_0 E^2,$$ (13.79)

which modifies and extends Eq. 13.67 to include dielectric materials.

13.6.2 Example: Ionic Solubility and Membrane Permeability

One of the most important features of cell membranes is that they are essentially impermeable to ions. Remarkably, this property follows directly from the difference in dielectric constant between oil and water. In this example we are going to model oil as hexadecane (since 16 carbons is typical for a cell membrane lipid) and refer to Table 13.2 for the dielectric coefficients.

(a) To explain the physical basis of why cell membranes are impermeable to ions, first calculate the electrostatic energy of a spherical shell of charge Q and radius R in a medium of dielectric constant ϵ_r, either by direct calculation or by recalling that the electrostatic potential energy of a capacitor with change Q is $U = \frac{Q^2}{2C}$.

Answer: For a spherical shell of charge q and radius R, the electric field at radius r ($r > R$) is

$$E = \frac{q}{4\pi \epsilon_r \epsilon_0 r^2}. \tag{13.80}$$

The corresponding electrostatic potential at radius R is

$$V(R) = -\int_\infty^R E \, dr, \tag{13.81}$$

i.e.,

$$V(R) = -\int_\infty^R dr \frac{q}{4\pi \epsilon_r \epsilon_0 r^2} = \frac{q}{4\pi \epsilon_r \epsilon_0 R}, \tag{13.82}$$

where we took the electrostatic potential to be zero at $r = \infty$. To add an additional element of charge dq to a shell of charge q at radius R requires an energy

$$V(R)dq = \frac{q \, dq}{4\pi \epsilon_r \epsilon_0 R}. \tag{13.83}$$

It follows that to increase the charge from an initial value of $q = 0$ to a final value of $q = Q$

(continued)

requires a total energy

$$U = \int_0^Q \frac{q \, dq}{4\pi \epsilon_r \epsilon_0 R} = \frac{\frac{1}{2}Q^2}{4\pi \epsilon_r \epsilon_0 R}, \qquad (13.84)$$

which is the electrostatic potential energy of a spherical shell of charge Q of radius R within a medium of dielectric constant ϵ_r.
Alternatively, the capacitance of a spherical shell of radius R, in a medium of dielectric constant ϵ_r, is given by Eq. 13.36, evaluated for $R_1 = R$ and $R_2 = \infty$ and replacing ϵ_0 by $\epsilon_r \epsilon_0$:

$$C = 4\pi R \epsilon_r \epsilon_0. \qquad (13.85)$$

The corresponding energy is then

$$U = \frac{\frac{1}{2}Q^2}{4\pi \epsilon_r \epsilon_0 R}, \qquad (13.86)$$

which is identical to Eq. 13.84.

(b) The expression calculated in (a) represents a simple model for the energy of an ion of charge Q within a medium of dielectric constant ϵ_r. Given that the dielectric constant of water is about $\epsilon_r = 80$, while the dielectric constant of hexadecane is $\epsilon_r = 2.05$ (see Table 13.2), explain why the energy of an ion in water is smaller than the energy of the same ion in hexadecane, and estimate the ratio of these two energies.

Answer: Equation 13.84 informs us that the electrostatic energy of an ion in water is about 39 times smaller than the electrostatic energy of the same ion in oil.

(c) In Chap. 6. *Diffusion: Membrane Permeability and the Rate of Actin Polymerization*, we introduced the partition coefficient, B, which specifies the concentration ratio for a species of ion dissolved in two immiscible solvents in contact. Specifically, if n_1 is the concentration of the ion in oil and n_2 is its concentration in water, then when these solvents are in contact with each other, we have

$$\frac{n_1}{n_2} = B. \qquad (13.87)$$

(continued)

In Chap. 6. *Diffusion: Membrane Permeability and the Rate of Actin Polymerization*, we saw that B for oil–water is an important factor in determining the permeability of cell membranes because membrane interiors are essentially a layer of oil. Initially, we treated B as an empirical quantity. However, in Chap. 8. *Statistical Mechanics: Boltzmann Factors, PCR, and Brownian Ratchets*, we indicated that B is given by a Boltzmann factor:

$$B = e^{-(U_1 - U_2)/(k_B T)}, \qquad (13.88)$$

where U_2 and U_1 are the energies of the ion in oil and water, respectively. Develop a hypothesis that explains why lipid bilayer membranes are impermeable to ions.

Answer: For an ion in oil or water, we hypothesize that a very important contribution to its total energy (U_1 and U_2, respectively) is electrostatic. In this case, $(U_1 - U_2)/(k_B T)$ may be very large, leading to a tiny B and a corresponding tiny cell membrane permeability to ions, based on what we found in Chap. 6. *Diffusion: Membrane Permeability and the Rate of Actin Polymerization*. The key reason then that cell membranes are impermeable to ions is that an ion has a much higher electrostatic energy in oil than in water because of the relatively much smaller dielectric constant of oil than of water.

(d) By expressing the energy that you calculated in (a) in terms of an integral of the square of the electric field over the volume of space outside the sphere, calculate the energy density, u, associated with the electric field in this case.

Answer: Since for $r > R$, $E(r) = \frac{Q}{4\pi \epsilon_r \epsilon_0 r^2}$, the specified integral is

$$\int_R^\infty E(r)^2 4\pi r^2 dr = \int_R^\infty \frac{Q^2}{(4\pi \epsilon_r \epsilon_0 r^2)^2} 4\pi r^2 dr$$

$$= \frac{1}{\epsilon_r \epsilon_0} \int_R^\infty \frac{Q^2}{(4\pi \epsilon_r \epsilon_0 r^2)} dr = \frac{1}{\epsilon_r \epsilon_0} \frac{Q^2}{4\pi \epsilon_r \epsilon_0 R} = \frac{2}{\epsilon_r \epsilon_0} U.$$

$$(13.89)$$

(continued)

Thus, we may express

$$U = \frac{1}{2}\epsilon_r\epsilon_0 \int_R^\infty E(r)^2 4\pi r^2 dr. \tag{13.90}$$

It follows that the energy density of the electric field is

$$U = \frac{1}{2}\epsilon_r\epsilon_0 E(r)^2, \tag{13.91}$$

which may be compared to Eq. 13.79.

13.7 Electrostatics in Ionic Solutions

In solution, many key biological molecules—DNA, proteins, membrane lipids, etc.—are charged. For example, at physiological pH, DNA is negatively charged, while histone proteins, which DNA winds around in eukaryotic cell nucleii, are positively charged. Therefore, we may expect electrostatic interactions to be an important contributor to the interactions among biological molecules. However, all of these molecules are in ionic solutions, where the electrostatic interactions between charges are "screened," which means that the long-ranged Coulomb interaction is transformed into a short-ranged so-called screened Coulomb interaction. How this happens is illustrated schematically in Fig. 13.12, which shows a snapshot of a possible ionic configuration. Screening is very important in biology: indeed, the electrostatic interactions between proteins is described by a screened Coulomb potential. In this case, the electrostatic energies involved are comparable to $k_B T$, allowing thermal fluctuations to give rise to the charge separations illustrated in the figure.

But how is the configuration of charges, shown in Fig. 13.12, even possible? After all, the negative ions depicted in the solution are subject to an electric field and able to move. Why don't they move to the locations of the positive ions, which would both lower their electrostatic energy and eliminate the electric field in the solution? The answer is that the rule that there cannot be an electrostatic field in a conductor—ionic solutions are conductors—presupposes that the electrostatic energies involved are much larger than $k_B T$. In this section, this assumption is not valid, and positive and negative ions become separated by thermal fluctuations.

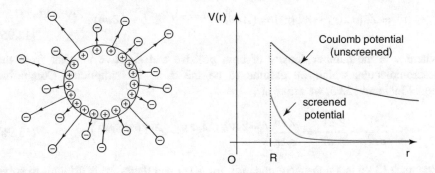

Fig. 13.12 Left side: a preponderance of negative charges in solution around a positively charged spherical macromolecule screen the positive charge's electric field. Right side: electrostatic potential plot as a function of distance from the macromolecule, the screened electrostatic potential, orange line at the bottom, decays much faster than the regular Coulomb potential (unscreened), top blue line

13.7.1 Spherically Symmetric Screening

To model the role of ions in solution, we will suppose that around a large spherical macromolecule of charge Q and radius R, there is a total charge density $\rho(r)$, as a result of positive ions, each of charge $+Ze$, and negative ions, each of charge $-Ze$. Applying Gauss's law (modified to account for the dielectric constant of the solution) to a concentric spherical shell between r and $r + dr$ (where $r > R$), we find

$$4\pi(r+dr)^2 E(r+dr) - 4\pi r^2 E(r) = \frac{4\pi r^2 dr \rho(r)}{\epsilon_r \epsilon_0}. \tag{13.92}$$

Dropping terms proportional to $(dr)^2$, using $(r+dr)^2 \simeq r^2 + 2rdr$ and $E(r+dr) \simeq E(r) + dr\frac{dE}{dr}$, and dividing by $4\pi r^2 dr$, we find

$$\frac{dE}{dr} + \frac{2}{r} E(r) = \frac{\rho(r)}{\epsilon_r \epsilon_0}. \tag{13.93}$$

Using a generalization to radial coordinate of Eq. 13.11, i.e., $E = -\frac{dV}{dr}$, we may rewrite Eq. 13.93 in terms of the potential $V(r)$:

$$-\frac{d^2 V}{dr^2} - \frac{2}{r}\frac{dV}{dr} = \frac{\rho(r)}{\epsilon_r \epsilon_0}. \tag{13.94}$$

The charge density, $\rho(r)$, is equal to the number density of positive charges $(+Ze)$ plus the number density of negative charges $(-Ze)$, where each number density depends on the electrostatic potential energy via the Boltzmann factor. Thus, the charge density is

$$\rho(r) = Ze[n_+(r) - n_-(r)] = (Zen)\, e^{-ZeV(r)/(k_BT)} - (Zen)\, e^{ZeV(r)/(k_BT)},$$
$$(13.95)$$

where n is the number density of both positive and negative ions far from the macromolecule, which we assume to be the same for simplicity. Combining Eqs. 13.94 and 13.95, we arrive at

$$-\frac{d^2V}{dr^2} - \frac{2}{r}\frac{dV}{dr} = \frac{Zen}{\epsilon_r\epsilon_0}\left(e^{-ZeV(r)/(k_BT)} - e^{ZeV(r)/(k_BT)}\right). \qquad (13.96)$$

Equation 13.96 is a non-linear equation for $V(r)$ and therefore is difficult to solve analytically. However, when $ZeV(r)/k_BT \ll 1$ (which will certainly be the case for large enough r), we may approximate

$$e^{\pm ZeV(r)/(k_BT)} \simeq 1 \pm \frac{ZeV(r)}{k_BT}. \qquad (13.97)$$

In this case,

$$\rho(r) \simeq -\frac{2(Ze)^2 n}{k_BT}V(r), \qquad (13.98)$$

and Eq. 13.96 becomes linear

$$\frac{d^2V}{dr^2} + \frac{2}{r}\frac{dV}{dr} \simeq \kappa^2 V(r), \qquad (13.99)$$

where

$$\kappa^2 = \frac{1}{\Lambda^2} = \frac{2(Ze)^2 n}{\epsilon_r\epsilon_0 k_BT}. \qquad (13.100)$$

$\Lambda = 1/\kappa$ is the so-called Debye screening length.[3] Equation 13.99 is a linear equation and so it is easier to find its solution. For $\kappa r \gg 1$, the solution for $V(r)$ is approximately given by

$$V(r) \simeq \frac{c_1}{r}e^{-\kappa r}. \qquad (13.101)$$

The functional dependence in Eq. 13.101 is important because it shows that the dependence on the distance r from the charge Q, source of the electrostatic potential, is not simply $1/r$ as for a charge in vacuum, but it is modified by an exponential decaying factor. This factor makes the potential to decay faster as a function of r, that is why we referred to this as a "screened" potential.

---

[3] http://en.wikipedia.org/wiki/Debye_length

To find the value of c_1 in Eq. 13.101, we need to specify a suitable boundary condition. To this end, we first use Gauss's law to determine the electric field at the surface of the spherical macromolecule. Then, we set that electric field equal to the electric field derived from Eq. 13.101, evaluated at R. The electric field from Eq. 13.101 is

$$E(r) = -\frac{dV}{dr} = -\frac{d}{dr}\left[\frac{c_1}{r}e^{-\kappa r}\right] = \frac{c_1(1+\kappa r)}{r^2}e^{-\kappa r}. \tag{13.102}$$

On the other hand, for a sphere of total charge Q and radius R, Gauss's law applied using a spherical Gaussian surface at (just outside) radius R gives

$$E(R) = \frac{Q}{4\pi\epsilon_r\epsilon_0 R^2}. \tag{13.103}$$

Equating Eq. 13.102, evaluated for $r = R$, to Eq. 13.103, we find

$$\frac{c_1(1+\kappa R)}{R^2}e^{-\kappa R} = \frac{Q}{4\pi\epsilon_r\epsilon_0 R^2}. \tag{13.104}$$

Thus,

$$c_1 = \frac{Q}{4\pi\epsilon_r\epsilon_0(1+\kappa R)}e^{\kappa R}, \tag{13.105}$$

and

$$V(r) \simeq \frac{Q}{4\pi\epsilon_r\epsilon_0(1+\kappa R)r}e^{-\kappa(r-R)}. \tag{13.106}$$

We can also calculate the corresponding charge density for $r > R$:

$$\rho(r) \simeq -\frac{2(Ze)^2 n}{k_B T} \times \frac{Q}{4\pi\epsilon_r\epsilon_0(1+\kappa R)r}e^{-\kappa(r-R)} = -\frac{Q}{4\pi\kappa^2(1+\kappa R)r}e^{-\kappa(r-R)}. \tag{13.107}$$

We see that the charge density in the solution surrounding the charge, Q, is opposite in sign. This net opposite charge "screens" the charge Q's electric field.

Equation 13.106 may be compared with the corresponding potential for a charge, Q, within a dielectric medium:

$$V(r) = \frac{Q}{4\pi\epsilon_r\epsilon_0 r}. \tag{13.108}$$

This comparison shows us that the potential in ionic solution is reduced from the Coulomb potential by an r-dependent exponential factor that becomes rapidly small for values of r greater than the Debye length, Λ (see the left side of Fig. 13.12).

This phenomenon is called "screening." If the Debye length were large compared to cellular scales, screening would not be especially important biologically. However, at a concentration of 154 mM NaCl, corresponding to normal saline, $\Lambda \simeq 0.8$ nm. Therefore, screening is important at the molecular scale. The result that electrostatic interactions in the ionic cellular milieu are much more short-ranged than in a dielectric solution without ions is fundamental in biology and biochemistry.

13.7.2 Example: Energy of a Charge Q in Ionic Solution

Calculate the energy of a charged spherical shell of radius R and charge Q in an ionic solution of Debye screening length $\Lambda = \kappa^{-1}$.

Answer: Equation 13.106 informs us that the electrostatic potential of a charged spherical shell of radius R and charge q in an ionic solution is

$$V = \frac{q}{4\pi \epsilon_r \epsilon_0 R(1 + \kappa R)}. \tag{13.109}$$

Therefore, electrostatic potential energy to increase the charge in the shell by dq is

$$dU = V\,dq = \frac{q\,dq}{4\pi \epsilon_r \epsilon_0 R(1 + \kappa R)}. \tag{13.110}$$

It follows that the total potential energy of the shell is

$$U = \frac{\frac{1}{2}Q^2}{4\pi \epsilon_r \epsilon_0 R(1 + \kappa R)}. \tag{13.111}$$

13.7.3 Viral Assembly

A nice application of Eq. 13.111 is to the assembly of a viral capsid, such as the capsid of hepatitis B virus (HBV) illustrated in the figure at the beginning of this chapter (Fig. 13.1) and in Fig. 13.13. HBV's capsid is a hollow shell with icosahedral symmetry. HBV is similar to many viruses with genomes based on double-stranded DNA in that it first self-assembles an empty viral capsid, and subsequently a molecular motor forces genomic material into the capsid. The hepatitis B capsid consists of 120 protein dimers, which self-assemble from solution to form the capsid.

Fig. 13.13 Sketch of the icosahedral hepatitis B viral capsid. Each red dot represents a hepatitis B http://en.wikipedia.org/wiki/Hepatitus_B viral capsid protein. The capsid is composed of a total of 240 of these proteins (120 protein dimers)

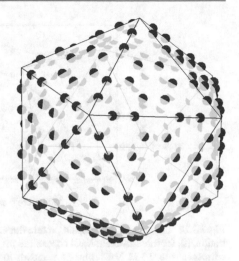

To discuss the electrostatics of the HBV capsid, although its shape is actually icosahedral, we will approximate it as being spherically symmetric. The capsid protein dimers are held together by hydrophobic forces, which operate between the contact surfaces of the dimers and make a negative contribution to the capsid binding energy. However because the dimers are charged in solution, there is also a positive electrostatic contribution to their total binding energy. At physiological conditions, $\Lambda \simeq 0.7$ nm. At the same time, for the hepatitis B virus, $R \simeq 42$ nm. It follows that $\kappa R = R/\Lambda \simeq 60$. Therefore, it is a good approximation to neglect 1 compared to κR in Eq. 13.111, so that, to a good approximation, the electrostatic part of the capsid's energy is

$$U \simeq \frac{\frac{1}{2}Q^2}{4\pi\epsilon_r\epsilon_0\kappa R^2} = \frac{\frac{1}{2}Q^2\Lambda}{4\pi\epsilon_r\epsilon_0 R^2}, \tag{13.112}$$

where Q is the total charge of the capsid. The total free energy of the capsid (G_C) is the sum of the electrostatic part plus a part from the protein–protein contacts which gives rise to binding, G_B, which is necessarily negative, i.e.,

$$G_C = \frac{\frac{1}{2}Q^2}{4\pi\epsilon_r\epsilon_0\kappa R^2} + G_B. \tag{13.113}$$

For dimers to assemble into a complete capsid, they effectively undergo the chemical reaction:

$$120\,\mathrm{d} \rightarrow \mathrm{c}, \tag{13.114}$$

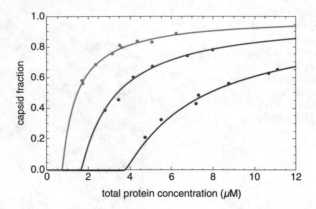

Fig. 13.14 Fraction of HBV capsid protein dimers in capsids, $\frac{[c]}{[t]}$, versus the total dimer concentration, [t], for three different NaCl concentrations [1]. Points are experimental measurements. Red corresponds to 0.3 M NaCl, blue corresponds to 0.5 M NaCl, and orange corresponds to 0.7 M NaCl. Lines correspond to Eq. 13.117 with $\log \frac{K}{K_\infty}$ =25 (red), 125 (blue), and 225 (orange)

where d denotes dimer and c denotes capsid. Chemistry's Law of Mass Action[4] then tells us that

$$\frac{[c]}{[d]^{120}} = K, \tag{13.115}$$

where [d] and [c] are the concentrations of dimer and capsid, respectively, and K is the equilibrium constant for capsid assembly. At the same time, the total dimer concentration, [t] in an experiment, is

$$[t] = [d] + 120[c] = [d] + 120K[d]^{120}. \tag{13.116}$$

The factor of 120 in this equation is because one capsid contains 120 dimers. Then, the fraction of dimers in capsids—the capsid fraction—is

$$\frac{120[c]}{[t]} = \frac{120[c]}{[d] + 120[c]} = \frac{120K[d]^{120}}{[d] + 120K[d]^{120}}. \tag{13.117}$$

Figure 13.14 shows a comparison between experimental measurements of the fraction of dimers in capsids versus the total dimer concentration from Ref. [1] to the predictions of Eq. 13.117 for three different NaCl concentrations. With appropriate values of K, it is clear that the model provides a good description of the measurements, giving us confidence that we can use the model to determine K for different [NaCl].

---

[4] https://en.wikipedia.org/wiki/Law_of_mass_action

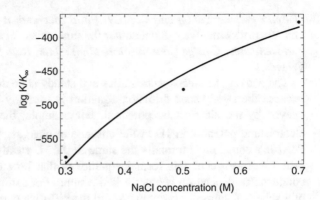

Fig. 13.15 Equilibrium constant for viral assembly as a function of NaCl concentration[1–3]. Circles are experimental values, determined from Fig. 13.14. The line is $\log \frac{K}{K_\infty} = -\frac{320}{\sqrt{[\text{NaCl}](M)}}$

When we discussed the cooperative binding of four oxygens by hemoglobin in Chap. 8. *Statistical Mechanics: Boltzmann Factors, PCR, and Brownian Ratchets*, for example, we found that the equilibrium constant K of such a reaction is proportional to a Boltzmann factor involving the total free energy. In analogy with that calculation, the equilibrium constant for capsid assembly is given by $K = \frac{e^{-G_C/(k_B T)}}{v^{120}}$, so that

$$\log K = -\frac{G_C}{k_B T} - 120 \log v = -\frac{G_B}{k_B T} - 120 \log v - \frac{\frac{1}{2}Q^2}{4\pi \epsilon_r \epsilon_0 \kappa R^2 k_B T}$$

$$= \log K_\infty - \frac{\frac{1}{2}Q^2}{4\pi R^2 Ze\sqrt{2\epsilon_r \epsilon_0 k_B T n}}, \tag{13.118}$$

where we introduced the equilibrium constant at high salt ($n = \infty$), $K_\infty = -\frac{G_B}{k_B T} - 120 \log v$. Evidently, $\log K - \log K_\infty = \log \frac{K}{K_\infty}$ is predicted to be proportional to $n^{-1/2}$. In Fig. 13.15, this prediction, shown as the line, is compared to the values of $\log \frac{K}{K_\infty}$ from Fig. 13.14, shown as the circles [1–3], revealing good agreement. As [NaCl] increases from 0.3 M to 0.7 M, $\log \frac{K}{K_\infty}$ increases by about 200. Because the difference in free energy of a capsid at finite [NaCl] and infinite [NaCl] is $\Delta G = -k_B T \log \frac{K}{K_\infty}$, this change in $\log \frac{K}{K_\infty}$ corresponds to a decrease in the free energy/electrostatic energy of the HBV capsid of about $200 k_B T$.

13.8 Chapter Outlook

This chapter completes this book's study of electrostatics, exploring the concepts of electrostatic potential energy, electric potential, and capacitance. We saw that the rules for combining capacitances in series or parallel are the same as the rules for combining diffusive conductance, flow conductance, and electrical conductance, which we encountered in Chap. 6. *Diffusion: Membrane Permeability and the Rate of Actin Polymerization* and Chap. 9. *Fluid Mechanics: Laminar Flow, Blushing, and Murray's Law*. We will encounter the same rules yet again in Chap. 14. *Circuits and Dendrites: Charge Conservation, Ohm's Law, Rate Equations, and Other Old Friends*.

The analogy between electrostatics and steady-state diffusion is especially deep because the steady-state diffusion equation is formally the same as the equation obeyed by the electrostatic potential. For example, the equation obeyed by the electrostatic potential in 1D in the absence of charges, $\frac{d^2 V}{dx^2} = 0$, which is called Poisson's equation, is formally the same as the 1D steady-state diffusion equation. One consequence of this formal equality is that how capacitance and diffusive conductance depend on geometry is the same. For example, both the capacitance of a spherical capacitor of radius R and the diffusive conductance between infinity and a spherical particle sink at radius R are linearly proportional to R, while the capacitance of a parallel plate capacitor with plates of area A and separation, d, and the diffusive conductance of a membrane of area A and thickness, d, are both proportional to $\frac{A}{d}$. As noted previously, this sort of analogy, where understanding of one system translates into understanding of another, physically quite different system, is beloved by physicists.

13.9 Problems

Problem 1: Facts to Know
(a) What is an insulator?
(b) What is a conductor?
(c) What is the value of the E-field inside a (perfect) conductor?
(d) How does the potential vary from place to place across the surface of a (perfect) conductor?
(e) What is capacitance?
(f) What is a capacitor?[5]
(g) What is a dielectric constant?

Problem 2: Electric Potential Energy of a Point Charge or of a System of Charges
Consider the following setup (similar to problem 2 of Chap. 12): a charge $Q_1 = Q$ is located at Cartesian coordinates $(-a, a, 0)$. A charge $Q_2 = Q$ is located at coordinates $(a, -a, 0)$. A charge $Q_3 = Q$ is located at coordinates $(-a, -a, 0)$. Then add a charge Q_0 at $(a, a, 0)$.

(a) Calculate the *total* potential energy *of the charge* Q_0 due to the interaction with the other three charges. (Hint: You need to consider the potential energy of Q_0 paired with each of the other charges, i.e., adding U_{01}, U_{02}, U_{03}, and U_{04}).

---

[5] http://en.wikipedia.org/wiki/Capacitor

(b) What happens to your answer to part (a) if the charge Q_3 at $(-a, -a, 0)$ is replaced by a charge $-2Q$?

Problem 3: Electrostatic Energies
(a) Calculate the electrostatic potential energy of a quadrupole, consisting of a charge $+2Q$ at the origin, $(0, 0, 0)$, a charge, $-Q$, at $(a, 0, 0)$, and a charge, $-Q$, at $(-a, 0, 0)$.
(b) Calculate the electrostatic energy of the following arrangement of charges: charges $+Q$ at $(0, 0, 0)$ and $(2a, 0, 0)$ and charges $-Q$ at $(-a, 0, 0)$ and $(a, 0, 0)$.

Problem 4: A Battery and Capacitor
A parallel plate capacitor with plate separation d is connected to a battery that provides potential difference ΔV_{bat}.

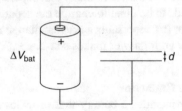

Without breaking any of the connections, insulating handles are used to increase the plate separation to $2d$.

(a) Does the capacitance change? If so, by how much? If not, why not?
(b) Does the potential difference across the capacitor change as the separation increases? If so, by how much? If not, why not?
(c) Does the capacitor charge Q change? If so, by how much? If not, why not?
(d) Does the magnitude of electric field between the plates of the capacitor change? If so, by how much? If not, why not?

Problem 5: An Isolated Capacitor with Charge
A capacitor with capacitance C is charged by an amount Q. It is isolated and not connected to a battery or any other electrical components.

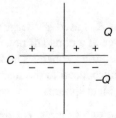

Insulating handles are used to increase the plate separation to $2d$.

(a) Does the capacitance change? If so, by how much? If not, why not?
(b) Does the capacitor charge Q change? If so, by how much? If not, why not?
(c) Does the potential difference across the capacitor change as the separation increases? If so, by how much? If not, why not?
(d) Does the magnitude of electric field between the plates of the capacitor change? If so, by how much? If not, why not?
(e) Compare how the voltage across the capacitor changes (or does not) in this problem and the last one. Does this make sense in terms of how the potential is defined, $V = -\int \mathbf{E} \cdot d\mathbf{l}$?

Problem 6: Work on a Capacitor
A capacitor has charge $+Q$ on one plate and $-Q$ on another plate, with these plates separated by a distance d.

(a) How much work needs to be done to make that separation $2d$?
(b) How much work needs to be done to change the separation from d to $d/2$?
(c) Which scenario, (a) or (b), represents a greater change in potential energy of the system? Explain why your answer makes sense.

Problem 7: Charge on a Capacitor
The circuit shown below includes a battery that provides a potential difference V and five capacitors that all have the same capacitance C.

(a) What is the charge on capacitor labeled C_1? Call it Q_1.
(b) What is the charge on capacitor labeled C_2 in terms of Q_1?

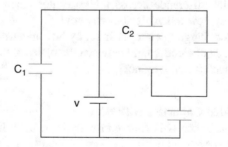

Problem 8: Charge Transfer
A conducting sphere A of radius R is used to charge a larger sphere of radius $3R$ by repeatedly touching it to a source at a constant potential V_0 that is located very far away and then touching it to the larger sphere. The larger sphere has a hole in it so that the smaller sphere can be introduced to the inside, but you should assume that the two conductors behave like ideal spheres.

(a) Find the capacitances of the two spheres, C_A and C_B, respectively, when they are far apart.

(b) The small sphere is touched once to the source and then touched to the outside of the larger sphere. What is the final potential of the larger sphere?

(c) The small sphere is touched once to the source and then touched to the inside of the larger sphere. What is the final potential of the larger sphere?

(d) Process (b) is repeated 100 times. For this many times, the potential of the larger sphere approaches its limiting value. Find that value.

(e) Process (c) is repeated 100 times. For this many times, the potential of the larger sphere approaches its limiting value. Find that value.

Problem 9: Forces on Dielectrics

(a) What is the total capacitance of two capacitors (C_1 and C_2) that are connected in series?

(b) What is the total capacitance of two capacitors (C_3 and C_4) that are connected in parallel?

(c) What is the energy of a capacitor of capacitance C, given that the total charge on the capacitors is Q?

(d) Consider a square parallel plate capacitor. The separation between plates is d and the dimensions of each plate is $L \times L$. What is the capacitance?

(e) Now, a block of a dielectric material of dielectric constant ϵ_r, width L, length L, and height d is partially slid between the plates. Over an area Lx the plates are separated by the dielectric, and over an area $L(L - x)$ the plates are separated by vacuum as shown in the diagram. How can the system be redrawn as two connected capacitors? What is the capacitance of this system? As usual, ignore ALL edge effects.

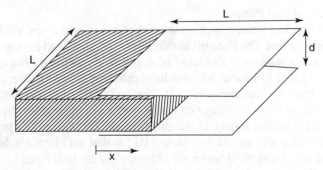

(f) If the capacitor's charge is fixed to be Q, what is the energy of the square capacitor with the partially slid-in dielectric? (Hint: Use your answer to (e).)

(g) If $\epsilon_r = 1$, what is the energy? Call this energy U_0. What is the physical interpretation of U_0 and $\epsilon_r = 1$?

(h) Rewrite the energy function you calculated in part (f) as a function of $\frac{x}{L}$. What is the physical interpretation of $\frac{x}{L}$? Use Wolfram Alpha to plot the potential energy as a function of $\frac{x}{L}$. In order to plot the function in Wolfram Alpha, let $U_0 = 1$ and $\epsilon_r \approx 80$, which is the dielectric constant for water. Record a drawing of the graph. Does the energy increase or decrease as x increases?

(i) Recall how to calculate the force from a potential energy function. Calculate the x-component of the force on the dielectric.
(j) Does this force pull the dielectric further into the capacitor or push it out? How can you tell the direction of the force directly from the graph you drew in (h)?

Problem 10: Spherical Capacitor Revisited

A spherical capacitor is composed of two concentric conducting spheres, one of radius a and the other of radius c $(c > a)$. In addition, between the two conductors, there is a spherical shell of dielectric material (relative permittivity/relative dielectric constant ϵ) with inner radius b $(c > b > a)$ and outer radius c. The charge on the inner conductor is $+Q$. The charge on the outer conductor is $-Q$.

(a) Make a sketch of the situation, indicating the relevant dimensions.
(b) Determine the magnitude of the electric field E at radius r for $a < r < b$.
(c) Determine the magnitude of the electric field E at radius r for $b < r < c$.
(d) What is the (induced) surface charge density on the inner surface of the dielectric.
(e) Sketch the radial component of the electric field versus r.
(f) Sketch the electrostatic potential versus r.
(g) Calculate the potential difference between the conductor at $r = a$ and that at $r = c$.
(h) What is the capacitance of this capacitor?

Problem 11: Parallel Plates

As shown in Fig. 13.16, three large conducting plates of area A and thickness d are parallel to one another. The plate in the middle is separated from the top and bottom plates—the outer plates—by distances of a and b, respectively. The outer plates are shorted together and therefore must have the same electrostatic potential. The middle plate carries a total charge of $+Q$. The goal of this problem is to determine Q_1 and Q_2, where Q_1 is the charge on the top surface of the middle plate and Q_2 is the charge on the bottom surface of the middle plate. Throughout this problem, you should ignore edge effects, and you should take it that the electric field above the top plate and the electric field below the bottom plate are both equal to zero. Given quantities in this problem are: Q, a, b, d, A, and ϵ_0.

(a) Make your own sketch, showing the plates, charges, and electric field lines. Show the direction of the electric fields on your sketch. Tip: Think about what is the electric field inside the conducting plates?
(b) What is the total charge on the outer surfaces of the middle plate? i.e., write down a relationship between Q and Q_1 and Q_2.
(c) What is the charge on the top surface of the top plate ?
(d) What is the charge on the bottom surface of the bottom plate (2 points)?

Fig. 13.16 Parallel
conducting plates

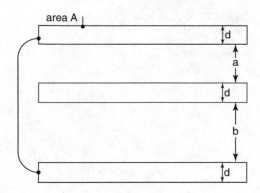

(e) Calculate the electric field between the bottom surface of the top plate and the top surface of the middle plate (E_1) in terms of the charge residing on the top surface of the middle plate, Q_1, and given/known quantities.

(f) Calculate the electric field between the top surface of the bottom plate and the bottom surface of the middle plate (E_2) in terms of the charge residing on the bottom surface of the middle plate, Q_2, and given/known quantities.

(g) Using your answer to (e), calculate the electrostatic potential differences between the top plate and the middle plate (V_1) in terms of Q_1 and given/known quantities.

(h) Using your answers to (f), calculate the electrostatic potential differences between the bottom plate and the middle plate (V_2) in terms of Q_2 and given/known quantities. How are V_1 and V_2 related?

(i) Using your results, solve for Q_1 and Q_2 in terms of given/known quantities.

Problem 12: Giant Sucking Sound
In this problem, you may find it useful to recall that the capacitance of a parallel plate capacitor (C) is given by $C = A\epsilon_r\epsilon_0/d$, where A is the area of the capacitor, ϵ is the dielectric constant of the material between the plates, and d is the separation between the plates

(a) Capacitor 1 is a rectangular parallel plate capacitor, with length z and width L, and the plate separation is d, and a fluid of dielectric constant ϵ fills the space between the plates. What is the capacitance, C_1, of capacitor 1?

(b) Capacitor 2 is a rectangular parallel plate capacitor, with length $L - z$ and width L, and the plate separation is d, and there is a vacuum in the space between the plates. What is the capacitance, C_2, of capacitor 2?

(c) Capacitor 1 and capacitor 2 are connected in parallel. What is the total capacitance of the two capacitors?

(d) If the capacitors hold a total charge Q between them, what is their total electrostatic energy in terms of the given quantities?

(e) In fact as shown in Fig. 13.17, capacitor 1 and capacitor 2 are simply two parts of a vertical square parallel plate capacitor of area L^2. The space between the

Fig. 13.17 Capacitor for
Problem 12(e)

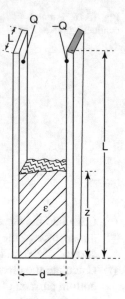

plates is partially filled with a liquid of dielectric constant ϵ up to a height z. Given your answer to (d), what is the electrostatic force on the liquid? Is the direction of this force upward or downward?

Problem 13: Screening Length

As any pre-medical student worth their salt should know—big groan—normal saline[6] is 0.154 M NaCl. (Groan.)

Estimate the Debye screening length for a normal saline solution.

Problem 14: Going Viral

Could you assemble a capsid in a solution of de-ionized water? Explain. (Hint: Make sure you study Sect. 13.7 *Electrostatics in Ionic Solutions*)

Problem 15: Spherical Capacitor Revisited

A spherical capacitor is composed of two concentric conducting spheres, one of radius a and the other of radius c $(c > a)$. In addition, between the two conductors, there is a spherical shell of dielectric material (relative permittivity/relative dielectric constant ϵ) with inner radius b $(c > b > a)$ and outer radius c. The charge on the inner conductor is $+Q$. The charge on the outer conductor is $-Q$.

---

[6] http://en.wikipedia.org/wiki/Saline_solution

(a) Make a sketch of the situation, indicating the relevant dimensions.
(b) Determine the magnitude of the electric field E at radius r for $a < r < b$.
(c) Determine the magnitude of the electric field E at radius r for $b < r < c$.
(d) What is the (induced) surface charge density on the inner surface of the dielectric.
(e) Sketch the radial component of the electric field versus r.
(f) Sketch the electrostatic potential versus r.
(g) Calculate the potential difference between the conductor at $r = a$ and that at $r = c$.
(h) What is the capacitance of this capacitor?

Problem 16: Another Giant Sucking Sound

In order to perform to specification, a large horizontal, parallel plate capacitor with plate separation d and plate area A must be filled up with liquid of dielectric constant ϵ and mass density ρ. Unfortunately, the liquid inlet is at the bottom of the capacitor. A student suggests that it should be possible to arrange for the capacitor to fill itself by charging it sufficiently. The goal of this problem is to investigate this proposal. Suppose that the charges on the capacitor plates are $\pm Q$ throughout this problem.

(a) What is (the magnitude of) the electric field between the plates before there is any fluid there? Call it E.
(b) Now suppose that the space between the plates has been filled to a depth z by the dielectric fluid of dielectric constant ϵ and mass density ρ. What is the electric field within the dielectric fluid? Call it E_1.
(c) What is the electrostatic potential energy, U_e, stored within the capacitor when the space between the plates has been filled to a depth z by the dielectric fluid of dielectric constant ϵ and mass density ρ?
(d) What is the gravitational potential energy, U_g, when the space between the plates has been filled to a depth z by the dielectric fluid of dielectric constant ϵ (relative to when the capacitor is empty)?
(e) What is the total potential energy, U, of the capacitor plus fluid system when the space between the plates has been filled to a depth z by the dielectric fluid of dielectric constant ϵ?
(f) What is the force on the fluid, F, when the space between the plates has been filled to a depth z by the dielectric fluid of dielectric constant ϵ?
(g) In light of your answer to (f), comment on whether the student's proposal is feasible or not and specify any conditions needed for feasibility.

Problem 17: Capacitor Design Project

After graduation, you are working for an electrical engineering company on a project to design a next-generation defibrillator. Your supervisor has asked you to investigate the possibility of using an infinitely long capacitance ladder in this device. To this end, you draw the capacitance ladder shown in Fig. 13.18. At point A, the voltage is V_A. At point B, the voltage is V_B with $V_B < V_A$. The "rungs" of your capacitance ladder have capacitance C and the "risers" have capacitance C_1,

Fig. 13.18 Flow circuit for
"Capacitor design project"

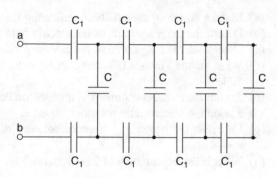

where C_1 is to be determined in terms of C in order to achieve the desired electrical
properties, which include a relatively large capacitance and a relatively long length
over which the charge decreases along the ladder. Because of these considerations,
you determine that the total capacitance of the ladder should be $8C$. The goal of this
problem is to determine how to pick C_1 and to determine what is the decay length
of the charge in the capacitance ladder.

(a) Draw a circuit that is equivalent to the infinite capacitance ladder of Fig. 13.18,
    but that replaces the circuit beyond the first rung by a single capacitor of the
    appropriate capacitance.
(b) Determine the required value of C_1 by first calculating the total capacitance of
    the circuit that you drew in (a) in terms of C and C_1 (or otherwise).
(c) What fraction of the total charge on the ladder is located on the first-rung
    capacitor, when C_1 has the value you determined in (b)?
(d) What fraction of the total charge on the ladder is located beyond the first rung?
(e) What fraction of the total charge on the ladder is located beyond the second
    rung?
(f) What fraction of the total charge on the ladder is located beyond the n-th rung?
(g) We can define the decay "length," N, of the capacitance ladder—'it is actually
    the number of rungs over which the charge decays—by asserting that the answer
    to (f) is equal to $e^{-n/N}$. Using this result and your answer to (f), calculate N.
    Next, use the result that for small x, $\log(1+x) \simeq x$, to simplify your expression
    for N.

References

1. P. Ceres, A. Zlotnik, Weak protein-protein interactions are sufficient to drive assembly of
   hepatitis B virus capsids. Biochemistry **41**, 11525–11531 (2002)
2. W.K. Kegel, P. van der Schoot, Competing hydrophobic and screened-Coulomb interactions in
   hepatitis B virus capsid assembly. Biophys. J. **86**, 3905–3913 (2004)
3. R. Phillips, J. Kondev, J. Theriot, *Physical Biology of the Cell* (Garland Science, 2008)

Circuits and Dendrites: Charge Conservation, Ohm's Law, Rate Equations, and Other Old Friends

14

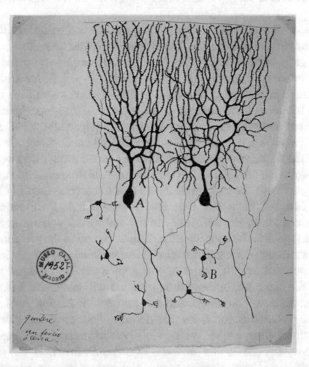

Fig. 14.1 Drawing of Purkinje cells (A) and granule cells (B) from pigeon cerebellum by Santiago Ram-n y Cajal, 1899; Instituto Cajal, Madrid, Spain. Public domain image from https://commons. wikimedia.org/wiki/File:PurkinjeCell.jpg

© The Author(s), under exclusive license to Springer Nature Switzerland AG 2023   669
S. Mochrie, C. De Grandi, *Introductory Physics for the Life Sciences*,
Undergraduate Texts in Physics, https://doi.org/10.1007/978-3-031-05808-0_14

14.1   Introduction

In Chap. 12. *Gauss's Law: Charges and Electric Fields* and Chap. 13. *Electric Potential, Capacitors, and Dielectrics*, we discussed only situations in electrostatic equilibrium. In other words, the charges were strictly stationary. In particular, inside conductors, we imagined the charges to have moved until the **E**-field was zero in the conductor, but that any motion that took place to create the zero field had already occurred before we started to analyze the situation. By contrast, in this chapter, we will consider electrical circuits in which there are charges in motion, as a result of electric fields in conductors. This situation is possible if the charges cannot accumulate at locations necessary to cancel the applied electric field. Essential concepts in the context of electrical circuits include current (usually I), voltage (usually V), and electrical resistance (usually R) or conductance (usually $G = \frac{1}{R}$). Current is the electrical current and is the amount of charge crossing a plane (usually the cross-section of a wire) per unit time. Voltage is what drives current through a resistance.

The motion of individual charges through a conducting medium may be described similarly to the way in which we described the motion of a sphere falling through a viscous fluid under the influence of gravity or the motion of a charged macromolecule moving through a gel in Chap. 2. *Force and Momentum: Newton's Laws and How to Apply Them*. In this chapter, we will see that there is a microscopic frictional force on each charge that gives rise to a linear relationship between current and voltage, i.e., $I = GV$, or equivalently $V = IR$, which is Ohm's law. We will see that electrical resistance and conductance are the electrical analogues of the diffusive resistance and diffusive conductance that we encountered in Chap. 6. *Diffusion: Membrane Permeability and the Rate of Actin Polymerization*, and of the fluid resistance and conductance that we encountered in Chap. 9. *Fluid Mechanics: Laminar Flow, Blushing, and Murray's Law*. Furthermore, electrical current is analogous to the diffusive current and to the volume flow rate, while voltage is analogous to the concentration difference and to the pressure difference.

We will also consider electrical circuits containing capacitors in addition to resistors. Because the rate of change of charge on a capacitor is equal to the current flowing into the capacitor, in these situations, we will encounter linear differential equations that describe how charge and current vary versus time, which are closely similar to the differential equations that describe how drug concentrations in the body vary versus time, which we previously encountered in Chap. 7. *Rates of Change: Drugs, Infections, and Weapons of Mass Destruction*. On this basis, we may anticipate that in such circuits charge and current relax exponentially toward their steady-state values.

In this chapter, therefore, we will encounter a number of "old friends" in a new context. The mathematical skills and physical intuition that we developed in the relevant earlier modules will all be applicable here. As noted previously, physicists love being able to exploit analogies in this way.

At the end of the chapter, we will discuss an important physiological application of electrical circuits involving resistance and capacitance, namely to the dendrites of neurons (Fig. 14.1), and therefore, we will discuss dendritic conduction, a model to describe how electric signals are propagated between neurons.

14.2 Your Learning Goals for This Chapter

By the end of this chapter, you should be able to:

- Utilize Ohm's law to describe the relationship between current, resistance, and voltage in a circuit.
- Apply Kirchhoff's current law, namely that the current into a node of an electrical circuit equals the current out of the node.
- Apply Kirchhoff's loop law, namely that the voltage drop between two locations in a circuit containing resistors and capacitors is independent of route.
- Calculate the voltage drops across resistors and capacitors.
- Calculate the current through resistors and capacitors.
- Show that charge and current in circuits containing resistors and capacitors relax exponentially to their steady-state values and be able to determine the time scale(s) for relaxation.

14.3 Current, Resistance, and Ohm's Law

To start to engage with these concepts, we consider a charge Q inside a conductor of length L and cross-sectional area A (see Fig. 14.2). We furthermore suppose that a voltage V is applied across the ends of the conductor and that as a result of this

Fig. 14.2 Quantities relevant to the resistance of a rectangular block

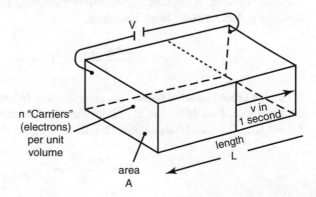

voltage, charges within the conductor experience a corresponding uniform electric field **E**. According to the relationships between voltage and electric field given in the previous chapter (see Eq. 13.4), we can determine the magnitude of the electric field to be $E = V/L$. Then, Newton's second law for the charge Q is

$$ma = QE - \zeta v = \frac{QV}{L} - \zeta v, \qquad (14.1)$$

where m is the mass of the charge, a is its acceleration, and $QE = \frac{QV}{L}$ is the electrostatic force on the charge, and we have added a frictional term, namely $-\zeta v$, which is of the same form as for a particle moving through a viscous fluid, with v the velocity of the charge and ζ is the friction coefficient. In Chap. 2. *Force and Momentum: Newton's Laws and How to Apply Them*, we found that, after a brief initial period of acceleration, the particle (charge) will reach a terminal velocity given by

$$v_t = \frac{QV}{\zeta L}. \qquad (14.2)$$

In our conductor, however, there is not just one charge; there are many. Let us suppose that there are n charges per unit volume, and then ask the question: How many charges cross a plane of cross-sectional area A in a time Δt? As we discussed in Chap. 2. *Force and Momentum: Newton's Laws and How to Apply Them*, the answer is

$$n A v_t \Delta t. \qquad (14.3)$$

Therefore, the total charge crossing a plane of cross-sectional area A in Δt is given by Eq. 14.3 multiplied by Q, namely,

$$Q n A v_t \Delta t = \frac{n A Q^2 V}{\zeta L} \Delta t, \qquad (14.4)$$

where in the second equation we have used the expression for terminal velocity, v_t, introduced above. By definition, the electric current, I, is the number of charges crossing a plane per unit time. Therefore,

$$I = \frac{n A Q^2 V}{\zeta L} = GV = \frac{V}{R}, \qquad (14.5)$$

where the second equality defines the conductance, G, and the third equality defines the resistance, $R = 1/G$. The SI unit for electric current is the Ampere or Amp (A). 1 A = 1 C/s. The conductance, G, is then defined as:

$$G = \frac{n A Q^2}{\zeta L} = \frac{1}{R}. \qquad (14.6)$$

Table 14.1 Linear relationships between quantities in different contexts. Table revisited from Table 13.1 already encountered in the previous chapter, adding one more linear relationship in the V is the voltage difference across the resistor R

Chapter/context	Quantity of interest	Quantity that changes	Proportionality	Relationship
Chapter 6. *Diffusion*	Diffusion current I	Concentration	Diffusive conductance G_D	$I = G_D(n_1 - n_2)$
Chapter 9. *Fluid Mechanics*	Volume flow rate Q_F	Pressure	Flow conductance G_F	$Q_F = G_F(P_1 - P_2)$
Chapter 13. *Electric potential*	Electric charge Q	Electric potential	Capacitance C	$Q = CV$
Chapter 14. *Circuits*	Electric current I	Electric potential	Conductance G	$I = GV = \frac{V}{R}$

Equation 14.5 reveals a linear relationship between voltage and current:

$$I = GV = \frac{V}{R}. \tag{14.7}$$

This equation constitutes the original Ohm's law, which is a key concept for understanding the behavior of moving charges, hence for being able to analyze electrical circuits. Although this is the first time we have discussed the original Ohm's law, we previously discussed Ohm's law for fluids in Chap. 9. *Fluid Mechanics: Laminar Flow, Blushing, and Murray's Law*, and all of the intuition you developed in the context of fluid flow, where there is a linear relation between fluid flow and pressure difference (or in the context of particle transport across membranes, where there is a linear relationship between particle current and concentration difference) is applicable here *mutatis mutandis*,[1] as also summarized in Table 14.1. Electrical resistance, R, is proportional to L and inversely proportional to A. It is also proportional to ζ, which is the "friction" coefficient of an individual charged particle. A proper explanation of the friction experienced by electrons within conductors is quantum mechanical and beyond the scope of this book. Suffice it to say that electrons in crystalline, metal conductors, such as copper, do not experience any obstruction, moving through a crystalline lattice. Rather resistance arises because of deviations from perfect periodicity, either as a result of thermal fluctuations or as a result of impurities in the conducting material. Irrespective of the microscopic mechanism, resistance is a frictional effect that causes electrostatic potential energy to be converted into heat. This is why, for example, a toaster is made with thin high-resistance wires, so that an electric current heats up the wires eventually causing them to glow red hot to make toast.

---

[1] http://en.wikipedia.org/wiki/Mutatis_mutandis

Voltage is measured in volts (V). Resistance is measured in Ohms (Ω), where $1\,\Omega = 1$ V/A. A resistor, which is a circuit element possessing resistance, is drawn in a circuit with the symbol: —\/\/\/—. According to Ohm's law, Eq. 14.7, for a given potential difference V, a material with low resistance—a good conductor—allows a high current to flow, while a material with very high resistance—a poor conductor—allows little current. An insulator has infinite resistance and no current flows. Ohm's law can be written equivalently as

$$V = IR, \tag{14.8}$$

which highlights that if a current, I, flows through a resistance R, the electric potential decreases correspondingly by $V = IR$. This result will help us understand the behavior of electrostatic potential around a circuit.

It is important to be clear that for the conductor described at the beginning of this section, we did not specify the sign of the charge Q. In general, electric current can be carried by either positive or negative charges. In conductors (e.g., metals), the charge carriers are electrons. In solutions, the charge carriers can be both positive and negative ions. Therefore, in order to avoid confusion, the convention is to define electric current as the flow of positive charges. In the case of a conductor, therefore, the current I as conventionally defined flows in the opposite direction to the direction in which electrons are actually moving, i.e., the opposite direction to the moving electrons' velocities.

14.4   Kirchhoff's Current Law: Conservation of Electric Charge

Just like conservation of particle number leads to the position-independence of particle current and liquid incompressibility leads to the requirement that the volume flow rates into and out of any location in a flow circuit are equal, electric charge conservation leads to the requirement that the total electrical current that flows into a junction (or node) in an electrical circuit must equal the total current flowing out from that junction. This requirement is called Kirchhoff's current law. For example, for a three-way junction with current I flowing in and currents I_1 and I_2 flowing out, then we must have $I = I_1 + I_2$ (Fig. 14.3). It is important to realize that the directions of the currents, i.e., the directions of the arrows in Fig. 14.3, are meaningful for Kirchhoff's law. Therefore, when solving circuit problems, you should always define the current directions at the beginning and keep them fixed subsequently.

Fig. 14.3 Kirchhoff's current
law

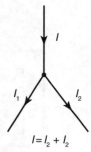

$$I = I_2 + I_2$$

14.5   Kirchhoff's Loop Law: Electrostatic Forces Are Conservative

Another important requirement for circuits involving resistors and capacitors follows from the conservative character of electrostatic forces. Since the integral of the electrostatic field, **E**, around a closed loop is zero,

$$\oint \mathbf{E} \cdot d\mathbf{s} = 0, \tag{14.9}$$

the electrostatic potential energy at a particular location depends only on that location and not on how a charge arrived at that location. These results manifest through what is called Kirchhoff's loop law: the sum of the potential differences around any loop (i.e., a closed path) in a circuit must be equal to zero.

$$\Delta V_{\text{loop}} = \sum_i \Delta V_i = 0. \tag{14.10}$$

In other words, if a charge starts at a point in a circuit, goes around a path, to come back to the same initial point, then the overall change in the charge's electrostatic potential is necessarily zero. This result is Kirchhoff's loop law. It is summarized in Fig. 14.4, where the rectangular yellow boxes are arbitrary circuit elements (e.g., batteries, resistors, capacitors, ...). The red arrow shows the direction (in this case clockwise) that is used to apply Kirchhoff's loop low. You can choose this direction as you like, provided that you pay attention to the sign of the potential difference for each element. For instance, by going from the negative to the positive side of a battery, the potential difference is positive ($\Delta V = V_0 > 0$), since a gain in potential energy happens through the battery; instead by going from the positive to the negative side of a battery, the potential difference will be negative ($\Delta V = -V_0 < 0$). Similarly, going along the direction of the current through a resistor, the potential decreases, and in particular, according to Ohm's law, the

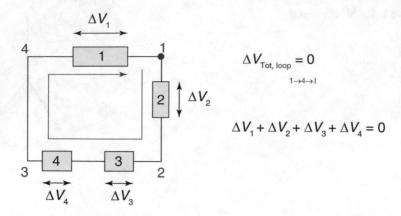

Fig. 14.4 Kirchhoff's loop law

Fig. 14.5 Kirchhoff's loop
law expresses that the
potential drop across circuit
elements in parallel must be
the same $\Delta V_{\text{left}} = \Delta V_{\text{right}}$

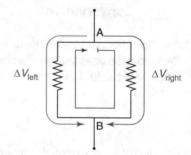

potential drop is given by $\Delta V = -IR$. Instead, going through a resistor against the direction of the current, the potential increases, and according to Ohm's law, the potential gain is given by $\Delta V = +IR$.

An important consequence of Kirchhoff's loop law is that the potential drop for circuit elements in parallel must always be the same. For instance, Fig. 14.5 shows part of a circuit that splits into two branches between points A and B; we say the two resistors are in parallel. Let us call ΔV_{right} the potential drop that occurs if the current goes from A to B through the right branch, and ΔV_{left} the potential drop that occurs if the current goes from A to B through the left branch. Applying Kirchhoff's loop law to the red clockwise path shown gives: $-\Delta V_{\text{right}} + \Delta V_{\text{left}} = 0$, which implies $\Delta V_{\text{left}} = \Delta V_{\text{right}}$. This is true in general for any parallel configuration connecting two points A and B, and the potential difference between A and B will be always the same independently from which path is taken:

$$\Delta V_{AB} = \Delta V_{\text{left}} = \Delta V_{\text{right}}$$

We saw an analogous result for pressure drops in Chap. 9. *Fluid Mechanics: Laminar Flow, Blushing, and Murray's Law.*

14.6   Resistors in Series and in Parallel

Resistances in series add, i.e., inverse conductances in series add, and inverse resistances in parallel add, i.e., conductances in parallel add, as expected based on the analogies with diffusion and viscous fluid flow. We can directly confirm these results as follows. For two resistors, R_1 and R_2, in series, the same current, I, flows through each. According to Ohm's Law (Eq. 14.8), we know that the voltage drop across R_1 is $I R_1$, and the voltage drop across R_2 is $I R_2$. If the voltage applied across both resistors is V, then we must have

$$V = I R_1 + I R_2 = I (R_1 + R_2). \qquad (14.11)$$

We see that the resistance of the two series resistors is

$$R_S = R_1 + R_2, \qquad (14.12)$$

or equivalently

$$\frac{1}{G_S} = \frac{1}{G_1} + \frac{1}{G_2}. \qquad (14.13)$$

For two resistors, R_1 and R_2, in parallel. The same voltage, V, is applied across both resistors. The current flow through each resistor is V/R_1 and V/R_2, respectively. The total current flow is therefore

$$I = \frac{V}{R_1} + \frac{V}{R_2} = V \left(\frac{1}{R_1} + \frac{1}{R_2} \right). \qquad (14.14)$$

We see that the resistance of the two parallel resistors is given by

$$\frac{1}{R_P} = \frac{1}{R_1} + \frac{1}{R_2}, \qquad (14.15)$$

or equivalently

$$G_P = G_1 + G_2. \qquad (14.16)$$

Thus, we find exactly the results that we expect on the basis of the analogy with diffusion and fluid flow.

14.6.1 Example: A Voltage Divider

Suppose that you have a power supply V, but you need a voltage $V_1 < V$. How can you create such a voltage? The answer is to use a voltage divider circuit, shown in Fig. 14.6. The goal of this example is to determine how to pick R_1 and R_2 to achieve the desired voltage V_1.

(a) What is the total current through the resistors?
    ```
    The total resistance is  R₁ + R₂. Therefore, the
    total current is
    ```

$$I = V/(R_1 + R_2). \tag{14.17}$$

(b) What is the voltage drop (V_1) across R_1?
    ```
    Using Ohm's law, the voltage drop across
    resistor R₁ is
    ```

$$V_1 = I R_1 = \frac{R_1}{R_1 + R_2} V = \frac{V}{1 + \frac{R_2}{R_1}}, \tag{14.18}$$

```
    as shown in Fig. 14.6.
```
(c) Thus, find the value of $\frac{R_2}{R_1}$ that gives the desired voltage, V_1.
    ```
    Solving for R₂/R₁, we find
    ```

$$\frac{R_2}{R_1} = \frac{V}{V_1} - 1. \tag{14.19}$$

Fig. 14.6 Voltage divider. The top "rail" is at a voltage V. The three different length lines on the bottom rail indicate that the rail is at ground, i.e., zero voltage

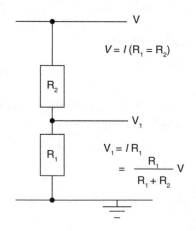

$V = I\,(R_1 = R_2)$

$V_1 = I R_1$

$\quad = \dfrac{R_1}{R_1 + R_2} V$

14.6.2 Example: An Infinite Resistance Ladder

(a) Calculate the total conductance, G_T, of the arrangement of resistors shown in Fig. 14.7.
   Answer: We first add in parallel the two conductances in the middle, $8G$ and $2G$; then we combine the result with the other two conductances in series, $5G$ at the top and $5G$ at the bottom. The desired total conductance is then found to be

$$\frac{1}{G_T} = \frac{1}{5G} + \frac{1}{8G + 2G} + \frac{1}{5G} = \frac{1}{2G}. \tag{14.20}$$

Therefore,

$$G_T = 2G. \tag{14.21}$$

(b) For the circuit of Fig. 14.7, the voltage at point A is V_A, and the voltage at point B is V_B. Calculate the voltage at point C (V_C) in this circuit in terms of V_A and V_B only.
   Answer: We have that the total current is

$$I = 2G(V_A - V_B). \tag{14.22}$$

We also have that

$$I = 5G(V_A - V_C). \tag{14.23}$$

(continued)

Fig. 14.7 First circuit for "An infinite resistance ladder".

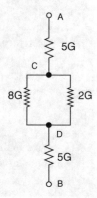

Equating these two equations for I, we find

$$V_C = V_A - \frac{2}{5}(V_A - V_B) = \frac{3}{5}V_A + \frac{2}{5}V_B, \qquad (14.24)$$

which is the desired answer.

(c) For the circuit in Fig. 14.7, calculate the voltage at point D (V_D) in terms of V_A and V_B only.

Answer: We have that the total current is

$$I = 2G(V_A - V_B). \qquad (14.25)$$

We also have that

$$I = 5G(V_D - V_B). \qquad (14.26)$$

Equating these two equations for I, we find

$$V_D = V_B + \frac{2}{5}(V_A - V_B) = \frac{2}{5}V_A + \frac{3}{5}V_B, \qquad (14.27)$$

which is the desired answer.

(d) The top of Fig. 14.8 shows an infinite resistance ladder, comprised of an infinite, periodic sequence of resistors. The conductance of this resistance ladder is G_T. Briefly explain how the conductance of this infinite ladder can be calculated from the circuit shown at the bottom of Fig. 14.8.

Answer: Because the resistance ladder after the first "rung" is identical to the entire resistance ladder, we have that G_T is given by

$$\frac{1}{G_T} = \frac{1}{5G} + \frac{1}{5G} + \frac{1}{8G + G_T}, \qquad (14.28)$$

which enables us to calculate G_T.

(e) Using your answer to (a), or otherwise, determine G_T for the circuit in (d) in terms of G.

Answer: The circuit of Fig. 14.7 and the circuit at the bottom of Fig. 14.8 are identical provided we pick

$$G_T = 2G. \qquad (14.29)$$

(f) What is the voltage difference between point D and point C in the circuit at the top of Fig. 14.8? i.e., what is $V_C - V_D$?

(continued)

Answer: Using the answers to (b) and (c), we
have

$$V_C - V_D = \frac{1}{5}(V_A - V_B).\tag{14.30}$$

(g) If the total current from point A to point C in the circuit at the top of
Fig. 14.8 is I, what is the current from point C to point E in terms of I?
Call this current I_1.

Answer: Because the voltage difference is the
same across the $8G$ conductance and the G_T
conductance, the ratio of the corresponding
currents equals the ratio of the conductances,
i.e.,

$$\frac{I_1}{I - I_1} = \frac{G_T}{8G} = \frac{2G}{8G} = \frac{1}{4}.\tag{14.31}$$

It follows that

$$I_1 = \frac{1}{5}I.\tag{14.32}$$

Alternatively, we have that

$$I = 2G(V_A - V_B)\tag{14.33}$$

and

$$I_1 = 2G \times \frac{1}{5}(V_A - V_B).\tag{14.34}$$

It again follows that

$$I_1 = \frac{1}{5}I.\tag{14.35}$$

(h) What is the current from point E to point F in terms of I? Call this
current I_2.

Answer:

$$I_2 = \frac{1}{5^2}I.\tag{14.36}$$

Extrapolating, we can see that the current
decreases exponentially along the ladder with
$I_n = \frac{1}{5^n}I$.

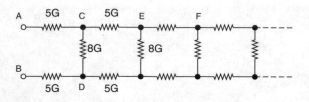

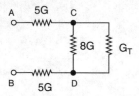

Fig. 14.8 Additional circuits for "An infinite resistance ladder".

14.7   Power Dissipation in a Resistor

When a current, I, flows through a resistor, the power dissipation, namely the rate
at which electrostatic potential energy decreases, is $P = VI$, where V is the
voltage drop across the resistor and I is the current flowing through the resistor.
This expression follows because V is the decrease in energy per charge when that
charge moves through the resistor and I is the charge per second passing through
the resistor. Thus, VI is the decrease in energy per second, namely the power
dissipation. Because $V = IR$, we can also write $P = I^2 R = \frac{V^2}{R} = \frac{I^2}{G} = GV^2$.
These expressions for power dissipation are just as we expect based on the analogy
between voltage and current on the one hand and pressure and volume flow rate on
the other, and mirror analogous expressions that we encountered in Chap. 9. *Fluid
Mechanics: Laminar Flow, Blushing, and Murray's Law.* Just as in that case, the
dissipated energy ends up as heat and the resistor heats up. In the electrical context,
this phenomenon is often termed "Joule heating."

14.8   Circuits with Resistors and Capacitors

Often it is useful to combine resistors and capacitors in the same circuit. In this case,
in addition to using the fact that the voltage drop across a capacitance C is $\frac{Q}{C}$, it is
also necessary to appreciate that the magnitude of the current flow onto a capacitor
plate, I, equals the rate of change of the charge, dQ/dt on that plate.

14.8.1 Example: Capacitor in Series with a Resistor

Consider the circuit shown in Fig. 14.9, consisting of a capacitor and resistor in series. The goal of this example is to determine the charge on the capacitor as a function of time, $t > 0$, given the initial condition that $Q(0) = 0$.

(a) Write an equation that expresses the relationship between the applied voltage, V, the charge on the capacitor, Q, and the current flowing through the resistor, I.
Using Kirchhoff's loop law, we can write the desired equation:

$$V = \frac{Q}{C} + IR. \qquad (14.37)$$

(b) Use the relationship between I and Q to eliminate I from your equation from (a) to obtain an equation involving Q only.
Using $I = \frac{dQ}{dt}$, we find

$$V = \frac{Q}{C} + IR = \frac{Q}{C} + R\frac{dQ}{dt}, \qquad (14.38)$$

which is a differential equation for Q.
(c) Solve your equation for Q by direct substitution, using an exponential trial solution.
As in Chap. 7. *Rates of Change: Drugs, Infections, and Weapons of Mass Destruction*, we solve this equation by direct substitution of a trial solution. The problem tells us to use an exponential trial solution, namely

$$Q = A + Be^{-\gamma t}, \qquad (14.39)$$

(continued)

Fig. 14.9 Resistor and capacitor in series

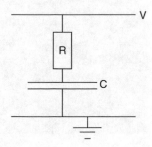

where A, B, and γ are to be determined. Given Eq. 14.39, we have that

$$\frac{dQ}{dt} = -\gamma B e^{-\gamma t}. \tag{14.40}$$

Substituting Eqs. 14.39 and 14.40 into Eq. 14.38, we find

$$V = \frac{A + B e^{-\gamma t}}{C} - R\gamma B e^{-\gamma t} = \frac{A}{C} + B e^{-\gamma t}\left(\frac{1}{C} - \gamma R\right). \tag{14.41}$$

Thus, we see that Eq. 14.39 is a solution to Eq. 14.38, provided we pick

$$\gamma = \frac{1}{RC} \tag{14.42}$$

and

$$A = CV, \tag{14.43}$$

so that

$$Q(t) = CV + B e^{-\frac{t}{RC}}. \tag{14.44}$$

The remaining unknown constant, B, is then determined by the initial conditions.

(d) What is the relaxation time of this circuit?
Answer: The product RC has dimensions of time and is often called the RC-time constant in the context of RC circuits.

(e) Apply the initial condition to find $Q(t)$.
Answer: Using the initial condition that $Q(0) = 0$, we have

$$0 = CV + B. \tag{14.45}$$

It follows that $B = -CV$, so that

$$Q(t) = CV(1 - e^{-\frac{t}{RC}}), \tag{14.46}$$

which is the desired solution in this case.

14.8.2 Example: Switched RC Circuit

Consider the RC circuit shown in Fig. 14.10, in which a switch can be used to apply a voltage of either V or 0 across a resistor, R, and capacitor, C, arranged in series. We analyzed this circuit in Sect. 14.8.1, where we showed that the general solution for the charge on the capacitor is

$$Q(t) = CV + Be^{-\frac{t}{RC}}, \tag{14.47}$$

where B is determined by the initial conditions. Equation 14.47 applies directly when the switch is arranged to include the battery in the circuit. Alternatively, when the switch is in its other configuration, Eq. 14.47 is applicable by setting $V = 0$. In fact, the switch alternates periodically between the two different configurations, switching every time step $RC \log 2$.

(a) Sketch the charge on the capacitor as a function of time under this voltage regime, assuming that a (dynamic) steady state has already been reached.
   Answer: The requested sketch is shown in
   Fig. 14.11, where we used that after a time
   period $\Delta t = RC \log 2$, $e^{-\frac{\Delta t}{RC}} = e^{-\log 2} = \frac{1}{2}$.
(b) Suppose, as shown in Fig. 14.11, that Q_B (Q_T) is the minimum (maximum) charge on the capacitor during this process. Calculate Q_B and Q_T.
   Answer: From the figure and the general
   solution, we have that

$$Q_B = CV + B, \tag{14.48}$$

$$Q_T = CV + \frac{B}{2}, \tag{14.49}$$

(continued)

Fig. 14.10 Circuit for "Switched RC-circuit"

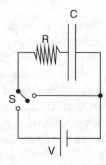

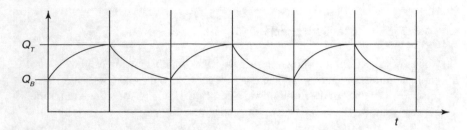

Fig. 14.11 Charge, $Q(t)$, versus time, t, for "Switched RC-circuit"

and

$$Q_B = \frac{1}{2}Q_T = \frac{1}{2}(Q_B + \frac{B}{2}),\qquad(14.50)$$

which are three equations for the three unknowns, Q_B, Q_T, and B, where is the same B as appears in Eq. 14.47. Solving, we find $Q_B = \frac{1}{3}CV$, $Q_T = \frac{2}{3}CV$, and $B = -\frac{2}{3}CV$.

(c) How does the voltage across the capacitor compare to the voltage across both the capacitor and the resistor?

Answer: The voltage across the resistor and the capacitor is the periodic "square-wave," described in the setup of this example. By contrast, the voltage across the capacitor alone is proportional to the charge shown in Fig. 14.11. Clearly, the voltage across the capacitor is significantly smoothed compared to the voltage across both resistor and capacitor.

14.9   Dendritic Conduction

There is an important physiological application of electrical circuits involving resistance and capacitance, namely to nerve impulses in the dendrites of neurons. Neurons are the principal cells of our nervous system. A schematic of a neuron, together with its associated Schwann cells, is shown in Fig. 14.12. Typically, a neuron consists of dendrites, an axon, and a cell body (or soma), containing the nucleus. Dendrites are multiple branched structures emanating from the neuron cell body. Generally, signal propagation proceeds from the axon of one neuron to a dendrite of another via a synapse. In many dendrites, electrical signal propagation can be understood on the basis of an appropriate RC circuit. By contrast, in axons, electrical signal propagation is more complicated and is beyond our scope.

Fig. 14.12 Schematic of a
neuron, showing dendrites, an
axon, and the cell body,
containing the nucleus. Also
shown are the neuron's
associated Schwann cells

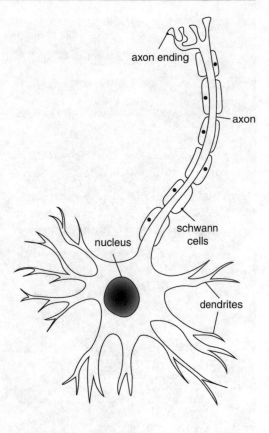

axon ending

axon

schwann
cells

nucleus

dendrites

We will model the electrical properties of a dendrite as a certain current ladder,
similar to the flow ladders that we encountered in Chap. 9. *Fluid Mechanics: Laminar Flow, Blushing, and Murray's Law* and the resistance ladder of Sect. 14.6.2,
but now including capacitors as well as resistors. Example 14.9.1 provides an
introduction to current ladders.

14.9.1 Example: A Current Ladder

Consider a circuit, consisting of N capacitors of capacitance C, and $2N$
resistors of resistance R, arranged as shown in Fig. 14.13. The goal of this
example is to be able to describe how the voltage across the capacitors varies
as a function of capacitor index number, n, and time, t. This first example will
help us then build a model for dendritic conduction.

(continued)

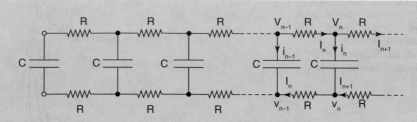

Fig. 14.13 The arrangement of capacitors (C) and resistors (R) for "A current ladder"

(a) Referring to Fig. 14.13, write the equations that express the relationship between V_n, V_{n-1}, and I_n, and between v_n, v_{n-1}, and I_n.
The desired equations are

$$V_{n-1} - V_n = I_n R, \qquad (14.51)$$

and

$$v_n - v_{n-1} = I_n R. \qquad (14.52)$$

(b) Referring to the current ladder in Fig. 14.13, write an equation that expresses the relationship between I_{n+1}, I_n, and i_n.
Answer: Kirchhoff's law immediately implies that

$$I_n = I_{n+1} + i_n. \qquad (14.53)$$

(c) Referring to Fig. 14.13 again, now write an equation that expresses the relationship between V_n, v_n, and q_n.
Answer: We have

$$q_n = C(V_n - v_n). \qquad (14.54)$$

Taking a derivative with respect to time ($i_n = \frac{dq_n}{dt}$), we find

$$i_n = C(\frac{dV_n}{dt} - \frac{dv_n}{dt}). \qquad (14.55)$$

(d) It turns out that it is convenient to focus on the voltage across the capacitors, namely $W_n = V_n - v_n$. Use your answers to (a)–(c), to find an equation that relates W_{n-1}, W_n, and W_{n+1} to $\frac{dW_n}{dt}$.
Adding Eqs. 14.51 and 14.52 implies that

$$W_{n-1} - W_n = 2I_n R. \qquad (14.56)$$

(continued)

Therefore,

$$W_{n-1} - 2W_n + W_{n+1} = 2(I_n - I_{n+1})R. \tag{14.57}$$

Now, using Eq. 14.53 and then Eq. 14.55, Eq. 14.57 becomes

$$W_{n-1} - 2W_n + W_{n+1} = 2RC\frac{dW_n}{dt}, \tag{14.58}$$

which is the desired equation.

(e) On the basis of the equation obtained in (d), describe the dynamics of voltage or charge on the current ladder.

Answer: The left-hand side of Eq. 14.58 is discrete version of the second derivative, i.e.,

$$W_{n+1} - 2W_n + W_{n-1} \simeq \frac{d^2W_n}{dn^2}. \tag{14.59}$$

It follows that we may conceive Eq. 14.58 as a diffusion equation,

$$\frac{d^2W_n}{dn^2} \simeq 2RC\frac{dW_n}{dt}, \tag{14.60}$$

which we previously encountered in Chap. 6. *Diffusion: Membrane Permeability and the Rate of Actin Polymerization*, and from which it is clear that the voltage (or charge) on the capacitors of the current ladder undergoes diffusion. If the capacitors are separated by a distance a, so that the position of capacitor n is $x = na$, then we may write

$$\frac{a^2}{2RC}\frac{d^2W_n}{dx^2} \simeq \frac{dW_n}{dt}, \tag{14.61}$$

permitting us to identify the diffusion coefficient as $D = \frac{1}{2(R/a)(C/a)} = \frac{a^2}{RC}$, where R/a and C/a are the resistance and capacitance per unit length.

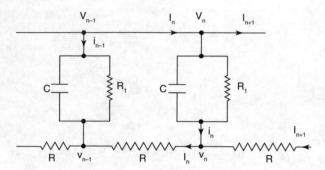

Fig. 14.14 The arrangement of capacitors (C), intracellular resistors (R), and transmembrane resistors (R_1), for modeling dendritic conduction

The current ladder of Sect. 14.9.1 is similar to a simple model for electrical conduction on a dendrite, except that to model a dendrite, the circuit of Example 14.9.1 is augmented with transmembrane resistances in parallel with the capacitors. In addition, the extracellular resistance is generally neglected because it is small compared to the intracellular resistance. Thus, the appropriate circuit to describe a dendrite is that shown in Fig. 14.14, where the top rail represents outside the neuron, the bottom rail represents inside the neuron, and the capacitors and resistors, which form the rungs of the ladder, represent the membrane capacitance and transmembrane resistance, respectively. Analysis of the circuit of Fig. 14.14 goes similarly to the analysis of Sect. 14.9.1 with two modifications: First, Eq. 14.51 is modified to read

$$V_{n-1} - V_n = 0. \tag{14.62}$$

It follows that Eq. 14.56 becomes

$$W_{n-1} - W_n = I_n R. \tag{14.63}$$

Second, the transmembrane resistors cause Eq. 14.55 to read

$$i_n = C(\frac{dV_n}{dt} - \frac{dv_n}{dt}) + \frac{1}{R_1}(V_n - v_n) = C\frac{dW_n}{dt} + \frac{W_n}{R_1}. \tag{14.64}$$

It follows that Eq. 14.58 is replaced by

$$W_{n-1} - 2W_n + W_{n+1} = RC\frac{dW_n}{dt} + \frac{R}{R_1}W_n. \tag{14.65}$$

Because a dendrite is a continuous object, it is necessary to go to a continuum description. We make the conversion by asserting that $x = na$ represents distance along the dendrite, where a is the separation between the rungs of the ladder. In this context, it is convenient to use the membrane capacitance per unit area, c, and the transmembrane resistance per unit area, r_1, and the intracellular resistivity, ρ. For the membrane capacitance per unit length and the transmembrane resistance per unit length, we write $C/a = 2\pi bc$ and $R_1/a = r_1/(2\pi b)$. For the intracellular resistance per unit length, corresponding to the resistance along the dendrite, we write $R/a = \rho/(\pi b^2)$. It follows that the voltage across the dendritic cell membrane is given by

$$\frac{d^2 W}{dx^2} = \frac{2\rho c}{b}\frac{dW}{dt} + \frac{2\rho}{br_1}W. \tag{14.66}$$

Equation 14.66 is called the cable equation. It differs from the diffusion equation by the presence of a term proportional to W. In the limit of small b, the $\frac{d^2 W}{dx^2}$-term is negligible, and we are left with an equation that has solutions that decay exponentially in time. Motivated by this observation, it is useful to introduce a trial solution for $W = W(x, t)$ that explicitly incorporates an exponential time dependence, that is, we try

$$W(x, t) = e^{-\frac{t}{r_1 c}}U(x, t), \tag{14.67}$$

where $U = U(x, t)$ is a function to be determined. Then, by direct substitution, we find

$$\frac{d^2 W}{dx^2} = e^{-t/(r_1 c)}\frac{d^2 U}{dx^2} \tag{14.68}$$

and

$$\frac{dW}{dt} = -\frac{1}{r_1 c}e^{-\frac{t}{r_1 c}}U + e^{-\frac{t}{r_1 c}}\frac{dU}{dt}. \tag{14.69}$$

Re-writing Eq. 14.66 in terms of U implies that

$$e^{-\frac{t}{r_1 c}}\frac{d^2 U}{dx^2} = e^{-\frac{t}{r_1 c}}\frac{2\rho c}{b}(\frac{dU}{dt} - \frac{1}{r_1 c}U) + \frac{2\rho}{br_1}e^{-\frac{t}{r_1 c}}U, \tag{14.70}$$

i.e.,

$$\frac{d^2 U}{dx^2} = \frac{2\rho c}{b}\frac{dU}{dt}, \tag{14.71}$$

which is a diffusion equation for U with diffusion coefficient $D = \frac{b}{2\rho c}$. Therefore, the behavior of the voltage, W, given by Eq. 14.67, is diffusive, but with an additional overall exponential decay. Interestingly, the time constant of the exponential decay, $r_1 c$, is independent of dendrite diameter. This exponential decay is important for how neurons function: Neuronal input signals arrive at distal dendritic locations and initiate a transmembrane voltage signal that diffuses to the cell body. If the total transmembrane voltage at the cell body exceeds a threshold value, an electrical signal that propagates along the axon is initiated, i.e., the neuron "fires." Dendritic input signals that arrive within a time $r_1 c$ of each other can add together to achieve the threshold for neuronal "firing," but signals that are not coincident within $r_1 c$ cannot. Every thought we have involves millions of our own RC circuits in an essential fashion.

We may also ask whether Eq. 14.66 might also be applicable to axons (Fig. 14.12)? We learned in Chap. 7 that the time for a voltage signal to diffuse a distance L is given by $\frac{L^2}{2D}$. On the other hand, the time that a voltage signal persists is $r_1 c$. If we set these two times equal to each other, we will discover an estimate of the maximum distance, L_{MAX}, that a voltage signal can travel before dissipating: $L_{MAX} \simeq \sqrt{2Dr_1 c} = \sqrt{\frac{r_1 b}{2\rho}}$. For realistic parameters, L_{MAX} is of the order of 1 mm. By contrast, the longest nerve in the human body—the sciatic nerve—is about 1 m long. On this basis, we can infer that signal propagation along axons requires physics beyond resistors and capacitors. In fact, signal transmission along an axon takes the form of propagating voltage spikes—called action potentials—which are very different from the exponentially decaying diffusive-type behavior that emerges from Eq. 14.66. The physics of action potentials depends critically on voltage-gated ion channels, which start in a closed, but active state, which then transition into an open state, when the transmembrane voltage exceeds a threshold value, and which subsequently transition into a closed, inactive state. Opening voltage-gated ion channels decreases the transmembrane resistance, leading to a non-linear elaboration of the cable equation (Eq. 14.66) that is coupled to equations describing ion channel state dynamics. These equations, which are a little too involved to include here, constitute the Hodgkin–Huxley model,[2] which well describe nerve impulse propagation along axons, and for which Alan Hodgkin and Andrew Huxley were awarded the 1963 Nobel Prize in Physiology or Medicine.

14.10 Chapter Outlook

A standout aspect of this chapter is how many of the concepts and methods recall material from previous chapters: Ohm's law recalls Ohm's law for fluids and the relationship between diffusion current, diffusive conductance, and concentration difference. Kirchhoff's current law recalls particle number conservation in Chap. 6. *Diffusion: Membrane Permeability and the Rate of Actin Polymerization* and volume conservation in incompressible fluids in Chap. 9. *Fluid Mechanics: Laminar Flow, Blushing, and Murray's Law*; the diffusion equation itself came back in the context of dendritic conduction; and the exponential decays in RC circuits recalled the exponential decays in serum concentration or the exponential decay in probability of remaining cancer-free, which we saw in Chap. 7. *Rates of Change: Drugs, Infections, and Weapons of Mass Destruction*. In fact, elements of the circulatory system can change volume (slightly) in response to pressure, i.e., they are (slightly)

---

[2] https://en.wikipedia.org/wiki/Hodgkin?Huxley_model

compliant and can effectively store a small blood volume, analogously to how capacitors store charge. Thus, what you have learned about RC circuits in this chapter also applies to the circulatory system with the appropriate translation: A compliant blood vessel is analogous to a resistor and capacitor in parallel.

Looking forward, we will see in Chap. 17. *Magnetic Fields and Ampere's Law* that electrical current gives rise to magnetic field, which is the focus of that chapter, and which is an essential partner to electric field in Maxwell's equations, as we will see in Chap. 18. *Faraday's Law and Electromagnetic Induction* and Chap. 19. *Maxwell's Equations and Then There Was Light*. In addition, our study of electric circuits in this chapter will be important in Chap. 18. *Faraday's Law and Electromagnetic Induction* because the observations that lead to Faraday's law were carried out in electrical circuits.

14.11 Problems

Problem 1: Combining Capacitances or Resistances
(a) Calculate the total capacitance (C_T) of the arrangement of four capacitors shown on the left of Fig. 14.15, each of capacitance C.
(b) Calculate the total resistance (R_T) of the arrangement of four resistors shown on the right, each of resistance R. What is the conductance, G_T?

Problem 2: Path of Least Resistance
Consider the circuit on the left side of Fig. 14.16, the currents I_a and I_b flow through points **a** and **b**, respectively. Answer the following questions in terms of the given quantities in the diagram, V_0 and R.

(a) What are the currents I_a and I_b? Which of the following is true: $I_a > I_b$, $I_a < I_b$, or $I_a = I_b$?
(b) A wire is used to connect point a to point b. Does a current I_{ab} flow through this wire? Would this current flow from $a \to b$, from $b \to a$, or does $I_{ab} = 0$? Explain your answer using physics concepts or equations (or both).

Fig. 14.15 Left: An arrangement of four capacitors, each of capacitance C. Right: An arrangement of four resistors, each of resistance R

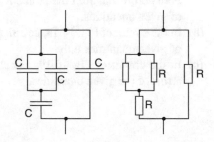

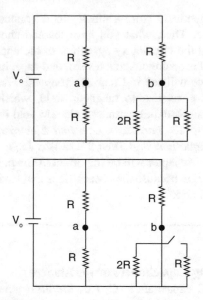

Fig. 14.16 Figures for Problem 2: Path of least resistance

(c) A switch is opened as shown on the right side of Fig. 14.16. Notice there is no wire connecting point a to b. Consider currents flowing through points **a** and **b** with the switch open. Which of the following is true: $I_a > I_b$, $I_a < I_b$, or $I_a = I_b$?

(d) A wire is used to connect point a to point b after the switch is opened. Does the current I_{ab} flow from $a \rightarrow b$, from $b \rightarrow a$, or does $I_{ab} = 0$? Explain your answer using physics concepts or equations (or both).

Problem 3: Circuits-a-Palooza

(a) In the circuit of Fig. 14.17, assuming steady state is reached in both cases, calculate the charge on and the energy stored in, each capacitor, C_1 and C_2 both before and after the points A and B are connected by a conductor in terms of given quantities.

(b) In the circuit of Fig. 14.18, calculate the power dissipated in resistor R_3 in terms of given quantities only.

(c) In the circuit of Fig. 14.19, calculate the energy stored on the capacitor C in terms of the given quantities,

Fig. 14.17
Circuits-a-Palooza (a)

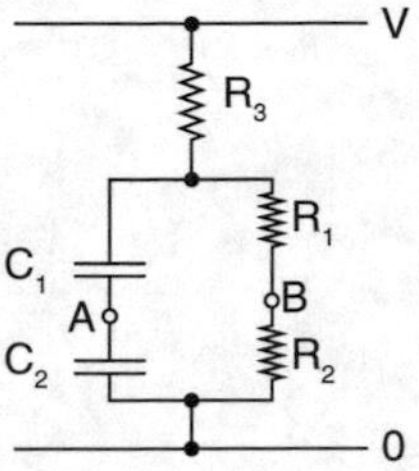

Fig. 14.18
Circuits-a-Palooza (b)

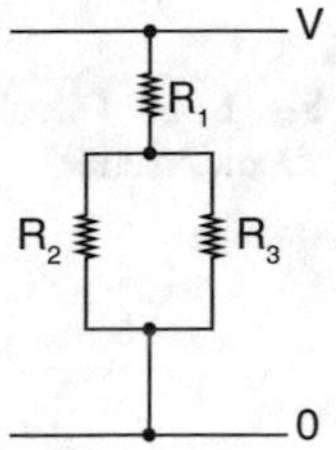

Fig. 14.19
Circuits-a-Palooza (c)

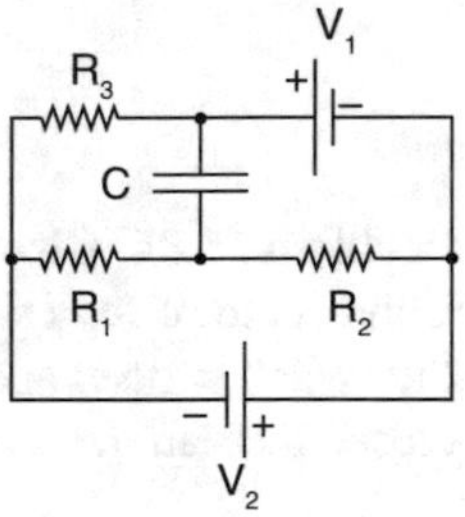

Problem 4: Sherlock R

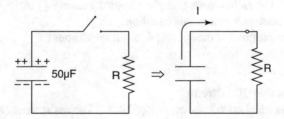

You are given an RC circuit with a capacitor with capacitance $C = 50\,\mu F$ and an unknown resistance R. The capacitor is initially fully charged, you close the switch, and let the capacitor discharge through the resistor. You can measure the current with a multimeter, and you get the graph below that represents the current as a function of time since the moment you closed the switch.

Extract the value of the unknown resistance.

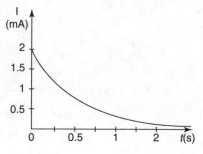

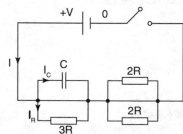

Fig. 14.20 Circuit for "RC circuit practice"

Problem 5: RC Circuit Practice

Consider the circuit shown in Fig. 14.20. V, C, and R are given quantities. For $t < 0$, the switch is open and no current flows. At $t = 0$, the switch is closed and remains closed thereafter.

(a) What is the resistance of the two resistors, each of resistance $2R$?
(b) What current flows through the switch at the steady state achieved at long times?
(c) Write down three independent equations relating the currents, I, $I_C = dQ/dt$, and I_R, and the charge on the capacitor, Q.
(d) Eliminate I and I_R to obtain a single equation relating Q and $\frac{dQ}{dt}$.
(e) Find the general solution to this equation.
(f) What is the solution for the specified initial conditions?

Problem 6: Another RC Circuit

Consider the electrical circuit shown in Fig. 14.21. The resistances (R_1 and R_2), the capacitor (C) and the voltage from the battery (V) are known. The currents (I and I_1) and the charge on the capacitor (Q) are unknown.

(a) What is the steady-state current flowing through the resistor R_1 at long times?
(b) What is the steady-state charge on the capacitor at long times?
(c) Write down three independent equations, relating I, I_1, and Q at an arbitrary time t.
(d) Eliminate I and I_1 to obtain a single equation relating Q and $\frac{dQ}{dt}$.
(e) What is the characteristic time constant for charging or discharging the capacitor?

Fig. 14.21 Circuit for "Another RC circuit"

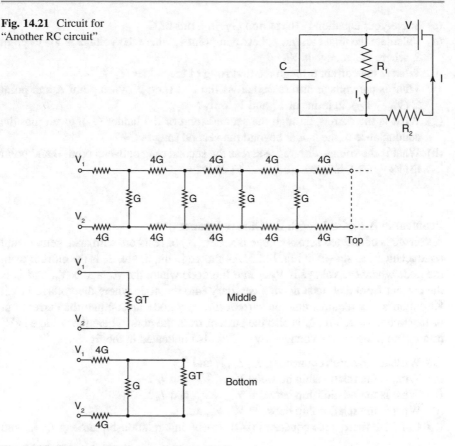

Fig. 14.22 Circuit for "Resistance ladder design"

Problem 7: Another Infinitely Long Resistance Ladder

In this problem, we will analyze the infinitely long resistance ladder shown in the top panel of Fig. 14.22, which is composed of infinitely many resistors of conductances $4G$ and G, arranged as shown. The total conductance of the resistance ladder is G_T, as indicated in the middle panel of Fig. 14.22. Notice, however, that the ladder after the first "rung" is the same as the entire ladder, as indicated in the bottom panel of Fig. 14.22. The ladder after the second rung is also the same as the entire ladder. These results are a consequence of the ladder being infinitely long. It follows that all three circuits shown in Fig. 14.22 have the same conductance, equal to the conductance of the entire ladder, G_T.

(a) Calculate the conductance of the circuit, conceived as shown in the bottom panel of Fig. 14.22, in terms of G and G_T.
(b) Given that the total conductance of the ladder, namely G_T, must be equal to your answer to (a) write down a single equation that establishes a relation between G_T and G.

(c) Solve your equation in (b) to find G_T in terms of G.
(d) Calculate the total current, I, between point 1, where the voltage is V_1, to point 2, where the voltage is V_2.
(e) What is the current through the first rung of the ladder (I_1)?
(f) What is the voltage difference across the first rung between point A and point C ($V_A - V_C$), in terms of V_1 and V_2 only?
(g) What is the current through the second rung of the ladder (I_2), given that the conductance of the ladder beyond the second rung is G_T?
(h) What is the voltage difference across the second rung between point B and point D ($V_B - V_D$), in terms of V_1 and V_2 only?

Problem 8: A Finite-Length Resistance Ladder

A voltage is connected across a large number (N) of resistors with arrangement and conductances, as shown in Fig. 14.23. As shown in the figure, I_n is the current from the node where the voltage is V_{n-1} into the node where the voltage is V_n, and i_n is the current from that node down a capillary into the node where the voltage is V'_n. Kirchhoff's law requires that the current into any node must equal the current out of that node. Therefore, I_n is also the current from the node where the voltage is V'_n into the node where the voltage is V'_{n-1}, as also indicated in the figure.

(a) Write an equation connecting I_n, I_{n+1}, and i_n.
(b) What is the relationship between V_{n-1}, V_n, and I_n?
(c) What is the relationship between V'_{n-1}, V'_n, and I_n?
(d) What is the relationship between V_n, V'_n, and i_n.
(e) Combine these three equations (b–d) to obtain a relationship among I_n, i_n, and i_{n-1}.
(f) Combine your answers to (a) and (e) to obtain a relationship among I_{n-1}, I_n, and I_{n+1}.
(g) Show by direct substitution that a solution to your equation in (f) is given by $I_n = Bx^n$, where B is a constant, independent of n, and determine the permitted values of x. Call those solutions x_+ and x_-.
(h) Both $I_n = Bx^n_\pm$ are solutions. Therefore, because we are dealing with a linear equation, in fact, the general solution for the current is given by the sum:

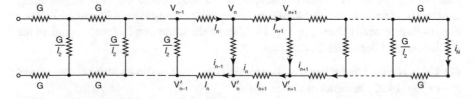

Fig. 14.23 A resistance ladder (top) and two equivalent circuits (middle and bottom)

$$I_n = B_+ x_+^n + B_- x_-^n. \tag{14.72}$$

In general, B_+ and B_- must be determined by the relevant boundary conditions. Specializing now to the case of an infinitely long resistance ladder ($n \to \infty$), what must be the boundary condition on I_n, given the physical situation that we are describing? How does this boundary condition affect B_+ and B_-? Given this boundary condition, how does I_n depend on n?

Problem 9: Resistance Ladder Design Project

Consider the resistance ladder shown in Fig. 14.24. The "rungs" of the ladder have conductance G, and the "risers" have conductance G_1, where G_1 is to be determined in terms of G in order to achieve the desired current properties, which include a relatively large conductance and a relatively long length over which the current decreases along the resistance ladder. Because of these considerations, it has been determined that the total conductance of the resistance ladder should be $5G$. The goal of this problem is to determine how to pick G_1 and to determine what is the decay length of the current in the current ladder.

(a) Draw a circuit that is equivalent to the infinite resistance ladder of Fig. 14.24, but that replaces the circuit beyond the first rung by a single resistor of the appropriate conductance.
(b) Determine the required value of G_1 by first calculating the total conductance of the circuit that you drew in (a) in terms of G and G_1.
(c) What fraction of the total current goes through the first rung when G_1 has the value you determined in (b)?
(d) What fraction of the total current goes through the resistance ladder beyond the first rung? i.e., what fraction of the current does not go through the first rung?
(e) What fraction of the total current goes through the resistance ladder beyond the second rung?
(f) What fraction of the total current goes through the resistance ladder beyond the n-th rung?
(g) We can define the decay "length," N, of the resistance ladder—it is actually the number of rungs over which the current decays—by asserting that the answer to (f) is equal to $e^{-n/N}$. Using this result and your answer to (f), calculate N.

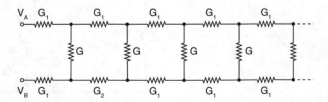

Fig. 14.24 A finite-length resistance ladder

Next, use the result that for small x, $\log(1+x) \simeq x$, to simplify your expression for N.

(h) Given that N is given by your simplified answer to (g), compare the total conductance of N conductances in parallel, each one of conductance G, to the total fluid conductance of the resistance ladder, namely $5G$.

Problem 10: One More RC Circuit

Consider the circuit shown in Fig. 14.25. At $t = 0$, the switch S is closed.

(a) There are four unknowns in this problem. What are they?
(b) Write four independent equations that can be used to determine the four unknowns as a function of time after the switch is closed.
(c) Give the steady-state potential difference between the points b and a on the circuit.
(d) Eliminate the other unknowns to obtain a single differential equation for Q.
(e) What is the relaxation time in this circuit?

Problem 11: Charge Relaxation

Consider the electrical circuit shown in Fig. 14.26. Initially, both switches are closed. However, at $t = 0$, they are both opened simultaneously. The goal of this problem is to determine the charge on the capacitors as a function of time.

(a) What is the charge on the capacitor on the left, Q_1, at $t = 0$?
(b) What is the charge on the capacitor on the right, Q_2, at $t = 0$?
(c) Determine equations that describe how Q_1 and Q_2 change in time for $t > 0$.
(d) Find the general solution to your equations.
(e) Apply appropriate initial conditions to determine $Q_1 = Q_1(t)$ and $Q_2 = Q_2(t)$ as functions of time.

Fig. 14.25 Circuit for "One More RC Circuit"

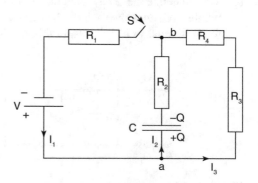

Fig. 14.26 Charge relaxation

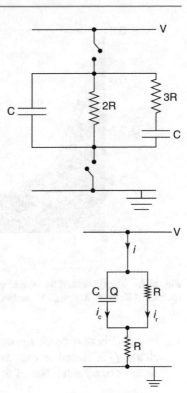

Fig. 14.27 Circuit for "Another RC circuit"

Problem 12: Yet Another RC Circuit

Consider the circuit shown in Fig. 14.27, in which a voltage V is applied at $t = 0$ to a resistor, R, and a capacitor, C, in parallel, which together are in series with another resistor, R. Denote the current through the first resistance as i_r and the charge on (current through) the capacitor as $Q = Q(t)$ (i_c).

(a) How is the current through the second resistor—call it i—related to i_r and i_c.
(b) Write an equation that relates V to i_R and i.
(c) Write an equation that relates V to Q and i.
(d) Write an equation that relates i_c to the charge on the capacitor, $Q(t)$, or its derivative, $Q'(t)$?
(e) Eliminate i_r, i, and i_c to find an equation relating $Q'(t)$ to $Q(t)$.
  Evidently, the RC-time constant of this circuit is $\tau = \frac{RC}{2}$.

Problem 13: Current Ladder Revisited

Consider a circuit similar to that shown in Fig. 14.13 consisting of N capacitors and $2N$ resistors. Capacitor N is short-circuited, so that the voltage across capacitor N is zero at all times, i.e., $W_N(t) = 0$.

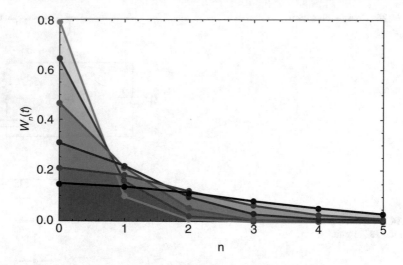

Fig. 14.28 $W_n(t)$ versus n for several values of t: $t = 0.25$ (orange), 0.5 (red), 1.0 (magenta), 2.0 (purple), 4.0 (blue), 8.0 (black). The lines are guides-to-the-eye

(a) Initially, there is no charge on any of the capacitors. At time $t = 0$, however, a charge Q is placed on capacitor 0, so that $W_0(0) = \frac{Q}{C} = V$. Explain why it is sensible to try a solution of the form

$$W_n(t) = W_0 e^{-\gamma t} \cos kn, \tag{14.73}$$

where k and γ are to be determined?

(b) What are the allowed values of k?

(c) Determine the relation between k and γ, and the corresponding allowed values of γ.

(d) The allowed values of γ specified in (c) constitute the eigenvalues for this problem. Determine the amplitude of each corresponding eigenmode, given the initial conditions in this case that $W_0(0) = V$, and $W_n(0) = 0$ for $n > 0$, and that

$$\Sigma_{m=0}^{N-1} \cos \frac{(2m+1)n\pi}{2N} = 0, \tag{14.74}$$

for integer n.[3] Plots of the resultant solution for various values of t are shown in Fig. 14.28.

---

[3] https://www.wolframalpha.com/input/?i=Sum+cos%28%282*m%2B1%29*pi*n%2F%282*N%29%29+from+m%3D0+to+N-1

Problem 14: Getting on My Nerves

The chapter discusses how electrical signals can propagate along the dendrites of neurons. Figure 14.29 shows the voltage at three positions—$x = 3$, $x = 4$, and $x = 5$—along a dendrite as a function of time, in response to an input signal (from another neuron's axon) consisting of a voltage spike occurring at $x = 0$ and $t = 0$.

(a) Notice that each curve appears to show an initial "lag time" when the signal is very small. Describe and explain how this initial period depends on position along the dendrite.

(b) Notice too that each curve decreases at large times. Describe and explain how this behavior depends on position along the dendrite.

(c) Explain which curve corresponds to which position along the dendrite.

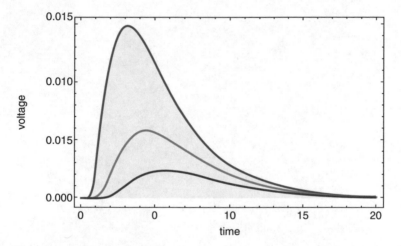

Fig. 14.29 Voltage at three positions—$x = 3$, $x = 4$, and $x = 5$—along a dendrite as a function of time, in response to an input signal (from another neuron's axon) consisting of a voltage spike occurring at $x = 0$ and $t = 0$

Optics: Refraction, Eyes, Lenses, Microscopes, and Telescopes

15

The eye of a human being is a microscope, which makes the world seem bigger than it really is.

Khalil Gibran

15.1   Introduction

The topic of this chapter is optics, including image formation by lenses and eyes, and microscopy which is the preeminent technique in modern cell biology, in particular. Optics is fundamental in experimental laboratory settings, when seeking to acquire an image of a sample of interest. Therefore, in this chapter, we will emphasize the practical aspects and what is important in the laboratory in the context of setting up or using a microscope for biological and biomedical research.

15.2   Your Learning Goals for This Chapter

By the end of this chapter you should be able to:

- Explain refraction and use Snell's law to relate angles of incidence and refraction and refractive indices either side of a planar interface.
- Apply the thin lens approximation to relate the focal length of a lens to its object and image locations, and relate these lengths to magnification.
- Explain simple optical arrangements, from the microscopy researcher's toolkit.
- Describe optical resolution.
- Deconstruct and design simple optical systems.

15.3   Refraction and Snell's Law

In Chap. 11. *Wave Equations: Strings and Wind*, we already learned a great deal about transmission and reflection of one-dimensional sound waves at an interface between acoustically different media. There, the reflection and transmission coefficients were determined by the requirement that the pressure and the displacement of media be the same in both media at the interface. Similarly, light waves undergo reflection and transmission at an interface between optically different media. Optical reflection and transmission coefficients are determined by the required boundary conditions on the electric and magnetic fields, $\mathbf{E}$ and $\mathbf{B}$. At perpendicular incidence, the resultant optical reflection and transmission coefficients are given by very similar expressions to those found for the reflection and transmission of sound waves, but with the *refractive index* playing the role of acoustic impedance. Table 15.1 shows the refractive indices of several media, relevant to biomedical microscopy.

To understand eyes, microscopes, telescopes, etc., we need to go beyond one-dimensional waves and understand an important aspect of the behavior of waves, propagating in two- or three-dimensions, that impinge on an interface at non-

Table 15.1 Refractive index
of several materials

Material	Refractive index
Air	1.0003
Water	1.33
Crown glass	1.52
Dense flint glass	1.65
Glycerol	1.47
40% sucrose solution	1.40
Immersion oil	1.51
Diamond	2.42

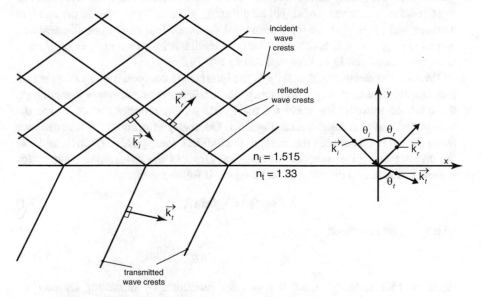

Fig. 15.1 Incident, reflected, and transmitted wave crests, showing the relevant wavevectors, $\mathbf{k}_i$, $\mathbf{k}_r$, and $\mathbf{k}_t$, and the angles, θ_i, θ_r, and θ_t. Left to right is towards positive x. Bottom to top is towards positive y

normal incidence, namely refraction. At non-normal incidence, transmitted waves experience the phenomenon of refraction, which is a change in the wave propagation direction after being transmitted through an interface into a different medium. Refraction at an interface between materials of different refractive index is the basis of optics and how our eyes work. But, of course, sound waves also undergo refraction when incident onto an inclined interface between acoustically different media.

To understand refraction, we consider light waves incident on an interface between two media of different refractive index, say glass and water. For an interface that lies in the plane $y = 0$, Fig. 15.1 sketches the locations of the crests of incident (i), reflected (r), and transmitted (t) waves in the xy-plane at one instant in time, together with the incident, reflected, and transmitted wavevectors, $\mathbf{k}_i$, $\mathbf{k}_r$, and $\mathbf{k}_t$, which are vectors, whose direction is perpendicular to the crests and troughs of the incident, reflected, and transmitted light waves, respectively, and whose lengths are equal to the respective wavevectors: $\frac{2\pi}{\lambda_i}$, $\frac{2\pi}{\lambda_i}$, and $\frac{2\pi}{\lambda_t}$, respectively, where λ_i, and λ_t are the incident and transmitted wavelengths. Also shown are the incident, reflected, and transmitted angles, θ_i, θ_r, and θ_t, respectively, which correspond to the angle between a vector perpendicular to the interface (interface normal vector) and the respective wavevectors. Importantly, the frequencies of the incident, reflected, and transmitted waves must all be the same, because the electric field exactly at the interface is composed of all three waves and oscillates with a single frequency, given by the frequency of the incident wave. On the other hand, the velocities of light in the two different media will be different, given by $c_i = c/n_i$ in the incident medium and $c_t = c/n_t$ in the transmitted medium. Just as for one-dimensional waves, $\omega = c_i k_i = c_t k_t$ is the same for both media. It follows that $k_i = n_i \omega/c$ and $k_t = n_t \omega/c$, and that $\lambda_i = \lambda/n_i$ and and $\lambda_t = \lambda/n_t$.

Because the electric field exactly at the interface is composed of all three waves and oscillates with a single frequency, the distance between wave crests along the interface must be the same for the incident (and reflected) wave as for the transmitted wave, as illustrated in Fig. 15.1. Geometry informs us that the distance along the interface between the crests of the incident wave yields $\lambda_i / \sin \theta_i$, and that the distance along the interface between the crests of the transmitted wave yields $\lambda_t / \sin \theta_t$. These two distances must be equal. It follows that

$$\lambda_i / \sin \theta_i = \lambda_t / \sin \theta_t, \tag{15.1}$$

which may be re-written

$$n_i \sin \theta_i = n_t \sin \theta_t. \tag{15.2}$$

Equation 15.2 is Snell's Law. It quantifies refraction by informing us how the direction of light propagation changes when the light goes from one material to another. Because the crests of the incident and reflected waves must also be equally spaced, we must also have $\theta_i = \theta_r$, i.e., the angle of incidence equals the angle of reflection. Although, we call it Snell's law today, after the seventeen century astronomer, Willebrord Snellius, in fact Snell's law was first correctly described by the Persian scientist, Ibn Sahl in 984, who furthermore employed it to design lenses.

15.3.1 Total Internal Reflection

Light, emanating from a point on an object into a particular direction is referred to as a light "ray." Consider a light ray, initially propagating in a high-refractive-index medium, that is incident at an angle, θ_i, onto a planar interface that separates the high-refractive-index medium (n_i) from a lower refractive index medium ($n_t < n_i$). Equation 15.2 specifies the transmitted angle, and as θ_i is increased, θ_t increases correspondingly. Because, in this case, $n_i > n_t$, from Snell's law Eq. 15.2, it must be $\theta_t > \theta_i$, so that the transmitted light ray refracts away from the normal. Eventually, however, the left-hand side of

$$\sin\theta_t = \frac{n_i}{n_t}\sin\theta_i \tag{15.3}$$

exceeds 1. In this case, when $\theta_i > \arcsin\frac{n_t}{n_i}$, there is no real solution for θ_t. Correspondingly, there is no transmitted wave in this case. Instead, the incident wave is completely reflected. This phenomenon is called total internal reflection (TIR). It is "internal," because the wave cannot exit from, e.g., high-refractive-index glass into air. As discussed in Problem 2 at the end of the chapter, in fact, under TIR conditions, electric and magnetic field intensities decay exponentially into the low-refractive-index medium with a decay length of about 0.2 μm. Total internal reflection fluorescence (TIRF) microscopy exploits this phenomenon to only excite fluorophores within about 0.2 μm of the surface of a glass cover slide, by directing fluorescence excitation light at a large angle to the cover slide.

Total internal reflection underlies the operation of transoceanic communications cables and flexible medical endoscopes, as well as the sparkle of diamond, which has a very large refractive index (Table 15.1). In cell biology, total internal reflection enables total internal reflection fluorescence (TIRF) microscopy, which yields wonderful images of cellular structures and behaviors, that are sufficiently close to a glass slide.

In TIRF microscopy,[1] light that will excite fluorescent probes in the sample under study is directed onto a glass-aqueous interface at an angle greater than the critical angle for total internal reflection. As a result, the light is not transmitted as a propagating wave into the "transmitted medium." Nevertheless, there is an electromagnetic disturbance in the transmitted medium, which instead of propagating cosinusoidally, in fact, decreases in amplitude exponentially with distance into the transmitted medium. What this means is that only fluorescent molecules close to the interface can be excited by the incident illumination beam. The upshot is clear fluorescent images from the molecules near the interface, without image "contamination" by fluorescence from fluorescent molecules that are further into the bulk of the sample.

15.4   Image Formation by a Spherical Surface and the Human Eye

Until now, we have considered how planar surfaces give rise to refraction of waves. We now turn to a discussion of image formation by *curved* surfaces between materials of different refractive index, which is an important application of Snell's law, and which is how our eye focusses light onto its retina (Fig. 15.2). In this context, we can envision the curve surfaces in question to be locally flat. Then light arriving at a particular location on the curved surface undergoes refraction, as given by Snell's law, with the incident angle determined by the angle between the

---

[1] http://en.wikipedia.org/wiki/Total_internal_reflection_fluorescence_microscopy

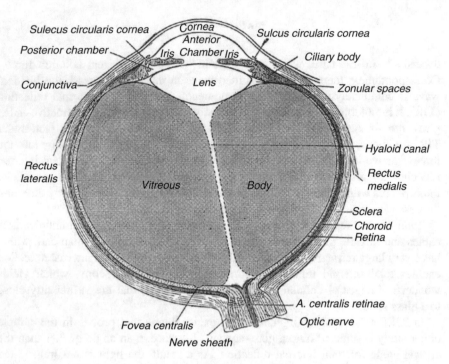

Fig. 15.2 Human eye. Anatomy of the human eye. The indices of refraction of the components of the eye are: 1.38, 1.33, 1.40, and 1.34 for the cornea, the aqueous humor, the lens, and the vitreous humor, respectively. Public domain image from H. Gray's Anatomy of the Human Body, 1918

direction of propagation and the local orientation of the curved interface. Focusing involves collecting the light rays emanating from an object in different directions and bringing all these rays from a given point on the object back to a corresponding point to form an image of the object at a different location. This is how our eyes work: The light rays from an object that we are looking at propagate to our eyes, where they are focused by our cornea and eye lens into an image on our retina, where the light that constitutes the image are converted into nerve impulses that proceed to the parts of our brain responsible for vision processing. In our eyes, the largest change in refractive index, and hence the greatest refraction occurs between air and the cornea. Nevertheless, refraction at the internal surfaces is essential for focusing. Changing focus is accomplished in humans via muscles that change the shape of the lens.

For simplicity, we will model image formation by the eye as image formation behind a single spherical surface—the surface of our retina is approximately spherical—that separates two media of different refractive index, when light from a point source in front of the surface propagates through the surface, undergoing refraction on the way, as shown in Fig. 15.3, which also specifies the relevant angles. First, we notice that the fact that the sum of the angles of a triangle must add up to π radians implies that

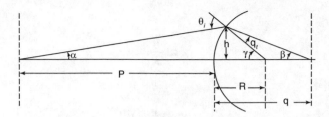

Fig. 15.3 Angles, distances, and geometry for image formation behind a spherical surface, which is a simple model for how our eyes focus. As sketched, the refractive index inside the sphere is larger than the refractive index outside the sphere, corresponding to the situation for our eyes

$$\gamma + \alpha + \pi - \theta_i = \pi \tag{15.4}$$

or

$$\theta_i = \gamma + \alpha \tag{15.5}$$

and

$$\theta_t + \beta + \pi - \gamma = \pi \tag{15.6}$$

or

$$\theta_t = \gamma - \beta. \tag{15.7}$$

When the object (p) and image (q) distances are large compared to the radius of the eye, which we will take to be the case throughout our discussion, all the relevant angles are small and we may use a small angle approximation for all trigonometric functions. Using this approximation, Snell's law implies

$$n_i \theta_i \simeq n_t \theta_t, \tag{15.8}$$

so that

$$n_i (\gamma + \alpha) \simeq n_t (\gamma - \beta), \tag{15.9}$$

or

$$n_i \alpha + n_t \beta \simeq (n_t - n_i)\gamma. \tag{15.10}$$

Furthermore, in a small angle approximation, we may express

$$\alpha \simeq \frac{h}{p}, \tag{15.11}$$

$$\beta \simeq \frac{h}{q}, \tag{15.12}$$

and

$$\gamma \simeq \frac{h}{R}. \tag{15.13}$$

It follows that

$$\frac{n_i}{p} + \frac{n_t}{q} \simeq \frac{n_t - n_i}{R}, \tag{15.14}$$

where we cancelled a factor of h everywhere. Remarkably, Eq. 15.14 is independent of α or h, implying that all the light from a point a distance p in front of the cornea is focused to a point q behind the cornea, irrespective of its initial propagation angle, α. Thus, we see that a spherical surface indeed gives rise to focusing. The image formed by a spherical surface in the way shown in Fig. 15.3 is said to be a *real* image, because you can actually put a light sensor—a retina in the case of an eye— at the location of the image and record an image of the object. Since the location of the retina is a fixed distance (q) behind the cornea, this calculation cannot be the whole story as far as our eyes are concerned. Otherwise we could only see objects in focus that are a definite distance (p) in front of our eyes. In fact, the lens and muscles of our eyes play an important role. The lens gives rise to additional focusing beyond that of the cornea alone, and because lens' radius of curvature is controlled by our eye muscles, we can focus on objects at a range of distances in front of our eyes.

15.4.1 Real and Virtual Images

We presented Fig. 15.3 in order to understand our eyes for which $n_t > n_i$. However, our calculation is also applicable in the case that $n_i > n_t$, but in that case the right-hand side of Eq. 15.14 is negative. The only way this equation can then be satisfied for fixed p is if q becomes negative, which implies that the image is now on the same side of the spherical surface as the object. Such an image is said to be a *virtual* image, because you cannot put a sensor at its location and record it. No light actually arrives at that location. By contrast, the light that arrives on our retinas forms a *real* image.

15.5   Image Formation by a Lens

A lens consists of two spherical surfaces, one after the other. Thus, focusing by a lens—in spectacles, microscopes, telescopes, etc.—may be understood on the basis of refraction at two sequential spherical surfaces. Focusing by two successive spherical surfaces is discussed in detail in Problem 3 "The thin lens formula" at the end of the chapter. The upshot of these calculations, which are built on the calculations of Sect. 15.4, is that there is a relationship between the distance from the object to the lens (p), and the distance from the lens to the image (q), namely

$$\frac{1}{p} + \frac{1}{q} = \frac{1}{f}, \tag{15.15}$$

where f is the focal length of the lens, which depends only on the properties of the lens. Equation 15.15 is called the *thin lens formula*, and is an important result. An important corollary of the thin lens formula is that the angular deviation that a light ray undergoes when it passes through a lens of focal length f at a distance h from the center of the lens is

$$\Delta\alpha = \frac{h}{f}. \tag{15.16}$$

As also discussed in Problem 3 "The thin lens formula," the "lens maker's formula," namely

$$\frac{1}{f} = (n-1)\left(\frac{1}{R_1} + \frac{1}{R_2}\right), \tag{15.17}$$

relates the focal length, f, to the refractive index of the lens, n, and the radius of curvature of the two spherical surfaces that constitute the lens. Converging lenses are usually either biconvex, with equal radii of curvature on both sides ($R_2 = R_1$), or plano-convex with one side curved and the other side flat (e.g., $\frac{1}{R_2} = 0$). In order to minimize aberrations, i.e., distortions in the image, the rule of thumb is to keep the refraction angles as small as possible. The locations of the object and the image are said to be conjugate planes.

15.6 Optical Applications of Lenses

15.6.1 Magnification by a Lens

Geometry furthermore informs us that the magnification, namely the height of the image divided by the height of the object is

$$M = -\frac{q}{p}, \tag{15.18}$$

as shown in Fig. 15.4, where the object is an arrow. The minus sign corresponds to the fact that the image is inverted relative to the object.

An important special case of Eq. 15.15 occurs for $p = f$ ($q = f$), so that the object (the image) is placed at the focal length of the lens. It follows according to Eq. 15.15 that then the image (object) is located at $q = \infty$ ($p = \infty$). In this case, light from a particular position across the object (image) is mapped, on the other side of the lens, into light propagating in a particular direction, as shown in Fig. 15.5. The relationship between the angle on one side of the lens and the position on the other side of the lens is determined by considering the special case of rays that go through

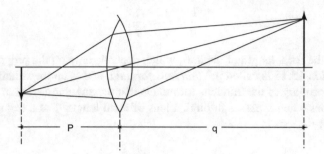

Fig. 15.4 Object and image formation by a lens

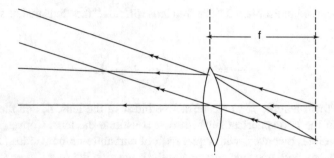

Fig. 15.5 Object located in the focal plane of the lens

the center of the lens. The propagation direction of these rays is unchanged by the lens, as shown in Fig. 15.6. From the geometry shown in this figure, we see that

$$\frac{x}{f} \simeq \theta, \tag{15.19}$$

where x is the positional displacement from the optical axis on one side, θ is the angular deviation from the direction of the optical axis on the other, and we again made a small angle approximation. Although we derived it by considering a special ray, Eq. 15.19 relates angle on one side of the lens to position on the other for all rays, irrespective of whether they go through the center of the lens or not.

15.6.2 Corrective Lenses

Many of us wear spectacles or contact lenses because otherwise we have difficulty focusing on distant objects (near-sightedness) or nearby objects (far-sightedness). In order for increasingly distant objects to remain in focus as p becomes larger and $\frac{1}{p}$ becomes smaller, it is necessary for $\frac{1}{f}$ to also become smaller and f to become larger in order to maintain the image at the location of the retina at fixed distance q. If a patient is short-sighted, they are unable to reduce the focal length of their

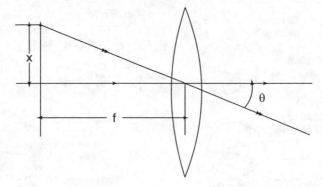

Fig. 15.6 Mapping from position to angle for a lens with the object at the focal length of the lens

eye lenses sufficiently. Therefore, to correct for near-sightedness, spectacles must increase the divergence of the light arriving at the eye. Concave lenses are necessary to achieve this result. Conversely, to correct for far-sightedness, spectacles must decrease the divergence of the light arriving at the eye, which requires convex lenses. These considerations are illustrated in Fig. 15.7.

15.6.3 Example: Magnification with Two Tandem Lenses and Infinity-Corrected Microscopes

Consider two lenses with focal lengths f_1 and f_2, respectively. Design an optical configuration of these two lenses to achieve a magnification $M = -\frac{f_2}{f_1}$.

Answer: Figure 15.8 shows a configuration with
the desired magnification, in which the object
plane is located a distance f_1 in front of the
f_1-focal-length lens and the f_2-focal length lens is
located a distance f_2 in front of the image plane.
The distance between the two lenses can take any
value in this application.
This configuration is the basic element within a
modern, so-called "infinity-corrected" microscope,
where the object-to-lens 1 distance equals the
focal length of the microscope objective--typically
effectively 2 mm for a ×100 objective--and the lens
2-to-image distance equals the focal length of the
"tube lens"--typically 200 mm. In this case, the
magnification is $\frac{f_2}{f_1} = \frac{200}{2} = 100$. Hence, it is a ×100

(continued)

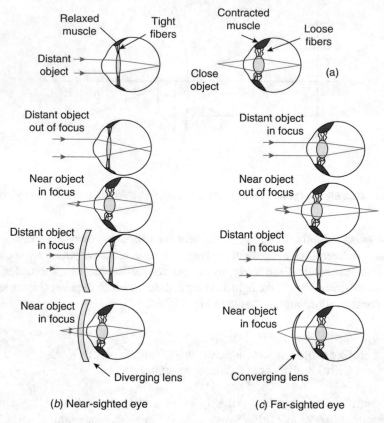

Fig. 15.7 (**a**) Changing the shape of the lens in our eye allows us to focus on distant and near objects. (**b**) Near-sighted patients cannot sufficiently flatten their lenses. Near-sightedness can be corrected with spectacles with diverging lenses. (**c**) Far-sighted patients cannot sufficiently curve their lenses. Far-sightedness can be corrected with spectacles with convergent lenses

objective, although the magnification depends on the focal length of the tube lens just as much as it depends on the effective focal length of the objective itself.

To provide a little more context, in a microscope, the image plane is arranged to coincidence with a sensor, such as a CCD ("charge-coupled-device"), which might typically be comprised of 1024×1024 square pixels, each $10\,\mu m$ on a side. Then, for a $\times 100$ objective, the image of a $0.5\,\mu m$-object becomes $50\,\mu m$ on the CCD, which is 5 pixels across.

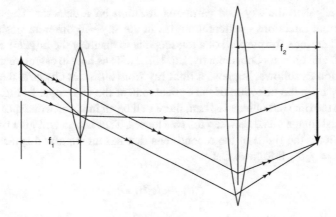

Fig. 15.8 Two tandem lenses (lens 1 on the left and lens 2 on the right) with object and image in the focal plane of each lens. The resultant magnification is $M = -\frac{f_2}{f_1}$

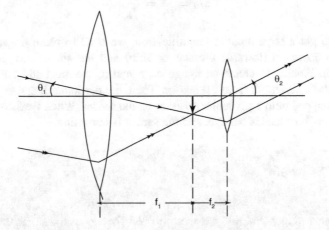

Fig. 15.9 A telescope arrangement of two tandem lenses separated by the sum of their focal lengths

15.6.4 Two Tandem Lenses as a Telescope: Magnifying Angles

Another useful arrangement of two tandem lenses constitutes a telescope. In a telescope, the two lenses have a common focal plane. Thus, in a telescope, the two lens of focal lengths f_1 and f_2, respectively, are separated from each other by a distance $f_1 + f_2$, as shown in Fig. 15.9.

How do telescopes work to examine distant objects? Stars are very, very away. The closest star, Proxima Centauri, is 9.4×10^{15} m (4.2 light-years) away. Therefore, the light from a particular distant star arrives at the Earth from a particular direction, i.e., from a particular (solid) angle. An essential aspect of a telescope is that it is designed to work with a third, lens that focuses light arriving from a particular angle in the sky to a particular location on a detector. This third lens maps different directions—different angles in the sky—to different positions on a detector. Therefore, the purpose of a telescope is to magnify the *angular separation* between distant objects. Conveniently, Eq. 15.19 tells us how to calculate the angular magnification as follows. Suppose a light ray from a distant object is displaced by an angle θ_1 from the direction of the optical axis of the telescope. In Fig. 15.9, this object is to the (far) left of lens 1. Then, there will be an image at a distance $x = f_1\theta_1$ at the mutual image plane between the two lenses. That image will give rise to a ray downstream (to the right) of the second lens that has an angular displacement, θ_2, given by $x = f_2\theta_2$. It follows that

$$f_1\theta_1 = f_2\theta_2, \tag{15.20}$$

i.e., the angular magnification is

$$M_\theta = \frac{\theta_2}{\theta_1} = \frac{f_1}{f_2}. \tag{15.21}$$

Note that to get a large angular magnification, we need to place the large-focal-length lens upstream (towards the star or ship) and the small focal length lens downstream. Finally, to create an image on a sensor, we must add the third lens located at its focal length from the sensor. Then, the particular angular position of the star is mapped onto a particular position on the sensor. When we look through a telescope, the third lens is our eye, and the sensor is our retina.

15.6.5 Example: Telescope as a Beam Expander

Although we often associate telescopes with observing distant objects, a telescopic arrangement of two lenses provides the means to change the cross-sectional area of a light beam. For example, in some applications, it can be highly desirable to focus a laser beam into the smallest possible focal spot in a microscope. As discussed in Sect. 15.7.2, achieving this goal requires that the back pupil of the microscope objective be fully illuminated by an illuminating laser beam. In turn, full illumination of the back pupil often requires that the diameter of the laser beam be expanded. Using two lenses of focal lengths f_1 and f_2, respectively, design a beam expander that increases the diameter of a laser beam by a factor $\frac{f_2}{f_1}$.

Answer: Inspection of the telescope geometry shown in Fig. 15.9 reveals that a telescope can serve as a beam expander by a factor f_2/f_1, provided light enters from the right, that is, provided

(continued)

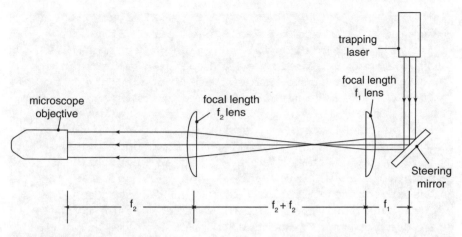

Fig. 15.10 "4f-configuration" of lenses for beam steering via a mirror. In this case, the first lens is separated from an upstream mirror by its focal length, f_1, and from the second lens by a distance equal to the sum of the focal lens of the two lenses, $f_1 + f_2$. The back pupil of the microscope objective is downstream of the second lens by a distance, f_2 equal to the focal length of that lens

the telescope is operated backwards. This is the
desired design. As well as expanding the beam,
a beam expander/backwards telescope necessarily
reduces angles within the incident light beam.
Therefore, the distribution of angles within
the transmitted beam will be narrower than the
distribution of angles within the incident beam.
Thus, a beam expander is also a beam collimator
(Fig. 15.10).

15.6.6 Example: Beam Steering

Figure 15.10 shows an optical setup—a so-called 4f-configuration—that permits a focused laser beam to be steered in the object plane via a mirror, while simultaneously ensuring that all of the beam passes unobstructed through the back pupil of the microscope objective, irrespective of the angle of the mirror! By considering planes that are conjugate to the mirror, explain how this arrangement ensures that all of the light reflected from the mirror makes it into the microscope objective.

(continued)

Answer: The location of the back pupil of the
microscope objective is conjugate to the location
of the mirror. Therefore, a hypothetical object
at the mirror is focused to its image at the back
aperture of the objective. Since rotating the
mirror does not change the position of a laser
spot at the mirror's location, rotating the mirror
does not change the position of that spot at the
back pupil of the objective. Rather what changes
both after the mirror and at the back pupil of
the objective is the angle of the laser beam.
Correspondingly, the position of the laser focus
in the focal plane of the objective moves as the
mirror angle is changed. Thus, it is possible to
steer the focus to any desired location, by means
of a mirror in a plane conjugate the back pupil of
the objective. The ability to steer a laser focus
is important in optical tweezers experiments and
in fluorescence recovery at photobleaching (FRAP)
experiments, for example.

15.6.7 Example: Effective Focal Length of Two Lenses

Microscope objectives generally consist of multiple lenses, which makes it
possible to largely eliminate optical aberrations—distortions in the images
that occur for individual lenses—but at a significant cost: The most expensive
$\times 100$-magnification, microscope objectives—which can cost up to $10,000 or
so—provide excellent, multicolor images over a relatively wide field of view.
Good $\times 100$ objectives can be purchased for $500. In order for these objectives
to function properly, the objective-to-cover slide distance typically must be
set to about 0.17 mm. Nevertheless, a $\times 100$-magnification, large numerical
aperture (large NA) objective can be modeled as a single lens with a focal
length of 2 mm.

In this example, we will see how two lenses with individual focal lengths,
f_1 and f_2, respectively, separated by a distance d behave as a single lens
with a focal length f that depends on f_1, f_2, and d, located at a position
that is different from the position of either of its two constituent lenses. Once
we can combine two lenses into a single effective lens, we can combine any
number of lenses—such as in a microscope objective—into a single effective

(continued)

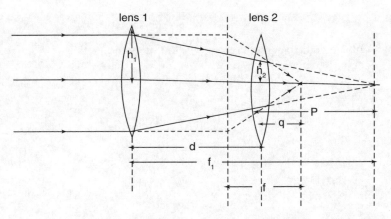

Fig. 15.11 Ray diagram for two tandem lenses (lens 1 on the left with focal length f_1 and lens 2 on the right with focal length f_2) that behave as a single effective lens with focal length f

lens. Thus, this example makes sense of how the multiple lenses that actually comprise a microscope objective can be considered as effectively a single lens and how the effective focal length of a microscope objective can be different from the objective to sample distance.

Consider, then, the ray diagram of Fig. 15.11, in which parallel light from the left of the two lenses is focused to a point to the right of the two lenses. The focal length of the single equivalent lens is defined naturally to be the distance between the focus and the plane at which the initial rays and their corresponding final rays intersect, as shown in the figure.

(a) Write an equation relating f_2, q, and p.

    Answer: The thin lens formula informs us that

$$\frac{1}{f_2} = \frac{1}{q} - \frac{1}{p}. \tag{15.22}$$

    The negative sign appears because the "object", at distance p, is on the same side of lens 2 as its image, at distance q.

(b) Write an equation relating d, p, and f_1.

    Answer: Inspection of the figure informs us that

$$f_1 = d + p. \tag{15.23}$$

(continued)

(c) Combine your answers to (a) and (b) to express q in terms of f_1, f_2, and d, only.

Answer: From (a), we have

$$\frac{1}{q} = \frac{1}{f_2} + \frac{1}{p}, \tag{15.24}$$

and from (b), we have

$$p = f_1 - d. \tag{15.25}$$

Thus,

$$\frac{1}{q} = \frac{1}{f_2} + \frac{1}{f_1 - d}, \tag{15.26}$$

or

$$q = \frac{f_2(f_1 - d)}{f_1 + f_2 - d}, \tag{15.27}$$

(d) Write an equation expressing the fact that the sum of the deflection angles (Eq. 15.16) of lens 1 and lens 2 equals the deflection of the single equivalent lens in terms of h_1, h_2, f, f_1, and f_2 only.

Answer: According to the figure, the deflection angle caused by lens 1 is $\frac{h_1}{f_1}$, the deflection angle caused by lens 2 is $\frac{h_2}{f_2}$, and the deflection angle caused by the single equivalent lens is $\frac{h_1}{f}$. Therefore,

$$\frac{h_2}{f_2} + \frac{h_1}{f_1} = \frac{h_1}{f}. \tag{15.28}$$

(e) Write a relationship between h_2, p, h_1, and f. Thus, eliminate h_2 and h_1 from your answer to (d).

Answer: According to the figure,

$$\frac{h_2}{q} = \frac{h_1}{f}. \tag{15.29}$$

(continued)

Therefore, $h_2 = \frac{q}{f} h_1$ and the answer to (d) becomes

$$\frac{q h_1}{f_2 f} + \frac{h_1}{f_1} = \frac{h_1}{f}. \tag{15.30}$$

Cancelling h_1 everywhere, we have

$$\frac{1}{f} = \frac{1}{f_1} + \frac{q}{f_2 f} \tag{15.31}$$

(f) Combine your answers to (c) and (e) to express f in terms of f_1, f_2, and d only.

Answer: Using Eq. 15.27 in Eq. 15.31, we find

$$\frac{1}{f} = \frac{1}{f_1} + \frac{1}{f} \frac{(f_1 - d)}{f_1 + f_2 - d}. \tag{15.32}$$

Rearranging Eq. 15.32 becomes

$$\frac{1}{f} \left(1 - \frac{(f_1 - d)}{f_1 + f_2 - d} \right) = \frac{1}{f_1} = \frac{1}{f} \frac{f_2}{f_1 + f_2 - d}, \tag{15.33}$$

i.e.,

$$\frac{1}{f} = \frac{f_1 + f_2 - d}{f_1 f_2} = \frac{1}{f_1} + \frac{1}{f_2} - \frac{d}{f_1 f_2}, \tag{15.34}$$

which is the desired result. Notice that when $d = f_1 + f_2$, $\frac{1}{f} = 0$, consistent with Example 15.6.3.

15.6.8 Kohler Illumination for Microscopy

Until now, we have not considered where the light emanating from the object of interest comes from. In fact, in addition to collecting light from an object of interest, microscopes must also provide the illumination. For example, in fluorescence microscopy experiments, fluorophores within the object under study must be excited by illumination of the appropriate wavelength. In bright-field experiments, the image is formed from light scattered from the object of interest.

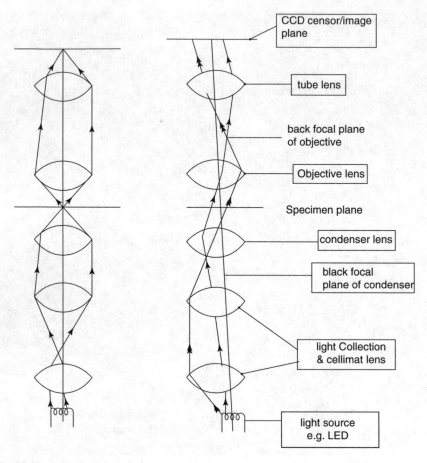

Fig. 15.12 Arrangement of lenses and two sets of conjugate planes in an "infinity-corrected" microscope with trans-illumination

Most microscopes employ so-called Kohler illumination,[2] illustrated in Fig. 15.12, which shows two sets of light paths in a microscope with bright-field Kohler illumination. It is the same light in both cases, just viewed differently. The light originates in a super-bright light emitting diode (LED) at the bottom of the figure. Light from the LED is collected by a collector lens and directed to the condenser lens, which sends the light to the sample of interest. Then, light scattered from the sample is collected by the objective and an image of the sample appears on the sensor.

First, let us look at the left-hand light path. This is the image-forming light path. This path has conjugate focal planes between the collector and the condenser, at

---
[2] http://en.wikipedia.org/wiki/Kohler_illumination

the specimen (a.k.a. sample, a.k.a. object) plane, and on the CCD sensor. (CCD is "charge-coupled device.")

Below the objective, the source of light is a LED. The light illuminating the sample in the image-forming path corresponds to parallel light rays from the LED, as illustrated in the sketch. Such rays, emerging in a definite direction from the LED, are focused to a definite location in the specimen plane. Since light is emitted equally in all directions from the LED, this implies that the illumination in the sample/object plane will be uniform, which is desirable.

The right-hand light path corresponds to the illumination light path. The illumination is a LED, and this path corresponds to light from a particular location on the LED. Planes conjugate to the LED are the so-called back-focal plane of the condenser and the back-focal plane of the objective. Back-focal planes are such that a focus there gives rise to parallel light on the sample side of the condenser. Importantly, this means that there is not an image of the LED in the sample/object plane, nor any of its conjugate planes.

As shown in Fig. 15.13, Similar considerations apply to so-called epi-illumination, where the illumination comes in through the objective, instead of from the other side, through a condenser, which is called trans-illumination. Epi-illumination is commonly used for fluorescence-based experiments with a bandpass filter acting as a mirror to reflect the excitation light towards the sample, but passing the emitted light towards the CCD.

15.7   Microscopy Beyond Lenses

In general, microscopy experiments involve several features beyond the lens optics that we have discussed so far in this chapter. In this section, we discuss two of the most important of these aspects, namely fluorescence and optical resolution, which will allow us to understand the overwhelmingly majority of biological microscopy setups and experiments.

15.7.1 Fluorescence Microscopy

Fluorescent molecules—fluorophores—transition into an excited state by absorbing light of one color, and then transition back to their unexcited state with the emission of fluorescent light of a different, longer-wavelength color. Fluorescence microscopy involves imaging a sample containing fluorophores, by imaging the emitted, fluorescent light. As an example, Fig. 15.14 shows the absorption (cyan) and emission (green) spectra of the fluorescent protein, EGFP. In a fluorescence microscopy experiment involving EGFP, the EGFP-containing sample is illuminated with excitation light with a wavelength near the peak of the EGFP emission spectrum, such as laser light of wavelength 488 nm. Then, excited EGFP molecules will emit light with a wavelength given according to the EGFP the emission spectrum (green in Fig. 15.14), which is used to make an image of the EGFP in

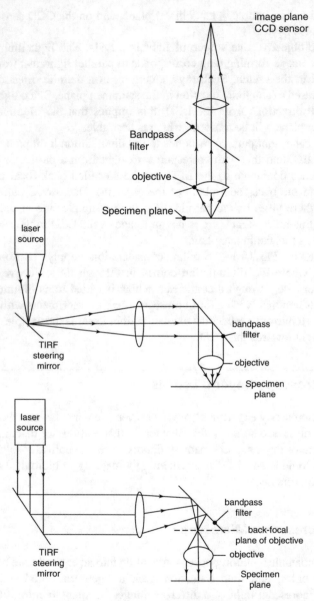

Fig. 15.13 Laser-based Kohler illumination for TIRF microscopy in an infinity-corrected micro-scope. The top panel shows the light path of fluorescence emission. The middle and bottom panels shows the light path of the excitation light from the laser. The middle panel shows that the steering mirror and the sample are in conjugate planes. The bottom panel shows that parallel light from the laser gives rise to sample illumination with parallel light. A bandpass filter reflects a narrow range of wavelengths about the laser, fluorescence excitation wavelength, but passes the fluorescence emission wavelength through into the CCD detector. In this case, the steering mirror steers the position of the focus in the back-focal plane of the objective, which allows for TIRF or not. Because the steering mirror is in a plane conjugate to the sample plane, changing the incidence angle for TIRF or not, does not change the location of the beam in the sample plane

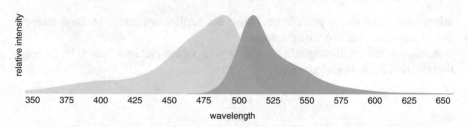

Fig. 15.14 Absorption spectrum, shown in cyan, and emission spectrum, shown in green, of the fluorescent protein, EGFP

the sample. Filters on the detection side ensure that no excitation light reaches the detector. The introduction and development of proteins consisting of a protein of interest (POI) genetically fused to a fluorescent protein, such as EGFP, has led to a revolution in cell biology over the last 25 years. Roger Tsien, Osamu Shimomura, and Martin Chalfie were awarded the 2008 Nobel Prize in Chemistry for the discovery and development of the green fluorescent protein (GFP) for biological microscopy

15.7.2 Optical Resolution

Why not study 5 nm-sized objects by using a ×10,000 microscope objective? The answer is that it is not possible because the wave character of light gives rise to a limit on the smallest object that can be resolved optically. A derivation of the optical resolution limit is beyond our scope, but in microscopy, it is important to know that a nm-size light source will appear to be much larger than nm-size, when observed optically by even the best microscope. Specifically, the smallest that any object will appear in a microscope is given by

$$\frac{0.61\lambda}{NA},\tag{15.35}$$

where λ is the wavelength of the light and NA is the numeral aperture of the microscope objective, given by $NA = n\sin\theta_{max}$, where n is the refractive index containing the object and θ_{max} is the largest angle collected by the objective. Thus, for example, an individual fluorescent protein molecule, that is 2 nm in size, emitting light with a wavelength of 510 nm, will appear as a spot with a full-width-at-half-maximum (FWHM) of about 0.22 μm (220 nm), which becomes 22 μm after magnification. This is the apparent size of a point source of light in a microscope.

However, it is possible to determine the location of a 220 nm spot to much better that a 220 nm precision. Localizing the center of such a spot with a precision of ±10–20 nm would be typical. This observation underlies super-resolution microscopy[3] and single particle tracking,[4] but further discussion of these state-of-the-art methods is also out of our scope.

Similarly, the smallest spot in the object plan, which can be achieved by focusing illumination light, is also given by

$$\frac{0.61\lambda}{NA}, \tag{15.36}$$

where NA is the numerical aperture of the focusing lens. Thus, the smallest spot that a lens can achieve is limited by the numerical aperture of the lens. For example, a 40 mm-diameter 200 mm-focal length lens in air has a numerical aperture of $NA = \frac{20}{200} = 0.1$. Therefore, for 510 nm-wavelength light, the smallest spot it is possible to achieve (by filling the entire area of the focusing lens with light) is 3 μm. To achieve the smallest possible focused spot in a microscope (0.22 μm) it is necessary to illuminate the entire back pupil of the microscope objective. Reducing the illuminated area of the back pupil by means of an aperture increases the spot size in the object plane correspondingly. Often, it is desirable to illuminate the entire microscope field of view (FOV), which is about 100 μm across. To achieve illumination over the entire FOV requires a beam diameter at the objective's back pupil of about 0.5 mm.

15.8   Chapter Outlook

The purpose of this chapter was first to introduce and then examine the consequences of the key feature of wave propagation in 2D and 3D, that does not occur for 1D wave propagation, namely refraction. Refraction leads directly to optics, lenses, focusing, magnification, etc. Optical imaging has long played a key role in biological research. Today this statement is truer than ever, because our ability to form images of sub-cellular structures and monitor processes inside cells, and the biological research enabled by these capabilities, has been revolutionized within the last twenty years, first, by the widespread adoption of fluorescent protein fusions, that now allow biologists to monitor specifically the behavior of particular proteins in living cells, and, second, by the invention of super-resolution microscopy methods, that now permit optical images of fluorescently labeled sub-cellular structures to be obtained at a resolution of tens of nanometers or less, far smaller than the wavelength of visible light. Both of these advances have been recognized by recent Nobel prizes: The 2008 Nobel Prize in Chemistry was awarded to Osamu Shimomura, Martin Chalfie, and Roger Tsien for "the discovery and development of green fluorescent protein, GFP," and the 2014 Nobel Prize in Chemistry was awarded to Eric Bertzig, Stefan Hell, and William Moerner for

---

[3] https://en.wikipedia.org/wiki/Super-resolution_microscopy

[4] https://en.wikipedia.org/wiki/Single-particle_tracking

"the development and super-resolved fluorescence microscopy." In this chapter, we discussed a number of lens arrangements—e.g., telescopes and 4f-configurations—that are very commonly employed in biological microscopy setups. We hope that this material will enable you to better understand optical systems that you may encounter in the future, irrespective of whether in a research or in a medical diagnostic setting (pathology).

15.9 Problems

Problem 1: Snell's Window
Snell's window is the phenomenon illustrated in Fig. 15.15 and more schematically in Fig. 15.16, in which an underwater viewer sees light from above the surface through a circular "window." Outside the window, no light is transmitted from the air side of the interface. The goal of this problem is to elucidate this phenomenon.

(a) What is the largest value of θ_i for which light can make it from the air side of the ocean's surface to the diver's eye?
(b) What is the corresponding largest value of θ_t (see Fig. 15.16) for which light can make it from the air side of the ocean's surface to the diver's eye? Give your answer in terms of the refractive index of water, $n = 1.33$. The refractive index of air is 1.0.
(c) Given that the diver's eye is at a depth d, what is the radius of the Snell's window that the diver sees?

Fig. 15.15 US Navy diver, Ryan Arnold, framed in Snell's window. From https://en.wikipedia. org/wiki/Snell%27s_window

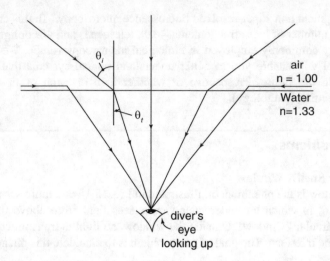

Fig. 15.16 Schematic for Snell's window problem. A diver's eye is at a vertical depth d under a perfectly calm, flat ocean surface. This figure shows a selection of light rays, propagating from above the surface of the water to the diver's eye

(d) Suppose that the diver has a laser pointer, what is the largest angle, relative to the surface normal, at which the diver can point the laser pointer and expect the laser beam to exit the water?

(e) In microscopy, in order to achieve the best resolution, it is imperative to collect light rays from the object under study from the largest angles possible. Explain why the highest-resolution microscope objectives, used in cell biology, are "oil-immersion" objectives that must be used with a special oil of refractive index 1.51 between the glass cover slide (refractive index 1.51) and the glass of the first lens of the microscope objective (refractive index 1.51).

Problem 2: Imaginary Sines and "Evanescent" Waves
In this problem, you will understand how total internal reflection comes about mathematically, and the character of the fields within the "external" medium.

(a) Write the transmitted wavevector, using the coordinate system of Fig. 15.1, in terms of $k = \frac{\omega}{c}$, θ_i, n_i, and n_t only.
Trigonometry tells us that

$$\mathbf{k}_t = k_x\mathbf{i} + k_y\mathbf{j} = n_t k \sin\theta_t \mathbf{i} - n_t k \cos\theta_t \mathbf{j}. \tag{15.37}$$

Next, using Snell's law, we find

$$\cos\theta_t = \pm\sqrt{1 - \frac{n_i^2}{n_t^2}\sin^2\theta_i}. \tag{15.38}$$

It follows that

$$\mathbf{k}_t = n_i k \sin \theta_i \mathbf{i} \pm n_t k \sqrt{1 - \frac{n_i^2}{n_t^2} \sin^2 \theta_i} \mathbf{j}. \tag{15.39}$$

Equation 15.39 is the transmitted wavevector expressed in terms of the refractive indices of the materials either side of the interface and the incident
angle. We will pick the $\pm$ sign depending on what makes
sense physically.

(b) Total internal reflection occurs when $\cos \theta_t$ becomes imaginary, which occurs when $\sin \theta_1 > \frac{n_t}{n_i}$. Show that in this situation the amplitude of the transmitted E-field decays exponentially versus y. This exponentially decaying E-field is what we called the evanescent wave.

When $\sin \theta_1 > \frac{n_t}{n_i}$, the argument of the square root in
Eq. 15.39 is negative, and therefore the square root
itself is imaginary. Thus, we may write

$$\mathbf{k}_t = n_i k \sin \theta_i \mathbf{i} + i k \alpha \mathbf{j}, \tag{15.40}$$

where

$$\alpha = \pm n_t \sqrt{\frac{n_i^2}{n_t^2} \sin^2 \theta_i - 1} \tag{15.41}$$

is real.

To determine the consequences of an imaginary wavevector, it is convenient to now represent a cosinusoidal wave as the real part of a complex exponential:

$$E_t(\mathbf{r}, t) = E_t \cos(\mathbf{k}_t \cdot \mathbf{r} - \omega t) = E_t Re[e^{i(\mathbf{k}_t \cdot \mathbf{r} - \omega t)}]$$

$$= E_t e^{-k\alpha y} Re[e^{i(n_i k \sin \theta_i x - \omega t)}] = E_t e^{-k\alpha y} \cos(n_i k \sin \theta_i x - \omega t). \tag{15.42}$$

Equation 15.42 informs us that when the incidence
angle is greater than $\sin^{-1} \frac{n_t}{n_i}$, and $k_y = i k \alpha$ is correspondingly imaginary, then the electric field in the tran-
smitted medium varies exponentially versus y. In the
context of Fig. 15.1, the electric field must go to
zero as y goes to negative infinity. Therefore, here,
we must pick the negative sign for α, in order that
the E- and B-fields do not grow exponentially as $y \to$
$-\infty$, which would be unphysical. On the contrary, with

the negative sign, the E- and B-fields decay exponentially within the transmitted medium.

(c) Calculate the penetration depth (Λ), which characterizes the decrease of the electric field within the transmitted medium.

Answer:

$$\Lambda = \frac{1}{k\alpha} = \frac{\lambda}{2\pi n_t \sqrt{\frac{n_i^2}{n_t^2} \sin^2 \theta_i - 1}}. \tag{15.43}$$

(d) Use WolframAlpha to plot your result for (c) in the case of green light ($\lambda \simeq$ 530 nm), incident from glass ($n_i = 1.515$) onto water ($n_t = 1.33$), and comment on the result.

Answer:　http://www.wolframalpha.com/input/?i=Plot%5B+Sqrt
%5B2*pi*530%2F1.33%2F%28%281.515%2F1.33%29%5E2*%28Sin%5Bx%
5D%29%5E2-1%29%5D+%2C%7Bx%2CArcSin%5B1.33%2F1.516%5D%2C1.5[5]

From this plot, it is apparent that the penetration
depth, Λ, is quite close to 100 nm--about 20% of the
wavelength of the light in question--except for angles near to the critical angle, where Λ becomes larger, eventually going to infinity right at the critical angle.

Problem 3: The Thin Lens Formula

The goal of this problem is to derive the "thin lens formula," which relates the location of an image, formed by a lens, to the location of the object and the focal length of the lens.

(a) Reproduce the left-hand portion Fig. 15.3, including rays and angles just to the right of the spherical surface. Now, add to the right-hand side of your sketch a near mirror image of the left-hand portion that you already sketched, which shows the path of light rays through a second nearby spherical surface. Indicate the new angles α', β', γ', θ_t' (inside the lens), and θ_i' (outside the lens), analogous to the angles already introduced.

(b) In the chapter, we found

$$n_i \alpha + n_t \beta = (n_t - n_i)\gamma \tag{15.44}$$

---

[5] http://www.wolframalpha.com/input/?i=Plot%5B+Sqrt%5B2*pi*530%2F1.33%2F%28%281.
515%2F1.33%29%5E2*%28Sin%5Bx%5D%29%5E2-1%29%5D+%2C%7Bx%2CArcSin
%5B1.33%2F1.516%5D%2C1.5

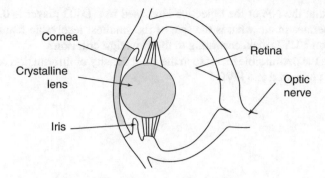

Fig. 15.17 Cross-section of a fish eye

for the first spherical surface. Following steps similar to those in the chapter, find a similar equation that relates the new angles α', β', and γ', and the refractive indices, n_i, and n_t.

(c) How is β' related to β?

(d) Combine Eq. 15.44, and your answers to (b) and (c) to find an equation that relates α, α', γ, γ', n_i, and n_t.

(e) Using a small angle approximation, as in the chapter, relate α, α', γ, and γ' to the object-to-lens distance, p, the lens-to-image distance, q, the radius of curvature, R, and h. (It is a thin lens, so $h' = h$.)

Problem 4: Comparative Anatomy
Why are fishes' eyes lenses (Fig. 15.17) much more spherical than ours?

Problem 5: Things to Know About Color
(a) What color is light of wavelength 470 nm, 505 nm, 530 nm, 590 nm, and 630 nm?

(b) What is the peak wavelength of the light emitted from green fluorescent protein (GFP)?

(c) What range of light wavelengths can the human eye see?

Problem 6: Blu-Ray Versus DVD?
(a) How does the resolution of a microscope depend on the numerical aperture (NA) of the microscope objective?

(b) How does the resolution of a microscope depend on the wavelength of the light used for the imaging?

(c) What wavelength of light is used in Blu-ray players?

(d) What wavelength of light is used in a DVD players?

(e) Given that the NA of the objective lens used in a Blu-ray player is 0.85 and the lens operates in air, what is the size of the smallest resolvable feature that can be put on a Blu-ray disk, according to the Rayleigh criterion?

(f) Given that the NA of the objective lens used in a DVD player is 0.60 and the lens operates in air, what is the size of the smallest resolvable feature that can be put on a DVD disk, according to the Rayleigh criterion?

(g) What is the permissible increase in the areal density of information on a Blu-ray disk as compared to a DVD?

Biologic: Genetic Circuits and Feedback 16

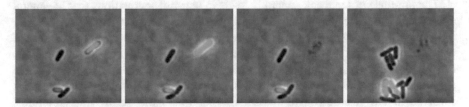

Fig. 16.1 Sequence of frames from the movie at http://www.youtube.com/watch?v= sLkZ9FPHJGM, associated with Ref. [6], used with permission. Increasing time runs from left to right. In the earliest, leftmost frame two *E. coli* cells are infected by phage lambda, as signaled by their green fluorescence. becomes infected and undergoes lysis, providing an example of the behavior in the lytic state. By contrast, in the lower left of the field of view, another *E. coli* becomes infected, but instead of undergoing lysis, this bacterium undergoes two rounds of cell division before the movie ends, resulting in four infected, fluorescent, progeny *E. coli* in the lysogenic state

16.1 Introduction

In this chapter, we will study two artificial genetic networks, namely the "genetic toggle switch" and the "repressilator." The genetic toggle switch was described and implemented in *E. coli* in Ref. [1], and the repressilator was described and implemented in *E. coli* in Ref. [2]. We present simple mathematical models of both

 Electronic Supplementary Material The online version of this article (https://doi.org/10.1007/ 978-3-031-05808-0_16) contains supplementary material, which is available to authorized users.

the genetic toggle switch and the repressilator that seek to predict the concentrations versus time of the protein products of the genes involved in the network. These examples provide an introduction to biological decision making, control and time-keeping, which are the subject matter of "Systems Biology" or "Synthetic Biology," which has emerged as a major subfield of Biology over the last two decades.

An essential concept for this biologic is "feedback." Feedback is exemplified by the protein product of a certain gene then going on to affect in some way its own expression. There are two types of feedback: positive feedback, in which case the protein tends to increase its own expression, for example, by being an activator for its own expression; and negative feedback, in which case the protein tends to decrease its own expression, for example, by being a repressor for its own expression. Feedback does not have to be direct. It can go through one or more intermediary genes/proteins. Beyond genetic networks, feedback is tremendously important in medicine—the human body involves countless feedback and control circuits, realized in many different ways.

The presence of feedback generally leads to non-linear equations that in general are difficulty to solve analytically. However, as we saw previously in Chap. 7. *Rates of Change: Drugs, Infections, and Weapons of Mass Destruction* in the context of our discussion of HIV infection within an individual patient, non-linear equations can have steady-state solutions. Accordingly, first we will look for steady-state solutions of the equations that describe the concentrations of the two proteins involved in the genetic toggle switch. Having found the steady-state solutions, next, we will examine what happens when the concentrations differ just a little from these steady-state values. This procedure leads us back to linear equations, which therefore may be solved using the eigenvalue-eigenvector approach, that we employed first in Chap. 7. *Rates of Change: Drugs, Infections, and Weapons of Mass Destruction* and then in Chap. 10. *Oscillations and Resonance.* Correspondingly, we will find exponential relaxation towards "stable" steady states, but exponential growth away from "unstable" steady states. Interestingly, for certain parameters, there are two stable steady states with two different sets of protein concentration. It is then possible to switch between the two stable states by changing initial conditions, for example. Hence, this is the genetic toggle switch.

We will then seek to follow a similar procedure for the equations that describe the repressilator. In this case, we will discover that for a certain range of model parameters. The putative steady state is invariably unstable, in that the real part of one of the corresponding eigenvalues is always positive. It turns out that this result is a manifestation of sustained oscillatory behavior for this range of parameter values.

Nature relies on multiple, interconnected gene circuits that are often more elaborate than the simple Frankensteinian examples that we will focus on. Nevertheless, analogous natural gene networks are ubiquitous. For example, the genetic toggle switch is closely analogous to the lytic-lysogenic switch in the life cycle of bacteriophage lambda[1] in infected *E. coli*, which realizes one of the most studied and best understood biological switches [3]. For infected bacteria, in the lysogenic state, the

---

[1] https://en.wikipedia.org/wiki/Lambda_phage

bacteriophage's genetic material is incorporated into the bacterial chromosome and is replicated along with the host's genetic material at cell division, but phage capsid proteins, for example, are not expressed. However, in the lytic state all the proteins required to form new phage are expressed, many copies of the phage assemble, and the host is caused to disintegrate (lyse), releasing many new bacteriophage particles, which are now free to infect a new host. These two different possible outcomes are visualized in the movie at http://www.youtube.com/watch?v=sLkZ9FPHJGM,[2] where phage infection is signaled by green fluorescence (Fig. 16.1) [4]. Early in the movie, in the top right of the field of view a replicating *E. coli* becomes infected and undergoes lysis, which is signaled in the movie by the disappearance of green fluorescence, and which provides an example of the behavior in the lytic state. By contrast, in the lower left of the field of view, another *E. coli* becomes infected, but instead of undergoing lysis, this bacterium undergoes two rounds of cell division before the movie ends, resulting in four infected, fluorescent, progeny *E. coli* in the lysogenic state. A version of the repressilator has recently been recognized in the natural gene circuit of *Bacillus subtilis*, a common soil bacterium [5].

16.2 Your Learning Goals for This Chapter

By the end of this chapter, you should be able to

- Write chemical rate equations appropriate for describing simple biologic circuits.
- Find the steady-state solutions of biologic chemical rate equations in simple cases.
- Carry out a linear stability analysis of biologic rate equations for small deviations of the concentrations from their steady-state values in simple cases.
- Determine the biologic "phase diagram" in simple cases.

16.3 Binary Biologic

For an *E. coli* bacterium infected by phage lambda, the existence of two possible outcomes, which is called bistability, depends on the interaction of two genes, *cI* and *cro*, and their protein products, cI and cro, each of which represses the other. The simplest way to rationalize the time evolution in an infected bacterium is to assume that both cI and cro are either *ON* or *OFF*. In the lysogenic state, cI is *ON*, that is, the concentration of cI is high. Therefore, because cI represses expression of cro, cro is *OFF*. Consequently, expression of cI continues at a high level, i.e., cI remains *ON*, which continues to repress cro, and so on. Alternatively, in the lytic

---

[2] http://www.youtube.com/watch?v=sLkZ9FPHJGM

Fig. 16.2 Schematic of the
biologic circuit involved in
the lysogenic-lytic switch of
E. coli, infected by phage
lambda

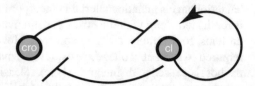

state, cro is *ON*, which represses cI, which therefore is *OFF* and unable to repress
cro so that cro remains *ON* and the lytic state persists, or would persist, except that
this state initiates a pathway to host cell lysis. Importantly, the overall feedback in
this case is positive, realized by the protein product of a certain gene then going on
to repress a gene whose protein product represses the original gene. In addition to
repressing cro, cI is also an activator for itself, realizing a second positive feedback
loop, which represents the "suspenders" in a "belt-and-suspenders" approach to
maintaining the lysogenic state. A graphical representation of the biologic of the
lysogenic-lytic switch is shown in Fig. 16.2, where the blunt arrows represent down-
regulation (repression) and the pointy arrow represents up-regulation (activation).

Binary biologic is simple and intuitive and can inform us about the network
topologies that may give rise to interesting behaviors, such as switching or oscillations. In particular, if binary biologic does not predict that bistability may occur
in a particular gene/protein network (for example), it is unlikely to occur. However,
binary biologic is too simple to reliably predict whether or not interesting behavior
actually occurs in a particular situation, because its underlying assumptions, namely
that genes are either *ON* or *OFF* and experience synchronous switching events, are
unrealistic. A better approach is to write and solve the appropriate chemical rate
equations, of the sort we first encountered in Chap. 7. *Rates of Change: Drugs,
Infections, and Weapons of Mass Destruction*. In the context of such models,
whether or not there is switching or oscillatory behavior depends on the structure of
the model and the values of the parameters of the model. As we will see, often these
parameters must lie within a certain range of values in order to realize bistability
or sustained oscillations within the context of a particular model. It follows that
if we want to predict, or even control, what goes on inside cells—perhaps to cure
disease—understanding the relevant equations and knowing their parameters will
be tremendously valuable. This is a vision for the medicine of the future. Even
chemical rate equations are not the whole story, because as we saw in Chap. 8.
Statistical Mechanics: Boltzmann Factors, PCR, and Brownian Ratchets, chemical
reactions can be viewed as random walks and, just like random walks, are subject to
fluctuations, which are likely to be important when the numbers of chemical species
are small, as can be the case inside a bacterium. Here, however, we consider the
chemical rate equation point of view only.

16.4   The Genetic Toggle Switch

The first biologic circuit we will examine mathematically is the so-called genetic
toggle switch [1], which consists of two genes each of which encodes for a protein
that represses the other's gene expression. Because of the even number (2) of

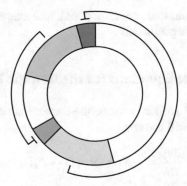

Fig. 16.3 Schematic of the genetic toggle switch plasmid. The upper orange region is the promoter for repressor 1. The lower green region is the promoter for repressor 2. The upper light orange region codes for repressor 1, which, as indicated, represses expression of repressor 2. The lower light green region codes for repressor 2, which, as indicated, represses promoter 1

components around the circuit, there is a net positive feedback, which can give rise to bistable behavior for certain parameter values. The design of the genetic toggle switch is shown in Fig. 16.3. The genetic toggle switch is similar to the lytic-lysogenic switch, except that in this case one of the states does not self-destruct, so here it is possible in principle to "toggle" back and forth between the two states.

As for the lysogenic-lytic switch, this is how the genetic toggle switch can realize bistability: Suppose that the concentration of repressor 1 is *ON*. This has the result of repressing expression of repressor 2. Since the concentration of repressor 2 is *OFF*, it is not able to repress the expression of repressor 1, which results in the expression of additional repressor 1, which therefore remains *ON*, which continues to repress repressor 2, which remains *OFF* and so on. The alternative state is that the concentration of repressor 1 remains low (repressor 1 is *OFF*), while the concentration of repressor 2 remains high (repressor 2 is *ON*). In fact, however, whether or not the genetic toggle switch actually realizes bistability depends on the physical (i.e., biochemical) properties of the proteins involved—their production and degradation rates, and details of the feedback. To elucidate these features, we will analyze the genetic toggle switch as follows:

- Write the chemical rate equations for the concentrations of the two repressors.
- Look for and find steady-state solutions to these equations.
- Examine which of the steady-state solutions are *unstable* and which are *stable*. What do we mean by "unstable" and "stable"? An unstable solution is analogous to a ball located at the top of a mountain—although ball will not roll down when it is perfectly balanced at the top, any small fluctuation that perturbs it from the exact balance condition will send the ball rolling down the mountain. A stable solution is like a ball located at the bottom of a valley. In this case, the ball returns to its stable equilibrium position after any fluctuations that take it away from the valley minimum.

- Interpret the stable solutions, which we find, and examine how these solutions depend on the model's parameters.

16.4.1 Chemical Rate Equations for the Genetic Toggle Switch

The chemical rate equations that describe the concentration of repressor 1 (c_1) and repressor 2 (c_2) that we will use are:

$$\frac{dc_1}{dt} = -K_1 c_1 + \gamma_1 (1 - P_1(c_2)) \tag{16.1}$$

and

$$\frac{dc_2}{dt} = -K_2 c_2 + \gamma_2 (1 - P_2(c_1)), \tag{16.2}$$

where

- K_1 is the degradation rate of repressor 1.
- K_2 is the degradation rate of repressor 2.
- γ_1 is the production rate of repressor 1 in the absence of any repression by repressor 2.
- γ_2 is the production rate of repressor 2 in the absence of any repression by repressor 1.
- $P_1(c_2)$ is the probability that repressor 2 binds promotor 1 thus repressing expression of repressor 1. See Fig. 16.3. Only that fraction of promotor 1 sites that are not occupied by repressor 2 can bind RNA polymerase and give rise to repressor 1 expression.
- $P_2(c_1)$ is the probability that repressor 1 binds promotor 2, thus repressing expression of repressor 2.

For analytic simplicity, we will take $K_1 = K_2$, $\gamma_1 = \gamma_2$, and $P_1 = P_2 = P$ with

$$P(c_2) = \frac{(c_2/c_0)^n}{1 + (c_2/c_0)^n}. \tag{16.3}$$

Equation 16.3 is a so-called Hill function. Where did this form come from? First, notice that Eq. 16.3 has the correct behavior for large and small c_2: for small c_2, Eq. 16.3 gives $P \simeq 0$, which is physically correct, and leads to little repression; for large c_2, Eq. 16.3 gives $P \simeq 1$, which is physically correct, and leads to near complete repression. But the values of c_0—which is a concentration—and n are parameters that are given to us by the cell, in the same way that K_1, γ_1, etc. are given to us by the cell. The Hill function for $n = 2$ and $n = 10$ is shown in Fig. 16.4.[3]

---

[3] http://www.wolframalpha.com/input/?i=Plot%5B%7Bx%5E2%2F%281%2Bx%5E2%29%2C+x%5E10%2F%281%2Bx%5E10%29%7D%2C%7Bx%2C0%2C4%7D%5D

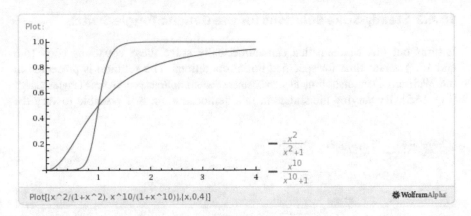

Fig. 16.4   Hill functions for $n = 2$ and $n = 10$

In both cases, we find a sigmoidal (S-shaped) curve. However, the curve transitions from 0 to 1 more suddenly for the larger value of n. This is generally true: the larger the value of n, the more sudden the transition.

A value of n greater than 1 is generally the result of "cooperativity." A well-known example of cooperativity occurs for oxygen binding by hemoglobin, which we encountered previously in Chap. 8. *Statistical Mechanics: Boltzmann Factors, PCR, and Brownian Ratchets*. In that case, we discovered that, if hemoglobin binds either zero or four oxygen molecules, a Hill function (Eq. 16.3) with $n = 4$ describes the probability that hemoglobin binds 4 oxygen molecules. Importantly, in order to realize bistable behavior in the context of this model some cooperativity ($n > 1$) is required.

With the simplifications described above, Eqs. 16.1 and 16.2 become

$$\frac{dc_1}{dt} = -Kc_1 + \frac{\gamma}{1 + (c_2/c_0)^n} \tag{16.4}$$

and

$$\frac{dc_2}{dt} = -Kc_2 + \frac{\gamma}{1 + (c_1/c_0)^n}. \tag{16.5}$$

Dividing through by Kc_0 and writing $x_1 = c_1/c_0$ and $x_2 = c_2/c_0$, these equations become

$$\frac{1}{K}\frac{dx_1}{dt} = -x_1 + \frac{a}{1 + x_2^n} \tag{16.6}$$

and

$$\frac{1}{K}\frac{dx_2}{dt} = -x_2 + \frac{a}{1 + x_1^n}, \tag{16.7}$$

where $a = \frac{\gamma}{Kc_0}$.

16.4.2 Steady-State Solutions for the Genetic Toggle Switch

It turns out that Mathematica can numerically solve these equations (Eqs. 16.6 and 16.7) versus time for specified initial conditions. The solution is presented in the Wolfram Demonstration at http://demonstrations.wolfram.com/GeneticToggleSwitch/[4] (Fig. 16.5). By varying the sliders in this demonstration, it is possible to vary the

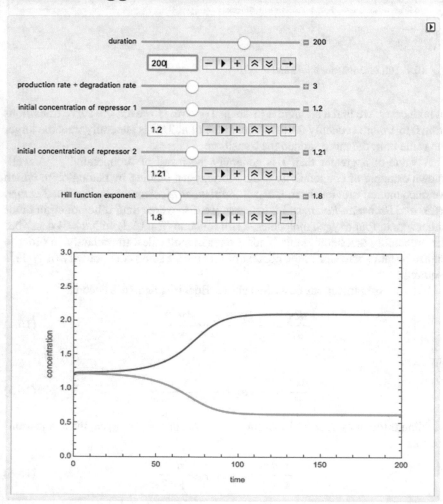

Fig. 16.5 Screenshot showing the numerical solution of Eqs. 16.6 and 16.7, implemented by the Wolfram Demonstration at http://demonstrations.wolfram.com/GeneticToggleSwitch/

---

[4] http://demonstrations.wolfram.com/GeneticToggleSwitch/

parameters of the model and the initial conditions. This exercise reveals that for some parameters there is a steady-state bistable solution, where x_1 (green) is large and x_2 (orange) is small, or vice versa, and that it is possible to switch between these two solutions by changing only the initial conditions. For other parameters, there is a steady-state solution with $x_1 = x_2$, irrespective of the initial conditions. For any parameters and initial conditions, the concentrations, x_1 and x_2, do approach constant values. These values, of course, are the steady-state solutions, defined via

$$- x_1 + \frac{a}{1 + x_2^n} = 0 \tag{16.8}$$

and

$$- x_2 + \frac{a}{1 + x_1^n} = 0. \tag{16.9}$$

In order to be able to proceed analytically, we will specialize to the case $n = 2$:

$$- x_1 + \frac{a}{1 + x_2^2} = 0 \tag{16.10}$$

and

$$- x_2 + \frac{a}{1 + x_1^2} = 0. \tag{16.11}$$

Conveniently, WolframAlpha can solve these equations. First, we look for solutions for which $x_1 = x_2$, which are not bistable. In this case, Eqs. 16.10 and 16.11 both become (with $x_1 = x_2 = x$):

$$- x + \frac{a}{1 + x^2} = 0, \tag{16.12}$$

whose solution[5] is shown in Fig. 16.6.[6] WolframAlpha can also solve Eqs. 16.10 and 16.11[7] in the more general case that $x_1 \neq x_2$. In this case, there are two additional solutions, which are given in Fig. 16.7, and which are real for $a > 2$

---

[5] http://www.wolframalpha.com/input/?i=Solve%5B-x%2Ba%2F%281%2Bx%5E2%29%3D%3D0%2Cx%5D

[6] http://www.wolframalpha.com/input/?i=Plot%5B+%28Sqrt%5B3%5D*Sqrt%5B27*a%5E2%2B4%5D%29%2B9*a%29%5E%281%2F3%29%2F%282%5E%281%2F3%29%29%2F%283%5E%282%2F3%29%29+-%28%282%2F3%29%5E%281%2F3%29%29%2F%28Sqrt%5B3%5D*Sqrt%5B27*a%5E2%2B4%5D%2B9*a%29%5E%281%2F3%29%2C%7Ba%2C0%2C4%7D+%5D

[7] http://www.wolframalpha.com/input/?i=Solve%5B%7B-x%2Ba%2F%281%2By%5E2%29%3D%3D0%2C-y%2Ba%2F%281%2Bx%5E2%29%3D%3D0%7D%2C%7Bx%2Cy%7D%5D

```
Solve[{-x+a/(1+x^2)==0},x]
```

Input interpretation:

solve	$-x + \dfrac{a}{1+x^2} = 0$	for	x

Results:

$$x = \frac{\sqrt[3]{\sqrt{3}\,\sqrt{27a^2+4}+9a}}{\sqrt[3]{2}\,3^{2/3}} - \frac{\sqrt[3]{\dfrac{2}{3}}}{\sqrt[3]{\sqrt{3}\,\sqrt{27a^2+4}+9a}}$$

Fig. 16.6 Solutions of Eq. 16.12 via WolframAlpha

```
Solve[{-x+a/(1+y^2)==0,-y+a/(1+x^2)==0},{x,y}]
```

Input interpretation:

solve	$-x + \dfrac{a}{1+y^2} = 0$ $-y + \dfrac{a}{1+x^2} = 0$	for	x, y

Results:

$y = 0$ and $a = 0$ and $x = 0$

$y = \dfrac{1}{2}\left(\sqrt{a^2-4}+a\right)$ and $x = \dfrac{1}{2}\left(a-\sqrt{a^2-4}\right)$ and $a \neq 0$

$y = \dfrac{1}{2}\left(a-\sqrt{a^2-4}\right)$ and $x = \dfrac{1}{2}\left(\sqrt{a^2-4}+a\right)$ and $a \neq 0$

Fig. 16.7 Solutions of Eqs. 16.10 and 16.11 via WolframAlpha

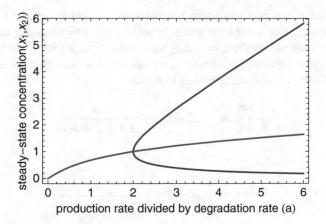

Fig. 16.8 The three real solutions of Eqs. 16.10 and 16.11 plotted versus the (normalized) repressor production rate divided by the repressor degradation rate [$a = \frac{\gamma}{K c_0}$]. The two bistable solutions are plotted in blue and red. The other solution is plotted in magenta only.

only. These two solutions are bistable: either x_1 is large and x_2 is small or x_2 is large and x_1 is small. This is what we mean by bistability.

The three steady-state solutions are plotted[8] all together in Fig. 16.8 in magenta, red and blue. The first solution, for which $x_1 = x_2$, is shown as the magenta line. For the second solution, x_1 is the red line and x_2 is the blue line. For the third solution, x_1 is the blue line and x_2 is the red line.

16.4.3 Stable or Unstable?

It will turn out that for $a < 2$, the magenta solution is "stable," but that, for $a > 2$, the magenta solution is "unstable," while the red and blue bistable solutions are "stable." To determine whether or not a particular solution is stable, we consider concentrations that are slightly different from the steady-state solutions that we just found, i.e., we set

$$x_1 = x_1^* + \delta_1 \tag{16.13}$$

and

$$x_2 = x_2^* + \delta_2, \tag{16.14}$$

---

[8] http://www.wolframalpha.com/input/?i=Plot%5B%281-Sqrt%5B1-4%2Fa%5E2%5D%29%2F%282%2Fa%29%2C%281%2BSqrt%5B1-4%2Fa%5E2%5D%29%2F%282%2Fa%29%7D%2C%7Ba%2C0%2C4%7D%5D

where x_1^* and x_2^* are the steady-state solutions that we just found and δ_1 and δ_2 are small deviations. For a stable solution, δ_1 and δ_2 will evolve to smaller values in time. For an unstable solution, δ_1 and δ_2 will evolve to larger values in time.

To determine the time evolution of δ_1 and δ_2, we substitute Eqs. 16.13 and 16.14 into Eqs. 16.6 and 16.7 (setting $n = 2$) to obtain

$$\frac{1}{K}\frac{d\delta_1}{dt} = -(x_1^* + \delta_1) + \frac{a}{1 + (x_2^* + \delta_2)^2} \tag{16.15}$$

and

$$\frac{1}{K}\frac{d\delta_2}{dt} = -(x_2^* + \delta_2) + \frac{a}{1 + (x_1^* + \delta_1)^2}. \tag{16.16}$$

We now envisage δ_1 and δ_2 to be very small, and carry out Taylor series expansions of the last term in Eqs. 16.15 and 16.16 to linear order in δ_2 and δ_1, respectively. In the case of Eq. 16.15, the linear expansion reads

$$f(x_2) = f(x_2^* + \delta_2) = f(x_2^*) + f'(x_2^*)\delta_2 + O(\delta_2^2), \tag{16.17}$$

with

$$f(x_2) = \frac{a}{1 + x_2^2}. \tag{16.18}$$

It follows that to linear order:

$$\frac{1}{K}\frac{d\delta_1}{dt} = -x_1^* - \delta_1 + \frac{a}{1 + (x_2^*)^2} - \frac{2ax_2^*}{[1 + (x_2^*)^2]^2}\delta_2 = -\delta_1 - \frac{2ax_2^*}{[1 + (x_2^*)^2]^2}\delta_2$$

$$= -\delta_1 - Y_2\delta_2, \tag{16.19}$$

where we introduced $Y_2 = \frac{2ax_2^*}{[1+(x_2^*)^2]^2}$ to save writing. Similarly,

$$\frac{1}{K}\frac{d\delta_2}{dt} = -\delta_2 - Y_1\delta_1. \tag{16.20}$$

These equations are reminiscent of similar equations, involving two coupled variables, that we encountered in Chap. 7. *Rates of Change: Drugs, Infections, and Weapons of Mass Destruction*. We may therefore expect to solve these equations similarly by finding their eigenvalues and eigenvectors. In fact, we are interested in whether δ_1 and δ_2 shrink or grow versus time. To answer this question, the key quantities that we need to find are the eigenvalues, which report upon the decay (or growth) rates of the system. To figure out the eigenvalues, we will

- assume an exponential decay for δ_1 and δ_2, i.e., assume that $\delta_1 = C_1 e^{\Lambda t}$ and $\delta_2 = C_2 e^{\Lambda t}$,
- substitute these guesses into Eqs. 16.19 and 16.20, and
- solve for Λ in terms of the parameters of the problem (given quantities),
- decide whether Λ is negative for decaying δ_1 and δ_2 and a stable solution, or whether Λ is positive for an unstable solution.

Following the first step of this procedure, we find in this case

$$\frac{\Lambda}{K}\delta_1 = -\delta_1 - Y_2\delta_2. \tag{16.21}$$

$$\frac{\Lambda}{K}\delta_2 = -\delta_2 - Y_1\delta_1, \tag{16.22}$$

or

$$(1 + \frac{\Lambda}{K})\delta_1 = -Y_2\delta_2. \tag{16.23}$$

$$(1 + \frac{\Lambda}{K})\delta_2 = -Y_1\delta_1. \tag{16.24}$$

These are the equations that we must solve. To this end, multiply Eq. 16.23 by Eq. 16.24 and divide both sides by $\delta_1\delta_2$ to find

$$(1 + \frac{\Lambda}{K})^2 = Y_1 Y_2, \tag{16.25}$$

i.e.,

$$\Lambda = -K(1 \pm \sqrt{Y_1 Y_2}). \tag{16.26}$$

Notice that Eq. 16.26 is applicable to all of the steady-state solutions that we found previously. We just have to use the appropriate values of Y_1 and Y_2 for the steady-state solution in question to find the effect of small deviations from this steady state.

Stability of the Bistable Solution
First, we examine the bistable solutions. Using the explicit expressions, given by Mathematica and shown blue and red in Fig. 16.8, we have

$$Y_1 = \frac{2x_1^* a}{[1 + (x_1^*)^2]^2} = \frac{4/a}{a + \sqrt{a^2 - 4}} \tag{16.27}$$

and

$$Y_2 = \frac{2x_2^* a}{[1 + (x_2^*)^2]^2} = \frac{4/a}{a - \sqrt{a^2 - 4}}. \tag{16.28}$$

It follows that

$$\sqrt{Y_1 Y_2} = 2/a \tag{16.29}$$

and thus

$$\Lambda = -K(1 \pm \frac{2}{a}). \tag{16.30}$$

Recall that these bistable solutions are real only for $a > 2$. Given that $a > 2$, Eq. 16.30 then informs us that the values of Λ corresponding to the two eigenmodes of the bistable solutions are both invariably negative. Consequently, they both correspond to decaying exponentials, and we see that the bistable solutions are stable, i.e., any deviation from the steady state will decay exponentially in time towards the steady state.

Stability of the Non-bistable Solution

Next, we examine the non-bistable solution, shown in Fig. 16.8 as the magenta line. In this case, $x_1^* = x_2^* = x^*$, so that $Y_1 = Y_2 = Y$, and

$$\sqrt{Y_1 Y_2} = Y = \frac{2x^* a}{[1 + (x^*)^2]^2}. \tag{16.31}$$

In this case therefore,

$$\Lambda = -K(1 \pm Y). \tag{16.32}$$

Using Mathematica again to do the algebra, we find

$$Y = \frac{(3^{1/3}2 - 2^{1/3}(-9a + \sqrt{12 + 81a^2})^{2/3})^3}{18a(-9a + \sqrt{12 + 81a^2})}, \tag{16.33}$$

which is plotted as the magenta line in Fig. 16.9. Evidently, Y is less than unity for $a < 2$ and greater than unity for $a > 2$. It follows that for $a < 2$, Λ is inevitably negative corresponding to a stable solution. However, for $a > 2$, in the case of the negative sign, Λ becomes positive, so that $e^{\Lambda t}$ grows exponentially as a function of time. We only need one of the eigenvalues to be positive to send us away from the steady-state solution. Therefore, this is indeed an unstable solution, and it is not realized in the cell, because any small fluctuation grows away from it. Such a fluctuation will, in fact, eventually approach one of the stable solutions, as you can see from the Wolfram Demonstration.

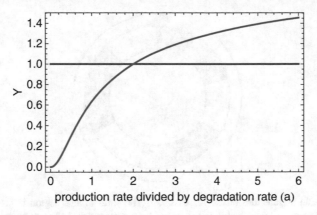

Fig. 16.9 Y versus a

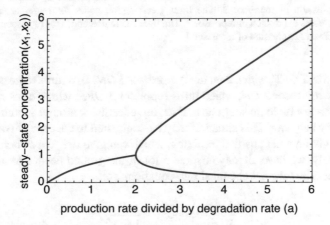

Fig. 16.10 Steady-state concentrations for the genetic toggle switch

Finally, then, we can report the "equation of state" of the genetic toggle switch in Fig. 16.10. For $a < 2$, there's a single solution with $c_1 = c_2$—no bistability. For $a > 2$, there are two bistable solutions with $x_1 > x_2$ or $x_2 > x_1$.

16.5 Repressilator

Figure 16.11 shows the architecture in the case of the so-called repressilator. This biologic circuit is composed of three genes and their gene products, each of which represses expression of the following gene's gene product [2]. This circuit uses the same basic element as the genetic toggle switch, but it uses three of these elements around the feedback loop, rather than two. In this case, therefore, going around the complete circuit gives an overall negative feedback. What does binary biologic suggest about the behavior of the repressilator? Suppose repressor 1 is *ON*. This then

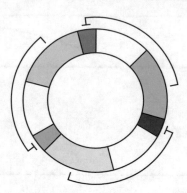

Fig. 16.11 Gene architecture of the repressilator. The solid orange region is the promoter for repressor 1. The solid green region is the promoter for repressor 2. The solid red region is the promoter for repressor 3. The light orange region codes for repressor 1, which, as indicated, represses expression of repressor 2. The light green region codes for repressor 2, which, as indicated, represses expression of repressor 3. The light red region codes for repressor 3, which, as indicated, represses expression of repressor 1

turns repressor 2 *OFF*, which then turns repressor 3 *ON*. This turns repressor 1 *OFF*, which turns repressor 2 *ON*, which turns repressor 3 *OFF*, which turns repressor 1 *ON*. Thus, binary biologic suggests that negative feedback around a circuit can lead to oscillatory behavior. This situation may be contrasted to the case of two negative feedbacks, giving a net positive feedback, which can give rise to a switch.

Fortunately, we have already covered a lot of the ground needed to analyze the repressilator. The relevant chemical rate equations are:

$$\frac{1}{K}\frac{dx_1}{dt} = -x_1 + \frac{a}{1+x_3^n},\tag{16.34}$$

$$\frac{1}{K}\frac{dx_2}{dt} = -x_2 + \frac{a}{1+x_1^n},\tag{16.35}$$

and

$$\frac{1}{K}\frac{dx_3}{dt} = -x_3 + \frac{a}{1+x_2^n}.\tag{16.36}$$

In this case, we will leave n as is. The reason is that this system does not realize sustained oscillations for $n = 2$ for any value of a. In fact, we will determine the "phase diagram" of the repressilator as function of a and n.

This system of equations (Eqs. 16.34, 16.35, and 16.36) can be solved numerically by Mathematica. A Wolfram Demonstration that does just this is http://demonstrations.wolfram.com/Repressilator/.[9] For certain parameter values, adjustable via the sliders in the Demonstration, the numerical solution approaches

---

[9] http://demonstrations.wolfram.com/Repressilator/

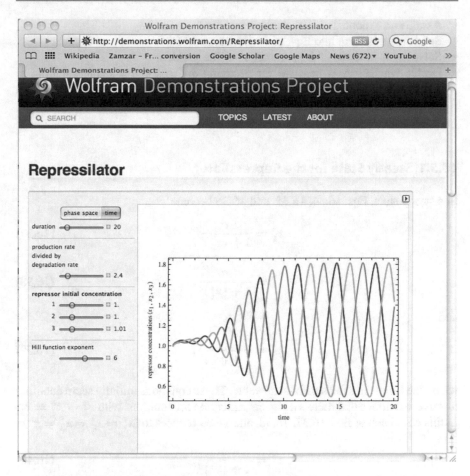

Fig. 16.12 Numerical solution of Eqs. 16.34, 16.35, and 16.36, implemented as the Wolfram Demonstration at http://demonstrations.wolfram.com/Repressilator/

a steady-state value for the concentrations after decaying oscillatory behavior at early times. For other parameter values, the numerical solutions show sustained, constant-amplitude oscillations (Fig. 16.12). These sustained oscillations are not a steady-state solution, but they are the interesting solution in this case, because they constitute a biologic clock, which might be a model for the clock that operates in cell division or establishes a circadian rhythm. Therefore, in this case, it is actually the unstable solutions that we are interested in, and the range of parameter values for which such unstable, oscillatory solutions are observed. Nevertheless, our procedure in this case will mirror the procedure that we followed in the case of the genetic toggle switch:

- Find the steady-state solutions.
- Derive the equations that describe how small deviations (δ_1, δ_2, and δ_3) in concentration from the steady-state values evolve in time

- Assume an exponential time dependence for δ_1, δ_2, and δ_3
- Substitute this form into Eqs. 16.34, 16.35, and 16.36.
- Solve for Λ in terms of the parameters of the problem
- Decide whether Λ is negative for decaying δ_1, etc., and a stable solution, or whether Λ is positive for an unstable solution. In this case, we are interested in the unstable solutions.

16.5.1 Steady State for the Repressilator

In a steady state, Eqs. 16.34, 16.35, and 16.36 become

$$x_1^* = \frac{a}{1 + (x_3^*)^n}, \tag{16.37}$$

$$x_2^* = \frac{a}{1 + (x_1^*)^n}, \tag{16.38}$$

and

$$x_3^* = \frac{a}{1 + (x_2^*)^n}, \tag{16.39}$$

where the * indicates the steady-state value. These equations initially seem daunting to solve, but notice that there's a big simplification for solutions with $x_1^* = x_2^* = x_3^*$. In this case, each of Eqs. 16.37, 16.38, and 16.39 reduces to ($x_1^* = x_2^* = x_3^* = x^*$):

$$x^* = \frac{a}{1 + (x^*)^n}. \tag{16.40}$$

Now, examination of the numerical solution presented in the Mathematica demonstration fails to reveal any steady-state solutions for which the concentrations (x_1, x_2, and x_3) are different from each other. Therefore, we will hypothesize that, in fact, all of the real solutions to Eqs. 16.34, 16.35, and 16.36 correspond to $x_1^* = x_2^* = x_3^*$. (Since concentrations must be real, we are interested solely in real solutions.)

In fact, it is possible to see graphically that Eq. 16.40 has only one real solution: plot $y = x^*$ and $y = \frac{a}{1+(x^*)^n}$. Where these two curves cross is the solution for x^* that we want. This graphical solution is illustrated schematically in the left of Fig. 16.13 for a value of a near 1. It is clear from this figure that irrespective of the value of a, there is always one real solution, i.e., there is one point only that the curves $y = x^*$ and $y = a/(1 + (x^*)^n)$ cross. Next, we examine the stability of this solution.

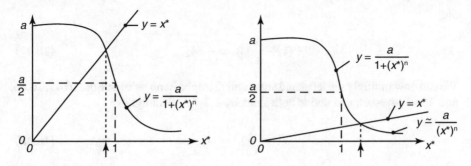

Fig. 16.13 Schematic graphical solution of Eq. 16.40 for a near unity (left) and for $a \gg 1$ (right). The arrows indicate the values of x^* that solve Eq. 16.40

16.5.2 Stability?

To examine the stability of this solution, we suppose that we know x^* and assume $x_1 = x^* + \delta_1$, $x_2 = x^* + \delta_2$, $x_3 = x^* + \delta_3$. Carrying out an analogous linear expansion to that carried out in our genetic toggle switch analysis, we find

$$\frac{1}{K}\frac{d\delta_1}{dt} = -\delta_1 - Y\delta_3, \tag{16.41}$$

$$\frac{1}{K}\frac{d\delta_2}{dt} = -\delta_2 - Y\delta_1, \tag{16.42}$$

and

$$\frac{1}{K}\frac{d\delta_3}{dt} = -\delta_3 - Y\delta_2, \tag{16.43}$$

where, here,

$$Y = \frac{n(x^*)^{n-1}a}{(1 + (x*)^n)^2} = \frac{n(x^*)^{n+1}}{a}, \tag{16.44}$$

where we used Eq. 16.40 in the last step.

Following the same steps as before, next, we assume that $\delta_1 = C_1 e^{\Lambda t}$, etc., with results that

$$(1 + \frac{\Lambda}{K})\delta_1 = -Y\delta_3, \tag{16.45}$$

$$(1 + \frac{\Lambda}{K})\delta_2 = -Y\delta_1, \tag{16.46}$$

and

$$(1 + \frac{\Lambda}{K})\delta_3 = -Y\delta_2. \tag{16.47}$$

We can now multiply the left-hand sides and the right-hand sides of Eqs. 16.45, 16.46, and 16.47 together, and divide both sides by $\delta_1\delta_2\delta_3$, yielding:

$$(1 + \frac{\Lambda}{K})^3 = -Y^3. \tag{16.48}$$

It follows that the eigenvalues are:

$$\Lambda = -K[1 - (-1)^{1/3}Y]. \tag{16.49}$$

Since three cube-roots of -1 are -1, $e^{i\pi/3} = 1/2 + i\sqrt{3}/2$, and $e^{-\pi i/3} = 1/2 - i\sqrt{3}/2$, the three eigenvalues may be written:

$$\Lambda = -K[1 + Y], \tag{16.50}$$

$$\Lambda = -K[1 - \frac{Y}{2} - i\frac{\sqrt{3}Y}{2}], \tag{16.51}$$

and

$$\Lambda = -K[1 - \frac{Y}{2} + i\frac{\sqrt{3}Y}{2}]. \tag{16.52}$$

We will see that Y is always positive. Therefore, Eq. 16.50 always corresponds to an exponential decay, and a stable solution. Remarkably, Eqs. 16.51 and 16.52 are imaginary, corresponding to oscillatory solutions! The real parts of Eqs. 16.51 and 16.52 are negative for $Y < 2$, corresponding to oscillatory solutions for the protein concentrations that decay exponentially to the steady-state values. These are stable solutions. By contrast, for $Y > 2$, the real parts of Eqs. 16.51 and 16.52 are positive, corresponding to oscillatory solutions whose amplitude grows exponentially in time. These solutions are unstable. Therefore, the condition to realize an unstable solution, which in fact corresponds to sustained oscillations, is that $Y > 2$.

How does Y depend on a and n? So far, we have not calculated an explicit value for Y, because we have not calculated an explicit value for x^*. The reason is that it is not possible to write the solution to Eq. 16.40 for arbitrary values of n. However, if we assume that $a \gg 1$, we can find a useful, approximate solution as follows. The right of Fig. 16.13 shows the graphical solution for $a \gg 1$. In this case, the intersection point of the two curves occurs for $x^* \gg 1$. In this case, we may first think to make the approximation

$$\frac{a}{1 + (x^*)^n} \simeq \frac{a}{(x^*)^n}.$$ (16.53)

In this approximation, we have that

$$x^* = \frac{a}{(x^*)^n}$$ (16.54)

and, therefore, that

$$Y = \frac{n(x^*)^{n+1}}{a} = n,$$ (16.55)

suggesting that sustained oscillations should occur whenever $n > 2$ for all values of a. However, the Wolfram Demonstration shows that, for small-enough values of a, there are not sustained oscillations, even for large values of n. We must conclude that Eq. 16.54 is too crude an approximation.

As an improvement over Eq. 16.54, we instead write

$$x^* = \frac{a}{(x^*)^n} \frac{1}{1 + (x^*)^{-n}} \simeq \frac{a}{(x^*)^n}(1 - (x^*)^{-n}),$$ (16.56)

where we used that $1/(1 + z) \simeq 1 - z$ for small $z = (x^*)^{-n}$. It follows that

$$(x^*)^{n+1} = a(1 - (x^*)^{-n}) \simeq a(1 - a^{-\frac{n}{n+1}}).$$ (16.57)

Combining Eq. 16.44 with Eq. 16.56, we arrive at

$$Y \simeq n(1 - a^{-\frac{n}{n+1}}),$$ (16.58)

which represents a useful approximate expression for Y that we can use to determine the stability of the steady-state solution. First, we notice that for $n = 2$, Y is always less than 2. This result survives an exact calculation, which is possible for $n = 2$. Therefore, there are no sustained oscillations for $n = 2$. It is a different story for $n > 2$. In this case, according to Eq. 16.52 the steady-state solution is unstable—i.e., $Y > 2$—for

$$n(1 - a^{-\frac{n}{n+1}}) > 2.$$ (16.59)

After a little algebra, we discover that the condition for the steady-state solution to be unstable is that

$$a > (1 - \frac{2}{n})^{-(1+\frac{1}{n})}.$$ (16.60)

Fig. 16.14 Approximate repressilator phase diagram, according to our model. The shaded region above the line corresponds to sustained oscillations—a so-called limit cycle. The non-shaded region below the line corresponds to a stable fixed point

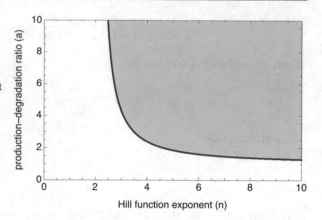

Equation 16.60 represents a "phase diagram" for the repressilator, specifying the region of $n - a$ "phase space" in which the steady-state solution is unstable and which therefore realizes sustained oscillations. This (approximate) repressilator phase diagram is shown in Fig. 16.14.The region above the red line corresponds to sustained oscillations—a limit cycle. The region below the red line corresponds to a stable fixed point. Figure 16.14 is approximate because Eq. 16.58 is approximate. Nevertheless, by running the repressilator Wolfram Demonstration, we can see that this graph is qualitatively correct. Specifically, you will find that to realize sustained oscillations requires larger values of a at smaller values of n and relatively smaller values of a at larger values of n.

Does the repressilator circuit actually give rise to oscillations when it is incorporated into living bacteria? Reference [2] shows that the fluorescence intensity from green fluorescent protein that is under the control of one of the circuit's repressors indeed shows oscillatory behavior, indicating that the concentration of the repressor itself also shows oscillatory behavior, qualitatively consistent with the model.

16.6   Chapter Outlook

Our bodies are comprised of many different cell types: neurons, liver cells, T and B cells, skin cells, cardiac cells, etc. As cells grow and divide during embryonic development, all of these different cell types emerge via a sequence of binary decisions, starting from pluripotent embryonic stem cells. Even as adults, we retain a number of different types of adult stem cells—hematopoietic stem cells (HSCs),[10] neural stem cells,[11] mesenchymal stem cells,[12] endothelial stem cells,[13] etc.—

---

[10] https://en.wikipedia.org/wiki/Hematopoietic_stem_cell

[11] https://en.wikipedia.org/wiki/Neural_stem_cell

[12] https://en.wikipedia.org/wiki/Mesenchymal_stem_cell

[13] https://en.wikipedia.org/wiki/Endothelial_stem_cell

which are generally multipotent, capable of renewing themselves and differentiating into a number of different, albeit related cell types. Each cell fate decision is controlled by a network of genes and proteins—a biologic circuit—that responds to internal and external cues to bring about a discrete change, i.e., a switch, in the pattern of gene expression within a cell, thus directing the cell to a particular tissue type. This discussion so-far envisions cell fate to be a unidirectional process with no escape from the final cell fate, i.e., a sequence of switches, each similar to the genetic toggle switch. However, this picture describes non-proliferative cells in multicellular organisms. Alternatively, many cells cyclically alternate between the different states of the cell cycle,[14] in which growing cells proceed from G_1 phase to S phase to G_2 phase to M phase and cytokinesis, each of which is a different cell fate. After cytokinesis, the progeny cells return to G_1 to start the cycle again, i.e., the cell cycle is an oscillator, like the repressilator. The 2001 Nobel Prize in Physiology or Medicine for 2001 was awarded to Leland Hartwell, Timothy Hunt, and Paul Nurse for their discoveries concerning the network of proteins involved in driving the cell cycle in yeast.

16.7   Problems

Problem 1: A Genetic Switch that Regulates Itself
Reference [7] informs us:

> Under conditions of nutrient deprivation, cells of *Bacillus subtilis* can undergo a process of development that leads to the formation of dormant, environmentally resistant spores. Under optimal conditions, most of the cells in a culture will each produce a spore. Sporulation takes approximately six to eight hours and involves extensive changes in gene expression and morphology. The hallmark of endospore formation is an asymmetric cell division that produces two cell types: a larger cell called the mother cell, and a smaller cell called the forespore. As development proceeds the forespore is engulfed by the mother cell, forming a cell within a cell. After engulfment, maturation of the spore proceeds and the resistance properties develop. Eventually, the mother cell lyses – via programmed cell death, which is another cell fate decision – releasing the mature spore. Spores can remain dormant for many years, but when appropriate nutrients are once again presented, the dormant spores germinate, lose their resistance properties, and resume vegetative growth.

Bacillus subtilis[15] is a Gram-positive bacterium commonly found in soil. It is known that *B. subtilis* expression of the factor ComK is limited to that fraction of any *B. subtilis* culture that is competent for genetic transformation—i.e., those bacteria that are not spores—and in fact ComK provides a master switch between spore (non-competent) and non-spore (competent) states. In this problem, we will consider a simple model for this *B. subtilis* switch. Similar behavior may also be relevant in

---

[14] https://en.wikipedia.org/wiki/Cell_cycle
[15] http://en.wikipedia.org/wiki/Bacillus_subtilis

cholera.[16] Interestingly, the *B. subtilis* switch is different from the genetic toggle switch considered in class.

In this case, the gene (comK) gives rise to a gene product (the protein, ComK) that *activates* its own gene. This can give rise to a bistable switch,[17] as we will now work out. First, we write a chemical rate equation for the concentration of gene product (x) as

$$\frac{dx}{dt} = -Kx + \gamma P(x). \tag{16.61}$$

The first term on the RHS (Right-Hand Side) is the ComK degradation rate. K is the fraction of ComK molecules that are degraded per second (K must be positive). The second term on the RHS is the ComK production rate. It is the product of two terms, $P(x)$ and γ. $P(x)$ is the probability that RNA polymerase[18] is bound to the comK promotor[19]. $P(x)$ depends on x because ComK is an activator[20] for itself. This IS positive feedback. γ is the number of ComK molecules produced per second (γ must also be positive), given that RNA polymerase binds the comK promotor. When the going is good for *B. subtilis*, we may expect γ to be relatively large. In tough times, γ will be relatively small.

(a) It is possible for Mathematica to numerically solve Eq. 16.61. A Mathematica demonstration that does just this, using Eq. 16.63 for $P(x)$, can be downloaded from BacillusSubtilis-Ch16.cdf. Run the demonstration. Note that the sliders allow you to vary the ratio of the ComK degradation rate to the ComK production rate ($A = \frac{K}{\gamma}$), the basal production rate ($\gamma\alpha$ by varying α), and the initial ComK concentration. The initial parameters are: $A = 0.5$, $\alpha = 0.07$, and an initial ComK concentration of 0.9. You can see bistability in action by slowly increasing the "Initial concentration of ComK" slider from 0.9 to 0.93 (without changing A and α). Do this. What happens? Contrast this behavior with what happens for $A = 0.2$ and $\alpha = 0.07$. For these parameter values, convince yourself that the competent state with a relatively high concentration of ComK is stable for all values of the initial ComK concentration. Contrast both of these behaviors with what happens for $A = 0.6$ and $\alpha = 0.07$. For these parameter values, convince yourself that the spore state with a relatively low concentration of ComK is stable for all values of the initial ComK concentration. Thus, we see that there's a finite range of values of A and α in which there's bistability. The numbers matter.

---

[16] http://www.ncbi.nlm.nih.gov/pubmed/20862321

[17] http://onlinelibrary.wiley.com/doi/10.1111/j.1365-2958.2005.04592.x/pdf

[18] http://en.wikipedia.org/wiki/RNA_polymerase

[19] http://en.wikipedia.org/wiki/Promoter_(biology)

[20] http://en.wikipedia.org/wiki/Activator_(genetics)

(b) Suppose that the concentration of ComK has reached a steady state, so that $\frac{dx}{dt} = 0$. Show that in a steady state Eq. 16.61 becomes

$$\frac{K}{\gamma}x = P(x). \qquad (16.62)$$

(c) Consider the functions $y = Ax$, where $A = \frac{K}{\gamma}$, and $y = P(x)$. At the value of x at which these two functions cross, it must be that $Ax = P(x)$. Therefore, we can solve Eq. 16.62 graphically by plotting $y = Ax$ and $y = P(x)$, and by finding the point at which they cross. In order to sketch the function $y = Ax$, I have to tell you a value for A. Plot $y = Ax$ for $A = 1$, $A = 0.5$ and $A = 0.25$.

(d) In order to sketch the function $y = P(x)$, you need an explicit functional form for $P(x)$. It turns out that in order to find an interesting, bistable solution, corresponding to a switch, it is necessary for $P(x)$ to be "sigmoidal" in shape— sort of an S-shaped curve. Below, we will discuss the physical origin of such a curve. To realize a simple sigmoidal curve, take

$$P(x) = \alpha + (1 - \alpha)\frac{(x/c)^4}{1 + (x/c)^4}, \qquad (16.63)$$

with $c = 1$ and $\alpha = 0.1$. Notice that $P(0) = \alpha$ and $P(\infty) = 1$. What is the meaning of α and c? When $x \ll c$, the second term in Eq. 16.63 is negligible, and we see that α is the probability that RNAP binds the promoter without any ComK activation. c is the concentration of ComK at which the probability that RNAP is bound to the promotor is $(1 + \alpha)/2$. Plot Eq. 16.63 on the same graph as your three straight lines.

(e) How many solutions are there for $A = 1$? Extract from your graph the approximate value for each solution.

(f) How many solutions are there for $A = 0.5$? Extract from your graph the approximate value for each solution.

(g) How many solutions are there for $A = 0.25$? Extract from your graph the approximate value for each solution.

(h) Find an approximate range of values of A for which there are three solutions.

(i) A number of the solutions you have found are "stable" and a number are "unstable." Suppose that x is near x_1, where x_1 is one of these solutions. What "stable" or "unstable" means is as follows. If x is a little different from x_1, then if x_1 is a stable solution, Eq. 16.61 will lead x to evolve in time so that it will return to x_1. On the other hand, if x is a little different from x_1 and x_1 is an unstable solutions, Eq. 16.61 will lead x to move away from x_1. An unstable solution is analogous to the case of a ball balanced on the very top of a hill; any little push will cause the ball to roll down the hill. A stable solution is like a ball at the bottom of a valley. In that case, after a little push, the ball will eventually return to the bottom of the valley. Unstable solutions do not correspond to the concentrations of ComK that can be realized experimentally. We are interested in the stable solutions.

To determine whether a solution is stable or not, we may carry out a linear expansion of the RHS of Eq. 16.61 for x near x_1 with the result that

$$\frac{dx}{dt} = \gamma \left(-A + \frac{dP(x_1)}{dx} \right) (x - x_1) = B(x - x_1), \tag{16.64}$$

where the last equality defines B. First, consider Eq. 16.64 in the case that B is positive. Then, if $x > x_1$, dx/dt is positive and x gets bigger, moving away from x_1. If $x < x_1$, dx/dt is negative and x gets smaller, again moving away from x_1. We must conclude that, if B is positive, x_1 is an unstable solution. Conversely, if B is negative, then, if $x > x_1$, dx/dt is negative and x gets smaller, moving towards x_1. If $x < x_1$, dx/dt is positive and x gets larger, again moving towards x_1. In this case, we must conclude that if B is negative, x_1 is a stable solution—one of the solutions we want. For B equal to a constant, we have encountered equations of the form of Eq. 16.64 a number of times before. What is the general solution to Eq. 16.64? What happens to this solution for B positive? What happens for B negative? What is the characteristic time for Eq. 16.64?

(j) Plot

$$B(x) = -A + \frac{dP(x)}{dx} \tag{16.65}$$

for $A = 1$, $A = 0.5$, and $A = 0.25$.

(k) Using your plots determine which of the graphical solutions you found are unstable solutions and which are stable solutions. Note that if $B(x_1)$ is negative, x_1 is a stable solution. If $B(x_1)$ is positive, x_1 is an unstable solution.

(l) Next, *sketch* the loci of the stable solutions as a function of A, i.e., for a given value of A, plot (approximately) the stable solutions for that value of A. Your answers to parts (e) to (h) should guide your graph. Repeat for a number of values of A between $A = 0.25$ and $A = 1$. (Recall that A represents the slope of the line you drew in part (c)). Notice the bistable behavior predicted by your calculation. Interpret the two branches.

(m) One loose end concerns the sigmoidal form of $P(x)$. To motivate where that (approximate) form comes from, suppose that four ComK molecules bind one RNAP, as suggested in Figure 1 of this paper.[21] This is a chemical reaction:

$$4\text{ComK} + \text{RNAP} \rightarrow \text{ComK}_4\text{RNAP}. \tag{16.66}$$

---

[21] http://onlinelibrary.wiley.com/doi/10.1111/j.1365-2958.2005.04592.x/pdf

Chemistry classes then inform us that

$$\frac{[\text{ComK}]^4[\text{RNAP}]}{[\text{ComK}_4\text{RNAP}]} = K_D, \tag{16.67}$$

where K_D is the dissociation constant for this reaction. At the same time, if this chemical reaction is the sole way that RNAP binds ComK, then we have that the probability that RNAP binds ComK is

$$\frac{[\text{ComK}_4\text{RNAP}]}{[\text{ComK}_4\text{RNAP}] + [\text{RNAP}]} = \frac{[\text{ComK}]^4}{[\text{ComK}]^4 + K_D} = \frac{(x/c)^4}{1 + (x/c)^4}, \tag{16.68}$$

since $x = [\text{ComK}]$ and where we defined $c^4 = K_D$. This is the factor that contributes to Eq. 16.63. The sigmoidal form is a signature of "cooperativity"—four ComK proteins cooperate to activate gene expression. We can show that a sigmoidal form is required in order to achieve bistability and switching behavior between the two bistable solutions. Show that the algebraic steps in Eq. 16.68 are correct.

Problem 2: Repressilator Supersized
In this problem, we consider a modified version of the repressilator—the supersized repressilator—that uses M gene-repressor pairs ($M > 3$), instead of the 3 of the original.

(a) Given that we are interested in the possibility of oscillations in this gene circuit, are there any restrictions on the possible values of M?
(b) How is Eq. 16.40 modified for this new version?
(c) How is Eq. 16.48 modified for this new version?
(d) Given your answer to (c), determine the condition for oscillatory solutions.
(e) How is Eq. 16.58 modified for the new version?
(f) Using your answer to (e) determine the phase diagram of the supersized repressilator. Is more or less cooperativity needed to achieve oscillations for $M > 3$?

Problem 3: Circadian Rhythms
Discuss the relevance of the material of this chapter to circadian rhythms.[22] Jeffrey C. Hall, Michael Rosbash, and Michael W. Young were awarded the 2017 Nobel Prize in Physiology or Medicine[23] "for their discoveries of molecular mechanisms controlling the circadian rhythm" in fruit flies.

---

[22] https://en.wikipedia.org/wiki/Circadian_rhythm

[23] https://www.nobelprize.org/prizes/medicine/2017/press-release/

References

1. T.S. Gardner, C.R. Cantor, J.J. Collins, Construction of a genetic toggle switch in *Escherichia coli*. Nature **403**, 339–342 (2000)
2. M.B. Elowitz, S. Leibler, A synthetic oscillatory network of transcriptional regulators. Nature **403**, 335–338 (2000)
3. M. Ptashne, *A Genetic Switch: Phage Lambda Revisited* (Cold Spring Harbor Press, Cold Spring Harbor, NY, 2004)
4. F. St-Pierre, D. Endy, Determination of cell fate selection during phase lambda infection. Proc. Natl. Acad. Sci. **105**, 20705–20710 (2008)
5. D. Schultz, P.G. Wolynes, E. Ben Jacob, J.N. Onuchic, Deciding fate in adverse times: Sporulation and competence in *Bacillus subtilis*. Proc. Natl. Acad. Sci. **106**, 21027–21034 (2009)
6. F. St-Pierre, D. Endy, Determination of cell fate selection during phage lambda infection. Proc. Natl. Acad. Sci. U. S. A. **105**, 20705–20710 (2008)
7. A.D. Grossman, Genetic networks controlling the initiation of sporulation and the development of genetic competence in *Bacillus subtilis*. Annu. Rev. Genetics **29**, 477 (1995)

Magnetic Fields and Ampere's Law 17

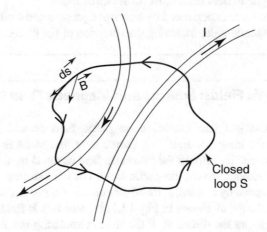

17.1 Introduction

In this chapter, we will advance our discussion of electromagnetism by introducing
the magnetic field, which is complementary to the electric field. Just as electric field
lines were an important tool for understanding electric fields, so too will magnetic
field lines be an important tool for understanding magnetic fields. Electric field lines
start on positive charges and end on negative electric charges. However, there are no
magnetic charges. Therefore, magnetic field lines must form closed loops. This idea
will lead us to "Gauss's law for magnetic fields", which is the second of Maxwell's
equations of electromagnetism. (Gauss's law is the first.) Rather than magnetic
charges being the source of magnetic field, we will discover that *moving charges*,

i.e., currents, are the sources of magnetic fields. The key tool for determining the magnetic field created by a current is Ampere's law, which will be the most technical new aspect that we will encounter in this chapter, and which is the third of Maxwell's equations.

17.2   Your Learning Goals for This Chapter

By the end of this chapter, you should be able to:

- Determine the magnetic field given a current distribution using Ampere's law and the right-hand rule, in situations with spherical, cylindrical, or planar symmetry.
- Sketch the magnetic field lines for a given current distribution, including the fact that magnetic field lines must form closed loops.
- Calculate the force experienced by a charged particle and a current-carrying wire in a magnetic field, including the direction of the force.

17.3   Magnetic Fields: Sources and Magnetic Field Lines

A magnet is an object that creates a magnetic field around itself. Like the electric field, $\mathbf{E}$, the magnetic field is a vector field that satisfies the principle of superposition. We usually refer to the magnetic field as the $\mathbf{B}$ field. Similarly to the electric field, we cannot see the magnetic field, but we can find out if there is a magnetic field in space by looking at its effects on other objects. For instance, using some small iron filings, as shown in Fig. 17.1, if there is a $\mathbf{B}$ field, the filings will align themselves along the direction of the field, visualizing the $\mathbf{B}$ field lines. The sources of magnetic fields are moving charges, i.e., currents. Indeed, if we place iron filings around a wire carrying current, Fig. 17.2, the filings will align in concentric circles, showing the $\mathbf{B}$ field created by the current in the wire. In Sect. 17.4, we will learn how to quantify the $\mathbf{B}$ field produced by a current I through Ampere's law. For now, let us keep in mind that as charges are sources of $\mathbf{E}$ field, moving charges, i.e., currents, are sources of $\mathbf{B}$ field.

17.3.1 Earth's Magnetic Field and Tesla

The Earth, because of its internal core made of iron, has a permanent magnetic field and, as shown in Fig. 17.3, can be thought of a gigantic magnet. The geographical north pole is a magnetic south pole. This is why the North pole of a compass is

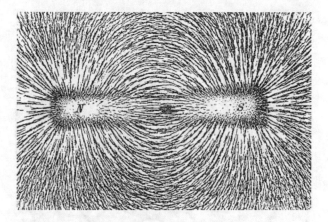

Fig. 17.1 Iron filings around a permanent magnet align along the magnetic field giving a good visualization of magnetic field lines. From Newton Henry Black, Harvey N. Davis (1913) Practical Physics, The MacMillan Co., USA, p. 242. Public domain image. https://commons.wikimedia.org/wiki/File:Magnet0873.png

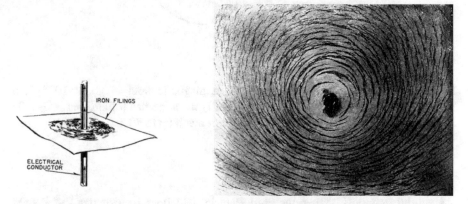

Fig. 17.2 Iron filings around a current-carrying wire align along the magnetic field giving a good visualization of magnetic field lines in this case. Public domain from Popular Science Monthly Volume 56. https://commons.wikimedia.org/wiki/File:PSM_V56_D0072_Magnetic_whirls_around_the_sending_wire.png

attracted toward the magnetic south pole, indicating where the North is. The Earth's magnetic field is all around us, but it is also very weak, and it is measured to be of the order of 10^{-5} Tesla. The Tesla (T) is the SI unit for magnetic field, and it is defined as

$$\text{Tesla} = \frac{\text{Newton}}{\text{Ampere} \times \text{meter}} = \frac{\text{kg}}{\text{C} \times \text{s}}. \tag{17.1}$$

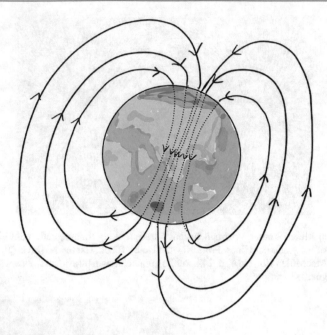

Fig. 17.3 The Earth's magnetic field

In comparison, a refrigerator magnet has a magnetic field of about 10^{-3} T, a diagnostic magnetic resonance imaging (MRI) machine has a field that is 1–5 T; high magnetic fields for scientific research can reach up to 40 T.

17.3.2 Magnetic Field Lines

As we did previously, when we used electric field lines to visualize the **E** field (Sect. 12.4.4), similarly we can introduce magnetic field lines—**B** field lines—to visualize the magnetic field vector. Examples of magnetic field lines for the case of a magnetic dipole are shown in Figs. 17.3 and 17.4 and similar to the pattern shown by the iron filings in Fig. 17.1. Magnetic field lines respect the following four rules:

1. Magnetic field lines are continuous curves, tangent to the magnetic field vector. Conversely, the magnetic field vector at any point is tangent to the field line at that point.
2. Magnetic field lines never cross.
3. The number of magnetic field lines per unit area through a surface perpendicular to the lines is proportional to the magnitude of the magnetic field in that region. Therefore, the field lines are close together where the magnetic field is strong and far apart where the field is weak.
4. Magnetic field lines are always closed, i.e., they must form loops. For a simple bar magnet (with a north and a south pole), the field lines are both inside and

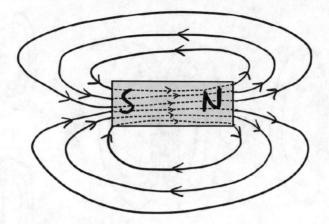

Fig. 17.4 Magnetic field lines of a dipole

outside the magnet: they come out from the north pole and curve back to re-
enter the magnet at the south pole (as shown in Fig. 17.4).

The last characteristic of **B**-field lines follows from the impossibility of having a
single magnetic monopole.

17.3.3 "Gauss's Law" for Magnetic Fields and the Impossibility of Having Magnetic Monopoles

Recall from Chap. 12. *Gauss's Law: Charges and Electric Fields:*

$$\oint_A \mathbf{E} \cdot d\mathbf{A} = \frac{Q}{\epsilon_0}, \tag{17.2}$$

which specifies how the electric field, **E**, is related to charge. In particular Gauss's
law states that the electric flux through the closed surface, A, namely $\Phi_E = \oint_A \mathbf{E} \cdot d\mathbf{A}$, is proportional to the charge Q enclosed within that closed surface. What would
be the corresponding "Gauss's law" for magnetic fields? How is the magnetic flux
related to the sources of magnetic fields?

Similarly to positive and negative charges, like magnetic poles repel each other,
and opposite magnetic poles attract each other. The fundamental difference though
is that it is never possible to have a single magnetic pole isolated on its own. For
instance, if we were to cut in half the magnet in Fig. 17.1, we will be left with two
smaller magnets each with a north and a south pole, as illustrated in Fig. 17.5. Thus,
every magnetic material behaves as a dipole, always possessing a north pole–south
pole structure.

Since there cannot exist single magnetic monopole, any closed surface must
contain north and south poles in pairs. The B field lines that come out of north

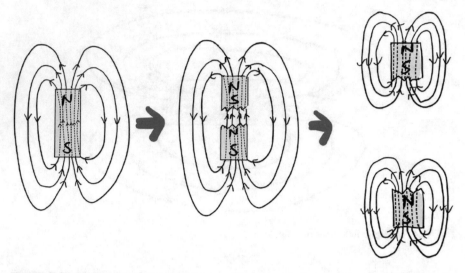

Fig. 17.5 By cutting a magnet in a half, you would get two smaller magnets each with a north and a south pole. It will never be possible to have a single magnetic monopole isolated on its own

poles inevitably go back into the south poles. As a result, the overall magnetic flux through a closed surface must always be zero. This observation leads to "Gauss's law" for magnetic fields":

$$\oint_A \mathbf{B} \cdot d\mathbf{A} = 0. \tag{17.3}$$

This equation constitutes the second of the four Maxwell's equations. The original Gauss's law (Eq. 17.2) is the first.

17.3.4 The Magnetic Field of a Wire Carrying Current

In 1820, the Danish scientist, Oersted, made an important observation: he noticed that a compass needle was deflected if placed nearby a wire carrying electric current. Oersted's discovery spurred further research to study the connection between electric and magnetic phenomena, which lead to the discovery that a wire carrying current, I, produces a magnetic field of magnitude:

$$B = \frac{\mu_0 I}{2\pi r}, \tag{17.4}$$

where r is the distance from the wire, and μ_0 is the vacuum permeability:

$$\mu_0 = 4\pi \times 10^{-7} \text{ T m/A}.$$

Both the magnetic field and the current that produce it have cylindrical symmetry centered at the wire. Correspondingly, the magnetic field lines are circles concentric with the wire, as shown in Fig. 17.2. The direction of the magnetic field is determined using the right-hand rule.

17.4 Ampere's Law

Now, we come to the third of Maxwell's equations, namely Ampere's law, which relates magnetic field to current. Ampere's law expresses that currents create magnetic fields, similarly to the way that charges create electric fields. Ampere's law reads

$$\oint_{s} \mathbf{B} \cdot d\mathbf{s} = \mu_0 I. \tag{17.5}$$

First, we discuss the left-hand side of Eq. 17.5. What does this mean?

- The integral on the left-hand side is a *line integral* around a closed loop (S). This means that the integral is carried over each point along the one-dimensional path, constituting the loop. In general, a line integral can be carried out along any path, but, in the case of Ampere's law, it is a closed path. The circle on the integral sign ($\oint_s$) indicates it is a closed-path line integral.
- $d\mathbf{s}$ is a vector element of the loop, whose direction is tangent to the loop and whose length is equal to the length of the loop element in question.
- $\mathbf{B}$ is the magnetic field at the loop element.
- $\mathbf{B} \cdot d\mathbf{s}$ is the scalar (dot) product of these two quantities, which then has to be integrated around the loop.

Actually, we encountered line integrals previously in Chap. 3. *Energy, Work, Geckos and ATP*, when we discussed work. There, we discussed that the work done by a force, $\mathbf{F}$, applied to a system moving from position i to position f is given by a line integral of the force along the path of the motion: $W = \int_i^f \mathbf{F} \cdot d\mathbf{l}$. Line integrals in general can be complicated to calculate. However, in the context of Ampere's law, by studying highly symmetric examples and exploiting the symmetry, we will be able to simplify and calculate the line integrals needed to apply Ampere's law. Just as we can use Gauss's Law to calculate the electric field from charges in sufficiently symmetric situations, we can use Ampere's law to calculate the magnetic field created by current in sufficiently symmetric situations.

Now, we turn to the right-hand side of Ampere's law.

- This involves μ_0, which is the vacuum permeability, a constant equal to $4\pi \times 10^{-7}$ T m/A.
- It also involves I, which is the *net* current that threads the closed loop.

17.4.1 The Right-Hand Rule and Counting Currents Threading a Loop

When applying Ampere's law, it is important to get the sign right. The direction of the current determines the direction of the magnetic field, as described by the right-hand rule in Fig. 17.6, as follows: align the thumb of your right hand with the direction of the current. Then, the positive direction of the magnetic field curls around with your fingers.

Once you have identified the direction of the B field created by a specific current then you know if the contribution of that current is positive or negative on the right-hand side of Ampere's law. More specifically, if the direction of the **B** field created by a current is aligned in the same direction along which you are doing the line integral, then the current counts as positive. Instead, if the direction of the **B** field created by a current is opposite to the direction along which you are doing the integration, then the current counts as negative. For the example in Fig. 17.7, the total/net current threading the loop is $I = I_2 - I_3 + I_4$.

Fig. 17.6 Right-hand rule to determine the direction of the **B** field created by a wire carrying current: align the thumb of your right hand with the direction of the current. Then, the positive direction of the magnetic field curls around with your fingers

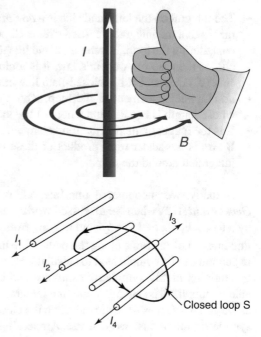

Fig. 17.7 Example of how to count currents threading a loop. In this case the total current threading the loop is $I_{tot} = I_2 - I_3 + I_4$

17.4.2 How to Apply Ampere's Law

A prescription for how to apply Ampere's law is as follows:

1. Make a sketch of your problem and assess the symmetry (e.g., whether it is cylindrical, planar). Similarly to Gauss's law, to solve a problem easily using Ampere's law requires that the symmetry/geometry is simple enough. If the geometry is not simple, then probably you cannot use Ampere's law. Add magnetic field lines to your sketch, making sure that every magnetic field line is a closed loop and that they respect the symmetry of the problem.

2. Select the appropriate Amperian loop, i.e., the closed loop along which you will do the line integral. In problems with cylindrical symmetry, this will be a (concentric) circle. In other case, the loop will be rectangular. Experience will guide you concerning how to draw an appropriate Amperian loop for the problem in hand.

   If your problem indeed has symmetry and if you picked your Amperian loop according to this symmetry, recognize that, on certain portions of your loop, **B** is parallel to the loop, and hence it is parallel to $d\mathbf{s}$. Recognize that on other portions of your surface, **B** is perpendicular to the loop and hence perpendicular to $d\mathbf{s}$.

3. Calculate the *left* side of Ampere's law (Eq. 17.5). For the portion(s) of your Amperian loop, for which **B** and $d\mathbf{s}$ are parallel (so that $\mathbf{B} \cdot d\mathbf{s}$ is not zero), recognize that often **B** has a constant magnitude and therefore can be brought outside the integral. For this portion of your Amperian loop,

$$\int \mathbf{B} \cdot d\mathbf{s} = Bs, \tag{17.6}$$

   where B is the magnitude of the electric field and s is the length of this portion of your Amperian loop. For cylindrically symmetric situations with circular loops, likely you will find $s = 2\pi r$ for an Amperian loop of radius r. Now, you know the left-hand side of Eq. 17.5.

4. Calculate the *right* side of Ampere's law. The right-hand side of Ampere's law depends on I, where I is the total net charge threading your Amperian loop.

5. Set your previous two answers equal to each other and solve for the magnetic field B.

17.4.3 Example: The Magnetic Field of a Current-Carrying Wire

We will now use Ampere's law to derive the previously introduced result (Eq. 17.4) for the magnetic field of a current-carrying wire. In the cylindrically symmetric situation shown end-on in Fig. 17.8, current, I, is flowing in a wire, directed out of the paper, as shown. Therefore, the magnetic field runs in counterclockwise circles (as shown).

(a) Explain why the magnetic field lines run in circular loops and how to verify that the direction of the magnetic field shown in Fig. 17.8 is correct.

```
Answer: The cylindrical symmetry of the wire
dictates that the pattern of magnetic field
lines also has cylindrical symmetry. Application
of the right-hand rule demonstrates that
the direction of the magnetic field is shown
correctly.
```

(b) Evaluate the right-hand side of Ampere's law using the Amperian loop shown.

```
Answer: The example sensibly specifies a
circular, concentric Amperian loop at radius r.
In this case, ds is parallel to B. In addition,
the magnitude of the magnetic field, B, is the
same everywhere around the circle. Therefore,
the right side of Ampere's law is
```

$$\int \mathbf{B} \cdot d\mathbf{s} = B2\pi r. \qquad (17.7)$$

(continued)

Fig. 17.8 A current-carrying wire, shown end-on. The dot at the center of the wire indicates that the direction of the current is out of the page. Three unequally spaced, circular magnetic field lines are shown as the continuous lines with arrows. An appropriate Amperian loop is shown as the dashed line

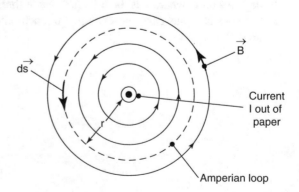

(c) Thus, determine how the magnetic field varies as a function of distance from the wire, r.

Answer: The right side is $\mu_0 I$. Thus, solving for B, we have

$$B = \frac{\mu_0 I}{2\pi r}. \tag{17.8}$$

We see that the magnetic field falls as $1/r$.

17.5 Magnetic Field of a Coil

Currents are sources of magnetic fields. We have seen that a current-carrying wire produces a magnetic field, the direction of which can be found using the right-hand rule as explained in Fig. 17.6 and the magnitude of which can be found using Ampere's law, as shown in Example 17.4.3. What if now we bend this wire into a circular loop? What is the magnetic field of a circular current-carrying loop, i.e., of a coil? The magnetic field produced by a coil is quite similar to the magnetic field of a permanent magnet, like the one shown in Figs. 17.1 and 17.4, with a north and a south pole. To identify which side is the north pole and which side is the south pole, we again use the right-hand rule as shown on the left side of Fig. 17.9: by curling your fingers in the direction of the current flowing into the coil, your thumb gives the direction of the magnetic field and therefore identifies the north pole.

To understand intuitively why the magnetic field of a coil resembles the field of a magnetic dipole), we consider applying the right-hand rule to each segment of the wire along the loop. Each segment produces a B field around it, so that we can envision many small B field lines circles around the coil as shown on the left side of Fig. 17.10. Since the B field obeys the principle of superposition, the total B field will be the sum of all the B field created by each segment, and so as shown in two steps in the middle and right of Fig. 17.10, the B field will be stronger in the middle of the coil pointing straight up and the bending as moving toward the side and curving downward as a dipole field.

17.5.1 Example: The Magnetic Field of a Long Solenoid

A long cylindrical solenoid consists of a conducting wire, wound into a tight helix, of radius R as shown in Fig. 17.12. There are n helical turns per unit

(continued)

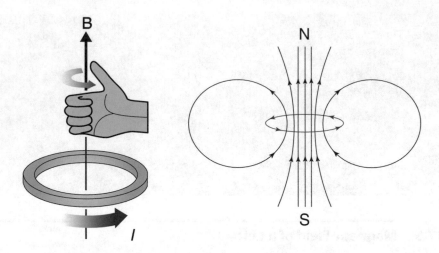

Fig. 17.9 *Left* Right-hand rule to determine the north pole of a coil currying current. *Right* B field lines for the magnetic field of a coil

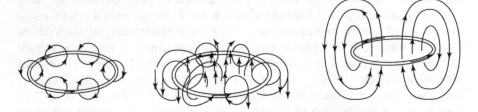

Fig. 17.10 Sketches to visualize intuitively, from left to right, how the individual B field of each segment of the coil adds up to create an overall B field that has the shape of a magnetic dipole

length. The wire carries a current I. The goal in this example is to determine the magnetic field of such a solenoid using Ampere's law.

(a) Sketch the magnetic field lines.

```
Answer: The magnetic field of a current-carrying
solenoid can be thought as the sum of the B
fields of many circular coils next to each
other with the current in each flowing in
the same direction, as shown in Fig. 17.11.
Applying the right-hand rule, we see that the
field lines must run along the axis of the
solenoid, as shown in Fig. 17.12. Since magnetic
```

(continued)

Fig. 17.11 The B fields of each single coil add up in a solenoid creating a uniform B field pointing along the axis of the solenoid

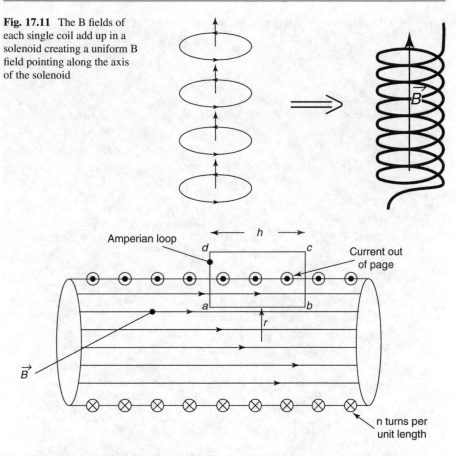

Fig. 17.12   The magnetic field of a cylindrical solenoid

field lines must form closed loops, actually there must be an equal number of magnetic field lines outside the solenoid as inside the solenoid. However, since the cross-sectional area inside the solenoid is much smaller than the cross-sectional area outside, the magnitude of the magnetic field inside is much larger than the magnetic field outside. Consequently, we approximate the field outside the solenoid to be zero, which is why no field lines are shown outside the solenoid in Fig. 17.12.

(continued)

(b) Pick a suitable Amperian loop and sketch it on the figure.

Answer: We choose the rectangular Amperian loop of length h, shown in Fig. 17.12, that spans the wires of the solenoid and whose inner side is at a distance r from the central axis of the solenoid. Since there is no field outside the solenoid, where the outer side is located does not matter in this case.

(c) Evaluate the left side of Ampere's law.

Answer: We label the corners of the loop: a, b, c, and d. From b to c and from d to a, $d\mathbf{s}$ and $\mathbf{B}$ are perpendicular and so contribute zero to the left side of Ampere's law. From c to d, outside the solenoid, the magnitude of the magnetic field is approximately zero, so that side also makes a negligible contribution to the left side of Ampere's law. Finally, from a to b, $d\mathbf{s}$ and $\mathbf{B}$ are parallel, B is the same everywhere along the line from a to b and therefore can be moved outside of the integral. It follows that the left side of Ampere's law is

$$Bh. \tag{17.9}$$

(d) Evaluate the right side of Ampere's law.

Answer: For the right side, each turn of the solenoid adds I to the current threading the loop. If there are n turns per unit length, there are nh turns threading the loop. Therefore, the total current threading through the loop is

$$nhI. \tag{17.10}$$

(e) Thus, determine the magnetic field inside the solenoid.

Answer: Equating Eqs. 17.9 and 17.10, and solving for B, we find

$$B = \mu_0 nI. \tag{17.11}$$

(continued)

Notice that this expression does not depend
on r. Thus, we see that the field inside the
solenoid is uniform. This is reminiscent of the
electric field in a parallel plate capacitor,
which is also a uniform field (provided that we
neglect edge effects). The solenoid therefore
can be seen as the counterpart of the capacitor.

17.5.2 Example: The Magnetic Field of a Toroidal Solenoid

A toroidal solenoid consists of a conducting wire, wound into a tight helix,
with the helix brought around to close on itself as shown in Fig. 17.13. There
are a total of N helical turns in the toroid. The wire carries a current I. The
goal of this example is to determine the magnetic field of such a solenoid
using Ampere's law.

(a) Sketch the magnetic field lines.

Answer: Based on what we know from the previous
cylindrical solenoid example, and guided by the
right-hand rule, we deduce that the magnetic

(continued)

Fig. 17.13 A toroidal
solenoid of N turns in total

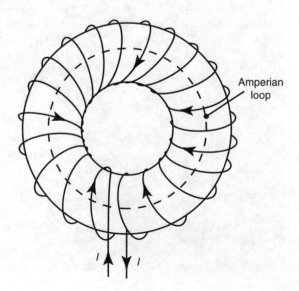

field lines run in circles around the inside
of the toroid clockwise. Outside the toroidal
solenoid the magnetic field is zero. For
clarity, we have not sketched the field lines
on Fig. 17.13.

(b) Choose and sketch an appropriate Amperian loop.

Answer: An appropriate Amperian loop is the
circular loop, shown as the dashed line in
Fig. 17.13.

(c) Calculate the left side of Ampere's law.

Answer: By symmetry, the magnitude of magnetic
field is the same everywhere around the circle,
and the direction of the magnetic field is
everywhere parallel to $d\mathbf{s}$. (We pick $d\mathbf{s}$ to go
around clockwise.) It follows that the left side
of Ampere's law is

$$B2\pi r. \tag{17.12}$$

(d) Calculate the right side of Ampere's law.

Answer: For the right side, each turn of the
solenoid adds I to the current threading the
loop. Given that there are N turns in total,
there are N turns threading the loop. Therefore,
the total current threading through the loop is

$$NI. \tag{17.13}$$

(e) Determine the magnetic field.

Using the answers to (c) and (d), Ampere's law
yields

$$B = \frac{\mu_0 NI}{2\pi r}, \tag{17.14}$$

which depends on r, but not on the shape of the
solenoid's cross-section.

17.6   Force on a Moving Charge and Another Version of the Right-Hand Rule

In view of the fact that an electric field, **E** causes a force on a charge Q, namely $\mathbf{F} = Q\mathbf{E}$, perhaps it is not surprising that a magnetic field also causes a force on a charge. In this case, however, there is a force only if the charge is moving:

$$\mathbf{F} = Q\mathbf{v} \times \mathbf{B}, \tag{17.15}$$

where **v** is the velocity of the charge. Equation 17.15 contains the vector product of **v** and **B**. The directions of the vectors in this equation are specified by a second version of the right-hand rule shown in Fig. 17.14, namely: align your thumb in the direction of the velocity vector, align your index finger in the direction of the **B** field, then coming out perpendicularly to your palm the middle finger will give you the direction of the force. As per definition of vector cross product, the force **F** is perpendicular to both **v** and **B**. It is important to notice that according to this prescription the order of **v** and **B** in the vector product is meaningful because $\mathbf{v} \times \mathbf{B}$ has the opposite direction to $\mathbf{B} \times \mathbf{v}$.

Fig. 17.14 Right-hand rule to determine the direction of the magnetic force: align your thumb in the direction of the velocity vector, align your index finger in the direction of the **B** field, then coming out perpendicularly to your palm the middle finger will give you the direction of the force

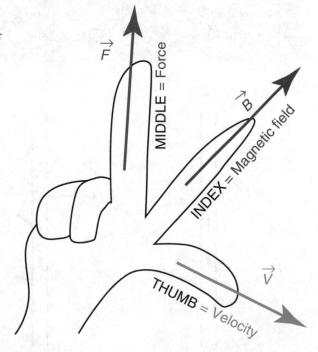

17.6.1 Example: The Force Between Two Parallel Wires Carrying Current

The goal of this example is to calculate the force that causes two parallel wires each carrying current to either attract or repel each other. The current in each wire is i. They are each a length ℓ long, and are separated from each other by a distance, R.

(a) Explain why there is a force between the wires shown in Fig. 17.15.

   Answer: The current in one wire creates a magnetic field at the location of the second wire. The current in the second wire corresponds to moving charges. Therefore, Eq. 17.15 implies that there will be a force on those charges and hence on the wire that carries those charges.

(b) Assuming that the currents (i) are parallel and equal, as shown in Fig. 17.15, what is the direction of the force between the wires?

   The current of wire 1 creates a magnetic field at the location of wire 2 that is directed out of the page via the first version of the

(continued)

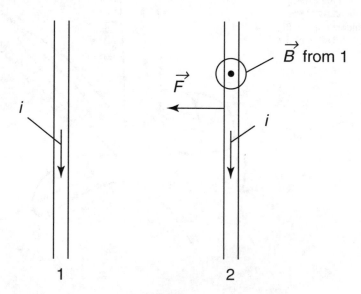

Fig. 17.15 Forces on parallel current-carrying wires

right-hand rule. i is parallel to $Q\mathbf{v}$. Therefore, the direction of the force on wire 2, using the second version of the right-hand rule, is toward wire 1. It is an attractive force for parallel currents.

(c) Determine the magnitude of this force.

Ampere's law informs us that the magnetic field created by wire 1 at wire 2 is $B = \frac{\mu_0 i}{2\pi R}$. Equation 17.15 tells us the force on a single charge, Q, moving with velocity $\mathbf{v}$. In our wire of length ℓ, there are many (N) moving charges. Therefore, the total force on the wire will be $NQvB$.

We discussed how to relate i to $Q\mathbf{v}$ in Chap. 14. *Circuits and Dendrites: Charge Conservation, Ohm's Law, Rate Equations, and Other Old Friends.* In brief, all charges "upstream" of a given reference plane within a length $v\Delta t$ cross the reference plane in a time Δt. If n is the number of charges per unit volume, then the total charge crossing the reference plane in Δt is $QnAv\Delta t$, where A is the cross-sectional area of the wire carrying the current. The total charge crossing the reference plane in Δt also equals $i\Delta t$. It follows that $i = nAQv$. On the other hand, the total number of charges in our wire of length ℓ is $N = nA\ell$. Therefore, $i\ell = NQv$. It follows that the magnitude of the force is

$$F = i\ell B = \frac{\mu_0 i^2 \ell}{2\pi R}, \tag{17.16}$$

which is the desired answer.

17.6.2 Force Between Two Coils

In the previous example (17.6.1), we saw that two parallel wires carrying current in the same direction will attract each other. Similarly, one can prove that two parallel

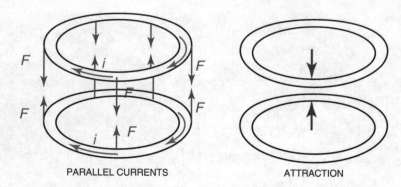

PARALLEL CURRENTS   ATTRACTION

Fig. 17.16 Two parallel coils carrying current in the same direction attract each other

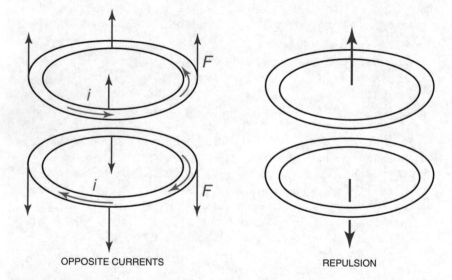

OPPOSITE CURRENTS   REPULSION

Fig. 17.17 Two parallel coils carrying current in the opposite direction repel each other

wires carrying currents in opposite direction will repel. What are the implications of these results for two coils carrying current?

Let us consider first the case of two parallel coils, one on top of the other, both carrying currents in the same direction as show in Fig. 17.16. This configuration is similar to the one in Example 17.6.1 of two parallel wires with the only difference that the wires are now bent to form a closed loop. From the point of view of analyzing the force between them though the analysis is the same, at each point along the coil there will be an attractive force between the two wires and so effectively the two coils will be attracting each other. Instead if we consider two parallel coils carrying currents in the opposite direction as show in Fig. 17.17, with a similar analysis we can conclude that the two coils repel each other.

If we think of each coil as a small magnet with a north and south pole as we have identified before through the right-hand rule (see Fig. 17.9), then these results make sense because they agree with the experimental evidence that opposite magnetic poles attract each other and same magnetic poles repel each other.

17.7   Magnetic Materials

To understand the magnetism of materials, we need to look at what happens microscopically. In a simplified picture (see Fig. 17.18), each of the electrons in a material gives rise to a tiny magnetic field. In many materials, the electrons are paired, so that the net magnetic field of the pair is zero. In other materials, however, there are unpaired electrons, resulting in a non-zero density of microscopic magnetic fields within a material. The magnitude of these tiny magnetic fields is specified by the corresponding magnetic moment, denoted μ, whose value depends on the specific orbitals from which the unpaired electrons derive. Magnetic moment has SI units of Am^2 or $J\ T^{-1}$. In most cases, the microscopic magnetic moments are randomly oriented, so that, on macroscopic scales, the vector sum of microscopic magnetic fields is zero and there are no macroscopic magnetic effects. (See the right side of Fig. 17.18.) Ferromagnetic materials, like iron for example, are exceptions. In ferromagnetic materials, the microscopic magnetic fields are all parallel with each other, as shown in Fig. 17.19. The superposition of all these microscopic magnetic fields is a non-zero macroscopic magnetic field. When placed in an external applied magnetic field, the microscopic magnets all align with the external magnetic field and remain in that orientation, even when the external field is removed, i.e., the material is a permanent magnet.

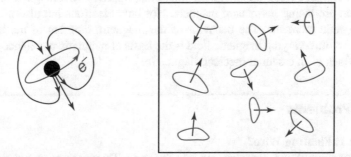

Fig. 17.18 Simplified microscopic description of microscopic magnetic fields and magnetic materials

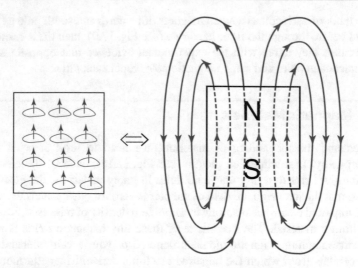

Fig. 17.19 Simplified microscopic description of a ferromagnetic material

17.8   Chapter Outlook

In this chapter, we introduced and discussed the magnetic field, **B**, and took two essential steps toward the complete set of Maxwell's equations with Gauss's law for magnetic fields and Ampere's law, which specifies how magnetic fields are created by electric currents. We are now ready to discover in the next chapter, Chap. 18. *Faraday's Law and Electromagnetic Induction*, that a changing magnetic field creates an electric field, which is the essence of Faraday's law. In this chapter, we also introduced electronic magnetic moments, which in ferromagnets align with each other, producing permanent magnets. Not only electrons but also nucleii can have magnetic moments. The behavior of the magnetic moment of the hydrogen nucleus in a time-varying magnetic field is the basis of magnetic resonance imaging (MRI), which is an essential medical diagnostic.

17.9   Problems

Problem 1: Floating Wire?
Three long straight wires are directed into the page. They each carry equal currents in the directions shown in the figure. The lower two wires are attached to a table. The upper wire is in the middle above them as shown, gravity is acting on it vertically. Is it possible for this wire to "float" so to form a triangle (as shown) with the lower wires? Explain your reasoning in details applying the appropriate physics principle and/or laws.

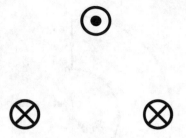

Problem 2: Amperian Loops

In this chapter, we have studied that the magnetic field from a current-carrying wire is $B = \frac{\mu_0 I}{2\pi r}$ where the field lines are concentric circles centered on the wire and the direction of the field is given by the right-hand rule. Consider the Amperian loop shown in the figure.

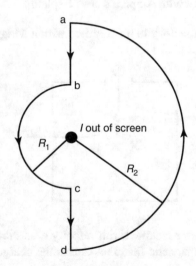

For each of the following, show or explain your work.

(a) What is $\int_a^b \mathbf{B} \cdot d\mathbf{s}$?

(b) What is $\int_b^c \mathbf{B} \cdot d\mathbf{s}$?

(c) What is $\int_c^d \mathbf{B} \cdot d\mathbf{s}$?

(d) What is $\int_d^a \mathbf{B} \cdot d\mathbf{s}$?

(e) Combining your answers to parts (a) through (d), what is $\oint \mathbf{B} \cdot d\mathbf{s}$ around the entire loop?

(f) Now do parts (a) through (e) for the following loop.

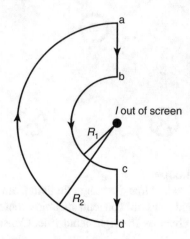

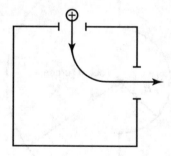

(g) Do your results agree with Ampere's law? Explain.

Problem 3: Charge Trajectory in a Chamber with a Magnetic Field

A positive charge q enters a chamber with velocity $\mathbf{v}$, as shown in the figure. Inside the chamber there is a magnetic field that causes the charge to take the trajectory shown in the figure.

(a) In what direction does the magnetic field point? Explain your reasoning.
(b) Explain why the trajectory is curved as shown.

Problem 4: Motion of a Charged Particle in a Magnetic Field
Carefully analyze and describe the motion of a particle of charge Q that moves in the xy plane only, in a magnetic field $\mathbf{B} = B\mathbf{k}$. The initial velocity of the particle is $v\mathbf{i}$.

Problem 5: Current Sheet
Consider a conducting sheet of thickness a lying in the xy plane, through which there flows a current of uniform current density (current per unit area) J in the direction that is directly toward you (Fig. 17.20).

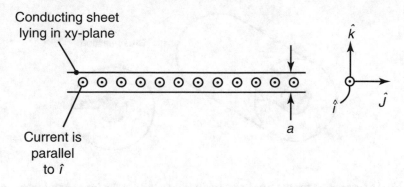

Fig. 17.20 Current sheet

(a) What is the direction of the magnetic field that is produced by this current, above the sheet ($z > 0$, and below the sheet ($z < 0$)?

(b) In preparation for a calculation using Ampere's law, sketch an appropriate Amperian loop or loops that will enable you to calculate the magnetic field.

(c) Carry out such a calculation, i.e., apply Ampere's law to calculate the magnetic field as a function of position.

(d) Now suppose that we add another current sheet, also of thickness a located a distance d above the first, in which the current flows in the opposite direction. What is the magnetic field now as a function of position?

Problem 6: Force on Parallel Wires

(a) What is the magnetic field a radial distance r from a long straight wire carrying a current i_1?

(b) What is the force per unit length on a long straight wire, carrying a current i_2, that runs parallel to a second long straight wire a distance d away and carries a current i_1.

(c) Do same-direction currents repel, or attract?

(d) Now, consider a current-carrying slinky—that is, a helical conductor or solenoid—for which the separation between successive turns of the helix (d) is very much less than the radius of the helix (R). Use your results from (b) to calculate the force per turn between neighboring turns of the slinky. Is that force attractive or repulsive?

Problem 7: Coaxial Cable

The coaxial cable shown in the figure consists of a solid inner conductor of radius R_1 surrounded by a hollow, very thin outer conductor of radius R_2. The two carry equal currents I, but in opposite directions. The current density is uniformly distributed over each conductor.

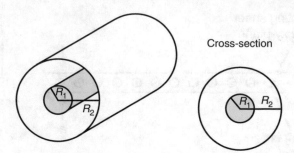

(a) Find expressions for the three magnetic fields: within the inner conductor, in the space between the conductors, and outside the outer conductor.
(b) Draw the corresponding B field lines for each region of part (a).
(c) Draw a graph of the magnetic field B versus r from $r = 0$ to infinity.

Problem 8: Toroidal Solenoid
The *toroidal* solenoid in the figure is a coil of wire wrapped around a doughnut-shaped ring (a *torus*) made of nonconducting material.

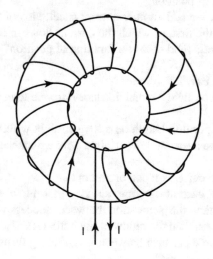

(a) From symmetry, what must be the *shape* of the magnetic field in this toroidal solenoid? Explain.
(b) Sketch the magnetic field lines.
(c) Consider a toroidal solenoid with N closely spaced coils carrying current I. Use Ampere's law to find an expression for the magnetic field strength at a point inside the torus at a distance r from the axis. Show the steps of your calculations.
(d) Is the magnetic field of a toroidal solenoid a uniform field? Explain.

(e) What is the magnetic field at $r = 0$?
(f) Assume a positive charge q is released in the middle of the solenoid with a velocity v as shown in the figure. What is its trajectory going to be? Is the charge going to go around a circle inside the solenoid? Explain.

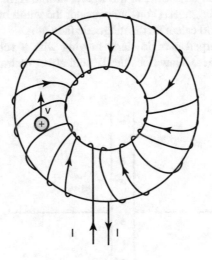

Problem 9: Parallel Infinities

An infinite sheet of current lying in the yz plane carries a surface current of linear density J_s. The current is in the positive z direction (i.e., out of the page in the figure below), and J_s represents the current per unit of length measured along the y axis. The figure below is an edge view of the sheet.

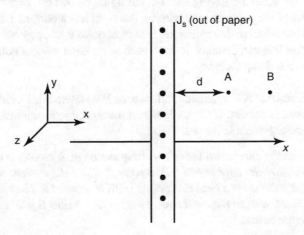

(a) By reflecting on the symmetry of the current distribution, determine what the B field lines look like outside the sheet of current. Explain in words their shape and with a simple sketch. Make sure to also show the direction of the B field lines.

(b) Consider the point A, shown in the figure, on the right side of the sheet at a distance d from the sheet. (You can consider the width of the sheet negligible). Explain how to and calculate the magnetic field at A.

(c) A negatively charged particle starts moving with a velocity $\mathbf{v} = v\hat{\mathbf{j}}$ on the left side of the sheet. How does the B field affect its trajectory? Explain your reasoning.

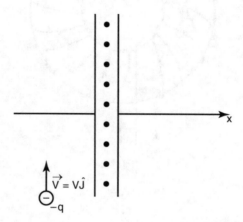

Problem 10: Between Two Currents

Consider the two current-carrying wires in the figure below: the wire on the left has a current I out of the page and the wire on the right has a current I into the page. Two students are asked to determine the B field at point P midway between the two wires. The point P is at a distance R from each wire. Each student comes up with a possible answer and explanation.

Evaluate both student A's statement and student B's statement. Decide whether or not the statement is correct. If the student explanation contains incorrect reasoning, explain how the argument is flawed.

Student A: *Since the current on the left is out of the page, it creates a counterclockwise field. Therefore the field points UP at point P. Since the current on the right is into the page, it will create a field that points DOWN at point P. The two B fields are equal and opposite and so they will cancel out, giving a zero B field at point P. See Student A's figure below.*

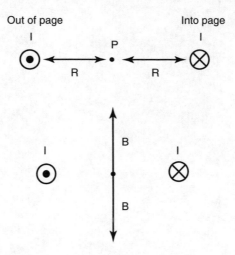

Student B: *I will take as Amperian loop a circle S centered at P with a radius bigger than R so it will include both wires. Ampere's law tells me that the integral is zero because the currents cancel. Therefore the B field in the center at P must be zero.* See Student B's figure below.

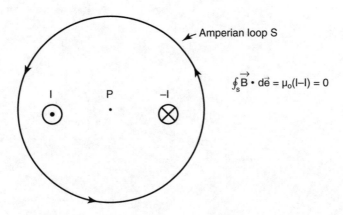

Faraday's Law and Electromagnetic Induction

18

18.1 Introduction

In 1831, Faraday observed surprisingly that in some situations a transient current could flow around a circuit in the absence of a battery. His experiments are summarized in Fig. 18.1: The top panel shows that opening or closing a switch in a circuit with a battery induces a momentary current in another nearby circuit. The middle panel shows that pushing a coil into a magnetic field or pulling the coil out of the magnetic field induces a momentary current in a circuit containing the coil. The bottom panel illustrates that pushing a magnet into a coil or pushing the magnet out of the coil induces a momentary current in a circuit containing the coil. What all three of these situations have in common is a time-varying magnetic field at the location of the circuit in which the transient current appears. In this chapter, we will see that what matters is a change in time of magnetic flux (analogous to the electric flux, which we encountered in Chap. 12. *Gauss's Law: Charges and Electric Fields*). Specifically, the rate of change of magnetic flux through a loop gives rise to an electric field around that loop. When the loop corresponds to a conducting circuit, this electric field causes current to flow around the circuit. The transient current is called an *induced* current because it is not caused by a battery, but it is induced as a consequence of the changing magnetic flux. In particular, for the situation, depicted at the bottom of Fig. 18.1, if we imagine pushing the magnet inside the coil, then waiting a few seconds, and then pulling it out, the induced current measured in the coil would behave as a function of time as shown in Fig. 18.2: The current is non-zero only while moving the magnet in or out. The current is zero when the magnet is not moving (either outside or outside the coil). Additionally, the direction of the current when pulling out the magnet is opposite to the direction of the current when pushing the magnet in. In either case, the direction of the induced current is such that the corresponding induced magnetic field tends to oppose the change in the magnetic flux through the loop.

Fig. 18.1 Three experiments
that show an induced current

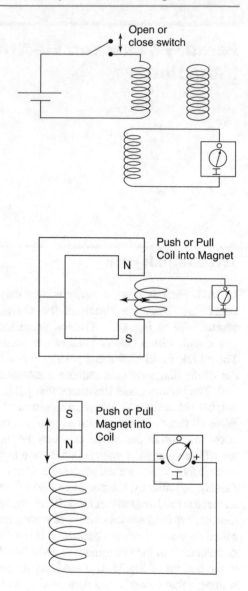

In this chapter, first, we will summarize Faraday's observations mathematically via Faraday's law, which we will then apply to a number of examples. Then, we will introduce inductance, which is the ratio of the rate of change of magnetic flux to the rate of change of current. Like capacitance, inductance depends only on the geometry of the circuit in question. Next, we will examine electrical circuits containing inductors, where we will again encounter linear differential equations that describe how current varies versus time, similar to equations that we encountered previously in Chap. 7. *Rates of Change: Epidemics, Infections, and*

Fig. 18.2 Induced current in the coil as a function of time while pushing the magnet in and then out of the coil

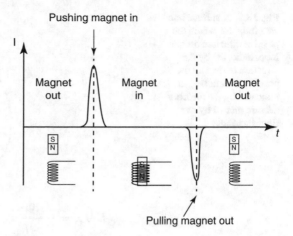

Weapons of Mass Destruction, Chap. 10. *Oscillations and Resonance*, and Chap. 14. *Circuits and Dendrites: Charge Conservation, Ohm's Law, Rate Equations, and Other Old Friends*.

18.2   Your Learning Goals for This Chapter

By the end of this chapter, you should be able to:

- Calculate the induced electromotive force around a circuit through which the magnetic flux is changing.
- Determine the direction of the induced current in a circuit through which the magnetic flux is changing.
- Calculate the inductance of a inductor with cylindrical, planar, or circular symmetry.
- Describe the behavior as a function of time of the current in a RL, RC, and RLC circuit.

18.3   Faraday's Law

Faraday's law was formulated to explain the observations described in Sect. 18.1, in which a changing magnetic flux through a conducting circuit gives rise to an induced current in that circuit. If there is an induced current, then the charges that are flowing must experience an electric field that causes their motion. Faraday's law quantifies this electric field by specifying the induced *electromotive force* or induced EMF (pronounced E-M-F). The EMF acts like a voltage in that it drives current around a circuit and is measured in Volts. Faraday's law states that the induced EMF is equal to the negative of the time rate of change of the magnetic flux:

Fig. 18.3 A surface bounded
by a loop, for which the
positive direction around the
loop, indicated by the
direction of ds, and the
positive direction of the area,
indicated by the direction of
dA, are related by the
right-hand rule

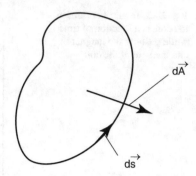

$$EMF = -\frac{d\Phi_B}{dt}. \tag{18.1}$$

It follows that for a circuit with total resistance R, using Ohm's law, the induced
current is related to the EMF and rate of change of magnetic flux via

$$I_{\text{ind}} = \frac{EMF}{R} = -\frac{1}{R}\frac{d\Phi_B}{dt}. \tag{18.2}$$

To better understand Faraday's law, we first need to understand the magnetic
flux, Φ_B. Analogously to the electric flux, the magnetic flux quantifies how many
magnetic field lines pass through a given surface. If we consider a close loop that
bounds a surface of area A, then the magnetic flux through this loop is

$$\Phi_B = \int_A \mathbf{B} \cdot d\mathbf{A}, \tag{18.3}$$

where $d\mathbf{A}$ is a vector element of area on the surface, whose direction is perpendicular
to the element of area in question. Importantly in this case, the flux is not the flux
through a closed surface, in contrast to the flux for "Gauss's Law for magnetic
fields" (Eq. 19.2). In applications of Faraday's law, A is often the surface enclosed
by a conducting loop. According to Eq. 18.1, there will be an induced current in that
conducting loop only if the flux through the loop, Φ_B, changes as a function of time.
How the change in flux comes about does not matter. It could be that the magnetic
field, $\mathbf{B}$, itself is changing, or the area of the loop A is changing, or it could be that
the angle between the magnetic field $\mathbf{B}$ and the area of the loop is changing, so that
their scalar (dot) product is changing. Examples and problems in this chapter will
realize each of these possibilities.

To determine the direction of the induced EMF and the corresponding induced
current, the right-hand rule is again essential. Shown in Fig. 18.3 is a surface
bounded by a loop, for which the positive direction around the loop, indicated by the
direction of $d\mathbf{s}$, and the positive direction of the area, indicated by the direction of
$d\mathbf{A}$, are related by the right-hand rule, i.e., that curling the fingers of your right hand

around the loop along $d\mathbf{s}$, and putting the direction of your thumb along the positive direction of the area vector of the surface $d\mathbf{A}$, you can check with your right hand that the right-hand rule is followed for Fig. 18.3. Consequently, if the magnetic field is parallel to $d\mathbf{A}$, then Φ_B is positive; if the induced current flows in the direction of $d\mathbf{s}$, then the induced current is positive. For example, suppose that the magnetic field is parallel to $d\mathbf{A}$ in Fig. 18.3. Then, Φ_B is positive. Further, suppose that the magnetic field is increasing versus time and that the loop is conducting. It follows that $\frac{d\Phi_B}{dt}$ is then positive. Faraday's law then informs us that the EMF around the loop is negative, as a result of the negative sign in Eq. 18.1. Correspondingly, the induced current around the loop is negative, which means that the direction of the current is opposite to the direction of $d\mathbf{s}$.

The induced current creates its own induced magnetic field, via Ampere's law. The direction of this induced magnetic field is also given by the right-hand rule. For a negative current around the loop of Fig. 18.3, the right-hand rule informs us that the direction of the induced magnetic field, near the center of the loop, is negative, that is, it is antiparallel to $d\mathbf{A}$. Thus, the induced magnetic field opposes the change in magnetic field creating it. This behavior is called Lenz's law. Lenz's law is the consequence of the negative sign in Eq. 18.1. It is important to note that the induced magnetic field opposes the *change* in the external magnetic field, not the external field itself. For example, if the external magnetic field is pointing upward and decreasing, the induced magnetic field will also point upward so as to oppose the decrease of the external field; if the external field is pointing upward and increasing, the induced magnetic field will be pointing downward to oppose the increase of the external field.

Although Eq. 18.1 is all that is needed in many applications, in order to arrive at the corresponding Maxwell's equation (Eq. 19.4) and then electromagnetic waves, it is necessary to re-write Eq. 18.1 in a form that explicitly contains the **E**- and **B**-fields. The existence of an induced EMF implies that there is an *induced electric field* that is pushing charges around the loop ($\mathbf{F} = q\mathbf{E}$), thereby giving rise to the induced current. The induced electric field does work on charges moving around the loop. The total work done on a charge q moving around the loop, s, is

$$W_{\text{loop}} = \oint_s \mathbf{F} \cdot d\mathbf{s} = \oint_s q\mathbf{E} \cdot d\mathbf{s} = q \oint_s \mathbf{E} \cdot d\mathbf{s}, \qquad (18.4)$$

and the circle on the integral ($\oint$) indicates that this is an integration over a closed loop. The induced EMF is the work done per unit of charge:

$$EMF = \frac{W_{\text{loop}}}{q} = \oint_s \mathbf{E} \cdot d\mathbf{s}. \qquad (18.5)$$

This expression relates the EMF to the induced electric field. Concerning the magnetic flux on the right-hand side of Eq. 18.1, we already know that we can write this as

$$\Phi_B = \int_A \mathbf{B} \cdot d\mathbf{A}. \tag{18.6}$$

Finally, substituting these last two expressions into Eq. 18.1, we obtain the more general expression of Faraday's law, which we will need:

$$\frac{d}{dt} \int_A \mathbf{B} \cdot d\mathbf{A} = - \oint_s \mathbf{E} \cdot d\mathbf{s}, \tag{18.7}$$

which corresponds to the fourth of Maxwell's equations (Eq. 19.4).

Equation 18.7 implies that a changing magnetic field gives rise to an electric field without the direct involvement of any charges! Such an induced electric field though, despite being able to exert forces on charges, differs from an electric field produced by static charges. First, while the electric field lines associated with charges start on positive charges and end on negative charges, the electric field lines, associated with a changing magnetic field via Faraday's Law, form closed loops. Second, for an induced electric field, we cannot define an electric potential, as we did for an electric field produced by charges. Indeed, expression Eq. 18.5 could never be true for an electric field that is the derivative of an electric potential, for which we know that

$$\oint_s \mathbf{E} \cdot d\mathbf{s} = 0, \tag{18.8}$$

as previously discussed in Chap. 12. *Gauss's Law: Charges and Electric Fields*. To be clear, an induced electric field cannot be obtained by taking the derivative of a potential function.

18.3.1 Example: Changing the Magnetic Flux by Changing Area

Shown in Fig. 18.4 is a sliding wire of resistance R moving at velocity v to the right on top of another wire of zero resistance, formed into the shape of a sideways "U" in a magnetic field of magnitude B, pointing out of the page (toward you, the reader). The two wires remain in electrical contact throughout, so current can flow around the circuit formed by the two wires. The base of the sideways U has height ℓ. The goal of this example is to determine the force on the moving wire.

(a) Calculate the rate of change of magnetic flux through the loop.

(continued)

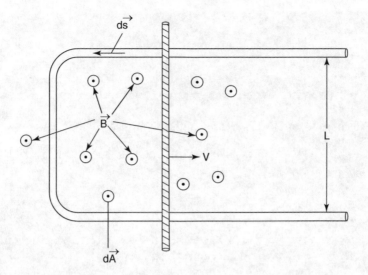

Fig. 18.4 Changing the flux by changing area

Answer: The magnetic flux through the loop formed by the U and the other wire is

$$\Phi_B = BA(t), \tag{18.9}$$

where $A(t)$ is the area at time t. What happened to the vector dot product? We chose the direction of **A** to be parallel to the direction of **B**, that is, out of the page. The positive direction of the loop is then anticlockwise, given by the first version of the right-hand rule, putting your thumb parallel to **A** out of the paper. It follows that

$$\frac{d\Phi_B}{dt} = B\frac{dA}{dt}. \tag{18.10}$$

But dA/dt is the height, ℓ, multiplied by the rate of change of length. But the rate of change of the length is simply v. Therefore, we have that

(continued)

$$\frac{d\Phi_B}{dt} = B\ell v, \qquad (18.11)$$

which is the desired expression for the rate of change of flux, in terms of given quantities.

(b) What is the resultant EMF around the circuit?

Answer: Faraday's law informs us that there is an EMF around the loop given by

$$EMF = -B\ell v. \qquad (18.12)$$

The minus sign implies that the direction of the EMF is clockwise around the loop.

(c) What is the resultant current, I, flowing around the loop?

Answer: As a result of this EMF, a current flows around the loop of magnitude given, according to Ohm's Law, by

$$I = \frac{EMF}{R} = -\frac{B\ell v}{R}. \qquad (18.13)$$

The direction of the current is along the EMF, i.e., clockwise. The minus sign implies that the direction of the current is clockwise around the loop.

(d) Why there is a force on the wire?

Answer: There is a changing magnetic flux because the area of the loop is changing. It follows that there is an EMF around the circuit. In turn, there is a current around the circuit. Since this current is in a magnetic field, there is a force on the wire, carrying the current. We just have to figure out what it is.

(e) What is the force on the sliding wire?

(continued)

Answer: The force (F) on a wire of length ℓ carrying current I in a magnetic field B is

$$F = \ell I B = -\frac{\ell^2 B^2 v}{R}. \qquad (18.14)$$

Applying the second version of the right-hand rule shows that the force is directed to the left, in the opposite direction to the velocity, and tends therefore to decrease the velocity, which slows the rate of change of area and flux, in accordance with Lenz's law.

18.4 Inductance and Inductors

According to Faraday's law, there are induced electric fields, and hence, currents, in circuits where there is a changing magnetic flux. Some circuits are more or less likely to experience induced current depending on their geometry; we define a new quantity—the inductance—to quantify how well a circuit element is able to experience an induced current. Recall that when we discussed Ampere's Law, we saw that there is a linear relationship between magnetic field and current. We also know there is a linear relationship between magnetic field and magnetic flux, and it follows that there is a linear relationship between current (I) and magnetic flux (Φ_B), and as a consequence between the rate of change of current (dI/dt) and the rate of change of magnetic flux $(d\Phi_B/dt)$. The constant of proportionality between these two quantities is called inductance, usually denoted by L:

$$\frac{d\Phi_B}{dt} = L\frac{dI}{dt}. \qquad (18.15)$$

Similarly to capacitance, inductance depends only on geometry. A circuit element possessing inductance is called an inductor. To calculate inductance:

1. Calculate B using Ampere's Law (usually).
2. Calculate Φ_B. In the simplest case, this is the product of B and the area of the Faraday's law loop. In more complicated situations in which the magnetic field changes as a function of position across the area of the Faraday's law loop, it is necessary to calculate the magnetic flux via an integral. It is always necessary to take into account the number of loops.
3. Use

$$\frac{d\Phi_B}{dt} = L\frac{dI}{dt} \qquad (18.16)$$

to identify the inductance.

The SI unit of inductance is the henry (H): $1H = \frac{T\,m^2}{A}$.

18.4.1 Example: Inductance of a Long Solenoid

Calculate the inductance of a solenoid of the total length ℓ and the total number of turns N.

Previously, we found, using Ampere's law, that the magnetic field inside such a solenoid runs parallel to the axis of the solenoid and has magnitude

$$B = \frac{N}{\ell}\mu_0 I. \tag{18.17}$$

The magnetic flux through one loop of the solenoid is therefore

$$BA = \frac{N}{\ell}\mu_0 I A, \tag{18.18}$$

where A is the cross-sectional area of the solenoid. The magnetic flux through all N loops of the solenoid is therefore

$$\Phi_B = NBA = \frac{N^2\mu_0 I A}{\ell}. \tag{18.19}$$

Faraday's law then informs us that

$$EMF = -\frac{d\Phi_B}{dt} = -\frac{d(NBA)}{dt} = -\frac{N^2\mu_0 A}{\ell}\frac{dI}{dt}, \tag{18.20}$$

since I is the only quantity that can change for a solenoid of fixed construction. By comparing Eq. 18.20 to the definition of inductance, namely Eq. 18.15, we see that the inductance of our solenoid is

$$L = \frac{N^2\mu_0 A}{\ell}. \tag{18.21}$$

18.4.2 Example: Inductance of a Coaxial Cable

A so-called coaxial cable is sketched on the left of Fig. 18.5. It consists of a central conductor of radius a and an outer conductor of radius b, separated by an insulator in-between. This is the "cable" in "cable TV." If you have cable TV, alternating current, carrying your TV signal, flows back and forth along the central conductor to your TV set, with the outer conductor completing the circuit. It turns out that at the frequencies relevant to cable TV, the current, I, is confined to thin surface layers at the outside of the inner conductor and the inside of the outer conductor. Consequently, there is zero magnetic field inside the conductors. To determine the inductance of a length ℓ of coaxial cable:

(a) Calculate B in terms of the current I.

Answer: Ampere's law informs us that the **B**-field consists of concentric circles in the region between the two conductors of magnitude

$$B = \frac{\mu_0 I}{2\pi r} \tag{18.22}$$

for $a < r < b$ and zero otherwise.

(b) Pick a suitable Faraday's law loop with which to calculate the magnetic flux and sketch it.

Answer: An appropriate loop is shown as the rectangle outlined as a thick line in the right panel of Fig. 18.5, which shows a vertical

(continued)

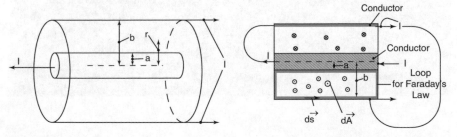

Fig. 18.5 Left: Sketch of the geometry of a coaxial cable. Right: A vertical "cut" through the center of the coaxial cable, illustrating the application of Faraday's law. The small circles indicate the presence and direction of the magnetic field

"cut" through the center of the coaxial cable.
In this view, the magnetic field is directed
directly out of the paper below the center
conductor (depicted as the points of arrows)
and is directed directly into the paper above
the center conductor (depicted as the crossed
fletchings of arrows). The Faraday's law loop
has length ℓ. Its top side is at r $=$ a and its
bottom side is at $r = b$.

(c) Using your Faraday's law loop from (b), calculate the magnetic flux.

Answer: In this case, we cannot simply multiply
the area of the loop by the magnetic field
because the magnetic field varies across the
loop--it varies versus r. Therefore in this
case, it is necessary to do an integral to
evaluate the magnetic flux. Specifically, we
calculate the flux through that part of the area
that lies between r and $r + dr$. We then add up all
these contributions for r between $r = a$ and $r = b$,
i.e., we integrate with respect to r from a to b:

$$\Phi_B = \int_a^b \ell \, dr \, B(r) = \int_a^b \ell \, dr \frac{\mu_0 I}{2\pi r} = \frac{\mu_0 I \ell}{2\pi} \log \frac{b}{a}. \tag{18.23}$$

(d) Calculate the inductance of a length ℓ of coaxial cable.

Answer: Given Eq. 18.23, we have

$$\frac{d\Phi_B}{dt} = \frac{\mu_0 \ell}{2\pi} \log \frac{b}{a} \frac{dI}{dt}. \tag{18.24}$$

It follows that the inductance of the cable is

$$L = \frac{\mu_0 \ell}{2\pi} \log \frac{b}{a}. \tag{18.25}$$

18.4.3 Inductors in Circuits

Similarly to having a resistance, every circuit component has some amount of inductance; therefore to do a complete circuit analysis, we need to take into account for the presence of inductors. Inductors are represented with the symbol —ᴍᴍᴍ— and with an associated inductance L as defined in the previous section.

If we consider a circuit consisting only of a battery and a resistor, we know that the battery creates a voltage, V, which drives current, I, around the circuit. The relationship between V and I is $V = IR$, where R is the resistance of the resistor in the circuit. If we now add an inductor of inductance L to the circuit, as shown in Fig. 18.6, the total EMF in the circuit is the sum of the EMF from the battery, namely its voltage V, and the EMF from the inductor, which we denote EMF_L. Thus, we have

$$V + EMF_L = IR, \tag{18.26}$$

which we can re-write

$$V = IR - EMF_L. \tag{18.27}$$

From Faraday's law and the definition of inductance in Eq. 18.15, the induced EMF in a circuit with an inductor is given by

$$EMF_L = -LdI/dt. \tag{18.28}$$

Therefore, we can re-write Eq. 18.27 as

$$V = IR + L\frac{dI}{dt}, \tag{18.29}$$

which suggests that we can treat a circuit containing an inductor as though there is a voltage drop (or gain) of $+L\frac{dI}{dt}$ across the inductor, similarly to the voltage drop IR across a resistor. It turns out that it is always possible to treat circuits with inductors as though there is a voltage drop of $+LdI/dt$ across the inductor, even though Faraday's law informs us that in fact the induced EMF corresponds to

Fig. 18.6 Circuit with a battery, a resistor, and an inductor

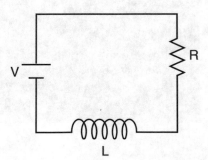

closed-loop electric field lines that extend all around the circuit. This approach also allows us to continue to use Kirchhoff's loop law in circuits involving inductors.

18.4.4 Example: An Inductor in Series with a Resistor Gives Rise to a Familiar Equation

Consider the circuit shown in Fig. 18.6, consisting of an inductor and a resistor in series with a battery providing a voltage V. The goal of this example is to determine the current in the circuit as a function of time, given the initial condition that $I(0) = 0$.

(a) Apply Kirchhoff's loop law to write an equation that expresses the relationship between the voltage V provided by the battery and the voltage drop across the inductor and resistor.
   Answer:

$$V - IR - L\frac{dI}{dt} = 0. \tag{18.30}$$

(b) Write your equation from part (a) in a form like:

$$\frac{dI}{dt} = \alpha - \beta I \tag{18.31}$$

and give the value of the constants α and β in terms of the given quantities (V, L, R).

   Answer: Rewriting the previous equation in the form

$$L\frac{dI}{dt} = V - IR \tag{18.32}$$

$$\frac{dI}{dt} = \frac{V}{L} - I\frac{R}{L}, \tag{18.33}$$

   we identify

$$\alpha = \frac{V}{L} \quad \text{and} \quad \beta = \frac{R}{L}. \tag{18.34}$$

(c) You should be quite familiar by now with a differential equation of the form of Eq. (18.31). We have seen the same equation in Chap. 7. *Rates of Change: Drugs, Infections, and Weapons of Mass Destruction*, (Sect. 7.4 for instance) when discussing the concentration of a drug in the human

(continued)

body. In that case I was replaced with N the number of molecules in the bloodstream. We have also encountered the same equation in Chap. 13, when discussing charging and discharging of a capacitor, in that case I was replaced with Q the charge on the capacitor plates. We know that a general solution to such differential equation is of the form:

$$I(t) = \frac{\alpha}{\beta} + A_0 e^{-\beta t}, \tag{18.35}$$

where A_0 is some constant to be determined by the initial condition. By direct substitution, prove that Eq. (18.35) is a solution of the differential equation Eq. (18.31).

Answer: First take the derivative of $I(t)$

$$\frac{dI}{dt} = -A_0 \beta e^{-\beta t}. \tag{18.36}$$

Substituting this and the function $I(t)$ into the differential equation and collecting terms,

$$\frac{dI}{dt} = \frac{V}{L} - I\frac{R}{L} \tag{18.37}$$

$$-A_0 \beta e^{-\beta t} = \frac{V}{L} - \frac{R}{L}\left(\frac{\alpha}{\beta} + A_0 e^{-\beta t}\right) \tag{18.38}$$

$$\left(\frac{R}{L} - \beta\right) A_0 e^{-\beta t} = \frac{V}{L} - \frac{R}{L}\frac{\alpha}{\beta}. \tag{18.39}$$

We see for $\beta = \frac{R}{L}$ and $\alpha = \frac{V}{L}$, and we get

$$\left(\frac{R}{L} - \frac{R}{L}\right) A_0 e^{-\beta t} = \frac{V}{L} - \frac{R}{L}\left(\frac{V}{L}\right)\left(\frac{L}{R}\right) \tag{18.40}$$

$$0 = 0. \tag{18.41}$$

This proves that Eq. (18.35) is a solution to the differential equation.

(d) Consider this circuit in the case that a switch, completing the circuit, is closed at time $t = 0$ What is the initial condition for the current in this case? In other words, what is $I(0)$?

Answer: $I(0) = 0$

(continued)

Fig. 18.7 Current versus
time for a circuit with a
battery, a resistor, and an
inductor in which a switch
completes the circuit at $t = 0$

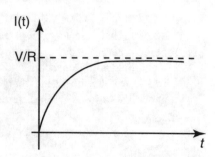

(e) Using your initial condition from part (d), determine the constant A_0 in
Eq. (18.35) in terms of V, L, and R.

Answer: Using the values for α and β, we can
write the current as

$$I(t) = \frac{\alpha}{\beta} + A_0 e^{-\beta t} = \frac{V}{R} + A_0 e^{-\frac{R}{L}t}. \tag{18.42}$$

Using the initial condition gives

$$I(0) = \frac{V}{R} + A_0 = 0 \qquad \rightarrow \qquad A_0 = -\frac{V}{R}. \tag{18.43}$$

(f) Having determined A_0, write the particular solution for $I(t)$ and sketch
it.

Answer: $I(t) = \frac{V}{R}\left(1 - e^{-\frac{R}{L}t}\right)$. The required sketch is
shown in Fig. 18.7.

(g) What is the time constant for the exponential behavior of $I(t)$?

Answer: The time constant is $\tau = \frac{L}{R}$.

18.4.5 Example: Inductors and Eigenvalues

In this example, we will consider the circuit shown in Fig. 18.8. Analyzing the
current flows in this more complicated circuit leads to equations, involving

(continued)

Fig. 18.8 Circuit for "Example: Inductors and eigenvalues"

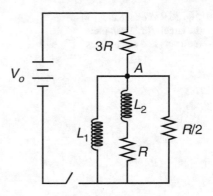

two variables, that are similar to equations that we encountered previously in Chap. 7. *Rates of Change: Drugs, Infections, and Weapons of Mass Destruction* and that are solved using an eigenvalue–eigenvector approach.

(a) Initially, the switch is closed and steady state has been reached. What then is the current through the resistor $3R$?

Answer: At steady state, the inductors act like wires (i.e., the current has reached its maximum). Because L_1 acts like a wire, the voltage drop across $3R$ is V_0.

$$I_{3R} = \frac{V_0}{3R}. \tag{18.44}$$

(b) During steady state, what is the current through the resistor $\frac{R}{2}$?

Answer: Because $\frac{R}{2}$ is in parallel with L_1, no current will flow through it. At steady state, L_1 acts like a wire so all the current will flow through the inductor.

$$I_{\frac{R}{2}} = 0. \tag{18.45}$$

(c) During steady state, what is the current through the resistor R?

Answer: For the same reason as (b), the current through R is zero.

$$I_R = 0. \tag{18.46}$$

(continued)

Fig. 18.9 Voltage loops in the circuit for "Example: Inductors and eigenvalues"

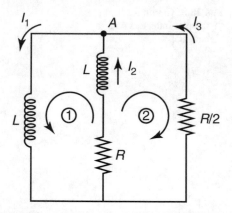

(d) During steady state, what is the current through inductor L_1?

   Answer: The current that flows through $3R$ is the same current that flows through L_1.

$$I_{L_1} = \frac{V_0}{3R}. \qquad (18.47)$$

(e) During steady state, what is the current through inductor L_2?

   Answer: The current that flows through R is the same current that flows through L_2.

$$I_{L_2} = 0. \qquad (18.48)$$

(f) At time $t = 0$, the switch is opened. The current through $3R$ stops immediately, and current flows as shown in Fig. 18.9. For the following questions, assume $L_1 = L_2 = L$. Write an equation for how the currents relate to one another at point A.

   Answer: Using Kirchhoff's rule for currents at a junction,

$$I_1 = I_2 + I_3. \qquad (18.49)$$

(g) We can write two equations that describe how I_1, I_2, and I_3 change in time for $t > 0$:

(continued)

$$\frac{dI_1}{dt} + \frac{dI_2}{dt} + \frac{R}{L}I_2 = 0$$

$$\frac{dI_2}{dt} + \frac{R}{L}I_2 - \frac{R}{2L}I_3 = 0.$$

Show that these are correct equations by analyzing the circuit and writing the appropriate Kirchhoff's law. Indicate on the circuit diagram where the equations came from and in what direction you took your loop.

Answer: Using Kirchhoff's loop rule around the loop labeled 1 in Fig. 18.9 and noticing for both currents I_1 and I_2, the loop goes in the same direction giving a negative sign to each term.

$$-I_2 R - L\frac{dI_2}{dt} - L\frac{dI_1}{dt} = 0. \tag{18.50}$$

Diving through by $-L$ gives the first equation,

$$\frac{dI_1}{dt} + \frac{dI_2}{dt} + \frac{R}{L}I_2 = 0. \tag{18.51}$$

Using Kirchhoff's loop rule around the loop labeled 2, the following current I_1 but opposing I_3 gives

$$-I_2 R - L\frac{dI_2}{dt} + I_3\frac{R}{2} = 0. \tag{18.52}$$

Diving through by $-L$ gives the second equation,

$$\frac{dI_2}{dt} + \frac{R}{L}I_2 - \frac{R}{2L}I_3 = 0. \tag{18.53}$$

(h) Use Kirchhoff's law applied to point A to eliminate I_3 from the differential equations.

Answer: Using $I_3 = I_1 - I_2$ in the second equation gives

$$\frac{dI_2}{dt} + \frac{R}{L}I_2 - \frac{R}{2L}(I_1 - I_2) = 0 \tag{18.54}$$

(continued)

$$\frac{dI_2}{dt} + \frac{3R}{2L}I_2 - \frac{R}{2L}I_1 = 0. \tag{18.55}$$

(i) Solve the differential equations to find the eigenvalues by assuming exponential trial solutions for I_1 and I_2.

Answer: Assume $I_1(t) = Ae^{\lambda t}$ and $I_2(t) = Be^{\lambda t}$. Taking the first derivatives

$$\frac{dI_1}{dt} = \lambda I_1 \tag{18.56}$$

$$\frac{dI_2}{dt} = \lambda I_2. \tag{18.57}$$

Using these in the first differential equations gives

$$\lambda I_1 + \lambda I_2 + \frac{R}{L}I_2 = 0 \tag{18.58}$$

$$\lambda I_1 = -\left(\lambda + \frac{R}{L}\right) I_2, \tag{18.59}$$

and in the second,

$$\lambda I_2 + \frac{3R}{2L}I_2 - \frac{R}{2L}I_1 = 0 \tag{18.60}$$

$$\frac{R}{2L}I_1 = \left(\lambda + \frac{3R}{2L}\right) I_2. \tag{18.61}$$

Dividing the first equation by the second gives the relationship

$$\frac{\lambda}{\frac{R}{2L}} = \frac{-\left(\lambda + \frac{R}{L}\right)}{\lambda + \frac{3R}{2L}}. \tag{18.62}$$

Solving for λ,

$$\lambda^2 + \frac{3R}{2L}\lambda = -\lambda\frac{R}{2L} - \frac{R^2}{2L^2} \tag{18.63}$$

$$\lambda^2 + \frac{2R}{L}\lambda + \frac{R^2}{2L^2} = 0 \tag{18.64}$$

(continued)

$$\lambda_{\pm} = \frac{1}{2}\left(\frac{-2R}{L} \pm \sqrt{\frac{4R^2}{L^2} - \frac{2R^2}{L^2}}\right) \qquad (18.65)$$

$$\lambda_{\pm} = \frac{-R}{L}\left(1 \mp \frac{\sqrt{2}}{2}\right). \qquad (18.66)$$

(j) Are the eigenvalues positive, negative, or zero? Does this make sense physically?

Answer: Because $\frac{\sqrt{2}}{2} < 1$, the term in parenthesis is positive and both eigenvalues are negative. This makes sense. Because of the presence of resistors, the current will decay to zero if there are no voltage sources in the circuit to keep the current flowing at long times.

(k) A long time after the switch is open, what is I_1, I_2, and I_3?

Answer: All three currents, I_1, I_2, and I_3, decay to zero.

18.4.6 RLC Circuits: Another Familiar Equation

If we now examine a circuit containing a resistance, inductance, and capacitance, namely an RLC circuit, such as that shown in Fig. 18.10, we find

$$V + EMF_L = IR + \frac{Q}{C} \qquad (18.67)$$

or

$$V = \frac{Q}{C} + IR - EMF_L = \frac{Q}{C} + IR + L\frac{dI}{dt}. \qquad (18.68)$$

Again, we can treat the circuit as if there is a voltage drop of $+LdI/dt$ across the inductor.

Furthermore, the charge on the capacitor in the circuit shown in Fig. 18.10 is related to the current that flows via

$$I = \frac{dQ}{dt}. \qquad (18.69)$$

Fig. 18.10 Series RLC
circuit

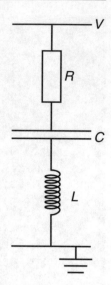

Using this equation, we may re-write Eq. 18.68

$$V = \frac{Q}{C} + \frac{dQ}{dt}R + L\frac{d^2Q}{dt^2}.$$
(18.70)

Remarkably, Eq. 18.70 is formally the same equation that describes driven, damped
simple harmonic motion, which we encountered in Chap. 10. *Oscillations and
Resonance*. Here, we have Q instead of x, L instead of m, $1/C$ instead of k, R
instead of ζ, and $V = V(t)$ instead of $F = F(t)$. It follows that RLC circuits must
behave in the same way as damped harmonic oscillators. In particular, driven RLC
circuits show resonance, just as driven damped simple harmonic oscillators. All of
the results obtained for resonance in the driven, damped harmonic oscillator case
apply to RLC circuits with the translation as given above. For example, the mean
power dissipation in a series RLC circuit ($\langle PD \rangle$) is

$$\langle PD \rangle = \frac{\frac{1}{2}V^2 R\omega^2}{(\frac{1}{C} - L\omega^2)^2 + (R\omega)^2},$$
(18.71)

Which shows a peak at a resonant frequency of $\omega_0 = 1/\sqrt{LC}$ with a full-width-at-
half-maximum of $FWHM = R/L$. An important example of resonance in an RLC
circuit is found in every radio tuner (Fig. 18.11).

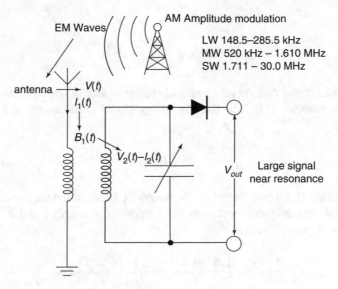

Fig. 18.11 AM radio tuner schematic

18.5 Magnetic Field Energy

Consider a circuit containing a resistor and a battery only. As we discussed in Chap. 14. *Circuits and Dendrites: Charge Conservation, Ohm's Law, Rate Equations, and Other Old Friends*, the power supplied by the battery is

$$P = IV = I^2 R = \frac{V^2}{R}. \tag{18.72}$$

If we now add an inductor to the circuit (as in Fig. 18.6), the rate at which energy is supplied by the battery remains $P = VI$. However, in this case, we have that $V = IR + L\frac{dI}{dt}$, so that

$$P = IV = I^2 R + LI\frac{dI}{dt}. \tag{18.73}$$

The first term on the right-hand side is Joule heating of the resistor, as before. To elucidate the second term of Eq. 18.73, it is convenient to re-write it using the chain rule, namely

$$\frac{d}{dt}\left(\frac{1}{2}LI^2\right) = LI\frac{dI}{dt}, \tag{18.74}$$

as

$$P = I^2 R + \frac{d}{dt}\left(\frac{1}{2}LI^2\right).$$

(18.75)

We interpret the second term as the rate of change of energy stored in the magnetic field of the inductor. It follows that the energy stored in the magnetic field of the inductor is

$$\frac{1}{2}LI^2.$$

(18.76)

It is interesting to express the magnetic energy in terms of the magnetic field. We will see how this program works out for a solenoid of length ℓ and N turns, for which $L = N^2\mu_0 A/\ell$, so that

$$\frac{1}{2}LI^2 = \frac{1}{2}\frac{N^2\mu_0 AI^2}{\ell} = \frac{1}{2\mu_0}\frac{N^2\mu_0^2 I^2}{\ell^2} A\ell.$$

(18.77)

For a solenoid, we also have that

$$B = \frac{N\mu_0 I}{\ell}.$$

(18.78)

It follows that

$$\frac{1}{2}LI^2 = \frac{1}{2\mu_0}B^2 A\ell.$$

(18.79)

Since $A\ell$ is the volume of the solenoid, interestingly we may identify

$$\frac{1}{2\mu_0}B^2$$

(18.80)

as the energy density of the magnetic field. In comparison, when we discussed capacitors, we found that the energy density of an electric field, E, is

$$\frac{1}{2}\epsilon_0 E^2.$$

(18.81)

Interestingly, the magnetic and electric energy densities mirror each other, when expressed in terms of their respective field strengths.

18.6   Chapter Outlook

We discovered in this chapter that a changing magnetic field creates an electric field, which is the essence of Faraday's law. The fact that magnetic field and electric field are inextricably linked becomes even clearer in Chap. 19. *Maxwell's Equations and Then There Was Light*, where we will see that Ampere's law, as discussed in Chap. 17. *Magnetic Fields and Ampere's Law*, is not complete, and that, in fact, in addition to current, a changing electric field also creates a magnetic field. In Chap. 19. *Maxwell's Equations and Then There Was Light*, we will also see that electromagnetic waves comprise both electric and magnetic fields in a definite ratio. In this chapter, we also introduced inductors and examined their role—i.e., the role of Faraday's law—in electrical circuits. In doing so, we encounter familiar differential equations yet again. The more that changes, the more that stays the same.

18.7   Problems

Problem 1: Motional EMF
In this problem, you will investigate how a moving conductor develops a voltage (emf) because of the work done by magnetic forces to separate the charges inside the conductor. This potential difference is called a *motional* emf, in contrast to the chemical emf generated in a battery, in which charges are separated by chemical reactions rather than magnetic forces.

Consider a conductor of length L that moves with velocity $\mathbf{v}$ through a perpendicular uniform magnetic field $\mathbf{B}$ into the page, as shown in the figure.

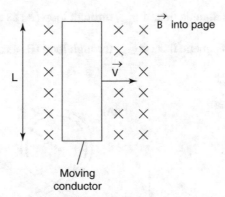

(a) The charges carried inside the conductor—assumed to be positive, each with charge q—also move with velocity $\mathbf{v}$. What is the magnetic force acting on each charge carrier? Specify the magnitude and the direction.
(b) This force causes the charge carriers to move, separating the positive and negative charges. The separated charges then create an electric field inside the

conductor. What is the direction of this electric field in the conductor? Make a sketch of the system and show the direction of the E-field.

(c) The charge carriers continue to separate until the magnetic force exactly balances the electric force generated by the newly created electric field. At this equilibrium condition, what is the strength of the electric field E?

(d) At this equilibrium condition, using the electric field strength you found in part (c), calculate the voltage difference between the top and the bottom of the conductor $\Delta V = V_{\text{top}} - V_{\text{bottom}}$.

(e) The result from part (d) is the *motional* emf that develops in the conductor due to the magnetic force. What happens to this emf if the bar stops moving?

Problem 2: Loop the Loop

Consider the two digital TV loop antenna designs shown in Fig. 18.12. Both consist of two circles, both of radius a, lying in the xy-plane. In one case (A), the wires connecting the circle cross. In the other case (B), the wires do not cross. You may ignore the small gap between the leads going to and coming from the loop, and the deviation from an exact circular shape near the connection between the circles. An electromagnetic (EM) wave, propagating in air along the x-axis, has a magnetic field at the location of the antenna directed along the z-axis and is given by

$$\mathbf{B}(t) = B\cos(\omega t)\mathbf{k}. \tag{18.82}$$

(This requires that the wavelength of the EM wave is much larger than the radius of the antenna, which you should assume is the case here.).

(a) What is the net magnetic flux, $\Phi_{B_{(A)}}$, through loop (A) as a function of B, t, ω, and a?

(b) What is the net magnetic flux, $\Phi_{B_{(B)}}$, through loop (B) as a function of B, t, ω, and a?

Fig. 18.12 Loop antennae

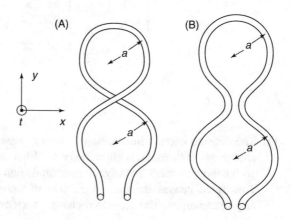

(c) What is the EMF, $V_{(A)}$, generated around loop (A)?
(d) What is the EMF, $V_{(B)}$, generated around loop (B)?
(e) Now, a resistor, R, is connected across the terminals of each loop. What is the current, $I_{(A)}$, around loop (A)?
(f) What is the current, $I_{(B)}$, around loop (B)? What is the direction of the current in this case?

Problem 3: Toroidal Solenoid

A toroidal solenoid of rectangular cross-section has inner radius a, outer radius b, height c, and has N turns. In this problem, $c \gg a$ and $c \gg b$. The current in the solenoid is $i(t)$ at time t.

(a) What is the magnetic field $B(r)$ inside the solenoid as a function of radial position r from the symmetry axis of the solenoid?
(b) What is the magnetic field outside the solenoid?
(c) The current is changing with time, so that the magnetic field changes with time. Sketch the solenoid and add electric field lines to your sketch.
(d) Carefully choose an appropriate loop with which to apply Faraday's law and indicate your loop on your sketch.
(e) Use your loop from (d) to calculate the electric field for values of the radial coordinate, r, for which $r < a$ in terms of $\frac{di}{dt}$. What is the direction of the electric field?
(f) Now, calculate the electric field for values of the radial coordinate, r, for which $a < r < b$ in terms of $\frac{di}{dt}$.

Problem 4: Coaxial Cable

Consider a coaxial cable composed of two cylindrical conducting shells of radius a and b of length ℓ, as illustrated at the top of Fig. 18.13.

(a) What is the inductance of the coaxial cable?
(b) What is the capacitance of the coaxial cable?
(c) Now, consider the circuit shown at the bottom of Fig. 18.13. One end of the coaxial cable is connected across a cosinusoidally varying emf ($V \cos \omega t$) in series with a resistor.
(d) What is the resonant frequency of this circuit?

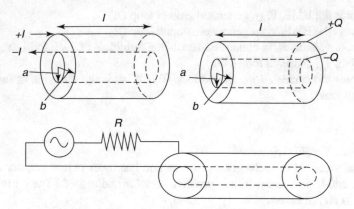

Fig. 18.13 Top: coaxial cable. Bottom: circuit including a coaxial cable

Problem 5: Rectangular Loop Entering a Magnetic Field Area

A conducting rectangular loop of length L and height h is moving with constant velocity v to the right, as shown in the figure. The loop enters from the left a region of total width L_0 with a constant and uniform B-field out of the page as shown and then exits from the right side. Assume $t = 0$ is when the right side of the loop starts entering the region with the magnetic field.

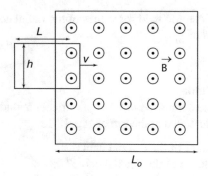

(a) What is the magnetic flux through the loop as a function of time? Make a sketch of $\Phi_B(t)$ from $t = 0$ until the time when the loop has exited the B-field region. Add to your plot as many quantitative features as you can in terms of the given variables.

(b) What is the induced emf in the loop as a function of time? Make a sketch of $\mathcal{E}(t)$ from $t = 0$ until the time when the loop has exited the B-field region. Add to your plot as many quantitative features as you can in terms of the given variables.

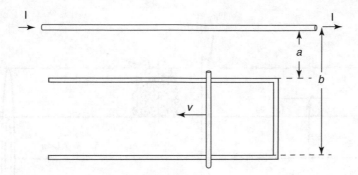

Fig. 18.14 Diagram for another wire on a slide

Problem 6: Another Wire on a Slide

(a) An infinitely long, straight conducting wire carries a current I as shown in (Fig. 18.14). Use Ampere's law to determine the magnetic field at a radial distance r from the center of the wire.

(b) As also shown in the figure, located near the first (straight) wire, there is another (second) wire bent into a sideways U-shape, on top of which a third wire slides at a constant velocity v. Indicate on a sketch of the situation, the direction of the magnetic field through the rectangular loop formed by the U of the second wire and the third wire.

(c) Calculate the magnetic flux through the rectangular loop formed by the U of the second wire and the third wire at time t, given that at $t = 0$, the third wire is on top of the bottom of the U. (Hint: Be sure to account for the fact that the magnetic field depends on r.)

(d) What is the EMF around the rectangular loop formed by the U of the second wire and the third wire at time t?

(e) Given that the resistance of the third wire is R and that the resistance of the U-shaped wire is zero, calculate the current in the third wire at time t.

Problem 7: Magnet Falling Through a Conducting Tube

Explain, in enough detail to be convincing, why a permanent magnet falls slowly through a conducting tube. Does the magnet have a terminal velocity?

Problem 8: Another Solenoid

A solenoid is constructed by winding a wire onto a dielectric cylinder of radius a and length d. The wire is wound around until the outer radius of the solenoid is b. Since $d \gg b$, end effects may be neglected throughout this problem. The cross-sectional area occupied by an individual strand of wire is A, so that a total of $N = (b - a)d/A$ turns of wire comprise the solenoid. You should assume that N is a very large number (Fig. 18.15).

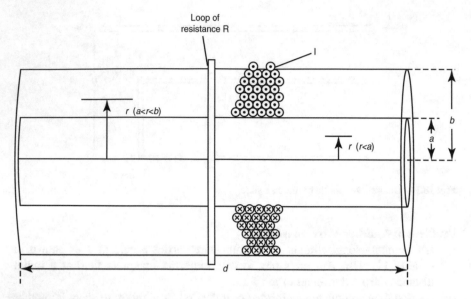

Fig. 18.15 Schematic of the solenoid for "Another solenoid", shown as a longitudinal cross-section

(a) A current I flows in the wire. Sketch your own longitudinal cross-section of the solenoid, like the figure, and indicate on your sketch two Amperian loops with which to apply Ampere's Law. Hint: your loops should be rectangles in this sketch, one to calculate the B-field for $r < a$ and the other to calculate the B-field for $a < r < b$.

(b) Given that the magnetic field outside the solenoid ($r > b$) is zero, find the magnetic field as a function of r, the radial distance from the center of the solenoid for $r < a$, and $a < r < b$.

(c) What is the energy density per unit volume stored in the magnetic field for $r < a$, $a < r < b$, and $r > b$?

(d) What is the total energy stored in the magnetic field? Hint: The answer in this case is the sum of two terms, is easier to write, is discussed in Sect. 18.5.

(e) A single loop of wire of resistance R is now wound around the outside of the solenoid. What is the magnetic flux, $\Phi_B = \int \mathbf{B} \cdot d\mathbf{A}$, through this loop? Hint: See the hint for part (d).

(f) The current in the solenoid varies as a function of time t as $I(t) = I_0 \cos \omega t$. What current flows around the loop of wire, introduced in part (e)?

Problem 9: Falling Loop

In a region of space containing a uniform magnetic field (of magnitude B, directed out of the paper), two conducting bars, each of length a, resistance R, and mass m, are connected to each other by means of two initially tightly coiled wires, also each

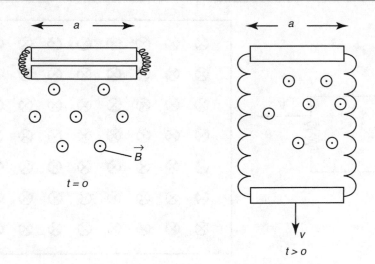

Fig. 18.16 Falling loop

of resistance R (Fig. 18.16). Initially, the bars are parallel and immediately adjacent to each other as shown on the left side of Fig. 18.16. At time $t = 0$, one of the bars is released and starts to fall under the influence of gravity. The acceleration due to gravity is g. Throughout its fall, the falling bar remains parallel to the stationary bar, and the coiled wires uncoil, creating a flat rectangular loop, with its vector area parallel to the magnetic field (right side of Fig. 18.16).

(a) Calculate the rate of change of magnetic flux, $\Phi_{B_{(A)}}$, through the loop.
(b) What current flows in the loop? Indicate its direction on a sketch of the loop.
(c) What is the force on the falling bar when its velocity is v? What is its direction?
(d) What is the value of the velocity at large times?

Problem 10: RC's Flying Circus
A square conducting loop of side L has a resistor R and a capacitor C. The loop is moving to the right with a constant velocity v approaching an area of uniform magnetic field B into the page, as shown in Fig. 18.17. Assume the B-field region extends infinitely to the right. Assume initially there is no charge on the capacitor and no current in the circuit, and that at time $t = 0$ the right edge of the loop enters the B-field region.

(a) Call the time it takes the loop to completely enter the region of magnetic field t_1. Give this time in terms of the given quantities.
(b) Make a sketch of the absolute value of the magnetic flux, $|\Phi_B|$, through the loop as a function of time for any $t \geq 0$. (You can neglect the size and shape of the resistor and capacitor and just consider the loop as a square.) Clearly, indicate the time t_1 on your sketch.

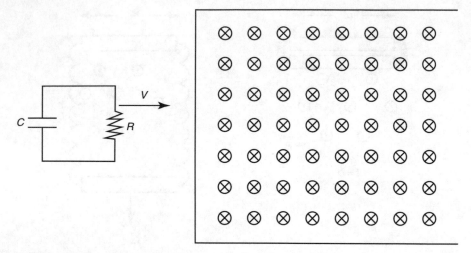

Fig. 18.17 Image for RC's Flying Circus problem

(c) Make a sketch of the absolute value of the induced emf, $|\mathcal{E}|$, around the loop as a function of time for any $t \geq 0$. Clearly indicate the times 0 and t_1 on your sketch.

(d) When the induced emf $\mathcal{E}$ is non-zero, what is its direction? Explain whether it is clockwise or counterclockwise.

(e) Make a sketch of the charge $Q(t)$ on the capacitor plate as a function of time for any $t \geq 0$. Clearly indicate the time t_1 on your sketch.

(f) What is the charge and the current in the circuit when $t \to \infty$? Explain your answer.

Problem 11: Alternating Current?

A rectangular loop of wire with area A rotates with angular velocity ω in a magnetic field as shown in Fig. 18.18. The angular velocity ω is constant.

(a) What is the induced emf in the coil as a function of time?

(b) If the coil has a total resistance R, what is the induced current into the coil?

(c) Make a sketch of the induced current as a function of time.

Problem 12: LR in Series

Answer the following questions in terms of the given quantities, V_0, L, and R.

- For L and R in series as shown in Fig. 18.19, a switch is closed at $t = 0$. What is the answer to the following questions immediately after the switch is closed.

(a) What is the current through the inductor? Explain.

Fig. 18.18 Image for
Alternating Current?

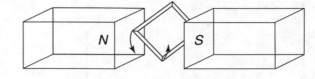

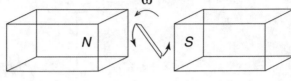

Fig. 18.19 Image for LR in
series

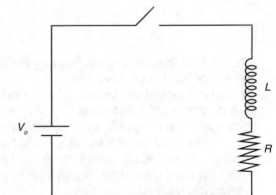

(b) What is the current through the resistor? Explain.
(c) What is the voltage across the inductor? Explain.
(d) What is the voltage across the resistor? Explain.
• Now imagine that the switch has been closed for a very long time and steady
   state has been reached. Answer questions (a) through (d) above. (Hint: You may
   find it easier to answer the (a) through (d) in a different order than asked.)

Problem 13: LR in Parallel

Answer the following questions in terms of the given quantities, V_0, L, and R_1 and
R_2.

For L and R_1 in parallel as shown in Fig. 18.20, a switch is closed at $t = 0$.

(a) Answer questions (a) through (d) for L and both R_1 and R_2 in this circuit
     immediately after the switch is closed.
(b) Now imagine that the switch has been closed for a very long time and steady
     state has been reached. Answer questions (a) through (d) above for L and both

Fig. 18.20 Image for LR in parallel

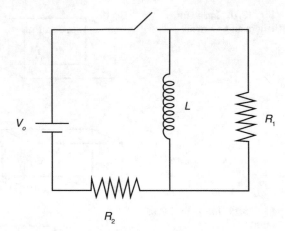

R_1 and R_2 in this circuit. (Hint: You may find it easier to answer the (a) through (d) in a different order than asked.)

(c) It is sometimes said that immediately after an inductor is connected to a steady voltage source, the inductor acts like an open switch or open circuit. Based on your answers to the questions above, how do you explain this statement?

(d) In a related statement, it is sometimes said that at long times after an inductor is connected to a steady voltage source, the inductor acts like an ordinary wire or a short circuit. Based on your answers to the questions above, how do you explain this statement?

(e) Compare this behavior with RC circuits. Specifically, if you replace the inductor in the above circuits with a capacitor, when can you say the capacitor acts like a wire and when can you say it acts like an open switch? Consider both early times and a long time after the battery has been connected.

Problem 14: LC Circuit

The switch in the circuit in Fig. 18.21 below has been in position 1 for a long time. It is changed to position 2 at time $t = 0$. Give your answers to the following questions in terms of the given quantities: R, L, C, and V_0.

(a) What is the maximum current through the inductor? (Hint: Recall how current is related to charge on the capacitor. Think about how you will determine $Q(t)$.)

(b) What is the first time at which the current is maximum?

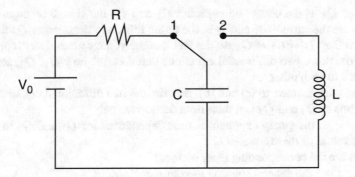

Fig. 18.21 Image for LC circuit

Problem 15: Flip-Flopper

Consider the circuit shown below. Initially, both capacitors are charged, as shown in the figure.

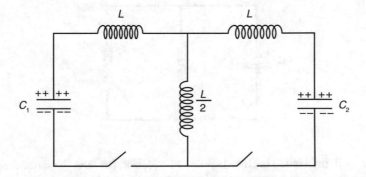

(a) At $t = 0$, both switches are closed simultaneously. The current in the circuit flows as shown in the diagram below. Write an equation for how the currents I_0, I_1, and I_2 relate to one another.

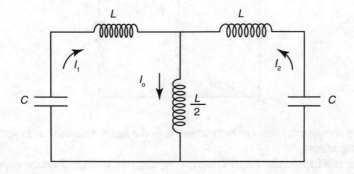

(b) Define Q_1 as the charge on capacitor C_1 and Q_2 the charge on capacitor C_2. How are the currents I_1 and I_2 in the figure related to the charges Q_1 and Q_2?

(c) Given that $C_1 = C_2 = C$, use the fact that the voltage around any loop is zero to write down two differential equations that describe how Q_1, Q_2, and I_0 are related to each other.

(d) Using your answers to (a) and (b), re-write the two differential equations in (c) in terms of Q_1 and Q_2 and their time derivatives only.

(e) Assume an imaginary exponential time dependence for Q_1 and Q_2 to find the eigenvalues of these equations.

(f) What are the corresponding eigenvectors?

(g) Write down the general solution to your equations.

Problem 16: Inductors in Parallel

Consider the circuit shown below. Initially, the capacitor is charged as shown on the figure.

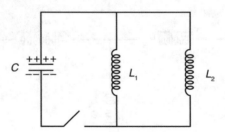

(a) At $t = 0$, the switch is closed. Write an equation for how the current I_0 in the figure relates to the currents flowing through L_1 and L_2.

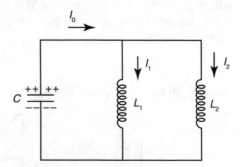

(b) Write an equation for how the current I_0 in the figure relates to the charge Q on the capacitor.

(c) Write two independent equations: one relating the charge Q on the capacitor to the current, I_1, through L_1, and one relating the charge Q on the capacitor to the current, I_2, through L_2. Indicate on your diagram where the equations came from and what law or rule you are using to determine these equations.

(d) Using the answers to the last three questions, write a single differential equation in terms of Q, $\frac{dQ}{dt}$ and $\frac{d^2Q}{dt^2}$ **only**.

(e) Solve the differential equation assuming a sinusoidal trial solution for Q.

(f) What is the oscillation frequency of the charge on the capacitor?

Maxwell's Equations and Then There Was Light

19

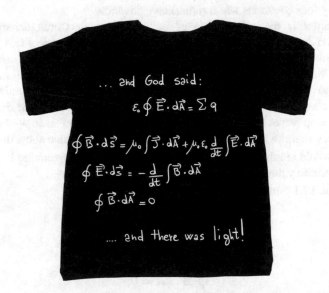

> "I have also a paper afloat, with an electromagnetic theory of light, which, till I am convinced to the contrary, I hold to be great guns" in the Scientific Letters and Papers of James Clerk Maxwell.

19.1 Introduction

When Fizeau[1] and Foucault[2] measured the speed of light in 1850, their result came in the context of an increasing understanding of electricity and magnetism, as a

---

[1] http://en.wikipedia.org/wiki/Hippolyte_Fizeau

[2] http://en.wikipedia.org/wiki/Leon_Foucault

© The Author(s), under exclusive license to Springer Nature Switzerland AG 2023
S. Mochrie, C. De Grandi, *Introductory Physics for the Life Sciences*,
Undergraduate Texts in Physics, https://doi.org/10.1007/978-3-031-05808-0_19

result of the earlier efforts of Gauss,[3] Ampere,[4] Faraday,[5] and others. In particular, the values of ϵ_0 and μ_0 were both known at this time from measurements on charges and currents. Thus, when Weber[6] and Kohlrausch[7] noticed that the experimentally measured value of the speed of light was close to the value of $\frac{1}{\sqrt{\epsilon_0\mu_0}}$, it was a striking and intriguing observation that eventually led James Clerk Maxwell, when he had formulated the complete set of his equations and shown that they give rise to waves, to write:

> We can scarcely avoid the conclusion that light consists in the transverse undulations of the same medium which is the cause of electric and magnetic phenomena.

In 1862, the new understanding that electrical and magnetic phenomena in circuits, on the one hand, and radio waves, light, and x-rays, on the other, are all manifestations of only four equations was a remarkable advance.

This chapter is the culmination of our study of electromagnetism that we embarked on in Chap. 12. *Gauss's Law: Charges and Electric Field*. First, we will complete Maxwell's equations by discussing the displacement current, which, together with Faraday's law, expresses how E- and B-fields are inextricably connected to each other. We will then show how Maxwell's equations in the absence of charges and currents give rise to a wave equation, which describes all electromagnetic waves—radio waves, microwaves, light, x-rays, etc.—that all must propagate at a definite speed $c = \frac{1}{\sqrt{\epsilon_0\mu_0}}$ in vacuum. We will also show that the ratio of E- and B-field amplitudes for an electromagnetic wave propagating in vacuum is fixed at a definite value, namely $\frac{E}{B} = c$.

Equations 19.1 through 19.2 are Maxwell's Equations[8]:

$$\oint_A \mathbf{E} \cdot d\mathbf{A} = \frac{Q}{\epsilon_0}, \tag{19.1}$$

$$\oint_A \mathbf{B} \cdot d\mathbf{A} = 0, \tag{19.2}$$

$$\oint_s \mathbf{B} \cdot d\mathbf{s} = \mu_0 I + \mu_0 \epsilon_0 \frac{d}{dt}\int_A \mathbf{E} \cdot d\mathbf{A}, \tag{19.3}$$

---

[3] http://en.wikipedia.org/wiki/Carl_Friedrich_Gauss

[4] http://en.wikipedia.org/wiki/Andre-Marie_Ampere

[5] http://en.wikipedia.org/wiki/Michael_Faraday

[6] http://en.wikipedia.org/wiki/Wilhelm_Eduard_Weber

[7] http://en.wikipedia.org/wiki/Rudolf_Kohlrausch

[8] http://en.wikipedia.org/wiki/James_Clerk_Maxwell

$$\frac{d}{dt} \int_A \mathbf{B} \cdot d\mathbf{A} = - \oint_s \mathbf{E} \cdot d\mathbf{s}. \qquad (19.4)$$

Gauss's law (Eq. 19.1) specifies how the electric field ($\mathbf{E}$) is related to charge, as we discussed in Chap. 12. *Gauss's Law: Charges and Electric Field*. The remaining Maxwell's equations involve not only $\mathbf{E}$, but also the magnetic field, $\mathbf{B}$. Equation 19.2 is often called "Gauss's law for magnetic fields" because its left-hand side involves an integral of the magnetic field over a closed surface, similarly to the left-hand side of Gauss's law (Eq. 19.1) that involves an integral of the electric field over a closed surface. However, the right-hand side of Eq. 19.2 is zero, instead of the enclosed electric charge, divided by ϵ_0, in the case of Eq. 19.1. As we discussed in Chap. 17. *Magnetic Fields and Ampere's Law*, this feature of Eq. 19.2 is because there are no magnetic charges—it is impossible to have a single magnetic pole. Equation 19.4 is Faraday's law, which we discussed in Chap. 18. *Faraday's Law and Electromagnetic Induction*. Without the second term on the right, Eq. 19.3 would be Ampere's law, as we discussed in Chap. 17. *Magnetic Fields and Ampere's Law*, relating the magnetic field to currents. In this chapter, we will explain why Ampere's law should be completed by the additional term, appearing in Eq. 19.3. This extra term is called the displacement current.

19.2   Your Learning Goals for This Chapter

By the end of this chapter, you should be able to:

- Explain the reasoning that leads to displacement current.
- Appreciate that Maxwell's equations in the absence of charges and currents possess wave solutions, with a definite value for the wave velocity equal to $c = \frac{1}{\sqrt{\epsilon_0 \mu_0}}$, which is the speed of light, and a definite value for the ratio of the E- and B-fields in the wave ($\frac{E}{B} = c$).

19.3   Ampere's Law Revisited: The Displacement Current

The version of Ampere's law, which we have used so far, is actually incomplete and is strictly valid only for situations in which the electric field is not time-dependent. Our goal in this section is to understand why and how Ampere's law should be extended to apply to situations involving time-varying electric fields. This discussion will reveal how electric and magnetic fields are inextricably connected to each other and will complete Maxwell's equations.

To understand why Ampere's law must be extended, it is convenient to consider the process of charging a capacitor with circular plates of radius R, as shown in Fig. 19.1. Suppose that during charging, a current I is flowing. Then, by picking a

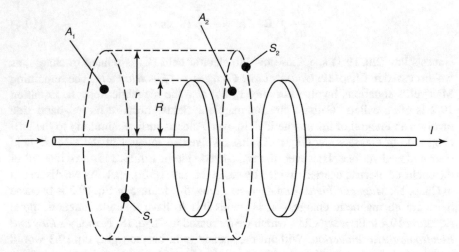

Fig. 19.1 Applying Ampere's law to a charging capacitor

circular Amperian loop, s_1, concentric with the wire, the left-hand side of Ampere's law evaluates to $2\pi r B$. This result is correct for any surface enclosed by s_1. Usually, we would think of that surface as being the circular area, A_1, in which case the net current through the surface—through A_1—is I, leading to $B = \frac{\mu_0 I}{2\pi r}$, as we have seen previously. However, an equally valid surface consists of the circular area A_2 plus the cylindrical sides between A_2 and A_1. This latter surface excludes A_1, but it is also bounded by s_1. In this case, however, the actual current through the surface is zero because there is no charge flowing through the capacitor. In this case, Ampere's law leads to $B = 0$. Thus, two correct applications of the original Ampere's law has resulted in two different answers. Further consideration of other Amperian loops, such as s_2, and the surfaces bounded by them suggests that $B = \frac{\mu I}{2\pi r}$ should be the same everywhere, including between the capacitor plates. Adding the new term in Eq. 19.3 ensures that this is the case.

To motivate the form of the new term, we write the relationship between the current, I, flowing onto the capacitor plates and the rate of change of electric flux between the plates, Φ_E:

$$I = \frac{dQ}{dt} = \frac{d}{dt}(CV) = \frac{d}{dt}\left(\frac{\epsilon_0 A}{d} \times Ed\right) = \epsilon_0 \frac{d}{dt}(AE) = \epsilon_0 \frac{d\Phi_E}{dt}. \qquad (19.5)$$

The right-hand side of Eq. 19.5 is called the displacement current, and was first introduced by Maxwell in 1861. Including the displacement current in Ampere's law, as in Eq. 19.3, ensures that the magnetic field is the same between capacitor plates as it is in the region of the current carrying wire.

To see the new term in action, we can apply Eq. 19.3 at a location that lies between the capacitor plates. We first pick a concentric circular Amperian loop, s_2, which encloses the surface, A_2. Then, the line integral on the left-hand side of

Eq. 19.3 evaluates to

$$\oint_s \mathbf{B} \cdot d\mathbf{s} = 2\pi r B. \tag{19.6}$$

On the right-hand side, the current is zero, while the area integral integrates to

$$\int_A \mathbf{E} \cdot d\mathbf{A} = EA = \pi R^2 E, \tag{19.7}$$

assuming that $r > R$, as in the figure, and that the electric field is constant within the cross-section of the capacitor and zero outside. Setting the two sides equal to each other then gives

$$2\pi r B = \mu_0 \epsilon_0 \frac{d}{dt}(\pi R^2 E), \tag{19.8}$$

i.e.,

$$B = \mu_0 \epsilon_0 \frac{1}{2\pi r} \pi R^2 \frac{dE}{dt} = \mu_0 \epsilon_0 \frac{1}{2\pi r} A \frac{dE}{dt} = \mu_0 \frac{1}{2\pi r} \frac{dQ}{dt} = \frac{\mu_0 I}{2\pi r}, \tag{19.9}$$

where we used that, between the plates $E = Q/(\epsilon_0 A)$, according to Gauss's law, and that $I = \frac{dQ}{dt}$. We see that (for $r > R$) the magnetic field in the region between the plates is indeed the same as in the regions where there is a current carrying wire. While Faraday informs us that a changing magnetic field gives rise to an electric field, the complete version of Ampere's law informs us that a changing electric field gives rise to a magnetic field, emphasizing that time-varying electric and magnetic fields are inseparable.

19.4 An Electromagnetic Wavefront

Without charges and currents, Maxwell's equations, including the displacement current, become

$$\oint_A \mathbf{E} \cdot d\mathbf{A} = 0, \tag{19.10}$$

$$\oint_A \mathbf{B} \cdot d\mathbf{A} = 0, \tag{19.11}$$

$$\oint_s \mathbf{B} \cdot d\mathbf{s} = \mu_0 \epsilon_0 \frac{d}{dt} \int_A \mathbf{E} \cdot d\mathbf{A}, \tag{19.12}$$

and

Fig. 19.2 Schematic of an
electromagnetic wavefront

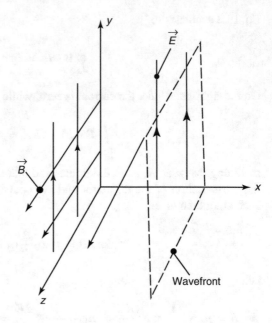

$$\frac{d}{dt}\int_A \mathbf{B}\cdot d\mathbf{A} = -\oint_s \mathbf{E}\cdot d\mathbf{s}. \tag{19.13}$$

Remarkably, these equations predict electromagnetic waves in the sense that they possess wave solutions.

To introduce electromagnetic wave propagation, we hypothesize an especially simple electromagnetic wave, namely an electromagnetic wavefront that is infinite along the y and z axes and that propagates along the x-axis at a velocity c. In front of this wavefront, the electric and magnetic (E- and B-) fields are zero. Behind the wavefront, that is, for $x - ct < 0$, the E-field is directed along the y-axis,

$$\mathbf{E} = E_y\hat{\mathbf{j}}, \tag{19.14}$$

and the B-field is directed along the z-axis,

$$\mathbf{B} = B_z\hat{\mathbf{k}}. \tag{19.15}$$

Such a wavefront is illustrated in Fig. 19.2. Importantly, the coordinate system in Fig. 19.2 is a right-handed coordinate system, so that $\hat{\mathbf{i}} \times \hat{\mathbf{j}} = \hat{\mathbf{k}}$, consistent with the rule to always rely on your right hand to decide on directions.

Since we have constructed both $\mathbf{E}$ and $\mathbf{B}$ to be functions of $x - ct$, we know from our discussion of waves in Chap. 11. *Wave Equations: Strings and Wind* that the wavefront's E- and B-fields both satisfy the wave equation. The question that we must now answer is whether $\mathbf{E}$ and $\mathbf{B}$ also satisfy Maxwell's equations? In fact, we

Fig. 19.3 A Gaussian
surface located to span the
electromagnetic wavefront

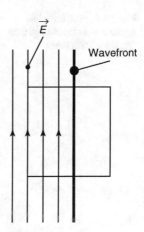

will show that **E** and **B** do satisfy Maxwell's equations, provided c has a certain value, which we will calculate, and provided E_y/B_z has a certain value, which we will also calculate. Under these conditions, then, this wavefront is a solution to Maxwell's equations.

19.4.1 Gauss's Law at the Wavefront

First, we consider a Gaussian surface in the form of a cube with faces along the x-, y-, and z-directions that spans the wavefront under consideration, as sketched in Fig. 19.3. We can check that Gauss's law (Eq. 19.10) is satisfied for this field configuration and this Gaussian surface. Interestingly, if the E-field lines were oriented away from the y-direction, they would then have to end on the wavefront. But the only way that field lines can end is on electric charges. But by assumption, there are no electric charges. Therefore, we may conclude that the E-field in an electromagnetic wave must point in a direction perpendicular to the direction of propagation of the wave, i.e., an electromagnetic wave is transverse as far as the E-field is concerned.

19.4.2 Gauss's Law for Magnetic Fields at the Wavefront

Equation 19.11 is satisfied similarly by the wavefront under consideration. In addition, since there are no magnetic charges, we may deduce that the B-field in an electromagnetic wave actually must also be perpendicular to the direction of propagation of the wave. Thus, electromagnetic waves are *transverse* waves.

Fig. 19.4 Faraday's law
applied to a closed loop than
spans the wavefront

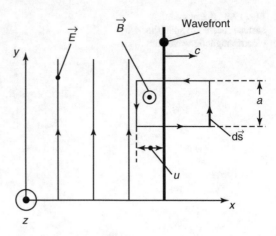

19.4.3 Faraday's Law at the Wavefront

Next we apply Faraday's law (Eq. 19.13) to the loop shown in Fig. 19.4. Applied to
this loop, Faraday's law becomes

$$-\frac{d}{dt}(B_z au) = -E_y a, \tag{19.16}$$

where a is the dimension of the loop along y and u is the dimension of the loop
along x behind the wavefront. As usual, we must use our right hand to determine
directions. Specifically, the way that $d\mathbf{s}$ is circulating in the figure implies that $d\mathbf{A}$
is directed out of the plane of the paper, parallel to $\mathbf{B}$. The way that $d\mathbf{s}$ is circulating
also implies that $d\mathbf{s}$ is antiparallel to $\mathbf{E}$ on left part of the loop, which leads to
the negative sign on the right side of Eq. 19.16. Since the wavefront is moving at
a velocity c, u is increasing with this same velocity, c. It follows that $\frac{du}{dt} = c$.
Therefore, canceling the common factor of $-a$ from both sides of Eq. 19.16, we
find that in order to satisfy Faraday's law, we must have that

$$B_z c = E_y. \tag{19.17}$$

Thus, we must conclude that in an electromagnetic wave the ratio of the amplitudes
of the E- and B-field must be given by the speed of the wave.

19.4.4 Ampere's Law at the Wavefront

The final Maxwell equation is Ampere's law, including the displacement current,
namely Eq. 19.12. Applied to the loop shown in Fig. 19.5, this equation implies that

Fig. 19.5 Ampere's law
(with displacement current)
applied to a closed loop than
spans the wavefront

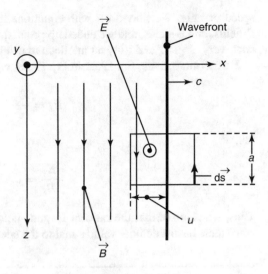

$$Ba = \mu_0 \epsilon_0 \frac{d}{dt}\left(E_y au\right) = \mu_0 \epsilon_0 E_y ac. \qquad (19.18)$$

Canceling a from both sides yields

$$B_z = \mu_0 \epsilon_0 c E_y. \qquad (19.19)$$

Therefore, Ampere's law (including displacement current) is satisfied if the ratio of E_y and B_z has the specific value given by Eq. 19.19.

19.5   The Speed of Light

Both Eqs. 19.19 and 19.17 specify the relative amplitude of E_y and B_z. In order to satisfy both equations simultaneously, the value of c must have a particular value. If we divide Eq. 19.19 by Eq. 19.17, we find that

$$\frac{1}{c} = \mu_0 \epsilon_0 c, \qquad (19.20)$$

that is,

$$c = \frac{1}{\sqrt{\epsilon_0 \mu_0}}. \qquad (19.21)$$

Equation 19.21 informs us that the speed of the wavefront must be $c = \frac{1}{\sqrt{\epsilon_0 \mu_0}} = 3 \times 10^8$ m s^{-1}, which is the speed of light. Thus, Maxwell's equations predict the speed of light. We started off with equations that told us about charges, currents, E-fields, and B-fields, and we ended up predicting light. This result is tremendously cool, very elegant, and truly an intellectual triumph.

Furthermore, using Eq. 19.21 in Eq. 19.19, we see that Eq. 19.19 becomes

$$B_z = \frac{E_y}{c}, \tag{19.22}$$

i.e.,

$$\frac{E_y}{B_z} = c. \tag{19.23}$$

Thus, we also see that the ratio of E- and -field amplitudes in an electromagnetic wave must have a definite value equal to the speed of light.

19.6   Wave Equation for Electromagnetic Waves

In the last section, we saw how a particular wave form satisfies Maxwell's equations. We now seek to generalize this result to show that that E- and B-fields obey the wave equation. For example, we may expect that

$$\frac{d^2 E_y}{dx^2} = \frac{1}{c^2} \frac{d^2 E_y}{dt^2}. \tag{19.24}$$

To derive Eq. 19.24, first, we take

$$E_y = E_y(x, t) \tag{19.25}$$

and

$$B_z = B_z(x, t), \tag{19.26}$$

so that both E_y and B_z vary along the x-direction. Then, we apply Faraday's law to the rectangular loop shown in Fig. 19.6, which starts at x and extends to $x + dx$ along the x-direction and has a dimension a in the y-direction. In this case, we find

$$E_y(x + dx)a - E_y(x)a = -\frac{d}{dt}(B_z dx a), \tag{19.27}$$

that is,

$$E_y(x)a + dx \frac{dE_y}{dx} a - E_y(x)a = -dx a \frac{dB_z}{dt}, \tag{19.28}$$

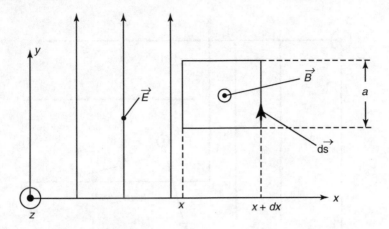

Fig. 19.6 Faraday's law applied to a loop for arbitrary $B_z(x,t)$ and $E_y(x,t)$

that is,

$$dx\frac{dE_y}{dx}a = -dxa\frac{dB_z}{dt}. \tag{19.29}$$

Now, canceling the common factor of dxa, we arrive at

$$\frac{dE_y}{dx} = -\frac{dB_z}{dt}. \tag{19.30}$$

Next, we apply Ampere's law including the displacement current to the loop shown in Fig. 19.7. In this case, we find

$$-B_z(x+dx)a + B_z(x)a = \epsilon_0\mu_0\frac{d}{dt}\left(E_y dxa\right), \tag{19.31}$$

that is,

$$-\frac{dB_z}{dx} = \epsilon_0\mu_0\frac{dE_y}{dt}, \tag{19.32}$$

similarly to Eq. 19.30. Now, we take the derivative of Eq. 19.30 with respect to x with the result that

$$\frac{d^2E_y}{dx^2} = -\frac{d^2B_z}{dxdt}, \tag{19.33}$$

and the derivative of Eq. 19.32 with respect to t with the result that

$$-\frac{d^2B_z}{dtdx} = \epsilon_0\mu_0\frac{d^2E_y}{d^2t}. \tag{19.34}$$

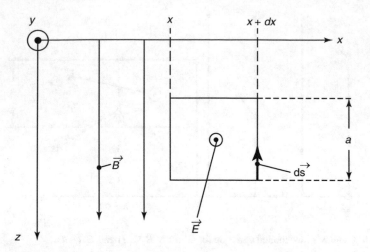

Fig. 19.7 Ampere's law, including the displacement current, applied to a loop for arbitrary $B_z(x, t)$ and $E_y(x, t)$

Because

$$\frac{d^2 B_z}{dt\,dx} = \frac{d^2 B_z}{dx\,dt},$$ (19.35)

we may combine Eqs. 19.33 and 19.34 to obtain the desired result, namely

$$\frac{d^2 E_y}{dx^2} = \epsilon_0 \mu_0 \frac{d^2 E_y}{dt^2},$$ (19.36)

which is the wave equation corresponding to a wave velocity $c = \frac{1}{\sqrt{\epsilon_0 \mu_0}}$, just as we found in our earlier wavefront example.

19.7 Chapter Outlook

Today, more than 150 years after they were first written down, Maxwell's equations, and the emergence from these equations of electromagnetic waves, represent a huge scientific triumph, and continue to set the standard for what physicists consider to be elegant and beautiful. Our modern world would be impossible without the understanding of electromagnetism provided by Maxwell's equations: no cars, no cellphones, no lasers, no computers, no internet, no solar panels..... The list is endless, but this book is not, and Maxwell's equations are a fitting high-note to end with.

19.8 Problem

Problem 1: Magnetic Field Inside a Capacitor

Apply Maxwell's version of Ampere's law (Eq. 19.3) to calculate the magnetic field between the plates of a circular capacitor of radius R for $r < R$, i.e., inside the capacitor, when a current I flows onto the capacitor.

Index

A

Acceleration, 16, 19
Actin polymerization, 238, 251, 308, 312, 410
Addition rule of probability, 190
Ampere's law, 769, 771, 833, 838
Amperian loop, 834
Atomic hypothesis, 240, 241
ATP hydrolysis, 163
Avogadro's number, 251

B

Bacterial growth, 221, 276, 374
Bat, 519
Bayes' theorem, 189
Beam steering, 719
Binding, 415
Binding energy, 160
Binomial distribution, 205
Boltzmann factor, 391
Boltzmann's constant, 251
Boundary conditions, 291
Brownian motion, 238, 240, 242
Brownian ratchet, 404, 410, 437, 440, 441
Buoyant force, 82, 450

C

Cancer, 199, 350, 382
Capacitor, 617, 636
Cell membrane, 649
Cell size, 307
Center of mass, 90
Central limit theorem, 214
Charge, 596
Circular motion, 18, 168
Circulatory system, 477
Complex numbers, 506, 535
complex numbers, 813

Conductor, 618, 619
Conservation law, 289
Conservation of energy, 136
Conservative forces, 144, 149, 151
Coulomb's law, 597
Coupled oscillators, 520, 524, 529, 532, 537

D

Damped simple harmonic motion, 509
Dielectrics, 644
Diffusion, 272
Diffusion coefficient, 249
Diffusion equation, 287, 689
Diffusion to capture, 302, 309, 328, 333, 334
Diffusive conductance, 295, 320
Displacement current, 833
Dissipation, 513
DNA zipping, 270, 400, 403
Donuts, 135

E

Echolocation, 519
Eigenmodes, 345, 520, 532
Eigenvalues, 345, 520
Eigenvectors, 345, 520
Einstein relation, 249, 437
Electric field, 598
Electric flux, 602, 834
Electromagnetic waves, 836, 842
Entropy, 419
Epidemic, 275, 276, 385
Escape velocity, 140
Euler's theorem, 508
Evolution, 258, 276
Exponential, 345
Exponential distribution, 212
Exponential relaxation, 85

© The Author(s), under exclusive license to Springer Nature Switzerland AG 2023
S. Mochrie, C. De Grandi, *Introductory Physics for the Life Sciences*,
Undergraduate Texts in Physics, https://doi.org/10.1007/978-3-031-05808-0

Printed in the United States
by Baker & Taylor Publisher Services